PROCEEDINGS OF SYMPOSIA
IN PURE MATHEMATICS
Volume 46, Part 1

Algebraic Geometry
Bowdoin 1985

Spencer J. Bloch, Editor

with the collaboration of H. Clemens,
D. Eisenbud, W. Fulton, D. Gieseker, J. Harris,
R. Hartshorne, and S. Mori

AMERICAN MATHEMATICAL SOCIETY
PROVIDENCE, RHODE ISLAND

PROCEEDINGS OF THE SUMMER RESEARCH INSTITUTE
ON ALGEBRAIC GEOMETRY
HELD AT BOWDOIN COLLEGE
BRUNSWICK, MAINE
JULY 8–26, 1985

with support from the National Science Foundation, Grant DMS-8415200

1980 *Mathematics Subject Classification* (1985 *Revision*):
Primary 14-06, 14-XX, 11G25, 11L40, 11Q25, 11R39, 13CXX, 13D05, 13E05, 13H15, 32JXX, 53C55.
Secondary 11G40, 13DXX, 13E15, 13F20, 13H10, 20C99, 22E55, 30F30, 32G13, 43A32, 55P62, 57D35.

Library of Congress Cataloging-in-Publication Data

Summer Research Institute on Algebraic Geometry (1985: Bowdoin College)
    Algebraic geometry: Bowdoin 1985/Spencer J. Bloch, editor; with the collaboration of H. Clemens. . . [et al.].

    (Proceedings of symposia in pure mathematics, 0082- 0717; v. 46)
    Proceedings of Summer Research Institute on Algebraic Geometry, Bowdoin College, Brunswick, Maine, July 8–26, 1985.
    Includes bibliographies.
    1. Geometry, Algebraic–Congresses. I. Bloch, Spencer. II. Clemens, C. Herbert (Charles Herbert), 1939–   . III. American Mathematical Society. IV. Title. V. Series.
QA564.S86   1985   516.3′5   87-12306
ISBN 0-8218-1476-1 (part 1)
ISBN 0-8218-1480-X (part 2)
ISBN 0-8218-1481-8 (set) (alk. paper)

Contents

PART 1

Algebraic Varieties; General Results

Curves

Surfaces

Threefolds

PART 2

Groups in Algebraic Geometry

Vector Bundles

Geometry in Characteristic p

Hodge Theory

Enumerative Geometry

Algebraic Cycles

Commutative Algebra

Algebraic Varieties; General Results

Proceedings of Symposia in Pure Mathematics
Volume **46** (1987)

On Varieties of Minimal Degree
(A Centennial Account)

DAVID EISENBUD AND JOE HARRIS

Abstract. This note contains a short tour through the folklore surrounding the rational normal scrolls, a general technique for finding such scrolls containing a given projective variety, and a new proof of the Del Pezzo–Bertini theorem classifying the varieties of minimal degree, which relies on a general description of the divisors on scrolls rather than on the usual enumeration of low-dimensional special cases and which works smoothly in all characteristics.

Introduction. Throughout, we work over an algebraically closed field k of arbitrary characteristic with subschemes $X \subset \mathbf{P}_k^r$. We say that X is a *variety* if it is reduced and irreducible, and that it is *nondegenerate* if it is not contained in a hyperplane. There is an elementary lower bound for the degree of such a variety:

PROPOSITION 0. *If $X \subset \mathbf{P}^r$ is a nondegenerate variety, then* $\deg X \geq 1 + \operatorname{codim} X$.

(PROOF. If $\operatorname{codim} X = 1$ the result is trivial. Else we project to $\mathbf{P}^{r-1}$ from a general point of X, reducing the degree by at least 1 and the codimension by 1, and are done by induction. □)

We say that $X \subset \mathbf{P}^r$ is a *variety of minimal degree* if X is nondegenerate and $\deg X = 1 + \operatorname{codim} X$. One hundred years ago Del Pezzo (1886) gave a remarkable classification for surfaces of minimal degree, and Bertini (1907) showed how to deduce a similar classification for varieties of any dimension. Of course the case of codimension 1 is trivial, X being then a quadric hypersurface, classified by its dimension and that of its singular locus. In other cases we may phrase the result as:

THEOREM 1. *If $X \subset \mathbf{P}^r$ is a variety of minimal degree, then X is a cone over a smooth such variety. If X is smooth and $\operatorname{codim} X > 1$, then $X \subset \mathbf{P}^r$ is either a rational normal scroll or the Veronese surface $\mathbf{P}^2 \subset \mathbf{P}^5$.*

1980 *Mathematics Subject Classification* (1985 *Revision*). Primary 14J26, 14J40; Secondary 14M20, 14M99.

(See §1 for the definition and some properties of rational normal scrolls.)

The purpose of this note is to give a short and direct proof of the Del Pezzo–Bertini theorem, valid in any characteristic. The proofs (Bertini (1907), Harris (1981), and Xambó (1981)) are all essentially similar: they treat first the cases of surfaces in general (which is also done in Nagata (1960) and Griffiths-Harris (1978)), and finally they reduce the case of arbitrary varieties to the case of surfaces, distinguishing according to whether the general 2-dimensional plane section of the given variety is a scroll or the Veronese surface. Instead, we base our discussion on the following general result (§2), which is useful in many other circumstances:

THEOREM 2. *Let $X \subset \mathbf{P}^r$ be a linearly normal variety, and $D \subset X$ a divisor. If D moves in a pencil $\{D_\lambda | \lambda \in \mathbf{P}^1\}$ of linearly equivalent divisors, then writing $\overline{D}_\lambda$ for the linear span of D_λ in $\mathbf{P}^r$, the variety*

$$S = \bigcup_\lambda \overline{D}_\lambda$$

is a rational normal scroll.

This allows us (in §3) to write an arbitrary variety X of minimal degree as a divisor on a scroll, and simple considerations of the geometry of scrolls then lead to the result.

1. Description of the varieties of minimal degree. We first explain some of the terms used in Theorems 1 and 2 above:

If $L \subset \mathbf{P}^{r+s+1}$ is a linear space of dimension s, $p_L : \mathbf{P}^{r+s+1} \to \mathbf{P}^r$ is the projection from L, and X is a variety in $\mathbf{P}^r$, then the cone over X is the closure of $p_L^{-1}X$. In equations, the cone is simply given by the same equations as X, written in the appropriate subset of the coordinates on $\mathbf{P}^{r+s+1}$. Thus a *cone in $\mathbf{P}^r$ over the Veronese surface $\mathbf{P}^2 \hookrightarrow \mathbf{P}^5$* may be defined as a variety given, with respect to suitable coordinates $x_0, \ldots, x_r$, by the (prime) ideal of 2×2 minors of the generic symmetric matrix:

$$\begin{pmatrix} X_0 & X_1 & X_2 \\ X_1 & X_3 & X_4 \\ X_2 & X_4 & X_5 \end{pmatrix}$$

(It is easy to see that a cone over any variety of minimal degree has minimal degree; our definition of rational normal scroll is such that the cone over a rational normal scroll is another rational normal scroll.)

Note that the Veronese surface contains no lines—indeed, any curve that lies on it must have even degree, as one sees by pulling back to $\mathbf{P}^2$—and thus a cone over the Veronese surface cannot contain a linear space of codimension 1. We shall see that this property separates the varieties of minimal degree which are cones over the Veronese surface from those that are scrolls.

We now describe rational normal scrolls in the terms necessary for Theorem 1. In our proof of the theorem we reduce rapidly to the case where X is a divisor on a scroll, and we shall describe these as well.

A rational normal scroll is a cone over a smooth linearly normal variety fibered over $\mathbf{P}^1$ by linear spaces; in particular, a rational normal scroll contains a pencil of linear spaces of codimension 1 (and these are the only linearly normal varieties with this property, as will follow from Proposition 2.1, below).

To be more explicit, think of $\mathbf{P}^r$ as the space of 1-quotients of k^{r+1}, so that a d-plane in $\mathbf{P}^r$ corresponds to a $d+1$-quotient of k^{r+1}. A variety $X \subset \mathbf{P}^r$ with a map $\pi \colon X \to \mathbf{P}^1$ whose fibers are d-planes is thus the projectivization of a rank $d+1$ vector bundle on $\mathbf{P}^1$ which is a quotient of

$$k^{r+1} \otimes_k \mathcal{O}_{\mathbf{P}^1} = \mathcal{O}_{\mathbf{P}^1}^{r+1}.$$

Slightly more generally, let

$$\mathcal{E} = \bigoplus_0^d \mathcal{O}_{\mathbf{P}^1}(a_i)$$

be a vector bundle on $\mathbf{P}^1$, and assume

$$0 \leq a_0 \leq \cdots \leq a_d, \quad \text{with } a_d > 0,$$

so that $\mathcal{E}$ is generated by $\sum a_i + d + 1$ global sections. Write $\mathbf{P}(\mathcal{E})$, or alternately $\mathbf{P}(a_0, \ldots, a_d)$, for the projectivized vector bundle

$$\mathbf{P}(\mathcal{E}) = \operatorname{Proj} \operatorname{Sym} \mathcal{E} \xrightarrow{\pi} \mathbf{P}^1$$

(whose points over $\lambda \in \mathbf{P}^1$ are quotients $\mathcal{E}_\lambda \to k(\lambda)$), and let $\mathcal{O}_{\mathbf{P}(\mathcal{E})}(1)$ be the tautological line bundle. Because the a_i are ≥ 0, $\mathcal{O}_{\mathbf{P}(\mathcal{E})}(1)$ is generated by its global sections (see the computation below) and defines a "tautological" map

$$\mathbf{P}(\mathcal{E}) \to \mathbf{P}^{\sum a_i + d}.$$

This map is birational because $a_d > 0$. We write $S(\mathcal{E})$ or $S(a_0, \ldots, a_d)$ for the image of this map, which, as we shall see, is a variety of dimension $d+1$ and degree $\sum a_i$, so that it is a variety of minimal degree. A *rational normal scroll* is simply one of the varieties $S(\mathcal{E})$. Note that $\mathcal{O}_{\mathbf{P}(\mathcal{E})}(1)$ induces $\mathcal{O}_{\mathbf{P}^d}(1)$ on each fiber $F \cong \mathbf{P}^d$ of $\mathbf{P}(\mathcal{E}) \to P^1$, so F is mapped isomorphically to a d-plane in $S(\mathcal{E})$.

The most familiar examples of rational normal scrolls are probably

(o) $\mathbf{P}^d$, which is $S(0, \ldots, 0, 1)$,

(i) the rational normal curve of degree a in $\mathbf{P}^a$, which is $S(a)$,

(ii) the cone over a plane conic, $S(0, 2) \subset \mathbf{P}^3$,

(iii) the nonsingular quadric in $\mathbf{P}^3$, $S(1, 1)$,

(iv) the projective plane blown up at one point, embedded as a surface of degree 3 in $\mathbf{P}^4$ by the series of conics in the plane passing through the point; this is $S(1, 2)$.

There is a pretty geometric description of $S(a_0, \ldots, a_d)$ from which the name "scroll" derives, and from which the equivalence of the two definitions above may be deduced:

The projection

$$\mathcal{E} = \bigoplus_0^d \mathcal{O}(a_i) \to \mathcal{O}(a_i)$$

defines a section $\mathbf{P}^1 \cong \mathbf{P}(\mathcal{O}(a_i)) \hookrightarrow \mathbf{P}(\mathcal{E})$, and

$$\mathcal{O}_{\mathbf{P}(\mathcal{E})}(1)|_{\mathbf{P}(\mathcal{O}(a_i))} = \mathcal{O}_{\mathbf{P}(\mathcal{O}(a_i))}(1) = \mathcal{O}_{\mathbf{P}^1}(a_i),$$

so this section is mapped to a rational normal curve of degree a_i in the $\mathbf{P}^{a_i} \subset \mathbf{P}^{\sum a_i + d}$ corresponding to the quotient $H^0(\mathcal{E}) \to H^0(\mathcal{O}_{\mathbf{P}^1}(a_i))$. (Of course if $a_i = 0$, the "rational normal curve of degree a_i" is a point $\subset \mathbf{P}^0$!) Thus we may construct the rational normal scroll $S(a_0, \dots, a_d) \subset \mathbf{P}^{\sum a_i + d}$ by considering the parametrized rational normal curves

$$\mathbf{P}^1 \overset{\phi_i}{\to} C_{a_i} \subset \mathbf{P}^{a_i} \subset \mathbf{P}^{\sum a_i + d}$$

corresponding to the decomposition

$$k^{\sum a_i + d + 1} = \bigoplus_0^d k^{a_i + 1},$$

and letting $S(a_0, \dots, a_d)$ be the union over $\lambda \subset \mathbf{P}^1$ of the d-planes spanned by $\phi_0(\lambda), \dots, \phi_d(\lambda)$. In particular, we see that the cone in $\mathbf{P}^{\sum a_i + d + s}$ over $S(a_0, \dots, a_d)$ is

$$S(\underbrace{0, \dots, 0}_{s}, a_0, \dots, a_d).$$

Also, $S(a_0, \dots, a_d)$ is nonsingular iff $(a_0, \dots, a_d) = (0, \dots, 0, 1)$ or $a_i > 0$ for all i.

We note that this description is convenient for giving the homogeneous ideal of $S(a_0, \dots, a_d)$. As is well known, the homogeneous ideal of a rational normal curve $S(a) \subset \mathbf{P}^a$ may be written as the ideal of 2×2 minors

$$\det_2 \begin{pmatrix} X_0, X_1, \dots, X_{a-1} \\ X_1, X_2, \dots, X_a \end{pmatrix},$$

and this expression gives the parametrization sending $(s, t) \in \mathbf{P}^1$ to the point of $\mathbf{P}^a$ where the linear forms

$$sX_0 + tX_1, \dots, sX_{a-1} + tX_a$$

all vanish. (This is s times the first row of the given matrix plus t times the second row.) It follows at once that $S(a_0, \dots, a_d)$ is at least set-theoretically the locus where the minors of a matrix of the form

$$\begin{pmatrix} X_{0,0} X_{0,1}, \dots, X_{0,a_0-1} & | & X_{1,0}, \dots, X_{1,a_1-1} & | & \dots X_{d,a_d-1} \\ & | & & | & \dots \\ X_{0,1}, X_{0,2}, \dots, X_{0,a_0} & | & X_{1,1}, \dots, X_{1,a_1} & | & \dots X_{d,a_d} \end{pmatrix}$$

all vanish. That these minors generate the whole homogeneous ideal follows easily as in the proof of Lemma 2.1 below.

The divisor class group of a projectivized vector bundle $\mathbf{P}(\mathcal{E})$ over $\mathbf{P}^1$ is easy to describe (Hartshorne (1977), Chapter II, exc. 7.9): Writing H for a divisor in the class determined by $\mathcal{O}_{\mathbf{P}(\mathcal{E})}(1)$, and F for the fiber of $\mathbf{P}(\mathcal{E}) \to \mathbf{P}^1$, the divisor class group may be written (confusing divisors and their classes systematically)

$$\mathbf{Z}H + \mathbf{Z}F.$$

Moreover, the chow ring is given by

$$\mathbf{Z}[F, H] / \left(F^2, H^{d+2}, H^{d+1}F, H^{d+1} - \left(\sum a_i\right) H^d F \right).$$

We shall only need a numerical part of this, giving the degree of a scroll:

$$\text{degree } S(a_0, \ldots, a_d) = H^{d+1} = \sum_0^d a_i.$$

The simplest way to understand this is perhaps from the geometric description given above: In $\mathbf{P}^{\sum_0^d a_i + d}$ we may take a hyperplane containing the natural copy of $\mathbf{P}^{\sum_1^d a_i + d - 1}$ and meeting $C_{a_0} \subset \mathbf{P}^{a_0}$ transversely. The hyperplane section is then the union of $S(a_1, \ldots, a_d)$ with a_0 copies of F (which is embedded as a d-plane).

It is also easy to compute the cohomology of the line bundles on $\mathbf{P}(\mathcal{E})$. In particular, the tautological map

$$\pi^*\mathcal{E} \to \mathcal{O}_{\mathbf{P}(\mathcal{E})}(1)$$

induces for any integer a a map

$$\text{Sym}_a \, \mathcal{E} = \pi_* \text{Sym}_a \, \pi^*\mathcal{E} \to \pi_* \mathcal{O}_{\mathbf{P}\mathcal{E}}(a)$$

and thus for every a, b a map

$$\mathcal{O}_{\mathbf{P}^1}(b) \otimes \text{Sym}_a \, \mathcal{E} \to \mathcal{O}_{\mathbf{P}^1}(b) \otimes \pi_* \mathcal{O}_{\mathbf{P}\mathcal{E}}(a) \cong \pi_* (\pi^* \mathcal{O}_{\mathbf{P}^1}(b) \otimes \mathcal{O}_{\mathbf{P}\mathcal{E}}(a)),$$

which is an isomorphism, as one easily checks locally. Since π is surjective, π_* induces an isomorphism on global sections, and we see that an element

$$\sigma \in H^0(\pi^* \mathcal{O}_{\mathbf{P}^1}(b) \otimes \mathcal{O}_{\mathbf{P}\mathcal{E}}(a))$$

may be represented as an element of

$$H^0(\mathcal{O}_{\mathbf{P}^1}(b) \otimes \text{Sym}_a \, \mathcal{E}) = H^0\left(\mathcal{O}_{\mathbf{P}^1}(b) \otimes \sum_{|I|=a} \mathcal{O}_{\mathbf{P}^1}\left(\sum_{i \in I} a_i\right)\right)$$

$$= \sum_{|I|=a} H^0\left(\mathcal{O}_{\mathbf{P}}\left(b + \sum_{i \in I} a_i\right)\right),$$

where the notation $\sum_{|I|=a}$ indicates summation over all collections I consisting of a elements (with repetitions) from $\{0, \ldots, d\}$.

From this we may derive a useful representation of divisors in $\mathbf{P}\mathcal{E}$, generalizing the idea of "bihomogeneous forms" in the case of $\mathbf{P}(\mathcal{O}_{\mathbf{P}^1}^{d+1}) = \mathbf{P}^1 \times \mathbf{P}^d$. If we let

$$x_i \in H^0(\pi^* \mathcal{O}_{\mathbf{P}^1}(-a_i) \otimes \mathcal{O}_{\mathbf{P}\mathcal{E}}(1)) = H^0 \mathcal{E}(-a_i)$$

be an element corresponding to a generator of the ith summand

$$\mathcal{O}_{\mathbf{P}^1}(a_i - a_i) = \mathcal{O}_{\mathbf{P}^1} \subset \mathcal{E}(-a_i),$$

and write

$$x^I := \prod_{i \in I} x_i \in H^0 \left[(\mathrm{Sym}_a \, \mathcal{E}) \left(-\sum_{i \in I} a_i \right) \right]$$

$$= H^0 \left(\pi^* \mathcal{O}_{\mathbf{P}^1} \left(-\sum_{i \in I} a_i \right) \otimes \mathcal{O}_{\mathbf{P}\mathcal{E}}(a) \right)$$

for the product, then we may represent σ conveniently as a "polynomial":

$$\sigma = \sum_{|I|=a} \alpha_I(s,t) x^I,$$

where s, t are homogeneous coordinates on $\mathbf{P}^1$ and where $\alpha_I(s,t)$ is a homogeneous form of degree

$$\deg \alpha_I(s,t) = b + \sum_{i \in |I|} a_i.$$

This representation is convenient because the "variables"

$$x_i \in H^0(\pi^* \mathcal{O}_{\mathbf{P}^1}(-a_i) \otimes \mathcal{O}_{\mathbf{P}\mathcal{E}}(1))$$

restrict to a basis of the linear forms on each fiber of $\mathbf{P}\mathcal{E} \to \mathbf{P}^1$, and the divisor D of σ meets the $\mathbf{P}^d \cong F_{(u,v)}$ over $(u,v) \in \mathbf{P}^1$ in the hypersurface with equation $\sum_{|I|=a} \alpha_I(u,v) x^I$.

In practice, we wish to use this idea on a Weil divisor X of a scroll $S(\mathcal{E})$. Since $S(\mathcal{E})$ is normal and $\mathbf{P}\mathcal{E} \to S\mathcal{E}$ is birational, we may do this by defining $\tilde{X} \subset \mathbf{P}\mathcal{E}$ to be the "strict transform" of X—that is, for an irreducible subvariety X of codimension 1, $\tilde{X}$ is the closure of the image in $\mathbf{P}\mathcal{E}$ of the complement, in X, of the fundamental locus of the inverse rational map, $S(\mathcal{E}) \to \mathbf{P}\mathcal{E}$. Then $\tilde{X}$ occupies a well-defined divisor class on $\mathbf{P}\mathcal{E}$, and we may apply the above technique to it.

2. Rational normal scrolls in the wild. The proof of Theorem 2 rests on a technique of constructing scrolls from their determinantal equations, as follows:

We say that a map of k-vector spaces

$$\phi : U \otimes V \to W$$

is *nondegenerate* if $\phi(u \otimes v) \neq 0$ whenever $u, v \neq 0$, or equivalently if each map $\phi_u : u \otimes V \to W$ is a monomorphism. The typical example, for our purposes, comes from a (reduced, irreducible) variety X and a pair of line bundles $\mathcal{L}$, $\mathcal{M}$; if $U = H^0(\mathcal{L})$, $V = H^(\mathcal{M})$, and $W = H^0(\mathcal{L} \otimes \mathcal{M})$, then the multiplication map is obviously nondegenerate in the above sense. In our application, X will be embedded linearly normally in $\mathbf{P}^r$ by $\mathcal{L} \otimes \mathcal{M}$, so we may identify $H^0(\mathcal{L} \otimes \mathcal{M})$ with $H^0(\mathcal{O}_{\mathbf{P}^r}(1))$.

In general, given any map

$$k^\gamma \otimes k^\delta \to H^0 \mathcal{O}_{\mathbf{P}^r}(1),$$

we define an associated map of sheaves

$$A_\phi : \mathcal{O}_{\mathbf{P}^r}^\delta(-1) \to \mathcal{O}_{\mathbf{P}^r}^\gamma.$$

by twisting the obvious map

$$k^\delta \otimes \mathcal{O}_{\mathbf{P}^r} \to k^{\gamma^*} \otimes \mathcal{O}_{\mathbf{P}^r}(1)$$

by $\mathcal{O}_{\mathbf{P}^r}(-1)$. Taking $\gamma = 2$, we have

LEMMA 2.1. *If $\phi: k^2 \otimes k^\delta \to H^0 \mathcal{O}_{\mathbf{P}^r}(1)$ is a nondegenerate pairing, then the ideal of 2×2 minors $\det_2 A_\phi$ is prime, and $V(\det_2 A_\phi)$ is a rational normal scroll of degree δ.*

PROOF. If the image of ϕ is a proper subspace of $H^0 \mathcal{O}_{\mathbf{P}^r}(1)$, then $V(\det_2 A_\phi)$ is a cone. Since the cone over a scroll is a scroll, we may by reducing r assume that ϕ is an epimorphism, so that the rank of A_ϕ never drops to 0 on $\mathbf{P}^r$. It follows that $\mathcal{L} = \operatorname{Coker} A_\phi$ is a line bundle on $S = V(\det_2 A_\phi)$, generated by the image of $V = k^{2^*}$. The linear series $(\mathcal{L}, V)$ defines a map $\pi: S \to \mathbf{P}^1$. If $(s, t) \in \mathbf{P}^1$, then the fiber F of π over (s, t) is the scheme defined by the vanishing of the composite map

$$\mathcal{O}_{\mathbf{P}^r}^\delta(-1) \to \mathcal{O}_{\mathbf{P}^r}^2 \overset{(s,t)}{\to} \mathcal{O}_{\mathbf{P}^r};$$

and this scheme is, by our nondegeneracy hypothesis, given by the vanishing of δ linearly independent linear forms, so F is a plane of codimension δ. By the general formula for the maximum codimension of (any component of) a determinantal variety we have $\operatorname{codim} S \le \delta - 1$, so the map $S \to \mathbf{P}^1$ is onto, and since the fibers are smooth and irreducible, and the map is proper, S is smooth and irreducible of codimension $\delta - 1$.

Since $\det_2 A_\phi$ thus has height $\delta - 1$ in the homogeneous coordinate ring of $\mathbf{P}^r$, it is perfect, and in particular unmixed (Arbarello et al. (1984), Chapter II, 4.1; note that the characteristic 0 hypothesis there is irrelevant). Thus $\det_2 A_\phi$ is the entire homogeneous ideal of S, and since $\det_2 A_\phi$ is perfect, S is arithmetically Cohen-Macaulay, so in particular S is linearly normal.

The fibers of π, being linear spaces in $\mathbf{P}^r$, correspond to quotients of k^{r+1}, and this defines a vector bundle on $\mathbf{P}^1$ of rank $r - \delta + 1$ such that $S \to \mathbf{P}^1$ is the associated projective space bundle; thus S is a rational normal scroll as claimed. $\square$

REMARK. Using the same ideas, one sees that the height of $\det_2(A_\phi)$ is $\delta - 1$ iff the rank of ϕ_u never drops by more than 1; then $X = V(\det_2 A_\phi)$ is a "crown", that is, the union of a scroll of codimension $\delta - 1$ and some linear spaces of codimension $\delta - 1$ which intersect the scroll along linear spaces of codimension δ (fibers of π)—see Xambò (1981).

With this result in hand, it is easy to complete the proof of Theorem 2:

PROOF OF THEOREM 2. Let $k^2 \cong V \subset H^0 \mathcal{O}_X(D)$ be the vector space of sections corresponding to the pencil D_λ, and let H be the hyperplane section of X. The natural multiplication map

$$V \otimes H^0 \mathcal{O}_X(H - D) \to H^0 \mathcal{O}_X(H) = H^0 \mathcal{O}_{\mathbf{P}^r}(1)$$

is nondegenerate, and thus gives rise to a scroll S containing all the D_λ, and thus X. The linear space $\overline{D}_\lambda$ is the intersection of all the hyperplanes containing

D_λ, which correspond to elements of $H^0 \mathcal{O}_X(H - D)$, so $\overline{D}_\lambda$ is the fiber over λ of $S \to \mathbf{P}^1$, as desired. $\square$

EXAMPLES. (i) Let C be a hyperelliptic (or elliptic) curve, $C \subset \mathbf{P}^r$ an embedding by a complete series of degree d. C is a divisor on the variety S which is the union of the secants corresponding to the $\mathfrak{g}_2^1$ on C (or, if C is elliptic, any $\mathfrak{g}_2^1$ on C). This variety is a rational normal scroll $S(\mathcal{E})$ and $\tilde{C} \sim 2H + (d - 2r + 2)F$ on $\mathbf{P}(\mathcal{E})$. More generally, a linearly normal curve $C \subset \mathbf{P}^r$ which possesses a $\mathfrak{g}_d^1$ lies on a scroll of dimension $\leq d$; if $C \subset \mathbf{P}^r$ is the canonical embedding, then this scroll is of dimension $\leq d - 1$, so in particular the canonical image of any trigonal curve is a divisor on a 2-dimensional scroll $S(\mathcal{E})$, and $\tilde{C} \sim 3H + (4-g)F$ on $\mathbf{P}(\mathcal{E})$. See Schreyer (1986) for a study of canonical curves using this idea.

(ii) A $K3$ surface, embedded linearly normally in any projective space, is a divisor on a 3-dimensional scroll if it contains an elliptic cubic (which then moves in a nontrivial linear series). See for example Saint-Donat (1974).

3. The classification theorem. Before giving our proof of the Del Pezzo–Bertini Theorem, we record three elementary observations about projections:

(1) If X is a variety of minimal degree, then X is linearly normal. (*Proof*: If X were the isomorphic projection of a nondegenerate variety X' in $\mathbf{P}^{r+1}$, then X' would have degree less than that allowed by Proposition 0.)

(2) If $X \subset \mathbf{P}^r$ is a variety of minimal degree and $p \in X$, then the projection $\pi_p X \subset \mathbf{P}^{r-1}$ is a variety of minimal degree, the map $X - p \to \pi_p X$ is separable, and if p is singular then X is a cone with vertex p. (*Proof*: Indeed, $\pi_p X$ is obviously nondegenerate. If X is a cone with vertex p, the result is obvious. Else $\dim \pi_p X = \dim X$ but $\deg \pi_p X \leq \deg X - 1$. The inequality must actually be an equality by Proposition 0, which shows in particular that p is a nonsingular point, and $\pi_p \colon X - p \to \pi_p X$ is birational.)

(3) If $p \in X \subset \mathbf{P}^r$ is any point on any variety, E_X the exceptional fiber of the blow-up of p in X, and $E_{\mathbf{P}^r} \cong \mathbf{P}^{r-1}$ the exceptional fiber of the blow-up of $\mathbf{P}^r$ at p, then E_X is naturally embedded in $E_{\mathbf{P}^r}$, which is mapped isomorphically to $\mathbf{P}^{r-1}$ by the map induced by π_p. Thus $E_X \subset \pi_p(X) \subset \mathbf{P}^{r-1}$. In particular, if p is a nonsingular point on X, so that E_X is a linear subspace of $\mathbf{P}^{r-1}$, then the "image of p" under $\pi_p \colon X \to \pi_p(X) \subset \mathbf{P}^{r-1}$ is a linear subspace of $\mathbf{P}^{r-1}$ which is a divisor on $\pi_p(X)$. More naively, this is the image of the tangent plane to X at p.

In view of observation (3) it will be useful to begin with the following result, which "recognizes" scrolls:

PROPOSITION 3.1. *If $X \subset \mathbf{P}^r$ is a variety of minimal degree, and X contains a linear subspace of $\mathbf{P}^r$ as a subspace of codimension 1, then X is a scroll.*

PROOF. By Proposition 2.1 it suffices to show that X contains a pencil of linear divisors, though the given subspace itself may not move.

Let $F \subset X$ be the given linear subspace. We may assume (by projecting, if necessary) that X is smooth along F. Let $H \subset \mathbf{P}^r$ be a general hyperplane containing F, and let $S = H \cap X - F$. Let π_F be projection from F. We distinguish two cases:

Case 1. $\dim \pi_F(X) \geq 2$. By Bertini's Theorem and observation (2) above, S is then a reduced and irreducible variety, of degree and dimension one less than that of X. Thus by Proposition 0, S is degenerate in H, so $F = H \cdot X - S$ moves in (at least) a pencil of linear spaces, and we are done.

Case 2. $\dim \pi_F(X) = 1$. By observation (1), $\pi_F(X)$ is a curve of minimal degree, say of degree s in $\mathbf{P}^s$. Projecting $\pi_F(X)$ from $s - 1$ general points on it gives a birational map to $\mathbf{P}^1$, so $\pi_F(X) \cong \mathbf{P}^1$. Further, the cone on S with vertex p is a union of s planes, the spans of F with the points of a general hyperplane section of $\pi_F(X)$, so S has s components. But $s = r - \dim F - 1 = \operatorname{codim} X = \deg X - 1 = \deg S$, so S is the union of s planes, and these are linearly equivalent to each other since the points of $\pi_F(X)$ are. Thus a component of S is a linear space moving in a pencil as desired. $\square$

PROOF OF THEOREM 1. Let $X \subset \mathbf{P}^r$ be a variety of minimal degree. We may assume that the codimension c of X is ≥ 2 and that X is not a cone. By Proposition 3.1 we may as well also assume that X contains no linear space of codimension 1, so that in particular the dimension d of X is ≥ 2, and we must prove that under these hypotheses X is the Veronese surface $\mathbf{P}^2 \subset \mathbf{P}^5$. In fact, it suffices to prove that $X \cong \mathbf{P}^2$; for the embedding of $\mathbf{P}^2$ by the complete series of curves of degree d gives a surface of degree d^2 and codimension

$$\binom{2 + d}{2} - 3,$$

which is $< d^2 - 1$ for $d \geq 3$.

Let $p \in X$ be any point. By observation (3) and Proposition 3.1, $\pi_p(X)$ is a scroll, so the cone $S \subset \mathbf{P}^r$ with vertex p over $\pi_p(X)$ (or over X) is a scroll, say $S = S(\mathcal{E})$, with $\mathcal{E} = \bigoplus_0^d \mathcal{O}_{\mathbf{P}^1}(a_i)$ and $0 \leq a_0 \leq \cdots \leq a_d$. X is a divisor on S.

Consider the strict transform $\tilde{X} \subset \mathbf{P}(\mathcal{E})$ of X under the desingularization $\mathbf{P}(\mathcal{E}) \to S(\mathcal{E}) = S$, and let its divisor class be $aH - bF$. We will prove under the hypotheses above that $a = 2$ and X is a surface. (Along the way we will see numerically that $b = 4$, $(a_0, a_1, a_2) = (0, 1, 2)$, so $c = 3$ and $X \subset \mathbf{P}^5$ as befits the Veronese, but we will not use this directly.)

First, because the degree $c + 1$ of X is 1 more than that of S, and on the other hand is $H^{d-1} \cdot (aH - bF)$, we get $b = (a - 1)c - 1$.

To bound a, first note that X must meet every fiber of $\mathbf{P}\mathcal{E} \to \mathbf{P}^1$, so $aH - bF|_F = aH|_F > 0$, and $a \geq 1$. If a were 1, then $\tilde{X}$ would meet each fiber F in a linear space of dimension $d - 1$. Since each fiber F is mapped isomorphically to a d-plane in $\mathbf{P}^r$ under $\mathbf{P}(\mathcal{E}) \to S(\mathcal{E})$, X would contain linear spaces of dimension $d - 1$, contrary to our hypothesis. Thus $a \geq 2$.

As in §2, $\tilde{X}$ may be represented by an equation $g = 0$ with g of the form:

$$g = \sum_{|I| = a} \alpha_I(s, t) x^I,$$

with

$$\deg \alpha_I = \left(\sum_{i \in I} a_i \right) - b = \sum_{i \in I} a_i - (a-1)c + 1.$$

If the variable x_0 did not occur in g, then $\tilde{X}$ would meet each fiber F in a cone over the preimage of p, and X itself would be a cone contrary to hypothesis. But for x_0 to occur we must have

$$0 \le \deg \alpha_{0,d,\dots,d} = a_0 + (a-1)a_d - (a-1)c + 1.$$

Since S is a cone we have $a_0 = 0$, and we derive

$$(*) \qquad\qquad\qquad a_d \ge c - 1/(a-1).$$

If x_d occurred in every nonzero term of g, then for every fiber F, $\tilde{X} \cap F$ would contain the $(d-1)$-plane $x_d = 0$, and again X would contain a $(d-1)$-plane, contradicting our hypotheses. Thus

$$(**) \qquad\qquad 0 \le \deg \alpha_{d-1,d-1,\dots,d-1} = aa_{d-1} - (a-1)c + 1.$$

Now if $a \ge 3$, then $a_d = c$ by $(*)$; but $c = \deg X - 1 = \deg S = \sum_{i=0}^{d} a_i$, so this implies $a_{d-1} = 0$, and $(**)$ gives a contradiction. Thus $a = 2$ as claimed, and $a_d \le c - 1$. Condition $(*)$ now gives $a_d = c - 1$, so $a_{d-1} = 1$ and $a_0 = \cdots = a_{d-2} = 0$. Applying $(**)$ again we get $a_d = 1$ or $a_d = 2$.

In the first case $(a_0, \dots, a_d) = (0, \dots, 0, 1, 1)$, so S is a cone over $\mathbf{P}^1 \times \mathbf{P}^1 \subset \mathbf{P}^3$. A suitable hypersurface section of S will consist of the union of two planes F_1 and F_2, the cones over the rulings of $\mathbf{P}^1 \times \mathbf{P}^1$. Since each of these rulings sweeps out all of $\mathbf{P}^1 \times \mathbf{P}^1$, X must meet each of F_1 and F_2 in codimension 1. Because $c = 2$ we have $\deg X = 3$, so either $X \cap F_1$ or $X \cap F_2$ must be a linear space, contradicting our assumption on X.

We thus see that $a = 2$, $c = 3$, and $(a_0, \dots, a_d) = (0, \dots, 0, 1, 2)$. Under these circumstances the sum of the terms of g involving $x_0, \dots, x_{d-2}$ may be written

$$\left(\sum_0^{d-2} \alpha_{i,d} x_i \right) x_d,$$

with $\alpha_{i,d}$ constant. Thus if $d \ge 3$ the locus $g = 0$ in each fiber F is a cone with vertex the $(d-3)$-dimensional linear space given by

$$x_d = x_{d-1} = \sum_0^{d-2} \alpha_{i,d} x_i = 0.$$

Of course S is itself a cone with $(d-2)$-dimensional vertex L, say. The $(d-2)$-dimensional subspaces of the fibers F given by $x_d = x_{d-1} = 0$ are all mapped isomorphically to L under $\mathbf{P}(\mathcal{E}) \to S$, and the restrictions of the coordinates $x_0, \dots, x_{d-2}$ are all identified, and become coordinates on L. Thus X meets the image of each fiber in a cone with vertex given in L by $\sum_0^{d-2} \alpha_{i,d} x_i = 0$, so X is a cone, contradicting our assumption. This shows $d = 2$.

We have now shown that $a = 2$ and X is a surface. In this case, for every fiber $F \cong \mathbf{P}^2$ of $\mathbf{P}(\mathcal{E})$, $F \cap \tilde{X}$ is a conic, necessarily nonsingular since else X would contain a line. Thus $\tilde{X}$ is a rational ruled surface. But the preimage in X of p is a line, so $\tilde{X}$ is the blow-up of X at p, and is not a minimal surface. This is only possible if $\tilde{X} \cong \mathbf{P}(\mathcal{O}_{\mathbf{P}^1} \oplus \mathcal{O}_{\mathbf{P}^1}(1))$ and $X \cong \mathbf{P}^2$, as required. $\square$

REFERENCES

1. E. Arbarello, M. Cornalba, P. A. Griffiths, and J. Harris, *Geometry of algebraic curves*, vol. I, Springer-Verlag, New York, 1984.

2. E. Bertini, *Introduzione alla geometria proiettiva degli iperspazi*, Enrico Spoerri, Pisa, 1907.

3. Del Pezzo, *Sulle superficie di ordine n immerse nello spazio di n + 1 dimensioni*, Rend. Circ. Mat. Palermo **1** 1886.

4. P. Griffiths and J. Harris, *Principles of algebraic geometry*, Wiley, New York, 1978.

5. J. Harris, *A bound on the geometric genus of projective varieties*, Ann. Scuola Norm. Sup. Pisa Cl. Sci. (4) **8** (1981), 35–68.

6. R. Hartshorne, *Algebraic geometry*, Springer-Verlag, New York, 1977.

7. M. Nagata, *On rational surfaces. I: Irreducible curves of arithmetic genus 0 or 1*, Mem. Coll. Sci. Univ. Kyoto, Ser. A, **32** (1960), 351–370.

8. B. Saint-Donat, *Projective models of K-3 surfaces*, Amer. J. Math. **96** (1974), 602–639.

9. F. O. Schreyer, *Syzygies of curves with special pencils*, Thesis, Brandeis University (1983); Math. Ann. **275** (1986), 105–137.

10. S. Xambò, *On projective varieties of minimal degree*, Collect. Math. **32** (1981), 149–163.

BRANDEIS UNIVERSITY

BROWN UNIVERSITY

Proceedings of Symposia in Pure Mathematics
Volume **46** (1987)

On the Topology of Algebraic Varieties

WILLIAM FULTON

Topology and algebraic geometry have been nourishing each other for more than a century. Even in periods with little interaction—such as the 1930s when topologists and algebraic geometers independently discovered characteristic classes for varieties—their growths have been intertwined. The two subjects have probably never been more active together than now.

To survey recent progress in this vast area would be a large task. These topics would certainly be included: the role of *topological invariants* in the classification of algebraic varieties, and the homeomorphism and diffeomorphism types of various classes of varieties; *Hodge theory*, including mixed Hodge structures, degeneration of Hodge structures, and rational homotopy type; *monodromy* and differential equations; *singularities*; *intersection homology*. Fortunately, many of these topics have been considered in seminars of this institute, while the summer institute of 1981 was entirely devoted to singularities.

These lectures have a much more modest aim: to discuss some recent work related to the homotopy groups π_i of algebraic varieties, with particular attention to π_0 (connectivity) and π_1 (fundamental groups):

I. *Introduction.* An attempt to unify classical theorems of Bézout, Bertini, Lefschetz, Zariski, and Barth leads to a general connectedness principle, which provides the theme of these lectures.

II. *Morse theory on analytic spaces.* A bit of the history, and an introduction to the work of Goresky and MacPherson [**54**].

III. *Lefschetz and connectedness theorems.* From this Morse theory one proves a version of Lefschetz's theorem, from which a general connectedness theorem follows.

IV. π_0 *and varieties of small codimension.* Some remarkable applications by Zak and others to problems in classical projective geometry.

1980 *Mathematics Subject Classification* (1985 *Revision*). Primary 14F45; Secondary 14F35, 14E20, 14M07.

Research partially supported by National Science Foundation Grant DMS 84-02209.

V. π_1 *and branched coverings.* Applications to fundamental groups and branched coverings of projective space, with an introduction to the work of Nori [**106**].

Some of this work has been discussed in other survey lectures: [**40, 88, 37, 35, 42, 36, 6, 7, 60, 65**]. One purpose of these talks is to bring the survey [**40**] up to date. The bibliography includes many papers and recent preprints with related and overlapping results and other points of view, as well as those papers explicitly referred to in the lectures.

Notation. V^r denotes an r-dimensional irreducible complex projective variety; $\mathbf{P}^n$ is complex projective n-space; D^r is a closed ball in real r-space, with ∂D^r its bounding $(r-1)$-sphere.

Base points are omitted from fundamental groups and homotopy groups of connected spaces; all assertions are independent of base point chosen.

Chapter I
Introduction

1. Five classical theorems. Our first goal is to compare five classical (pre-Arcata!) results, looking for a common thread:

(1) "Bézout": *If A^a and B^b are subvarieties of $\mathbf{P}^m$, with $a + b \geq m$, then $A \cap B \neq \varnothing$.*

(2) "Bertini": *If $f: V^n \to \mathbf{P}^m$ is a morphism, with $\dim f(V) \geq 2$, and $L^{m-1} \subset \mathbf{P}^m$ is a hyperplane, then $f^{-1}(L)$ is connected.*

(3) "Lefschetz": *If V^n is a subvariety of $\mathbf{P}^m$, and L^{m-d} is a linear subspace of $\mathbf{P}^m$ which contains the singular locus of V, then*

$$\pi_i(V, V \cap L) = 0 \quad for \ i \leq n - d.$$

Equivalently,

$$\pi_i(V \cap L) \to \pi_i(V) \quad \begin{cases} \cong & i < n - d, \\ \twoheadrightarrow & i = n - d. \end{cases}$$

(4) "Zariski": *If $H \subset \mathbf{P}^n$ is any hypersurface, and L^{n-d} is a generic linear subspace of $\mathbf{P}^n$, then*

$$\pi_i(L - L \cap H) \to \pi_i(\mathbf{P}^n - H) \quad \begin{cases} \cong & i < n - d, \\ \twoheadrightarrow & i = n - d. \end{cases}$$

(5) "Barth": *If V^n is a nonsingular subvariety of $\mathbf{P}^m$, then*

$$\pi_i(\mathbf{P}^m, V) = 0 \quad for \ i \leq 2n - m + 1.$$

The five names are those usually associated with earliest special cases of these results, although the versions stated were all known at least ten years ago. The truly classical theorems were all about *generic* intersections of hyperplane sections. For example, the usual classical version of Bézout's theorem gives the degree of $A \cap B$ (as a cycle) when A and B meet properly. Bertini's theorem stated that for generic L, $f^{-1}(L)$ is irreducible; the assertion (2) follows by applying the Enriques-Zariski principle that limits of connected projective varieties remain connected.

In Lefschetz's version [90] of (3), V was nonsingular, L generic, and the conclusion was only for homology groups. The modern version of (3) came from the first use of Morse theory in algebraic geometry by Thom, Bott [9], Andreotti and Frankel [1], cf. Milnor [98].

Zariski's original theorem stated (3) for the fundamental group only. Hamm and Lê [62] gave a modern proof with the generalization to higher homotopy.

Barth's original theorem [3] was also only for homology groups; it was extended to homotopy groups by Larsen [85, 6]. Note that for applications of (5), we have

$$\pi_i(\mathbf{P}^m) = \begin{cases} \mathbf{Z}, & i = 0, 2, \\ 0, & i = 1, \ 3 \le i \le 2m \end{cases}$$

(as follows from the fibration $\mathbf{C}^{m+1} - \{0\} \to \mathbf{P}^m$), so the knowledge of $\pi_i(\mathbf{P}^m, V)$ implies a good knowledge of the homotopy and homology of V.

Clearly (1)–(5) are related. Looking for a common unification forces us to state better theorems. For example, (1) indicates that the theorem should allow arbitrary *singular* varieties and intersection with *nonlinear* subvarieties of $\mathbf{P}^m$; note that the conclusion can be stated in the form

$$\pi_0(A \cap B) \twoheadrightarrow \pi_0(A \times B).$$

Similarly (2) indicates that one should allow *mappings* to $\mathbf{P}^m$ as well as subvarieties; its conclusion is that

$$\pi_0(f^{-1}(L)) \xrightarrow{\cong} \pi_0(V).$$

Certainly (3) asks to include *higher homotopy* and homology in the story, and (4) indicates that there should be versions for *noncompact* varieties such as $\mathbf{P}^n - H$.

The relation of Barth's theorem to the others was for a time less evident, although (5) does follow from (3) when V is a complete intersection. We shall see how Barth's theorem can be regarded as the self-intersection case of a Lefschetz theorem.

It is also natural to look for an underlying reason for these theorems. They all have to do with varieties in (or mapping to) projective space, which was known to have an *ample* tangent bundle. For example, Frankel [30] showed that if A^a and B^b were submanifolds of an m-dimensional manifold with suitably positive tangent bundle, and $a + b \ge m$, then A must meet B.

Remarkably Mori [103] and Siu and Yau [122] proved the conjecture of Frankel and Hartshorne that $\mathbf{P}^m$ is the *only* projective manifold with ample (positive) tangent bundle.

2. The connectedness principle. The general principle which has come from musings such as those in §1 can be stated as the following theme:

Given a "suitably positive" imbedding $M \hookrightarrow P$ of codimension d and a proper morphism $f: X^n \to P$,

$$
\begin{array}{ccc}
f^{-1}(M) & \hookrightarrow & X \\
\downarrow & & \downarrow f \\
M & \hookrightarrow & P
\end{array}
$$

we should have

$$\pi_i(X, f^{-1}(M)) \xrightarrow{\cong} \pi_i(P, M), \qquad i \leq n - d - \text{"defect"}.$$

This defect should be measured by (a) lack of positivity of M in P; (b) singularities of X; (c) dimension of fibres of f. Usually $\pi_i(P, M)$ is zero (or simple) for small i, so the conclusion is that, as regards connectivity, $f^{-1}(M)$ must look like X. If the defect is zero, we deduce that

$$
\begin{aligned}
f^{-1}(M) &\neq \varnothing & &\text{if } n \geq d, \\
f^{-1}(M) \text{ is connected and} & \Bigg\} & &\text{if } n > d. \\
\pi_1(f^{-1}(M)) \twoheadrightarrow \pi_1(X) &
\end{aligned}
$$

If f is not proper, one expects the same to hold only after replacing M by an ε-neighborhood, cf. Deligne [20].

The most basic case is with $P = \mathbf{P}^m$, and $M = L^{m-d}$ a linear subspace. In this case the principle gives the theorems of Bertini and Lefschetz (setting $X = V$), and Zariski ($X = \mathbf{P}^n - H$), as stated in §1, (2)–(4).

The case which allows one to include all the classical theorems is with $P = \mathbf{P}^m \times \mathbf{P}^m$, and $M = \mathbf{P}^m$ diagonally imbedded in P. In this case $X = A \times B$, mapped by the product imbedding to P, yields the Bézout claim (1), while (2)–(4) are recovered by setting $X = V \times L$ for L a linear subspace. We shall see that the Barth theorem (5) follows by setting $X = V \times V$.

The idea of considering the diagonal imbedding of $\mathbf{P}^m$ in $\mathbf{P}^m \times \mathbf{P}^m$ in these terms is due to J. Hansen, who saw it as the key to the results of [38]. The extension to π_1, to noncompact varieties, and the conjectured extension to higher homotopy were given by Deligne [20].

Unfortunately there is not yet one general connectedness theorem which includes all the known theorems that follow this general principle. When $\mathbf{P}^n$ is replaced by other homogeneous manifolds such as a flag manifold, one can measure the defect of positivity of its tangent bundle, and one expects the principle to hold with this defect. For results in this direction see the papers of Barth [4], Faltings [29], Goldstein [48, 49], Hansen [66], Okonek [109], Peternell [113], and Sommese [124–129].

There are also some local theorems which imply some of the results on $\mathbf{P}^m$, cf. Faltings [27], Goresky and MacPherson [54], Hamm and Lê [64], Lê and Saito [89].

Most of the results stated here for π_0 and π_1 are valid, when suitably interpreted, over arbitrary algebraically closed fields, cf. [**38**, **40**].

Why should one expect this connectedness principle to be valid? In some cases one can define a Morse function which measures distance from M. Positivity should imply that all the Morse indices of this function are at least $n - d - 1$ (perhaps minus a defect). Then one can construct X from $f^{-1}(M)$ by adding only cells of dimension at least $n - d - 1$, which yields the required vanishing of relative homotopy groups. (In practice we proceed this way for special $M \hookrightarrow P$, and use some simple geometric constructions to deduce the theorem for more general $M \hookrightarrow P$.)

Chapter II
Morse Theory on Projective Varieties

1. Classical Morse theory. Like so much of our subject, the origins can be traced to Riemann: if $f\colon X \to \mathbf{P}^1$ is a meromorphic function on a smooth projective curve X, with n sheets and w simple branch points, the genus (so the topology) g of X is given by the formula $w = 2g + 2n - 2$.

Zeuthen and Segre gave an analogue for a smooth projective surface X: for a pencil of curves on X whose generic member is smooth of genus g, all meeting transversally at a base points, with d of the curves singular, each having one ordinary node, they saw that the *Zeuthen-Segre invariant* $I = d - 4g - a$ is independent of choice of pencil. In fact, $I + 4$ is the topological Euler characteristic of X. If $\tilde{X}$ is the blow-up of X at the a base points, the curves of the pencil are the fibres of a morphism $f\colon \tilde{X} \to \mathbf{P}^1$ (see Figure 1). The topology of $\tilde{X}$ is studied via the singular fibration f. This program was carried forward by Poincaré and particularly Lefschetz [**90**], who used such "Lefschetz pencils" of hypersurfaces to analyze n-dimensional varieties.

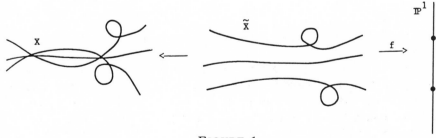

FIGURE 1

In Morse theory the topology of a real compact manifold M is studied via a suitably general C^∞ real-valued function $f\colon M \to \mathbf{R}$ (see Figure 2). Setting $M_v = f^{-1}(-\infty, v]$, the topology of M_v changes only when one crosses a critical value, i.e., the image of a point P with $df_P = 0$. Then for small ε we have

$$M_{v+\varepsilon} \underset{\text{homeo}}{\approx} M_{v-\varepsilon} \cup_B A;$$

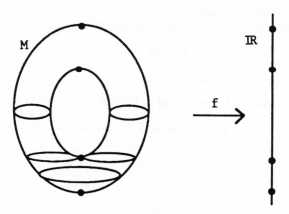

$$\textsc{Figure}\ 2$$

i.e., $M_{v+\varepsilon}$, is homeomorphic to the space obtained from $M_{v-\varepsilon}$ by gluing on the space A along the subspace B. Here

$$(A, B) = Morse\ Data = (D^\lambda \times D^{n-\lambda}, \partial D^\lambda \times D^{n-\lambda})\underset{\text{homotopy}}{\sim}(D^\lambda, \partial D^\lambda),$$

where $\lambda = \text{Index}_P(f)$ is the *Morse index*: at the critical point P, the *Hessian* H_f is a quadratic form on the tangent space $T_P M$ given in local coordinates by

$$H_f(v) = \sum_{i,j} \frac{\partial^2 f}{\partial x_i \partial x_j}(P)v_i v_j,$$

and λ is the number of negative terms in the diagonalization of H_f. (One assumes H_f is nondegenerate, and that v is the image of only one critical point.) It follows that

$$\pi_i(M_{v+\varepsilon}, M_{v-\varepsilon}) = 0 \quad \text{for } i < \lambda.$$

For this beautiful story we refer of course to Milnor [**98**].

2. Local complex application. It is a fundamental fact about complex varieties that their local topology is largely controlled by the number of equations needed to define them. From this point of view, a local complete intersection looks nonsingular. The proof is quite typical of the way (real) Morse theory is used to study complex varieties.

Let $X \subset \mathbf{C}^N$ be a complex algebraic set. Assume that 0 is an isolated singular point of X. Let $M = \partial B_\varepsilon(0)$ be the boundary of a small ball around 0 in $\mathbf{C}^N$. For small ε, M meets X transversally and $M \cap X$ is independent[1] of ε; $M \cap X$ is called the *real link* of X at 0.

---

[1] We skip over all such technical points, which are proved by methods developed in Milnor [**98**].

THEOREM (HAMM [**57, 58**]). *If X can be defined (set-theoretically) by r holomorphic equations in a neighborhood of 0 in $\mathbf{C}^N$, then*

$$\pi_i(M, M \cap X) = 0 \quad \text{for } i < N - r.$$

Thus if X is n-dimensional, and requires $(N-n)+q$ equations (q is a "defect"),

$$\pi_{i-1}(M \cap X) = \pi_i(M, M \cap X) = 0 \quad \text{for } i < n - q.$$

PROOF. Let $h_1, \ldots, h_r$ define X near 0, and define a real-valued function f on M by

$$f(z) = \sum_{j=1}^{r} |h_j(z)|^2 (1 - \|z\|^2).$$

Then $X \cap M = M_0 \subset M_\delta = f^{-1}(-\infty, \delta] \subset M$, and for δ very small compared to ε, the inclusion of M_0 in M_δ is a deformation retract. To show that $\pi_i(M, M_\delta) = 0$ for $i < N - r$, it suffices to show that the Morse indices of f at all critical points of M are at least $N - r$. (Actually f may not have nondegenerate critical points with distinct values, but a slight perturbation of f will, and the indices of the perturbation will not be lower than those of f.)

Given $P \in M$, consider the *Levi form* L_f, a Hermitian form on $T_p\mathbf{C}^N$ defined by

$$L_f(v) = \sum_{i,j} \frac{\partial^2 f}{\partial z_i \partial \bar{z}_j} v_i \bar{v}_j.$$

This is a complex analogue of the Hessian H_f, but better behaved with regard to restriction to subspaces and composition of f by log (see an appendix to [**54**, §II.4]). They are related by the identity

$$L_f(v) = \tfrac{1}{4}[H_f(v) + H_f(iv)].$$

LEMMA. *Let W be a complex subspace of T_PM on which L_f is negative definite. Then*

$$\text{Index}_P(f) \geq \dim_{\mathbf{C}}(W).$$

The proof is linear algebra, cf. Barth [**6**, p. 311].

One applies this lemma to the space

$$W = \{v \in T_PM | \partial f_P(v) = 0 \text{ and } \exists c \in \mathbf{C} \ni: (dh_j)_P(v) = ch_j(P) \text{ for } j = 1, \ldots, r\}.$$

Then $\dim_{\mathbf{C}} W \geq N - r$, and for $v \in W$, $v \neq 0$, one calculates (cf. Hamm [**58**, p. 175]) that

$$L_f(v) = -\sum_{j=1}^{r} |h_j(P)|^2 (|c|^2(1 - \varepsilon^2) + \|v\|^2) < 0,$$

which concludes the proof.

3. Morse theory on singular real spaces. To develop a Morse theory on a real (compact) analytic space X, imbedded in a real analytic manifold M, we choose a real analytic stratification of M into a disjoint union of locally closed submanifolds (the strata), so that X is a union of strata. Then for a dense open set of "generic" C^∞ functions $f: M \to \mathbf{R}$, we have (Pignoni [**115**]):

(i) For any stratum S, the restriction of f to S has only nondegenerate critical points.

(ii) If a stratum S is contained in the closure of another stratum S', and a sequence of points P_t in S' approaches $P \in S$ such that the tangent spaces $T_{P_t} S'$ approach a limiting space T, then $df_P(T) \neq 0$.

(iii) Distinct critical points have distinct values.

Such f will be called a *Morse function*. For example, if X is the cuspidal curve $y^2 = x^3$ in $M = \mathbf{R}^2$, with strata $\{0\}$, $X - \{0\}$, $M - X$, and f is projection to the y-axis, condition (ii) is not satisfied; the projection to the x-axis is a Morse function.

Let P be a point in a stratum S in X, critical for the restriction of the Morse function f to S, and set $v = f(P)$. Let k be the (real) dimension of S, and let λ be the Morse index of the restriction of f to S. Define *Tangential Morse Data* TMD by

$$\text{TMD} = (D^\lambda \times D^{k-\lambda}, \partial D^\lambda \times D^{k-\lambda}).$$

Let N be a normal slice to S at P, which is a submanifold of M of codimension k meeting S transversally at P. Let $B_\delta(P)$ be a small closed ball around P. The *Normal Morse Data* NMD is defined by

$$\text{NMD} = X \cap N \cap B_\delta(P) \cap (f^{-1}[-\varepsilon, \varepsilon], f^{-1}(-\varepsilon)),$$

where ε is a positive number chosen small in comparison with δ.

THEOREM (GORESKY-MACPHERSON [**51, 52, 54**]). *The product of tangential and normal Morse data gives Morse data for f :*

$$X_{v+\varepsilon} \underset{\text{homeo}}{\approx} X_{v-\varepsilon} \cup_B A,$$

where $(A, B) = \text{TMD} \times \text{NMD}$.

We refer to [**52, 54**] for the proof, which requires a good deal of stratification technique.

Consider the example shown in Figure 3 of a singular surface X in $\mathbf{R}^3$, which could be obtained by naïvely trying to turn a sphere inside out. We stratify X with a point stratum Q, a curve stratum S; the rest of X has three surface strata. If f is height, there are 5 critical points. At the higher critical point on S, we have $k = 1$, $\lambda = 1$, and

$$\text{TMD} = (\,\bullet\!\!-\!\!-\!\!-\!\!\bullet \ , \ \bullet \qquad \bullet \,)$$

$$\text{NMD} = \left(\times \ , \ \begin{matrix} \bullet \\ \bullet \end{matrix} \right)$$

so

$$\text{Morse Data} = \left(\ \boxtimes \ , \ \boxtimes \ \right)$$

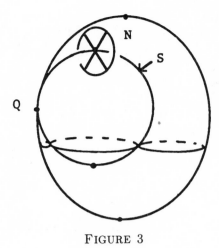

FIGURE 3

At the lower critical point on S, NMD is the same, but $\lambda = 0$, so TMD is $(\bullet\!-\!\!\bullet\,,\quad)$ and

$$\text{Morse Data} = \left(\ \text{\includegraphics{membrane}}\ ,\ \underline{\qquad\qquad}\ \right)$$

Note that the Normal Morse Data at these two critical points is the same; in other words, the Normal Morse Data for f and for $-f$ are the same at each critical point. However, if the surface had only one membrane across the center instead of two, this would no longer be true: one NMD would be $(\curlyvee\ ,\ \bullet\,)$, the other $(\curlywedge\ ,\ \bullet\ \bullet\ )$.

4. Morse theory on singular complex spaces. When X is a complex space, one may take the strata to be complex manifolds. In this case one can describe the Normal Morse Data at a critical point P in terms of the *complex link* of X at P. In particular the Normal Morse Data at P is independent of choice of Morse function (cf. the examples in §3).

To define the complex link $\mathcal{L}$ of X at P, cut X by a (complex) normal slice N, thus reducing to the case where $X = X \cap N$ has an isolated singularity at P. Let $B_\delta(P)$ be a small neighborhood of P. Then $\mathcal{L}$ is the intersection of $X \cap B_\delta(P)$ with a nearby complex hyperplane. More precisely, let F be a holomorphic function on M near P (so that $\text{Re}(F)$ is a Morse function near P), and set

$$\mathcal{L} = \text{complex link} = X \cap B_\delta(P) \cap F^{-1}(\varepsilon e^{i\theta})$$

for ε small compared with δ, any $0 \le \theta \le 2\pi$. This generalizes the familiar Milnor fibration; $F^{-1}(0)$ is contractible, and, locally, X can be recovered from $\mathcal{L}$ and the monodromy (see Figure 4).

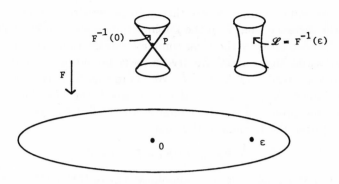

$$\text{F}^{-1}(0) \qquad \text{P} \qquad \qquad \mathcal{L} = \text{F}^{-1}(\varepsilon)$$

$$\text{F} \downarrow$$

$$\bullet\, 0 \qquad\qquad \bullet\, \varepsilon$$

FIGURE 4

THEOREM (GORESKY-MACPHERSON [**51, 53, 54**]). *The Normal Morse Data has the homotopy type of the pair* $(C\mathcal{L}, \mathcal{L})$:

$$\text{NMD} \underset{\text{homotopy}}{\sim} (C\mathcal{L}, \mathcal{L})$$

where $C\mathcal{L}$ *is the cone over* $\mathcal{L}$.

For a proof, as well as the precise description of NMD up to homeomorphism, we refer to the cited works of Goresky and MacPherson.

COROLLARY. *Suppose* X *is an* n-*dimensional complex analytic space which is a local complete intersection at* P. *Let* k *be the* (*complex*) *dimension of the stratum containing* P. *Then*

$$\pi_i(C\mathcal{L}, \mathcal{L}) = 0 \quad \text{for } i < n - k.$$

PROOF. Note that the normal slice of X is also a local complete intersection, so one is reduced to the case where $k = 0$. The theorem of Hamm from §2 then applies.

If X requires q more equations at P than its codimension in M, then q provides a "defect":

$$\pi_i(C\mathcal{L}, \mathcal{L}) = 0 \quad \text{for } i < n - k - q.$$

Chapter III
Lefschetz and Connectedness Theorems

1. Lefschetz theorem for complex projective varieties. The basic Lefschetz theorem, extended by replacing "nonsingular" by "local complete intersection," reads:

THEOREM. *Let* X^n *be a subvariety of* $\mathbf{P}^m$, L^{m-d} *a linear subspace, such that* X *is a local complete intersection at each point outside* L. *Then*

$$\pi_i(X, X \cap L) = 0 \quad \text{for } i \leq n - d.$$

PROOF. We want to apply Morse theory to (a perturbation of) the function f which is the distance from L in the Fubini-Study metric on $\mathbf{P}^m$. The Normal Morse Data is controlled by the local complete intersection hypothesis. To control the Tangential Morse Data, we need a lower bound on the index of $f|S$ at a point P in a stratum S of X, $P \notin L$. We claim that $T_P S$ contains a complex subspace of dimension at least $k - d + 1$ on which the Levi form is negative definite. For the proof see the discussion at the end of this section. Granting this it follows from the Lemma in II.2 that

$$\lambda = \text{Index}_P(f|S) \geq k - d + 1.$$

The Tangential Morse Data has the homotopy type of $(D^\lambda, \partial D^\lambda)$, so the relative homotopy groups vanish through dimension $k - d$. Together we have

$$\pi_i(\text{TMD}) = 0 \quad \text{for } i \leq k - d,$$
$$\pi_i(\text{NMD}) = 0 \quad \text{for } i \leq n - k - 1,$$

the latter by the corollary in II.4. By a basic fact about the homotopy groups of a product of pairs (cf. Blakers-Massey [8]) we get

$$\pi_i(\text{TMD} \times \text{NMD}) = 0 \quad \text{for } i \leq (k - d) + (n - k - 1) + 1,$$

so $\pi_i(MD) = 0$ for $i \leq n - d$.

Note that $X \cap L$ is a deformation retract of $X \cap L_\varepsilon$, for L_ε an ε-neighborhood of L. We can apply Morse theory to go from $X \cap L_\varepsilon$ to X. At each crossing of a critical value the relative homotopy groups don't change through degree $n - d$, so

$$\pi_i(X, X \cap L) = 0 \quad \text{for } i \leq n - d,$$

as required.

There are several ways this theorem can be improved:

(1) If X is locally defined by at most $m - n + q$ equations at points outside L, q is a "defect":

$$\pi_i(X, X \cap L) = 0 \quad \text{for } i \leq n - d - q.$$

(2) If, instead of X^n imbedded in $\mathbf{P}^m$, we have a finite (proper) morphism $f: X^n \to \mathbf{P}^m$, or a finite morphism $f: X^n \to U$, where U is a Zariski neighborhood of L in $\mathbf{P}^m$, the analogous result holds:

$$\pi_i(X, f^{-1}(L)) = 0 \quad \text{for } i \leq n - d.$$

Since $\pi_i(\mathbf{P}^m, L^{m-d}) = 0$ for $i \leq n - d$ as well, this Lefschetz theorem is a special case of the connectedness principle stated in I.2.

(3) If $f: X \to \mathbf{P}^m$ is only quasi-finite (finite fibres), the theorem is still true provided L is replaced by an ε-neighborhood L_ε:

$$\pi_i(X, f^{-1}(L_\varepsilon)) = 0 \quad \text{for } i \leq n - d.$$

For noncompact X one needs to consider strata and critical points "at infinity" (outside X). This is a whole theory which is dual to the singular, compact theory sketched in Chapter II; we refer to [**51, 54**] for this.

(4) For arbitrary singular X, control on the complex links gives a Lefschetz theorem with corresponding defect.

In Chapters IV and V we will need only the following special case:

PROPOSITION. *Let $f: X \to U$ be a finite morphism from an n-dimensional irreducible variety X to a Zariski neighborhood U of a linear subspace L^{m-d} in $\mathbf{P}^m$. Then*

(a) $n \geq d \Rightarrow f^{-1}(L) \neq \varnothing$;

(b) $n > d \Rightarrow f^{-1}(L)$ *is connected;*

(c) $n > d$, X *locally analytically irreducible* $\Rightarrow \pi_1(f^{-1}L) \twoheadrightarrow \pi_1(X)$.

PROOF. The reasoning is the same as in the theorem; for π_0 and π_1 the singularities of X have little effect. For example, to prove (c) it suffices to see that all the Morse data (A, B) have B connected. Since all the complex links $\mathcal{L}$ are assumed to be connected, the only case that could cause a problem is for a maximal stratum, when $\mathcal{L}$ is empty. But in this case the Morse data has the homotopy type of $(D^\lambda, \partial D^\lambda)$, with $\lambda \geq n - d + 1 \geq 2$, so ∂D^λ is connected.

This proposition can also be proved without Morse theory. For a complete elementary discussion see [**40**]. Note that the hypothesis on X in (c) is necessary [**40**, p. 56].

The claim left unproved in the proof of the Lefschetz theorem has appeared in some guise in most treatments of the subject, cf. [**1, 9, 55, 98, 126**]. Instead of repeating a proof, we prove the following lemma, which suffices for the theorem, and has other useful applications.

LEMMA. *Let E be a holomorphic vector bundle of rank d on a complex manifold X, with a Hermitian metric which is positive in the sense of Griffiths [**55**]. Let S be a k-dimensional locally closed complex submanifold of $E - \{0\}$, let f be the restriction of the metric to S, and let $P \in S$ be a critical point of f. Then the tangent space $T_P S$ contains a complex subspace W with*

$$\dim(W) \geq k - d + 1$$

on which the Levi form $L_f(P)$ is negative definite.

Special cases of this lemma occur in the literature, starting with [**55**], although we have not found precisely this version. We thank T. Garrity for showing us a proof.

Griffith's notion of positivity can be described as follows. Let E be a holomorphic bundle of rank d on a complex manifold X with a Hermitian metric h. We write $\phi = \phi_h$ for the corresponding (nonnegative) real-valued function on $E: \phi(P) = h(P, P)$. For all $P \in E - \{0\}$, we have the Levi form $L_\phi(P)$ on the tangent space $T_P E$. Let x be the image of P in X. Any trivialization of E near x, $E|U \cong U \times \mathbf{C}^d$, determines a splitting

$$(1) \qquad\qquad T_P E = T_x X \oplus E_x,$$

where $E_x = \mathbf{C}^d$ is the fibre of E at x. If $z_1, \ldots, z_n$ are local coordinates at x, with $z_i(x) = 0$, the Hermitian metric is then given by a $d \times d$ Hermitian matrix $H(z)$ of functions near $z = 0$. One can find such a local frame so that H has the simple form

$$(2). \qquad\qquad H(z) = I + O(|z|^2),$$

where I is the identity matrix. For the existence of such a "normal trivialization" see [55, p. 195] or [136, p. 80]. It follows that the curvature matrix is

$$(3) \qquad\qquad -\partial\bar{\partial}H(0) = -\left(\frac{\partial^2\phi}{\partial z_i \partial \bar{z}_j}(P)\right).$$

Griffiths positivity is that the curvature matrix be positive definite, or, equivalently, that the *Levi form* $L_\phi(P)$ *be negative definite on the subspace* $T_x X \oplus 0$ *of* $T_P E$, *for a normal trivialization of* E *at* P.

PROOF OF THE LEMMA. Choose a normal trivialization of E at P as above, and let $t_1, \ldots, t_d$ be coordinates on $\mathbf{C}^d = E_x$, so $z_1, \ldots, z_n, t_1, \ldots, t_d$ are coordinates for E near P. The normalization (2) implies that

$$(4) \qquad \frac{\partial\phi}{\partial z_i}(P) = 0, \quad 1 \le i \le n; \qquad \frac{\partial\phi}{\partial t_i}(P) = \bar{t}_i(0), \quad 1 \le i \le d.$$

From $T_P S \subset T_P E = T_x X \oplus E_x$ the projections induce linear mappings

$$\alpha: T_P S \to T_x X, \qquad \beta: T_P S \to E_x.$$

The required subspace W is defined to be the kernel of β.

Since f is the restriction of ϕ to a complex submanifold S of E, a chain rule calculation shows that $L_f(P) = L_\phi(P)$ on $T_P S \subset T_P E$. Therefore if v is a nonzero vector in W,

$$L_f(P)(v) = L_\phi(P)(v) = L_\phi(P)(\alpha(v) \oplus 0),$$

and this is negative by the assumption of Griffiths positivity. To conclude the proof we must show that $\dim(W) \ge k - d + 1$, or $\operatorname{rank}(\beta) < d$. Letting $u_1, \ldots, u_k$ be local coordinates for S at P, since P is a critical point for $\phi|S$ we have, for all j,

$$0 = \frac{\partial\phi}{\partial u_j}(P) = \sum_{i=1}^{n} \frac{\partial z_i}{\partial u_j}(P)\frac{\partial\phi}{\phi z_i}(P) + \sum_{i=1}^{d} \frac{\partial t_i}{\partial u_j}(P)\frac{\partial\phi}{\partial t_i}(P)$$

$$= \sum_{i=1}^{d} \frac{\partial t_i}{\partial u_j}(P)\bar{t}_i(P)$$

by (4). But if $\operatorname{rank}\beta = \operatorname{rank}(\partial t_i(P)/\partial u_j) = d$, this would force all $t_i(P)$ to be zero, contradicting the fact that P is not in the zero section.

PROBLEM. Is this lemma true for a bundle which is known only to be ample? This would allow a useful extension of connectivity theorems to such bundles. It seems optimistic to hope for a positive Hermitian metric on every ample bundle, although this question of Griffiths' [55] remains open. Perhaps a Finsler metric (cf. Kobayashi [82]) could be used instead. When E is generated by its sections, see Sommese [126], and Goresky-MacPherson [54].

To use the lemma to complete the proof of the Lefschetz theorem one can proceed as follows. If $n + d \leq m$, one can choose a linear subspace M^{d-1} of $\mathbf{P}^m$ disjoint from X. Projection from M gives a morphism $\mathbf{P}^m - M \to L$ which realizes $\mathbf{P}^m - M$ as the bundle $\mathcal{O}(1) \oplus \cdots \oplus \mathcal{O}(1)$ (d copies) over L. Since this bundle is Hermitian positive, the lemma gives the required bounds on the indices. In general one can reduce to this case, in fact to the case where $d = 1$, by choosing a chain of linear spaces

$$L = L_0 \subset L_1 \subset \cdots \subset L_d = \mathbf{P}^m$$

and using the case $d = 1$ to bound the homotopy of the pairs $(X \cap L_i, X \cap L_{i-1})$.

2. The connectedness theorem. Since the normal bundle to the diagonal imbedding

$$\Delta = \mathbf{P}^m \hookrightarrow \mathbf{P}^m \times \mathbf{P}^m$$

is the tangent bundle to $\mathbf{P}^m$, which is positive, it is not unreasonable to expect the connectedness principle to be valid for this diagonal imbedding. In fact, a simple construction of Deligne shows how to deduce it from the case of a linear subspace $L^m \hookrightarrow \mathbf{P}^{2m+1}$.

THEOREM. *Let X^n be a projective variety, $f: X \to \mathbf{P}^m \times \mathbf{P}^m$ a finite morphism. Then*
(a) $n \geq m \Rightarrow f^{-1}(\Delta)$ *is nonempty;*
(b) $n > m \Rightarrow f^{-1}(\Delta)$ *is connected;*
(c) $n > m$, X *locally analytically irreducible* $\Rightarrow \pi_1(f^{-1}(\Delta)) \twoheadrightarrow \pi_1(X)$;
(d) X *a local complete intersection at each point not in* $f^{-1}(L) \Rightarrow$

$$\pi_i(X, f^{-1}(\Delta)) \xrightarrow{\sim} \pi_i(\mathbf{P}^m \times \mathbf{P}^m, \Delta) \quad \text{for } i \leq n - m.$$

Note that $\pi_i(\mathbf{P}^m \times \mathbf{P}^m, \Delta)$ is $\mathbf{Z}$ if $i = 2$, and 0 otherwise in this range ($i \leq 2m$).

PROOF. Let $(z_0, \ldots, z_m, w_0, \ldots, w_m)$ be homogeneous coordinates on $\mathbf{P}^{2m+1}$, and define linear subspaces L_0^m, L_∞^m, L^m of $\mathbf{P}^{2m+1}$ by

$$L_0 = \{(z_0 : \ldots : z_m : 0 : \ldots : 0)\},$$
$$L_\infty = \{(0 : \ldots : 0 : w_0 : \ldots : w_m)\},$$
$$L = \{(z_0 : \ldots : z_m : z_0 : \ldots : z_m)\}.$$

The projection $\pi: \mathbf{P}^{2m+1} - L_0 \cup L_\infty \to \mathbf{P}^m \times \mathbf{P}^m$ defined by

$$\pi(z_0 : \ldots : z_m : w_0 : \ldots : w_m) = (z_0 : \ldots : z_n) \times (w_0 : \ldots : w_m)$$

is a $\mathbf{C}^*$ bundle. Consider the diagram:

$$
\begin{array}{ccccc}
\tilde{X} & \xrightarrow{\tilde{f}} & \mathbf{P}^{2m+1} - L_0 \cup L_\infty & \hookleftarrow & L \\
\downarrow & \square & \downarrow \pi & & \downarrow \wr\wr \\
X & \xrightarrow{f} & \mathbf{P}^m \times \mathbf{P}^m & \hookleftarrow & \Delta
\end{array}
$$

where the left square is a fibre (Cartesian) square. Then $\tilde{f}^{-1}(L)$ maps isomorphically to $f^{-1}(\Delta)$, and $\tilde{X} \to X$ is a $\mathbf{C}^*$ bundle. This means that $\tilde{X}$ and X have the same homotopy groups except for $\pi_1(\mathbf{C}^*) = \mathbf{Z}$. Applying the theorem of III.1 to $\tilde{f}$ and L (see the improvement (2) stated in that section), and keeping track of the extra "$\mathbf{Z}$," the result follows. For details, see [40, §9].

The connectedness principle is also valid for the diagonal imbedding

$$
\Delta = \mathbf{P}^m \hookrightarrow \mathbf{P}^m \times \cdots \times \mathbf{P}^m
$$

for more than two factors; the proof is essentially the same.

Note that if $X = A \times B$, with A^a and B^b subvarieties of $\mathbf{P}^m$, then (a) gives the "Bézout" assertion of I.1. In addition, (b) implies that

$$
A \cap B \text{ is connected if } a + b > m.
$$

The "Bertini" and "Lefschetz" assertions follow by setting $X = V \times L$. More interestingly, setting $X = V \times V$ yields the following strengthening of the Barth-Larsen theorem, which was conjectured at Arcata [6]:

COROLLARY. *If* $V^n \subset \mathbf{P}^m$ *is a local complete intersection, then*

$$
\pi_i(\mathbf{P}^m, V) = 0 \quad \text{for } i \le 2n - m + 1.
$$

PROOF. The point here is that the connectedness theorem forces surjections

$$
\pi_i(V \times V \cap \Delta) \twoheadrightarrow \pi_i(V \times V),
$$

but the diagonal map $\pi_i(V) \to \pi_i(V) \times \pi_i(V)$ can't be surjective unless $\pi_i(V) = 0$. (Again there is one copy of $\mathbf{Z}$ to worry about, but this can be avoided, cf. [40, §9].)

More generally, we see that the connectedness principle is valid for any imbedding $Y^{m-d} \hookrightarrow \mathbf{P}^m$ of a local complete intersection Y in $\mathbf{P}^m$. In particular:

COROLLARY. *With* Y *as above, and* $f: X^n \to \mathbf{P}^m$, X *a local complete intersection, we have*

$$
\pi_i(X, f^{-1}(Y)) \to \pi_i(\mathbf{P}^m, Y) \quad
\begin{cases}
\cong & i \le n - d, \\
\twoheadrightarrow & i = n - d + 1.
\end{cases}
$$

For $Y = L$ one recovers the Lefschetz theorem, while for $Y = X$ one sees the Barth theorem again. For the proof, which is similar, see [40, §9].

Note that when $i \le m - 2d + 1$, then $\pi_i(\mathbf{P}^n, Y) = 0$ by the Barth-Larsen theorem, and the corollary reduces to

$$
\pi_i(X, f^{-1}(Y)) = 0 \quad \text{for } i \le \min(n - d, m - 2d + 1).
$$

Results like this, with more smoothness assumptions, had been proved by Barth [2] and Larsen [85] and generalized to other homogeneous manifolds, by Sommese [124, 128, 129], Goldstein [48], and Sommese–Van de Ven [130].

At this conference we learned of a new theorem of Okonek [109], which contains all these results. Sommese [125] has defined a notion of k-ample for a bundle, 0-ample being ample, and k a sort of defect from ampleness. This defect is easily calculated for the tangent bundle of familiar homogeneous spaces, cf. [48].

THEOREM (OKONEK [109]). *Let $P = G/H$, where G is a simply connected complex Lie group, H a closed subgroup, and let Y be a local complete intersection of codimension d in P. Let X^n be a compact complex space which is a local complete intersection, $f: X \to P$ a morphism. If f^*T_P is k-ample, then*

$$\pi_i(X, f^{-1}(Y)) \to \pi_i(P, Y) \quad \begin{cases} \cong & i \leq n - d - k, \\ \twoheadrightarrow & i = n - d - k + 1. \end{cases}$$

In addition, there are corrections for the case when X has worse singularities. The proof uses Barth's "transplanting" technique [2], as well as the Goresky-MacPherson stratified Morse theory.

In this discussion we have omitted closely related questions of q-convexity of complements of subvarieties, and vanishing of sheaf cohomology groups. For some of the extensive work in this direction see Diederich and Fornaess [21, 22], Faltings [28, 29], Fritzsche [31, 32], Okonek [110, 111], Peternell [112, 113, 114]. In the context of the preceding theorem, the connectivity bounds that come out are of the form

$$\pi_i(X, X \cap Y) = 0 \quad \text{for } i \leq [n/(d + k)] - 1.$$

Nori [106] has proved another important theorem of Lefschetz type, which will be discussed in Chapter V.

The assertions (a) and (b) of the connectedness theorem have elementary proofs (cf. [38, 40, 27, 71]). For recent results in this direction see Brodmann [10, 11].

Chapter IV
π_0 and Varieties of Small Codimension

1. Projections and hyperplane sections. A typical application of the connectedness theorem to projective geometry follows the following pattern, discovered by Hansen (cf. [65]).

Given a morphism $f: X \to Y$, and a closed subvariety V of X, consider the product mapping

$$V \times X \xrightarrow{F} Y \times Y$$

and the inclusion

$$\Delta_V \subset F^{-1}(\Delta_Y) = \{(v, x) | f(v) = f(x)\}.$$

If connectedness applies, for example, if $Y = \mathbf{P}^m$ and $\dim V + \dim X > m$, then $F^{-1}(\Delta_Y)$ will be connected. This implies that either $\Delta_V = F^{-1}(\Delta_Y)$, or some degenerations are forced: there is a sequence (v_t, x_t) converging to (v_0, v_0) with $v_t \in V$, $x_t \in X$, $f(v_t) = f(x_t)$, $v_t \neq x_t$. For $V = X$, $Y = \mathbf{P}^m$, this implies:

PROPOSITION [38]. *If $f: X^n \to \mathbf{P}^m$ is unramified, and $2n > m$, then f must be an imbedding.*

In particular, any irreducible subvariety of $\mathbf{P}^m$ of more than half the dimension cannot have étale coverings. In fact the following version of Barth's theorem is true, with no assumptions about singularities of X.

COROLLARY. *For $X^n \subset \mathbf{P}^m$, $n > 2m$, $\pi_1(X) = 0$.*

If X is normal, the connectedness theorem (c) of §III.2 for the product imbedding of $X \times X$ in $\mathbf{P}^m \times \mathbf{P}^m$ gives

$$\pi_1(\Delta_X) \twoheadrightarrow \pi_1(X \times X)$$

which forces $\pi_1(X) = 0$. In the general case one proceeds similarly for the normalization of X, combined with a use of the proposition, cf. [40, §5].

THEOREM (ZAK [138, 40]). *Let $X^n \subset \mathbf{P}^m$ be nonsingular, not contained in any hyperplane, and let $L^k \subset \mathbf{P}^m$ be any linear subspace, $n \leq k \leq m - 1$. Then*

$$\dim\{x \in X | T_x X \subset L\} \leq k - n.$$

Here $T_x X$ denotes the imbedded projective tangent space to X at x. For $k = n$ we have:

COROLLARY 1. *The Gauss map $X \to \mathrm{Grass}(n, \mathbf{P}^m)$, $x \mapsto T_x X$, is always finite.*

The fact that the Gauss map is generically finite had been known (Piene, Griffiths-Harris). Recently Ran [118] showed how to deduce finiteness from generic finiteness, also in the case of submanifolds of abelian varieties. In the latter case, Smyth and Sommese [123] have bounded the degree of the Gauss mapping.

The case $k = m - 1$ of Zak's theorem gives three striking corollaries about arbitrary hyperplane sections:

COROLLARY 2. $\dim \mathrm{Sing}(X \cap L^{m-1}) \leq m - n - 1$.

Indeed, $\mathrm{Sing}(X \cap L) = \{x \in X | T_x X \subset L\}$.

COROLLARY 3. $m < 2n \Rightarrow X \cap L^{m-1}$ *is reduced.*

COROLLARY 4. $m < 2n - 1 \Rightarrow X \cap L^{m-1}$ *is normal.*

In addition, if X^* denotes the dual variety,

COROLLARY 5. $\dim X^* \geq n$.

Following an idea of Landman, further restrictions on the dimension of X^* have been given by Ein [25].

PROOF OF THEOREM. For a subvariety $V^r \subset X$, set

$$\text{Tan}_V X = \bigcup \{T_v X | v \in V\},$$

$$\text{Sec}_V X = \text{Closure} \left(\bigcup \{\text{line } \overline{vx} | v \in V, \ x \in X, \ v \neq x\} \right)$$

We have $\text{Tan}_V X \subset \text{Sec}_V X$; the maximum, and expected, dimensions of these varieties are $r+n$ and $r+n+1$ respectively. The claim is that these two varieties both have their expected dimensions, or they are equal.

If not, say $t = \dim(\text{Tan}_V X) < r + n$. A generic L^{m-t-1} will hit $\text{Sec}_V X$, but not $\text{Tan}_V X$. Let $f: X \to \mathbf{P}^t$ be induced by projection from L. Applying connectedness to

$$V \times X \to \mathbf{P}^t \times \mathbf{P}^t$$

gives a sequence (v_t, x_t) approaching (v_0, v_0), such that the lines $\overline{v_t x_t}$ all meet L. This would force a limiting line to meet L, but by assumption $T_{v_0} V$ does not meet L! (For $V = X$, this was proved in [38].)

Now apply this to an irreducible component V^t of $\{x \in X | T_x X \subset L\}$, with $t > k - n$. Since $\text{Tan}_V X \subset L$, it is smaller than expected, so it must be $\text{Sec}_V X$. But then $X \subset L$, contradicting our original assumption.

2. Proof of a conjecture of Hartshorne. There has been a good deal of work on the general conjecture that nonsingular varieties X^n of small codimension in $\mathbf{P}^m$ should be complete intersections, although the question is still open in codimension 2, $n \geq 4$. Hartshorne [69, 70] conjectured that this is true whenever $n > \frac{2}{3}m$, and pointed out connections with Lefschetz and Barth theorems. Since any complete intersection is linearly normal—it is not the projection of a subvariety of $\mathbf{P}^{m+1}$ which is not contained in a hyperplane—one expects subvarieties of small codimension to be linearly normal. Equivalently subvarieties of small codimension, not contained in hyperplanes, should not project isomorphically.

The following classical examples were known:

(1) $X^2 = \mathbf{P}^2 \hookrightarrow \mathbf{P}^5$, the Veronese imbedding, which projects to $\mathbf{P}^4$.

(2) $X^4 = \mathbf{P}^2 \times \mathbf{P}^2 \hookrightarrow \mathbf{P}^8$, the Segre imbedding, which projects to $\mathbf{P}^7$.

(3) $X^8 = \text{Grass}(1, \mathbf{P}^5) \hookrightarrow \mathbf{P}^{14}$, the Plücker imbedding, which projects to $\mathbf{P}^{13}$.

For each such $X^n \hookrightarrow \mathbf{P}^m$, $n = \frac{2}{3}(m - 2)$. Although there were no asymptotic classes of examples suggesting this $\frac{2}{3}$—and there are still none—Hartshorne made the bold conjecture:

If $X^n \subset \mathbf{P}^m$ is nonsingular, not contained in a hyperplane, and $n > \frac{2}{3}(m-2)$, then X does not project isomorphically into $\mathbf{P}^{m-1}$.

Zak [138] has proved this conjecture. In fact, Lazarsfeld [40] found another proof,[2] which also uses the connectedness theorem, and also gives precisely the bound conjectured by Hartshorne. Indeed, Hartshorne's numbers hint at how to apply connectedness: if $3n > 2(m-2)$, then connectedness will imply that for a finite morphism

$$f \times f \times f \colon X^n \times X^n \times X^n \to \mathbf{P}^{m-2} \times \mathbf{P}^{m-2} \times \mathbf{P}^{m-2}$$

the inverse image of the diagonal must be connected. Lazarsfeld takes f to be projection from a carefully chosen line in $\mathbf{P}^m$; for details, see [40].

A letter of Zak claimed to show that the three classical examples are the only three on the boundary line $3n = 2(m-2)$. However, Lazarsfeld discovered a fourth,

(4) $X^{16} \hookrightarrow \mathbf{P}^{26}$

which can be realized as a projectivized orbit of a representation of the exceptional group E_6.

Zak then proved the amazing theorem that these 4 are the only examples! Partial results in these directions had also been obtained by T. Fujita and J. Roberts. Lazarsfeld and Van de Ven have outlined Zak's proof [88].[3]

Note that this leaves the asymptotic situation as mysterious as ever. Few nontrivial examples of large dimension are known.

3. Vector bundles. If E is an ample vector bundle of rank d on a variety X, any section $s\colon X \to E$ should be a "positive" imbedding, in that we may expect the connectedness principle to be valid. In particular, if V^n is a closed subvariety of E, one expects:

(a) $n \geq d \Rightarrow s^{-1}(V)$ is not empty, and

(b) $n > d \Rightarrow s^{-1}(V)$ is connected.

For example, if E has a Hermitian metric which is Griffiths positive, these assertions follow from the Goresky-MacPherson theory, together with the lemma of §III.1. Unfortunately, many ample bundles that arise in nature are not known to have such a metric.

Nevertheless, (a) is always true; the proof uses intersection theory [41, 42]. Although (b) is open, the following special case is known:

THEOREM [43]. *Let* $\sigma\colon A \to B$ *be a homomorphism of vector bundles of ranks* a *and* b *on a variety* X^n. *Assume* $A^\vee \otimes B$ *is ample. Let*

$$D_k(\sigma) = \{x \in X \mid \operatorname{rank} \sigma(x) \leq k\}.$$

If $(a-k)(b-k) \leq n$, *then* $D_k(\sigma)$ *is nonempty, and if* $(a-k)(b-k) < n$, *then* $D_k(\sigma)$ *is connected.*

These would follow from (a) and (b), setting $E = \operatorname{Hom}(A, B)$, V the universal degeneracy locus in E, and s the section determined by σ.

---

[2]For other proofs see Th. Peternell, J. Le Potier, and M. Schneider, *Vanishing theorems, linear and quadratic normality*, Invent. Math. **87** (1987), 573–586.

[3]Zak's paper, *Severi varieties*, has appeared in Math. USSR-Sb. **54** (1986), 113–127.

This can be applied to show that the loci W_d^r of special divisors on an arbitrary smooth curve of genus g is connected if

$$\rho = g - (r+1)(g-d+r) > 0.$$

Similarly [39], one can show that

$$\dim W_{d-1}^r \geq \dim W_d^r - (r+1).$$

Analogues of this theorem are expected for symmetric and skew-symmetric bundle maps; they would also follow from (b). Some cases of this have been verified by Tu [134].

For sufficiently ample E on a local complete intersection X^n it has been proved that

$$\pi_i(X, Z(s)) = 0 \quad \text{for } i \leq n - d,$$

as expected from the connectedness principle. The strongest theorem is Okonek's [109]: if E is generated by its sections and k-ample, then

$$\pi_i(X, Z(s)) = 0 \quad \text{for } i \leq n - d - k.$$

In the next chapter we will use a special case due to Lazarsfeld [86].

4. A counterexample. The positivity of the normal bundle to an imbedding $M \hookrightarrow P$ is not enough for the validity of the connectedness principle. Hansen pointed out that examples used by Hironaka and Hartshorne also give counterexamples in this context. Take $X = \mathbf{P}^3$, with a linear action of a finite group G on X such that there is a line not meeting the fixed point locus. Then if $P = X/G$, M is the image of the line in P, f the projection of X to P, the normal bundle to M in P is $\mathcal{O}(1) \oplus \mathcal{O}(1)$ on $M \cong \mathbf{P}^1$, but $f^{-1}(M)$ is not connected.

Chapter V
π_1 and Branched Coverings

1. Normal branched coverings of $\mathbf{P}^n$. The first application of the connectedness theorem to coverings was found by Gaffney and Lazarsfeld [44]. Consider a finite morphism $f \colon X^n \to \mathbf{P}^n$, with X normal, and d sheets. One would expect, "in general", $n+1$ sheets (or all d sheets if $d \leq n+1$) to come together at some point of X. Part of the connectedness philosophy is that what is expected "in general" may be true always in the presence of positivity.

THEOREM [44]. *At least* $\min(d, n+1)$ *sheets must come together at some point of* X.

If R_l is the set where more than l sheets come together, they show that

$$\operatorname{codim}(R_l, X) \leq l \quad \text{for } l \leq \min(d-1, n).$$

The essential case, achieved by cutting by hyperplanes, is to show that R_n is not empty if $d \geq n + 1$. By induction one knows that R_{n-1} contains a curve V. Applying connectedness to

$$V \times X \to \mathbf{P}^n \times \mathbf{P}^n$$

gives the result: if (v_t, x_t) converges to (v_0, v_0), and at least n sheets come together at v_t, and $v_t \neq x_t$, then at least $n + 1$ sheets come together at v_0.

COROLLARY. *If $d \leq n$, then $\pi_1(X) = 0$.*

If $X' \to X$ were a finite étale covering, cut by hyperplanes to get to the case $d = n$. Then $X' \to \mathbf{P}^n$ contradicts the theorem. This shows that

$$\pi_1^{\mathrm{alg}}(X) = \varprojlim \pi_1(X)/N = 0,$$

where the inverse limit is over all normal subgroups N of finite index. The old question of whether the canonical map

$$\pi_1(X) \to \pi_1^{\mathrm{alg}}(X)$$

must always be an injection for any algebraic variety X is still open, so a separate argument must be given for the topological fundamental group [**40**, §6].

A similar useful application of connectedness has been given by Lazarsfeld [**36**]: with $f: X^n \to \mathbf{P}^n$ as above, if there is any positive-dimensional locally closed subvariety V of $\mathbf{P}^n$ such that the mapping $f^{-1}(V) \to V$ is set-theoretically one-to-one, then $\pi_1(X) = 0$. This applies, for example, to any cyclic covering. It can also be applied to some of the Hirzebruch surfaces described in Barth's lectures.

2. Nonsingular coverings of $\mathbf{P}^n$. More can be said for a branched covering $f: X^n \to \mathbf{P}^n$ when X is nonsingular. In this case (since X is Cohen-Macaulay) we may write

$$f_*(\mathcal{O}_X) = \mathcal{O}_{\mathbf{P}^n} \oplus E^{\vee}$$

where $E^{\vee}$ is the kernel of trace: $f_*\mathcal{O}_X \to \mathcal{O}_{\mathbf{P}^n}$. The surjection of $\mathrm{Sym}(E^{\vee})$ to $f_*\mathcal{O}_X$ factors f:

where p is the projection. Lazarsfeld [**87**] proved that E is always an ample vector bundle—that in fact, $E(-1)$ is generated by its sections; the proof uses Kodaira vanishing on X.

THEOREM (LAZARSFELD [**86**]). *For such $X^n \to \mathbf{P}^n$,*

$$\pi_i(X) \to \pi_i(\mathbf{P}^n) \qquad \begin{cases} \cong & \text{for } i \leq n - (d - 1), \\ \twoheadrightarrow & \text{for } i = n - (d - 1) + 1. \end{cases}$$

In fact for any E of rank e on $\mathbf{P}^n$, with $E(-1)$ generated by sections, and any compact local complete intersection $X^n \subset E$, Lazarsfeld proves that

$$\pi_i(E, X) = 0 \quad \text{for } i \leq n - e + 1.$$

For a sketch of the proof, consider the diagram

$$
\begin{array}{ccccc}
X & \hookrightarrow & Y & \hookrightarrow & X \times X \\
\downarrow & & \downarrow & \square & \downarrow \\
E & \overset{\text{diag}}{\hookrightarrow} & E \oplus E & \hookrightarrow & E \times E \\
\downarrow & & \downarrow & \square & \downarrow \\
& & \mathbf{P}^n & \overset{\text{diag}}{\hookrightarrow} & \mathbf{P}^n \times \mathbf{P}^n
\end{array}
$$

The connectedness theorem controls the groups $\pi_i(X \times X, Y)$. And X is the zero of a section of the pull-back of E to Y, which controls $\pi_i(Y, X)$.

Cornalba [19] showed that if $f \colon X^n \to Y^n$ is a cyclic covering of nonsingular projective varieties, with branch divisor B ample on Y, then

$$
\pi_i(X) \to \pi_i(Y) \text{ is } \begin{cases} \cong & i < n, \\ \twoheadrightarrow & i = n. \end{cases}
$$

In this case f factors through an ample line bundle.

Ein [24] has proved that for any branched covering $X^n \to \mathbf{P}^n$ with X nonsingular, the ramification divisor $R = R_1$ is always an ample divisor. Special cases had been proved by Landman (unpublished) and Lanteri [84].

3. Generic projections. Moishezon and Teicher [101] consider a generic projection $f \colon X^2 \to \mathbf{P}^2$ of the surface $X = \mathbf{P}^1 \times \mathbf{P}^1$ imbedded in projective space by a linear system $|al_1 + bl_2|$, where l_1 and l_2 are the two rulings. Then f has degree $2ab$, and the least Galois covering $\tilde{f} \colon \tilde{X} \to \mathbf{P}^2$ dominating f has degree $(2ab)!$, and $\tilde{X}$ is smooth. Miyaoka had showed that such surfaces can have positive index. In [101] it is shown that for a, b sufficiently large and relatively prime, $\tilde{X}$ is a surface of general type with positive index and with $\pi_1(\tilde{X}) = 0$, thus providing a counterexample to a conjecture of Bogomolov. Unlike the simple proofs of triviality of fundamental groups of branched coverings of $\mathbf{P}^n$ described in the preceding chapter, the proof in [101] requires elaborate braid calculations.

Let X^2 be nonsingular, $f \colon X \to \mathbf{P}^2$ a finite morphism of degree n which locally looks like a generic projection: the ramification curve $R \subset X$ is nonsingular, the branch curve $B \subset \mathbf{P}^2$ has only ordinary nodes and cusps, and the map $R \to B$ is birational. A basic question, studied by Chisini [17, 18] and Segre [119], is:

To what extent does B determine $f \colon X \to \mathbf{P}^2$?

Note that the analogous question for $X^1 \to \mathbf{P}^1$ is completely understood: if the degree n is fixed, there are a finite number of such "simple" coverings, whose number was counted by Hurwitz. In higher dimensions the situation is much more rigid: one expects B to determine the degree, as well as the covering, uniquely.

Chisini [17] gave an example where $B \subset \mathbf{P}^2$ does not determine the covering: the projection of the Veronese has degree 4 and branches on the dual curve B to a smooth cubic D. The incidence variety

$$
\{(P, l) \in \mathbf{P}^2 \times \mathbf{P}^{2^\vee} | P \in D, \ P \in l\}
$$

projects to $\mathbf{P}^{2\vee}$, branching along $B = D^\vee$, with degree $= 3$. Catanese [13] constructed two more 3-sheeted coverings with the same branch curve. All four coverings have different normal bundles $N_{R/X}$ to R in X.

In general choosing local coordinates at the cusps of B trivializes $N_{R/X}$ at the points over the cusps. Catanese [13] shows that *this marked normal bundle* $N_{R/X}$ *determines the entire covering.* The idea is that the marked normal bundle determines the covering in a tubular neighborhood of R. The fact that $X - R$ is Stein is used to extend an isomorphism to the whole surface—an idea I heard from Washnitzer in 1964.

It follows from Catanese's theorem that there are only finitely many coverings with the same branch curve. In fact, Nori recently pointed out an upper bound for the degree of the covering:

$$\deg(f) \le 2d^2/(d^2 - 2r),$$

where $d = \deg(B)$, and r is the number of singular points of B. This inequality is sharp, as shown by the generic projection of a smooth surface in $\mathbf{P}^3$. To prove it Nori computes the intersection matrix of R and f^*B on X:

$$\begin{pmatrix} (d^2 - 2r)/2 & d^2 \\ d^2 & d^2 \deg(f) \end{pmatrix}$$

By the Hodge index theorem, the determinant of this matrix is nonpositive, which gives the result.

4. The Zariski problem. Let $D \subset \mathbf{P}^2$ be a curve—irreducible for simplicity —with only (ordinary) nodes as singularities. Zariski's problem [139] was to prove that

$$\pi_1(\mathbf{P}^2 - D) \cong \mathbf{Z}/d\mathbf{Z}.$$

Equivalently, if $f(x,y) = 0$ defines D, then the branched covering of $\mathbf{P}^2$ defined by $z^d = f(x,y)$ gives the universal covering of $\mathbf{P}^2 - D$. It suffices to show $\pi_1(\mathbf{P}^2 - D)$ is abelian, since it is easy to calculate $H_1(\mathbf{P}^2 - D)$. For the algebraic fundamental group, it suffices to show that a Galois covering of $\mathbf{P}^2$ branched along D has abelian Galois group.

There are now at least five proofs of Zariski's assertion.

Given a Galois covering $f: X \to \mathbf{P}^2$ branching along D, Abhyankar showed that the Galois group G is abelian if one can show that any two irreducible components of $f^{-1}(D)$ must meet. This is because the cyclic inertia groups of the irreducible components generate G, and two such subgroups commute whenever their components intersect. In fact, $f^{-1}(D)$ is irreducible, as follows by applying the connectedness theorem to

$$\tilde{D} \times X \xrightarrow{F} \mathbf{P}^2 \times \mathbf{P}^2,$$

where $\tilde{D} \to D \subset \mathbf{P}^2$ is the normalization of D. Indeed, $F^{-1}(\Delta)$ is nonsingular, as well as connected, and it dominates D, cf. [34]. Deligne [20] developed the π_1 version of the connectedness theorem, and applied it to

$$\tilde{D} \times (\mathbf{P}^2 - D) \to \mathbf{P}^2 \times \mathbf{P}^2$$

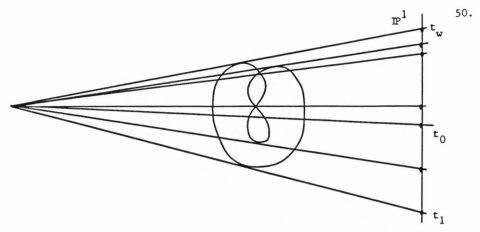

$$\text{FIGURE 5}$$

to prove that the topological group $\pi_1(\mathbf{P}^2 - D)$ is abelian. (See §5 below for similar techniques.)

Lê and Saito have proved a local version of this theorem. Applying it to the cone over D in C^3 near 0 gives a second proof [89]. Two proofs were given by Nori [106], one of which will be discussed in the next section.

Recently Harris [67], by proving Severi's assertion that the family of all irreducible nodal curves of given degree and genus is irreducible, certified Zariski's original proof. This assertion implies that D degenerates to a curve D_0 which consists of d general lines; then

$$\pi_1(\mathbf{P}^2 - D_0) \twoheadrightarrow \pi_1(\mathbf{P}^2 - D),$$

and $\pi_1(\mathbf{P}^2 - D_0)$ is abelian.

In fact, this approach gives more information than just the fundamental group. Recall the approach of Zariski, Enriques, and Van Kampen to studying $\pi_1(\mathbf{P}^2 - D)$ for any plane curve D. Take a generic pencil of lines l_t in $\mathbf{P}^2$, parametrized by $t \in \mathbf{P}^1$ (see Figure 5). Choose general $t_0 \in \mathbf{P}^1$, and standard generators γ_i for $\pi_1(\mathbf{P}^1 - \{t_1, \ldots, t_w\}, t_0)$, where $t_1, \ldots, t_w$ are the points for which l_{t_i} meets D in fewer than d points:

Then generators for $\pi_1(l_{t_0} - l_{t_0} \cap D, t_0)$ give generators for $\pi_1(\mathbf{P}^2 - D, t_0)$ (by I.1(4)), and relations are obtained by carrying these generators around by the w loops γ_i (see Cheniot [14] for a modern proof).

Chisini [15] encoded this information in a "braid diagram":

which describes how the points of $l_t \cap D$ move around as t moves around the successive γ_i. He showed that for D nonsingular one could achieve the canonical form indicated:

$$(12)(12)(13)(13)\cdots(1d)(1d)(23)(23)\cdots(2d)(2d)\cdots(d-1d)(d-1d).$$

Note that there are $d(d-1)/2$ pairs of transpositions, one transposition for each tangency.

THEOREM (CHISINI-HARRIS). *If D is a nodal curve of degree d with δ nodes, it has a braid description which can be obtained from the above description by removing δ pairs.*

Chisini proves this [16] by degenerating D to d general lines, and considering a small deformation of the lines. This program was carried out for several classes of curves by Chisini and Tibiletti [132, 133]. It should be rewarding to take another look at their methods and results.

There has been recent work on $\pi_1(\mathbf{P}^2 - D)$, when D has cusps as well as nodes. Libgober [92, 94] uses Deligne's techniques to get control over the Alexander polynomial associated to the canonical surjection

$$\pi_1(\mathbf{P}^2 - D \cup L) \to \mathbf{Z}$$

when L is a line, in terms of singularities.

Moishezon [100] showed that if $X^2 \subset \mathbf{P}^3$ is a nonsingular hypersurface of degree n, and $C \subset \mathbf{P}^2$ is the branch curve of a generic projection from X to $\mathbf{P}^2$, then

$$\pi_1(\mathbf{P}^2 - C) = B_n/\text{Center}(B_n),$$

where B_n is the braid group for n points. When $n = 3$ this had been proved by Zariski.

Dolgachev and Libgober [23] compute some examples of $\pi_1(\mathbf{P}^n - V)$, where V is the dual of a smooth variety in $(\mathbf{P}^n)^\vee$.

Although we have discussed $\pi_1(\mathbf{P}^2 - D)$, note that by Zariski's theorem (§I.1(4)), if V is any algebraic subset of $\mathbf{P}^n$, then

$$\pi_1(\mathbf{P}^n - V) = \pi_1(\mathbf{P}^2 - \mathbf{P}^2 \cap V)$$

for a generic plane $\mathbf{P}^2 \subset \mathbf{P}^n$, so the case of plane curves is the crucial one for fundamental groups.

Libgober has begun an investigation of higher homotopy groups of complements of hypersurfaces [95, 96, 97].

One of Zariski's aims in studying $\pi_1(\mathbf{P}^2 - D)$ was to relate invariants of branched coverings $X^2 \to \mathbf{P}^2$ (e.g., the first betti number of X for cyclic coverings) to subtle invariants of D, cf. Lehr [91]. For some recent work on this, see Libgober [92], Khashin [79], Buium [12].

5. Nori's work on fundamental groups. The first of Nori's proofs of Zariski's conjecture uses an action of $Sl(3)$ on $\mathbf{P}^2$; this corresponds to moving the diagonal inside $\mathbf{P}^2 \times \mathbf{P}^2$, which is another reflection of the positivity of $\mathbf{P}^2$, cf. [106, §4].

Nori's second approach gives a vast generalization, which includes strong conclusions when $\mathbf{P}^2$ is replaced by an arbitrary nonsingular surface X. For simplicity we assume X is projective, and D is an irreducible nodal curve on X, with δ nodes. Define N to be the kernel:

$$0 \to N \to \pi_1(X - D) \to \pi_1(X) \to 0.$$

Then N is the normal subgroup generated by loops γ_D which go from the base point to a point near a smooth point of D, go once around D, and return to the base point (see Figure 6); such γ_D is well defined up to conjugation.

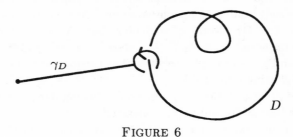

FIGURE 6

Nori asks:

(1) When is N a central subgroup of $\pi_1(X - D)$?

(2) When is N abelian?

The easy case is when D is nonsingular and ample, i.e., the self-intersection D^2 is positive. Let U be a tubular neighborhood of D. The noncompact version of the Lefschetz theorem implies that $\pi_1(U-D) \twoheadrightarrow \pi_1(X-D)$. Since $U - D \to D$ is a $\mathbf{C}^*$ bundle, it follows that the loop γ_D is central in $\pi_1(U - D)$. It follows that γ_D is central in $X - D$, so N is central.

When X is singular, one can apply the preceding to the proper transform of D in the blow-up of X at the singular points of D. This gives the following result, which generalizes previous results of Abhyankar and Prill.

PROPOSITION. *If $D^2 > 4\delta$, then N is central.*

Nori gives examples with $D^2 = 4\delta$ and N not central.

The deeper result, which implies Zariski's conjecture, is:

THEOREM. *If $D^2 > 2\delta$, then N is abelian.*

This too is sharp; Nori gives examples with $D^2 = 2\delta$ and N not abelian. There are two ingredients to Nori's proof:

(A) A "WEAK LEFSCHETZ THEOREM". *Let $\tilde{D} \to D$ be the normalization, and extend this to a tubular neighborhood*

$$\begin{array}{ccc} U & \xrightarrow{f} & X \\ \cup & & \cup \\ \tilde{D} & \to & D \end{array}$$

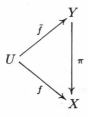

Then

$$G_D = \text{Image}(\pi_1(U - f^{-1}(D)) \to \pi_1(X - D))$$

has finite index in $\pi_1(X - D)$.

(B) *If $\pi: Y \to X$ is a finite covering, branched along D, then any two irreducible components of $\pi^{-1}(D)$ must meet.*

Note that the degree of the normal bundle $N_{\tilde{D}/U}$ is $D^2 - 2\delta$, so the assumption is precisely that this bundle is ample. Nori shows that G_D may be smaller than $\pi_1(X - D)$. Note in particular that (A) forces $\text{Image}(\pi_1(\tilde{D}) \to \pi_1(X))$ to have finite index. In fact, Nori shows that this index is bounded by $D^2/(D^2 - 2\delta)$, and gives examples to show that this bound too is sharp. Nori is led to ask the following remarkable

Question. Let D be any effective divisor on a nonsingular projective surface X with $D^2 > 0$. Is the normal subgroup generated by all $\text{Image}(\pi_1(\tilde{C}) \to \pi_1(X))$, where $\tilde{C}$ is the normalization of an irreducible component C of D, of finite index in $\pi_1(X)$? A positive answer for some elliptic surfaces has been given by Gurjar and Shastri [**56**].

To see how (A) and (B) imply the theorem, let $\pi: Y \to X$ be the covering associated to the subgroup G_D; we have

$$\begin{array}{ccc} & & Y \\ & \overset{\tilde{f}}{\nearrow} & \uparrow \pi \\ U & & \\ & \underset{f}{\searrow} & \downarrow \\ & & X \end{array}$$

and $Y \to X$ is unramified over $\tilde{f}(U)$. Now (B) implies that all components of $\pi^{-1}(D)$ must meet $\tilde{f}(\tilde{D})$. This forces π to be unramified, so $N \subset G_D$. But since γ_D is central in G_D, γ_D is central in N, so N is abelian.

Nori's proof of (B)—which implies Zariski's conjecture for the algebraic fundamental group—is remarkably elementary, and uses none of the recent Lefschetz or connectivity theorems. Assuming $\pi: Y \to X$ is Galois, he shows that any irreducible component of $\pi^{-1}(D)$ supports a Cartier divisor with positive self-intersection. Then the Hodge index theorem (another form of "positivity"!) implies that any two components must meet. For more details see [106] or [36].

The proof of (A) is more difficult. It is sketched in notes of Libgober prepared for these proceedings.

BIBLIOGRAPHY

In addition to the articles referred to in the lectures, this bibliography includes references published since other surveys, particularly [40] and [30]. It also contains the references, numbered [11]–[20], which were omitted from the printing of the Ravello survey [36]; they are references numbered [81, 92, 105, 106, 116, 120, 131, 135, 139, 140], respectively in the following list.

1. A. Andreotti and T. Frankel, *The Lefschetz theorem on hyperplane sections*, Ann. of Math. (2) **69** (1959), 713–717.

2. W. Barth, *Transplanting cohomology classes in complex projective space*, Amer. J. Math. **92** (1970), 951–967.

3. ____, *Der Abstand von einer algebraischen Mannigfaltigkeit im komplex-projektiven Raum*, Math. Ann. **187** (1970), 150–162.

4. ____, *Verallgemeinerung des Bertinischen Theorems in Abelschen Mannigfaltigkeiten*, Ann. Scuola Norm. Sup. Pisa Cl. Sci. (4) **23** (1969), 317–330.

5. ____, *Über die analytische Cohomologiegruppe $H^{n-1}(P_n \setminus A, F)$*, Invent. Math. **9** (1970), 135–144.

6. ____, *Larsen's theorem on the homotopy groups of projective manifolds of small embedding codimension*, Proc. Sympos. Pure Math., vol 29, Amer. Math. Soc., Providence, R.I., 1975, pp. 307–313.

7. ____, *Submanifolds of low codimensions in projective space*, Proc. Internat. Congr. Math. (Vancouver, 1974), Vol. 1, Canad. Math. Congress, Montreal, 1975, pp. 409–413.

8. A. L. Blakers and W. S. Massey, *The homotopy groups of a triad. II*, Ann. of Math. (2) **55** (1952), 192–201.

9. R. Bott, *On a theorem of Lefschetz*, Michigan Math. J. **6** (1959), 211–216.

10. M. Brodmann, *A few remarks on blowing-up and connectedness*, J. Reine Angew. Math. **370** (1986), 52–60.

11. M. Brodmann and J. Rung, *Local cohomology and the connectedness dimension in algebraic varieties*, Comment. Math. Helv. **61** (1986), 481–490.

12. A. Buium, *Sur le nombre de Picard des revêtements doubles des surfaces algébriques*, C. R. Acad. Sci. Paris **296** (1983), 361–364.

13. F. Catanese, *On a problem of Chisini*, Duke J. Math. **53** (1986), 33–42.

14. D. Cheniot, *Unre demonstration du théorème de Zariski sur les sections hyperplanes d'une hypersurface projective et du théorème de Van Kampen sur le groupe fondamental du complémentaire d'une courbe projective plane*, Compositio Math. **27** (1973), 141–158.

15. O. Chisini, *Una suggestiva rappresentazione reale per le curve algebriche piane*, Rend. Istit. Lombardo **66** (1933), 1141–1155.

16. ____, *Forme canoniche per il fascio caratteristico rappresentativo do una curva algebraica piana*, Rend. Istit. Lombardo **70** (1937), 49–61.

17. ____, *Sulla identità birazionale delle funzioni algebriche di due variabili dotate di una medesima curva di diramazione*, Rend. Istit. Lombardo **77** (1944), 339–356.

18. ____, *Courbes de diramation des plans multiples et tresses algébriques*, Liege (Proc. 2nd Colloq. Algebraic Geom., Liege, 1952), Masson, Paris, 1952, pp. 11–27.

19. M. Cornalba, *Una osservazione sulla topologia dei rivestimenti ciclici di varietà algebriche*, Boll. Un. Math. Ital. A (5) **18** (1981), 323–328.

20. P. Deligne, *Le groupe fondamental du complément d'une courbe plane n'ayant que des points doubles ordinaires est abélien*, Sém. Bourbaki, no. 543, 1979/80, Lecture Notes in Math., vol. 842, Springer-Verlag, 1981, pp. 1–10.

21. K. Diederich and J. E. Fornaess, *Smoothing q-convex functions and vanishing theorems*, Invent. Math. **82** (1985), 291–305.

22. ____, *Smoothing q-convex functions in the singular case*, Math. Ann. **273** (1986), 665–671.

23. I. Dolgachev and A. Libgober, *On the fundamental group of the complement to a discriminant variety*, Lecture Notes in Math., vol. 862, Springer-Verlag, 1981, pp. 1–25.

24. L. Ein, *The ramification divisors for branched coverings of P^n*, Math. Ann. **261** (1982), 483–485.

25. ____, *Varieties with small dual varieties*. I, Invent. Math. **86** (1986), 63–74; II, Duke J. Math. **52** (1985), 895–907.

26. G. Faltings, *Über die Annullatoren lokaler Kohomologiegruppen*, Archiv der Math. **30** (1978), 473–476.

27. ____, *Algebraisation of some formal vector-bundles*, Ann. of Math. (2) **110** (1979), 501–514.

28. ____, *Über lokale Kohomologiegruppen hoher Ordnung*, J. Reine Angew. Math. **313** (1980), 43–51.

29. ____, *Formale Geometrie und homogene Räume*, Invent. Math. **64** (1981), 123–165.

30. T. Frankel, *Manifolds with positive curvature*, Pacific J. Math. **11** (1961), 165–174.

31. K. Fritzsche, *q-konvexe Restmengen in kompakten komplexen Mannigfaltigkeiten*, Math. Ann. **221** (1976), 251–273.

32. ____, *Pseudoconvexity properties of complements of analytic subvarieties*, Math. Ann. **230** (1977), 107–122.

33. T. Fujita and J. Roberts, *Varieties with small secant varieties: The extremal case*, Amer. J. Math. **193** (1981), 953–976.

34. W. Fulton, *On the fundamental group of the complement of a node curve*, Ann. of Math. **111** (1980), 407–409.

35. ____, *On the fundamental group of the complement of a plane curve*, Proc. 18th Scandanavian Congr. Math. 1980, Birkhäuser, 1981, pp. 38–55.

36. ____, *On nodal curves*, Lecture Notes in Math., vol. 997, Springer-Verlag, 1983, pp. 146–155.

37. ____, *Some aspects of positivity in algebraic geometry*, Proc. Internat. Congr. Math. (Warsaw, 1983), Part 1, pp. 711–718.

38. W. Fulton and J. Hansen, *A connectedness theorem for projective varieties, with applications to intersections and singularities of mappings*, Ann. of Math. (2) **110** (1979), 159–166.

39. W. Fulton, J. Harris, and R. Lazarsfeld, *Excess linear series on an algebraic curve*, Proc. Amer. Math. Soc. **92** (1984), 320–322.

40. W. Fulton and R. Lazarsfeld, *Connectivity and its applications in algebraic geometry*, Proc. Midwest Algebraic Geom. Conf. 1980, Lecture Notes in Math., vol. 862, Springer-Verlag, 1981, pp. 26–92.

41. ____, *Positive polynomials for ample vector bundles*, Ann. of Math. (2) **118** (1983), 35–60.

42. ____, *Positivity and excess intersections*, Enumerative Geom. and Classical Algebraic Geom., Progr. Math., vol. 24, Birkhäuser, 1982, pp. 97–105.

43. ____, *On the connectedness of degeneracy loci and special divisors*, Acta Math. **146** (1981), 271–283.

44. T. Gaffney and R. Lazarsfeld, *On the ramification of branched coverings of P^n*, Invent. Math. **15** (1972), 67–71.

45. O. Gerstner and L. Kaup, *Homotopiegruppen von Hyperflachenschnitten*, Math. Ann. **204** (1973), 105–130.

46. A. Gimigliano, *Intersezioni complete in prodotti di spazi proiettivi*, Boll. Un. Mat. Ital. D (6) **1** (1982), 229–245.

47. N. Goldstein, *A second Lefschetz for general manifold sections in complex projective space*, Math. Ann. **246** (1979), 41–68.

48. ____, *Ampleness and connectedness in complex G/P*, Trans. Amer. Math. Soc. **274** (1982), 361–373.

49. ____, *Ampleness in complex homogeneous spaces and a second Lefschetz theorem*, Pacific J. Math. **106** (1983), 271–291.

50. M. Goresky and R. MacPherson, *Morse theory and intersection homology*, Asterisque, no. 101, Soc. Math. France, Paris, 1983, pp. 135–192.

51. ____, *Stratified Morse theory*, Proc. Sympos. Pure Math., vol. 40, pt. 1, Amer. Math. Soc., Providence, R.I., 1983, pp. 517–533.

52. ____, *Stratified Morse theory.* II: *Morse theory of Whitney stratified spaces* (to appear as Part 1 of [**54**]).

53. ____, *Stratified Morse theory.* III: *Connectivity theorems for complex analytic varieties* (to appear as part 2 of [**54**]).

54. ____, *Stratified Morse theory, The book*, to appear.

55. P. Griffiths, *Hermitian differential geometry, Chern classes, and positive vector bundles*, Global Analysis, Univ. Tokyo Press, Tokyo, 1969, pp. 195–251.

56. R. V. Gurjar and A. R. Shastri, *Covering spaces of an elliptic surface*, Compositio Math. **54** (1985), 95–104.

57. H. Hamm, *Lokale topologische Eigenschaften komplexer Räume*, Math. Ann. **191** (1971), 235–252.

58. ____, *On the vanishing of local homotopy groups for isolated singularities of complex spaces*, J. Reine Angew. Math. **323** (1981), 172–176.

59. ____, *Zum Homotopietyp Steinscher Räume*, J. Reine Angew. Math. **338** (1983), 121–135.

60. ____, *Lefschetz theorems for singular varieties*, Proc. Sympos. Pure Math., vol. 40, pt. 1, Amer. Math. Soc., Providence, R.I., 1983, pp. 547–558.

61. ____, *Zum Homotopietyp q-vollständiger Räume*, J. Reine Angew. Math. **364** (1986), 1–9.

62. H. Hamm and Lê Dũng Tráng, *Un théorème de Zariski du type de Lefschetz*, Ann. Sci. École Norm. Sup. **6** (1973), 317–366.

63. ____, *Lefschetz theorems on quasi-projective varieties*, Bull. Soc. Math. France **113** (1985), 123–142.

64. ____, *Local generalizations of Lefschetz-Zariski theorems*, Preprint.

65. J. Hansen, *Connectedness theorems in algebraic geometry*, Proc. 18th Scandanavian Congr. Math. 1980, Birkhäuser, 1981, pp. 336–346.

66. ____, *A connectedness theorem for flagmanifolds and Grassmannians*, Amer. J. Math. **105** (1983), 633–639.

67. J. Harris, *On a problem of Severi*, Invent. Math. **84** (1986), 445–461.

68. J. Harris and D. Mumford, *On the Kodaira dimension of the moduli space of curves*, Invent. Math. **67** (1982), 23–86.

69. R. Hartshorne, *Varieties of small codimension of projective space*, Bull. Amer. Math. Soc. **80** (1974), 1017–1032.

70. ____, *Equivalence relations on algebraic cycles and subvarieties of small codimension*, Proc. Sympos. Pure Math., vol. 29, Amer. Math. Soc., Providence, R.I., 1975, pp. 129–164.

71. J.-P. Jouanolou, *Théorèmes de Bertini et applications*, Progr. Math., vol. 42, Birkhäuser, 1983.

72. K. K. Karchyauskas, *A generalized Lefschetz theorem*, Functional Anal. Appl. **11** (1978), 312–313.

73. ____, *Homotopy properties of complex algebraic sets*, J. Soviet Math. **19** (1982), 1253–1257.

74. M. Kato, *Topology of k-regular spaces and algebraic sets*, Manifolds—Tokyo 1973, Univ. Tokyo Press, Tokyo, 1975, pp. 153–159.

75. ____, *Partial Poincaré duality for k-regular spaces and complex algebraic sets*, Topology **16** (1977), 33–50.

76. M. Kato and Y. Matsumoto, *On the connectivity of the Milnor fiber of a holomorphic function at a critical point*, Manifolds—Tokyo 1973, Univ. Tokyo Press, Tokyo, 1975, pp. 131–136.

77. L. Kaup, *Zur Homologie projektiv algebraischer Varietäten*, Ann. Scuola Norm. Sup. Pisa Cl. Sci. (4) **26** (1972), 479–513.

78. S. Kawai, *On finite ramified covering spaces of* P^n, Comm. Math. Univ. Sancti Pauli **30** (1981), 87–103.

79. S. I. Khashin, *The irregularity of double surfaces*, Math. Notes **33** (1983), 233–235.

80. F. Kirwan, *Cohomology of quotients in symplectic and algebraic geometry*, Math. Notes, no. 31, Princeton Univ. Press, Princeton, N.J., 1984.

81. S. L. Kleiman, *Théorème de l'indice de Hodge*, Lecture Notes in Math., vol. 225, Springer-Verlag, 1971, pp. 662–666.

82. S. Kobayashi, *Negative vector bundles and complex projective manifolds*, Nagoya Math. J. **57** (1975), 153–166.

83. K. Lamotke, *The topology of complex projective varieties after S. Lefschetz*, Topology **20** (1981), 15–51.

84. A. Lanteri, *Su un teorema di Chisini*, Rend. Atti Acc. Lincei **66** (1979), 523–532.

85. M. E. Larsen, *On the topology of complex projective manifolds*, Invent. Math. **19** (1973), 251–260.

86. R. Lazarsfeld, *Some applications of the theory of positive vector bundles*, Lecture Notes in Math., vol. 1092, Springer-Verlag, 1984, pp. 29–61.

87. _____, *A Barth-type theorem for branched coverings of projective space*, Math. Ann. **249** (1980), 153–162.

88. R. Lazarsfeld and A. Van de Ven, *Topics in the geometry of projective space, Recent work of F. L. Zak*, DMV Seminar, Band 4, Birkhäuser, 1984.

89. Lê Dũng Tráng and K. Saito, *The local* π_1 *of the complement of a hypersurface with normal crossings in codimension 1 is abelian*, Ark. Mat. **22** (1984), 1–24.

90. S. Lefschetz, *L'analysis situs et la géométrie algébrique*, Gauthier-Villars, Paris, 1924.

91. M. Lehr, *Regular linear systems of curves with the singularities of a given curve as base points*, Amer. J. Math. **54** (1932), 471–488.

92. A. Libgober, *Alexander polynomial of plane algebraic curves and cyclic multiple planes*, Duke Math. J. **49** (1982), 833–851.

93. _____, *Alexander modules of plane algebraic curves*, Contemp. Math., vol. 20, Amer. Math. Soc., Providence, R.I., 1983, pp. 231–247.

94. _____, *Alexander invariants of plane algebraic curves*, Proc. Sympos. Pure Math., vol. 40, pt. 2, Amer. Math. Soc., Providence, R.I., 1983, pp. 135–143.

95. _____, *Homotopy groups of the complements to singular hypersurfaces*, Bull. Amer. Math. Soc. (N.S.) **13** (1985), 49–51.

96. _____, *On* π_2 *of the complements to hypersurfaces which are generic projections*, Preprint.

97. _____, *On the homotopy type of the complement to plane algebraic curves*, J. Reine Angew. Math. **367** (1986), 103–114.

98. J. Milnor, *Morse theory*, Ann. of Math. Stud., vol. 51, Princeton Univ. Press, Princeton, N.J., 1963.

99. _____, *Singular points of complex hypersurfaces*, Ann. of Math. Stud., vol. 61, Princeton Univ. Press, Princeton, N.J., 1968.

100. B. Moishezon, *Stable branch curves and braid monodromies*, Lecture Notes in Math., vol. 862, Springer-Verlag, 1981, pp. 107–192.

101. B. Moishezon and M. Teicher, *Simply-connected algebraic surfaces of positive index*, Invent. Math. (to appear).

102. J. Morgan, *The algebraic topology of smooth algebraic varieties*, Inst. Hautes Études Sci. Publ. Math. **48** (1978), 137–204.

103. S. Mori, *Projective manifolds with ample tangent bundles*, Ann. of Math. (2) **110** (1979), 593–606.

104. M. Namba, *Lectures on branched coverings and algebraic families*, Preprint.

105. A. Nobile, *On families of singular plane projective curves*, Ann. Mat. Pura Appl. (4) (1984), 341–378.

106. M. V. Nori, *Zariski's conjecture and related problems*, Ann. Sci. École Norm. Sup. (4) **16** (1983), 305–344.

107. M. Oka, *On the cohomology structure of projective varieties*, Manifolds—Tokyo 1973, Univ. Tokyo Press, Tokyo, 1975, pp. 137–143.

108. C. Okonek, *A comparison theorem*, Math. Ann. **273** (1986), 271–276.

109. _____, *Barth-Lefschetz theorems for singular spaces*, J. Reine Angew. Math. **374** (1987), 24–38.

110. _____, *Subvarieties in homogeneous manifolds*, Lecture Notes in Math., vol. 1194, Springer-Verlag, 1986, pp. 127–132.

111. _____, *Concavity, convexity and complements in complex spaces*, Lecture Notes in Math., vol. 1194, Springer-Verlag, 1986, pp. 104–126.

112. M. Peternell, *Ein Lefschetz-Satz für Schnitte in projektiv algebraischen Mannigfaltigkeiten*, Math. Ann. (2) **264** (1983), 361–388.

113. _____, *Homotopie in homogenen komplexen Mannigfaltigkeiten*, Math. Z. **188** (1985), 271–278.

114. _____, *Continuous q-convex exhaustion functions*, Invent. Math. **85** (1986), 249–262.

115. R. Pignoni, *Density and stability of Morse functions on a stratified space*, Ann. Scuola Norm. Sup. Pisa Cl. Sci. (4) **4** (1979), 592–608.

116. D. Prill, *The fundamental group of the complement of an algebraic curve*, Manuscripta Math. **14** (1974), 163–172.

117. Z. Ran, *On projective varieties of codimension 2*, Invent. Math. **73** (1983), 333–336.

118. _____, *The structure of Gauss-like maps*, Compositio Math. **52** (1984), 171–177.

119. B. Segre, *Sulla caratterizzazione delle curve di diramazione dei piani multipli generale*, R. Accad. d'Ital. Cl. di Sci. Fis. Mat. et Nat., Memorie **1** (1930), 71–97.

120. F. Severi, *Vorlesungen über algebraische Geometrie*, Anhang. F, Teubner, Leipzig, 1921.

121. B. Shiffman and A. Sommese, *Vanishing theorems on complex manifolds*, Progr. Math., vol. 56, Birkhäuser, 1985.

122. Y.-T. Siu and S.-T. Yau, *Compact Kähler manifolds of positive bisectional curvature*, Invent. Math. **59** (1980), 189–204.

123. B. Smyth and A. Sommese, *On the degree of the Gauss mapping of a submanifold of an Abelian variety*, Comment. Math. Helv. **59** (1984), 341–346.

124. A. Sommese, *Theorems of Barth-Lefschetz type for complex subspaces of homogeneous complex manifolds*, Proc. Nat. Acad. Sci. U.S.A. **74** (1977), 1332–1333.

125. _____, *Subvarieties of Abelian varieties*, Math. Ann. **233** (1978), 229–256.

126. _____, *Concavity theorems*, Math. Ann. **235** (1978), 37–53.

127. _____, *Complex subspaces of homogeneous complex manifolds. I: Transplanting theorems*, Duke Math. J. **46** (1979), 527–548.

128. _____, *Complex subspaces of homogeneous complex manifolds. II: Homotopy results*, Nagoya Math. J. **86** (1982), 101–129.

129. _____, *Concavity theorems. II*, Preprint.

130. A. Sommese and A. Van de Ven, *Homotopy groups of pullbacks of varieties*, Nagoya Math. J. **102** (1986), 79–90.

131. A. Tannenbaum, *Families of algebraic curves with nodes*, Compositio Math. **41** (1980), 107–126.

132. C. Marchionna Tibiletti, *Trecce relative a forme canoniche del gruppo di monodromia*, Rend. Istit. Lombardo **19** (1955), 25–40.

133. _____, *Trecce algebriche di curve di diramazione: costruzione ed applicazioni*, Sem. Mat. Rend. Padova Univ. **23-24** (1955), 183–214.

134. L. Tu, *A note on the connectedness of symmetric degeneracy loci*, Preprint.

135. B. L. Van der Waerden, *Zur algebraischen Geometrie. XI: Projektive und birationale Äquivalenz und Moduln von ebenen Kurven*, Math. Ann. (2) **114** (1937), 683–699.

136. R. O. Wells, *Differential analysis on complex manifolds*, Springer-Verlag, 1980.

137. S. Yamamoto, *Covering spaces of P^2 branched along two non-singular curves with normal crossings*, Comm. Math. Univ. Sancti Pauli **33** (1984), 163–190.

138. F. L. Zak, *Projections of algebraic varieties*, Mat. Sb. (N.S.) **116** (1981), 593–602; English transl. in Math. USSR-Sb. **44** (1983), 535–544.

139. O. Zariski, *On the problem of existence of algebraic functions of two variables possessing a given branch curve*, Amer. J. Math. **51** (1929), 305–328.

140. _____, *On the irregularity of cyclic multiple planes*, Ann. of Math. (2) **32** (1931), 485–511.

BROWN UNIVERSITY

Curves

Proceedings of Symposia in Pure Mathematics
Volume 46 (1987)

Fay's Trisecant Formula
and a Characterization of Jacobian Varieties

ENRICO ARBARELLO

As is well known the space of all principally polarized abelian varieties of dimension g is the generalized Siegel upper-half plane $\mathcal{H}_g$, that is, the space of $g \times g$ complex symmetric matrices with positive imaginary part. To each such matrix τ one associates the complex torus

$$X_\tau = \mathbf{C}^g / \mathbf{Z}^g + \tau \mathbf{Z}^g,$$

and the Riemann theta function

$$\theta(z) = \theta(z, \tau) = \sum_{p \in \mathbf{Z}^g} \exp 2\pi i \left\{ \tfrac{1}{2} {}^t p \tau p + {}^t p z \right\}, \qquad z \in \mathbf{C}^g.$$

This function is quasiperiodic with respect to the lattice $\Lambda_\tau = \mathbf{Z}^g + \tau \mathbf{Z}^g$:

$$\theta(z + n + {}^t m \tau) = \exp 2\pi i \left\{ -\tfrac{1}{2} {}^t m \tau m - {}^t m z \right\} \theta(z),$$

for $n, m \in \mathbf{Z}^g$. Therefore the zero locus of the Riemann theta function defines a divisor $\Theta_\tau \subset X_\tau$ which is a principal polarization of X_τ in the sense that Θ_τ is ample and the linear system $|\Theta_\tau|$ is 0-dimensional.

The geometry of the linear system $|2\Theta_\tau|$ is particularly interesting. It can be described in terms of second-order theta functions. Given any half integer vector $n \in \tfrac{1}{2} \mathbf{Z}^g / \mathbf{Z}^g$, one defines the second-order theta function with characteristics $(n, 0)$ as follows:

$$\theta[n](z, \tau) = \sum_{p \in \mathbf{Z}^g} \exp 2\pi i \left\{ \tfrac{1}{2} {}^t (p + n) \tau (p + n) + {}^t (p + n) z \right\}.$$

Using the Riemann-Roch theorem and a direct computation, one can see that the 2^g functions

$$\hat{\theta}[n](z, \tau) = \theta[n](2z, 2\tau), \qquad n \in \tfrac{1}{2} \mathbf{Z}^g / \mathbf{Z}^g,$$

form a basis for the vector space of sections of $\mathcal{O}(\Theta_\tau)$. The morphism defined by the linear system $|2\Theta_\tau|$ is then given by

$$\begin{aligned} \vec{\theta} : \quad X_\tau &\to \quad \mathbf{P}^N, \qquad N = 2^g - 1, \\ \varsigma &\to \quad \vec{\theta}(\varsigma) = [\dots, \hat{\theta}[n](z, \tau), \dots]. \end{aligned}$$

1980 *Mathematics Subject Classification* (1985 *Revision*). Primary 14K25, 14H40.

This is a two-to-one morphism. Its image is the so-called Kummer variety $K(X_\tau)$. The Kummer variety has degree $2^g g!$ and it is smooth except at the images of the points of order two of X_τ, where it is singular.

Many identities involving the theta function can be geometrically interpreted in terms of the Kummer variety by means of a fundamental identity discovered by Riemann:

$$(1) \qquad \theta(z+\varsigma)\theta(z-\varsigma) = \sum_{n\in(1/2)\mathbf{Z}^g/\mathbf{Z}^g} \hat\theta[n](z)\hat\theta[n](z).$$

Let us now consider a compact Riemann surface C of genus $g > 1$ and its Jacobian

$$J(C) = H^{1,0}(C)^*/H_1(C,\mathbf{Z}).$$

The principally polarized abelian variety $J(C)$ corresponds to the period matrix

$$\tau = \left(\int_{b_j} \omega_i \right), \qquad i,j = 1,\ldots,g,$$

where $a_1,\ldots,a_g,b_1,\ldots,b_g$ is a symplectic basis of $H_1(C,\mathbf{Z})$ and $\omega_1,\ldots,\omega_g$ a basis of the vector space of holomorphic differentials on C normalized in such a way that $\int_{a_j}\omega_i = \delta_{ij}$. We set $\vec\omega = (\omega_1,\ldots,\omega_g)$. Fixing a point $p_0 \in C$ the Abel-Jacobi map $C \to J(C)$ is defined by $p \mapsto \int_{p_0}^P \vec\omega$. We shall denote by Γ the (isomorphic) image of C under the Abel-Jacobi map.

We now come to Fay's trisecant formula. *Given any three points α,β,γ on Γ and any point ς on $\frac{1}{2}(\Gamma - \alpha - \beta - \gamma)$* (here $\frac{1}{2}$ denotes the inverse of the multiplication by 2 map on $J(C)$), *then there exist constants c_1,c_2 such that*

$$(2) \quad \theta(z-\alpha)\theta(z+2\varsigma+\alpha) = c_1\theta(z-\beta)\theta(z+2\varsigma+\beta) + c_2\theta(z-\gamma)\theta(z+2\varsigma+\gamma).$$

For a proof of this formula one can see [**FY**, p. 34], [**M3**, p. 214], or [**F**]. Using Riemann's relation (1) the trisecant formula (2) becomes

$$\sum_{n\in(1/2)\mathbf{Z}^g/\mathbf{Z}^g} (\hat\theta[n](\varsigma+\alpha) - c_1\hat\theta[n](\varsigma+\beta) - c_2\hat\theta[n](\varsigma+\gamma)) \cdot \hat\theta[n](z+\varsigma) = 0.$$

Therefore by the linear independence of the $\hat\theta[n](z)$'s, the trisecant formula is equivalent to the system

$$(3) \qquad \hat\theta[n](\varsigma+\alpha) - c_1\hat\theta[n](\varsigma+\beta) - c_2\hat\theta[n](\varsigma+\gamma) = 0,$$
$$\varsigma \in \tfrac{1}{2}(\Gamma - \alpha - \beta - \gamma), \ n \in \tfrac{1}{2}\mathbf{Z}^g/\mathbf{Z}^g.$$

We can write this system in vector notation as

$$(4) \qquad \vec\theta(\varsigma+\alpha) = c_1\vec\theta(\varsigma+\beta) + c_2\vec\theta(\varsigma+\gamma) = 0, \qquad \varsigma \in \tfrac{1}{2}(\Gamma - \alpha - \beta - \gamma)$$

or, equivalently, as

$$(5) \qquad \vec\theta(\varsigma+\alpha) \wedge \vec\theta(\varsigma+\beta) \wedge \vec\theta(\varsigma+\gamma) = 0, \qquad \varsigma \in \tfrac{1}{2}(\Gamma - \alpha - \beta - \gamma).$$

We can therefore say that for any choice of points α,β,γ on Γ the curve $\frac{1}{2}(\Gamma - \alpha - \beta - \gamma)$ parametrizes trisecant lines to the Kummer variety of $J(C)$; in

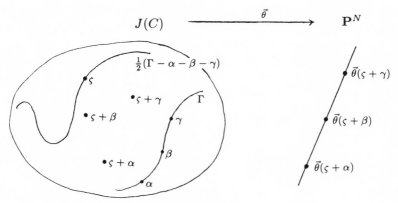

FIGURE 1

in fact, (5) says that for any $\varsigma \in \frac{1}{2}(\Gamma - \alpha - \beta - \gamma)$ the three points $\vec{\theta}(\varsigma + \alpha)$, $\vec{\theta}(\varsigma + \beta), \vec{\theta}(\varsigma + \gamma)$ are collinear (see Figure 1).

Gunning in [**G**] showed that the trisecant formula can be efficiently used to characterize Jacobians among all abelian varieties. He shows that *an irreducible principally polarized abelian variety X* (irreducible means that the theta divisor of X is irreducible) *is a Jacobian if and only if there exists an irreducible curve Γ in X such that for general points α, β, γ on Γ and for any point ς on $\frac{1}{2}(\Gamma - \alpha - \beta - \gamma)$ the three points $\vec{\theta}(\varsigma + \alpha), \vec{\theta}(\varsigma + \beta)$, and $\vec{\theta}(\varsigma + \gamma)$ are collinear.*

Gunning introduces the subvariety $\tilde{V}_{\alpha,\beta,\gamma} \subset X$ defined by

$$\tilde{V}_{\alpha,\beta,\gamma} = \{\varsigma \in X : \vec{\theta}(\varsigma + \gamma) \wedge \vec{\theta}(\varsigma + \beta) \wedge \vec{\theta}(\varsigma + \gamma) = 0\}$$

and, using Matsusaka's criterion, shows that the existence of Γ is equivalent to the following conditions:

(6) $\dim_{-\alpha-\beta} 2\tilde{V}_{\alpha,\beta,\gamma} > 0$ for some α, β, γ on X;

(7)   there is no complex multiplication on X mapping $\beta - \alpha$ and
         $\gamma - \alpha$ into 0.

Therefore conditions (6) and (7) are necessary and sufficient for X to be the Jacobian of a curve. In fact if (6) and (7) are satisfied, then $2\tilde{V}_{\alpha,\beta,\gamma}$ turns out to be a smooth curve Γ and X is the Jacobian of Γ.

Let us go back to the trisecant formula (3). As Mumford notices in [**M4**] and as was more implicitly observed in [**FY**], when the points α, β, γ on Γ tend to $0 \in \Gamma$, one then gets the Kodomcev-Petviashvili equation
(8)
$$D_1^4\theta(z) \cdot \theta(z) - 4D_1^3\theta(z) \cdot D_1\theta(z) + 3(D_1^2\theta(z))^2 - 3(D_2\theta(z))^2$$
$$+ 3D_2^2\theta(z) \cdot \theta(z) + 3D_1\theta(z) \cdot D_3\theta(z) - 3D_1 D_3\theta(z) \cdot \theta(z) + d_4\theta(z) \cdot \theta(z) = 0,$$

where d_4 is a suitable constant and where D_1, D_2, and D_3 are constant vector fields in $\mathbf{C}^g$ defined as follows. Write $\vec{\omega}$ in local coordinates near $p_0 \in C$ as

$\vec{\omega} = 2(W^{(1)} + W^{(2)}\varepsilon + W^{(3)}\varepsilon^2 + \cdots)d\varepsilon$, where $W^{(i)} = (W_1^{(i)}, \ldots, W_g^{(i)}) \in \mathbf{C}^g$,
and then set

$$D_i = \sum_{j=1}^{g} W_j^{(i)} \frac{\partial}{\partial \varsigma_j}.$$

The equation (8) is actually equivalent to the original K.P. equation, which is the following:

$$\frac{\partial}{\partial x}(2u_{xxx} + 3uu_x - u_t) + 3u_{yy} = 0,$$

where

$$u(x, y, t; z) = \frac{\partial^2}{\partial x^2} lg\theta(xW^{(1)} + yW^{(2)} + tW^{(3)} + z) + c_1.$$

The fact that the theta function of a Riemann surface satisfies the K.P. equation was already known. Novikov, Krichever, and Dubrovin (see, for instance, [**DU1, K, N**]) studied extensively θ-function solutions of the K.d.V. and K.P. equations. It was Novikov who first conjectured that, among all abelian varieties, Jacobians are characterized by the fact that their theta function satisfies the K.P. equation, thus providing an analytical characterization of the "Jacobian locus" $\mathcal{J}_g \subset \mathcal{H}_g$. In [**DU2**] Dubrovin gives good evidence of the validity of this conjecture. He proves a statement which is only a little bit weaker than the following: the Jacobian locus is characterized by the K.P. equation up to irreducible components. The first proof of the validity of Novikov's conjecture is due to T. Shiota [**S**] and follows the point of view first taken by Krichever, then analyzed by Mumford [**M2**], and finally enriched by the contributions of the Japanese school (see, for instance, [**D**]).

Here we take the "trisecant point view," which also leads to a proof of Novikov's conjecture (see [**W1, W3, AD1, AD2**]).

Motivated by Gunning's criterion and by the relation between the trisecant formula and the K.P. equation, Welters gives an infinitesimal version of Gunning's criterion which is, at the same time, a drastic improvement of it. The criterion is the following:

Let X be an irreducible principally polarized abelian variety (i.p.p.a.v.) of dimension g. Then X is a Jacobian if and only if there exist constant vector fields $D_1 \neq 0$, D_2 such that

(9) $\dim_0 \tilde{V}_{D_1, D_2} > 0,$

where

(10) $\tilde{V}_{D_1, D_2} = \{\varsigma \in X : \vec{\theta}(\varsigma) \wedge D_1\vec{\theta}(\varsigma) \wedge (D_1^2 + D_2)\vec{\theta}(\varsigma) = 0\}.$

Moreover if (9) holds, then $\tilde{V}_{D_1, D_2}$ is a smooth curve and X is the Jacobian of $\Gamma = 2\tilde{V}_{D_1, D_2}$.

One can easily see that, in this formulation, trisecant lines are substituted with inflectionary lines, in the sense that the images, under $\vec{\theta}$, of the points of $\tilde{V}_{D_1, D_2}$ are inflection points of the Kummer variety, that is, points p for which there exists a line in $\mathbf{P}^N$ having intersection multiplicity with the Kummer at p

which is greater than or equal to three. So $\vec{\theta}(\tilde{V}_{D_1,D_2})$ is a curve of flexes of the Kummer. Another thing is also evident. Let us look at the second-order germ of curve

$$\bar{\zeta}(\varepsilon) = \varepsilon W^{(1)} + \varepsilon^2 W^{(2)},$$

where $W^{(i)} = (W_1^{(i)}, \ldots, W_g^{(1)}) \in \mathbf{C}^g$ and $D_i = \sum_{j=1}^g W_j^{(i)} \partial/\partial \varsigma_j$, $i = 1, 2$. It is then easy to check that $\tilde{V}_{D_1,D_2}$ coincides, up to second order, with this germ; so that, in particular,

(11) $$\vec{\theta}(\bar{\zeta}(\varepsilon)) \wedge D_1 \vec{\theta}(\bar{\zeta}(\varepsilon)) \wedge (D_1^2 + D_2)\vec{\theta}(\bar{\zeta}(\varepsilon)) = O(\varepsilon^3).$$

We shall now start, following [AD1], to translate Welters's criterion into analytical terms. Clearly condition (9) is equivalent to the existence of an infinite order germ of curve

$$\varsigma(\varepsilon) = \varepsilon W^{(1)} + \varepsilon^2 W^{(2)} + \varepsilon^{(3)} W^{(3)} + \cdots$$

contained in $\tilde{V}_{D_1,D_2}$. The fact that, up to second order, this germ coincides with $\bar{\zeta}(\varepsilon)$ is dictated by the remark we just made. Therefore condition (9) can now be expressed in the equivalent form

(12) $$\vec{\theta}(\varsigma(\varepsilon)) \wedge D_1 \vec{\theta}(\varsigma(\varepsilon)) \wedge \bar{\Delta}_2 \vec{\theta}(\varsigma(\varepsilon)) = 0,$$

where $\bar{\Delta}_2 = D_1^2 + D_2$.

We now make a general remark. Let $W^{(1)}, W^{(2)} W^{(3)}, \ldots$ be as above. Let $W^{(i)} = (W_1^{(i)}, \ldots, W_g^{(i)})$ and set $D_i = \sum_{j=1}^g W_j^{(i)} \partial/\partial \varsigma_j$. Also set

$$\Delta_s = \Delta_s(D_1, \ldots, D_s) = \sum_{i_1 + 2i_2 + \cdots + si_s = s} \frac{1}{i_1! \cdots i_s!} D_1^{i_1} \cdots D_s^{i_s}.$$

Then it is easy to check that for a C^∞ function $f(z)$ on $\mathbf{C}^g$ one has

(13) $$f(\varsigma + \varsigma(\varepsilon)) = \sum_{s \geq 0} \Delta_s(f)\varepsilon^s.$$

We also set $D(\varepsilon) = \sum_{i \geq 1} D_i \varepsilon^i$ so that

(14) $$e^{D(\varepsilon)} = \sum_{s \geq 0} \Delta_s \varepsilon^s.$$

The equation (12) can now be written as

(15) $$e^{D(\varepsilon)}(\vec{\theta}(\varsigma) \wedge D_1 \vec{\theta}(\varsigma) \wedge \bar{\Delta}_2 \vec{\theta}(\varsigma))\big|_{\varsigma=0} = 0.$$

This, in turn, is equivalent to the existence of three relatively prime elements $d(\varepsilon), c(\varepsilon), b(\varepsilon)$ in $\mathbf{C}[[\varepsilon]]$ such that

(16) $$(d(\varepsilon)e^{D(\varepsilon)}\vec{\theta} + c(\varepsilon)e^{D(\varepsilon)}D_1\vec{\theta} + b(\varepsilon)e^{D(\varepsilon)}\bar{\Delta}_2\vec{\theta})\big|_{\varsigma=0} \equiv 0.$$

Computing at $\varepsilon = 0$ we get (recall that $\hat{\theta}[n]$ is an even function)

$$(d(0)\vec{\theta}(0) + b(0)\bar{\Delta}_2\vec{\theta}(0))\big|_{\varsigma=0} = 0.$$

Since the theta divisor Θ is irreducible we get $d(0) = b(0) = 0$. In particular, since d, c, b are relatively prime, this means that c is invertible so that we may assume $c = 1$. We now compute the coefficient of ε on the left-hand side of (16); this is $\Delta_1^2\vec\theta(0) + b_1\Delta_1^2\vec\theta(0)$, where $b(\varepsilon) = \sum_{i\geq 1} b_i\varepsilon^i$. Thus $b_1 = -1$. Changing variables we can assume $\varepsilon = -1$. This amounts to changing each D_i by adding to it a linear combination of $D_1, \ldots, D_{i-1}$, which is of course an allowable operation. Now set $d(\varepsilon) = \sum_{i=1}^\infty d_{i+1}\varepsilon^i$. The equation (16) modulo ε^3 is $\varepsilon\vec\theta(0)(d_2+\varepsilon d_3) = 0$. Hence $d_2 = d_3 = 0$. Now equation (16) becomes

$$(d(\varepsilon)e^{D(\varepsilon)}\vec\theta + e^{D(\varepsilon)}\Delta_1\vec\theta - \varepsilon e^{D(\varepsilon)}\bar\Delta_2\vec\theta)\big|_{\varsigma=0} = 0.$$

Looking at the coefficient of ε^s, we get

(17) $$\left[\Delta_s\Delta_1 - \Delta_{s-1}\bar\Delta_2 + \sum_{i=3}^s d_{i+1}\Delta_{s-i}\right]\vec\theta(\varsigma)\Big|_{\varsigma=0} = 0, \qquad s \geq 1.$$

Using Riemann's relation (1) and following the same method that showed how (2) and (4) are equivalent, we can see that (17) is equivalent to

(18) $$\left[\Delta_s\Delta_1 - \Delta_{s-1}\bar\Delta_2 + \sum_{i=\varsigma}^s d_{i+1}\Delta_{s-i}\right]\theta(z+\varsigma)\theta(z-\varsigma)\Big|_{\varsigma=0} = 0, \qquad s \geq 1.$$

We shall use the following notation

(19)
$$P_s(z) = P_s(D_1, \ldots, D_s, d_4, \ldots, d_{s+1})(z)$$
$$= \left[\Delta_s\Delta_1 - \Delta_{s-1}\bar\Delta_2 + \sum_{i=\varsigma}^s d_{i+1}\Delta_{s-i}\right]\theta(z+\varsigma)\theta(z-\varsigma)\Big|_{\varsigma=0}.$$

As we suggest in the notation, $P_s(z)$ only depends on the knowledge of $D_1, \ldots, D_s$ and $d_4, \ldots, d_{s+1}$.

We can therefore conclude with the following analytical translation of Welters's criterion.

Let X be an i.p.p.a.v. Then X is a Jacobian if and only if there exist constant vector fields $D_1 \neq 0$, $D_2, D_3, \ldots$ and complex numbers $d_4, d_5, \ldots$ such that $P_s(z) = 0$, $s \geq 1$.

An easy computation shows that $P_1(z)$ and $P_2(z)$ are identically zero while the equation $P_3(z) = 0$ is exactly the K.P. equation (8). Therefore in order to prove Novikov's conjecture, we must show that

(20) $$P_3(z) = 0 \Rightarrow \exists D_4, D_5, \ldots, d_5, d_6, \ldots \text{ such that } P_s(z) = 0, \ s \geq 4.$$

This is proved in [AD2]. Here we shall present a procedure that points towards a completely algebraic proof of (20) and which works, at present, only under some additional hypotheses on X.

We start by making a geometrical consideration. Consider the theta divisor $\Theta \subset X$. The function $D_1\theta$, when restricted to Θ, can be considered as a section of the line bundle $\Theta|_\Theta$. Its zero locus defines a divisor on Θ which we denote by $D_1\Theta$, so that

$$D_1\Theta = \{\varsigma \in X : \theta(\varsigma) = D_1\theta(\varsigma) = 0\}.$$

Let us look at the K.P. equation (8) and let us restrict it to $D_1\Theta$. We get

$$[(D_1^2\theta)^2 - (D_2\theta)^2]\big|_{D_1\Theta} = 0$$

or, equivalently,

(21) $$[(D_1^2 + D_2)\theta][(D_1^2 - D_2)\theta]\big|_{D_1\Theta} = 0.$$

We then see that if the K.P. equation is satisfied, then

(22) $$\{\theta = 0\} \cap \{D_1\theta = 0\} \subset \{(D_1^2 + D_2)\theta = 0\} \cup \{(D_1^2 - D_2)\theta = 0\}.$$

This "splitting" of $D_1\Theta$ is not surprising. Let us go back to the trisecant formula (2). One sees that

(23) $$\Theta_\beta \cap \Theta_\gamma \subset \Theta_\alpha \cup \Theta_{-\alpha+2\varsigma},$$

where $\Theta_x = \Theta + x$. The importance of this formula has been pointed out by Mumford on several occasions [**M1, M3, M4**]. Then, in [**W2**], Welters shows that, under some additional hypotheses, this splitting indeed characterizes Jacobian varieties among all abelian varieties. Now, as one easily sees, the statement (22) is nothing other than the infinitesimal version of (23) when $\alpha, \beta, \gamma,$ and 2ς tend to 0. The splitting of Θ has been recently studied from a new point of view by A. Beauville and O. Debarre [**BD**].

What is perhaps more surprising about formula (22), or equivalently about (21), is how simple becomes the K.P. equation when restricted to $D_1\Theta$, especially if one observes that in (21) the vector field D_3 and the constant d_4 have disappeared. It is then very natural to analyze what happens to the equations $P_s(z) = 0$ when restricted to $D_1\Theta$.

First of all observe that

$$\Delta_s\Delta_1 - \Delta_{s-1}\bar{\Delta}_2 + \sum_{i=\varsigma}^{s} d_{i+1}\Delta_{s-1}$$

is a differential polynomial in $D_1, \ldots, D_s$ in which the only term involving D_s is the term D_sD_1 coming from $\Delta_s\Delta_1$. Therefore the only term in $P_s(z)$ involving D_s and d_{s+1} is

$$2D_1D_s\theta(z) \cdot \theta(z) - 2D_1\theta(z) \cdot D_s\theta(z) + d_{s+1}\theta(z) \cdot \theta(z).$$

Now this term vanishes when restricted to $D_1\Theta$. Therefore we observe for $P_s(z)$ the same phenomenon happening to $P_3(z)$. Let us formalize this by setting

$$P'_s(z) = P_s(D_1, \ldots, D_{s-1}, 0, d_4, \ldots, d_s, 0)(z).$$

Then the observation we just made can be written as follows:

$$P'_s\big|_{D_1\Theta} = P_s\big|_{D_1\Theta}.$$

We are now going to prove that (see [**AD2**])

(24) $$P'_s(z)|_{D_1\Theta} = 0 \Leftrightarrow \exists D_s, d_{s+1} \text{ such that } P_s(z) = 0.$$

So that to solve the equation $P_s(z) = 0$ on X it will suffice to solve it on $D_1\Theta$; in particular, we see that (22) is *equivalent* to the K.P. equation (8). One implication in (24) is trivial. To prove the other we first observe that the quasiperiodicity factor of $\theta(z + \varsigma)\theta(z - \varsigma)$ is an exponential which only depends on z. On the other hand, the D_i's are derivations in the ς variables, and, using this, one easily checks that $P_s(z)$ *is a well-defined section of* 2Θ. We now look at the exact sequences

(25)
$$0 \longrightarrow \mathcal{O}_X(\Theta) \xrightarrow{\cdot\theta} \mathcal{O}_X(2\Theta) \longrightarrow \mathcal{O}_\Theta(2\Theta) \longrightarrow 0$$
$$0 \longrightarrow \mathcal{O}_\Theta(\Theta) \xrightarrow{\cdot D_1\theta} \mathcal{O}_\Theta(2\Theta) \longrightarrow \mathcal{O}_{D_1\Theta}(2\Theta) \longrightarrow 0$$

We are assuming that $P'_s(z)|_{D_1\Theta} = 0$. Now we recall (see, for instance, [**I**]) that $H^0(\Theta, \mathcal{O}_\Theta(\Theta))$ is generated by $\partial\theta/\partial z_1, \ldots, \partial\theta/\partial z_g$. Looking at the second exact sequence in (25), it follows that there exists a constant vector field D_s such that

$$P'_s(z)\big|_\Theta = 2D_1\theta \cdot D_s\theta\big|_\Theta.$$

Let us then look at the section of $\mathcal{O}_X(\Theta)$ given by

$$P'_s(z) + D_1 D_s\theta(z + \varsigma)\theta(z - \varsigma)|_{\varsigma=0} = P'_s(z) - 2D_1\theta \cdot D_s\theta + 2D_1 D_s\theta \cdot \theta.$$

Its restriction to Θ is zero. Looking at the first exact sequence in (25), we can then find a complex number d_{s+1} such that

$$P'_s(z) - 2D_1\theta \cdot D_s\theta + 2D_1 D_s\theta \cdot \theta = -d_{s+1}\theta \cdot \theta,$$

proving that there exist D_s and d_{s+1} such that $P_s(z) = 0$.

We can now conclude that in order to prove (20) it suffices to prove that

(26) $$P_3(z) = \cdots = P_{s-1}(z) \Rightarrow 36 P_s(z)\Big|_{D_1\Theta} = 0.$$

We are now going to show that, exactly as it happened for $P_3(z)$ (the K.P. equation), $P_s(z)$ also simplifies a great deal when restricted to $D_1\Theta$ and when we assume that $P_3 = \cdots = P_{s-1} = 0$. (Working with P_s modulo $P_3, \ldots, P_{s-1}$ will be very useful. In fact, when we assume that $P_3 = \cdots = P_{s-1} = 0$, we are also tacitly assuming that all the z-derivatives of: $P_3, \ldots, P_{s-1}$ vanish.) We have

$$P(z, \varepsilon) =: \sum_{s \geq 1} P_s(z)\varepsilon^s$$

$$= \sum_{s \geq 1} \left[\Delta_s\Delta_1 - \Delta_{s-1}\bar\Delta_2 + \sum_{i=3}^{s} d_{i+1}\Delta_{s-i}\right] \theta(z + \varsigma)\theta(z - \varsigma)\big|_{\varsigma=0}\varepsilon^s$$

$$= \left[\Delta_1 - \varepsilon\bar\Delta_2 + \sum_{i=3}^{\infty} d_{i+1}\varepsilon^i\right] \theta(z + \varsigma(\varepsilon) + \varsigma)\theta(z - \varsigma(\varepsilon) - \varsigma)\big|_{\varsigma=0},$$

$$R(z,\varepsilon) =: P(z + \varsigma(\varepsilon), \varepsilon) = e^{\sum D_i \varepsilon^i} P(z, \varepsilon)$$

$$= \sum_{s \geq 0} \sum_{i=0}^{s} \Delta_i P_{s-i} \qquad (P_0 = P_1 = P_2 = 0).$$

(Here we consider the D_i's as operators in the z-variables.)

If we set

$$R(z, \varepsilon) = \sum_{s \geq 0} R_s(z) \varepsilon^s,$$

we then get

$$R_s(z) = P_s(z) + \Delta_1 P_{s-1}(z) + \cdots + \Delta_s P_0(z).$$

Therefore working with P_s modulo $P_3, \ldots, P_{s-1}$ is like working with R_s modulo $R_3, \ldots, R_{s-1}$ meaning that

(27) $$P_3 = \cdots = P_{s-1} = 0 \Leftrightarrow R_3 = \cdots = R_{s-1} = 0.$$

In particular (26), which is what we must prove, is equivalent to

(28) $$R_3(z) = \cdots = R_{s-1}(z) = 0 \Rightarrow R_s(z)\big|_{D_1 \Theta} = 0.$$

We now compute $R_s(z)$. We have

$$R(z,\varepsilon) = \left[\Delta_1 - \varepsilon \bar{\Delta}_2 + \sum_{i=3}^{\infty} d_{i+1}\varepsilon^i \right] \theta(z + 2\varsigma(\varepsilon) + \varsigma)\theta(z - \varsigma)\big|_{\varsigma=0}$$

$$= \left[\Delta_1 - \varepsilon\bar{\Delta}_2 + \sum_{i=3}^{\infty} d_{i+1}\varepsilon^i \right] \cdot \left[\left(\sum_{k \geq 0} \tilde{\Delta}_k \theta(z + \varsigma)\varepsilon^k \right) \theta(z - \varsigma) \right] \Big|_{\varsigma=0},$$

where $\tilde{\Delta}_k(D_1, \ldots, D_k) = \Delta_k(2D_1, \ldots, 2D_k)$. Hence
(29)
$$R_k(z) = \Delta_1(\tilde{\Delta}_k\theta(z + \varsigma) \cdot \theta(z - \varsigma))\big|_{\varsigma=0} - \bar{\Delta}_2(\tilde{\Delta}_{k-1}\theta(z + \varsigma) \cdot \theta(z - \varsigma))\big|_{\varsigma=0}$$

$$+ \sum_{i=3}^{k} d_{i+1}\tilde{\Delta}_{k-i}\theta(z) \cdot \theta(z)$$

$$= \Delta_1\tilde{\Delta}_k\theta(z) \cdot \theta(z) - \tilde{\Delta}_k\theta(z) \cdot \Delta_1\theta(z) - D_1^2\tilde{\Delta}_{k-1}\theta(z) \cdot \theta(z)$$

$$+ 2D_1\tilde{\Delta}_{k-1}\theta(z) \cdot D_1\theta(z) - \tilde{\Delta}_{k-1}\theta(z) \cdot D_1^2\theta(z) - D_2\tilde{\Delta}_{k-1}\theta(z) \cdot \theta(z)$$

$$+ \tilde{\Delta}_{k-1}\theta(z) \cdot D_2\theta(z) + \sum_{i=3}^{k} d_{i+1}\tilde{\Delta}_{k-i}\theta(z) \cdot \theta(z).$$

This is the formula which best generalizes the expression (8) for the K.P. equation. One also sees that

(30) $$R_s(z)\big|_{D_1\Theta} \equiv -\tilde{\Delta}_{k-1}\theta(z) \cdot (D_1^2 - D_2)\theta(z)\big|_{D_1\Theta}.$$

This formula directly generalizes (21). We are now going to prove (28). We shall do so under two additional hypotheses. The first one is easily stated and concerns the singular locus of the theta divisor; namely, we shall assume

(31) $$\dim \Theta_{\text{sing}} \leq g - 3.$$

The second one is going to be stated in terms of a subscheme of X which plays an important role in both [**S**] and [**AD2**]. We define

$$\Sigma = \{ \varsigma \in X : D_1^n \theta(z) = 0 \ \forall n \geq 0 \}.$$

Our second assumption is

(32) $$\dim \Sigma \leq g - 3.$$

Under these assumptions we are now going to prove (28). We therefore suppose that

$$R_3 = \cdots = R_{s-1} = 0,$$

and we try to show that

(33) $$\tilde{\Delta}_{s-1}\theta \cdot (D_1^2 - D_2)\theta \big|_{D_1\Theta} = 0.$$

Then let V be a component of $D_1\Theta$. We want to show that $\tilde{\Delta}_{s-1}\theta \cdot (D_1^2 - D_2)\theta$ vanishes on V. If V is reduced, this is easy. Observe in fact that, by hypothesis, $P_3 = 0$, so that

$$(D_1^2 + D_2)\theta \cdot (D_1^2 - D_2)\theta = 0$$

on V. Since V is reduced, then either $(D_1^2 + D_2)\theta$ or $(D_1^2 - D_2)\theta$ vanishes on V. If $(D_1^2 - D_2)\theta$ vanishes on V, then (33) is trivially satisfied. We can therefore assume that $(D_1^2 + D_2)\theta$ vanishes on V and that $(D_1^2 - D_2)\theta$ does not. Now, since by hypothesis $R_{s-1}(z) = 0$, we also have that $\Delta_1 R_{s-1}(z) = 0$, and a straightforward computation, based on the expression (29), shows that

(34) $$\Delta_1 R_{s-1}\big|_{D_1\Theta} = [-\tilde{\Delta}_{s-1}\theta \cdot D_1^2\theta + D_1\tilde{\Delta}_{s-2}\theta \cdot D_1^2\theta - \tilde{\Delta}_{s-2}\theta \cdot D_1^3\theta$$
$$+ \Delta_1\tilde{\Delta}_{s-2}\theta \cdot D_2\theta + \tilde{\Delta}_{s-2}\theta \cdot D_1 D_2\theta]\big|_{D_1\Theta} = 0.$$

On the other hand, since $R_{s-1}|_{D_1\Theta} = 0$ and since we are assuming that $(D_1^2 - D_2)\theta \neq 0$ on V, we can conclude that $\tilde{\Delta}_{s-2}\theta$ vanishes on V so that from (34) we get

(35)
$$\Delta_1 R_{s-1}\big|_V = [-\tilde{\Delta}_{s-1}\theta \cdot D_1^2\theta + D_1\tilde{\Delta}_{s-1}\theta \cdot (D_1^2 + D_2)\theta]\big|_V$$
$$= -\tilde{\Delta}_{s-1}\theta \cdot D_1^2\theta\big|_V = 0.$$

But now the assumption that $(D_1^2 - D_2)\theta|_V \neq 0$ and $(D_1^2 + D_2)\theta|_V = 0$ implies that $D_1^2\theta|_V \neq 0$, showing that $\tilde{\Delta}_{s-1}\theta$ vanishes on V and therefore proving that

$$R_{s-1}|_V = \tilde{\Delta}_{s-1}\theta \cdot (D_1^2 - D_2)\theta|_V = 0.$$

We are now going to treat the case in which V is nonreduced. Let x be a general point in V. By our assumption on the singular locus of Θ, x is a smooth point of Θ, and therefore there is an irreducible element $h \in \mathcal{O}_{X,x}$ such that the ideal of V in x is of the form (h^m, θ) for some $m \geq 1$. Since the reduced case has already been treated, we may assume that $m \geq 2$. Also because of the assumption (32) we may assume that h does not divide $D_1(h)$, otherwise V

would be invariant under the D_1-flow and would then be a $(g-2)$-dimensional subscheme of Σ. We have

(36)
$$
\begin{aligned}
D_1\theta &\equiv h^m + a\theta, \\
D_1^2\theta &\equiv mh^{m-1}D_1(h) \quad (\mathrm{mod}(\theta, h^m)), \\
D_2\theta &\equiv ch^s \quad (\mathrm{mod}\,\theta), \\
D_3\theta &\equiv \gamma h^t \quad (\mathrm{mod}\,\theta), \\
D_1^3\theta &\equiv m(m-1)D_1(h)^2 h^{m-2} \quad (\mathrm{mod}(\theta, h^m)),
\end{aligned}
$$

where a is an element in $\mathcal{O}_{X,x}$ and c, γ are invertible elements of $\mathcal{O}_{X,x}$.

Observe that $s \geq 1$. In fact since $(D_1^2 - D_2)\theta \cdot (D_1^2 + D_2)\theta$ vanishes on V, then either $(D_1^2 - D_2)\theta$ vanishes on V_{red} or $(D_1^2 + D_2)\theta$ does so that $D_1^2\theta = \pm D_2\theta$ on V_{red}. But since $m \geq 2$, we get $s \geq 1$. Actually more is true. In fact we have

(37)
$$
\begin{aligned}
D_2 D_1\theta &\equiv mD_2(h)h^{m-1} + ach^s \quad (\mathrm{mod}(h^m, \theta)), \\
D_1 D_2\theta &\equiv scD_1(h)h^{s-1} + D_1(c)h^s \quad (\mathrm{mod}(h^m, \theta)),
\end{aligned}
$$

and since h does not divide $D_1(h)$ we must have $s \geq m$. We are now going to prove that $m = 2$. (For this proof we follow a suggestion of De Concini.) Plug (36) into the K.P. equation (8) and consider it modulo (h^{2m-1}, θ). We get

$$
-4m(m-1)D_1(h)^2 h^{2m-2} + 3m^2 D_1(h)^2 h^{2m-2} + 3\gamma h^{m+t} \equiv 0 \quad (\mathrm{mod}(h^{2m-1}, \theta)).
$$

If $t = 0$, then we must have $m = 2$ and we are done. If $t > 0$, then proceeding as in (37) we see that t must be greater than or equal to m, and therefore, since h does not divide $D_1(h)$, we must have $4m(m-1) - 3m^2 = 0$, which gives $m = 4$. Now take the D_1-derivative of the K.P. equation (8) and restrict it to Θ. We get

$$
\begin{aligned}
[D_1^4\theta \cdot D_1\theta - 4D_1^4\theta \cdot D_1\theta - 4D_1^3\theta \cdot D_1^2\theta + 6D_1^3\theta \cdot D_1^2\theta + 3D_2^2\theta \cdot D_1\theta \\
- 6D_2\theta \cdot D_1 D_2\theta - 3D_1 D_3\theta \cdot D_1\theta + 3D_1^2\theta \cdot D_3\theta + 3D_1\theta \cdot D_1 D_3\theta]\big|_\Theta \equiv 0.
\end{aligned}
$$

Now plug (36) into this equation and look at it modulo (h^6). We get

$$
24D_1(h)^3 h^5 \equiv 0 \quad (\mathrm{mod}(h^6, \theta)),
$$

which is absurd since h does not divide $D_1(h)$. We can then assume that $m = 2$.

It is in fact worth noticing that the case in which $D_1\Theta$ is not reduced does happen. It happens in the hyperelliptic case where $D_1\Theta$ has multiplicity two and, a posteriori, this is the only case.

In view of (29) and the inductive hypothesis, we have

(38) $\quad 0 = R_{s-1}\big|_\Theta = -\tilde{\Delta}_{s-1}\theta \cdot D_1\theta + 2D_1\tilde{\Delta}_{s-2}\theta \cdot D_1\theta - \tilde{\Delta}_{s-2}\theta \cdot D_1^2\theta + \tilde{\Delta}_{s-2}\ddot{\,}D_2\theta$

and

(39)
$$
0 = \tilde{\Delta}_{s-2}\theta \cdot (D_1^2 - D_2)\theta\big|_V.
$$

On the other hand we have, in (36), $s \geq m = 2$, so that (39) implies that, on V,

$$
\tilde{\Delta}_{s-2}\theta \equiv \alpha h \quad (\mathrm{mod}(h^2, \theta)).
$$

From (38) we then get

$$-2\alpha D_1(h)h^2 + 2\alpha D_1(h)h^2 + \tilde{\Delta}_{s-1}\theta \cdot h^2 \equiv 0 \quad (\mathrm{mod}(h^3, \theta)).$$

This means that $\tilde{\Delta}_{s-1}\theta \in (h, \theta)$ so that $\tilde{\Delta}_{s-1}\theta \cdot (D_1^2 - D_2)\theta \in (h^2, \theta)$, proving that R_{s-1} vanishes on V.

This concludes the proof of (20). We summarize the result in the following

THEOREM. *Let X be an i.p.p.a.v. of dimension $g > 1$. Let $\Theta \subset X$ be its theta divisor. Assume there exist constant vector fields $D_1 \neq 0$, D_2, and D_3 on X and a complex number d_4 such that the K.P. equation (8) is satisfied. Assume furthermore that*

(i) $\dim \Theta_{\mathrm{sing}} \leq g - 3$ *and*

(ii) $\dim(\Sigma = \{\varsigma \in X : D_1^n \theta(\varsigma) = 0, \forall n\}) \leq g - 3$.

Then X is a Jacobian.

As we already mentioned, conditions (i) and (ii) are not necessary and proofs of the stronger theorem can be found in [S] and [AD2]. The possible advantage of the present approach is that it is purely algebraic.

It is with pleasure that I thank both Corrado De Concini and Gerald Welters for the many conversations on the subject of this note. Communicating with Welters was essential for the understanding of (24). On the other hand, many of the ideas in this note are already contained in [AD2] and are the result of a joint work with De Concini.

REFERENCES

[AD1] E. Arbarello and C. De Concini, *On a set of equations characterizing Riemann matrices*, Ann. of Math. (2) **120** (1984), 119–140.

[AD2] _____, *Another proof of a conjecture of S. P. Novikov on periods of abelian integrals on Riemann surfaces*, Preprint.

[BD] A. Beauville and O. Debarre, *Une rélation entre deux approches du problème de Schottky*, Preprint.

[D] E. Date, M. Jimbo, M. Kashiwara, and T. Miwa, *Transformation groups for soliton equations*, Proceedings RIMS Sympos. Non-Linear Integrable Systems, Classical Theory and Quantum Theory, World Scientific, Singapore, 1983.

[DU1] B. A. Dubrovin, *Theta functions and non-linear equations*, Russian Math. Surveys **36** (1981), 11–92.

[DU2] _____, *The Kadomcev-Petviasvili equation and the relations between the periods of holomorphic differentials on Riemann surfaces*, Math. USSR Izv. **19** (1982), 285–296.

[F] H. M. Farkas, *On Fay's trisecant formula*, J. Analyse Math. **44** (1984), 205–217.

[FY] J. D. Fay, *Theta functions on Riemann surfaces*, Lectures Notes in Math., vol. 352, Springer-Verlag, 1973.

[G] R. C. Gunning, *Some curves in Abelian varieties*, Invent. Math. **66** (1982), 377–389.

[K] I. M. Krichever, *Methods of algebraic geometry in the theory of nonlinear equations*, Russian Math. Surveys **32** (1977), 185–213.

[I] J. Igusa, *Theta functions*, Grundlehren Math. Wiss., vol. 194, Springer Verlag, 1972.

[M] M. Mulase, *Cohomological structure in soliton equations and Jacobian varieties*, J. Differential Geom. **19** (1984), 403–430.

[M1] D. Mumford, *Curves and their Jacobians*, University of Michigan Press, Ann. Arbor, Mich., 1975.

[M2] ———, *An algebro-geometric construction of commuting operators and of solutions to the Toda lattice equation, Korteweg-de Vries equation and related non-linear equations*, (Proc. Internat. Sympos. Alg. Geometry, Kyoto, 1977), Kinokuniya Book Store, Tokyo, 1978, pp. 115–153.

[M3] ———, *Tata lectures on theta*. II, Birkhaüser, Boston, 1983.

[M4] D. Mumford and J. Fogarty, *Geometric invariant theory*, Ergeb. Math. Grenzgeb., vol. 34, Springer-Verlag, 1982.

[N] S. P. Novikov, *The periodic problem for the Korteweg-de Vries equation*, Functional Anal. Appl. **8** (1974), 236–246.

[S] T. Shiota, *Characterization of Jacobian varieties in terms of soliton equations*, Invent. Math. **83** (1986), 333–382.

[W1] G. E. Welters, *A criterion for Jacobian varieties*, Ann. of Math. (2) **120** (1984), 497–504.

[W2] ———, *A characterization of non-hyperelliptic Jacobi varieties*, Invent. Math. **74** (1983), 437–440.

[W3] ———, *On flexes of the Kummer variety*, Nederl. Akad. Wetensch. Proc. Ser. A 86 **45** (1983), 501–520.

ISTITUTO "G. CASTELNUOVO", UNIVERSITÀ "LA SAPIENZA", ROMA, ITALY

Proceedings of Symposia in Pure Mathematics
Volume **46** (1987)

Deformations and Smoothing
of Complete Linear Systems on Reducible Curves

MEI-CHU CHANG AND ZIV RAN

In this paper we take up the following questions. Consider a nodal curve $C = C_1 \cup C_2$, where C_1 and C_2 are nonsingular curves meeting transversely, and let $|L|$ be a complete linear system on C, where L is a line bundle. For $i = 1, 2$, let (L_i, V_i) be the induced linear system on C_i, i.e., $L_i = L|_{C_i}$ and $V_i = \operatorname{im}(H^0(L) \to H^0(L_i))$. Then

(a) suppose (L_1, V_1) and (L_2, V_2) both have unobstructed deformations, i.e., their deformation spaces are smooth and of the expected dimension; then is the same true of $|L|$? More generally, how, if at all, is the deformation theory of $|L|$ determined by those of (L_1, V_1) and (L_2, V_2)?

(b) Give some useful criteria, especially in cases where (a) has a good answer, to ensure that $|L|$, or subsystems thereof, be smoothable, i.e., a limit of linear systems on smooth curves.

As regards question (a), one cannot, as we shall explain below, expect results in the general case. In this paper we shall, however, present such results in some special cases. The precise hypotheses are too complicated to reproduce here (see §1), but they include mainly

(i) maximal rank of the restriction maps $\rho_i \colon H^0(L) \to H^0(L_i)$, and

(ii) assumptions on the number of intersection points $C_1 \cap C_2$ as compared to the dimensions involved.

As is well known, for $|L|$ to have unobstructed deformations is equivalent to the μ_0-mapping

$$\mu_0 = \mu_{0,L} \colon H^0(L) \otimes H^0(\omega_c(-L)) \to H^0(\omega_c)$$

having maximal rank. Thus some of our results can be interpreted as asserting maximal rank for $\mu_{0,L}$ under the assumption of maximal rank for μ_{0,L_1,V_1} and μ_{0,L_2,V_2}. In this form, results of this type go back to Sernesi [8]. Sernesi considers the case where $C_1 \subset \mathbf{P}^r$ is linearly normal and has μ_0 of maximal rank,

1980 *Mathematics Subject Classification* (1985 *Revision*). Primary 14H12.

The first author was partially supported by the Rackham Foundation, and the second author by the National Science Foundation.

and C_2 is a rational normal curve in a general hyperplane $H \subset \mathbf{P}^r$, meeting C_1 in $r + 1$ points; he proves in this case that C has μ_0 of maximal rank. The deformation-theoretic significance of μ_0 plays no explicit role in Sernesi's argument. This paper grew out of an attempt to understand Sernesi's result on μ_0 from a deformation-theoretic viewpoint. As a result, we obtain a transparent (to us)[1] proof of this result (in char. 0) and various generalizations of it. These results are presented in §1. We note, incidentally, that the μ_0 mapping for curves in $\mathbf{P}^3$ was recently found to have applications to the equations defining these curves; cf. [2, 3].

A linear system on a reducible curve is, from our point of view, hardly interesting unless it is smoothable. Consequently we present, in §2, a number of smoothing criteria (applicable to incomplete linear systems as well). These include injectivity of μ_0 and generic dimension for the locally trivial deformations, the latter being a recent result of Sernesi [9]. Finally we give 2 new smoothing criteria, in terms of normal bundles, for reducible space curves; the proof of these involves, at least implicitly, the recently developed theory of generalized linear systems [4]. As a consequence we show the existence of components of the Hilbert scheme $H_{d,g,r}$ of smooth curves in $\mathbf{P}^r$ which are smooth of the expected dimension and expected number of moduli and such that the ratio d/g approaches, as $d \to \infty$, either $5/18$ (for $r = 3$) or $(3r - 3)/(5r - 4)$ (for $r \geq 4$), improving results of Sernesi [8].

As indicated above, this paper owes a great deal to the work of Sernesi. In addition, we are grateful to Professor Sernesi for helpful conversations.

Notations and conventions. In this paper we work over the complex numbers. If E is a coherent sheaf on a noetherian scheme S and $x \in S$ is a closed point, then the fibre of E at x, $E(x)$, is $E \otimes k(x)$. Likewise, if $\varphi \colon E \to F$ is a map of coherent sheaves, then $\varphi(x) \colon E(x) \to F(x)$ denotes the induced map. If E and F are locally free, we say that φ is *fibrewise injective* at x if $\varphi(x)$ is injective. Note that in this case there is a neighborhood U of x over which φ is injective, $E/\varphi(F)$ is free and such that $\varphi(y)$ is injective for all $y \in U$. A *curve* is a reduced, purely 1-dimensional projective scheme, usually with only nodes as singularities. If C_1 and C_2 are smooth curves and A is a smooth divisor on C_1 identified with one on C_2, then $C = C_1 \cup C_2$ denotes the resulting nodal curve. If (V, L) is a linear system on C, we shall denote by $\mu_{0,V,L}$ the multiplication map $V \otimes H^0(\omega_C(-L)) \to H^0(\omega_C)$. If $V = H^0(L)$, we denote $\mu_{0,V,L}$ by $\mu_{0,L}$ or sometimes just μ_0. As usual, $\varphi_{L,V}$ or φ_L denotes the rational map to $\mathbf{P}^r$ induced by (L, V) or $|L|$.

1. Local analysis of W_d^r. In this we carry out our basic analysis of the deformation theory of complete linear systems, i.e., the local structure of W_d^r, for reducible curves. We begin with some notation, which will be fixed throughout our discussion. C_1 and C_2 are nonsingular curves of genus g_1, g_2 respectively, A is

---

[1]The referee has taken exception to this statement.

a smooth divisor of degree a on C_1 identified with one on C_2, and $C = C_1 \cup_A C_2$ is the resulting a-nodal curve of arithmetic genus $g = g_1 + g_2 + a - 1$.

Now let L be a line bundle of degree d on C and put $L_i = L|_{C_i}$, $r = h^0(L) - 1$, $r_i = \mathrm{rk}(H^0(L) \to H^0(L_i)) - 1$. Let P be the space parametrizing locally deformations of L (i.e., a neighborhood or germ of L in the Picard variety of C). We similarly define P_i, $i = 1, 2$. Note that we have a map $\alpha = (\alpha_1, \alpha_2)$: $P \to P_1 \times P_2$ which is smooth of relative dimension $a-1$, the α_i being restriction maps. Let $W \subset P$ be the locus of line bundles L' with $h^0(L') \geq r + 1$, and similarly define $W_i \subset P_i$, $i = 1, 2$. We note that W, as well as the W_i, are defined as schemes by the usual determinantal conditions, as in [1]: namely, take divisors $D_i \subset C_i$ sufficiently general of sufficiently large degree, and let $D = D_1 + D_2 \subset C$ and $\tilde{D} = D \times P \subset C \times P$. Then we have a complex of vector bundles over P

$$(1.1) \qquad p_{2*}(\mathcal{L}(\tilde{D})) \xrightarrow{\varphi} p_{2*}(\mathcal{L}(\tilde{D})|_{\tilde{D}})$$

where p_2: $C \times P \to P$ is the projection and $\mathcal{L}$ is the universal (Poincáre) bundle on $C \times P$, and W is just the $(n - r - 1)$st determinantal scheme of φ, where $n = d + \deg D - g + 1$. For simplicity, we denote the complex (1.1) by $K^0 \xrightarrow{\varphi} K^1$. Similarly, we have complexes $K_i^0 \xrightarrow{\varphi_i} K_i^1$ on P_i, as well as a natural commutative diagram

$$(1.2) \qquad \begin{array}{ccc} K^0 & \xrightarrow{\varphi} & K^1 \\ \downarrow & & \downarrow \\ \alpha^*(K_1^0 \oplus K_2^0) & \xrightarrow{\varphi_1 \oplus \varphi_2} & \alpha^*(K_1^1 \oplus K_2^1) \end{array}$$

Note that the vertical arrows, induced by restrictions, are *fibrewise* injective. Now let G_i be the canonical blow-up of W_i, i.e., the subscheme of the Grassmannian $G(r_i + 1, K_i^0)$ defined by the vanishing of the composite map $S_i \to q_i^* K_i^0 \to q_i^* K_i^1$, where q_i: $G(r_i + 1, K_i^0) \to P_i$ is the projection and S_i is the universal bundle. The restriction $E_i = S_i|_{G_i}$ is the universal bundle on G_i. Also, since $h^0(L) = r+1$, W itself carries a universal bundle E, i.e., there is an exact (locally split) sequence of vector bundles on W

$$(1.3) \qquad 0 \to E \to K^0|_W \to K^1|_W.$$

We now introduce a key condition on the complete linear system $|L|$ to make our analysis work.

DEFINITION. *A complete linear system $|L|$ on $C = C_1 \cup_A C_2$ is said to be trim if*

(i) *L has no base points on A;*

(ii) *the map φ_L: $C \to \mathbf{P}^r$ is an embedding in a neighborhood of A;*

(iii) *the restriction maps ρ_i: $H^0(L) \to H^0(L_i)$ both have maximal rank, i.e., each one is either injective or surjective.*

Of course, condition (iii) is the essential one. We can now state our main result on deformations of linear systems as follows.

THEOREM 1.1. *Using the above notation, let* $|L|$ *be a trim complete linear system on* C, *and suppose for* $i = 1, 2$ *that* G_i *is smooth of dimension* m_i *in a neighborhood of* (L_i, V_i). *Let* S *be the span of* $\varphi_L(A)$ *and* $s = \dim S$.

(i) *If* $s = a - 1$, *then* W *is smooth of dimension* $m_1 + m_2 + a - 1$, *near* L.

(ii) *If* $s < a - 1$, *and* $\varphi_L(A)$ *is in general position in* S, *i.e., any* $s + 1$ *points of it span* S, *we have*

$$m_1 + m_2 - (r_1 + r_2 - 2s)(a - s - 1) \leq \dim_L W \leq \dim T_L W \leq m_1 + m_2.$$

Now assume moreover $a = s + 2 = r + 1$, $\dim_L W_1 > 0$, *that* L_1 *is sufficiently general on* W_1, *and that* S *is a sufficiently general hyperplane with respect to* $\varphi_{L_1}(C_1)$. *Then*

(iii) *if* $r_2 = s$, *then* W *is smooth of dimension* $m_1 + m_2 - 1$ *at* L;

(iv) *if* $r_2 = r$ *and* L_2 *and* G_2 *satisfy similar hypotheses as* L_1 *and* G_1, *then* W *is smooth of dimension* $m_1 + m_2 - 2$ *at* L.

PROOF. The idea is to decompose P on $P_1 \times P_2 \times$ (gluing data) and analyze W in terms of this decomposition. More precisely, we shall construct and analyze a natural map $\lambda: W \to G_1 \times G_2$. To begin with, we construct maps $\lambda_i: W \to G_i$. Assume first that the restriction map ρ_i is injective. Comparing (1.2) and (1.3), we get a diagram

$$
\begin{array}{ccccccc}
0 & \to & E & \to & K^0|_W & \to & K^1|_W \\
& & \downarrow \rho & & \downarrow & & \downarrow \\
0 & \to & E' & \xrightarrow{j} & \alpha_i^* K_i^0|_W & \xrightarrow{\varphi_i} & \alpha_i^* K_i^1|_W
\end{array}
$$

with $E' = \ker \varphi_i = p_{2*} j_i^*(h)$ where $j_i: C_i \times W \hookrightarrow C \times W$ is the inclusion and $p_2: C_i \times W \to W$ is the projection. Note that the fibre of E (resp. E') at L may be identified with $H^0(L)$ (resp. $H^0(L_i)$): hence ρ induces an injection on fibres at L, and by construction (cf. [1, Chapter IV, §3]) so does j. Hence the composite map $\psi: E \to \alpha_i^* K_i^0|_W$ is fibrewise injective. Since the composite $\varphi_1 \circ \psi$ vanishes, this means that the projection $\alpha_i: W \to P_i$ factors through G_i, whence a map $\lambda_i: W \to G_i$.

Now assume instead that ρ_i is surjective. Using the above notation, this means that the fibre map $\varphi_1(L)$ has nullspace of dimension $r_i + 1$; hence E' has precisely $r_i + 1$ minimal generators. On the other hand, because the map $\psi(L)$ has rank $\geq r_i + 1$, it is possible to choose a partial set of minimal generators $e_0, \ldots, e_{r_i}$ of E such that $\psi(e_0), \ldots, \psi(e_{r_i})$ are minimal generators of $\alpha_i^* K_i^0|_W$. Thus $\rho(e_0), \ldots, \rho(e_{r_i})$ is a set, necessarily complete, of minimal generators of E' whose images under j are still minimal for $\alpha_i^* K_i^0|_W$. It follows that E' is locally free of rank $r_i + 1$, and $E \to E'$ is surjective; hence we again get a natural map $W \to G_i$ (of course in this case $G_i \xrightarrow{\sim} W_i$).

We may informally describe what we have seen so far by saying that as the complete linear system $|L|$ moves within W, the induced linear systems on C_1 and C_2 move "smoothly," i.e., they maintain constant dimension r_1 and r_2 respectively. We note that the trimness hypothesis on L (rather, part (iii) of it) seems essential for this: i.e., without it, it is presumably possible for the dimensions of the induced linear systems to jump upwards. Although it would still be possible to define a map $W \to G_1 \times G_2$ by choosing sufficiently general subsheaves of E of rank $r_1 + 1$, $r_2 + 1$, respectively, such a map would not be canonical, hence presumably of little use.

We now have a diagram

$$
\begin{array}{ccc}
W & \subseteq & P \\
\downarrow & & \downarrow \alpha \\
G_1 \times G_2 & \to & P_1 \times P_2.
\end{array}
$$

Let U be the group $\prod_{j=1}^{a} \mathbf{C}_j^* / (\text{diagonal})$ (or a small neighborhood of the identity in it). Things being local, we may trivialize the universal bundles $\mathcal{L}_i$ on $P_i \times C_i$ in a neighborhood of $P_i \times A$ and then P may be identified with $P_1 \times P_2 \times U$ by sending a triple $(M_1, M_2, (u_1, \ldots, u_a))$ to the line bundle M on C obtained by identifying $z_j \in M_1(A_j) \simeq \mathbf{C}$ with $u_j z_j \in M_2(A_j) \simeq \mathbf{C}$. Now put $L_0 = H^0(L|_A)$ and let F be the trivial bundle $L_0 \otimes \mathcal{O}$ on $G_1 \times G_2$. Then restriction gives us maps

$$
\beta_i \colon p_i^* E_i \to F
$$

which have rank precisely $s + 1$ at the point $(L_1, V_1) \times (L_2, V_2)$. Note that the evident map $W \to G_1 \times G_2 \times U$ is clearly an embedding. Let $\overline{W}$ be its image. We identify $\overline{W}$ as follows.

Claim 1. $\overline{W}$ coincides locally (scheme-theoretically) with the subscheme W' of $G_1 \times G_2 \times U$ defined by the conditions

(a) rank $\beta_1 = $ rank $\beta_2 = s + 1$, and

(b) $u . \operatorname{im} \beta_1 = \operatorname{im} \beta_2 \in G(s + 1, L_0)$ (with the obvious action of U on $G(s + 1, L_0)$).

Proof of claim. Note the exact sequence

$$
0 \to E \xrightarrow{i} \alpha^*(E_1 \oplus E_2) \xrightarrow{\beta} F.
$$

Since i is clearly fibrewise injective, the image of β is a locally split rank-$(s+1)$ subbundle, and this just means $\overline{W} \subset W'$. The opposite inclusion is proved similarly.

Note that if $a = s + 1$, then both conditions (a) and (b) above are vacuous, i.e., $\overline{W} = G_1 \times G_2 \times U$, which proves assertion (i) of the theorem. So from now on we assume $a > s + 1$.

Claim 2. The map $\alpha \colon \overline{W} \to G_1 \times G_2$ is an embedding.

Proof of claim. Note that a fibre of α may be identified with the subgroup $U_0 = \{u \in U : u . K = K\} \subseteq U$, for a suitable $K \in G(s + 1, L_0)$, so the claim is that U_0 is trivial (of course, being a subgroup, U_0 is automatically reduced). This follows from the following

LEMMA 1.2. *Let $L_0 = \bigoplus_1^a \mathbf{C} x_i$ be a vector space and $K \subset F$ a subspace of dimension $b < a$ such that the natural map $K \to \bigoplus_{j=1}^{b} \mathbf{C} x_{i_j}$ is surjective for* *all $i_1 < \cdots < i_b$. Let $U \subset \mathrm{PGL}(L_0)$ be the subgroup of matrices diagonal with respect to $x_1, \ldots, x_a$. Then the map*

$$g : U \to G(b, L_0), \qquad g(u) = u \cdot K,$$

is an embedding. Moreover if $b = 1$ or $a - 1$, g is open.

PROOF. Replacing L_0 by $\bigwedge^b L_0$ and K by $\bigwedge^b K$, we may assume $b = 1$, in which case the result is obvious.

Assertion (ii) of the theorem follows immediately from claims 1 and 2. So from now on we shall assume $a = s + 2$. Using Lemma 1.2 with $b = a - 1$, it clearly follows that the projection of $\overline{W} \subset G_1 \times G_2 \times U$ to $G_1 \times G_2$ is an embedding, with image $H_1 \times H_2$, where $H_i \subset G_i$ is the subscheme defined by the determinantal condition rank $\beta_i = s + 1$. Thus to complete the proof of assertions (iii) and (iv) of the theorem, it would suffice to prove that H_i is smooth of dimension $m_i - 1$, provided $s = r_i - 1$. To this end take a small neighborhood N of the point corresponding to A in the symmetric product $(C_i)_a$. Then restriction gives us a natural map over $G_i \times N$,

$$\beta_i \colon p_1^* E_i \to L_0 \otimes \mathcal{O}.$$

Let $\tilde{H}_i$ denote the determinantal locus $\{\mathrm{rank}\ \tilde{\beta}_i \le s + 1\}$. The key point is that the projection map $p_1 \colon \tilde{H}_i \to G_i$ is smooth near $((L_i, V_i), A)$: indeed the fibre of this map over (L_i', V_i') consists of the divisors A' such that $\varphi_{V_i}(A')$ lies on a hyperplane in $\mathbf{P}^r$, and since A' and the hyperplane determine each other locally, p_1 may be realized as an open submorphism of a projective bundle, hence, in particular, is smooth. (We leave it to the fastidious reader to translate the foregoing argument into schematese.)

Now since both G_i and the map p_1 are smooth locally, so is $\tilde{H}_i$, and clearly $\dim \tilde{H}_i = m_j + r_i$. We claim next that $p_2(\tilde{H}_i)$ has dimension $r_i + 1 = a$, i.e., that p_2 is surjective locally. Suppose not. Since clearly $\dim p_2(p_1^{-1}(L_i, V_i)) = r_i$, it follows that $p_2(p_1^{-1}(L_i, V_i)) = p_2(\tilde{H}_i)$. Using the hypotheses of (iii), it follows that there exists a 1-parameter family $\{(L_t, V_t)\} \subset G_i$ continuing (L_i, V_i), with L_t a *nontrivial* family of line bundles, such that for small t and any small perturbation of the initial hyperplane S to a V_t-hyperplane, the corresponding "analytic continuation" of A actually lies on a V_i-hyperplane. Now using analytic continuation plus the fact that the monodromy acts as the full symmetric group on the points of a V_t-hyperplane section, and the fact that $a > r_i$, it follows that an entire V_t-hyperplane section must be on a V_1-hyperplane, which contradicts the nontriviality of the family $\{L_t\}$.

Now using the Bertini-Sard theorem it follows that a general fibre of p_2 is smooth of dimension $m_i - 1$. Since H_i is such a fibre, this finally completes the proof of Theorem 1.1.

We proceed now to formulate some of our results in terms of μ_0-mappings. Recall first that if C is any nodal curve of arithmetic genus g, and (L, V) is a g_d^r on C, then as in [1] the space G parametrizing deformations of (L, V) has expected dimension $\rho = g - (r+1)(g-d+r)$ (the Brill-Noether number), and moreover G is smooth at (L,V) of dimension ρ (resp. 0) iff the map

$$\mu_0 \colon V \otimes H^0(\omega_C(-L)) \to H^0(\omega_C)$$

is injective (resp. surjective).

THEOREM 1.3. *In the situation of Theorem 1.1, let μ_0 and $\mu_{0,i}$ denote the μ_0-mapping associated with L and (L_i, V_i) respectively, $i = 1, 2$. Then we have the following conclusions in the various cases.*

(i) *If $\mu_{0,1}$ and $\mu_{0,2}$ are both injective, then so is μ_0.*

(ii) *If $\mu_{0,1}$ and $\mu_{0,2}$ are both surjective, then so is μ_0.*

(iii) *If $\mu_{0,1}$ is injective but not surjective, and $\mu_{0,2}$ is injective, then μ_0 is injective.*

(iv) *If both $\mu_{0,1}$ and $\mu_{0,2}$ are injective and nonsurjective, then μ_0 is injective.*

PROOF. This follows directly from Theorem 1.1, modulo checking that the various Brill-Noether numbers add up right.

Next we compare our results on the μ_0-mapping with those of Sernesi [8]. Sernesi's main result on μ_0 is essentially the special case of Theorem 1.3 above, cases (ii) and (iii) where (L_1, V_1) is a complete very ample linear system (without, however, any generality hypothesis) and C_2 is a rational normal curve in a general hyperplane in $\mathbf{P}^r$ meeting (the image of) C_1 in $r+1$ points. Sernesi's proof proceeds by a direct projective-geometric analysis of the μ_0-mapping, with no explicit reference to its deformation-theoretic significance. The main point is to prove, in case (iii), that the multiplication map

$$\tau \;\colon\; H^0(L_1) \otimes H^0(\omega_{C_1}(-L_1 + A)) \to H^0(\omega_{C_1}(A))$$

is injective. On the other hand, perhaps the main point in our proof was the fact that the scheme H_i constructed above was smooth of dimension $m_i - 1$. It is not difficult to check a priori that this is equivalent to the condition that the image of

$$\nu \;\colon\; H^0(L_1 - A) \otimes H^0(\omega_{C_1}(-L_1 + A)) \to H^0(\omega_{C_1})$$

is not contained in $\operatorname{im}\mu_{0,1}$. (Since $H^0(L_1 - A) = 1$, $H^0(\omega_{C_1}(-L_1 + A)) = h^0(\omega_{C_1}(-L_1)) + 1$, we automatically have $\dim(\operatorname{im}\mu_{0,1} + \operatorname{im}\nu) \le \dim \operatorname{im}\mu_{0,1} + 1$.) But it is straightforward to see that the latter condition is equivalent to injectivity of $\tau_{V_1} \;\colon\; V_1 \otimes H^0(\omega_{C_1}(-L_1 + A)) \to H^0(\omega_{C_1}(A))$. Thus our proof is somewhat similar to Sernesi's except that rather than analyze τ_{V_1} directly, we have appealed to Bertini (at the cost of introducing a generality hypothesis plus of course char. 0).

Perhaps the most glaring omission from Theorems 1.1 and 1.3 is the range where $a > s + 2$. In this range, however, we cannot in general expect the same sort of results; that is, the deformation theory of the complete linear system $|L|$ is not determined just by those of the induced linear systems (L_1, V_1) and (L_2, V_2) separately, but also by the way these fit together. We proceed to analyze this situation in a special case.

We say that a trim linear system $|L|$ on $C = C_1 \cup_A C_2$ is *very trim* if the restriction maps $H^0(L) \to H^0(L_i)$ are both surjective, $i = 1, 2$. Let us fix such a system, and note the exact sequence

$$H^0(L_1 - A) \oplus H^0(L_2 - A) \to H^0(L) \xrightarrow{\psi} H^0(L|A)$$
$$\to H^1(L_1 - A) \oplus H^1(L_2 - A) \to H^1(L) \to 0$$

and its dual

$$0 \to H^0(\omega_C(-L)) \to H^0(\omega_{C_1}(-L_1 + A)) \otimes H^0(\omega_{C_2}(-L_2 + A))$$
$$\to H^0(-L|A) \to H^0(L)^v \cdots.$$

From the hypothesis that L is very trim, it follows that the maps $H^0(\omega_C(-L)) \to H^0(-L|A)$ and $H^0(\omega_{C_i}(-L_i + A)) \to H^0(-L|A)$, $i = 1, 2$, all have the same image, namely $(\text{im}\,\psi)^\perp$; hence

$$X := \frac{H^0(\omega_C(-L))}{H^0(\omega_{C_1}(-L_1 + A)) \oplus H^0(\omega_{C_2}(-L_2 + A))}$$
$$\simeq \frac{H^0(\omega_{C_i}(-L_i + A))}{H^0(\omega_{C_i}(-L_i))}, \qquad i = 1, 2.$$

Thus we can map by multiplication

$$\frac{H^0(L)}{H^0(L_1 - A) \oplus H^0(L_2 - A)} \otimes X \xrightarrow{\nu_i} \frac{H^0(\omega_{C_i}(A))}{\text{im}\,\mu_{0,L_i} + \text{im}\,\mu_{0,L_i-A}}$$

Let

$$R_i = \frac{H^0(\omega_{C_i})}{\text{im}\,\mu_{0,L_i} + \text{im}\,\mu_{0,L_i-A}} \quad \text{and} \quad T = \nu_i^{-1}(R_i)$$

which is independent of i.

PROPOSITION 1.4. *In the above situation, $\mu_{0,L}$ is injective (resp. surjective) iff so is $\nu_1 \oplus \nu_2 : T \to R_1 \oplus R_2$.*

PROOF. We analyze the situation as in the proof of Theorem 1.1, using the same notations. Note that $H^0(L)/(H^0(L_1 - A) \oplus H^0(L_2 - A))$ may be identified with the cotangent space to $G(s + 1, L_0)$ at K, T with the subspace of vectors perpendicular to the orbit of U, and R_i with the tangent space to H_i at $|L_i|$. Also, a standard type of computation (compare [1]) identifies ν_i with the codifferential of the natural map $H_i \to G(s + 1, L_0)$. From this and the earlier analysis the proposition follows.

The condition of the proposition seems difficult to analyze in general, so we specialize further.

COROLLARY 1.5. *Using notations as above, assume $|L|$ is very trim and moreover*

(i) *μ_{0,L_1} is surjective;*

(ii) *the multiplication map $H^0(L_2) \otimes H^0(\omega_{C_2}(-L_2 + A)) \to H^0(\omega_{C_2}(A))$ is surjective.*

Then $\mu_{0,L}$ is surjective.

PROOF. Our assumptions imply that $R_1 = 0$ and that ν_2 is surjective; hence so is its restriction $T \xrightarrow{\nu_2} R_2$, hence the conclusion.

COROLLARY 1.6. *Using notations as above, assume $|L|$ is very trim and morover*

(i) μ_{0,L_2} *is surjective*;

(ii) $a \geq 2g_2 + 4$.

Then $\mu_{0,L}$ is surjective.

PROOF. This follows from the previous corollary plus a result of Green [5, Corollary 4.e.4], which implies that in our case condition (ii) of that corollary is satisfied.

2. Smoothing criteria and applications. In this section we give some smoothing criteria for linear systems on nodal curves, especially of the type considered in §1. We begin with a relatively trivial case.

PROPOSITION 2.1. *Let (L, V) be a linear system on a nodal (or even just abstractly smoothable) curve C, and suppose that the map $\mu_{0,L,V}$ is injective. Then (L, V) is smoothable.*

PROOF. By general nonsense (compare [1]) there exists a space $\mathcal{G}$ parametrizing deformations of the triple (C, L, V), which is purely of dimension $\geq 3g - 3 + \rho$ (where g is the arithmetic genus of C and $\rho = g - (\dim V)h^1(L)$ is the Brill-Noether number) and maps to a versal deformation space M of C. By hypothesis, the fiber of $\mathcal{G}$ over C is smooth of dimension ρ and since M is smooth of dimension $3g - 3$, it follows that $\mathcal{G}$ maps onto M and in particular (L, V) is smoothable.

Next, we state in a form convenient for our purposes, a smoothing criterion discovered recently by Sernesi [9], which, while simple, is actually rather powerful and implies a number of earlier smoothing criteria (e.g., those of Hartshorne and Hirschowitz [6]).

Fix integers d, g, r and put $\chi = \chi(d, g, r) = (r+1)d + (r-3)(1-g)$, the expected dimension of the Hilbert scheme $H_{d,g,r}$ of (smooth) curves of degree d and genus g in $\mathbf{P}^r$. Let C be an abstract nodal curve with δ nodes and (L, V) a g_d^r on C. Fix a basis $v_0, \ldots, v_r$ of V up to scalar multiple. Let H be the space parametrizing local deformations of $(C, L, V, \overline{v}_0, \ldots, \overline{v}_r)$; thus if V has no base points and $\varphi_{L,V}$ is an embedding, then H is a local germ of $H_{d,g,r}$. Let $H' \subset H$ be the subspace parametrizing those deformations for which the underlying deformation of C is locally trivial.

THEOREM 2.2 (SERNESI). *In the above situation, if $\dim H' \leq \chi - \delta$, then (C, L, V) is smoothable.*

For completeness' sake, we sketch the proof. Using the well-known fact that $\dim H \geq \chi$ one sees easily that it suffices to prove that "each node imposes at most 1 condition on H." To see this let P be one of the nodes of C and $U \subset C$ a small neighborhood of it. Then the deformation space A of U may be identified with the germ of $\mathbf{A}^1$ at the origin. Considering the natural map $H \to A$, we see that the subspace of H parametrizing deformations locally trivial near P is a Cartier divisor, hence has codimension ≤ 1, as claimed.

Further smoothing criteria, applicable to space curves with a *degenerate* component, can be obtained by using the theory of generalized linear systems [7]. We shall return to this topic in greater detail elsewhere, but meanwhile we shall sketch the proof of a sample result. As its statement is of necessity somewhat complicated, the reader may wish, by way of motivation, to first consult Example 2.7 below.

THEOREM 2.3. *Let C_1 be a nondegenerate nonsingular curve of degree d_1 and genus g_1 in $\mathbf{P}^r$, $r \geq 3$, and $\overline{C}_2$ a nondegenerate nodal curve of degree d_2 and geometric genus g_2 in a general hyperplane $H \subset \mathbf{P}^r$, meeting C_1 transversely in a divisor A of degree a, and with normalization C_2. Let N_1 be the normal bundle of $C_1 \hookrightarrow \mathbf{P}^r$ and N_2 that of $C_2 \to H$. Assume either*

(i) $H^1(N_1)$, $H^1(N_2(-A))$, *and* $H^1(\mathcal{O}_{C_2}(1+A))$ *all vanish*, $\mathcal{O}_{C_2}(1+A)$ *is very ample, and* $a \geq g_2 - d_2 + r$; *or*

(ii) $H^1(N_1(-A))$, $H^1(N_2)$, *and* $H^1(\mathcal{O}_{C_2}(1 + 2A))$ *all vanish*, $\mathcal{O}_{C_2}(1 + A)$ *is very ample, and* $a \leq \min(g_2 - d_2 + r, d_2)$.

Then the map $C_1 \cup_A C_2 \to \mathbf{P}^r$ can be deformed to an embedding of a smooth curve in $\mathbf{P}^r$, whose image X' has $H^1(N_{X'}) = 0$.

(ii)$'$ *Additionally, in case* (ii), *if moreover $H^1(N_2(-1)) = H^1(\mathcal{O}_{C_2}(2A)) = 0$, then for any set A' of at most d_2 points from a general hyperplane section of X', we have $H^1(N_{X'}(-A')) = 0$.*

REMARKS 2.4. (1) Note that $a \geq 2g_2 - d_2 + 1$ implies that $\mathcal{O}_{C_2}(1 + A)$ is nonspecial and very ample.

(2) The hypothesis $H^1(N_i(-A)) = 0$ implies that, by taking C_i general in the unique component of the Hilbert scheme of $\mathbf{P}^r$ or H to which it belongs, A becomes a general a-tuple of points on H. In particular, given, for $i = 1, 2$, components of the Hilbert scheme whose general curve C_i satisfies appropriate hypotheses as in case (ii) of the theorem (with A either a subset of a general hyperplane section of C_1 or a general a-tuple on C_2), then a curve $C_1 \cup_A C_2$ can always be constructed.

(3) Assertion (ii)$'$ of the theorem shows, in the case considered, that X' satisfies similar hypotheses as C does, so that the operation of attaching C_2 can be iterated. That the same is also true in case (i) is clear.

(4) If $\overline{C}_2$ does have nodes (e.g., for $r = 3$), then as an embedding of a smooth curve degenerates to $C_1 \cup C_2 \to \mathbf{P}^r$, the limit of the image has embedded points at the nodes of $\overline{C}_2$.

PROOF OF THEOREM 2.3. Let $\mathbf{P}_2^r$ be another copy of $\mathbf{P}^r$ and $\tilde{\mathbf{P}}_2^r$ the blow-up of $\mathbf{P}_2^r$ at a point with projection $\pi: \tilde{\mathbf{P}}_2^r \to \mathbf{P}^{r-1}$ and exceptional divisor E. Let V be the union of $\mathbf{P}^r$ and $\tilde{\mathbf{P}}_2^r$ glued via an identification of E with H. By hypothesis, we may choose a very ample r-dimensional subsystem of $|\mathcal{O}_{C_2}(1+A)|$ and lift the corresponding embedding $C_2 \hookrightarrow \mathbf{P}_2^r$ to an embedding $C_2 \hookrightarrow \tilde{\mathbf{P}}_2^r$ which projects to the given map $C_2 \to H \cong \mathbf{P}^{r-1}$. Gluing $C_2 \hookrightarrow \tilde{\mathbf{P}}_2^r$ with $C_1 \hookrightarrow \mathbf{P}^r$, we obtain an embedding of $C_1 \cup_A C_2$ in V which, under the natural map $V \to \mathbf{P}^r$, projects to the given map $C_1 \cup C_2 \to \mathbf{P}^r$. Now attach fibres L_i of π at all points of $C_1 \cap H \backslash A$ to obtain a locally complete intersection curve $X \subset V$. Let N be the normal bundle of X in V and N_2' and N_{L_i} those of C_2 and L_i in $\tilde{\mathbf{P}}_2^r$. Note that $N_{L_i} = (r-1)\mathcal{O}_{L_i}$. We have exact sequences

$$0 \to N_2'(-A) \oplus \bigoplus_i N_{L_i} \to N \to N_1 \to 0,$$

$$0 \to M \to N \to N_2' \to 0 \quad \text{(this defines } M\text{)},$$

$$0 \to \bigoplus_i N_{L_i}(-1) \to M \to N_1(A) \to 0,$$

$$0 \to \mathcal{O}_{C_2}(1+2A) \to N_2' \to N_2 \to 0.$$

(The latter comes from $0 \to \pi^*\mathcal{O}_{\mathbf{P}^{r-1}}(1)(2E) \to T_{\tilde{\mathbf{P}}_2^r} \to \pi^*T_{\mathbf{P}^{r-1}} \to 0$.) Using these sequences and our vanishing hypotheses, we see in either case (i) or case (ii) that $H^1(N) = 0$. This implies that deformations of X in V are unobstructed of dimension $\chi(N)$. On the other hand, there is by general principles a deformation space for the triple $(X, V, X \to V)$ which has dimension $\geq \chi(N) + 1$. This implies that there is a deformation of $(X, V, X \to V)$ in which V deforms nontrivially. But the only nontrivial deformation of V is a smoothing to $\mathbf{P}^r$, and the only kind of deformation of $X \to V$ which can go along with this is an embedding of a smooth curve $X' \hookrightarrow \mathbf{P}^r$. Now underlying the deformation of $X' \hookrightarrow \mathbf{P}^r$ to $X \to V$ is one of $X' \hookrightarrow \mathbf{P}^r$ to $C_1 \cup_A C_2 \to \mathbf{P}^r$, and hence the latter is smoothable. Since the normal bundle $N_{X'}$ degenerates to N, we have $H^1(N_{X'}) = 0$. Finally, assertion (ii)' is proved similarly, using the fact that a hyperplane section of C_2 in $\mathbf{P}_2^r$ is part of the limit of a hyperplane section of X' in $\mathbf{P}^r$.

REMARK. Some of the deformation-theoretic facts alluded to above, while simple variants of standard facts, are somewhat hard to locate exactly in the literature. But for helpful hints, see Artin's Tata notes or the second edition of *Geometric Invariant Theory*. Hopefully all these facts will be contained in a forthcoming set of notes by Sernesi [10].

REMARK 2.5. It is interesting to note that similar considerations also lead to the following nonsmoothing result (to be discussed in more detail elsewhere): if $C = C_1 \cup_A \overline{C}_2 \subset \mathbf{P}^r$, $\overline{C}_2 \subset H$ are as above, and moreover C is linearly normal and $h^0(\mathcal{O}_{\overline{C}_2}(1+A)) = r$, then C is not smoothable. For a smoothing of C can be modified to a map into a degeneration of $\mathbf{P}^r$ with V as special fibre, as above, but this contradicts $h^0(\mathcal{O}_{\overline{C}_2}(1+A)) = r$. Note that to have $h^0(\mathcal{O}_{\overline{C}_2}(1+A)) = r$ it suffices, for instance, that A be general on C_2 and $a \leq h^1(\mathcal{O}_{C_2}(1))$.

REMARK 2.6. Note that in cases where Theorem 2.3 is applicable, Theorem 2.2 usually is not: i.e., in such cases, Theorem 2.2 *may* be applicable only if $d_2 = g_2 + r - 1$.

We proceed now to construct some examples of good families of space curves. In the construction we shall use some relatively simple cases of the following result whose proof will be presented elsewhere [3a]. It would be difficult, albeit tedious, to give an ad hoc proof of the cases needed below.

THEOREM. *Let $C \subset \mathbf{P}^r$ be a general, noncanonical, embedding of a general curve, $r \geq 3$, with normal bundle N. Then $H^1(N(-1)) = 0$.*

EXAMPLES 2.7. (1) Let $C_1 \subset \mathbf{P}^3$ be a nonsingular curve with, say, $H^1(N_{C_1}(-1)) = 0$ and μ_0 surjective, and attach to it an elliptic cubic C_2 at a points, $5 \leq a \leq 9$. Then by Theorems 1.3 and 2.2, $C_1 \cup C_2$ has μ_0 surjective and is smoothable (since $h^0(\mathcal{I}_{A,\mathbf{P}^2}(3)) = 10 - a$). But it is not clear if this operation can be iterated.

(2) Now let $C_1 \subset \mathbf{P}^3$ be a nonsingular curve with $H^1(N_{C_1}) = 0$ and μ_0 surjective (e.g., a general curve of degree 9 and genus 8), and attach to it a plane curve $\overline{C}_2$ of degree d_2 and geometric genus g_2. Since $N_2 = \omega_{C_2}(3)$, the hypotheses of Theorem 2.3, case (ii)$'$ will be satisfied if

$$\max(2g_2 - d_2, \, g_2 - d_2 + 3) \leq a \leq 3d - 1,$$

and those of Corollary 1.6 if $a \geq 2g_2 + 4$, and the operation of attaching C_2 can be iterated. For instance, taking $d_2 = 5$, $g_2 = 5$, $a = 14$ yields families of smooth curves with μ_0 surjective and $H^1(N) = 0$ whose ratio degree/genus is asymptotically $5/18$. This improves, in part, a result of Sernesi [8] who obtains a ratio of $1/3$ (cf. also [4]), although Sernesi's result is more precise than just an asymptotic one. Presumably our result too could be made more precise with more careful and varied choices of the initial curve C_1.

(3) Now let $C_1 \subset \mathbf{P}^r$ be a curve with $H^1(N_{C_1}(-1)) = 0$ and μ_0 surjective (e.g., a general embedding of degree $3r$ of a general curve of genus $2(r+1)$), and attach to it a curve $C_2 \subset H$ of general moduli, genus $2r$, and degree $3(r - 1)$ at $a = 3(r-1)$ points. Then $H^1(\mathcal{O}_{C_2}(2A)) = 0$ follows from injectivity of $\mu_{0,A}$ (Gieseker-Petri) and $H^1(N_{C_2}(-1)) = 0$ as noted above. Hence the hypotheses of Theorem 2.3, (ii)$'$ are satisfied. Moreover, since C_1 and C_2 actually have μ_0 bijective, it follows from Theorem 1.3, (ii) that $C_1 \cup C_2$ has μ_0 surjective. Iterating this procedure yields families of smooth curves with μ_0 surjective, $H^1(N) = 0$ and asymptotic degree/genus ratio of $(3r - 3)/(5r - 4)$. This again improves a result of Sernesi [8], who obtains the ratio $(r - 1)/r$.

REFERENCES

1. E. Arbarello et al., *Geometry of algebraic curves*, Springer-Verlag, New York, 1984.
2. M.-C. Chang and Z. Ran, *Unirationality of the moduli spaces of curves of genus 11, 13, (and 12)*, Invent. Math. **76** (1984), 41–54.
3. ____, *The Kodaira dimensions of the moduli space of curves of genus 15*, J. Differential Geom. **23** (1986).
3a. ____, *The Kodaira dimensions of the moduli space of curves of genus 15*, J. Differential Geom. **24** (1986), 205–220.
4. D. Gieseker, *A construction of special space curves*, Algebraic Geometry: Proceedings, Ann Arbor, 1982, Lecture Notes in Math., vol. 1008, Springer-Verlag, 1983.
5. M. Green, *Koszul cohomology and the geometry of projective varieties*, J. Differential Geom. **19** (1984), 125–171.

6. R. Hartshorne and A. Hirschowitz, *Smoothing algebraic space curves*, Algebraic Geometry: Proceedings, Sitges 1983, Lecture Notes in Math., vol. 1124, Springer-Verlag, 1985, pp. 98–131.

7. Z. Ran, *Degeneration of linear systems*, Preprint.

8. E. Sernesi, *On the existence of certain families of curves*, Invent. Math. **75** (1984), 125–171.

9. ____, private communication (to appear).

10. ____, *Notes on the Hilbert scheme and deformations*, (to appear).

UNIVERSITY OF CALIFORNIA, RIVERSIDE

Proceedings of Symposia in Pure Mathematics
Volume **46** (1987)

Complete Subvarieties of the Moduli Space
of Smooth Curves

STEVEN DIAZ

As usual let M_g be the moduli space of smooth curves of genus g. By definition M_g consists of one point for each isomorphism class of smooth curve of genus g. M_g can be made into a quasiprojective variety of dimension $3g - 3$ if $g \geq 2$, 1 if $g = 1$, and 0 if $g = 0$. When you have a quasiprojective variety, some natural questions to ask about it are: Is it affine? projective? and if the answers to both of those questions are no, what kind of projective subvarieties does it have? We will discuss these questions for M_g.

The purpose of M_g is to parametrize families of curves. A family of smooth curves of genus g is a flat, proper, surjective morphism $\pi: X \to Y$ such that all the fibers are smooth curves of genus g. The basic properties of M_g are:

(1)   Given a family as just defined, the map $f_\pi: Y \to M_g$ sending $y \in Y$ to the point representing the isomorphism class of $\pi^{-1}(y)$ is a morphism.

(2)   Given a morphism $h: Z \to M_g$ there exists a finite cover $p: Z' \to Z$ and a family of smooth curves of genus g, $\pi: X \to Z'$, such that the morphism f_π defined in (1) is equal to $h \circ p$. (Ideally you would like to be able to do this without the finite cover p; however, p is often needed to do some "unwinding" around curves with automorphisms.)

Terminology. A family $\pi: X \to Y$ is complete if Y is complete. It is nondegenerate if the morphism f_π to M_g has finite fibers and generically nondegenerate if there exists a dense open subset $U \subset Y$ such that the family $\pi: \pi^{-1}(U) \to U$ is nondegenerate.

From (1) and (2) we see that looking for projective subvarieties of M_g is almost the same as looking for complete (generically) nondegenerate families of smooth curves of genus g.

1980 *Mathematics Subject Classification* (1985 *Revision*). Primary 14H10; Secondary 14D20.

For $g \geq 1$, $\mathcal{M}_g$ is not projective. This is because smooth curves can degenerate to singular curves in an essential way. That is, one can construct a family of curves of genus g where the general fiber is smooth but some special fibers are singular and no amount of base changing or blowing up and down will get rid of the singular fibers.

$\mathcal{M}_0$ is a point.

$\mathcal{M}_1 \cong \mathbf{A}^1$ (the j line).

$\mathcal{M}_2$ is an affine three-fold (see Igusa).

However for $g \geq 3$, $\mathcal{M}_g$ is not affine because of the following result.

THEOREM. *Given any finite number of points contained in $\mathcal{M}_g$, $g \geq 3$, there exists a projective curve contained in $\mathcal{M}_g$ through those points.*

PROOF. (I don't know whom to attribute this theorem to.) There exists a projective compactification $\tilde{\mathcal{M}}_g$ of $\mathcal{M}_g$ called the Satake compactification in which $\tilde{\mathcal{M}}_g - \mathcal{M}_g$ has codimension 2. (It is obtained by compactifying the locus of Jacobians in the moduli space of Abelian varieties. See Satake.) The theorem follows from this and elementary properties of projective varieties. $\square$

REMARK. This theorem shows that projective curves in $\mathcal{M}_g$ are very numerous. Yet they seem to be very hard to find explicitly. For example, to my knowledge no one has ever explicitly written down a complete nondegenerate family of smooth curves of genus 3 whose base has dimension 1.

I know of only one way of constructing complete families of smooth curves. I have been told it goes back at least to Kodaira. It gives the following result.

THEOREM. *Given any positive integer k there exists a genus g such that $\mathcal{M}_g$ contains a complete subvariety of dimension k.*

PROOF. (Sketch only. See Miller for more details.) Start with a smooth curve C of genus $g_0 > 1$. Let $\pi': C' \to C$ be an unramified double cover. In $C' \times C$ define $\Gamma = \{(p, q): \pi'(p) = q\}$. If we could take a double cover of $C' \times C$ branched over Γ, this would give a complete family of smooth curves with base C. This, however, may be impossible due to monodromy problems. If g' is the genus of C', for any two points $x, y \in C'$ there exist $2^{2g'}$ different double covers of C' branched over x and y. This problem can be overcome by pulling back to a $2^{2g'}$-sheeted unramified cover $\tilde{C}$ of C. One gets a family $\pi: \tilde{X} \to \tilde{C}$. Letting each fiber of π take the place of C one can repeat the process fiber by fiber on this family to obtain a family with a base of dimension 2 and continue until the base has as large a dimension as required. $\square$

REMARK. A simple application of the Riemann-Hurwitz formula shows that for given k the g obtained in this way is $4^k g_0 - \frac{2}{3}(4^k - 1)$.

In the previous theorem we found a complete curve through arbitrary points. In this theorem the complete subvarieties constructed always lie in the locus of points corresponding to curves with automorphisms. Whether there exists a complete subvariety larger than a curve through a general point of $\mathcal{M}_g$ is unknown.

The two theorems we have discussed so far gave lower bounds on the dimension of the largest projective subvariety of M_g. There are also upper bounds. Unfortunately as yet the lower bounds do not equal the upper bounds. The following are currently the strongest known results.

THEOREM (HARER). *There do not exist complete subvarieties of M_g $(g \geq 1)$ of dimension $2g - 1$ or greater.*

PROOF. He shows that $H_k(M_g, Q) = 0$ for $k > 4g - 3$ and any complete subvariety would give rise to nonzero homology classes.   □

THEOREM (DIAZ). *In M_3 the largest complete subvariety is a curve.*

PROOF. We have already seen that complete curves do exist.

The locus H of points corresponding to hyperelliptic curves is a divisor on M_3. A computation of its class shows that H is ample. Any complete surface in M_3 would therefore have to meet H in a complete curve, but it is well known that H is affine and contains no complete curves.   □

THEOREM (DIAZ). *If $Z \subset M_g$, $g \geq 3$, is a complete subvariety, then dimension $Z \leq g - 2$.*

PROOF. The proof uses a stratification of M_g which was inspired by the stratification studied by Arbarello. Much of the idea of how to use such a stratification is also due to Arbarello. As Arbarello's stratification is much simpler than the stratification the actual proof uses, there still remains the interesting question of whether Arbarello's stratification can also provide a proof. Therefore we will start with a discussion of Arbarello's stratification.

For $l \leq g$ define $A_l = \{[C] \in M_g$: there exists a finite morphism $f: C \to \mathbf{P}^1$ of degree $\leq l$ with one point of total ramification$\}$. The point of total ramification is a Weierstrass point. A_2 is the locus of hyperelliptic curves. Arbarello shows that A_l is irreducible of dimension $2g - 3 + l$. The A_l stratify M_g.

$$M_g = A_g \supsetneq A_{g-1} \supsetneq \cdots \supsetneq A_2 \supsetneq A_1 = \varnothing.$$

To complete the proof of the theorem all one would need is the following lemma.

(3) $A_l - A_{l-1}$ contains no complete curve.

If this were true, then a k-dimensional complete subvariety of A_l would meet A_{l-1} in at least a $(k-1)$-dimensional complete subvariety. Counting the steps in the stratification would finish the proof. The truth of (3) is still an open question. An almost-proof proceeds as follows.

Suppose $B \subset A_l - A_{l-1}$ is a complete curve. Then for some finite cover $\tilde{B}$ of B we have the following commutative diagram.

(4)
$$
\begin{array}{ccc}
X & \longrightarrow & E \\
 & \searrow \quad \swarrow & \\
 & \tilde{B} &
\end{array}
\quad s_i \atop i=1,2,\ldots
$$

$X \to \tilde{B}$ is a family of smooth curves of genus g.

$E \to \tilde{B}$ is a $\mathbf{P}^1$ bundle. The s_i are sections.

$X \to E$ is finite and branched exactly over the s_i with a total ramification point over s_1. For $b \in \tilde{B}$, let X_b and E_b be the corresponding fibers; then the restricted map $X_b \to E_b$ expresses X_b as an l-sheeted cover of $\mathbf{P}^1$ with one point of total ramification. Arbarello shows that if any of the s_i for $i > 1$ meet s_1 (that is, if any ramification points run into the total ramification point), the resulting fiber X_b would have either to be singular or to represent a point in A_{l-1}, contrary to assumption. The problem is that a $\mathbf{P}^1$ bundle can have disjoint sections, so all the s_i, $i > 1$, might be disjoint from s_1.

A remedy is to use a slightly larger stratification. Define $H_g(i,j) = \{[C] \in \mathcal{M}_g$: there exists a morphism $f: C \to \mathbf{P}^1$ such that degree $f \leq i$ and the number of points in $f^{-1}\{0, \infty\}$, not counting multiplicity, is $\leq j\}$. For any fixed $k \geq g$, we have $\mathcal{M}_g = H_g(k, g) \supset H_g(k, g-1) \supset \cdots \supset H_g(k, 2) \supset H_g(k, 1) = \varnothing$. A_l is a component of $H_g(g, l)$. In general $H_g(i, j)$ can have many components. Now we can prove

LEMMA. $H_g(i, j) - H_g(i, j-1)$ does not contain a complete curve.

(In the same way that (3) would imply the theorem, this lemma also implies the theorem.)

PROOF. Suppose $B \subset H_g(i, j) - H_g(i, j-1)$ is a complete curve. In the same way that we obtained diagram (4), this gives a diagram like (4), where now s_1 and s_2 correspond to $0, \infty$ in the definition of $H_g(i, j)$. Using the compactification of the Hurwitz scheme in Harris-Mumford and Diaz, one shows that if s_1 or s_2 meet each other or any other s_i the resulting fiber X_b will either be singular or represent a point in $H_g(i, j-1)$, contrary to assumption. But if s_1, s_2 never meet each other or any other s_i, then s_1, s_2, s_3 are three disjoint sections of the $\mathbf{P}^1$ bundle. This shows that the bundle is trivial and all sections are constant which means the family $X \to \tilde{B}$ is trivial—a contradiction. □

COROLLARY. If $Z \subset \overline{\mathcal{M}}_g$ (the Deligne-Mumford compactification of $\mathcal{M}_g$) is a complete subvariety of dimension $\geq 2g - 1$, then Z must meet δ_0.

PROOF. By the previous theorem Z must meet some δ_i. If $i \neq 0$ apply the theorem to the induced families of curves of genera i and $g - i$ to show there is a further degeneration. Repeat the process until you force a degeneration to δ_0. □

Why would one want a theorem like this? Many of the theorems which have recently been proven about curves have been proven by degenerating smooth curves to singular curves. This theorem shows that any complete subvariety of $\overline{\mathcal{M}}_g$ of dimension greater than $g - 2$ must contain points corresponding to singular curves. One could then at least hope to study the curves this subvariety represents by the technique of degenerating to singular curves.

References

1. E. Arbarello, *Weierstrass points and moduli of curves*, Compositio Math. **29** (1974), 325–342.

2. S. Diaz, *Exceptional Weierstrass points and the divisor on moduli space that they define*, Mem. Amer. Math. Soc. No. 327 (1985).

3. ____, *A bound on the dimension of complete subvarieties of* M_g, Duke Math. J. **51** (1984), 405–408.

4. J. Harer, *The second homology group of the mapping class group of an orientable surface*, Invent. Math. **72** (1983), 221–239.

5. ____, *Stability of the homology of the mapping class groups of orientable surfaces*, Ann. of Math. (2) **121** (1985), 215–249.

6. ____, *The virtual cohomological dimension of the mapping class group of an orientable surface*, Invent. Math. (to appear).

7. J. Harris and D. Mumford, *On the Kodaira dimension of the moduli space of curves*, Invent. Math. **67** (1982), 23–86.

8. J. I. Igusa, *Arithmetic variety of moduli for genus two*, Ann. of Math. (2) **72** (1960), 612–649.

9. E. Miller, *The homology of the mapping class group*, J. Differential Geom. **24** (1986), 1–14..

10. I. Satake, *On the compactification of the Siegel space*, J. Indian Math. Soc. (N.S.) **20** (1956), 259–281.

BRANDEIS UNIVERSITY

Current address: University of Pennsylvania, Philadelphia

Proceedings of Symposia in Pure Mathematics
Volume **46** (1987)

The Irreducibility of the Hilbert Scheme
of Smooth Space Curves

LAWRENCE EIN

Denote by $H_{d,g,n}$ the open subscheme of the Hilbert scheme parametrizing the smooth irreducible curves of degree d and genus g in $\mathbf{P}^n$.

In [**2**], we showed that $H_{d,g,3}$ is irreducible if $d \geq g+3$. This result answered a question asked by Hartshorne and Hirschowitz [**4**, p. 14]. More generally, Severi has asserted that $H_{d,g,n}$ is irreducible if $d \geq g + n$ with an incomplete proof [**8**, p. 370]. Using a theorem of Kaji (Lemma 1) and a refinement of the method in [**2**] enables us to show that $H_{d,g,4}$ is irreducible if $d \geq g + 4$. I should mention that G. Sacchiero informed me that he had a proof for Lemma 1. Furthermore, independently he had also observed that Lemma 1 will give a refinement for the results in [**2**]. Following a suggestion of Joe Harris, we demonstrated by an example that Severi's assertion is false for $n \geq 6$. Throughout the paper we shall work over the complex numbers.

I would like to thank R. Hartshorne, R. Lazarsfeld, and G. Sacchiero for helpful discussions.

Let C be a smooth irreducible genus g curve, and $\mathcal{L}$ a degree d line bundle on C. Let V be an $(r+1)$-dimensional subspace of $H^0(\mathcal{L})$. Assume that V induces a birational unramified map $f\colon C \to C_1 = f(C) \subseteq \mathbf{P}(V)$. Consider the following diagram (where $P^1(\mathcal{L})$ is the first principal part of $\mathcal{L}$ and $F = \ker(df)$):

$$
\begin{array}{ccc}
0 & & 0 \\
\downarrow & & \downarrow \\
F & = & F \\
\downarrow & & \downarrow \\
0 \to f^*\Omega^1_{\mathbf{P}^r} \otimes \mathcal{L} \to V \otimes \mathcal{O}_C \to \mathcal{L} \to 0 \\
\downarrow df \qquad\quad \downarrow \qquad\quad \| \\
0 \to \Omega^1_C \otimes \mathcal{L} \to P^1(\mathcal{L}) \to \mathcal{L} \to 0 \\
\downarrow \qquad\quad \downarrow \\
0 \qquad\quad 0
\end{array}
$$

1980 *Mathematics Subject Classification* (1985 *Revision*). Primary 14–02.
Partially supported by an NSF Grant.

Let $X = \mathbf{P}(P^1(\mathcal{L}))$ and $\pi\colon X \to C$ be the projection map. Let $h\colon X \to h(X) = Y \subseteq P(V)$ be the natural birational map. Then Y is the tangent surface of X. Let $C_0 \subseteq X$ be the section induced by $P^1(\mathcal{L}) \to \mathcal{L}$. Denote by U the tautological bundle of X. Then $U|_{C_0} = \mathcal{L}$.

LEMMA 1 (KAJI, SACCHIERO). *Let C and $\mathcal{L}$ be as above. Let p be a general point on C. Suppose l_p is the tangent line of C_1 at $f(p)$. If $r \geq 3$, then*

(a) $f^{-1}(l_p) = 2p$.

(b) *Suppose $g_1\colon C_1 \to \mathbf{P}^{r-1}$ is the projection map from $f(p)$. Then $g_p = g_1 \circ f$ is an unramified birational map from C to its image.*

PROOF. See [**7**, Theorem 3.1 and Remark 3.8].

COROLLARY 2. *Let $Z = h^{-1}h(C_0)$. Then the ideal sheaf of Z in X is of the form $I_{Z/X} = \mathcal{O}_X(-2C_0) \otimes I_{Z_0}$, where I_{Z_0} is the ideal sheaf of a zero-dimensional scheme.*

LEMMA 3. *Let $p_1, p_2, \ldots, p_{r-2}$ be $r - 2$ general points on C. Then $f(p_1), f(p_2), \ldots, f(p_{r-2})$ span an $(r - 3)$-plane L in $\mathbf{P}^r$. Then*

(a) $f^{-1}(L) = \{p_1, p_2, \ldots, p_{r-2}\}$.

(b) *Suppose that $Q = h^{-1}(L)$. Then $Q = \coprod_{i=1}^{r-2} q_i$, where q_i is a length 2 0-dimensional subscheme of X and $\pi_* \mathcal{O}_{q_i} \cong \mathcal{O}_C/\mathcal{O}_C(-2p_i)$.*

PROOF. (a) This follows from the uniform position lemma of [**1**, p. 112].

(b) This follows from (a) and Corollary 2.

LEMMA 4. *Let $F = \ker(V \otimes \mathcal{O}_C \to P^1(\mathcal{O}_C(1)))$. Then there is the following exact sequence:*

$$(4.1) \qquad 0 \to K_C^{-1} \otimes \mathcal{L}^{-2} \otimes \mathcal{O}_C \left(\sum_1^{r-2} 2p_j \right) \to F \to \sum_1^{r-2} \mathcal{O}_C(-2p_j) \to 0.$$

PROOF. Let $W = H^0(I_{L/\mathbf{P}^r}(1))$. Then $\dim W = 3$. There is the exact sequence $W \to I_{L/\mathbf{P}^r}(1) \to 0$.

Observe that there is the following diagram:

$$(4.2)$$

$$
\begin{array}{ccccccccc}
 & & 0 & & 0 & & 0 & & \\
 & & \downarrow & & \downarrow & & \downarrow & & \\
0 \to & & M' & \to & W \otimes \mathcal{O}_X & \overset{\alpha}{\to} & I_Q \otimes U & \to & 0 \\
 & & \downarrow & & \downarrow & & \downarrow & & \\
0 \to & & M & \to & V \otimes \mathcal{O}_X & \to & U & \to & 0 \\
 & & \downarrow{\scriptstyle\beta} & & \downarrow & & \downarrow & & \\
0 \to & & \sum_{i=1}^{r-2} I_{q_i} & \to & V/W \otimes \mathcal{O}_X & \to & U|_Q & \to & 0 \\
 & & \downarrow & & \downarrow & & \downarrow & & \\
 & & 0 & & 0 & & 0 & &
\end{array}
$$

where $M = \ker(V \otimes \mathcal{O}_X \to U)$ and $M' = \ker(W \otimes \mathcal{O}_X \to I_Q \otimes U)$. Observe that M' is a rank 2 locally free sheaf. Since $h^{-1}(L) = Q$, α is surjective. It follows from the snake lemma that β is also surjective. Set $f_i = \pi^{-1}(p_i) \cong \mathbf{P}^1$. Denote by $\bar{q}_i$ the point supporting Q_i. Observe that

(4.3) $\dim I_Q \otimes U \otimes k(\bar{q}_i) = 2,$ where $k(\bar{q}_i)$ is the residue field of $\bar{q}_i$.

(4.4)   The restriction map $H^0(W \otimes \mathcal{O}_X) \twoheadrightarrow I_Q \otimes U \otimes k(\bar{q}_i)$ is surjective.

There is also the exact sequence:

$$0 \to \text{tor}_1(I_Q \otimes U, \mathcal{O}_{f_i}) \to M \otimes \mathcal{O}_{f_i} \to W \otimes \mathcal{O}_{f_i} \to I_Q \otimes U \otimes \mathcal{O}_{f_i} \to 0.$$

Since M' is locally free, $\text{tor}_1(I_Q \otimes U, \mathcal{O}_{f_i}) = 0$. Observe that $h^0(I_Q \otimes U \otimes \mathcal{O}_{f_i}) = 2$. It follows from (4.3) and (4.4) that $H^0(W \otimes \mathcal{O}_{f_i}) \to H^0(I_Q \otimes U \otimes \mathcal{O}_{f_i})$ is surjective. It follows that $h^0(M' \otimes \mathcal{O}_{f_i}) = 1$ and $h^1(M' \otimes \mathcal{O}_{f_i}) = 0$. So $\pi_* M'$ is an invertible sheaf and $R^1 \pi_* M' = 0$. Observe that $\pi_* M = F$. It follows from (4.2) that there is the following exact sequence:

$$0 \to K_C^{-1} \otimes \mathcal{L}^{-2} \otimes \mathcal{O}_C \left(\sum_1^{r-2} 2p_i \right) \to F \to \sum_1^{r-2} \mathcal{O}_C(-2p_i) \to 0.$$

This completes the proof of Lemma 4.

The fine moduli space of smooth irreducible genus g curves with level m structure is denoted by $\mathcal{M}_{g,m}$. Suppose that $\mathcal{C} \to \mathcal{M}_{g,m}$ is the universal family of curves. Let $\mathcal{P}\text{ic}\,\mathcal{C}$ be the relative Picard scheme. Set

$$W_d^r = \{(\mathcal{L}, C) \in \mathcal{P}\text{ic}\,\mathcal{C} | \deg \mathcal{L} = d \text{ and } h^0(\mathcal{L}) \geq r + 1\}.$$

For the rest of the paper we shall use the following notations. We shall assume that $\mathcal{L}$ is a degree d very ample line bundle on C. Suppose that $h^0(\mathcal{L}) = r + 1$ and $h^1(\mathcal{L}) = \delta > 0$. Denote by f the natural map $f \colon C \to \mathbf{P}(H^0(\mathcal{L})) = \mathbf{P}^r$. Let N^* be the conormal sheaf of C in $\mathbf{P}^r$. It follows from diagram (1.1) there is the following exact sequence:

(5.1) $0 \to N^* \otimes K \to H^0(\mathcal{L}) \otimes K \otimes \mathcal{L}^{-1} \to P^1(\mathcal{L}) \otimes K \otimes \mathcal{L}^{-1} \to 0.$

Consider the map

$$\mu \colon H^0(\mathcal{L}) \otimes H^0(K \otimes \mathcal{L}^{-1}) \to H^0(P^1(\mathcal{L}) \otimes K \otimes \mathcal{L}^{-1}).$$

It is well known that $H^0(P^1(\mathcal{L}) \otimes K \otimes \mathcal{L}^{-1})$ is naturally isomorphic to the cotangent space of $\mathcal{P}\text{ic}\,\mathcal{C}$ at $(\mathcal{L}, C)$. The image of μ is the annihilator of the Zariski tangent space of W_d^r at $(\mathcal{L}, C)$.

THEOREM 5. *Assume $r \geq 3$. Let N be the normal sheaf of C in $\mathbf{P}(H^0(\mathcal{L}))$. Then*
  (a) *If $\delta \leq 2$, then $h^0(N^* \otimes K) = h^1(N) = 0$.*
  (b) *If $\delta \geq 2$, then $h^0(N^* \otimes K) = h^1(N) \leq (r-2)(\delta-2)$.*

PROOF. It follows from Lemma 4 that there is the following exact sequence:

(5.2) $0 \to \mathcal{L}^{-3} \left(\sum_1^{r-2} 2p_i \right) \to N^* \otimes K \to \sum_1^{r-2} K \otimes \mathcal{L}^{-1}(-2p_i) \to 0.$

It follows that

$$h^1(N) = h^0(N^* \otimes K) \le \sum_1^{r-2} h^0(K \otimes \mathcal{L}^{-1}(-2p_i)).$$

THEOREM 6. *Assume $r \ge 3$. Suppose Y is an irreducible component of $\mathcal{W}_d^r$ containing $(\mathcal{L}, C)$. Then*

(a) *If $\delta \le 2$, then $\operatorname{rank} \mu = (r+1)\delta$. Then $\mathcal{W}_d^r$ is smooth at $(\mathcal{L}, C)$ and $\dim Y = 4g - 3 - (r+1)\delta$.*

(b) *If $\delta \ge 2$, then $\operatorname{rank} \mu \ge 5\delta + 2(d-g) - 4$. Furthermore, $\dim Y \le 6g - 5\delta - 2d + 1$.*

PROOF. This follows immediately from the exact sequence (5.1) and Theorem 5.

(*)   If $C \in H_{d,g,n}$, then $\chi(N_{C/\mathbf{P}^n}) = (n+1)d + (n-3)(1-g)$.

As in [6], we can conclude that the dimension of each irreducible component of $H_{d,g,n}$ is greater than or equal to $(n+1)d + (n-3)(1-g)$.

THEOREM 7. *$H_{d,g,4}$ is irreducible if $d \ge g + 4$.*

PROOF. There is a unique irreducible open set of $H_{d,g,4}$ corresponding to nonspecial curves ($h^1(\mathcal{O}_C(1)) = 0$). Suppose for contradiction $H_{d,g,4}$ is reducible. Then there is an irreducible component W of $H_{d,g,4}$ such that the general curves in the family W satisfy $h^0(\mathcal{O}_C(1)) = r + 1$ and $h^1(\mathcal{O}_C(1)) = \delta > 0$. We denote by $H_{d,g,4}^m$ the Hilbert scheme of degree d genus g smooth irreducible curves in $\mathbf{P}^4$ with level m structure. Let W_m be an irreducible component of $H_{d,g,4}^m$ corresponding to W. Then $\dim W = \dim W_m$. There is a natural map $h\colon W_m \to \mathcal{W}_d^r \subseteq \operatorname{Pic} \mathcal{C}$. Let Y be an irreducible component of $\mathcal{W}_d^r$ containing $h(W_m)$. Let x be a general point of $h(W_m)$. Then

$$\dim h^{-1}h(x) = \dim G(5, r+1) + \dim(\operatorname{Aut} \mathbf{P}^4) = 5d + 5 + 5\delta - 5g.$$

It follows from Theorem 6,

$$\dim W = \dim W_m = \dim h^{-1}h(x) + \dim h(W_m) < 5d + (1-g).$$

This contradicts (*).

THEOREM 8. *Assume that $n \ge 5$. If*

$$d > \frac{2n-2}{n+2}g + \frac{n+8}{n+2},$$

then $H_{d,g,n}$ is irreducible.

PROOF. The proof is similar to Theorem 8 in [2]. We shall omit the details here.

The following example is suggested by Joe Harris. This example will show the bound that we obtained in Theorem 8 is not too far from being sharp.

PROPOSITION 9. *Assume that $n \geq 6$. Then $H_{16n-35,8n+6,n}$ is reducible.*

PROOF. Note that there is a unique irreducible open set W of $H_{16n-35,\ 8n+6,\ n}$ corresponding to nonspecial curves. Furthermore,

$$\dim W = (n+1)d + (n-3)(1-g) = 8n^2 - 20.$$

Now consider a generic trigonal curve C of genus $8n + 6$. C can be imbedded as a curve of type $(3, 4n + 4)$ in a quadric surface Q. Let $\mathcal{L} = \mathcal{O}_Q(1, 4n - 13)|_C$. Then $h^0(\mathcal{L}) = 8n - 24$ and $h^1(\mathcal{L}) \neq 0$. We can embed C into $\mathbf{P}^n$ by an $(n+1)$-dimensional subspace of $H^0(\mathcal{L})$. In this fashion, we can construct a family E of trigonal curves in $\mathbf{P}^n$. Furthermore,

$$\dim E \geq \dim(\text{trigonal curves of genus } 8n + 6) + \dim(\text{Aut } \mathbf{P}^n)$$
$$+ \dim(G(n + 1, 8n - 24))$$
$$= (16n + 13) + (n+1)^2 - 1 + (7n - 25)(n+1) = 8n^2 + 12 > \dim W.$$

This proves that $H_{16n-35,\ 8n+6,\ n}$ is reducible.

REFERENCES

1. E. Arbarello, M. Cornalba, P. Griffiths, and J. Harris, *Geometry of algebraic curves.* Volume I, Springer-Verlag, 1984.
2. L. Ein, *Hilbert scheme of smooth space curves*, Ann. École Nat. Sup. Méc. Nantes (to appear).
3. J. Harris, *Curves in projective spaces*, Sem. Math. Sup., no. 85, Presses Univ. Montréal, Montréal, 1982.
4. R. Hartshorne and A. Hirschowitz, *Smoothing algebraic space curves*, Algebraic geometry, (Sitges, 1983), Lecture Notes in Math., vol. 1124, Springer-Verlag, 1985, pp. 98–131.
5. L. Gruson, R. Lazarsfeld, and C. Peskine, *On a theorem of Castelnuovo and equations defining space curves*, Invent. Math. **72** (1983), 491–506.
6. Hajime Kaji, *On the tangentially degenerate curves*, J. London Math. Soc. (to appear).
7. S. Mori, *Projective manifolds with ample tangent bundles*, Ann. of Math. (2) **110** (1979), 593–606.
8. F. Severi, *Vorlesungen über algebraische Geometrie*, (E. Löffler Übersetzung) Leipzig, 1921.

UNIVERSITY OF ILLINOIS, CHICAGO

Proceedings of Symposia in Pure Mathematics
Volume **46** (1987)

On Theta Functions for Jacobi Varieties

R. C. GUNNING

Jacobi varieties are a quite special class of principally polarized Abelian varieties, and their particular nature is reflected in a number of special properties and identities satisfied by the theta functions associated to them. The investigation of the latter has long been pursued, probably the first major result being Riemann's singularity theorem (also known as Riemann's vanishing theorem). This pursuit has been especially active in recent years, but despite the extensive results that have been accumulated since Riemann's time and the flurry of current activity, the topic is far from being exhausted; indeed in many areas present knowledge remains rather meagre. In the hope of enticing others to consider looking into those areas further, this article is a survey of some work and problems involving descriptions of the subvarieties of special positive divisors in terms of second-order theta series.

1. Suppose then that $J = J(\Omega)$ is the Jacobi variety associated to the period matrix Ω of a marked compact Riemann surface M of genus $g > 0$; thus $J = \mathbf{C}^g/L$ for the lattice subgroup $L = (I\Omega)\mathbf{Z}^{2g} \subset \mathbf{C}^g$, and the canonical Abelian integrals $w_1, \ldots, w_g$ on M are the coordinate functions of a holomorphic mapping from the universal covering space $\tilde{M}$ of M into $\mathbf{C}^g$ inducing a biholomorphic mapping $w\colon M \to W_1$ between M and a nonsingular curve $W_1 \subseteq J$. It is convenient to normalize these Abelian integrals to vanish at the base point $z_0 \in \tilde{M}$, so that z_0 is mapped to the origin $0 \in \mathbf{C}^g$, and to view the image of the origin $0 \in \mathbf{C}^g$ as the zero element in the group J. The theta series with characteristic $[\nu|\tau] \in \mathbf{C}^g \times \mathbf{C}^g$ has the form

$$(1) \quad \theta[\nu|\tau](w;\Omega) = \sum_{n\in\mathbf{Z}^g} \exp 2\pi i \left[\frac{1}{2}{}^t(n+\nu)\Omega(n+\nu) + {}^t(n+\nu)(w+\tau)\right],$$

and the second-order theta series with characteristic $[\nu|\tau]$ will be taken in the form

$$(2) \quad \theta_2[\nu|\tau](w;\Omega) = \theta\left[\frac{\nu}{2}\middle|\tau\right](2w;2\Omega).$$

1980 *Mathematics Subject Classification* (1985 *Revision*). Primary 14K25; Secondary 14H40.
Research supported in part by NSF Grant DMS-8401273.

These are entire functions of the variable $w \in \mathbf{C}^g$, and as ν varies over $\mathbf{Z}^g/2\mathbf{Z}^g$ the 2^g functions (2) describe a basis for the vector space of entire functions on $\mathbf{C}^g$ transforming by the factor of automorphy $\rho_{-\tau}\xi^2$ under the action of the lattice subgroup L; here $\rho_\tau \in \mathrm{Hom}(L, \mathbf{C}^*)$ is the representation

$$(3) \qquad \rho_\tau(\delta_j) = 1, \qquad \rho_\tau(\Omega\delta_j) = \exp 2\pi i \tau_j,$$

where ${}^t\delta_j = (0, \dots, 0, 1, 0, \dots, 0)$ has 1 in the jth place, and the factor of automorphy ξ is defined by

$$(4) \qquad \xi(\delta_j, w) = 1, \qquad \xi(\Omega\delta_j, w) = \exp -2\pi i(w_j + \varepsilon_j),$$

where 2ε is the diagonal vector of the period matrix Ω. It is convenient to view these 2^g second-order theta functions as forming a single vector-valued function

$$(5) \qquad \boldsymbol{\theta}_2[\tau](w) = \{\theta_2[\nu|\tau](w; \Omega): \nu \in \mathbf{Z}^g/2\mathbf{Z}^g\}$$

and just to set $\boldsymbol{\theta}_2(w) = \boldsymbol{\theta}_2[0](w)$, noting that $\boldsymbol{\theta}_2[\tau](w) = \boldsymbol{\theta}_2(w + \frac{1}{2}\tau)$.

The composition of the second-order theta functions (2) for $\nu \in \mathbf{Z}^g/2\mathbf{Z}^g$ and the holomorphic mapping $w\colon \tilde{M} \to \mathbf{C}^g$ defined by the Abelian integrals on M, or loosely speaking the restrictions of these theta functions to the curve $W_1 \subseteq J(M)$, are the Riemannian theta functions $\theta_2[\nu|\tau](w(z); \Omega)$; they are holomorphic functions on $\tilde{M}$ that transform under the action of the covering group Γ, the group of automorphisms of $\tilde{M}$ for which $M = \tilde{M}/\Gamma$, by a factor of automorphy for Γ induced by the factor of automorphy $\rho_{-\tau}\xi^2$ for L. It is possible to choose a factor of automorphy ς for the action of Γ on M such that ς represents the line bundle over M associated to the base point $z_0 \in M$ and that the factor of automorphy induced by ξ has the form

$$(6) \qquad \xi(T, z) = \rho_r(T) \cdot \varsigma(T, z)^g,$$

where $\rho_\tau \in \mathrm{Hom}(\Gamma, \mathbf{C}^*)$ is induced by (3) and $r \in \mathbf{C}^g$ is essentially the vector of Riemann constants. The Riemannian theta functions $\theta_2[\nu|\tau](w(z); \Omega)$ thus transform by the factor of automorphy $\rho_{2r-\tau}\varsigma^{2g}$. These functions as ν varies over $\mathbf{Z}^g/2\mathbf{Z}^g$ span the $(g + 1)$-dimensional vector space $\Gamma(M, \mathcal{O}(\rho_{2r-\tau}\varsigma^{2g}))$ of holomorphic functions that transform by this factor of automorphy. These observations are easily proved, and details can be found in [9]. Since $2^g > g + 1$ whenever $g > 1$, there are for $g > 1$ some linear relations among the Riemannian theta functions. The explicit description of these linear relations appears to be a very difficult problem indeed; it is of interest in that it is an approach to the problem of providing an explicit description of factors of automorphy for L describing the Picard bundle over M, and thus of developing an appropriate explicit theory of vector-valued theta functions paralleling the classical theory of scalar-valued theta functions. These problems are discussed in [8] and [13], where further references can also be found.

Just using the observation that the Riemannian second-order theta functions span $\Gamma(M, \mathcal{O}(\rho_{2r-\tau}\varsigma^{2g}))$, though, it is a simple matter to write down explicit descriptions of the spaces of special positive divisors in terms of these functions. As is no doubt quite familiar, these subvarieties can be described as

(7) $$W_r^\nu = \{t \in J; \gamma(\rho_t \varsigma^r) > \nu\},$$

where $\gamma(\rho_t \varsigma^r) = \dim \Gamma(M, \mathcal{O}(\rho_t \varsigma^r))$. Fix a positive divisor $\mathfrak{d} = k_1 z_1 + k_2 z_2 + \cdots$ of degree r, where $z_1, z_2, \ldots$ are distinct points of M, and introduce the $2^g \times r$ matrix

(8) $$\Theta_2(\mathfrak{d}; t) = \left\{ \boldsymbol{\theta}_2[t](w(z_1)), \frac{\partial}{\partial z_1} \boldsymbol{\theta}_2[t](w(z_1)), \ldots, \frac{\partial^{k_1-1}}{\partial z_1^{k_1-1}} \boldsymbol{\theta}_2[t](w(z_1)), \right.$$
$$\left. \boldsymbol{\theta}_2[t](w(z_2)), \cdots \right\}.$$

In these terms

(9) $$W_{r-2}^\nu = \{t \in J : \operatorname{rank} \Theta_2(\mathfrak{d}; t - w(\mathfrak{d})) < r - \nu\},$$

where as usual $w(\mathfrak{d}) = k_1 w(z_1) + k_2 w(z_2) + \cdots \in J$. The proof is a straightforward matter of expressing the dimension $\gamma(\rho_{t-w(\mathfrak{d})} \varsigma^{2g-r})$ in terms of the rank of the matrix $\Theta_2(\mathfrak{d}; t - w(\mathfrak{d}))$ and using the definition (7) and the Riemann-Roch theorem; details can be found in [9]. In particular if z_1, z_2, z_3 are any three distinct points of M, then

(10) $$W_1 = \{t \in J \colon \boldsymbol{\theta}_2[t - w(\mathfrak{d})](w(z_1)), \boldsymbol{\theta}_2[t - w(\mathfrak{d})](w(z_2)),$$
$$\boldsymbol{\theta}_2[t - w(\mathfrak{d})](w(z_3)) \text{ are linearly dependent}\};$$

alternatively set $\boldsymbol{\theta}_2^{(\nu)}[\tau](w(z)) = (\partial^\nu/\partial z^\nu)\boldsymbol{\theta}_2[\tau](w(z))$ for $\nu = 0, 1, 2, \ldots$, and in terms of the divisor $\mathfrak{d} = 3z$

(11) $$W_1 = \{t \in J \colon \boldsymbol{\theta}_2[t - 3w(z)](w(z)), \boldsymbol{\theta}_2^{(1)}[t - 3w(z)](w(z)),$$
$$\boldsymbol{\theta}_2^{(2)}[t - 3w(z)](w(z)) \text{ are linearly dependent}\}.$$

These are quite explicit descriptions of the curve W_1, in terms just of the period matrix Ω and any three distinct points $w(z_1), w(z_2), w(z_3)$ of W_1, or essentially equivalent information. In general (8) is an explicit description of the subvarieties W_r^ν, in terms just of the period matrix Ω and r distinct points $w(z_1), \ldots, w(z_r)$ of W_1, or essentially equivalent information.

The existence of the array of subvarieties W_r^ν satisfying the known interrelations is a special property of Jacobi varieties, characterizing them among all principally polarized Abelian varieties and providing a finer classification of Jacobi varieties according to further properties of these arrays. That leads of course to a very interesting class of problems: how little of this array of subvarieties with their special properties suffices to characterize Jacobi varieties, and how effective a characterization is it? More generally, some properties of these arrays characterize special classes of Jacobi varieties or more general classes of principally polarized Abelian varieties: what are such classes, their characteristic properties, and their interrelationships? A well-known example of results in this direction is due to Andreotti and Mayer [1], and involves just the portion $W_{g-1}^1 \subset W_{g-1}$

of this array together with the properties that W_{g-1} has dimension $g - 1$ and carries a known homology class and that W_{g-1}^1 is of dimension $g - 3$ or $g - 4$ and is the singular locus of W_{g-1}. This describes a subvariety of the variety of principally polarized Abelian varieties containing the set of Jacobi varieties as a dense subset of a component; it is now known that this is generally a larger class than just the Jacobi varieties, and references to subsequent work can be found in [2]. This part of the array can be described quite explicitly in terms of the period matrix alone, since Riemann's theorem shows that up to a translation W_{g-1} is just the zero locus of the first-order theta function $\theta[0|0](w; \Omega)$. A somewhat simpler approach is to consider just the part W_1 of this array, since as Matsusaka has shown [14] the presence of an irreducible curve in J carrying the appropriate homology class characterizes Jacobi varieties. It turns out that (10) and (11) can be used to describe this curve W_1 fairly explicitly just in terms of the period matrix alone; indeed I showed in [10] that if t_i are any three distinct points of a principally polarized Abelian variety for which W_{g-1} is irreducible, then the locus

$$(12) \qquad \{t \in J: \ \boldsymbol{\theta}_2[t](t_1), \boldsymbol{\theta}_2[t](t_2), \boldsymbol{\theta}_2[t](t_3) \text{ are linearly dependent}\}$$

is either a finite set of points or a nonsingular irreducible curve in J, and in the latter case it is a translate of $\pm W_1$ and J is a Jacobi variety, and Welters showed [17] the analogous result for (11). This characterization of Jacobi varieties has subsequently been much improved and extended, and made much more explicitly a property of the period matrix Ω, by Arbarello–De Concini [3] and Welters [18].

The linear dependence relation in (12) is of a rather contrived form, only natural in the context of a general result such as (8). It would appear more natural to consider for instance the locus

$$(13) \qquad \{(t_1, t_2, t_3) \in J^3: \ \boldsymbol{\theta}_2(t_1), \boldsymbol{\theta}_2(t_2), \boldsymbol{\theta}_2(t_3) \text{ are linearly dependent}\}.$$

This locus of course contains the $2g$-dimensional subvarieties $t_i = \pm t_j$, $i \neq j$, and if J is a Jacobi variety it follows readily from (10) that it contains the 4-dimensional subvariety

$$(14) \ \{(t_1, t_2, t_3): \ t_1 = \tfrac{1}{2}w(z_1 + z_2 - z_3 - z_4), \ t_2 = \tfrac{1}{2}w(z_1 + z_3 - z_2 - z_4),$$
$$t_3 = \tfrac{1}{2}w(z_1 + z_4 - z_2 - z_3), \ z_j \in M\};$$

it is unclear whether it can generally contain any other points, and it would be interesting to know to what extent the presence of at least some 4-dimensional piece outside any of the diagonals characterizes Jacobi varieties.

2. The relations (9) lead to a variety of identities among Abelian integrals expressible in terms of theta series, and thus to further special properties of the theta series of Jacobi varieties. The simplest relations correspond to the cases $\nu = 0$, $1 \leq r - 2 \leq g$, where the locus $W_{r-2}^0 = W_{r-2}$ is just the set of points $w(x_1) + \cdots + w(x_{r-2})$ as $x_1, \ldots, x_{r-2}$ vary over $\tilde{M}$, and (9) amounts to the identity

$$(15) \qquad \operatorname{rank} \Theta_2(k_1 z_1 + \cdots; w(x_1) + \cdots + w(x_{r-2}) - w(z_1) - \cdots) < r.$$

There is generally a single nontrivial linear relation among the r columns of the matrix in (15), depending on the points x_j, z_k. The functional equations satisfied by the theta functions imply further functional equations for the coefficients expressing this linear dependence, and the latter are easily seen to be sufficient to determine these coefficients quite explicitly, as I have shown in [11].

To describe these relations it is convenient to introduce the elementary function $q(z_1, z_2)$ of the Riemann surface, equivalent to Klein's prime form. The function $q(z_1, z_2)$ can be characterized as a holomorphic function on $\tilde{M} \times \tilde{M}$ such that

$$(16) \qquad q(Tz_1, z_2) = \rho_{w(z_2)}(T)\varsigma(T, z_1)q(z_1, z_2) \quad \text{for any } T \in \Gamma$$

and

$$(17) \qquad q(z_2, z_1) = -q(z_1, z_2);$$

it is determined uniquely up to a constant factor, and as a function of z_1 is just the standard section of the point bundle $\varsigma_{z_2} = \rho_{w(z_2)}\varsigma$ associated to the point z_2, having divisor z_2 on M. It is further convenient to introduce a related standard coordinatization of the Riemann surface M, using the function

$$(18) \qquad z = \int_{z_0} \partial_1 q(t, t)\, dt$$

as a local coordinate, where

$$(19) \qquad \partial_1 q(t, t) = \partial q(z_1, z_2)/\partial z_1|_{z_1 = z_2 = t}$$

in terms of any local coordinate t; since $q(z_1, z_2)$ has a simple zero at $z_1 = z_2$ this does describe a valid local coordinatization of $\tilde{M}$. It has the advantage that the factor of automorphy for Γ describing the canonical bundle of $M = \tilde{M}/\Gamma$ has the simple form

$$(20) \qquad \kappa(T, z) = (dT(z)/dz)^{-1} = \rho_{2r}(T)\varsigma(T, z)^{2g-2};$$

arbitrary derivatives in terms of these canonical coordinates thus have relatively simple transformation properties. Furthermore in terms of the canonical coordinates the elementary function has the series expansion

$$(21) \qquad q(z_1, z_2) = (z_1 - z_2) + (z_1 - z_2)^3 \tilde{q}(z_1, z_2)$$

near the diagonal $z_1 = z_2$, where $\tilde{q}$ is holomorphic near the diagonal and $\tilde{q}(z_2, z_1) = \tilde{q}(z_1, z_2)$. Some properties of this coordinatization are discussed in [11], but little else seems known about it.

In these terms the simplest nontrivial explicit identity arising from (15), that for the case $r = 3$, is Fay's trisecant identity:

$$
\begin{aligned}
0 = {} & q(z_1, z_2)q(z_3, z_4)\boldsymbol{\theta}_2(w(z_1 + z_2 - z_3 - z_4)/2) \\
(22) \qquad & - q(z_1, z_3)q(z_2, z_4)\boldsymbol{\theta}_2(w(z_1 + z_3 - z_2 - z_4)/2) \\
& + q(z_1, z_4)q(z_2, z_3)\boldsymbol{\theta}_2(w(z_1 + z_4 - z_2 - z_3)/2).
\end{aligned}
$$

Although this was probably essentially known to Riemann, and closely related results have appeared in the work of Frobenius and others, it seems to have been isolated clearly for the first time by John Fay [5]; other recent proofs can be found in [4] and [15]. The formulas for general r are quite similar, but somewhat more complicated. They can all be proved in the same way indicated before; details and explicit formulas can be found in [11]. Alternatively as observed by Fay, all these formulas can be deduced from the trisecant formula by a straightforward induction, when they are rewritten in terms of first-order theta functions by means of Weierstrass's addition theorem

$$(23) \qquad {}^t\boldsymbol{\theta}_2(u) \cdot \boldsymbol{\theta}_2(v) = \theta(u+v)\theta(u-v),$$

where the left-hand side is the scalar product of the vector-valued second-order theta functions and $\theta(u) = \theta[0|0](u; \Omega)$ is the standard first-order theta function.

For the case $\nu = 0$, $r = g+2$ the relation (9) can be written alternatively as the identity

$$(24) \qquad \operatorname{rank} \Theta_2(\mathfrak{d}; t - w(\mathfrak{d})) < g+2 \quad \text{for all } t \in J,$$

and again there is generally a single nontrivial linear relation among the $g+2$ columns of the matrix in (24), depending on the parameters t, z_j. Arguing as in [11] it is quite easy to deduce from the functional equations what the coefficients of this relation are, and thus to deduce an identity of the form

$$(25) \quad 0 = \sum_{j=1}^{g+2} \left[\prod_{k \neq j} q(z_j, z_k)^{-1} \right] \theta(r + w(z_j) - t)\boldsymbol{\theta}_2(t/2 - w(z_1 + \cdots + z_{g+2} - 2z_j)/2)$$

for arbitrary points $z_j \in \tilde{M}$, $t \in \mathbf{C}^g$, or alternatively

$$(26) \quad 0 = \sum_{j=1}^{g+2} \left[\prod_{k \neq j} q(z_j, z_k)^{-1} \right] \theta(2t + r - w(z_1 + \cdots + z_{g+2} - z_j)\boldsymbol{\theta}_2(t - w(z_j))$$

for arbitrary points $z_j \in \tilde{M}$, $t \in \mathbf{C}^g$. This can be reduced to the other form of the generalized trisecant identity by an application of the further easily derived identity

$$(27) \qquad \left[\prod_{1 \leq j < k \leq g} q(x_j, x_k) \right] \theta(r + w(z) - w(x_1 + \cdots + x_g))$$

$$= c \det\{w_j'(x_k)\} \cdot \left[\prod_{1 \leq j \leq g} q(z, x_j) \right],$$

where $x_1, \ldots, x_g$, $z \in \tilde{M}$, and $\det\{w_j'(x_k)\}$ is the determinant of the $g \times g$ matrix formed of the derivatives with respect to the canonical coordinates on $\tilde{M}$ of the g basic Abelian integrals evaluated at the g points x_j. Here c is a constant factor that cannot really be evaluated, since the elementary function $q(z_1, z_2)$ and hence the canonical coordinate function on $\tilde{M}$ are only determined up to a constant factor. However it is always possible to determine this factor so that $c = 1$; that further specifies the canonical coordinates on $\tilde{M}$.

The identities that arise for general values of ν, r can be derived similarly, or alternatively can be reduced to the identities already discussed; they reflect further special properties of the Riemann surface. Only one example will be considered here, perhaps the simplest example, but one that illustrates the sort of identities that arise and the interest that may reside in them. The subvariety W_2^1 is nonempty only for a hyperelliptic Riemann surface, and there it consists of a single point, the hyperelliptic point $e \in J$; if $\tau \colon M \to M$ is the hyperelliptic involution, then $w(z) + w(\tau z) = e$ for any point $z \in M$. The identity in this case amounts to the following:

$$
\begin{aligned}
0 = {} & q(z_1, \tau z_4) q(z_2, z_3) \boldsymbol{\theta}_2(e/2 + w(z_1 - z_2 - z_3 - z_4)/2) \\
& - q(z_2, \tau z_4) q(z_1, z_3) \boldsymbol{\theta}_2(e/2 + w(z_2 - z_1 - z_3 - z_4)/2) \\
& + q(q_3, \tau z_4) q(z_1, z_2) \boldsymbol{\theta}_2(e/2 + w(z_3 - z_1 - z_2 - z_4)/2).
\end{aligned}
$$
(28)

It is actually an immediate consequence of the trisecant identity (22), since $e - w(z_4) = w(\tau z_4)$.

A number of further identities can be derived quite easily from the trisecant formula and its extensions by considering coincidences between the points appearing in these formulas. Actually these formulas generally reduce to inanities when there are coincidences, all terms tending to zero; but suitable differentiation will lead to interesting and quite nontrivial results. For instance applying the differential operator $\partial^2/\partial z_1 \partial z_3$, interpreted as differentiation in the canonical coordinates on $\tilde{M}$, and setting $z_2 = z_1$, $z_4 = z_3$ leads quite easily to the identity

$$
\begin{aligned}
\boldsymbol{\theta}_2(w(z_1 - z_3)) = {} & \boldsymbol{\theta}_2(0) \cdot q(z_1, z_3)^2 \partial^2 \log q(z_1, z_3)/\partial z_1 \partial z_3 \\
& + \sum_j \partial_{j_1 j_2} \boldsymbol{\theta}_2(0) \cdot \tfrac{1}{2} q(z_1, z_3)^2 w'_{j_1}(z_1) w'_{j_2}(z_3),
\end{aligned}
$$
(29)

where $\partial_{j_1} \partial_{j_2} \boldsymbol{\theta}_2(0) = \partial^2 \boldsymbol{\theta}_2(w)/\partial w_{j_1} \partial w_{j_2}|_{w_1 = \cdots = w_g = 0}$ and the derivatives $w'_j(z)$ are again in terms of the canonical local coordinates on $\tilde{M}$. An abundance of equations ensues similarly, including the explicit solutions of the KP equations in terms of theta series on Jacobi varieties; more detailed discussion of these constructions can be found in [5] and [11], the latter including some formulas derived from the extended versions of the trisecant formula.

These various identities, the full tabulation of which remains possibly rather remote and also possibly only of interest to a computer, are another array of special properties of Jacobi varieties, and just as with the array of subvarieties of special positive divisors there naturally arise the questions: which such identities characterize Jacobi varieties, and how effective is such a characterization? More generally again, some identities serve to characterize subclasses of Jacobi varieties or more extended classes of principally polarized Abelian varieties: what are such classes, their characteristic properties, and their interrelationships? The Novikov conjecture recently solved by Shiota [16] is the prime example of such a result at present. Further very interesting conjectures and results along this line can be found in [7], and Welters has recently solved some of these conjectures closely related to the formula (29) in another lovely paper [19].

3. The systematic organization of the various identities that can be derived by the indicated methods presents other interesting problems. The consideration of (29) and similar results suggests viewing as basic identities expressions of the second-order theta functions on various standard subvarieties of J as linear combinations of canonical Abelian differentials and other such functions on the Riemann surface, with coefficients that are reasonably simply determined in terms of the theta functions. Indeed (29) can be viewed as expressing the function $q(z_1, z_3)^{-2}\theta_2(w(z_1 - z_3))$ as a linear combination of the Abelian differentials $w'_j(z)$ and the Abelian differentials of the second kind $w'_a(z) = \partial^2 \log q(z, a)/\partial z \partial a$ having a double pole at $z = a$; actually of course these latter functions are more properly just the coefficients of these differentials in terms of the canonical coordinates on $\tilde{M}$, the coordinatization chosen to reflect the same transformation properties as the theta functions. The coefficients are the constant vectors $\theta_2(0)$ and $\partial_{j_1 j_2}\theta_2(0)$; these $\binom{g+1}{2} + 1$ vectors for $j_1 \leq j_2$ are known to be linearly independent for Jacobi varieties. The result is a reasonably simple expression of the second-order theta functions on the subvariety $W_1 - W_1$ in terms of standard forms associated to the given Riemann surface. A number of further identities can then readily be deduced from this by differentiating repeatedly and equating the variables. There is a natural family of identities generalizing (29) in this same spirit. They can be derived from the trisecant identity and its extensions in much the same way as (29) was derived; the next case was derived that way in [11]. However with the general principle in mind it is considerably easier and more systematic just to use the functional equations to establish the general form of the identities, and then to analyze the resulting forms more closely to evaluate the coefficients appearing in them as in [12]. Further details can be found there, but there may be some interest here in at least a sketch of the next case that arises.

A first approximation to this and the other identities, rather simpler and for some purposes even more interesting than the general form, can be obtained by considering only the leading or most singular terms in the expression of the theta functions. For that purpose let $S \subseteq \mathbf{C}^{2^g}$ be the linear subspace spanned by the vectors $\theta_2(0)$ and $\partial_{jk}\theta_2(0)$, and let P be a linear projection of the space $\mathbf{C}^{2^g}$ onto a subspace complementary to S, so that P annihilates precisely the subspace S. Applying the operator P to the vector of second-order theta functions has the effect of ignoring that portion of this set of theta functions spanned by vectors in S, which will turn out to be the least singular part in the expression of the theta functions in terms of Abelian differentials on M. Even when applied to the formula (29) this is of some interest, for that identity then takes on the particularly simple form $P\theta_2(w(z_1 - z_3)) = 0$. This can be interpreted as exhibiting a subset $P\theta_2$ of the second-order theta functions vanishing on the subvariety $W_1 - W_1 \subseteq J$; Welters has shown in [19] that for $g > 4$ this subset of theta functions vanishes precisely at $W_1 - W_1$, and in special cases it is moreover known that these functions generate the local ideal of this subvariety at most points. The further identities that can be derived from (29) by differentiating repeatedly and equating the variables have correspondingly simpler forms. For example applying the operator $\partial^4/\partial z_3^4$ and setting $z_3 = z_1$ leads to the further identity

$$(31) \qquad \sum_{j_1 \cdots j_4} \partial_{j_1 \cdots j_4} P\theta_2(0) w'_{j_1}(z_1) w'_{j_2}(z_1) w'_{j_3}(z_1) w'_{j_4}(z_1) = 0.$$

When the terms in the kernel of P_1 are taken into account as well, this is the by now familiar expression leading to an explicit solution of the KP equation in terms of theta functions of Jacobi varieties as discussed for instance in [15]. In the simpler form here (31) can be interpreted as a collection of quartic polynomials

$$(32) \qquad \sum_{j_1 \cdots j_4} \partial_{j_1 \cdots j_4} P\theta_2(0) x_{j_1} x_{j_2} x_{j_3} x_{j_4}$$

vanishing on the canonical curve associated to the Riemann surface M, the imbedding of M in $\mathbf{P}^{g-1}$ defined by the g canonical Abelian differentials; at least in the standard general case (in which M is neither hyperelliptic nor trigonal nor a nonsingular plane quintic) the polynomials (32) describe precisely the canonical curve.

In these terms then the extension of (29) to the next case in this list of formulas is the identity

$$q(z_1, z_2)^2 q(a_1, a_2)^2 \left[\prod_{\mu, \nu = 1,2} q(z_\mu, a_\nu) \right]^{-2} \cdot P\theta_2(w(z_1 + z_2 - a_1 - a_2))$$

$$(33) \qquad = \sum_k \alpha_{k_3}^{k_1 k_2} w'_{k_1}(a_1) w'_{k_2}(a_2) [w'_{a_1 a_2}(z_1) w'_{k_3}(z_2) + w'_{a_1 a_2}(z_2) w'_{k_3}(z_1)]$$

$$+ \sum_k \beta_{k_1 k_2}(a_1, a_2) w'_{k_1}(a_1) w'_{k_2}(a_2),$$

where $\alpha_{k_3}^{k_1 k_2}$ are some constant vectors that are skew-symmetric in the indices k_1, k_2, k_3 and $\beta_{k_1 k_2}(a_1, a_2)$ are some holomorphic vector-valued functions of the parameters a_1, a_2 that are symmetric in the indices k_1, k_2. Here w'_j are as before the coefficients of the canonical holomorphic Abelian differentials when expressed in terms of the canonical coordinates on $\tilde{M}$, or equivalently the derivatives with respect to the canonical coordinates on $\tilde{M}$ of the canonical holomorphic Abelian integrals, and $w'_{a_1 a_2}$ are in the same sense the Abelian differentials of the third kind having simple poles at a_1, a_2 with residues respectively $+1, -1$ there. Under the further projection P' annihilating the span of S and the vectors $\alpha_{k_3}^{k_1 k_2}$ this formula implies further that $P\alpha_{k_3}^{k_1 k_2} = 0$ and

$$(34) \qquad P\beta_{k_1 k_2}(a_1, a_2) = \sum_j \beta_{k_1 k_2}^{j_1 j_2} w'_{j_1}(a_1) w'_{j_2}(a_2)$$

for some constant vectors $\beta_{k_1 k_2}^{j_1 j_2}$ that are symmetric in j_1, j_2, in k_1, k_2, and under the interchange of the pairs (j_1, j_2) and (k_1, k_2), and are such that

$$(35) \qquad \sum_k \beta_{k_1 k_2}^{j_1 j_2} w'_{k_1}(z) w'_{k_2}(z) = 0 \quad \text{for all } j_1, j_2.$$

The determination of these auxiliary vectors α, β in terms of the theta functions is a considerably more complicated and interesting matter than what arose in the earlier case (29), involving further properties of the Riemann surface M rather than simple expressions independent of M; the vectors α can be expressed in terms of the vectors $\partial_{j_1 \ldots j_4} P_1 \boldsymbol{\theta}_2(0)$, while the vectors β involve sixth derivatives as well.

References

1. A. Andreotti and A. Mayer, *On period relations for Abelian integrals on algebraic curves*, Ann. Scuola Norm. Sup. Pisa **21** (1967), 189–238.

2. E. Arbarello, M. Cornalba, P. A. Griffiths, and J. Harris, *Geometry of algebraic curves*, I. Springer-Verlag, New York, Berlin, Heidelberg, Tokyo, 1985.

3. E. Arbarello and C. DeConcini, *On a set of equations characterizing Riemann matrices*, Ann. of Math. (2) **120** (1984), 119–140.

4. H. M. Farkas, *On Fay's trisecant formula*, J. Analyse Math. **44** (1985), 205–217.

5. J. D. Fay, *Theta functions on Riemann surfaces*, Lecture Notes in Math., vol. 352, Springer-Verlag, Berlin, Heidelberg, New York, 1973.

6. ____, *On the even-order vanishing of Jacobian theta functions*, Duke Math. J. **51** (1984), 109–172.

7. B. van Geemen and G. van der Geer, *Kummer varieties and the moduli spaces of Abelian varieties*, Amer. J. Math. **108** (1986), 615–642.

8. R. C. Gunning, *Riemann surfaces and generalized theta functions*, Springer-Verlag, Berlin, Heidelberg, New York, 1976.

9. ____, *On generalized theta functions*, Amer. J. Math. **104** (1982), 183–208.

10. ____, *Some curves in Abelian varieties*, Invent. Math. **66** (1982), 377–389.

11. ____, *Some identities for Abelian integrals*, Amer. J. Math. **108** (1986), 39–74.

12. ____, *Lectures on Riemann surfaces and theta functions*, Princeton Univ. Press (to appear).

13. G. Kempf, *Inversion of Abelian integrals*, Bull. Amer. Math. Soc. (N.S.) **6** (1982), 25–32.

14. T. Matsusaka, *On a characterization of a Jacobi variety*, Mem. Coll. Sci. Univ. Kyoto **32** (1959), 1–19.

15. D. Mumford, *Tata lectures on theta*. II, Birkhauser, Boston, Basel, Stuttgart, 1984.

16. T. Shiota, *Characterization of Jacobian varieties in terms of soliton equations*, Invent. Math. **83** (1986), 333–382.

17. G. E. Welters, *On flexes of the Kummer variety*, Indag. Math. **45** (1983), 501–520.

18. ____, *A criterion for Jacobi varieties*, Ann. of Math. (2) **120** (1984), 497–504.

19. ____, *The surface $C - C$ on Jacobi varieties and second order theta functions*.

PRINCETON UNIVERSITY

Proceedings of Symposia in Pure Mathematics
Volume **46** (1987)

Curves and Their Moduli

JOE HARRIS

Introduction. It has been suggested that the fundamental problem of algebraic geometry is classification: the description of the set of isomorphism classes of algebraic varieties. Indeed, the analogous statement could well be made about many areas of mathematics, but a special circumstance holds in algebraic geometry: in a broad range of cases, the set of isomorphism classes of algebro-geometric objects itself possesses the structure of an algebraic variety, and it is the geometry of these varieties—these moduli or parameter spaces—that we undertake to study.

Accepting this point of view, there might still be some differences as to how to interpret the terms "algebraic variety" and "isomorphism" in this context. To most nineteenth-century geometers, the most natural interpretation of these terms would be "projective algebraic variety" and "projective isomorphism"; and the problem of describing the isomorphism classes of curves, for example, would be interpreted as meaning the study of the Hilbert scheme $\mathcal{H}_{d,g,r}$ parametrizing curves of degree d and genus g in $\mathbf{P}^r$ (or a suitably defined object in case $r = 1$ or 2). The techniques used for the study of the varieties $\mathcal{H}$ in the nineteenth century were, moreover, almost entirely extrinsic, dealing exclusively with the projective geometry of the embedded curves.

Our twentieth-century viewpoint is completely different: these days, when we use the word "variety" without further specification we usually mean "abstract algebraic variety." To understand curves in projective space, we would say, let us understand both abstract curves and the ways in which an abstract curve may be mapped to projective space. The classification problem for curves would be accordingly interpreted as meaning the study of the moduli spaces $\mathcal{M}_g$ parametrizing isomorphism classes of abstract curves of genus g, and of the natural map

$$\varphi \colon \mathcal{H}_{d,g,r} \to \mathcal{M}_g$$

defined on the open subset of $\mathcal{H}_{d,g,r}$ parametrizing smooth curves.

1980 *Mathematics Subject Classification* (1985 *Revision*). Primary 14H10; Secondary 14H15.

We have thus three objects of study: $\mathcal{H}$, $\mathcal{M}$ and φ; and in these lectures I hope to give an overview of what we know about these objects (at least over the complex numbers, to which we will restrict ourselves), devoting one lecture to each. Specifically, in the first lecture I would like to talk about the moduli space itself, and in the second the fibers of the map φ, at least over a general point of $\mathcal{M}$. In the third lecture, rather than try to catalogue what we know about $\mathcal{H}$ I will focus on an example—that of plane curves—that I hope may illustrate how the study of the varieties $\mathcal{H}$ may be approached from this point of view. There is also an appendix to the first lecture, illustrating some of the techniques we have for dealing with the geometry of the moduli space. One note: there will be varying amounts of content in these lectures, ranging from the first lecture, which will be almost entirely a litany of statements and conjectures, to the third lecture and the appendix, where I will attempt to at least sketch the proofs of the statements made.

The basic thesis of the lectures is that, while very different techniques are used in the study of the spaces $\mathcal{M}$ and $\mathcal{H}$ and the map φ, our understanding of each contributes to the knowledge of the others. This is true not just in the general sense that having more than one way to view an object always helps us, but also—thanks largely to the notion of semistable reduction and the corresponding stable compactification of the moduli space of curves—in specific applications: for example, in the second and third lectures we use degenerations in $\mathcal{M}_g$ to study both the fibers of φ and the global structure of $\mathcal{H}$; and in the first lecture we see that in the study of $\mathcal{M}_g$ itself the images of the maps φ play a crucial role. It is this interweaving of ideas from intrinsic and extrinsic geometry in the study of curves that I hope to bring out in what follows.

Lecture 1

The Moduli Space of Curves

1. Constructing the moduli space. To begin with, the moduli space $\mathcal{M}_g = \{$isomorphism classes of smooth complete curves of genus $g\}$ is simply a set, and the first thing we have to do is to give it the structure of an algebraic variety. Now, there is a natural condition we would want any such structure to satisfy: namely, for any family $\pi: \mathfrak{X} \to B$ of smooth curves (that is, any morphism of varieties whose fibers $\mathfrak{X}_b = \pi^{-1}\{b\}$ are smooth complete curves of genus g) the set map $\varphi: B \to \mathcal{M}_g$ induced by sending $b \in B$ to $\varphi(b) = [\mathfrak{X}_b]$ should be a regular map. It's not hard to see that in fact this condition uniquely determines the structure of variety on $\mathcal{M}_g$; what we really have to do is to show that one such structure exists. (In fact, a generalization of this condition determines a scheme structure on $\mathcal{M}_g$: we want there to be a natural transformation of functors

$$\{\text{families of curves over the scheme } B\} \to \{\text{morphisms from } B \text{ to } \mathcal{M}_g\}.)$$

There are three principal approaches to constructing a space $\mathcal{M}_g$ with this property, each of which has something to say about the geometry of $\mathcal{M}_g$.

1. *As a quotient of Teichmüller space.* In this approach, we consider the data of a Riemann surface C of genus g together with an isotopy class $[\varphi]$ of diffeomorphisms $\varphi \colon C \to C_0$ of C with a fixed compact orientable 2-manifold C_0 of genus g. The basic theorem of Teichmüller asserts that the set of such pairs $(C, [\varphi])$ is naturally parametrized by an open subset Σ_g in $\mathbf{C}^{3g-3}$ homeomorphic to a ball. The moduli space $\mathfrak{M}_g$ is thus realized as the quotient of Σ_g by the *Teichmüller modular group* T_g of isotopy classes of C^∞ automorphisms of C_0.

This approach has been valuable especially as a way of describing the topology of $\mathfrak{M}_g$; for example, as we shall see, in low degrees the cohomology of $\mathfrak{M}_g$ is just the cohomology of the Teichmüller modular group, which in turn can be dealt with purely topologically.

2. *As a subvariety of a quotient of the Siegel upper half-space.* In the previous case the data we looked at was equivalent to a curve C and a set of generators $\gamma_1, \ldots, \gamma_{2g}$ for the fundamental group $\pi_1(C)$ isotopic to the standard one. Here we look at pairs $(C; \delta_1, \ldots, \delta_{2g})$ where the δ_i are a basis for the homology group $H_1(C, \mathbf{Z})$ normalized with respect to the intersection pairing. Such pairs may be represented by their period matrices Z, which are points in the Siegel upper half-space $\mathcal{H}_g$ of symmetric $g \times g$ complex matrices with positive definite imaginary part; two such period matrices will come from the same curve C iff they differ only in the choice of basis δ_i, that is, by the action of the integral symplectic group $\mathrm{Sp}\,(2g, \mathbf{Z})$. The moduli space $\mathfrak{M}_g$ is thus realized as a subvariety of the quotient $\mathcal{A}_g = \mathcal{H}_g / \mathrm{Sp}\,(2g, \mathbf{Z})$.

The problem with this approach is of course that for $g \geq 4$ the moduli space is a proper subvariety of $\mathcal{A}_g$, so we don't get as much information about $\mathfrak{M}_g$ as we'd like. The whole question of describing the image of $\mathfrak{M}_g$ in $\mathcal{A}_g$ is a long-standing one, on which there has been spectacular progress in the last few years, and which is the subject of a seminar here in its own right.

Observe that each of the first two approaches is a natural generalization of the construction of the moduli $\mathfrak{M}_1$ of elliptic curves (the j-line) as a quotient of the upper half-plane by $T_1 = \mathrm{SL}(2, \mathbf{Z}) = \mathrm{Sp}\,(2, \mathbf{Z})$.

3. *As a quotient of a Hilbert scheme.* The basic point here is that, compared to the construction of a moduli space, it is a relatively straightforward matter to construct a parameter space for curves in projective space; we thus look at pairs (C, φ) consisting of a curve C and a projective embedding of C and parametrize those by the points of a Hilbert scheme. To minimize the choices involved, we choose φ to be given by sections of a multiple of the canonical bundle/dualizing sheaf on C, i.e.,

$$\varphi \colon C \to \mathbf{P}^N = \mathbf{P}^{(2n-1)(g-1)-1},$$
$$p \mapsto [\omega_1(p), \ldots, \omega_{(2n-1)(g-1)}],$$

where the ω_i are a basis for the n-fold differentials on C and $n \geq 3$. If $\mathcal{H}$ is then the Hilbert scheme parametrizing the images (strictly speaking, it will be an open subset of the subvariety of the Hilbert scheme $\mathcal{H}'$ of curves of degree

$2n(g-1)$ and genus g in $\mathbf{P}^N$ parametrizing n-canonical curves), we can realize the moduli space $\mathcal{M}_g$ as the quotient of $\mathcal{H}$ by $\mathrm{PGL}(N+1, \mathbf{C})$.

Of course, since we are here taking the quotient of a variety by a positive-dimensional group, we have to worry a bit about the existence of a good quotient. What allows us to see that a quotient exists is Mumford's geometric invariant theory, as carried out for curves in, e.g., [**39**].

What is most important about this approach to constructing $\mathcal{M}_g$ is that it gives us a wonderful compactification $\overline{\mathcal{M}}_g$ of $\mathcal{M}_g$. This arises by enlarging the variety $\mathcal{H}$ to the locally closed variety $\overline{\mathcal{H}}$ of *stable n-canonical curves* in $\mathbf{P}^N$ (here stable means reduced, connected, having only nodes as singularities and having finite automorphism group) and then taking the quotient. This has the following marvelous properties:

(i) The points $\overline{\mathcal{M}}_g$ correspond to isomorphism classes of stable curves (and the basic property above characterizing the structure of variety on the set $\mathcal{M}_g$ continues to hold on $\overline{\mathcal{M}}_g$);

(ii) The local structure of $\overline{\mathcal{M}}_g$ at the boundary is as well understood as in the interior—e.g., we know where it is smooth and where it is singular, and have a natural description of its tangent spaces in the former case; and

(iii) $\overline{\mathcal{M}}_g$ is a projective variety.

The boundary $\Delta = \overline{\mathcal{M}}_g - \mathcal{M}_g$ can, by the above, be described completely. For example, the locus of curves having exactly δ nodes has pure codimension δ in $\overline{\mathcal{M}}_g$; in particular, Δ itself is a union of irreducible divisors $\Delta_0, \Delta_1, \ldots, \Delta_{[g/2]}$ in $\overline{\mathcal{M}}_g$, where Δ_0 is the closure of the locus of irreducible curves with one node and Δ_i the closure of the locus of curves consisting of curves of genus i and $g-i$ identified at one point.

It should be noted that the second construction also gives rise to a compactification of $\mathcal{M}_g$: we can take $\overline{\mathcal{M}}_g^s$ to be the closure of $\mathcal{M}_g$ in the Satake compactification $\overline{\mathcal{A}}_g$ of $\mathcal{A}_g$. This is often referred to as the Satake compactification of $\mathcal{M}_g$; it is not as commonly used as $\overline{\mathcal{M}}_g$ because it does not share properties (i) and (ii) above, but it does tell us something about $\mathcal{M}_g$: for example, the fact that the complement of $\mathcal{M}_g$ in the Satake compactification has codimension 2 when $g \geq 3$ implies that *there are no nonconstant holomorphic functions on $\mathcal{M}_g$ for $g \geq 3$.*

One further remark to make is that all three of the above constructions generalize to give constructions of the moduli space $\mathcal{M}_{g,k}$ of curves C with k distinct marked points $p_1, \ldots, p_k \in C$. In particular, the space $\mathcal{M}_{g,1}$ is often called the *universal curve* over $\mathcal{M}_g$ (and sometimes written $\mathcal{C}_g$) since it is, via the natural map $\mathcal{C}_g \to \mathcal{M}_g$, the family of curves whose fiber over a point C in the open subset $\mathcal{M}_g^0 \subset \mathcal{M}_g$ of curves without automorphisms is isomorphic to C (of course, its fiber over any point $C \in \mathcal{M}_g$ is just the quotient of C by its automorphism group).

2. The topology of the moduli space. There have been a number of striking results on the topology of $\mathcal{M}_g$ proved in the last few years. To begin with,

John Harer [**24**] succeeded in calculating directly the first two cohomology groups of the Teichmüller modular group (and hence the corresponding cohomology groups for $\mathcal{M}_g$, since the fixed points of elements of T_g correspond to curves with automorphisms and so have codimension $g - 2$ in $\mathcal{M}_g$); he finds that

$$H^1(\mathcal{M}_g, \mathbf{Z}) = 0 \quad \text{for } g > 2,$$

and

$$H^2(\mathcal{M}_g, \mathbf{Z}) = \mathbf{Z}, \qquad g > 4.$$

This says in particular that $\text{Pic}(\mathcal{M}_g) = \mathbf{Z}$, and it's not hard to locate a generator, at least over $\mathbf{Q}$: the first chern class λ of the *Hodge bundle* E will do (loosely, the bundle whose fiber over any point $C \in \mathcal{M}_g$ is the space of regular differentials on C; a more precise definition will be given below). In fact, it has recently been shown by Arbarello and Cornalba [**2**] that λ generates over $\mathbf{Z}$. It also implies that the Picard group of compactification $\overline{\mathcal{M}}_g$ is $\mathbf{Z}^{[g/2]+2}$, since the components of the boundary are readily seen to be independent. Harer was also able to show that the homology groups $H_k(\mathcal{M}_g, \mathbf{Z})$ vanish for $k > 4g - 3$; it's not known if this is sharp.

Another theorem of Harer's is the *stability theorem* [**25**]: this says that we have isomorphisms

$$H^k(\mathcal{M}_g, \mathbf{Z}) \cong H^k(\mathcal{M}_{g+1}, \mathbf{Z})$$

whenever $g \geq 3k + 1$; since these isomorphisms respect cup product, this allows us to define the *stable cohomology ring* $H^*(\mathcal{M}, \mathbf{Z})$ by simply setting $H^k(\mathcal{M}, \mathbf{Z}) = H^k(\mathcal{M}_g, \mathbf{Z})$ for $g \gg 0$.

What does this mean? There is a a conjectured geometric interpretation of these isomorphisms, as follows. Basically, a cohomology class on any space can be thought of (ignoring torsion for the moment) as simply a linear function on cycles that measures their nontriviality. Since a cycle in $\mathcal{M}_g$ is a family of curves, a cohomology class on $\mathcal{M}_g$ may thus be thought of as a device that measures the nontriviality of a family $\pi\colon \mathfrak{X} \to B$ of smooth curves. This suggests a couple of ways of defining cohomology classes on $\mathcal{M}_g$.

(i) *By the twisting of the Hodge bundle.* To a family $\pi\colon \mathfrak{X} \to B$ we can associate the Hodge bundle E on B, whose fiber over $b \in B$ is just the space of differentials $H^0(\mathfrak{X}_b, K_{\mathfrak{X}_b})$; precisely, E is the direct image of the relative dualizing sheaf $\omega_{\mathfrak{X}/B}$. One way of measuring the variation in a family, then, is to look at the chern classes $c_i(E)$. In particular, if we apply this construction to the universal curve $\mathcal{C}_g^0 = \pi^{-1}(\mathcal{M}_g^0) \to \mathcal{M}_g^0$ we can define cohomology classes $c_i(E)$ on $\mathcal{M}_g^0$; since the codimension of $\mathcal{M}_g - \mathcal{M}_g^0$ in $\mathcal{M}_g$ is $g - 2$, for $i < g - 2$ these will extend to classes $c_i(E)$ in $\mathcal{M}_g$ that pull back to the classes $c_i(E) \in H^{2i}(B, \mathbf{Z})$ on any family $\mathfrak{X} \to B$.

(ii) *By the twisting of the relative dualizing sheaf.* For any family $\pi\colon \mathfrak{X} \to B$ of smooth curves of genus $g \geq 2$, let $\omega = c_1(\omega_{\mathfrak{X}/B})$ be the first chern class of the relative dualizing sheaf of the family. Of course, if $\mathfrak{X}$ is just a product $B \times C$, ω will be the pullback to $\mathfrak{X}$ of the canonical bundle on C, and so will have square zero in $H^*(\mathfrak{X}, \mathbf{Z})$; conversely, for example, it can be seen that if B is complete and $\omega^2 = 0$ then the family is trivial. This suggests that we can measure the variation in such a family by associating to it the push-forward to B of the various powers of the class ω on $\mathfrak{X}$, i.e., the classes

$$\kappa_i = \pi_*(\omega^{i+1}).$$

Again, for $i \leq g - 3$ we can define classes $\kappa_i \in H^{2i}(\mathcal{M}_g, \mathbf{Z})$ by making this construction on the universal curve. The classes κ_i are called the *tautological classes* of the family.

The first thing to observe about these classes is that the κ_i give a more complete set of invariants, in the sense that the $c_i(E)$ are expressible as polynomials in the κ_i. This is an immediate consequence of the Grothendieck-Riemann-Roch formula: since $R^1\pi_*\omega_{\mathfrak{X}/B} \cong \mathcal{O}$ is trivial for any family, we have

$$\begin{aligned}
\mathrm{ch}(E) = \mathrm{ch}(\pi_!\omega_{\mathfrak{X}/B}) &= \pi_*(\mathrm{ch}(\omega_{\mathfrak{X}/B}) \cdot \mathrm{Td}(\omega^*_{\mathfrak{X}/B})) \\
&= \pi_*(e^\omega \cdot (-\omega/(1 - e^\omega))) \\
&= \pi_*((1 + \omega + \omega^2/2 + \cdots) \cdot (1 - \omega/2 + \omega^2/12 + \cdots))
\end{aligned}$$

so that, e.g., $c_1(E) = \pi_*(\omega^2/12) = \kappa_1/12$.

In these terms, we can rephrase Harer's first theorem as saying that $\mathrm{Pic}(\mathcal{M}_g) \otimes \mathbf{Q} = H^2(\mathcal{M}_g, \mathbf{Q}) = \mathbf{Q} \cdot c_1(E) = \mathbf{Q} \cdot \kappa_1$; as such, it is the first case of the principal

CONJECTURE. The rational stable homology of the moduli space is freely generated by the classes κ_i; i.e.,

$$H^*(\mathcal{M}, \mathbf{Q}) = \mathbf{Q}[\kappa_1, \kappa_2, \ldots].$$

One half of this is known: Ed Miller [**38**] has shown that the monomials in the classes κ_i are independent in $H^*(\mathcal{M})$. The question of whether they generate remains a very mysterious one. Essentially the only evidence for it is the fact that every subvariety of $\mathcal{M}_g$ of low codimension that has been explicitly described— k-gonal curves, curves with exceptional linear series in general, curves with funny Weierstrass points, curves with theta-nulls vanishing to certain orders, etc.—can be seen to have fundamental class in the subring of $H^*(\mathcal{M}_g)$ generated by the κ_i (the appendix gives an indication of how this goes).

There is one further and highly spectacular result on the topology of $\mathcal{M}_g$ that has recently been proved by Harer and Zagier [**26**]: they have succeeded in calculating the Euler characteristic, as V-manifold, of $\mathcal{M}_g$. Briefly, a V-manifold is a space where each point p has an open neighborhood given as a quotient of a ball by a finite group G_p; $\mathcal{M}_g$ is just such a gadget, where the atlas around $C \in \mathcal{M}_g$ is given by the map of the Kuranishi space of C to $\mathcal{M}_g$ and the group in question is just the automorphism group of C. By the Euler characteristic of a V-manifold we mean either: take a finite triangulation of X in which the group G_p is a constant G_σ along each open simplex σ, if one exists (as it does in case of $\mathcal{M}_g$) and set

$$\chi_V(X) = \sum (-1)^{\dim \sigma} \cdot |G_\sigma|^{-1};$$

or, find a Galois cover $\varphi: Y \to X$ with stabilizer group G_p over $p \in X$, and take

$$\chi_V(x) = \chi(Y)/\deg(\varphi).$$

Thus, for example, the Euler characteristic of $\mathcal{M}_1$ as V-manifold is given by

$$\chi_V(\mathcal{M}_1) = \chi(\mathbf{C} - \{0, 1728\})/2 + \chi(\{0\})/4 + \chi(\{1728\})/6$$
$$= -1/2 + 1/4 + 1/6 = -1/12.$$

The theorem of Harer and Zagier is that in general

$$\chi_V(\mathcal{M}_g) = \varsigma(1 - 2g)$$

where ς is a Riemann's zeta function.

3. Complete subvarieties of the moduli space. Actually, this is part of a more general question: is the moduli space $\mathcal{M}_g$ more like an affine variety or a projective one? This would include such questions as what is the cohomological dimension of $\mathcal{M}_g$, either topologically (as in Harer's theorem quoted in §2 above) or algebraically. We will focus here, however, on the specific issue of the existence, dimension, and location of complete subvarieties of $\mathcal{M}_g$. What is known so far is this:

(i) For $g \geq 3$, there exists a complete curve through a general point (or, for that matter, through any given finite collection of points) in $\mathcal{M}_g$; in particular, there are no nonconstant regular functions on $\mathcal{M}_g$. This statement follows from the fact that the Satake compactification $\overline{\mathcal{M}}_g^s$ is a projective variety and that the complement of $\mathcal{M}_g$ in $\overline{\mathcal{M}}_g^s$ has codimension 2. We can thus embed $\mathcal{M}_g^s$ in projective space and take its intersection with a general collection of $3g - 4$ hypersurfaces through a given set of points to arrive at such a curve. It should be mentioned that these curves are not very explicit; I don't believe anyone has ever written down a family of quartic plane curves, for example, corresponding to a complete curve in $\mathcal{M}_3$ (I expect the degree, as a curve in the space $\mathbf{P}^{14}$ of plane quartics, would be quite high).

(ii) For any k, there exist g such that $\mathcal{M}_g$ contains complete, k-dimensional subvarieties. This follows from a construction of Kodaira's: given a complete family $\pi: X \to B$ of smooth curves of genus g, we can look at the family of all curves of genus $3g - 1$ that are triple covers of fibers of π totally ramified over one point (in other words, after making a base change $Z \to X$ there will exist a triple cover of $Z \times_B X$ totally ramified over the inverse image of the diagonal and we can view this triple cover as a family of curves over Z); we get a complete family over a base of dimension $\dim B + 1$. This construction gives rise to complete subvarieties of $\mathcal{M}_g$ of dimension on the order of $\log_3(g)$, which is about the best we can do at present. It should be pointed out also that the subvarieties we construct in this way all lie in the relatively small subvariety of $\mathcal{M}_g$ of curves admitting maps of degree 2 or more to curves of positive genus.

(iii) Finally, the best known upper bound is due to Steve Diaz, who in [11] shows that there do not exist complete subvarieties of $\mathcal{M}_g$ of dimension $g - 1$ or greater. Whether or not this is sharp is unclear; it is still an open question, for example, whether there exist complete surfaces in $\mathcal{M}_4$.

Clearly, a large gap remains between the dimension of the largest known complete subvarieties of $\mathcal{M}_g$ and Diaz's bound, and it seems a fascinating question to know where in that range the actual maximum occurs. It would be, to me, equally interesting to know what is the largest dimension of a complete subvariety *through a general point* C of $\mathcal{M}_g$; at present, we do not even know if there exists a complete surface containing a general C.

It is worth mentioning here a further (conjectured) corollary of Diaz's theorem. In every example that has been analyzed, it has been the case that the image in $\mathcal{M}_g$ of the open subset of any component $\mathcal{H}$ of the Hilbert scheme parametrizing smooth curves of degree d and genus g in $\mathbf{P}^r$ has dimension at least g; the family of smooth complete intersection curves in $\mathbf{P}^r$ seems to come closest, having dimension on the order of g plus a multiple of $\sqrt{g}$. It seems reasonable to guess that this is the case in general; in which case, of course, Diaz's theorem would imply that *every component of the Hilbert scheme contains families of curves that degenerate in moduli.* Whether this is actually true or not remains open.

4. Divisors and line bundles on the moduli space.

The Deligne-Mumford compactification of $\mathcal{M}_g$ allows us to do projective geometry on the moduli space, and the first questions we get into naturally relate to the divisor theory of $\overline{\mathcal{M}}_g$. Specifically, knowing that the Picard group $\mathrm{Pic}(\overline{\mathcal{M}}_g)$ is generated by the classes λ and δ_i, we can ask for the cones of ample and of effective divisors. In regard to both these questions, we do not have complete answers, but some intriguing partial results, which we shall try to describe. One note: it has often proved advantageous, both for the sake of being able to draw pictures and for the sake of being able to prove a theorem, to restrict attention to the sublattice Λ of $\mathrm{Pic}(\overline{\mathcal{M}}_g)$ generated by λ and "the class of the boundary," that is, $\delta = \sum \delta_\alpha$.

We start with the ample cone. The fundamental deep result here is of course that there *are* ample divisor classes on $\overline{\mathcal{M}}_g$; this was first shown by Knudsen [**35**], and then by Morrison and Mumford [**39**]; in the latter paper it is actually shown that the divisor classes $a\lambda - b\delta$ are ample for a and b positive and $a > 11.2 \cdot b$.

Given this, the question of determining the ample cone exactly amounts to answering the question: what linear combinations of λ and δ have positive degree on every one-parameter family of stable curves? There are, to begin with, some obvious necessary conditions: since there are, as we said, complete curves in $\mathcal{M}_g$, on which the degree of λ will be positive and the degree of δ zero, we must have $a > 0$. Secondly, since there are lots of curves contained in the boundary of $\overline{\mathcal{M}}_g$ on which the degree of λ is zero and the degree of δ negative (for example, take any fixed curves B and C of genera α and $g - \alpha$, and identify a fixed point of B to a variable point of C; an example of this is worked out in the appendix), we must also have $b > 0$; so we need look only at the first quadrant of the a, b-plane. Next, an example in [**39**] shows that μ must be at least 11: if we take a fixed curve C of genus $g - 1$ and a fixed point $p \in C$, and a pencil $(B_t)_{t \in \mathbf{P}^1}$ of plane cubic curves with a base point q, the family of stable curves (D_t) we get by identifying $p \in C$ with $q \in B_t$ has intersection number 11 with δ (12 with δ_0 and -1 with δ_1) and 1 with λ. The picture of the ample cone thus looks like this:

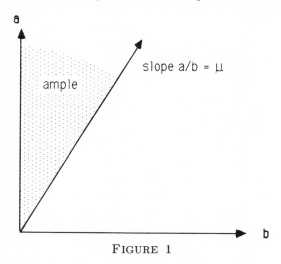

FIGURE 1

where μ is some constant less than or equal to 11.2 but greater than or equal to 11.

To nail down the exact value of μ we need, as indicated, an inequality on the degrees of λ and δ on a curve in $\overline{\mathcal{M}}_g$. This is provided by the basic estimate [9]: if $\Gamma \subset \overline{\mathcal{M}}_g$ is any complete curve not contained in the boundary, then

$$(*) \qquad \deg_\Gamma(\delta) \le (8 + 4/g) \cdot \deg_\Gamma(\lambda).$$

We note that this estimate is sharp, being attained, for example, by a general family of hyperelliptic curves $\{C_t = (y^2 - \prod(x - \lambda_i(t)))\}$. We can also use it to study families of singular curves, by applying it to their normalizations and using Arakelov's estimate that the self-intersection of a section of a family of generically smooth curves of genus $g \ge 2$ is nonpositive. In this way, it is easy to see that the ratio $\deg(\delta)/\deg(\lambda) = 11$ achieved by the family described above is the highest possible; we deduce that the constant μ is exactly 11.

The ample cone in all of $\mathrm{Pic}(\overline{\mathcal{M}}_g)$ is not known, nor does the outlook appear promising. We need, first of all, a refined analogue of $(*)$ that takes into account the degrees of the various divisors δ_α on Γ; and while some refinements of $(*)$ have been obtained, they do not appear to be sharp. Secondly, and perhaps more dauntingly, to analyze curves contained in the boundary in this detail Arakelov's inequality is not enough: we would need similar estimates on the cone of ample divisor classes in the universal curve $\overline{\mathcal{C}}_g$ and more generally in the moduli $\overline{\mathcal{M}}_{g,k}$ of stable k-pointed curves. An investigation in this direction has been begun by Cukierman [10], but we are at present far from a clear answer.

The problem of determining the effective cone in $\mathrm{Pic}(\overline{\mathcal{M}}_g) \otimes \mathbf{Q}$ is more elusive, even if we restrict to the sublattice generated by λ and δ. Once more, in the

a, b-plane it is just the first quadrant that is unsettled (here the lower half-plane is clearly not effective; the second quadrant is clearly effective); again, the locus of effective divisor classes in the first quadrant will be the sector above a ray of some slope s_g. There is here an additional incentive for finding the exact value of the slope s_g: by a standard calculation with the Grothendieck-Riemann-Roch formula (cf. [29]) we find that the canonical class of the moduli functor is

$$K_{\overline{\mathcal{M}}_g} = 13 \cdot \lambda - 2 \cdot \delta.$$

(On the moduli space the coefficient of δ_1 has to be adjusted; cf. the appendix for a discussion of the difference.) This, together with an analysis of the singularities of the moduli space, implies that *the Kodaira dimension of* $\mathcal{M}_g$ *is nonnegative if and only if* $s_g \leq 6\frac{1}{2}$; *and* $\mathcal{M}_g$ *is of general type if* $s_g < 6\frac{1}{2}$.

There is only one way so far used to prove a bound $s_g \leq s$ on s_g from above, and that is to exhibit an effective divisor $D \sim a\lambda - b\delta$ with slope $a/b = s$. There are a number of naturally defined divisor classes whose slope has been computed: for example, there is the divisor Θ_g of curves with a vanishing Θ-null (that is, a semicanonical pencil); the slope of this divisor tends to 8 as g gets large. Diaz in his thesis [12] analyzes the divisor D_{g-1} of curves possessing a Weierstrass point with first nongap $g - 1$; he finds the slope approaches 9 as $g \to \infty$. The curves of genus g whose canonical model lies on a quadric of rank 3 also form a divisor Q_g, whose slope approaches 7 when g gets large. Finally, in [29] (for $r = 1$) and in [15] (in general) it is shown that when the Brill-Noether number $\rho = g - (r + 1)(g - d + r) = -1$ (see Lecture 2), the locus $\Sigma_{d,r}$ in $\overline{\mathcal{M}}_g$ of curves possessing a linear series of degree d and dimension r (that is, for $r \geq 3$, the image in $\overline{\mathcal{M}}_g$ of the components of the Hilbert scheme $\mathcal{H}_{d,g,r}$ whose general members are smooth) is a divisor, and its slope, irrespective of r and d, is $s = 6 + 12/(g + 1)$. This is the lowest slope known to be attained by an effective divisor at present, and so when $g + 1$ is composite our picture of the effective cone looks like

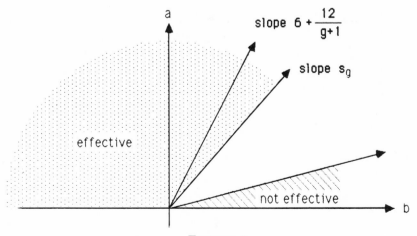

FIGURE 2

It will be noticed that the divisor Q_g is in fact reducible: it consists of the union of the divisors $Q_{g,d}$ of curves C possessing a pencil $V \subset H^0(C, \mathcal{L})$ of degree d with $H^1(C, \mathcal{L}^2) \neq 0$, where d runs from $g - 1$ down to the greatest integer $[(g + 2)/2]$; the component $Q_{g,g-1}$ is just the divisor Θ_g above, while at the other extreme if g is odd then for $d = (g + 1)/2$ the component $Q_{g,d}$ is the divisor $\Sigma_{d,1}$. In view of the fact that the slope of the sum Q_g approaches 7 while the slope of the component Θ_g approaches 8, this naturally suggests looking at the divisor $\Sigma_{d,1}$ to minimize the slope, which is what led to the study of $\Sigma_{d,1}$. In case $g + 1$ is prime, so that there are no r and d for which $\rho = -1$, it also suggests looking at the divisor $Q_{g,(g+2)/2}$; this divisor is studied in [15] and found to have slope

$$6g^2 + 28g + 4/g^2 + 2g,$$

which is still the lowest known for g with $g + 1$ prime.

The upshot of this is that we can say that $\mathcal{M}_g$ *is of general type for $g \geq 24$, and has positive Kodaira dimension for $g = 23$.* By contrast, it was known classically ([44]; cf. also [3]) that $\mathcal{M}_g$ is unirational for $g \leq 10$; in the last ten years it has been discovered that $\mathcal{M}_g$ is unirational for $g = 12$ [43], 11 and 13 [7], and of negative Kodaira dimension for $g = 15$ [8]. The cases of genus 14 and 16–22 are still unknown.

As for bounding from below the slope s_g of the effective cone, there is likewise only one way known to do this: exhibit a curve Γ in $\overline{\mathcal{M}}_g$ whose deformations fill up an open subset of $\overline{\mathcal{M}}_g$. Any effective divisor must then have nonnegative intersection number with Γ, so that we will have

$$s_g \geq \deg_\Gamma(\delta)/\deg_\Gamma(\lambda).$$

Just as the problem of determining the ample cone came down to a question of what we can say about the degrees of λ and δ on an arbitrary curve in $\overline{\mathcal{M}}_g$, this raises the question: what can we say about those degrees on a curve *through a general point* in $\overline{\mathcal{M}}_g$? The only example of this that has been worked out is the family of curves obtained by taking, for some $d \geq (g + 2)/2$ and $b = 2g - 2 + 2d$, a one-parameter family B_t of b-tuples of points on $\mathbf{P}^1$ and considering the curve in $\overline{\mathcal{M}}_g$ formed by all d-sheeted covers of $\mathbf{P}^1$ branched over one of the B_t. The degrees of λ and δ on this curve were determined in [28], explicitly for small g and in terms of values of characters of symmetric groups in general. In particular, it was found in case $g = 3$ that $s_3 = 9$; in case $g = 4$, that

$$s_4 \geq 46759680/5550633 \sim 8.4242\ldots$$

(showing that $6 + 12/(g + 1)$ may not be achieved if $g + 1$ is not composite); and in case $g = 5$ we have $7.65 \leq s_5 \leq 8$, narrowing down the range somewhat. Unfortunately, the bound obtained in [28] tends to zero as g gets large; asymptotically, we appear to get

$$s_g \geq 576/5g + o(g^{-1})$$

so that for example it is not even known if the limiting value $s = \liminf_{g \to \infty} s_g$ is six, or zero, or somewhere in between.

Lecture 2

Brill-Noether Theory

1. The basic set-up. Recall that we are dealing with the basic map $\varphi\colon \mathcal{H} = \mathcal{H}_{d,g,r} \to \mathcal{M}_g$, where $\mathcal{H}$ is, roughly, the Hilbert scheme parametrizing curves of degree d and genus g in $\mathbf{P}^r$. We say roughly because we will want to modify this to fit the situation: certainly in general we would want to consider only the "geometric Hilbert scheme," that is, the union of the components of the Hilbert scheme whose general members are nondegenerate (i.e., not lying in a hyperplane) and reduced of pure dimension 1. Sometimes we will want to restrict ourselves to the open subset of smooth curves; other times, such as when $r = 2$, we will want to look at the locally closed subvariety of $\mathcal{H}_{d,g,r}$ consisting of curves of geometric genus g. Of course, in case $r = 1$ we want to take $\mathcal{H}$ to be the Hurwitz scheme of branched covers of $\mathbf{P}^1$; and in general we may prefer to take $\mathcal{H}$ to be the space parametrizing maps $f\colon C \to \mathbf{P}^r$ rather than their images.

With all these caveats, our goal here is to say what we can about the fibers of the map φ. This question, it should be said at the outset, breaks up into two parts: to describe the fiber of φ over a general point of $\mathcal{M}_g$, and to see what sort of exceptional behavior goes on over subvarieties of $\mathcal{M}_g$. The first of these is one that, in the last 15 years or so, we have come to understand reasonably well; the second seems naturally much more difficult and not susceptible of a general solution. In this discussion we will focus almost entirely on the first. Thus, after a brief discussion in §1 giving the definition of the relevant objects, we will try to give a survey of the known results on the geometry of the general fiber φ_C in §2, together with at least a hint of the techniques involved. Finally, in §3, we will discuss in more detail one of these techniques, that of *limit linear series*, and use this to give a reasonably complete proof of the Brill-Noether theorem.

In the interests of uniformity, we will take $\mathcal{H}$ here to be the space parametrizing nondegenerate maps $f\colon C \to \mathbf{P}^r$ (that is, ones whose images do not lie in proper linear subspaces of $\mathbf{P}^r$); we will see that as long as we stick to a general point of $\mathcal{M}_g$ this will not differ too much from the Hilbert scheme, except of course in the cases $r = 1$ or 2. We then have

$$\varphi_C = \{\text{maps } f\colon C \to \mathbf{P}^r\} = \{\text{triples } (L, V, \Sigma)\},$$

where

$L \in \operatorname{Pic}^d(C)$ is a line bundle of degree d on C
$\qquad (L = f^*\mathcal{O}_{\mathbf{P}^r}(1));$
$V \subset H^0(C, L)$ is an $(r + 1)$-dimensional vector space
$\qquad$ of sections $(V = f^*H^0(\mathbf{P}^r, \mathcal{O}_{\mathbf{P}^r}(1)));$ and
$\Sigma = (\sigma_0, \sigma_1, \ldots, \sigma_r)$ is a basis for V $(\sigma_i = f^*(X_i),$
$\qquad$ where the X_i are homogeneous coordinates on $\mathbf{P}^r)$.

Of course, given L and V all Σ's are equivalent mod PGL_{r+1}; so the fiber φ_C is just a PGL_{r+1}-bundle over the variety $G_d^r(C)$ parametrizing pairs (L, V) (or "g_d^r's", as they are called) and it is to this variety that we turn our attention. We can in fact go further, and look at the variety parametrizing L's; that is, at the image

$$W_d^r(C) = \{L \colon H^0(C, L) \geq r + 1\} \subset \mathrm{Pic}^d(C)$$

of $G_d^r(C)$ in $\mathrm{Pic}^d(C)$. Either way, our basic problem is to describe the varieties $G_d^r(C)$ and $W_d^r(C)$ for a general curve C.

This may seem like a fairly complicated way to parametrize maps of a given curve C to $\mathbf{P}^r$, which after all can be given just by $(r+1)$-tuples of rational functions on C. To see the advantages of this approach, look at the relatively trivial case, where $d > 2g-2$ (and, of necessity, $r \leq d-g$). Here $W_d^r(C)$ is clearly all of $\mathrm{Pic}^d(C)$ for any curve C; $G_d^r(C)$ is a Grassmannian $\mathbf{G}(r, d-g)$-bundle over $W_d^r(C)$; and φ_C as we said is a PGL_{r+1}-bundle over $G_d^r(C)$; we may conclude that in this case the Hilbert scheme $\mathcal{H}_{d,g,r}$ is irreducible of dimension

$$\dim \mathcal{H}_{d,g,r} = \dim \mathcal{M}_g + \dim \mathrm{Pic}^d(C) + \dim \mathbf{G}(r, d-g) + \dim \mathrm{PGL}_{r+1}$$
$$= 3g - 3 + g + (r+1)(d-g+r) + (r+1)^2 - 1.$$

What would we like to know about $G_d^r(C)$ and $W_d^r(C)$? Just the usual things we ask about any variety: their dimensions (and in particular when are they nonempty), their irreducible components, and their smoothness or singularities. As it happens, we can answer all of these questions for general curves C.

2. A survey of results. To begin with, to estimate the dimension of $W_d^r(C)$ we introduce the variety C_d of effective divisors of degree d on C (i.e., the quotient of the ordinary d-fold product C^d by the symmetric group on d letters), and the subvariety $C_d^r \subset C^d$ of divisors D such that $h^0(C, \mathcal{O}(D)) \geq r + 1$. C_d maps to $\mathrm{Pic}^d(C)$ (and C_d^r to $W_d^r(C)$) by $D \mapsto L = \mathcal{O}(D)$; and as long as we are in the case $r \geq d-g$, for a general point $L \in W_d^r(C)$ we will have $h^0(C, L) = r + 1$, so the fiber of C_d^r over L will just be a $\mathbf{P}^r$ (cf. [**5**, p. 163]). To describe C_d^r, recall that by the Riemann-Roch formula, a divisor $D = p_1 + p_2 + \cdots + p_d$ will have $h^0(C, \mathcal{O}(D)) \geq r + 1$ if and only if it imposes $d - r$ or fewer conditions on the canonical series of C; that is (assuming the points p_i are distinct, which they will be if D is general in the fiber of C_d^r over a general point of $W_d^r(C)$), letting $\omega_1, \ldots, \omega_g$ be a basis for the space of holomorphic differentials on C and z_i a local coordinate on C in a neighborhood U_i of p_i, and writing $\omega_\alpha = f_{\alpha,i}(z_i)$ in U_i, if and only if the matrix

$$B = \begin{pmatrix} f_{1,1}(z_1) & \cdots & f_{g,1}(z_1) \\ \vdots & & \vdots \\ f_{1,g}(z_d) & \cdots & f_{d,g}(z_d) \end{pmatrix}$$

has rank at most $d - r$. We have thus defined a map η from a neighborhood of D in C_d to the space M of $g \times d$ matrices such that if $M' \subset M$ denotes the subvariety of matrices of rank $d - r$ or less, then $C_d^r = \eta^{-1}(M')$. Since M is

smooth and M' is irreducible of codimension $r(g - d + r)$ in M, we deduce that
every component of C_d^r has codimension at most $r(g-d+r)$ in C_d (cf. [**5**, p. 159]).
It follows that, if C_d^r is nonempty, then

$$\dim C_d^r \geq d - r(g - d + r),$$

and hence

$$\dim W_d^r(C) \geq d - r(g - d + r) - r = g - (r + 1)(g - d + r),$$

a number called the *Brill-Noether number* and usually denoted ρ. (The inequality
$\dim W_d^r(C) \geq \rho$ if $W_d^r(C) \neq \varnothing$ was first obtained by Brill and Noether.) Note
that the same argument can be made for a divisor varying on a family of curves
to deduce that any component of $\mathcal{H}$ has dimension at least $\rho+3g-3+(r+1)^2-1$.

Given this, the natural question to ask is whether the dimension of $W_d^r(C)$ is
indeed equal to ρ for general C. The first major step in this direction was taken,
independently, by Kempf [**30**] and Kleiman-Laksov [**32**, **33**], who proved that for
any curve C the variety $W_d^r(C)$ is nonempty (and hence of dimension at least ρ)
when ρ is nonnegative. The idea of all three proofs was to interpret the matrix
(1), or a related one, globally as a map between vector bundles. Specifically,
the matrix (1) can be viewed as a map between the bundles $\tilde{E}$ and $\tilde{F}$, where
$\tilde{E} = H^0(C, \omega_C) \otimes \mathcal{O}_{C_d}$ is the trivial bundle and $\tilde{F}$ the bundle whose fiber over
a point $D \in C_d$ is the space $H^0(C, \omega_C)/H^0(C, \omega_C(-D))$ of differentials on C
modulo those vanishing on D—i.e., the bundle

$$\tilde{F} = \pi_{2*}(\pi_1^* \omega_C \otimes \mathcal{O}_{\mathcal{D}}),$$

where $\mathcal{D}$ is the universal divisor of degree d in $C \times C_d$ and π_1, π_2 are the
projection maps on $C \times C_d$. The matrix (1) is then just the expression, in terms
of frames for $\tilde{E}$ and $\tilde{F}$, of the natural evaluation map from $\tilde{E}$ to $\tilde{F}$. We may then
compute the chern classes of the bundle $\tilde{F}$ and apply the formula of Porteous
for the fundamental class of the degeneracy locus of a bundle map, to deduce
that the locus C_d^r is nonempty.

An alternate approach is to consider a related vector bundle map on the
variety $\text{Pic}^d(C)$ (cf. [**5**, §§VII.2 and VII.4]). Fix a divisor D_0 of some suitably
large degree m and define a vector bundle E on $\text{Pic}^d(C)$ whose fiber over a point
$\mathcal{L} \in \text{Pic}^d(C)$ will be the $(d + m - g + 1)$-dimensional vector space of sections of
the vector bundle $\mathcal{L} \otimes \mathcal{O}(D_0)$; let F be the bundle of rank m whose fiber over
$\mathcal{L}$ is the vector space of sections of $\mathcal{L} \otimes \mathcal{O}(D_0)$ modulo those vanishing on D_0.
There is an obvious evaluation map $\psi: E \to F$, and the locus where this map
has kernel of dimension $r + 1$ or more is exactly the locus $W_d^r(C)$. As before,
we deduce from this without further computation that any component of the
variety $W_d^r(C)$ has codimension at most ρ; and after finding the chern classes
of E and F and applying Porteous we may conclude that $W_d^r(C)$ is nonempty
when $\rho \geq 0$ (and determine its fundamental class in case it has dimension ρ).

This approach was taken further by Fulton and Lazarsfeld in [**21**]. In this
paper they observe that the bundle E on $\text{Pic}^d(C)$ constructed above is ample,

and the bundle F algebraically equivalent to the trivial bundle; they are then able to deduce from their general theorems about positive vector bundles that $W_d^r(C)$ is nonempty when $\rho \geq 0$ and that it is connected when $\rho > 0$.

It remains now to say whether the dimension of $W_d^r(C)$ on a general curve C is exactly ρ, or strictly greater; and whether in particular a general curve may possess any g_d^r's with negative ρ. We will say that a curve C satisfies the *Brill-Noether condition* if we have $\dim W_d^r(C) = \rho$; the statement that a general curve C satisfies the Brill-Noether condition is called the *Brill-Noether theorem*. It is distinguished from the previous theorems in that statements like "$\dim W_d^r(C) = \rho$" are clearly not true of every curve of genus g; at most they hold on an open dense subset of $\mathcal{M}_g$. It follows that their proofs must be of a different character from those above, which took place on an arbitrary fixed curve of genus g; some variational element must be present. (Of course, since the statement "$\dim W_d^r(C) = \rho$" is an open condition in $\mathcal{M}_g$ it will suffice to exhibit a single smooth curve satisfying this condition for each g, d, and r; but this is not as much help as it would seem: at least for large g, all the curves we can write down explicitly, such as hyperelliptic, trigonal, or plane curves, etc., in fact tend to violate the Brill-Noether condition.) One natural approach would be to go to a particular curve C with a g_d^r and apply deformation theory to determine when a g_d^r on C could be extended to a given first-order deformation of C, in the hope of showing for example that a g_d^r with $\rho < 0$ would not extend to every first-order deformation. In general, this just reduces to the *Petri condition* discussed below, which seems to be harder to verify than the Brill-Noether condition itself; except in cases where $r = 1$ where it does give extra information about pencils with special properties as for example in Diaz [13].

In fact, the way in which a variational element might be introduced was first suggested by Castelnuovo almost a century ago. In [6], Castelnuovo proposes to consider a family of curves C_λ tending to the stable curve C_0 obtained from $\mathbf{P}^1$ by identifying g pairs of points p_i and q_i—that is, a rational curve with g nodes. Assuming we could come up with a reasonable notion of g_d^r on C_0, we would hope to have the statement

(2) $$\dim G_d^r(C_\lambda) \leq \dim G_d^r(C_0)$$

for general λ; and a g_d^r on C_0 would be given by a g_d^r on $\mathbf{P}^1$ with the property that every divisor D of the g_d^r that contained the point p_i would contain the point q_i as well and vice versa. Embedding $\mathbf{P}^1$ in $\mathbf{P}^d$ by the complete linear series $|\mathcal{O}_{\mathbf{P}^1}(d)|$, so that a g_d^r on $\mathbf{P}^1$ will be cut out by the hyperplanes containing a $(d-r-1)$-plane $\Lambda \subset \mathbf{P}^d$, these in turn correspond to $(d-r-1)$-planes meeting each of the g chords $L_i = \overline{p_i q_i}$. We would thus have

(3) $$G_d^r(C_0) \cong \{\Lambda \in \mathbf{G}(d-r-1, d) \colon \Lambda \cap \overline{p_i q_i} \neq \varnothing\};$$

and since the Schubert cycle $\Sigma_r(L_i)$ of planes meeting the line L_i has codimension r in the $(r+1)(d-r)$-dimensional Grassmannian $\mathbf{G}(d-r-1, d)$, we would have

(4) $$\dim G_d^r(C_0) = (r+1)(d-r) - rg = \rho.$$

This line of argument as it stands clearly needs a lot of work. (Castelnuovo, it should be said, was not trying to prove the Brill-Noether theorem, which was at the time considered known, but only to calculate, in case $\rho = 0$, the actual number of g_d^r's on a general curve; he did find the correct number.) First, there is the question of the notion of g_d^r on a singular curve: if we simply parrot the definition on a smooth curve (a line bundle with a vector space of sections) the family of varieties $\{G_d^r(C_\lambda)\}$ will not be proper at $\lambda = 0$ and so we will not be able to apply the upper-semicontinuity of dimension (2); if we try to compactify $G_d^r(C_0)$, we have to be careful about what happens to the description (3) of g_d^r's on C_0. These problems were first solved by Kleiman [**31**], using his theory of r-special subschemes.

The second difficulty with the proposed argument is the assumption that the g cycles $\Sigma_r(L_i)$ of codimension r in the Grassmannian will in fact intersect properly. To illustrate, consider the statement that *not every curve of genus three is hyperelliptic*, corresponding to the fact that $\rho = g - (r+1)(g-d+r) = 3 - 2 \cdot 2 = -1$. To establish this, we would specialize to the curve C_0 obtained from $\tilde{C} = \mathbf{P}^1$ by identifying p_i with q_i for $i = 1, 2, 3$, then argue that a degree 2 map of C_0 to $\mathbf{P}^1$ would be a degree 2 map of $\tilde{C}$ to $\mathbf{P}^1$ carrying p_i and q_i to the same point—that is, if we realize $\tilde{C}$ as a plane conic,

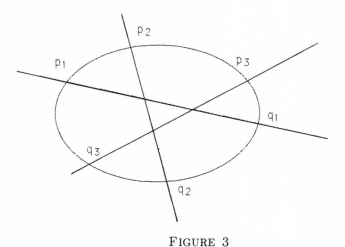

FIGURE 3

the projection of $\tilde{C}$ to a line from a point lying on each of the three chords $\overline{p_i q_i}$. Of course we can see in Figure 3 that no such point exists, and deduce the statement. But what if we had chosen p_i and q_i as in Figure 4? This diagram shows that the Schubert cycles $\Sigma_r(L_i)$ may in fact not intersect properly in some cases—i.e., in this case that there do exist rational curves with 3 nodes that are limits of hyperelliptic curves of genus 3, and in general that it is true of g-nodal curves as of smooth ones that while the general such curve may satisfy the Brill-Noether condition, some special ones will violate it. This suggests further specialization is required—in other words, that we let the configuration (p_i, q_i) vary and analyze the limit of the intersection of the corresponding Schubert cycles. This was carried out in [23] by letting all the points involved come together in sequence, completing, in combination with all the above, the proof of the Brill-Noether theorem.

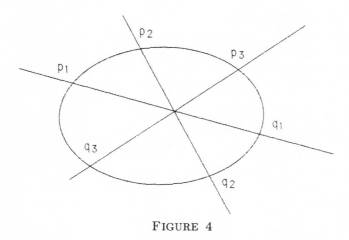

FIGURE 4

An alternate approach to this argument was given in [16], where it was observed that if the chords $L_i = \overline{p_i q_i}$ specialize to tangent lines to the rational normal curve $\mathbf{P}^1 \subset \mathbf{P}^r$ at the points p_i, then (at least over $\mathbf{C}$) the corresponding Schubert cycles will intersect properly (though not necessarily transversely), *regardless of the position of the p_i*. This amounts in the specific case of genus 3 curves discussed above to the observation that while three chords to a plane conic may be concurrent, three tangent lines can never be; in general it says that while a g-nodal curve may violate Brill-Noether (i.e., may be the limit of smooth curves of genus g that do), g-cuspidal curves will never. Another consequence of the generally good behavior of g-cuspidal curves was the result that for C general of genus g and any $r \geq 3$, *the general g_d^r on C gives an embedding of C in $\mathbf{P}^r$*. (In particular, we have the reassurance that components of the Hilbert scheme $\mathcal{H}_{d,g,r}$ do correspond to components of the variety parametrizing maps $f \colon C \to \mathbf{P}^r$, at least for those components dominating $\mathcal{M}_g$.)

The next major advance in this direction was Gieseker's proof of the Petri conjecture. K. Petri had conjectured that for a general curve C and *any* line bundle $\mathcal{L}$ on C, the natural multiplication map

$$\mu_0 \colon H^0(C, \mathcal{L}) \otimes H^0(C, \omega_C \otimes \mathcal{L}^{-1}) \to H^0(C, \omega_C)$$

is injective. Petri had made the statement in connection with the homogeneous coordinate ring of the curve C as embedded in projective space by the linear system $|\mathcal{L}|$. Later it was realized [1] that more was at stake: in fact the obstructions to deforming the line bundle $\mathcal{L}$ together with all its sections onto first-order deformations of C lived naturally in the kernel of μ_0. Consequences of Petri's statement would thus include the smoothness of $W_d^r(C) - W_d^{r+1}(C)$ and the smoothness of $G_d^r(C)$ for general curves C (and, thereby, the Brill-Noether theorem as well).

Giesker proved the Petri conjecture by introducing a new idea. Previous arguments for the Brill-Noether statement proceeded by specializing to a g-nodal curve $C_0 = \mathbf{P}^1/p_i \sim q_i$, translating the Brill-Noether statement into a question of intersections of Schubert cycles associated to the points $p_i, q_i \in \mathbf{P}^1$ (in effect, getting rid of the curve C_0), and then specializing the points. In keeping with the idea that one should take advantage of the existence of an explicit compactification of moduli of curves, Gieseker proposed instead to follow the curve C_0 to its stable limit as the points were specialized and ask what a g_d^r on that limiting curve looked like. This he did: he said that if you let the points p_i and q_i all come together and take the stable limit of the curve C_0 (and then make whatever base changes and blow-ups are necessary to ensure that a given line bundle on the general member of the family extends to a line bundle on the special fiber) you get a curve that looks like

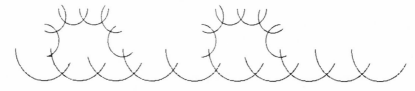

FIGURE 5

The scary part of this picture is simply the fact that the limiting curve is now reducible, which (as we will see in §3 below) gives the analysis of limits of g_d^r's a very different character. Gieseker was the first to attempt such an analysis; and, overcoming formidable technical obstacles, he succeeded in [**22**] in establishing the Petri statement.

Note that applying Gieseker's philosophy to the refinement of the original Brill-Noether argument using cuspidal rather than nodal curves suggests that we try following to their limit a family of curves degenerating to a g-cuspidal one; doing this, we arrive at a curve looking like

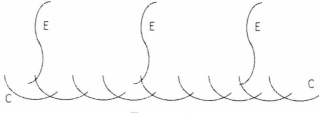

FIGURE 6

where E denotes an elliptic curve and C a rational one. An analysis of linear series on a family of curves degenerating to one of this type was carried out in [**48**], which resulted in a technical simplification of Gieseker's proof.

More recently, Rob Lazarsfeld [37] has come up with a proof of the Petri statement that involves no degeneration at all, specializing instead to curves on a K-3 surface and using the geometry of vector bundles on that surface to obtain the result.

To conclude our discussion of basic results, Fulton and Lazarsfeld point out in [21] that their connectedness statement, together with Gieseker's smoothness statement (and the basic estimates on the dimensions of the $W_d^r(C)$'s) imply that *for a general curve C the varieties $W_d^r(C)$ are irreducible when $\rho > 0$*. This in turn tells us, for $r \geq 3$, that *there is a unique component of the geometric Hilbert scheme $\mathcal{H}_{d,g,r}$ dominating $\mathcal{M}_g$ when $\rho > 0$*. (It is seen in [20] that this is also true when $\rho = 0$.) The other results tell us that *this component has dimension $3g - 3 + \rho + (r+1)^2 - 1$, and is smooth over an open subset of $\mathcal{M}_g$*; and that conversely *no such component exists when $\rho < 0$*.

This more or less concludes our summary of what is known about the varieties $G_d^r(C)$ and $W_d^r(C)$ for general C. What is known about them for arbitrary C is another story altogether. The first basic result about them is H. Marten's theorem, which says that for $d < g + r$, $\dim W_d^r(C) \leq d - 2r$ for any curve C, with equality holding if and only if C is hyperelliptic. This was extended by Mumford to the theorem that $\dim W_d^r(C) \leq d - 2r - 1$ for C nonhyperelliptic, with equality holding if and only if C is trigonal (and $r = 1$, $d = 3$), a two-sheeted cover of an elliptic curve ($r = 1$, $d = 4$), or a plane quintic ($r = 2$, $d = 5$); further progress in this direction has been made by Keem and G. Martens (see [5] for references). Of course, the borderline cases here are all comprised of linear series that map the curve C multiply to one to a curve of lower genus, and so do not actually arise as fibers of maps φ, suggesting that something better might be true in the case of components of $W_d^r(C)$ whose general members were birationally very ample, but so far nothing stronger than the estimate $\dim W_d^r(C) \leq d - 3r$, which is not expected to be sharp, is known in this case [5].

3. Limit linear series and a proof of the Brill-Noether theorem.

In this section we would like to carry out, in reasonable detail, an analysis of limits of linear series on reducible curves, as suggested above; we will obtain as a result a simple proof of the basic Brill-Noether statement.

We consider the following situation: we have a family $\pi \colon \mathcal{C} \to B$ of curves over a smooth, one-dimensional, suitably small base B (i.e., a disc, spec of a DVR, or whatever), with smooth total space $\mathcal{C}$; we assume the fibers C_λ of π over points $\lambda \neq 0 \in B$ are smooth, while the fiber C_0 consists of two smooth, reduced curves Y and Z meeting transversely in a point p as in Figure 7. We suppose that we are given a g_d^r on the general fiber C_λ of our family: that is, a line bundle $\mathcal{L}'$ on $\mathcal{C} - C_0$ of degree d on each fiber, and an $(r + 1)$-dimensional vector space of sections $V_\lambda \subset H^0(C_\lambda, \mathcal{L}')$ (more precisely, a locally free subsheaf of rank $r + 1$ in $\pi_* \mathcal{L}'$); we ask what we get in the limit as $\lambda \to 0$.

To begin with, because the space $\mathcal{C}$ is smooth, the line bundle $\mathcal{L}'$ will extend to a line bundle $\mathcal{L}$ on all of $\mathcal{C}$, and we will get a well-defined limiting vector space

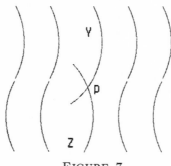

FIGURE 7

$V_0 = \lim V_\lambda \subset H^0(C_0, \mathcal{L})$. The problem—or rather the additional potential—is that there is more than one way to extend $\mathcal{L}'$ to $\mathcal{L}$: we could replace a given extension $\mathcal{L}$, for example, with the twist $\mathcal{L}(Y) = \mathcal{L} \otimes \mathcal{O}_\mathcal{C}(Y)$, whose restriction to Z will have degree one greater than $\mathcal{L}$ and whose restriction to Y degree one less. In fact, given any integer α, we see that there is a unique extension $\mathcal{L}^\alpha$ of $\mathcal{L}'$ to $\mathcal{C}$ having degree α on Y and $d - \alpha$ on Z. (Note that since B is taken to be small, $\mathcal{O}_\mathcal{C}(C_0) \cong \mathcal{O}$, and so twisting by Y is the same as twisting by $-Z$.) We thus get a sequence of line bundles and spaces of sections on C_0, no one of which will in general contain all the information we want. Luckily, two do: we focus our attention on the particular extensions, denoted $\mathcal{L}^Y$ and $\mathcal{L}^Z$, whose restrictions to Y have degrees d and 0 respectively, and on the corresponding limiting vector spaces of sections $V_Y \subset H^0(C_0, \mathcal{L}^Y)$ and $V_Z \subset H^0(C_0, \mathcal{L}^Z)$. Since $\mathcal{L}^Y$ has degree 0 on Z, any section $\sigma \in V_Y$ of its restriction to C_0 is determined by its values on Y; so we can think of V_Y as a g_d^r on Y, and likewise of V_Z as a g_d^r on Z. We thus get, as the limit of our family of g_d^r's, a g_d^r each on Y and Z; we now ask what in general will be the relation between them.

To answer this question, we have to introduce the ramification and vanishing sequences of a linear series on a curve C at a point $p \in C$. This is simple: if $(\mathcal{L}, V \subset H^0(C, \mathcal{L}))$ is a g_d^r on C, we can write

$$\{\operatorname{ord}_p(\sigma)\}_{\sigma \in V} = \{a_0, a_1, \ldots, a_r\},$$

where $a_0 < a_1 < \cdots < a_r$; the sequence of numbers $a_i = a_i(V, p)$ is called the *vanishing* sequence of (L, V) at p. We similarly define the *ramification* sequence of (L, V) at p to be the nondecreasing sequence $\alpha_i = a_i - i$. By the *ramification index* of V at p, we mean the sum $\alpha(V, p) = \sum_i \alpha_i(V, p)$, and by the *total ramification* of V we mean the sum of all the ramification indices $\alpha(V) = \sum_p \alpha(V, p)$ (all but finitely many of the terms of this sum will be zero). The ramification index of V at p can also be realized as the order of vanishing of the determinant

$$\begin{vmatrix} \sigma_0(z) & \cdots & \sigma_r(z) \\ \dfrac{d}{dz}\sigma_0(z) & \cdots & \dfrac{d}{dz}\sigma_r(z) \\ \vdots & & \vdots \\ \dfrac{d^r}{dz^r}\sigma(z) & \cdots & \dfrac{d^r}{dz^r}\sigma_r(z) \end{vmatrix}$$

where $\sigma_0, \ldots, \sigma_r$ is a basis for V and z a local coordinate on C near p. Since this determinant transforms like a section of the line bundle $\mathcal{L}^{r+1} \otimes \omega_C^{r(r+1)/2}$, which has degree $(r+1)d + r(r+1)(g-1)$, we arrive thereby at the *Plücker formula*:

$$\alpha(V) = (r+1)d + r(r+1)(g-1).$$

We can now express the relation between the two g_d^r's $|V^Y|$ and $|V^Z|$ above: it is simply that their vanishing sequences at the point p must be complementary in d; that is, we must have

(5) $$a_i(V_Y, p) + a_{r-i}(V_Z, p) \geq d$$

for all i. To see why this is so, consider a divisor D^Y of the g_d^r $|V^Y|$ on Y; it will be cut out by a divisor D of the linear system $|\mathcal{L}^Y|$ on $\mathcal{C}$ not containing Y. Let α be the order of D along Z, and note that we have thereby $\mathrm{ord}_p(D^Y) \geq \alpha$. Now observe that since $\mathcal{L}^Z \cong \mathcal{L}^Y(d \cdot Y)$, the divisor $E' = D + d \cdot Y$ is in the linear system $|\mathcal{L}^Z|$; and since E contains the central fiber C_0 with multiplicity α, so is the divisor $E = E' - \alpha \cdot C_0 = D + (d-\alpha) \cdot Y - \alpha \cdot Z$. Moreover, E does not contain Z, and so restricts to give a divisor D^Z of $|V^Z|$; since E does not contain Y with multiplicity $d - \alpha$, we must have $\mathrm{ord}_p(D^Z) \geq d - \alpha$. We have thus associated to a divisor $D^Y \in |V^Y|$ a divisor $D^Z \in |V^Z|$ with $\mathrm{ord}_p(D^Y) + \mathrm{ord}_p(D^Z) \geq d$. It is a straightforward linear algebra exercise to see that we can find a set of generators D_i^Y for $|V^Y|$ with $\mathrm{ord}_p(D_i^Y, p) = a_i(V^Y, p)$ such that the corresponding divisors D_i^Z generated $|V^Z|$ and satisfy $\mathrm{ord}_p(D_i^Z, p) = a_i(V^Z, p)$; the basic relation (5) follows.

Essentially the same analysis can be carried out for a family of g_d^r's on a one-parameter family of curves $\pi \colon \mathcal{C} \to B$ degenerating (with smooth total space) to any *curve of compact type*, that is, a semistable curve C_0 whose dual graph is a tree. Specifically, for each component Y of C_0 there will be a unique extension of the line bundle $\mathcal{L}'$ on $\mathcal{C}$ having degree d on Y and degree 0 on all other components of C_0, so that we get a collection of linear systems $|V^Y|$ on the various components; and we may see in the same way that for each node $p = Y \cap Z$ of C_0 the linear systems $|V^Y|$ and $|V^Z|$ satisfy the relation (5). (It is also true that after possibly some further base change and blowing up, we can ensure that equality holds throughout (5).) Such a collection of data is called a *limit linear series*, and the various g_d^r's $|V^Y|$ its *aspects*; for more details and applications, see [17].

To see how this sort of analysis goes in practice, let us use it to prove, as promised above, the Brill-Noether theorem. Start by taking C_0 to be any semistable curve of genus g consisting of a tree of N rational curves Y_i, to which g elliptic curves E_i are attached, each E_i attached at one point p_i (such as the curve in Figure 6). Let $\pi \colon \mathcal{C} \to B$ be any smoothing of the curve C_0; we claim that *for general $\lambda \in B$, the fiber C_λ satisfies the Brill-Noether condition*.

To prove this, suppose we have a family of g_d^r's on the smooth fibers of $\mathcal{C}$, given as above, and consider the limit linear series associated to it. (If the general fiber C_λ has a g_d^r, we can always assume we have such a family after making a base change and blowing up; the new central fiber will have more components but will still fit the description above.) To begin with, by the Plücker formula, the total ramification of each of the aspects $|V^{Y_i}|$ of our limit g_d^r on the rational components of C_0 is $(r+1)(d-r)$. On the other hand, since the elliptic curves E_i do not have any rational functions with only one pole, the vanishing sequence of the aspect $|V^{E_i}|$ at the point p_i will be at most

$$(6) \qquad \underline{a}(V^{E_i}, p_i) \le (d-r-1, d-r, \ldots, d-3, d-2, d)$$

and the ramification index of $|V^{E_i}|$ at p_i correspondingly at most $(r+1)(d-r)-r$. Thus the sum of the ramification indices of the aspects of the limit g_d^r at the nodes of C_0 is at most $(N+g)(r+1)(d-r) - rg$.

But the curve C_0 has $N+g-1$ nodes, and at each of those nodes the sum of the ramification indices of the relevant aspects must be at least $(r+1)(d-r)$ by the relation (5). We must therefore have

$$(N+g)(r+1)(d-r) - rg \ge (N+g-1)(r+1)(d-r),$$

i.e., $(r+1)(d-r) \le rg$; or, equivalently, $\rho \ge 0$. This proves that *the general curve C_λ possesses no g_d^r's with $\rho < 0$.*

To complete the proof, observe that equality can hold in the inequality (6) above only if the line bundle $\mathcal{L}^{E_i}$ restricted to E_i is isomorphic to $\mathcal{O}_{E_i}(d \cdot p_i)$, or, equivalently, if for any rational component Y of C_0 the line bundle $\mathcal{L}^Y$ restricted to E_i is trivial. It follows that if our family of g_d^r's had Brill-Noether number ρ, then *the line bundle $\mathcal{L}^Y$ will have to be trivial on at least $g - \rho$ of the curves E_i.* In terms of the identification of the Jacobian of C_0 with the product of the elliptic curves E_i, this says that the limit in $J(C_0)$ of the varieties $W_d^r(C_\lambda)$ will be supported on the union of the ρ-dimensional coordinate planes, and hence that $\dim W_d^r(C_\lambda) \le \rho$.

Lecture 3

Plane Curves

1. Preliminaries: families of plane curves. In the first two lectures we have discussed two of the three objects mentioned in the introduction: the moduli space $\mathcal{M}_g$ of abstract curves of genus g, and the fibers of the natural map φ from the space $\mathcal{H} = \mathcal{H}_{d,g,r}$ of curves of degree d and genus g in $\mathbf{P}^r$ to $\mathcal{M}_g$. The time, obviously, has come to talk about $\mathcal{H}$ itself.

There is, however, a problem. Both $\mathcal{M}_g$ and the (general) fibers of φ are reasonably well behaved, the geometry of $\mathcal{M}_g$ changing gradually as g goes to infinity, and that of $W_d^r(C)$ for general C almost uniform with respect to varying r, d, and g. The behavior of the varieties $\mathcal{H}_{d,g,r}$ is, by contract, unspeakable: with the exception of the cases where $r = 1$ or 2, or where d is large with respect to g (as was discussed in the last lecture), virtually no general statement can be made about them. $\mathcal{H}_{d,g,r}$ may be reducible, with components of arbitrarily many different dimensions; no one has any notion of what these dimensions might be (except for the inequality $\dim \mathcal{H}_{d,g,r} \ge 3g - 3 + \rho + (r+1)^2 - 1$ when $\mathcal{H}_{d,g,r} \ne \varnothing$);

there will in general be nonreduced components of $\mathcal{H}_{d,g,r}$; there exist components
of $\mathcal{H}_{d,g,r}$ whose general point corresponds to a singular curve, and so on. Finally,
all of these phenomena occur only in certain ranges of d, g, and r; but no one
knows exactly what those ranges are.

In sum, we really don't have much of a general picture of the varieties $\mathcal{H}_{d,g,r}$,
quite possibly because there simply isn't one to be had (the place to look for one
would be Robin Hartshorne's lectures on space curves). We have, however, been
able to analyze the structure of $\mathcal{H}$ in some particular cases. In this lecture, we
will focus on one such case, that of plane curves, and concern ourselves with a
proof of the

THEOREM. *The space of reduced and irreducible plane curves of degree d and
geometric genus g is irreducible.*

We hope that this will serve to illustrate at least some of the possible approaches to the study of the $\mathcal{H}_{d,g,r}$ in general. The proof given here is a hybrid
of the proof in [**27**] and one given by M. Nori [**41**]; another proof has been given
by Ziv Ran [**42**]. We will discuss briefly the various approaches at the end.

We begin by introducing some of the relevant varieties and recounting some of
the basic facts about them. First, we let $\mathbf{P}^N$, $N = d(d+3)/2$, be the projective
space parametrizing all plane curves of degree d. We then introduce locally
closed subvarieties as follows: we let

$$
\begin{array}{ccc}
U^{d,\delta} & \hookrightarrow\ U_{d,g} & \hookrightarrow\ \mathbf{P}^N \\
\updownarrow & \updownarrow & \\
V^{d,\delta} & \hookrightarrow\ V_{d,g} &
\end{array}
$$

$V_{d,g}$ is the subset of $\mathbf{P}^N$ consisting of reduced and irreducible curves of geometric genus g;

$V^{d,\delta}$ is the subset of $V_{d,g}$ consisting of curves with $\delta = \frac{1}{2}(d-1)(d-2) - g$
nodes as singularities;

$U_{d,g}$ is the subset of $\mathbf{P}^N$ consisting of reduced, but not necessarily irreducible,
curves of geometric genus g; and, analogously,

$U^{d,\delta}$ is the subset of $U_{d,g}$ consisting of curves with δ nodes as singularities.

(Recall here that by the geometric genus of any reduced curve C we mean the
arithmetic genus of its normalization; thus if C is the union of k components C_i
of geometric genera g_i, the geometric genus of C will be $\sum g_i - k + 1$. For the
remainder of this lecture, we will by genus mean geometric genus; the notion will
be applied only to reduced curves.)

What do we know about these varieties? Some facts are immediate: $U^{d,\delta}$ and
$V^{d,\delta}$ are open subsets of $U_{d,g}$ and $V_{d,g}$ respectively; $V_{d,g}$ is a union of connected
components of $U_{d,g}$, and $V^{d,\delta}$ is likewise a union of connected components of
$U^{d,\delta}$. On a slightly less elementary level, we know from the last lecture that the
dimension of all four is at least $3g - 3 + \rho + (r+1)^2 - 1 = 3d + g - 1$, where ρ is the
Brill-Noether number; observe that this is equal to $N - \delta$. To go much beyond
this we really should introduce some of the theory of plane curve singularities;
we will take a moment out here and do this. (The application of deformation

theory to the geometry of the varieties $V_{d,g}$ was carried out first by Arbarello and Cornalba [1, 4] and then, independently and via a different approach, by Zariski [47]; we will follow Zariski's approach here. A reference for the latter results quoted is [27]; cf. also [14] for an extension of this approach.)

Basically, the deformation theory says this: suppose that C is a plane curve, and $p \in C$ a singular point. If the plane curve C is given near p by the polynomial $f(x, y)$, then, firstly, a deformation of C

$$f(x, y) + \varepsilon^m g(x, y) = [\varepsilon^{m+1}]$$

can preserve the genus of C only if the polynomial g lies in the *adjoint*, or *conductor* ideal in the local ring of C at p. (More precisely, in terms of the identification of the base of the étale versal deformation space of the singularity (C, p) with a neighborhood of 0 in the quotient of the local ring $\mathcal{O}_{C,p}$ by the Jacobian ideal, we can say that the reduced tangent cone to the locus of deformations preserving the genus is the adjoint ideal.) Secondly, the deformation can be equisingular (i.e., topologically trivial) only if g lies in an ideal, called the *equisingular ideal*, contained in the conductor. Moreover, while the equisingular ideal is not in general easy to describe, we know that it is equal to the conductor only in the case of a node, where both coincide with the maximal ideal. (It also follows from a theorem of Wahl [45], and can be checked directly in the case of a node, that a first-order deformation $f + \varepsilon g$ of C is equisingular if and only if g lies in the equisingular ideal.)

Applying this to the varieties of plane curves, we see first that the Zariski tangent space to $U^{d,\delta}$ at a point C, viewed, via the inclusion of $U^{d,\delta}$ in $\mathbf{P}^N$ as a linear system of curves of degree d mod C, lies in the linear system of curves of degree d containing the nodes of the curve C. Since the adjoint ideals of any reduced plane curve impose independent conditions on curves of degree $d - 2$ in the plane, we conclude that the tangent space to $U^{d,\delta}$ has dimension at most $N - \delta$; since this coincides with the lower bound on the dimension of $U^{d,\delta}$ above, we deduce that $U^{d,\delta}$ *is smooth of dimension* $N - \delta$.

More generally, if C is a general point of any component of $U_{d,g}$, then every deformation of C in $U_{d,g}$ is necessarily equisingular; we conclude that the reduced tangent cone to $U_{d,g}$ at C is contained in the linear system of curves of degree d satisfying the adjoint conditions of C, with equality only if C is nodal (since $d > d - 2$, any ideal strictly contained in the conductor would impose strictly more conditions). It follows then by the same dimension estimate that C must indeed be nodal, i.e., that $U^{d,\delta}$ *is dense in* $U_{d,g}$.

The last statement has a "parametric" analogue, which will be useful: if $V \subset \mathbf{P}^N$ is any locus whose general point C is the image of an abstract curve $\tilde{C}$ of geometric genus g via a map π not constant on any connected component of $\tilde{C}$, then the dimension of V is at most $3d + g - 1$; and if the dimension of V is exactly $3d + g - 1$ then C must be nodal. This is just the above statement if C is reduced (i.e., π has degree one over every irreducible component of C); in general we just have to check that even though the geometric genus of C_{red}

may be larger than g (if, for example, several rational components $\tilde{C}$ map to the same component of C), this will be more than compensated for by the fact that the degree of C_{red} will have to be correspondingly less than d.

We can take advantage of the basic "margin of error" in the estimate $d > d - 2$ to prove a little more; we will need these results later on. Let D be any divisor of degree m on a line $L \subset \mathbf{P}^2$, and let $W_D \subset V_{d,g} \subset \mathbf{P}^N$ be the locus of reduced curves C of degree d and geometric genus g not containing the line L but containing the divisor D (i.e, such that as schemes, $D \subset C \cap L$). Applying the deformation theory to the blow-up of $\mathbf{P}^2$ at the points of D (including infinitely near ones), it follows from the fact that the adjoint conditions impose independent conditions on curves of degree $d - 1$ passing through D that W_D is nonempty of dimension $3d + g - 1 - m$. We may use this in turn to deduce that if $W_k \subset V_{d,g} \subset \mathbf{P}^N$ is the locus of curves meeting L in k or fewer distinct points, then W_k has dimension $2d + g - 1 + k$: just let $\Sigma \subset L^{(d)}$ be the locus of divisors of degree d on L supported at k or fewer points and $\Theta \subset \Sigma \times V_{d,g}$ the locus of pairs (D, C) such that $C\,|_L \geq D$, count dimensions, and remark that a general point of Θ is certainly a general point of its fiber over Σ. We can also combine these two versions of the lemma: for any divisor D of degree m on L, letting $W_{D,k} \subset V_{d,g} \subset \mathbf{P}^N$ be the locus of curves C containing the divisor D and meeting L in at most k distinct points outside of D, we see that $W_{D,k}$ is nonempty of dimension $3d + g - 1 + k$.

Finally, a parametric version of the last statement: if, for given D and k as above, $W \subset \mathbf{P}^N$ is any locus whose general point C is the image of a curve $\tilde{C}$ of geometric genus g via a map π that is not constant on any connected component of $\tilde{C}$, and such that $\pi^{-1}(L)$ contains k or fewer points outside of $\pi^{-1}(D)$, then we have again that

$$(1) \qquad\qquad \dim(W) \leq 2d + g - 1 + k;$$

and if equality holds in (1) then C is nodal, and smooth at its points of intersection with L.

2. Severi's argument and corollaries.

2. Severi's argument and corollaries. The first question to be asked in trying to prove the theorem is why it should be true in the first place. It is a classical theorem that the space $\mathcal{H}_{d,g,1}$—that is, the Hurwitz space of branched covers of $\mathbf{P}^1$—is irreducible for every d and g; but on the other hand as we have said the spaces $\mathcal{H}_{d,g,r}$ for $r \geq 3$ tend to be very, very reducible. In fact, as far as I can tell the only a priori reason to suspect that $V_{d,g}$ is irreducible in general is that given by Severi in his famous argument [**44**], which also forms the basis for the proof below.

Severi suggests first looking at the variety $V_{d,0}$ of rational curves, and at the variety $V_{d,g}$ in a neighborhood of $V_{d,0}$. The variety $V_{d,0}$ is itself visibly irreducible, being dominated by the space of triples $[F_0, F_1, F_2]$ of homogeneous polynomials of degree d on $\mathbf{P}^1$. (This is of course just a special case of the argument in Lecture 2 for the irreducibility of $\mathcal{H}_{d,g,r}$ when d is large relative to g and r.) Moreover we see exactly what $V_{d,g}$ looks like locally around a general point C_0 of $V_{d,g}$: the curve C_0 will have as singularities exactly $\delta = (d-1)(d-2)/2$ nodes, and $V_{d,g}$ will locally be the union of smooth sheets corresponding to subsets of g of these nodes (Figure 8). That these sheets are smooth and in fact intersect transversely again follows from the description of the deformations of plane curves with nodes, which says that the variety of deformations of C_0 that smooth a given subset of g of the nodes and are locally trivial around the other $\delta - g$ is smooth with tangent space given by the system of plane curves of degree d through the $\delta - g$ preserved nodes.

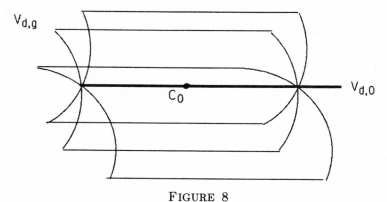

FIGURE 8

Clearly, $V_{d,g}$ is locally reducible near the point $C_0 \in V_{d,0}$. But now let C_0 vary in the space $V^{d,\delta}$ of nodal rational curves, returning eventually to C_0; as it does so, the monodromy will permute the δ nodes of C_0 and correspondingly the various sheets of $V_{d,g}$ near C_0. We claim that *the monodromy acts transitively on these sheets*, at least over $\mathbf{C}$. To see this, observe that any nodal rational curve C_0 in the plane may be realized as the projection of a rational normal curve $\tilde{C} \subset \mathbf{P}^d$ from a $(d-3)$-plane $\Lambda \subset \mathbf{P}^d$; the nodes of C_0 will correspond to the points of intersection of Λ with the chordal threefold X of $\tilde{C}$. By a standard theorem [49], the monodromy action on the points of intersection of an irreducible projective variety with a general linear subspace of complementary dimension is the full symmetric group on those points. In the present circumstances, this says that as Λ varies in $\mathbf{G}(d-3,d)$ the monodromy action on its intersection with X, and hence on the nodes of C_0, is the symmetric group on $(d-1)(d-2)/2$ letters, and so acts transitively on subsets of g of those nodes; the claim follows.

We see from the above that there is a unique component of $V_{d,g}$ containing $V_{d,0}$ in its closure; we thus have the

First reduction. To prove the theorem it is sufficient to show that every component of $V_{d,g}$ contains $V_{d,0}$ in its closure.

This can be broadened quite a bit: to begin with, by induction it will suffice to show that every component of $V_{d,g}$ contains in its closure a component of $V_{d',g'}$ for some $g' < g$. Then, too, we can look at the locus $U^{d,d(d-1)/2}$ of d-tuples of lines meeting pairwise transversely: this lies in the closure of $V_{d,0}$ and is, in a neighborhood of a general point $C_0 \in V_{d,0}$, the intersection of sheets of $V_{d,g}$. If we can show that any component of $V_{d,g}$ contains $U_{d,1-d}$ in its closure, then we will be done; and now it follows by induction that *it suffices to show that any component of $V_{d,g}$ contains in its closure a component of $U_{d',g'}$ with $d' < d$ or $g' < g$*. Finally, we observe that by the picture of $V_{d,g}$ in the neighborhood of any nodal curve, we see that if a component of $V_{d,g}$ just contains one point C_0 of $U^{d,\delta'}$ in its closure, it will contain the entire component of $U^{d,\delta'}$ containing C_0. We summarize these in the

Second reduction. To prove the theorem it will suffice to show that any component of $V_{d,g}$ contains in its closure a nodal curve of geometric genus $g' < g$.

This reduction certainly makes the theorem quite plausible: all we need to do is to show that every component of $V_{d,g}$ admits the simplest kind of degenerations. Now, it's not hard to see that every component Ω of $V_{d,g}$ contains degenerations in its closure: for example, we can just choose any $C \in \Omega$ meeting the line $Z_0 = 0$ transversely, and take the limit as t goes to zero of the curves C_t obtained from C by applying the linear transformation $\psi_t(Z_0, Z_1, Z_2) = (tZ_0, Z_1, Z_2)$ to get a curve in $\overline{\Omega}$ consisting of d concurrent lines (or, assuming C does not contain the point $Z_0 = Z_1 = 0$, take the limit as t goes to ∞ to get a d-fold line). (In fact, Severi tried to use the presence of these degenerations to prove the theorem; but there does not seem to be any way to make his arguments precise). Another way to exhibit degenerations is to point out that for any point $p \in \mathbf{P}^2$ the locus of curves containing p is a hyperplane in the space $\mathbf{P}^N$ of plane curves of degree d; if we take $p_1, p_2, \ldots, p_{d+1}$ to lie on a line $L \subset \mathbf{P}^2$ we deduce that $\overline{\Omega}$ contains curves containing all the p_i and hence containing L. The problem here is that the presence of these possibly wild degenerations does not guarantee that Ω admits the milder degenerations to nodal curves we need.

Another way of showing that every component of $V_{d,g}$ admitted degenerations was found by Diaz, who in [**11**] proved that the moduli space $\mathcal{M}_g$ of curves of genus g does not contain any complete subvarieties of dimension $g - 1$. Looking at the rational map

$$\varphi \colon V_{d,g} \to \mathcal{M}_g$$

and observing that the fibers of φ cannot have dimension greater than $d + 2$ (from the theorem that the dimension of any component of $W_d^r(C)$ whose general member is birationally very ample has dimension at most $d - 3r$; cf. [**5**]) and that the dimension of $V_{d,g}$ is $3d + g - 1$, Diaz concluded that the closure in $\overline{\mathcal{M}}_g$ of the image of φ on any component Ω of $V_{d,g}$ necessarily met the boundary $\Delta \subset \overline{\mathcal{M}}_g$. An arc in Ω whose image in $\overline{\mathcal{M}}_g$ tended to Δ would thus be a family of curves degenerating in moduli, whose existence had not previously been known; but here again the problem arose that there was no way to control the singularities of the limiting curve in $\mathbf{P}^2$: for example, it was a priori possible that any such arc in Ω would tend to a nonreduced curve.

These attempts made clear that what we really need here is not just to degenerate, but to have some control over the degeneration, i.e., to exhibit reduced curves in the closure of any Ω having lower geometric genus but still reasonably bounded singularities. The question is, how do we exert such control? For example, we have seen that in $\overline{\Omega}$ there are curves that contain a line L, but how do we know that not every such curve in $\overline{\Omega}$ is just a d-tuple of concurrent lines?

The answer is *by keeping track of dimensions*, or degrees of freedom, in our families. Thus, to answer the last question, we know that the dimension of Ω is $3d + g - 1$, and so the locus in $\overline{\Omega}$ of curves containing L is, by the argument above, of dimension at least $\dim \Omega - (d+1) = 2d + g - 2$; but the family of d-tuples of concurrent lines including L has dimension just d.

As another example, recall that the deformation theory result above says that any locus $\Omega' \subset \mathbf{P}^N$ consisting of curves of degree d and genus $g' < g$ and having dimension $3d + g - 2$ must be open in a component of $U_{d,g-1}$. We conclude that *in order to prove the theorem it will suffice to show just that $\overline{\Omega} - \Omega$ has codimension one*, i.e., that $\overline{\Omega}$ contains a locus of codimension one consisting of curves other than reduced curves of geometric genus g.

Now, clearly the first two constructions of degenerations given above will not produce such loci. Diaz's construction, on the other hand, seems a much better bet: after all, all we have to show is that the inverse image, under the map φ, of the boundary $\Delta \subset \overline{\mathcal{M}}_g$ has codimension one in $\overline{\Omega}$; and since Δ has codimension one in $\overline{\mathcal{M}}_g$ (and is in fact the support of a Cartier divisor), things are looking good. The problem is that φ is not a regular map, only a rational one. When we talk about the inverse image of Δ, we mean $\varphi_1(\varphi_2^{-1}\Delta)$, where Γ is the graph of φ and φ_1, φ_2 the projection maps to $\overline{\Omega}$ and $\overline{\mathcal{M}}_g$ respectively; and while $\varphi_2^{-1}(\Delta)$ will necessarily have codimension 1 in Γ, the fibers of φ_1 on $\varphi_2^{-1}\Delta$ may be positive-dimensional.

To see how this may occur, consider a family of plane curves over a two-dimensional base Ω, generically smooth in a neighborhood of a point $p \in \mathbf{P}^2$ and specializing to a curve C_0 with a cusp at p—e.g., the family given, in terms of local coordinates x, y on $\mathbf{P}^2$ near p and a, b on Ω, by the equation $y^2 = x^3 + ax + b$ (see Figure 9). Assuming the curves C_λ are well-behaved away from p, the map $\varphi \colon \Omega \to \overline{\mathcal{M}}_g$ will be defined everywhere except at the origin $a = b = 0$ in Ω. There, however, it is undefined, and we will have to blow up twice—once at the origin and then at the intersection of the first exceptional divisor with the proper transform on the a-axis—to resolve it. The map will then blow down the first exceptional divisor, so the graph Γ of φ will look like Figure 10. This is the basic example of a situation where a divisor in $\overline{\mathcal{M}}_g$—in this case Δ_1—may have inverse image of codimension greather than one in $\overline{\Omega}$.

As an alternate example, not involving singularities, consider the map $\varphi \colon \Omega = \mathbf{P}^{14} \to \overline{\mathcal{M}}_3$ from the space of plane quartics to their moduli. In $\overline{\mathcal{M}}_3$, the locus H of hyperelliptic curves is a divisor; but anytime we have a family of smooth curves of genus 3 approaching a hyperelliptic one, the canonical models will tend to a double conic, and the locus Σ in $\mathbf{P}^{14}$ of double conics is of course 5-dimensional. What's going on here is again simple to describe, at least over a general point of H: the map φ is blowing up the locus Σ, replacing points $2C \in \Sigma$ with pairs $(2C, D)$ where D is a normal direction to H in $\mathbf{P}^{14}$, represented by a divisor

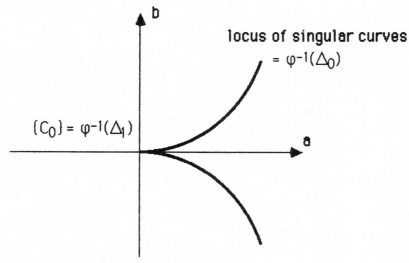

FIGURE 9

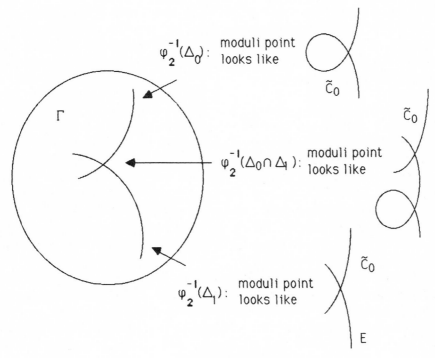

FIGURE 10

$D \in |\mathcal{O}_{2C}(4)|$, and mapping the pair $(2C, D)$ to the curve of genus 3 given as the double cover of $C \cong \mathbf{P}^1$ branched over the 8 points of D. (By the way, the inverse image $\varphi^{-1}(H)$ does not have codimension 9 in $\mathbf{P}^{14}$, but codimension 3: the locus in $\overline{\mathcal{M}}_3$ of curves consisting of two elliptic curves meeting at two points lies in H, and its inverse image in $\mathbf{P}^{14}$ contains the locus of quartics with a tacnode.)

In any event, it is clear that to find a locus of codimension one in $\overline{\Omega}$ consisting of curves of lower geometric genus, we need to find a point $C \in \varphi^{-1}(\Delta) \subset \overline{\Omega}$ such that the fiber of $\varphi_2^{-1}(\Delta) \subset \Gamma$ over C is finite (or, for that matter, contains an isolated point). Now, one circumstance in which a point $(C, D) \in \Gamma$ will be an isolated point of $\varphi_2^{-1}(\Delta) \cap \varphi_1^{-1}(\{C\})$ is when the stable curve D is obtained by contracting rational components of a partial normalization of the plane curve C, since there are only finitely many nodal partial normalizations of a given curve. We have thus arrived at our final reduction:

Third reduction. It suffices, to prove the theorem, to exhibit in any component Ω of $V_{d,g}$ an arc $B \subset \overline{\Omega}$ with $B - \{0\} \subset \Omega$ such that the stable limit as $\lambda \to 0$ of the normalizations of the curves $\{C_\lambda\}_{\lambda \in B - \{0\}}$ is obtained from a partial normalization of the curve C_0 by contracting rational components.

(Note that under the hypotheses of this last statement, the stable limit X_0 of the normalizations X_λ of the C_λ does not have to lie in the boundary $\Delta \subset \overline{\mathcal{M}}_g$. If indeed X_0 is smooth, then letting $\mathcal{L}_\lambda$ be the pullback of $\mathcal{O}_{\mathbf{P}^2}(1)$ to X_λ and $\mathcal{L}_0$ the limit on X_0 of the $\mathcal{L}_\lambda$, the limit $V_0 \subset H^0(X_0, \mathcal{L}_0)$ of the linear series $V_\lambda \subset H^0(X_\lambda, \mathcal{L}_\lambda)$ giving the maps from X_λ to $\mathbf{P}^2$ will have a base point. It's not hard to see that the corresponding locus in $\overline{V}_{d,g}$ has pure codimension one, and hence consists generically of nodal curves with $\delta + 1$ nodes, so in this case the result follows immediately.)

3. Analyzing a degeneration. It remains now only to find an arc B such as described in the last reduction on any given component Ω of $V_{d,g}$. Actually, this turns out to be relatively easy: almost any arc that is given in reasonably explicit form will serve. For example, consider the argument above that every component $\overline{\Omega}$ of the closure of $V_{d,g}$ contains curves C containing a line L, since we can force a curve in $\overline{\Omega}$ to pass through $d + 1$ fixed points $p_i \in L$. We go back to that construction and ask: suppose we take a suitably general arc $\{C_\lambda\}$ in the locus of curves passing through d fixed points p_i, tending to a curve C_0 containing L. What will the curve C_0 look like, and what will be the limiting moduli point in $\overline{\mathcal{M}}_g$ of this family?

To set this up, take $C \in \Omega$ general, and let $p_1, \ldots, p_d$ be the intersection of C with L, so that the intersection of Ω with the locus of curves passing through the p_i contains a component Ω_D whose general member is indeed a nodal curve of genus g not containing L. Let B be an arc in the closure $\overline{\Omega}_D$ tending to a general point C_0 in a component of the locus $\overline{\Omega}_L$ of curves $C \in \overline{\Omega}_D$ containing L. (For example, we can take B open in a general $(2d + g - 2)$-fold hyperplane section of $\overline{\Omega}_D$ in $\mathbf{P}^N$.) Take a sufficiently small neighborhood of $0 \in B$ that all the curves in this neighborhood except for C_0 itself are nice nodal curves of genus g. Let X_λ be the normalization of C_λ for $\lambda \neq 0$; the curves X_λ form the fibers of a family $\pi : \tilde{\mathfrak{X}} \to \Delta - \{0\}$ over the punctured disc, and the normalization maps $\eta_\lambda : X_\lambda \to C_\lambda$ string together to form a map $\tilde{\eta} : \tilde{\mathfrak{X}} \to \mathbf{P}^2$. Applying semistable reduction, we can (after base change) complete this family to a family $\pi \colon \mathfrak{X} \to B$ of nodal curves, proper over B and satisfying the conditions that

(i) the total space of $\mathfrak{X}$ is smooth;

(ii) the map $\tilde{\eta}$ extends to a regular map $\eta \colon \mathfrak{X} \to \mathbf{P}^2$; and

(iii) $\mathfrak{X}$ is minimal with respect to these properties: i.e., there are no rational components of the central fiber X_0 meeting the rest of the central fiber only once and on which the map η is constant.

We have, of course, no a priori idea of what the central fiber X_0 of this family looks like, or how the map $\eta_0 = \eta \,|\, X_0$ behaves; but we can use the conditions we do have—that η_0 has degree d, that X_0 has arithmetic genus g, and that η_0 is the limit of maps η_λ whose images contain the points p_i—together with the basic dimension estimates above to describe X_0 completely.

We start by introducing some notation. By construction, the image of η_0 contains the line L; let $Y_0 \subset X_0$ be the union of the components of X_0 mapping to L, let Y_1 be the union of the remaining components, and let $q_1, \ldots, q_k$ be the points of intersection of Y_0 and Y_1 (see Figure 11). Denote by α the degree of η_0 on the curve Y_0, so that we can write

$$C_0 = \alpha \cdot L + C_1,$$

where C_1 is the image of Y_1 under η_0; clearly a crucial part of our analysis must be to control α.

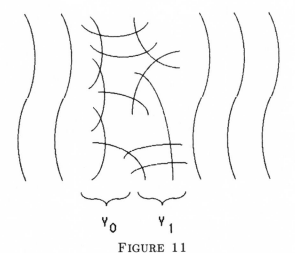

$$Y_0 \qquad Y_1$$

FIGURE 11

As indicated, we need to use the fact that η_0 is the limit of maps whose images contain the points p_i; to express this in the current setting let $\tilde{\Gamma}_i$ be the inverse image of the point p_i in $\tilde{\mathfrak{X}}$, and let Γ_i be its closure. Observe that since the total space $\mathfrak{X}$ is smooth, the sections Γ_i of π must meet the central fiber X_0 in smooth points of X_0. We can now write

$$\eta^* L = \sum \Gamma_i + M,$$

where M is a divisor whose support is exactly the curve Y_0. In particular, observe that any point of Y_1 lying over a point of L other than one of the p_i must be one of the points $q_1, \ldots, q_k$ of intersection of Y_1 with Y_0. In fact, we can refine this a little: of the k points q_i, say that β lie on connected components of Y_0 on which η is not constant, and $\gamma = k - \beta$ on connected components of Y_0 on which η is constant. A connected component Y of Y_0 on which η is constant must meet one of the sections Γ_i and hence map to one of the points p_i: if Y were disjoint from all Γ_i, then the part M_Y of the divisor M supported on Y would have self-intersection $(M_Y \cdot M_Y) = (M_Y \cdot \eta^* L) = 0$, contradicting the fact that any divisor supported on a proper subset of a fiber of π must have negative self-intersection. Thus, C_1 *can contain at most β points of L other than the p_i.* (Note that η can't be constant on any connected component of Y_1, since every connected component of Y_1 must meet Y_0.)

The remaining question to ask is: what is the geometric genus g_1 of Y_1? To estimate this, we use the fact that the arithmetic genus of the whole fiber X_0 is g; it follows that

$$p_a(Y_0) + p_a(Y_1) + k - 1 = g.$$

Now, since every connected component of Y_0 must meet Y_1, Y_0 can have at most k connected components. In fact, we can do a little better: there are at most γ connected components of Y_0 on which η is constant, and at most α connected components on which it is nonconstant. Thus

$$p_a(Y_0) \geq 1 - \gamma - \alpha = 1 - k + \beta - \alpha$$

and hence

$$g(Y_1) \leq p_a(Y_1) = g + 1 - k - p_a(Y_0) \leq g + \alpha - \beta.$$

But now we have a curve $C_1 = \eta(Y_1)$ that is the image of a nodal curve of geometric genus at most $g + \alpha - \beta$ via a map not constant on any connected component of Y_1, of degree $d - \alpha$, and such that at most β points of Y_1 map to points of L other than $p_1, \ldots, p_d$; and the curve C_1 moves in a family of plane curves of dimension

$$\dim(\Omega_1) = \dim(\Omega_D) - 1 = \dim(\Omega) - d - 1 = 2d + g - 2.$$

We conclude that

$$2d + g - 2 \leq 2(d - \alpha) + (g + \alpha - \beta) + \beta - 1$$
$$= 2d + g - 1 - \alpha,$$

from which it follows that $\alpha = 1$. Moreover, from the fact that we have equality in the last equation, we see that

(i) *every connected component of Y_0 has arithmetic genus zero.* But now any connected component of Y_0 on which η is constant can meet Y_1 in only one point (equality above implies that the k points of $Y_0 \cap Y_1$ map via η to distinct points of L); since each connected component of Y_0 is a tree, the minimality of

$\mathfrak{X}$ (condition (iii) above) implies that there are no connected components of Y_0 on which η is constant, i.e.,

(ii) $\gamma = 0$, or, equivalently, $\beta = k$. Since $\alpha = 1$, there is at most one component of Y_0 on which η is nonconstant; thus

(iii) Y_0 *is connected*: it consists of a tree with one irreducible component Z mapping isomorphically to L, plus chains of rational curves joining Z to the points $q_1, \ldots, q_k$ on Y_1.

Note that since the k points $q_1, \ldots, q_k$ do not map to points p_i, the points of Y_1 that do map to points p_i are points of intersection of Y_1 with the sections Γ_i; in particular, since each Γ_i occurs with multiplicity one in $\eta^* L$, we see that whenever C_1 meets L at a point p_i, it does so transversely. Finally, from equality in the last equation we deduce further that

(iv) Y_1 *is smooth of genus* $g + 1 - k$, and its image is a nodal curve C_1 of degree $d - 1$ (thus having $\delta - d + k + 1$ nodes), meeting L transversely at a subset of l of the points p_i and at k further points $r_i = \eta(q_i)$ that are smooth points of C_1. It follows then that

(v) *the stable reduction of X_0 is the union $\tilde{X}_0$ of Y_1 and $Z \approx \mathbf{P}^1$, joined at the k points* $q_1, \ldots, q_k$ (or the curve we get from this by contracting Z if $k = 1$ or 2).

In sum, the picture of the degeneration is this: as C_λ tends to $C_0 = L \cup C_1$, we see a node tending to each of the l points p_i that C_1 passes through, and $m_i - 1$ nodes tending to the point r_i of intersection of C_1 with L away from the points p_i, where m_i is the order of contact of C_1 with L at r_i. Note that l may be zero—i.e., C_1 may miss all of the points p_i—but k cannot be; C_1 must contain at least one point of L other than the p_i. Also, observe that there is only one other restriction on the numbers k and l: l must be at least δ. As for the multiplicities m_i of intersection of C_1 with L at the points r_i, they have to satisfy only the restriction that $\sum(m_i - 1) \leq \delta - l$. The components of the variety Ω_L are thus classified by the sequences $(l, m_1, \ldots, m_k)$ satisfying these inequalities.

For example, consider a family (C_λ) of curves of degree 8 and genus 16 meeting L in fixed points $p_1, \ldots, p_8$, specializing to a curve C_0 containing L as above— that is, with C_0 a general point of a component of the locus of curves in the closure of $V_{8,16}$ containing L. The limiting curve will then consist of the line L plus an irreducible nodal septic C_1 meeting L at smooth points. One possibility, by what we have seen, is that C_1 will pass through p_1 but none of the points $p_2, \ldots, p_8$, and will meet L three times away from p_1, once transversely, once simply tangent, and once with contact of order 3. The picture of C_0 is as in Figure 12.

In this case, the central fiber C_0 of the family $\pi \colon \mathcal{C} \to B$ constructed above will look like Figure 13, where Y_1 is the normalization of C_1, all the components of Y_0 are rational, and the numbers next to components of Y_0 are the multiplicities with which each appears in M; these are dictated by the requirements that η be constant on all but one component, so that the degree of $\eta^* L = \sum \Gamma_i + M$ on all other components of Y_0 is zero, and that M meet Y_1 with multiplicities $1, 2$,

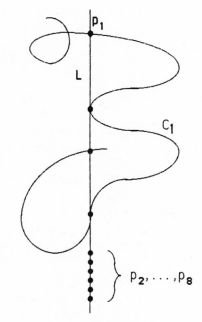

FIGURE 12

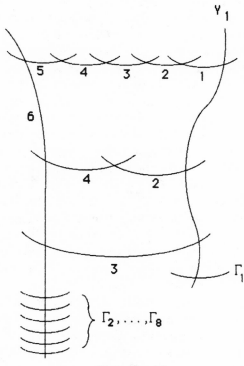

FIGURE 13

and 3. The stable reduction of C_0 thus is the union of the curve Y_1 with a copy of $\mathbf{P}^1$ meeting it at the three points q_1, q_2, and q_3 (Figure 14), or in other words the partial normalization of the curve C_0 in which the two branches at p_1 are separated, and the remaining singularities of C_0 reduced to nodes.

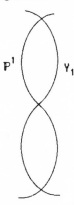

$\mathbf{P}^1$ Y_1

FIGURE 14

What we would actually see, if we carried out this experiment, is one of the five nodes of the general member C_λ of our family tending to the point p_1, one tending to the point $\eta(q_2)$ where C_1 is simply tangent to L, two tending to the point $\eta(q_3)$ where C_1 has contact of order 3 with L, and of course one staying away from L and arriving at the (unique) node of C_1.

To conclude our argument, we observe that in the general case the family $B = \{C_\lambda\}$ does indeed satisfy the conditions (whether $k > 1$ and the stable limit $\tilde{X}_0$ of the curves X_λ is singular, or $k = 1$ and X_0 is smooth) of our third reduction, and we are done.

Finally, we should remark that there is nothing really special about the family of curves $B = \{C_\lambda\}$ used in this proof. The proof in [27], for example, looks at families of curves having high order of contact with a line at one (unspecified) point; the proof in [42] looks at curves specializing to one with a multiple point. There is a basic idea common to all three: that to understand what happens to a family of curves in projective space, we have to look both at the family in the projective space, and in moduli. It is this idea, more than anything else, that this argument is intended to illustrate.

Appendix

Computing Divisor Classes in $\overline{\mathcal{M}}_g$

In this appendix we would like to illustrate by example how the classes of divisors in the moduli space of stable curves may be computed. The general situation is that we are given a divisor D in moduli space characterized by some geometric property of the curves parametrized—consisting, for example, of curves with a linear system of certain degree and dimension, or possessing a special Weierstrass point. Of course, we know that the Picard group of $\overline{\mathcal{M}}_g$ is generated by the classes λ of the Hodge bundle, and $\delta_0, \delta_1, \delta_2, \ldots$ of the components of the boundary. We thus know that there exists a relation

$$(1) \qquad\qquad D \sim a\lambda + b_0\delta_0 + b_1\delta_1 + \cdots$$

and the problem is to determine the coefficients $a, b_0, b_1, \ldots$, etc.

We will consider here a very concrete, relatively low-dimensional example of such a problem: in the moduli space of curves of genus three, the locus of hyperelliptic curves is a divisor, and we will compute its class. While this particular computation has no consequences by itself, most of the techniques that are commonly used to solve such problems in general can be seen clearly in this case.

One word of warning before we start: all the computations in §§1, 2, and 3 below will be made not on the moduli space but on the *moduli functor*. What this means is this: to an arbitrary family of stable curves $\pi \colon \mathfrak{X} \to B$ we can associate divisor classes λ and δ_i and D, where λ is the chern class of the Hodge bundle $\pi_*\omega_{\mathfrak{X}/B}$ and δ_i and D are the classes of the loci in B of points whose fibers are singular or hyperelliptic, counted with the appropriate multiplicity (as described in [**5**, vol. 2] or below), at least in case B is not contained in any of these loci. A relation (1) then holds among these divisors, and the numbers a and b_i we compute are the coefficients of this relation. (We say that λ, δ_i, and D are elements of the *Picard group of the moduli functor*, and the relation (1) holds in this group; cf. [**50**] for a beautiful discussion of moduli functors.) This is not the same thing as computing a relation among the corresponding divisors in the *space* $\overline{\mathcal{M}}_3$, as we will indicate in §4; but in many ways the easiest thing to do is to make the computation first in the setting of the moduli functor and then to translate the result into terms of the moduli space. This is what we will do; we warn the reader now, however, that until we get to §4 and clean up the mess, no further mention will be made of this.

1. The direct approach. The most natural thing to do to compute the class of the divisor D in $\overline{\mathcal{M}}_3$ would be to characterize hyperelliptic curves by some geometric property and to use this characterization to determine the class of D. There are many ways to characterize hyperelliptic curves, of course, but the best one for our present purposes is to say that a curve C of genus 3 is hyperelliptic if it possesses a hyperelliptic Weierstrass point—that is, a point $p \in C$ such that $h^0(C, \mathcal{O}(2p)) = 2$; or, equivalently (and more usefully for us), such that $h^0(C, \omega_C(-2p)) = 2$. (Note that this characterization is not valid when C is reducible, so we already have to restrict ourselves to families of irreducible curves.)

Suppose now we have a family $\pi \colon \mathfrak{X} \to B$ of curves of genus 3; for convenience we will denote both the relative dualizing sheaf $\omega_{\mathfrak{X}/B}$ of the family and its first chern class simply by ω. To locate the hyperelliptic Weierstrass points of the fibers of π, we introduce two vector bundles on $\mathfrak{X}$. First, we have the pullback of the Hodge bundle:

$$E = \pi^*\pi_*\omega,$$

the bundle of rank 3 whose fiber at a point p of $\mathfrak{X}$ is just the space of regular differentials on the fiber $\mathfrak{X}_{\pi(p)}$ of $\mathfrak{X}$ through p. Secondly, we have the bundle F whose fiber at a point p of $\mathfrak{X}$ will be the space of differentials on $\mathfrak{X}_{\pi(p)}$ modulo those vanishing to order 2 at p. To define F precisely, let Δ be the diagonal in the fiber product of $\mathfrak{X}$ with itself over B, and let π_1 and π_2 be the projection maps from $\mathfrak{X} \times_B \mathfrak{X}$ to $\mathfrak{X}$. We then set

$$F = \pi_{2*}((\pi_1^*\omega) \otimes (\mathcal{O}/\mathcal{O}(-2\Delta))).$$

Note that this does not do what we want it to at the points $p \in \mathfrak{X}$ that are singular points of the fibers. Thus, unless we are willing to deal in the sequel with coherent, rather than locally free sheaves, we are forced to restrict ourselves further to families of smooth curves.

We have now a natural vector bundle map $\varphi \colon E \to F$ obtained simply by evaluating regular forms on the fibers of π (or, less simply but more accurately, by pushing forward from $\mathfrak{X} \times_B \mathfrak{X}$ to $\mathfrak{X}$ the quotient map from $\pi_1^*\omega$ to $(\pi_1^*\omega) \otimes (\mathcal{O}/\mathcal{O}(-2\Delta)))$; and, visibly, a point p of $\mathfrak{X}$ will be a hyperelliptic Weierstrass point if and only if the rank of this map at p is 1. By general vector bundle theory, the locus where a map φ from a vector bundle of rank 3 to one of rank 2 has rank 1 will have codimension at most 2 if it is nonempty; and if it does have codimension 2 the class of the locus where it has rank 1 will be given by Porteous's formula:

$$(2) \qquad [\{p\colon \operatorname{rank} \varphi_p \le 1\}] = [c(E^* - F^*)]_2,$$

where $c(E^* - F^*)$ denotes the chern class $c(E^*)/c(F^*)$ of the virtual bundle $E^* - F^*$.

We can compute the right-hand side of (2) readily enough: to begin with, E^* being just the dual of the Hodge bundle, its chern class begins

$$(3) \qquad c(E^*) = 1 - \lambda + \cdots$$

(we will use the symbol λ to denote both the first chern class of the Hodge bundle on B and its pullback to $\mathfrak{X}$); the remaining terms will not affect our result. Secondly, we see that the bundle F naturally surjects onto the line bundle ω by evaluation, and that the kernel, which is the line bundle whose fiber at a point $p \in \mathfrak{X}$ is the space of differentials on $\mathfrak{X}_{\pi(p)}$ vanishing at p modulo those vanishing to order 2 at p, is isomorphic to $\omega^{\otimes 2}$—this is just the push-forward to $\mathfrak{X}$ of the tensor product of $\pi_1^*\omega$ with the exact sequence

$$0 \to \mathcal{O}(-\Delta)/\mathcal{O}(-2\Delta) \to \mathcal{O}/\mathcal{O}(-2\Delta) \to \mathcal{O}/\mathcal{O}(-\Delta) \to 0$$

on $\mathfrak{X} \times_B \mathfrak{X}$. From the exact sequence

$$0 \to \omega^2 \to F \to \omega \to 0$$

we see that

$$c(F) = (1+\omega)(1+2\omega) = 1 + 3\omega + 2\omega^2$$

and hence

$$(4) \qquad c(-F^*) = (1 - 3\omega + 2\omega^2)^{-1} = 1 + 3\omega + 7\omega^2.$$

Multiplying (3) and (4), we have

$$[c(E^* - F^*)]_2 = -3\omega\lambda + 7\omega^2.$$

This tells us the class in $\mathfrak{X}$ of the locus of hyperelliptic Weierstrass points; since this locus covers the divisor D of hyperelliptic curves 8 times, the class of D will just be one-eighth of the push forward to B of this class. To evaluate this, note that λ is the pullback to $\mathfrak{X}$ of a divisor class of B, and ω is a divisor class of degree 4 over B, so

$$\pi_*(-3\omega\lambda) = -12\lambda.$$

On the other hand, by the standard formula derived in the first lecture, $\pi_*\omega^2 = 12\lambda$; so altogether

$$\pi_*([c(E^* - F^*)]_2) = 72\lambda$$

and we have $D \sim 9\lambda$ in the moduli functor $\mathcal{M}_3$ of smooth curves of genus 3. Since $\lambda \neq 0$ and the δ_i are of course zero in $\mathcal{M}_3$, we deduce from this that the coefficient a in the relation (1) is equal to 9.

As indicated, we run into problems of two sorts when we try to extend this sort of computation over all of $\overline{\mathcal{M}}_3$: first, our characterization of hyperelliptic Weierstrass points fails; and secondly, the bundle F used extends to a coherent sheaf, but not locally a free one, in families of stable curves. The first of these problems is one that can be dealt with, for example by using the theory of limit linear series (discussed below) or admissible covers to characterize Weierstrass points on singular curves (cf. [**18**, **19**, **29**, or **36**, and **46**]); indeed, F. Cukierman has used this approach to determine the class of the divisor of all Weierstrass points in the moduli space $\overline{\mathcal{C}}_g$ of stable pointed curves (cf. [**7**]). The second problem, however, is much deeper; a good calculus of degeneracy loci for maps between coherent sheaves would be extremely useful but also seems at present largely intractable.

2. The first indirect approach. Given that we cannot simply calculate the class of D directly, there is a standard alternate approach to try. We know that a relation of the form (1) exists; to obtain information about its coefficients we can take any curve we like in $\overline{\mathcal{M}}_3$ and intersect both sides of (1) with it. This gives a linear relation among the coefficients; and with enough such relations we may hope to determine them all.

For example, consider a pencil of plane quartic curves, given by a polynomial $F_\mu(X_0, X_1, X_2)$ of degree four in X varying linearly with a parameter $\mu = [\mu_0, \mu_1] \in \mathbf{P}^1$—i.e., a bihomogeneous polynomial of bidegree (1,4). We think of this as a one-parameter family of curves of (arithmetic) genus three; since a general such pencil will contain only smooth curves and irreducible nodal ones, all the members of the family will in particular be stable, so we actually get a map from $\mathbf{P}^1$ to $\overline{\mathcal{M}}_3$.

We ask now for the degrees of the various divisors $\lambda, \delta_0, \delta_1$, and D on this curve $\mathbf{P}^1 \subset \overline{\mathcal{M}}_3$. Two of these are immediate: as we said, every curve C_μ in the pencil will be irreducible with at most nodes; in particular none will lie in Δ_1 so the degree of δ_1 on our family is zero. Likewise, the dualizing sheaf $\omega_{C_\mu} = \mathcal{O}_{C_\mu}(1)$ is very ample on each element of our family, so none are in the (closure of the) locus of hyperelliptic curves; thus the degree of D in our family is also zero.

To find the degree of δ_0 we ask how many elements of our pencil are singular. This can be done in many ways; probably the simplest is to say that the curve C_μ will be singular if the three partial derivatives $F_i = \partial F / \partial X_i$ have a common zero in X. The singular elements of the pencil thus correspond to intersections of the three divisors $F_i(\mu, X) = 0$ in the product $\mathbf{P}^1 \times \mathbf{P}^2$; since F_i is bihomogeneous of bidegree $(1,3)$ in μ and X, if we denote by η_1 and η_2 the pullbacks to $\mathbf{P}^1 \times \mathbf{P}^2$ of the generators of $H^2(\mathbf{P}^1, \mathbf{Z})$ and $H^2(\mathbf{P}^2, \mathbf{Z})$ respectively, the intersection number is given by the product

$$(\eta_1 + 3\eta_2)^3 = 3 \cdot \eta_1 \cdot (3\eta_2)^2 = 27.$$

Thus the degree of δ_0 in our family is 27. Of course, we have to check both that our curve $\mathbf{P}^1 \subset \overline{\mathcal{M}}_3$ intersects Δ_0 transversely, and that the divisors $F_i = 0$ intersect transversely in $\mathbf{P}^1 \times \mathbf{P}^2$ (or, more generally, that the intersection multiplicity of $\mathbf{P}^1$ with Δ_0 at a point corresponding to an irreducible curve C_μ is equal to the sum of the intersection multiplicities of the divisors $F_i = 0$ at the points (μ, X) corresponding to the singularities of C_μ); but we will omit such verifications here.

It remains to calculate the degree of the Hodge bundle on our family. Again, there are many ways to approach this; probably the most natural way to do this is to actually determine the Hodge bundle by writing down sections. To do this, let x and y be euclidean coordinates on $\mathbf{P}^2$, μ a euclidean coordinate on $\mathbf{P}^1$, and $f_\mu(x, y)$ the equation of our pencil of curves. For any $\mu \neq \infty$, we can write down a basis for $H^0(C_\mu, \omega_{C_\mu})$ by setting

$$\varphi = \frac{dx}{df_\mu / dy}$$

and taking as our basis the forms $\varphi_0 = \varphi$, $\varphi_1 = x\varphi$, and $\varphi_2 = y\varphi$. Since this is true for all finite μ, viewing the φ_i as sections of the Hodge bundle E of our family of curves we obtain a frame for E over the open set $(\mu \neq \infty)$ in $\mathbf{P}^1$. At the point $\mu = \infty$, moreover, we see that each of the sections φ_i of E has a simple zero (since the coefficients of f will in general have simple poles in μ); but after multiplying each by $1/\mu$ we again get a frame for the Hodge bundle in the open set $\mu \neq 0$. We thus have an identification

$$E \cong \mathcal{O}_{\mathbf{P}^1}(1)^{\otimes 3}$$

and in particular we see that the degree of the Hodge bundle is 3.

Now, restricting to our family $\mathbf{P}^1 \subset \overline{\mathcal{M}}_3$ the relation (1), we obtain

$$3a - 27b_0 = 0;$$

and since we already have $a = 9$ we conclude that $b_0 = 1$.

We could continue with this approach and finish the computation of the class of D in $\overline{\mathcal{M}}_3$ without difficulty. All we need is another family of stable curves of genus three whose intersection numbers with the various divisors in (1) we can compute, and that has nonzero intersection with the boundary component Δ_1; for example, we could take a generic pencil of plane quartics including a quartic with a cusp, and make a base change to arrive at a family meeting Δ_1. If we try to take the same approach to problems involving the moduli of curves of higher genus, however, we quickly run into trouble, for a simple reason: it is extremely difficult to write down "general" one-parameter families of curves of genus g when g becomes large. In effect, the ready availability of families of curves of genus 3 is a reflection of the fact that $\overline{\mathcal{M}}_3$ is unirational; for higher g about the only curve passing through a general point of $\overline{\mathcal{M}}_g$ whose numerical invariants have been calculated is the one worked out by Harris and Morrison in [**28**].

One possible solution to this problem would be to look for families among the more special curves of genus g (for example, the locus of hyperelliptic curves of genus g is unirational for any g, and it is correspondingly easy to write down explicitly one-parameter families of these) but it very often happens that these loci do not intersect properly with the divisors whose class we are trying to determine. (For example, in the computation above we could have tried to use the family of hyperelliptic curves $y_2 = f_\mu(x)$ where f is linear in μ and of degree 8 in x; but it would have been difficult then to evaluate the intersection number of this curve in $\overline{\mathcal{M}}_3$ with the divisor D of hyperelliptic curves.) This is not to say that this approach is ruled out (to carry our example one step further, we could describe the normal bundle to the hyperelliptic locus and calculate the intersection number of our family with D that way) but another idea will help.

3. The second indirect approach. One solution to the problem posed by the lack of curves through the interior of $\overline{\mathcal{M}}_g$—by far the most commonly used one to date—is to look at curves in the boundary of $\overline{\mathcal{M}}_g$, that is, one-parameter families of singular stable curves. There is certainly no shortage of these: we could take a fixed curve C of genus $g-1$ and identify two variable points p and q on C, for example; or take fixed curves C_1 and C_2 of genera i and $g-i$ and attach them at variable points $p \in C_1$, $q \in C_2$, and so on. The difficulties here lie in evaluating the degrees of the various divisors in $\overline{\mathcal{M}}_g$ on our curves, and are twofold. First, since our curve in $\overline{\mathcal{M}}_g$ will lie in one or more of the components of the boundary of $\overline{\mathcal{M}}_g$, it will obviously fail to meet some of them properly. This is not a major problem: all we need to determine the degrees of the δ_i on our curve is a little deformation theory to enable us to describe the normal bundle of the boundary component Δ_i in $\overline{\mathcal{M}}_g$, as illustrated below. Secondly, in order to describe the intersection of our curve in $\overline{\mathcal{M}}_g$ with a divisor D given as the closure in $\overline{\mathcal{M}}_g$ of the locus of smooth curves with some linear series, Weierstrass point, vanishing theta-null, or whatever, *we have to be able to say exactly which stable* *curves are limits of smooth curves with such a linear series.* This question has been answered in the case of pencils by the theory of *admissible covers*, and for arbitrary linear series when the stable curve in question has compact Jacobian (i.e., its dual graph is a tree) by the theory of *limit linear series* discussed in the second lecture. At least one of these is thus a necessary ingredient in any computation at the boundary of $\overline{M}_g$, and we will assume for the following that the reader is familiar with the discussion of limit linear series in the second lecture.

Back to $\overline{M}_3$. As indicated, we have a lot of options in choosing a curve contained in the boundary of $\overline{M}_3$; probably the simplest is just to take a fixed elliptic curve E with fixed point $p \in E$, and attach to it a fixed curve B of genus 2 by identifying p with a variable point $q \in B$—that is, take the family

$$\{C_q = B \cup E/q \sim p\}_{q \in B}\,.$$

To construct explicitly the corresponding family of curves, we just take the products $B \times B$ and $B \times E$ and identify the diagonal $\Delta \subset B \times B$ with the section $B \times \{p\} \subset B \times E$ to obtain a family

$$\pi = \pi_1 \colon \mathfrak{X} \to B.$$

As usual, in calculating the degrees of the divisors λ, δ_i, and D on the curve $B \subset \overline{M}_3$, we start with the easy ones. First, there are visibly no nonseparating nodes on any of the curves C_q, so the degree of δ_0 on our family is zero. Secondly, the space of sections of the dualizing sheaf of any of the curves C_q is naturally just the direct sum of the spaces of regular forms on B and E —that is,

$$\pi_*(\omega_{\mathfrak{X}/B}) = (H^0(E, \omega_E) \oplus H^0(B, \omega_B)) \otimes \mathcal{O}_B$$

—and in particular the degree of λ on B is zero.

Next, to evaluate the degree of δ_1 on B, we observe that since B is contained in the smooth locus of Δ_1 the restriction to B of the divisor class δ_1 is just the class of the normal bundle of Δ_1 in $\overline{M}_3$ restricted to B. (We are working on the moduli functor, which is smooth.) Now, the cotangent space to $\overline{M}_3$ at a point C is naturally identified with the space

$$H^0(C, \omega_C \otimes \Omega_C),$$

where ω_C is the dualizing sheaf of C and Ω_C the cotangent sheaf; in these terms the conormal space to Δ_1 in $\overline{M}_3$ is naturally identified with the one-dimensional vector space of torsion sections of $\omega_C \otimes \Omega_C$, and this may in turn be identified with the tensor product of the cotangent spaces to the branches of C at its node. We thus have an identification

$$N_{\Delta_1/\overline{M}_3}(C_q) = T_p E \otimes T_q B$$

that globalizes to an isomorphism of bundles on B:

$$N_{\Delta_1/\overline{M}_3} \otimes \mathcal{O}_B \cong N_{B \times \{p\}/B \times E} \otimes N_{\Delta/B \times B}.$$

Of course, the first factor on the right-hand side above is trivial; the second factor, on the other hand, is isomorphic to the tangent bundle of B and in particular has degree -2. We thus have $\deg_B(\delta_1) = -2$.

It remains to determine the degree of the divisor D on B, and, as we said, in order to do this we need to know when a curve C_q is a limit of smooth hyperelliptic curves. We claim that in fact *a curve C_q is a limit of smooth hyperelliptic curves if and only if the point $q \in B$ is a Weierstrass point of B.*

To see this, we invoke the theory of limit linear series, though here the theory of admissible covers would be equally effective. The theory of limit linear series tells us that C_q will be in D only if there are

(i) a g_2^1 on B with vanishing sequence (a_0, a_1) at q, and

(ii) a g_2^1 on E with vanishing sequence (b_0, b_1) at p for some a_0, a_1, b_0, b_1 satisfying $a_0 + b_1 \geq 2$ and $a_1 + b_0 \geq 2$. Our claim (in one direction, at any rate) follows immediately: since neither E nor B are rational, a g_2^1 on either cannot have a base point, so we must have a_0 and b_0 both 0; we must then have $a_1 = b_1 = 2$, and in particular the point q must be a branch point of the (unique) g_2^1 on B. Finally, since the locus of curves satisfying this condition has the (correct) codimension 1 in Δ_1, the basic theorems of [**17**] allow us to deduce that any such curve is indeed the limit of smooth hyperelliptics.

We see from this that the curve $B \subset \overline{\mathcal{M}}_3$ meets D in exactly six points, corresponding to the Weierstrass points of B. In fact it does so transversely, a fact whose checking would require a more detailed account of either admissible covers or limit linear series than we want to go into here. Given this, we have that the degree of D on the curve B is 6, and we deduce that in the relation (1) we have $b_1 = 3$. Putting together our three computations, we have the relation

$$(5) \qquad\qquad D \sim 9\lambda - \delta_0 - 3\delta_1$$

on the moduli functor $\overline{\mathcal{M}}_3$.

4. From the moduli functor to the moduli space. As we promised, we now want to take the relation (5) and translate it into a relation among divisors on the moduli space $\overline{\mathcal{M}}_3$. To begin with, let us say what these divisor classes are. First, D, Δ_0, and Δ_1 are loci in the space $\overline{\mathcal{M}}_3$, and we will denote by $[D]$, $[\Delta_0]$, and $[\Delta_1]$ their classes in $\mathrm{Pic}(\overline{\mathcal{M}}_3)$. The divisor class λ is not defined on $\overline{\mathcal{M}}_3$ per se, but may be defined as an element of $\mathrm{Pic}(\overline{\mathcal{M}}_3) \otimes \mathbf{Q}$ by going to a branched cover $\varphi \colon \mathcal{N} \to \overline{\mathcal{M}}_3$ over which there is a universal curve (at least in codimension 1), such as the moduli of curves with level n structure, taking the chern class of the Hodge bundle on $\mathcal{N}$, pushing it forward via φ, and dividing by the degree of the covering φ; we will denote the resulting class by $[\lambda]$ to distinguish it from the class λ on the moduli functor.

Now suppose we have a family of stable curves $\pi \colon \mathcal{X} \to B$ over a one-dimensional base B, general in the sense that any singular fibers of π are general members of their boundary components Δ_i and that the curve B is transverse to the discriminant locus in the versal deformation space of each singular fiber.

We have then a map $\psi\colon B \to \overline{\mathcal{M}}_3$; denote the image by $\overline{B}$. We know the relation (5) holds among the divisor classes λ, δ_i, and D on B; we want a relation among the intersections of $\overline{B}$ with the divisor classes $[\lambda]$, $[\delta_i]$, and $[D]$.

To begin with, as indicated above, the degree of $[\lambda]$ on $\overline{B}$ is the degree of λ on B. Secondly, a general point of the boundary component Δ_0 corresponds to a curve C with no automorphisms, so the versal deformation space of C maps isomorphically to $\overline{\mathcal{M}}_3$; thus the curve $\overline{B}$ will be transverse to Δ_0. On the other hand, a general member C of Δ_1 is a curve with an automorphism group of order 2 (if $C = E \cup B/p \sim q$, the involution on the elliptic curve E fixing p gives an automorphism of C), and so the versal deformation space of C maps 2-to-1 to $\overline{\mathcal{M}}_3$, the branch divisor of the covering being just the discriminant locus in the deformation space. If B is transverse to the discriminant locus, then the curve $\overline{B}$ will be simply tangent to Δ_1 at C (note that by this description $\overline{\mathcal{M}}_3$ is smooth at C in this case). It follows that the intersection number of $[\Delta_1]$ with $\overline{B}$ will be one-half the degree of the divisor class δ_1 on B. A similar argument shows that, since the general member of the divisor D is a curve with an automorphism group of order 2, the intersection number of $\overline{B}$ with $[D]$ is half the degree of D on B.

Note that instead of starting with a one-parameter family of stable curves and taking the image in $\overline{\mathcal{M}}_3$, we could have started with a curve $\overline{B}$ in $\overline{\mathcal{M}}_3$, transverse to the loci D and Δ_i. In this case there would not exist a family of stable curves over $\overline{B}$, but there would exist one over a finite cover $\psi\colon B \to \overline{B}$; the cover ψ would of necessity be ramified over the points of intersection of $\overline{B}$ with D and Δ_i, and the computation would work out the same. Another way of saying all this would be to say that "the moduli functor maps to the moduli space by a map that has degree one and is ramified over the divisors D and Δ_i," but that might be confusing.

However one wants to view the relation between the moduli space and functor. We see from the above that if the relation (5) holds on the curve B, the corresponding relations hold among the intersection numbers of the corresponding divisor classes on $\overline{\mathcal{M}}_3$ with $\overline{B}$ after dividing the $[D]$ and $[\Delta_1]$ terms by 2; we thus have in $\mathrm{Pic}(\overline{\mathcal{M}}_3)$ the relation

$$[D] \sim 18[\lambda] - 2[\Delta_0] - 3[\Delta_1].$$

References

1. E. Arbarello and M. Cornalba, *Su una congettura di Petri*, Comment. Math. Helv. **56**(1981), 1–38.

2. _____, *The Picard groups of the moduli spaces of curves*, Topology (to appear).

3. _____, *Footnotes to a paper of Beniamino Serge*, Math. Ann. **256** (1981), 341–362.

4. _____, *A few remarks about the variety of irreducible plane curves of given degree and genus*, Ann. Sci. École Norm. Sup.(4) **16** (1983), 467–488.

5. E. Arbarello, M. Cornalba, P. Griffiths, and J. Harris, *Geometry of algebraic curves*, Springer-Verlag, 1984.

6. G. Castelnuovo, *Numero delle involuzioni razionali giacenti sopra una curva di dato genere*, Rend. R. Accad. Lincei **5** (1889).

7. M. Chang and Z. Ran, *Unirationality of the moduli spaces of curves of genus* 11, 13 (*and* 12), Invent. Math. **76** (1984),

8. ____, *The Kodaira dimension of the moduli space of curves of genus* 15, J. Differential Geom. (to appear).

9. M. Cornalba and J. Harris, *An inequality for families of stable varieties*, (in preparation).

10. F. Cukierman, Ph.D. thesis, Brown University, Providence, R.I., 1987.

11. S. Diaz, *A bound on the dimensions of complete subvarieties of* M_g, Duke Math J. **51** (1984), 405–408.

12. ____, *Exceptional Weierstrass points and the divisor on moduli space that they define*, Mem. Amer. Math. Soc. No. 327 (1985).

13. ____, *Tangent spaces to moduli via deformations, with applications to Weierstrass points*, Duke Math J. **51** (1984), 905–922.

14. S. Diaz and J. Harris, *Ideals associated to deformations of singular plane curves*, Preprint.

15. D. Eisenbud and J. Harris, *The Kodaira dimension of the moduli space of curves of genus* ≥ 23, Invent. Math. (to appear).

16. ____, *Divisors on general curves and cuspidal rational curves*, Invent. Math. **74** (1983), 317–418.

17. ____, *Limit linear series: Basic theory*, Invent. Math. (to appear).

18. ____, *Existence, decomposition and limits of some Weierstrass points*, Invent. Math. (to appear).

19. ____, *The monodromy of Weierstrass points*, Invent. Math. (to appear).

20. ____, *Irreducibility and monodromy of some families of linear series*, Preprint.

21. W. Fulton and R. Lazarsfeld, *On the connectedness of degeneracy loci and special divisors*, Acta Math. **146** (1981), 271–283.

22. D. Gieseker, *Stable curves and special divisors*, Invent. Math. **66** (1982), 251–275.

23. P. Griffiths and J. Harris, *On the variety of special linear systems on a general curve*, Duke Math J. **47** (1980), 233–272.

24. J. Harer, *The second homology group of the mapping class group of an orientable surface*, Invent. Math. **72** (1983), 221–239.

25. ____, *Stability of the homology of the mapping class groups of orientable surfaces*, Ann. of Math. (2) **121** (1985).

$\sqrt{}$26. J. Harer and D. Zagier, *The Euler characteristic of the moduli space of curves*, Preprint.

27. J. Harris, *On the Severi problem*, Invent. Math. (to appear).

28. J. Harris and I. Morrison, *Slopes of effective divisors on moduli*, Preprint.

29. J. Harris and D. Mumford, *On the Kodaira dimension of the moduli space of curves*, Invent. Math. **67** (1982), 23–88.

30. G. Kempf, *Schubert methods with an application to algebraic curves*, Publ. Math. Centrum, Amsterdam, 1971.

31. S. Kleiman, *r-special subschemes and an argument of Severi's*, Adv. in Math. **22** (1976), 1–31.

32. S. Kleiman and D. Laksov, *On the existence of special divisors*, Amer. J. Math. **94** (1972), 431–436.

33. ____, *Another proof of the existence of special divisors*, Acta Math. **132** (1974), 163–176.

34. W. Kleinert, *Families of curves and line bundles on their moduli spaces*, Preprint, Humboldt Univ., East Berlin.

35. F. Knudsen, *The projectivity of the moduli spaces of curves*, Math Scand. **52** (1983).

36. R. Lax, *Weierstrass points on rational nodal curves*, Preprint.

37. R. Lazarsfeld, *Brill-Noether-Petri without degenerations*, Preprint.

38. E. Miller, *The homology of the moduli spaces and the mapping class group*, Preprint.

39. D. Mumford, *Stability of projective varieties*, Enseign. Math. (2) **23** (1977), 39–110.

40. ____, *Curves and their Jacobians*, Univ. of Mich. Press, Ann Arbor, Mich., 1976.

41. M. Nori, personal communication.

42. Z. Ran, *On nodal plane curves*, Invent. Math. (to appear).

43. E. Sernesi, *L'unirazionalità dei moduli delle curve di genere dodici*, Ann. Scuola Norm. Sup. Pisa Cl. Sci (4) **8** (1981), 405–439.

44. F. Severi, *Vorlesungen über algebraische Geometrie*, Teubner, Leipzig, 1921.

45. J. Wahl, *Equisingular deformations of plane algebroid curves*, Trans. Amer. Math. Soc. **193** (1974), 143–170.

46. C. Widland, *On Weierstrass points on Gorenstein curves*, Ph.D. thesis, Louisiana State University, 1984.

47. O. Zariski, *Algebraic systems of plane curves*, Amer. J. Math. **104** (1982).

48. D. Eisenbud and J. Harris, *A simpler proof of the Gieseker-Petri theorem on special divisors*, Invent. Math. **74** (1983), 269–280.

49. J. Harris, *The genus of space curves*, Math. Ann. **249** (1980), 191–204.

50. D. Mumford, *Picard groups of moduli problems*, Arithmetic Algebraic Geometry, O.F.G. Schilling, editor, Harper and Row, New York, 1965, pp. 33–81.

BROWN UNIVERSITY

Proceedings of Symposia in Pure Mathematics
Volume **46** (1987)

On the Classification of Algebraic Space Curves, II

ROBIN HARTSHORNE

0. Introduction. We consider irreducible nonsingular curves C in $\mathbf{P}_k^3$, the projective 3-space over an algebraically closed field k. If d and g denote the degree and genus of C, respectively, then one knows that d and g are constant in flat families, so it is natural to separate curves into classes according to these two invariants. The classification problem then falls into three parts:

(a) For which pairs of integers (d, g) do there exist such curves?

(b) For each d, g, let $H_{d,g}$ denote the set of all irreducible nonsingular curves in $\mathbf{P}^3$ with the given degree and genus. Give this set a natural algebraic structure, and discuss its properties: irreducible components, dimension, singular points, ...

(c) Discuss the properties of each particular curve C of degree d and genus g: least degree of a surface containing it, postulation, normal bundle, existence of special linear series, etc.

These questions existed and were already studied extensively in the late 19th century, especially in works of Halphen [**20**] and Noether [**40**], who shared the Steiner Prize in 1882. Then for a long time there was not much work on curves until recently when various statements which had long been accepted as true were brought into question again and provided with modern proofs. Since questions about abstract curves (existence of special linear systems, global character of the moduli space $\mathcal{M}_g$) are discussed in Joe Harris's lectures, we direct our attention here to curves embedded in projective space.

There are several areas of significant progress since the Arcata conference eleven years ago.

One is the theorem of Gruson and Peskine which determines the possible degree and genus of a curve in $\mathbf{P}^3$. This question was already answered in Halphen's paper, but in trying to understand that work, Gruson and Peskine found that his proof was incorrect. Their method is quite different from Halphen's, and it rests on some beautiful applications of the theory of integral quadratic forms. We give their result and some ideas of its proof in §1.

1980 *Mathematics Subject Classification* (1985 *Revision*). Primary 14H50; Secondary 14F05, 14H10, 14H45.

In any serious study of space curves it becomes necessary to consider the least degree s of a surface containing a given curve. Another important invariant is the largest integer e for which $H^1(\mathcal{O}_C(e)) \neq 0$. These integers behave in a semicontinuous manner in a flat family, so they may jump within an irreducible component of $H_{d,g}$. In §2 we study the relationship between the invariants d, g, s, e by focusing our attention on the problem of determining the maximum possible genus g of a curve of degree d not contained in a surface of degree $< k$, for any integer $k > 0$. A major contribution was made in an earlier paper of Gruson and Peskine [17], where they solved this problem, provided $d > k(k-1)$ (later improved to be $d \geq (k-1)^2 + 1$ [19]). For smaller d, the problem is more complicated, and we can only give a conjectural answer and some partial results.

An important technique for studying these questions, which is entirely new since Arcata, is by means of vector bundles, and more generally reflexive sheaves, on the projective space. It turns out that the difficult "range B" of the maximum genus problem discussed in §2 corresponds exactly to the theory of stable rank 2 reflexive sheaves on $\mathbf{P}^3$. So in §3 we explain briefly this correspondence, and how some recent results about reflexive sheaves give information about curves.

In §4 we survey some other kinds of recent work concerning space curves. There have been many papers discussing particular properties of curves or of the Hilbert scheme $H_{d,g}$ which parametrizes them. Some of these illustrate bad things which may happen, starting with Mumford's example of an irreducible component of $H_{14,24}$ which is generically nonreduced. Others speak of the existence of curves with nice properties such as having a stable normal bundle or being of maximal rank. It is clear from all this that the complete classification of space curves is extremely complicated.

To make some sense out of this chaos, I would like to suggest some principles with which to approach the classification problem. The first is to search for some nonempty open subset $U_{d,g} \subseteq H_{d,g}$ whose points correspond to "general" curves of degree d and genus g, and to regard the other curves as "special." The second is to try to explain bad behavior as resulting from the curve being contained in a surface of too low degree, either because it is "special" within $H_{d,g}$, or because all curves of $H_{d,g}$ are forced to lie on a surface of low degree. Turned around, this says "if a certain property P is not obviously contradicted by C lying on a surface of degree s, then it should hold on a nonempty open set of $H_{d,g}$."

These principles are necessarily vague in general, but in §4 we discuss some particular questions in this light, and use these principles to generate some explicit conjectures concerning the set of d, g for which certain properties should hold in general.

The title of this article alludes to an earlier article [24] in which I tried to describe what was known and not known about space curves in 1979. It is gratifying to note that most of the problems and conjectures stated there have been solved in the intervening six years. I would like to acknowledge my indebtedness to André Hirschowitz, who forced me to write that earlier article, and then provided a series of new ideas which have led to the solution of many of the problems stated there.

1. Degree and genus of curves in $\mathbf{P}^3$. In this section we will determine the set of pairs of integers (d, g) for which there exists a curve in $\mathbf{P}^3$. This result appears already in the work of Halphen [20], but his proof was incorrect. A new proof was recently given by Gruson and Peskine [18, 27].

THEOREM 1.1. *Let C be an irreducible nonsingular curve of degree d and genus g in $\mathbf{P}_k^3$ (k an algebraically closed field of arbitrary characteristic).*

(a) *If C is contained in a plane, then $g = \frac{1}{2}(d-1)(d-2)$. For any $d \geq 1$ there exist such curves.*

(b) *If C lies on an irreducible quadric surface Q, then there are integers $a, b \geq 0$ such that $d = a + b$ and $g = (a-1)(b-1)$. For any $a, b > 0$ there exist such curves on an irreducible nonsingular quadric surface.*

(c) *If C is not contained in a plane or a quadric surface, then*

$$0 \leq g \leq \tfrac{1}{6}d(d-3) + 1.$$

(d) *For every $d > 0$ and g satisfying the inequality of (c), there exists an irreducible nonsingular curve of degree d and genus g in $\mathbf{P}^3$.*

The complete proof of this theorem appears in the papers of Gruson and Peskine [17] and [18], and in the Séminaire Bourbaki talk [27], so here we will merely give a sketch of the main ideas of the proof.

Parts (a) and (b) of course are classical and elementary. The inequality (c) is due to Halphen, and proved by Gruson and Peskine [17]. The idea is to use the technique of Castelnuovo: cut the curve C with a general plane H and then study carefully the postulation of the finite set of points Z where C intersects H.

The most difficult part is the existence statement (d). This is also the place where Halphen's proof breaks down, because he tried to construct all such curves on cubic surfaces, and we know now that there are gaps, namely, values of (d, g) satisfying the hypotheses of (d), but for which there is no such curve on a cubic surface, not even on a singular cubic surface.

The existence is deduced as a consequence of the following two results.

PROPOSITION 1.2. *For any $d > 0$ and any g satisfying*

$$d^{3/2}/\sqrt{3} - d + 1 < g \leq \tfrac{1}{6}d(d-3) + 1,$$

there is an irreducible nonsingular curve C of degree d and genus g on a nonsingular cubic surface in $\mathbf{P}^3$.

PROPOSITION 1.3. *For any $d > 0$ and any g satisfying*

$$0 \leq g \leq \tfrac{1}{8}(d-1)^2,$$

there is an irreducible nonsingular curve C of degree d and genus g on a (singular) rational quartic surface in $\mathbf{P}^3$.

For the curves on the cubic surface X, we use the usual representation of X as a plane with six points blown up [**23**, V, §4]. Then Pic $X \cong \mathbf{Z}^7$, and we write any divisor class as $D = al - \sum b_i e_i$ with $a, b_1, \ldots, b_6 \in \mathbf{Z}$, where l is the total transform of a line in the plane, and $e_1, \ldots, e_6$ are the classes of the exceptional curves coming from the six blown-up points.

One knows that the divisor class $|D|$ contains an irreducible nonsingular curve if

$$a \geq b_1 + b_2 + b_3 \quad \text{and} \quad b_1 \geq b_2 \geq \cdots \geq b_6 \geq 0$$

(except for $(a; b_1, \ldots, b_6) = (n; n, 0, \ldots, 0)$ with $n > 1$). In this case the degree and genus of the curve are given by

$$d = 3a - \sum b_i, \qquad g = \frac{1}{2}\left(a^2 - \sum b_i^2 - d\right) + 1.$$

It is then a problem in number theory to find integers $(a; b_1, \ldots, b_6)$ satisfying the inequalities above, and giving all (d, g) in the statement of (1.2).

Without going into all the details, we mention only the key step, which is a result about sums of five squares of bounded integers.

LEMMA 1.4. *Let $k > 0$ be an integer. Then every positive integer $n < 3k^2 - 2k + 3$ is a sum of five squares $n = \sum_{i=1}^{5} x_i^2$ of integers x_i with $|x_i| \leq k$.*

The lemma is proved quite easily using the theorem of Gauss which says that a positive integer n is a sum of 3 squares if and only if it is not of the form $4^a(8b-1)$ for some $a, b \in \mathbf{Z}$. For example, the first step is the case $0 < n < (k+1)^2$. If n is a sum of 3 squares $n = x_1^2 + x_2^2 + x_3^2$, each $|x_i|$ is necessarily $\leq k$, so we are done. If n is not a sum of 3 squares, then $n = 4^a m$ with $m \equiv 7 \pmod 8$. So $m - 1$ is a sum of 3 squares, say $m - 1 = y_1^2 + y_2^2 + y_3^2$. Thus $n = 4^a(y_1^2 + y_2^2 + y_3^2 + 1)$ is a sum of 4 squares.

To prove Proposition 1.3, Gruson and Peskine consider the surface S obtained by blowing up 9 points in general position in $\mathbf{P}^2$. The quartic curves with a double point at P_1 and passing through $P_2, \ldots, P_9$ give a morphism of S to $\mathbf{P}^3$ whose image X is a rational quartic surface with a double line. The strategy is to construct the required curves on S in such a way that their images in $\mathbf{P}^3$ remain nonsingular.

Again we give only the main ideas of the proof. The Picard group Pic S is isomorphic to $\mathbf{Z}^{10}$ with basis $l, e_1, \ldots, e_9$, similar to the case of the cubic surface. The intersection form is diagonal of type $(1; -1, -1, \ldots, -1)$. The canonical class ω is given by $(-3; -1, \ldots, -1)$. We consider the group G of isometries of Pic S which leave ω fixed. The key lemma is the following.

LEMMA 1.5. *Let*

$$c = (1; 1, 0, \ldots, 0) \quad \text{and} \quad b = (0; 0, \ldots, 0, -1, -1)$$

in Pic S. Then for any $n \geq 0$, there exist $\sigma, \sigma' \in G$ such that $c \cdot \sigma(c) = n$ and $b \cdot \sigma'(c) = n$.

This lemma is proved by a careful study of the quadratic form on Pic S. Let $H = \omega^{\perp}$ be the orthogonal space of ω. Then $\omega \in H$ and $W = H/(\omega)$ is a negative definite even quadratic form of rank 8, isomorphic to $-\Gamma_8$ (in the notation of Serre's "Cours d'arithmétique" [50], for example). Then one uses the fact that Γ_8 represents all positive even integers. One also must use the theorem of Witt, applied to $W/2W$ with the quadratic form $\frac{1}{2}(x \cdot y) \pmod 2$, which says that the automorphisms of the quadratic form act transitively on isotropic elements. This enables one to show that the group G is big enough.

One further point is this: the group G acts only on Pic S, not on S. So to obtain curves, we apply G to get a divisor class with the right numerical properties, and then argue (under suitable conditions) that Bertini's theorem applies to show that this divisor class contains an irreducible nonsingular curve.

Thus one obtains a proof of Theorem 1.1. An alternative proof of existence of curves in the range $g < \frac{1}{8}d^2$ is given (over $\mathbf{C}$) by the theorem of Mori [37].

THEOREM 1.6 (MORI). $(k = \mathbf{C})$ *There exists an irreducible nonsingular curve C of degree d and genus g contained in a nonsingular quartic surface of $\mathbf{P}^3$ (depending on C), if and only if $d > 0$ and either*

(a) $g = \frac{1}{8}d^2 + 1$ *or*

(b) $g < \frac{1}{8}d^2$ *and* $(d, g) \neq (5, 3)$.

The idea of Mori's proof is this: starting with a product of two suitably chosen elliptic curves and dividing by an involution, one obtains a Kümmer surface S_0 with divisor classes H_0 and C_0 satisfying $H_0^2 = 4$, $C_0 \cdot H_0 = d$, and $C_0^2 = 2g - 2$. Now apply Kodaira's theory of deformations of $K3$ surfaces [36] to obtain a $K3$ surface S with Pic $S \cong \mathbf{Z}^2$ generated by divisor classes H and C with the same numerical properties as H_0 and C_0. Then Saint-Donat's study of divisors on $K3$ surfaces [49] applies to show that H is very ample, and hence it gives an embedding of S as a quartic surface in $\mathbf{P}^3$. Furthermore, the divisor class C contains an irreducible nonsingular curve of genus g, whose image in $\mathbf{P}^3$ has degree d.

Let us say just a few words about generalizations to higher-dimensional projective spaces. For any $n \geq 3$ one can ask for what pairs of integers (d, g) does there exist an irreducible nonsingular curve C of degree d and genus g in $\mathbf{P}^n$, not contained in any hyperplane $\mathbf{P}^{n-1}$ of the $\mathbf{P}^n$. Castelnuovo determined the maximum g for given d. Harris studied the general problem and proved a number of partial results in his Montreal notes [21]. A complete answer to this question for $n = 4, 5$ was recently given by Rathmann [45], but for $n \geq 6$ the problem remains open.

2. Relations between d, g, s, and e. Now that we have seen which degree and genus are possible for a curve in $\mathbf{P}^3$, we go on to study some more numerical invariants. As before, C denotes an irreducible nonsingular curve in $\mathbf{P}^3$ over an algebraically closed field k. We let

$$d = \text{degree of } C,$$

$$g = \text{genus of } C,$$

$$s = \text{least degree of a surface containing } C,$$

$$e = \text{least integer for which } H^1(\mathcal{O}_C(e)) \neq 0.$$

One could ask what possible 4-tuples (d, g, s, e) occur for curves in $\mathbf{P}^3$, but this question seems unmanageable at present. So we set ourselves a more limited task. Already from the discussion of d, g in the previous section, the principle emerges that curves of higher genus lie on surfaces of lower degree. So we seek to quantify this idea with the following.

PROBLEM 2.1. For given integers $d, k > 0$, determine the integer $G(d, k)$ which is the maximum genus of a curve C of degree d which is not contained in any surface of degree $< k$. (In other words, $s(C) \geq k$.)

For $k = 1, 2, 3$, the answer follows already from the study of d, g in the previous section:

$$G(d, 1) = \tfrac{1}{2}(d - 1)(d - 2)$$

is simply the maximum genus of a curve of degree d in $\mathbf{P}^3$.

$$G(d, 2) = \tfrac{1}{4}d^2 - d + 1 - \varepsilon,$$

where $d \geq 3$ and $\varepsilon = 0$ or 1 according to d even or odd. This is the maximum genus of a curve in $\mathbf{P}^3$ not contained in a plane. Similarly,

$$G(d, 3) = \tfrac{1}{6}d(d - 3) + 1 - \varepsilon,$$

where $d \geq 5$ and $\varepsilon = 0$ or $\tfrac{2}{3}$ according to $d \equiv 0 \pmod{3}$ or not.

For $k \geq 4$, the situation is more complicated. We begin with an easy result, which also brings into play the integer e.

PROPOSITION 2.2. Let C be a curve of degree d with $s(C) \geq k$.
(a) Then $d \geq \tfrac{1}{6}(k^2 + 4k + 6)$.
(b) If $e \leq k - 1$, then $g \leq d(k - 1) + 1 - \binom{k+2}{3}$.
(c) If $e \geq k - 1$, then $d \geq \tfrac{1}{3}(k^2 + 4k + 6)$.

PROOF. This result follows directly from the Riemann-Roch theorem and Clifford's theorem. If $e < k - 1$, then $h^1(\mathcal{O}_C(k - 1)) = 0$. So by Riemann-Roch,

$$h^0(\mathcal{O}_C(k - 1)) = d(k - 1) + 1 - g.$$

On the other hand, by hypothesis, $h^0(\mathcal{O}_C(k - 1)) = 0$, so $h^0(\mathcal{O}_C(k - 1)) \geq h^0(\mathcal{O}_{\mathbf{P}^3}(k - 1)) = \binom{k+2}{3}$. Combining inequalities gives (b). Writing $g \geq 0$ and combining with (b) gives (a).

On the other hand, if $e \geq k - 1$, then $\mathcal{O}_C(k - 1)$ is special, so by Clifford's theorem,

$$h^0(\mathcal{O}_C(k - 1)) - 1 \leq \tfrac{1}{2}d(k - 1).$$

Combining with $h^0(\mathcal{O}_C(k - 1)) \geq \binom{k+2}{3}$ gives (c).

This suggests we should divide our attention into ranges according to the relative size of d and k. We say that (d, k) belongs to

range A if $\frac{1}{6}(k^2 + 4k + 6) \leq d < \frac{1}{3}(k^2 + 4k + 6)$,

range B if $\frac{1}{3}(k^2 + 4k + 6) \leq d \leq k(k - 1)$,

range C if $d > k(k - 1)$.

Using this terminology, (2.2) gives a bound on g in range A, so we make the CONJECTURE 2.3. For $(d, k) \in$ range A,

$$G(d, k) = d(k - 1) + 1 - \binom{k + 2}{3}.$$

Note that to find $G(d, k)$ for any d, k requires proving a *bound*, namely, showing that $g \leq G(d, k)$ for any curve, and an *existence*, giving the existence of a curve with $g = G(d, k)$. So in range A, the bound is proved by Proposition 2.2b. The existence is still open, though it has been established in many cases by Ballico and Ellia [in preparation].

At the other extreme, in range C, the exact value of $G(d, k)$ has been determined by Gruson and Peskine [**17**].

THEOREM 2.4. *Assume that $d > k(k - 1)$. Then*

$$G(d, g) = d^2/2k + \tfrac{1}{2}d(k - 4) + 1 - \varepsilon,$$

where $\varepsilon \geq 0$ depends on the congruence of $d \pmod{k}$, namely,

$$\varepsilon = \tfrac{1}{2}f(k - f + 1 + f/k),$$

where $d \equiv f \pmod{k}$ and $0 \leq f < k$. Furthermore, for each d, k the maximum genus is attained by a curve lying on a surface of degree k and linked to a plane curve.

This result is the natural generalization of the results mentioned above for $k = 1, 2, 3$. It is proved by cutting the curve with a general plane and making a careful study of the postulation of the resulting finite set of points in the plane.

Now we come to the discussion of the most difficult range B. The statement of the expected result for $G(d, k)$ in this range has evolved over a number of years into a rather complicated formula. This makes it difficult to motivate, so we will simply state the result.

Given integers k, f, we define integers

$$A = A(k, f) = \lceil \tfrac{1}{3}(k^2 - kf + f^2 - 2k + 7f + 12) \rceil, \text{ resp. ditto } + 1$$
$$\text{if } f = 2k - 7 \text{ or } 2k - 9,$$

$$B = B(k, f) = \lceil \tfrac{1}{3}(k^2 - kf + f^2 + 6f + 11) \rceil, \text{ resp. ditto } + 1$$
$$\text{if } f = 2k - 8 \text{ or } 2k - 10.$$

Here $\lceil \ \rceil$ is the *round-up* of a rational number to the next higher integer.

CONJECTURE 2.5. Let k, d, f be integers such that $k \geq 5$, $k-1 \leq f \leq 2k-6$, and $A(k,f) \leq d < A(k, f+1)$. Then

$$G(d,k) = d(k+1) + 1 - \binom{k+2}{3} + \binom{f-k+4}{3} + h,$$

where

$$h = \begin{cases} 0 & \text{if } A(k,f) \leq d < B(k,f), \\ \frac{1}{2}(d-B)(d-B+1) & \text{if } B = B(k,f) \leq d < A(k, f+1). \end{cases}$$

The reader can verify that range B is empty for $k \leq 4$, and that for $k \geq 5$, as f varies from $k-1$ to $2k-6$, the intervals given for d fill up the range B. Now we state what is known so far about this conjecture.

THEOREM 2.5 (EXISTENCE) [29]. (char $k = 0$) For every (d,k) in range B, there is a curve with the conjectured genus $G(d,k)$.

THEOREM 2.7 (BOUND) [29]. Let C be an irreducible nonsingular curve of degree d, genus g, not contained in a surface of degree $< k$, with (d,k) in range B. Let $e = e(C)$.
(a) If $e < k-1$, then $g \leq$ conjectured $G(d,k)$ (follows from (2.2)).
(b) If $e \geq k-1$, then $d \geq A(k,e)$ and $e \leq 2k-6$.
(c) If $d < A(k, e+1)$, then $g \leq$ conjectured $G(d,k)$.
(d) In any case, the conjecture holds for $k \leq 10$, and all d.

Note that d and k determine an integer f, by the inequalities of Conjecture 2.5. Then part (b) of Theorem 2.7 says that $e \leq f$, and part (c) says that if $e = f$ then the bound holds. Thus the only case where the bound is not yet proved is if $e < f$, i.e., if $d \geq A(k, e+1)$.

The proofs of these results are achieved by using the theory of stable reflexive sheaves, which we explain in the next section.

3. Stable reflexive sheaves. In this section we will review the correspondence between curves and reflexive sheaves, and show how the theory of reflexive sheaves can be applied to problems of space curves. See [25, 26] as general references.

Let us start with a rank 2 vector bundle $\mathcal{E}$ on $\mathbf{P}^3$, regarded as a locally free sheaf of rank 2. If $s \in H^0(\mathcal{E})$ is a global section, we define the *zero-set* $Y = (s)_0$ as follows. The section s determines a map $s^\vee : \mathcal{E}^\vee \to \mathcal{O}$ whose image, by definition, is $\mathcal{I}_Y$. In good cases Y will be a curve, in the broad sense of being a locally complete intersection codimension 2 subscheme of $\mathbf{P}^3$. The construction just described gives rise to an exact sequence

$$0 \to \mathcal{O}(-c_1) \to \mathcal{E}^\vee \to \mathcal{I}_Y \to 0,$$

which expresses the relation between $\mathcal{E}$ and Y.

This exact sequence expresses $\mathcal{E}^\vee$ as an extension of $\mathcal{I}_Y$ by $\mathcal{O}(-c_1)$ and thus determines an element $\xi \in \mathrm{Ext}^1(\mathcal{I}_Y, \mathcal{O}(-c_1))$. By chasing some standard exact sequences one finds this Ext group is isomorphic to $H^0(\omega_Y(4-c_1))$. Since $\mathcal{E}$ is locally free, the element $\xi \in H^0(\omega_Y(4-c_1))$ is a nowhere-vanishing section of the sheaf $\omega_Y(4-c_1)$. In other words, it gives an isomorphism $\mathcal{O}_Y \xrightarrow{\sim} \omega_Y(4-c_1)$. The existence of this isomorphism imposes strong restrictions on Y. For example, its degree d divides $2g-2$. In particular, not every curve in $\mathbf{P}^3$ can arise in this way.

Now let us reverse this procedure, starting with an arbitrary curve Y in $\mathbf{P}^3$. Take an integer m and a section $\xi \in H^0(\omega_Y(m))$. As above, $H^0(\omega_Y(m))$ is isomorphic to $\mathrm{Ext}^1(\mathcal{I}_Y, \mathcal{O}(m-4))$, so we obtain an extension

$$0 \to \mathcal{O}(m-4) \to \mathcal{F} \to \mathcal{I}_Y \to 0$$

for some coherent sheaf $\mathcal{F}$. If we now assume that ξ generates the sheaf $\omega_Y(m)$ at all but a finite number of points, then $\mathcal{F}$ will be locally free of rank 2 except at those points. Furthermore, if Y is at least Cohen-Macaulay at those points, then $\mathcal{F}$ will have depth 2 there, and so $\mathcal{F}$ will be a rank 2 reflexive sheaf on $\mathbf{P}^3$.

If we regard reflexive sheaves as "vector bundles with singularities," then we get a dictionary between rank 2 reflexive sheaves on $\mathbf{P}^3$ and rather general curves in $\mathbf{P}^3$. We refer to [25] for a systematic development of this theory.

One point we would like to discuss here, however, is under what conditions a reflexive sheaf corresponds to a nonsingular curve. If $\mathcal{F}$ is a rank 2 reflexive sheaf on $\mathbf{P}^3$, the sheaf $\mathrm{Ext}^1(\mathcal{F}, \mathcal{O})$ is a measure of how non-locally-free $\mathcal{F}$ is. Indeed, this sheaf is supported at the set of points where $\mathcal{F}$ is not locally free. If $\mathcal{F}$ arises from a curve Y as above, then one easily sees that $\mathrm{Ext}^1(\mathcal{F}, \mathcal{O}_Y) \cong \omega_Y(m)/\xi \cdot \mathcal{O}_Y$. In particular, if Y is nonsingular, then ω_Y is invertible, and its quotient above is isomorphic to $\mathcal{O}_{Y,P}/(t_P^r)$ at each point $P \in Y$, for some integer r depending on P and nonzero only at the points where ξ does not generate $\omega_Y(m)$. In particular, there exists a set of regular parameters x, y, z in the local ring $\mathcal{O}_P$ of P on $\mathbf{P}^3$ such that $\mathrm{Ext}^1(\mathcal{F}, \mathcal{O})_P \cong \mathcal{O}_P/(x, y, z^r)$. So we make the

DEFINITION. A rank 2 reflexive sheaf $\mathcal{F}$ on $\mathbf{P}^3$ is *curvilinear* if at every non-locally-free point P of $\mathcal{F}$, there exists a set of regular parameters $x, y, z \in \mathcal{O}_P$ and an integer $r > 0$ such that $\mathrm{Ext}^1(\mathcal{F}, \mathcal{O})_P \cong \mathcal{O}_P/(x, y, z^r)$.

Now we have the

PROPOSITION 3.1. *Let $\mathcal{F}$ be a rank 2 reflexive sheaf on $\mathbf{P}^3$.*

(a) *If there is a section $s \in H^0(\mathcal{F}(n))$ for some n whose zero set $(s)_0$ is a nonsingular curve, then $\mathcal{F}$ is curvilinear.*

(b) (char $k = 0$). *Conversely, if $\mathcal{F}$ is curvilinear, then for any integer n such that $\mathcal{F}(n)$ is generated by global sections, and for any sufficiently general section $s \in H^0(\mathcal{F}(n))$, the zero set $Y = (s)_0$ is a nonsingular curve.*

PROOF [29]. We have already proved (a) above, since being curvilinear is a local property unaffected by the twist. To prove (b) one first makes a local calculation, showing that locally, the zero set of a general section is nonsingular in the neighborhood of a curvilinear singular point of $\mathcal{F}$. Then one considers the locus of pairs (s, P) with $s \in H^0(\mathcal{F}(n))$, $P \in \mathbf{P}^3$, $P \in (s)_0$ and proves the result by counting dimensions and using the theorem of generic smoothness.

Now let us look at the situation of the previous section, in range B, and see how to translate this into a question concerning reflexive sheaves.

We have an irreducible nonsingular curve C in $\mathbf{P}^3$ with invariants d, g, e as before, and $h^0(\mathcal{I}_C(k-1)) = 0$ for a certain k. Furthermore, to study Conjecture 2.5 we may restrict e to lie in the interval $k - 1 \leq e \leq 2k - 6$.

The first step is to choose ξ. The most efficient results come from taking the most negative twist of ω_C which has a nonzero section. By Serre duality on C, this is precisely $H^0(\omega_C(-e))$. So we take $\xi \in H^0(\omega_C(-e))$ and obtain a reflexive sheaf

$$0 \to \mathcal{O}(-e - 4) \to \mathcal{E}' \to \mathcal{I}_C \to 0.$$

Twisting by $k-1$ and using the hypothesis $h^0(\mathcal{I}_C(k-1)) = 0$, plus the hypothesis $e \geq k - 1$, we find that $H^0(\mathcal{E}'(k - 1)) = 0$. So we let $\mathcal{E} = \mathcal{E}'(k - 1)$ for ease of notation. Then $H^0(\mathcal{E}) = 0$, and its Chern classes can be expressed in terms of d, g, e, k:

$$c_1 = 2k - e - 6,$$

$$c_2 = d + (k - 1)(k - e - 5),$$

$$c_3 = 2g - 2 - de.$$

Note that again by our hypothesis on e, we find $c_1 \geq 0$. This, together with $H^0(\mathcal{E}) = 0$, implies that $\mathcal{E}$ is stable. So our problem is transformed into a problem of bounding the Chern classes of a stable rank 2 reflexive sheaf with $H^0(\mathcal{E}) = 0$. Note that as c_1 increases, the condition $H^0(\mathcal{E}) = 0$ expresses something stronger than just stable. For each c_1, then, we seek a lower bound for c_2 and an upper bound for c_3 in terms of c_1 and c_2. This will give a bound on g which we hope will prove Conjecture 2.5.

Note also from this construction how the curve being in range B gives rise to a stable reflexive sheaf. So we can say roughly that curves in range B correspond to *stable* reflexive sheaves, and this helps explain the interest of range B.

Now we describe the solution of the problem about reflexive sheaves mentioned above. First, for any integer c_1, define integers

$$A = \lceil \tfrac{1}{3}(c_1^2 + 2c_1 + 3) \rceil, \quad \text{resp. ditto } + 1 \text{ if } c_1 = 1, 3,$$

$$B = \lceil \tfrac{1}{3}(c_1^2 + 3c_1 + 8) \rceil, \quad \text{resp. ditto } + 1 \text{ if } c_1 = 2, 4.$$

THEOREM 3.2 (BOUNDS) [29]. *Let $\mathcal{E}$ be a rank 2 reflexive sheaf on $\mathbf{P}^3$ with $c_1 \geq -1$. Assume that $H^0(\mathcal{E}) = 0$. Then $c_2 \geq A$. Furthermore*
   (a) *if $A \leq c_2 \leq B$, then*

$$c_3 \leq (c_1 + 4)c_2 - 2\binom{c_1 + 3}{3} - 2,$$

and
   (b) *if $c_2 > B$, then*

$$c_3 \leq c_2^2 - c_2(2B - c_1 - 5) + B^2 - B - 2\binom{c_1 + 3}{3} - 2.$$

THEOREM 3.3 (EXISTENCE) [**29**]. (char $k = 0$) *The above statement is sharp, in the sense that for each $c_1 \geq -1$ and for each $c_2 \geq A$, there exists a stable curvilinear rank 2 reflexive sheaf $\mathcal{E}$ on $\mathbf{P}^3$ with $H^0(\mathcal{E}) = 0$ and whose c_3 is equal to the bound given in* (3.2).

The proof of the bound of Theorem 3.2, which generalizes the main theorem of [**26**], is a little complicated. One uses induction on c_1, the cases $c_1 = 0, -1$ being known already, for in that case the condition $H^0(\mathcal{E}) = 0$ simply says that $\mathcal{E}$ is stable, nothing more. Then one divides into two cases according to whether $H^2(\mathcal{E})$ is zero or not. If it is zero, one gets the bound (a) immediately from the Riemann-Roch theorem. If $H^2(\mathcal{E})$ is not zero, then its Serre dual $\operatorname{Ext}^1(\mathcal{E}, \mathcal{O}(-4))$ is also nonzero, so one can construct a nonsplit extension

$$0 \to \mathcal{O}(-4) \to \mathcal{M} \to \mathcal{E} \to 0,$$

where $\mathcal{M}$ is now a certain rank 3 reflexive sheaf on $\mathbf{P}^3$. If $\mathcal{M}$ is stable, one has bounds on the Chern classes of $\mathcal{M}$ which imply the desired bounds on the Chern classes of $\mathcal{E}$. If $\mathcal{M}$ is not stable, one deduces the existence of a *reduction step*, namely, an exact sequence

$$0 \to \mathcal{E}' \to \mathcal{E} \to \mathcal{I}_{Z,X}(n) \to 0,$$

where $\mathcal{E}'$ is another rank 2 reflexive sheaf on $\mathbf{P}^3$, and X is a surface, Z a curve on X, and n an integer. In this case one finds $c_1(\mathcal{E}') < c_1(\mathcal{E})$, so one can apply the induction hypothesis to $\mathcal{E}'$. Then some unpleasant calculations show that the result for $\mathcal{E}'$ implies the result for $\mathcal{E}$.

For the existence theorem (Theorem 3.3) one uses deformation theory. Starting with a torsion-free sheaf with suitable Chern classes, one seeks to deform it into a reflexive sheaf with the most general pattern of cohomology groups. We say a rank 2 torsion-free sheaf $\mathcal{F}$ on $\mathbf{P}^3$ has *seminatural cohomology* if for every $l \geq -\frac{1}{2}c_1 - 2$, at most one of the four groups $H^i(\mathcal{F}(l))$, $i = 0, 1, 2, 3$, is nonzero. In the range (a) of Theorem 3.2 one can prove the existence of reflexive sheaves with seminatural cohomology having $H^0(\mathcal{F}) = 0$ and the desired maximum value of c_3. From these one obtains examples of sheaves with the maximum c_3 in range (b) by using an *antireduction step*, namely, the reverse of the reduction step described above.

Returning to the conjecture about curves (Conjecture 2.5), one would hope that this satisfactory solution of the corresponding question about reflexive sheaves would settle the matter. Unfortunately, not so. The trouble is that given a curve C, we have no control over the invariant e. We have seen (Theorem 2.7) that $e \leq f$. If $e = f$, then indeed the bounds (Theorem 3.2) on c_3 give the desired bounds on g. If, however, $e < f$, then we obtain some reflexive sheaves which are not extremal for Theorem 3.2, and those bounds are not sufficient to give the bound we want for g.

On the other hand, using Proposition 3.1, the existence of sheaves with maximum c_3 given by Theorem 3.3 is enough to give the existence of curves with maximum genus (Theorem 2.6).

4. The search for a general space curve of given degree and genus.
Let us fix integers d, g for which there are curves of degree d and genus g in
$\mathbf{P}^3$, and consider the problem of classifying them. This is a difficult problem.
Almost any general statement one can think of is likely to be false. Halphen, for
example, considers the problem of prescribing numerical invariants which should
distinguish the irreducible components of $H_{d,g}$. Each time he adds another
invariant, there is an example where the corresponding family is reducible. This
philosophy that anything can happen is perhaps best exemplified by Mumford's
example in one of his "Pathologies" papers [**38**], of an irreducible component of
the Hilbert scheme which is generically nonreduced. One might think that the
classification problem is hopeless and therefore give up.

On the other hand, there have been a number of recent results indicating that
for certain classes of curves, good things happen. Typically if the degree is large
with respect to the genus or if the genus is large with respect to the degree, one
can say something. But there always remains that unknown range in the middle
where anything can happen.

The purpose of this talk is to suggest that for any d, g, even in the mysterious
middle range, one can identify a component or components of the Hilbert scheme
constituting the "general" curves of the given degree and genus, whose behavior
is predictable. In particular, unless there is an obvious reason preventing it, such
a curve should have any nice property one hopes for.

To be more precise, for any curve C, denote by $s(C)$ the least degree of a
surface containing C. Let

$$s(d, g) = \max_{C \in H_{d,g}} s(C).$$

A curve C with $s(C) = s(d, g)$ will be called *superficially general*. Thus for any
d, g, the superficially general curves form a nonempty open subset $H_{d,g}^0 \subseteq H_{d,g}$.
These superficially general curves are candidates for the general curves of given
degree and genus.

Furthermore, I would like to suggest the following *principle*: that nongeneral
behavior of these "general" curves should be a consequence of their being forced
to lie on a surface of degree $s(d, g)$. In other words, if the fact of C lying on
a surface of degree $s(d, g)$ does not automatically preclude it, C should have
whatever nice property we like.

So far there is very little evidence for such an assertion. The most one can
say is that it is consistent with known results, and that it stems from a belief
that the world is ordered by logical principles. In the following pages we will
examine how these ideas apply to some particular nice properties of curves.
One point worth mentioning, however, aside from the philosophical desirability
of having some guiding principles, is that these principles will generate some
explicit conjectures, which may have a unifying influence on the great quantity
of special results being discovered these days.

To begin with, let us list some properties which might hold for a "general" curve.

P_1: $H_{d,g}$ has the expected dimension $4d$ near C.

P_2: $H^1(\mathcal{N}) = 0$, where $\mathcal{N}$ is the normal bundle of C.

P_3: C corresponds to a smooth point $x_C \in H_{d,g}$—in this case we say C is *unobstructed*.

P_4: $H^1(\mathcal{N}(-2)) = 0$.

P_5: The mapping $\phi: H_{d,g} \to \mathcal{M}_g$ is dominant near C, in which case we say C has *general moduli*.

P_6: The normal bundle $\mathcal{N}$ of C is a stable (resp. semistable) vector bundle on C.

P_7: C has *maximal rank*, meaning for every $l \geq 0$ the restriction map

$$\rho(l): H^0(\mathbf{P}^3, \mathcal{O}(l)) \to H^0(\mathcal{O}_C(l))$$

is either injective or surjective.

P_8: C is *linearly normal*, meaning

$$\rho(1): H^0(\mathbf{P}^3, \mathcal{O}(1)) \to H^0(\mathcal{O}_C(1))$$

is surjective.

One could easily add to this list.

The work I suggest to do is to review what is known about each of these properties, to study the effect of C lying on a surface of small degree on this property, and thence to generate a conjectural characterization of the set (d, g) for which there exists a curve with the given property.

For values of (d, g) for which the nice property does not hold, one should generate a conjectural statement of how far off it is for general curves. So, for example, we should guess $\dim H_{d,g}$ for general curves, we should guess $\dim H^1(\mathcal{N})$, and so forth.

There is not time in these lectures to consider all these problems, so we will discuss only three, and then list a few specific open problems.

a. *Normal bundles of space curves.* This is a fairly recent topic historically, but is a good illustration of the ideas discussed above. It all began in 1979 when Van de Ven, in answer to a question of Grauert, gave an example of a space curve with an indecomposable normal bundle [**51**]. Since then there have been at least 15 papers published about normal bundles of space curves (see References).

Let me briefly review some of the results.

If $C \subseteq \mathbf{P}^3$ is a rational curve of degree d, then $C \cong \mathbf{P}^1$, so its normal bundle splits into a direct sum of two line bundles. The degree of $\mathcal{N}$ is $4d - 2$, so we can write

$$\mathcal{N} \cong \mathcal{O}_{\mathbf{P}^1}(2d - 1 - a) \oplus \mathcal{O}_{\mathbf{P}^1}(2d - 1 + a)$$

for some $a \geq 0$. The main results in this case are that if $d \geq 4$, then $0 \leq a \leq d-4$ and that all such a occur (assuming char $k = 0$). The general case is $a = 0$.

Furthermore, the variety of these rational curves with a given splitting type has been exhaustively studied [**9, 10, 11, 16, 46, 47**].

If C is an elliptic curve of degree d, then $\mathcal{N}$ has degree $4d$, so $\mathcal{N}(-2)$ has degree 0. Let m denote the maximum degree of a sub-line-bundle of $\mathcal{N}(-2)$. Then

$$m = 0 \quad \text{if } d = 4, 5,$$

$$0 \le m \le d - 3 \quad \text{if } d = 6, 7$$

$$0 \le m \le d - 4 \quad \text{if } d \ge 8,$$

and all such m occur. The general case is $m = 0$, in which case $\mathcal{N}$ is semistable [**13, 33**].

For curves with genus ≥ 2 there are partial results. First of all, there are some specific examples of curves with stable normal bundle, for $(d, g) = (6, 2)$ [**48**], $(6, 3)$ [**12**], $(7, 5)$ [**8**], $(9, 9)$ [**41**]. More generally, Ellingsrud and Hirschowitz [**14**] show for instance that for any $g \ge 3$ and $d \ge g + 3$, there exist curves C with $\mathcal{N}_C$ stable. They, and later Perrin [**43**], also investigate the integers $D_s(g)$ (resp. $D_{ss}(g)$) which is the least d such that there exists a curve C with $\mathcal{N}_C$ stable (resp. semistable).

Now let us make some speculations about the general situation. According to the general principle stated above, there should exist a curve C with given d, g having a stable normal bundle unless that is already contradicted by the fact of C lying on a surface of degree $s(d, g)$. We start with an easy observation.

PROPOSITION 4.1. *Let C be a nonsingular curve of degree d and genus g lying on a nonsingular surface S of degree s. Assume furthermore that $\mathcal{N}_C$ is stable (resp. semistable). Then*

$$g < d(s - 2) + 1 \quad (resp. \ \le).$$

PROOF. We consider the exact sequence of normal bundles

$$0 \to \mathcal{N}_{C/S} \to \mathcal{N}_{C/\mathbf{P}^3} \to \mathcal{N}_{S/\mathbf{P}^3} \otimes \mathcal{O}_C \to 0.$$

Note that $\mathcal{N}_{S/\mathbf{P}^3} \cong \mathcal{O}_S(s)$, so that the right-hand sheaf is isomorphic to $\mathcal{O}_C(s)$. If $\mathcal{N}_C$ is stable, then the degree of any quotient line bundle must be bigger than $\frac{1}{2} \deg \mathcal{N}_C$. This says $ds > \frac{1}{2}(4d + 2g - 2)$, whence the desired inequality, and similarly for semistable.

Our principle says if this obvious necessary condition is satisfied, a sufficiently general curve should have a stable normal bundle. So we obtain

CONJECTURE 4.2. Given d, g with $g \ge 2$ for which the Hilbert scheme $H_{d,g}$ is nonempty, let $s = s(d, g)$ as defined above, and suppose that $g < d(s - 2) + 1$ (resp. $\le$). Then there exists a curve $C \in H_{d,g}$ with stable (resp. semistable) normal bundle.

This conjecture is somewhat unsatisfactory because of the expression $s(d, g)$ which enters it. If one replaces it by its conjectural value coming from the conjectures of §2, namely, for $G(d, k) \ge g > G(d, k+1)$, $s(d, g)$ should be k, then one obtains the following more precise conjecture.

CONJECTURE 4.3. For each $d \geq 6$, choose k (uniquely) so that

$$k^2 - 2k + 2 \leq d \leq k^2.$$

Then there exists a curve C with semistable (resp. stable) normal bundle, with degree d and genus g if and only if $g \leq G(d, k)$ (resp. $<$ if $d = k^2 - 2k + 2$ or $d = k^2$, and furthermore, $g \geq 2$).

Note that the range of $G(d, k)$ which enters here is the top piece of the range B ($f = 2k - 6$) and the bottom piece of the range C.

According to the tables of Perrin [43], the conjecture (at least for "semistable") is true for $d \leq 20$, except for $d = 14, 17$, where it is not yet known.

b. *Vanishing of $H^1(\mathcal{N})$ or $H^1(\mathcal{N}(-2))$.* Suppose that the curve C is contained in a nonsingular surface S of degree s. Then from the exact sequence of normal sheaves

$$0 \to \mathcal{N}_{C/S} \to \mathcal{N} \to \mathcal{N}_S \otimes \mathcal{O}_C \to 0$$

using the fact that $\mathcal{N}_S \cong \mathcal{O}_S(s)$, one obtains surjective mappings

$$H^1(\mathcal{N}) \to H^1(\mathcal{O}_C(s)) \to 0$$

and

$$H^1(\mathcal{N}(-2)) \to H^1(\mathcal{O}_C(s - 2)) \to 0.$$

These give obvious necessary conditions for the vanishing of $H^1(\mathcal{N})$ or $H^1(\mathcal{N}(-2))$. Applying our general principle, we may suppose that if these necessary conditions are satisfied, then in general the corresponding group should vanish. So we formulate

CONJECTURE 4.4. Let C be a curve with invariants d, g, s, e. If $e < s$ (resp. $e < s - 2$), then there exists a curve $C \in H_{d,g}$ with $H^1(\mathcal{N}) = 0$ (resp. $H^1(\mathcal{N}(-2)) = 0$).

In order to get a conjecture involving only d and g, one must substitute for s, e the "expected" values of s, e. The problem falls into the A range and the beginning of the B range.

We do not yet have an exact conjectural bound for g in terms of d, s, e, so we cannot formulate an exact conjecture. We expect an answer with g asymptotically of the order of $\frac{1}{2}\sqrt{3}d^{3/2}$. Existence of curves with $H^1(\mathcal{N}) = 0$ or $H^1(\mathcal{N}(-2)) = 0$ is studied by Ellingsrud and Hirschowitz [14] and Perrin [44].

The importance of the condition $H^1(\mathcal{N}(-2)) = 0$ follows from the work of Perrin [44]. In particular, it relates to the previous problem about semistability of $\mathcal{N}$.

PROPOSITION 4.5. *Let C be a nonsingular curve in $\mathbf{P}^3$ with $H^1(\mathcal{N}(-2)) = 0$. Then $\mathcal{N}$ is semistable.*

PROOF. Since $\deg \mathcal{N}(-2) = 2g - 2$, by Riemann-Roch, $\chi(\mathcal{N}(-2)) = 0$, so our hypothesis is equivalent to $H^0(\mathcal{N}(-2)) = 0$. Now suppose $\mathcal{N}$, hence equivalently $\mathcal{N}(-2)$, is not semistable. Then there exists a sub-line-bundle $\mathcal{L} \subseteq \mathcal{N}(-2)$ with $\deg \mathcal{L} > g - 1$. Now Riemann-Roch applied to $\mathcal{L}$ gives $h^0(\mathcal{L}) > 0$, so $h^0(\mathcal{N}(-2)) > 0$, a contradiction.

c. *Curves of maximal rank.* An important classical problem for a space curve
C is to find the *postulation* of the curve, which amounts to knowing $h^0(\mathcal{I}_C(n))$
for all n. In some good cases it is easy to calculate. For complete intersections,
or more generally, projectively normal curves, the restriction map

$$\rho(n)\colon H^0(\mathcal{O}_{\mathbf{P}^3}(n)) \to H^0(\mathcal{O}_C(n))$$

is surjective, so one can compute $h^0(\mathcal{I}_C(n)) = \dim \ker \rho(n)$.

In general, however, the cokernel of $\rho(n)$, which is $H^1(\mathcal{I}_C(n))$, may be nonzero,
which makes the postulation more difficult to calculate. Following the general
principles mentioned above, one may hope that in general the rank of $\rho(n)$ is as
large as it could be. So we make the

DEFINITION. The curve C is of *maximal rank* if for every n, the map $\rho(n)$ is
of maximal rank, i.e., either injective or surjective.

The property of maximal rank is clearly an open condition on $H_{d,g}$. So one's
first hope might be that for any d, g, a sufficiently general curve has maximal
rank. This is not the case.

EXAMPLE. Consider curves of degree 8 and genus 8 in $\mathbf{P}^3$. We will show

(a) $H_{8,8}$ is irreducible, and any curve $C \in H_{8,8}$ is a curve of type $(3,5)$ on a
nonsingular quadric surface Q.

(b) No curve in $H_{8,8}$ has maximal rank.

First suppose $C \in H_{8,8}$. Then $\mathcal{O}_C(2)$ has degree 16, so is nonspecial, so by
Riemann-Roch, $h^0(\mathcal{O}_C(2)) = 9$. Since $h^0(\mathcal{O}_{\mathbf{P}^3}(2)) = 10$, it follows that
$h^0(\mathcal{I}_C(2)) \neq 0$. Thus C lies on a surface of degree 2. In a plane or on a quadric
cone there are no curves of this degree and genus, so C must lie on a nonsingular
quadric surface Q. Then the only possible bidegree is $(3,5)$. These curves exist,
and as Q varies in $\mathbf{P}^3$, they form an irreducible family $H_{8,8}$.

Now consider $H^0(\mathcal{I}_C(3))$. If $q \in H^0(\mathcal{I}_C(2))$ is the equation of Q, then
$x_0 q, x_1 q, x_2 q, x_3 q \in H^0(\mathcal{I}_C(3))$, so $h^0(\mathcal{I}_C(3)) \geq 4$. But $h^0(\mathcal{O}_{\mathbf{P}^3}(3)) = 20$ and
by Riemann-Roch, $h^0(\mathcal{O}_C(3)) = 17$. Thus $\rho(3)$ has rank ≤ 16, so C does not
have maximal rank.

In this example we can see what the trouble is. Since C lies on a surface of
degree 2, its equation $q \in H^0(\mathcal{I}_C(2))$ gives rise to many sections of $H^0(\mathcal{I}_C(n))$
for $n \geq 2$, and these contradict maximal rank. So by our general principle we
should say if this does not happen, then a general curve of the given degree and
genus should have maximal rank. This principle is hard to interpret in general,
but looking only at the case $e < s - 1$, and expressing the fact that $\rho(s)$ should
be surjective, we arrive at the following statement.

CONJECTURE 4.6. For each $d > 0$, choose t (uniquely) such that

$$\tfrac{1}{2}(t-1)(t+2) < d \leq \tfrac{1}{2}t(t+3).$$

Then for all g such that $0 \leq g \leq d(t-1) - \binom{t+2}{3} + 1$ there should exist a curve
of maximal rank.

This would answer part of the following

PROBLEM 4.7. Determine the set of (d, g) for which there exists a curve C of degree d and genus g in $\mathbf{P}^3$ of maximal rank.

Now let us review what is known about the existence of curves of maximal rank in $\mathbf{P}^3$.

The first case studied was for reducible curves, showing that a disjoint union of r lines in general position in $\mathbf{P}^3$ is a curve of maximal rank [26a]. Next was the result of Hirschowitz [30] dealing with a general rational curve in $\mathbf{P}^3$.

After that there is a series of papers proving the existence of curves of maximal rank in broader and broader ranges: first for curves of genus ≤ 3; then for any genus and d sufficiently large with respect to g; then for the nonspecial case $d \geq g+3$; and finally for curves of general moduli with $d \geq \frac{3}{4}g+3$, as conjectured by Harris (see References).

Then one has some sparse families of curves of large genus which have maximal rank: the projectively normal curves and the curves of maximal genus in the B range described in §2 above.

Let us say a few words about the proofs of these results. The most prevalent technique today is to construct singular or reducible curves which by some complicated inductive process can be shown to have maximal rank, and then smooth them. There is usually some rather messy arithmetic and various ad hoc methods to show that one can obtain all the desired pairs (d, g). To illustrate these techniques, we will give the complete proof of one special case (albeit a pivotal case) of the theorem on skew lines.

THEOREM 4.8. *A union of seven lines in general position in* $\mathbf{P}^3$ *is a curve of maximal rank.*

PROOF [26a]. First of all, maximal rank is an open property, so it is enough to find any set of seven lines, or even any scheme which is a specialization of a flat family of seven lines, which has maximal rank. Next, let us consider

$$\rho(4)\colon H^0(\mathcal{O}_{\mathbf{P}^3}(4)) \to H^0(\mathcal{O}_C(4)),$$

where $C = L_1 \cup \cdots \cup L_7$ is the union of seven lines. Note that both these spaces have dimension 35. If we show that $\rho(4)$ is an isomorphism, then clearly $h^0(\mathcal{I}_C(n)) = 0$ for $n \leq 4$. On the other hand, $h^1(\mathcal{I}_C(4)) = 0$, and in any case $h^2(\mathcal{I}_C(3)) = 0$, so by Castelnuovo's theorem, $\mathcal{I}_C$ is 5-regular, and $h^1(\mathcal{I}_C(n)) = 0$ for $n \geq 4$. Then C will have maximal rank.

Thus we reduce to showing $\rho(4)$ an isomorphism, for which it is sufficient to show $h^0(\mathcal{I}_C(4)) = 0$, i.e., C is not contained in any surface of degree 4.

Now we introduce a specialization of the scheme C. Let Q be a fixed nonsingular quadric surface, and degenerate C so that L_1, L_2, L_3 lie in one family of lines on Q. Further degenerate so that L_4 meets L_5 at a point x and L_5 meets L_6 at a point y, and so that x and y lie on Q. Note that the limit of C is a scheme C_0 with nilpotent elements at x and y. Note that $C_0 \cap Q$ is a scheme consisting of the lines L_1, L_2, L_3, plus the points x, y with scheme structure defined by $m_{x,Q}^2$

and $m_{y,Q}^2$, respectively, plus four other reduced points where L_4, L_5, L_6, L_7 meet Q.

If C_0 were contained in a surface X of degree 4, that surface would intersect Q in a curve of type $(4,4)$ containing $C_0 \cap Q$. One sees easily that this is not possible. Hence X contains Q, so $X = Q \cup Q'$, where Q' is another quadric surface containing $C_0 - C_0 \cap Q$, namely, $Q' \supseteq L_4 \cup L_5 \cup L_6 \cup L_7$. But this is also impossible. Done.

d. *A few problems.* We conclude this section with a few problems suggested by the general principles above.

PROBLEM 4d.1. Find the expected value of e for a general curve $C \in H_{d,g}$.

PROBLEM 4d.2. Find a conjectural value for the dimension of the variety $H_{d,g}$ of superficially general curves in $H_{d,g}$.

PROBLEM 4d.3. Show that a sufficiently general superficially general curve is unobstructed.

PROBLEM 4d.4. For any d, g with $d \leq g + 3$ for which $H_{d,g}$ is not empty, show that $H_{d,g}$ contains a linearly normal curve.

REFERENCES

1. E. Ballico and Ph. Ellia, *Generic curves of small genus in $\mathbf{P}^3$ are of maximal rank*, Math. Ann. **264** (1983), 211–225.

2. ____, *Generic curves of high degree in $\mathbf{P}^3$ are of maximal rank*.

3. ____, *On postulation of curves in $\mathbf{P}^4$*, Math. Z. **188** (1985), 215–223.

4. E. Ballico, *On the postulation of canonical curves in $\mathbf{P}^3$*, Ark. Mat. **22** (1984), 139–151.

5. E. Ballico and Ph. Ellia, *On postulation of curves: embeddings by complete linear series*, Arch. Math. (Basel) **43** (1984), 244–249.

6. ____, *The maximal rank conjecture for nonspecial curves in $\mathbf{P}^3$*, Invent. Math. **79** (1985), 541–555.

7. ____, *On projective curves embedded by complete linear systems*.

8. ____, *Some more examples of curves in $\mathbf{P}^3$ with stable normal bundle*, J. Reine Angew. Math. **350** (1984), 87–93.

9. E. Ballico, *On the rationality of the variety of smooth rational space curves with fixed degree and normal bundle*, Proc. Amer. Math. Soc. **91** (1984), 510–512.

10. D. Eisenbud and A. Van de Ven, *On the normal bundles of smooth rational space curves*, Math. Ann. **256** (1981), 453–463.

11. ____, *On the variety of smooth rational space curves with given degree and normal bundle*, Invent. Math. **67** (1982), 89–100.

12. Ph. Ellia, *Exemples de courbes de $\mathbf{P}^3$ à fibré normal semistable, stable*, Math. Ann. **264** (1983), 389–396.

13. G. Ellingsrud and D. Laksov, *The normal bundle of elliptic space curves of degree 5*, (Proc 18th Scand. Congr. Math., Aarhus, 1980), Progr. in Math., vol. 11, Birkhäuser, Boston, 1981, pp. 285–287.

14. G. Ellingsrud and A. Hirschowitz, *Sur le fibré normal des courbes gauches*, C. R. Acad. Sci. Paris Sér. I. Math. **299** (1984), 245–248.

15. G. Ellingsrud, L. Gruson, C. Peskine, and S. A. Strømme, *On the normal bundle of curves on smooth projective surfaces*.

16. F. Ghione and G. Sacchiero, *Normal bundles of rational curves in $\mathbf{P}^3$*, Manuscripta Math. **33** (1980/81), 111–128.

17. L. Gruson and C. Peskine, *Genre des courbes de l'espace projectif*, Algebraic Geometry, Tromsø 1977, Lecture Notes in Math, vol. 687, Springer-Verlag, 1978, pp. 31–59.

18. ____, *Genre des courbes de l'espace projectif*. II, Ann. Sci. École Norm. Sup. (4) **15** (1982), 401–418.

19. ____, *Postulation des courbes gauches*, Algebraic Geometry (Ravello), Lecture Notes in Math., vol. 997, 1983, pp. 218–227.

20. G. Halphen, *Mémoire sur la classification des courbes gauches algébriques*, J. École Polyt. **52** (1882), 1–200.

21. J. Harris (with the collaboration of D. Eisenbud), *Curves in projective space*, Sem. Math. Sup., Univ. Montreal, 1982.

22. J. Harris and K. Hulek, *On the normal bundle of curves on complete intersection surfaces*, Math. Ann. **264** (1983), 129–135.

23. R. Hartshorne, *Algebraic geometry*, Graduate Texts in Math., vol. 52, Springer-Verlag, New York, 1977.

24. ____, *On the classification of algebraic space curves*, Vector Bundles and Differential Equations (Nice, 1979), A. Hirschowitz, editor, Birkhäuser, Boston, 1980, pp. 83–112.

25. ____, *Stable reflexive sheaves*, Math. Ann. **254** (1980), 121–176.

26. ____, *Stable reflexive sheaves*. II, Invent. Math. **66** (1982), 165–190.

26a. R. Hartshorne and A. Hirschowitz, *Droites en position générale dans l'espace projectif*, Algebraic Geometry, (Proc. La Rábida, 1981), Lecture Notes in Math., vol. 961, Springer-Verlag, 1982, pp. 169–188.

27. R. Hartshorne, *Genre des courbes algébriques dans l'espace projectif (d'après L. Gruson et C. Peskine)*, Sém. Bourbaki **592** (1982), Astérisque **92–93** (1982), 301–313.

28. R. Hartshorne and A. Hirschowitz, *Courbes rationnelles et droites en position générale*, Ann. Inst. Fourier (Grenoble) **35** (1985), 39–58.

29. ____, *Nouvelles courbes de bon genre dans l'espace projectif*, (in preparation).

30. A. Hirschowitz, *Sur la postulation générique des courbes rationnelles*, Acta Math. **146** (1981), 209–230.

31. Masaaki Homma, *On projective normality and defining equations of a projective curve of genus three embedded by a complete linear system*, Tsukuba J. Math. **4** (1980), 269–279.

32. K. Hulek, *The normal bundle of a curve on a quadric*, Math. Ann. **258** (1981), 201–206.

33. K. Hulek and G. Sacchiero, *On the normal bundle of elliptic space curves*, Arch. Math. (Basel) **40** (1983), 61–68.

34. K. Hulek, *Complete intersection curves, the splitting of the normal bundle, and the Veronese surface*.

35. J. O. Kleppe, *The Hilbert-flag scheme, its properties and its connection with the Hilbert scheme. Applications to curves in 3-space*, Thesis, Oslo, 1981.

36. K. Kodaira, *On the structure of compact complex analytic surfaces*. I, Amer. J. Math. **86** (1964), 751–798.

37. S. Mori, *On degrees and genera of curves on smooth quartic surfaces in* $\mathbf{P}^3$, Nagoya Math. J. **96** (1984), 127–132.

38. D. Mumford, *Further pathologies in algebraic geometry*, Amer. J. Math. **84** (1962), 642–648.

39. ____, *Varieties defined by quartic equations* (with an appendix by G. Kempf), Questions on Algebraic Varieties (C.I.M.E., III Ciclo, Varenna, 1969), Cremonese, editor, Rome, 1970, pp. 29–100.

40. M. Noether, *Zur Grundlegung der Theorie der algebraischen Raumcurven*, Verl. König. Akad. Wiss., Berlin, 1883.

41. P. E. Newstead, *A space curve whose normal bundle is stable*, J. London Math. Soc. **28** (1983), 428–434.

42. D. Perrin, *Courbes passant par k points généraux de* $\mathbf{P}^3$; h^0-*stabilité*, C. R. Acad. Sci. Paris Sér. I. Math. **299** (1984), 879–882.

43. ____, *Courbes gauches, fibré normal et liaison*, C. R. Acad. Sci. Paris Sér. I. Math. **300** (1985), 39–42.

44. ____, *Courbes passant par n points généraux de* $\mathbf{P}^3$, Thèse, Univ. Paris VI, 1985.

45. J. Rathmann, *The genus of algebraic space curves*, Thesis, Univ. of Calif. at Berkeley, 1986.

46. G. Sacchiero, *Normal bundles of rational curves in projective space*, Ann. Univ. Ferrara Sez. VII (NS) **26** (1980), 33–40. (Italian)

47. _____, *On the varieties parametrizing rational space curves with fixed normal bundle*, Manuscripta Math. **37** (1982), 217–228.

48. _____, *Exemple de courbes de* $\mathbf{P}^3$ *de fibré normal stable*, Comm. in Alg. **18** (1983), 2115–2121.

49. B. Saint-Donat, *Projective models of K-3 surfaces*, Amer. J. Math. **96** (1974), 602–639.

50. J.-P. Serre, *A course in arithmetic*, Graduate Texts in Math., vol. 7, Springer-Verlag, 1973.

51. A. Van de Ven, *Le fibré normal d'une courbe dans* $\mathbf{P}_3$ *ne se décompose pas toujours*, C. R. Acad. Sci. Paris Ser. A–B **289** (1979), A111-A113.

52. C. Walter, (in preparation).

UNIVERSITY OF CALIFORNIA, BERKELEY

Proceedings of Symposia in Pure Mathematics
Volume **46** (1987)

The Rationality of Certain Spaces Associated to Trigonal Curves

N. I. SHEPHERD-BARRON

The object of this paper is to prove that the varieties M_4 (the moduli space for curves of genus four), T_g (the sublocus of M_g corresponding to trigonal curves), and $T_{g',1}$ (the locus of pointed trigonal curves of genus g') are rational when $g \equiv 2$ (4), $g \geq 6$, g' is odd, and $g' \geq 5$. (For the sake of clarity, we exclude the possibility of a curve being hyperelliptic when we speak of its being trigonal.) In §§1–3 we work over the complex numbers, while in the appendix we prove that in fact M_4 is rational over any field whose characteristic is neither 2 nor 3. I am very grateful to Professor S. Mori for suggesting to me that this might be true and for other valuable discussions on this subject.

1. THEOREM 1. *M_4 is rational.*

PROOF. Recall that the canonical model C of a generic curve of genus 4 is the complete intersection of a smooth quadric and a cubic in $\mathbf{P}^3$. The quadric Q containing C is unique, and so $M_4 \simeq \mathbf{P}(H^0(\mathcal{O}_Q(3)))/\operatorname{Aut} Q$, where $\simeq$ denotes birational equivalence.

Up to isogeny, $\operatorname{Aut} Q$ can be identified with the subgroup $Z \cdot O(4)$ of $GL(4)$, where Z is the center of $GL(4)$. Put $V = H^0(\mathcal{O}_Q(3))$; since Z acts trivially on $\mathbf{P}(V)$, it is enough to show that $\mathbf{P}(V)/O(4)$ is rational.

There is an exact sequence $0 \to H^0(\mathcal{O}_{\mathbf{P}^3}(1)) \to H^0(\mathcal{O}_{\mathbf{P}^3}(3)) \to V \to 0$ of $O(4)$-modules; it will be convenient to have an explicit splitting of this. This can be done by identifying V with the space of forms $f \in H^0(\mathcal{O}_{\mathbf{P}^3}(3))$ that are harmonic with respect to the differential operator $\Delta_Q = \sum_{i,j} q'_{ij} \, \partial^2/\partial x_i \, \partial x_j$, where $q'_{ij} = (q^{-1})_{ji}$ and the quadric Q is given by $Q = \sum q_{ij} x_i x_j$ with $q_{ij} = q_{ji}$. Equivalently, we can use the symbolical method; writing $f = a_x^3$ for $f \in H^0(\mathcal{O}_{\mathbf{P}^3}(3))$, where $a_x = a_1 x_1 + \cdots + a_4 x_4$, we can identify V with $\{f | (a,a) \cdot a_x = 0\}$, where $(a,a) = \sum q'_{ij} a_i a_j$ (remember that the coefficients a_i are contragredient to the variables x_i).

1980 *Mathematics Subject Classification* (1985 *Revision*). Primary 14H10.

Supported by National Science Foundation Grants Nos. MCS83-00863 and DMS85-03743.

PROPOSITION 2. *The generic form* $f \in H^0(\mathcal{O}_{\mathbf{P}^3}(3))$ *can be written uniquely as the sum of five cubes.*

This is an old and well-known result, due to Sylvester. We shall sketch a proof by Richmond [2].

Suppose that $b_1, \ldots, b_5$ are generic linear forms in $x_1, \ldots, x_4$, and that there are (possibly different) linear forms $c_1, \ldots, c_5$ such that $\sum b_i^3 = \sum c_i^3$. Denote the points in the dual space $(\mathbf{P}^3)^\vee$ corresponding to these forms by b_i^*, c_i^*. In $(\mathbf{P}^3)^\vee$, there is a web W of quadrics $\{Q_t\}_{t \in W}$ through $c_1^*, \ldots, c_5^*, b_5^*$. We can regard these quadrics as second-order differential operators Q_t^* which annihilate each of $c_1, \ldots, c_5, b_5$; application of each Q_t^* to the relation $\sum b_i^3 = \sum c_i^3$ yields a linear relation between $b_1, \ldots, b_4$. However, such relations must be trivial, and so the quadrics Q_t pass through each point b_i^*, c_j^*. Suppose that at least nine of the points b_i^*, c_j^* are distinct. Then the quadrics Q_t contain a common curve Γ. Necessarily $\deg \Gamma \leq 2$. If $\deg \Gamma = 2$, then Γ is either a plane conic or the union of two skew lines; in either case $\dim H^0(I_\Gamma(2)) = 4$, where I_Γ is the sheaf of ideals defining Γ, and Γ is the intersection of the quadrics containing it. So every point b_i^*, c_i^* lies on Γ, which contradicts the genericity of $b_1, \ldots, b_5$. Then suppose that Γ is a line. Choose distinct quadrics Q_1, Q_2 in the web W, so that $Q_1 \cap Q_2 = \Gamma \cap \Delta$, where Δ is a reduced curve of degree 3 meeting Γ in at least one point. Now the quadrics in W that contain Δ are just those in the pencil spanned by $\{Q_1, Q_2\}$, and so W cuts out a pencil Π on Δ. However, Γ passes through at most four of the points b_i^*, c_i^*; taking into account the point(s) $\Gamma \cap \Delta$, there are at least six fixed points of the pencil Π, which is absurd. Hence the b_i^*, c_j^* are at most eight in number; i.e., after reordering the forms b_i, c_j if necessary, there are constants λ_4, λ_5 such that $c_4 = \lambda_4 \cdot b_4$, $c_5 = \lambda_5 \cdot b_5$. Then

$$\sum_1^3 b_i^3 + (1 - \lambda_4^3)b_4^3 + (1 - \lambda_5^3)b_5^3 = \sum_1^3 c_j^3.$$

Applying the first-order operator corresponding to a plane in $(\mathbf{P}^3)^\vee$ passing through c_1^*, c_2^*, and c_3^* gives a linear relation between $b_1^2, \ldots, b_5^2$, which must be trivial. Hence $\lambda_4^3 = \lambda_5^3 = 1$ and the points $b_1^*, b_2^*, b_3^*, c_1^*, c_2^*, c_3^*$ are coplanar. The system of quadrics in $(\mathbf{P}^3)^\vee$ through c_1^*, c_2^*, c_3^* has no unassigned base points, and so (after reordering if necessary) we have $b_i^* = c_i^*$ for all i. Say $b_i = \lambda_i c_i$; then $\sum(1 - \lambda_i^3)b_i^3 = 0$, and so each $\lambda_i^3 = 1$. Q.E.D.

Proposition 2 can be restated as follows: the group $H = \mu_3^5 \rtimes S_5$ acts in the obvious way on $\mathbf{P}((\mathbf{C}^4)^5)$, and the rational map $\psi \colon (\mathbf{P}(\mathbf{C}^4)^5) \to \mathbf{P}(H^0(\mathcal{O}_{\mathbf{P}^3}(3)))$ given by $\psi(\alpha_1, \ldots, \alpha_5) = \sum_i(\sum_j \alpha_{ij}x_j)^3$ induces a birational equivalence

$$\varphi \colon \mathbf{P}((\mathbf{C}^4)^5)/H \to \mathbf{P}(H^0(\mathcal{O}_{\mathbf{P}^3}(3))).$$

COROLLARY 3. *The generic form* $f \in V$ *can be written uniquely as the sum of five cubes.*

PROOF. For any symmetric 4×4 matrix $Q' = (q'_{ij})$, let $\Delta_{Q'}^*$ denote the operator $\sum q'_{ij} \partial^2/\partial x_i \partial x_j$. Let $S \cong \mathbf{P}^9$ be the variety parametrizing the nonzero such Q, modulo scalars, and let $D \subset S$ be the locus where $\det Q' = 0$. Put $Y = \mathbf{P}(H^0(\mathcal{O}_{\mathbf{P}^3}(3)))$, and consider the incidence relation

$$Z = \{(f, Q') \in Y \times S | \Delta_{Q'}^*(f) = 0\}.$$

Let $p_1 \colon Z \to Y$, $p_2 \colon Z \to S$ be the projections. For all $f \in Y$, $p_1^{-1}(f) \cong \mathbf{P}^{15}$. Let $U \subset Y$ be a dense open set over which φ is an isomorphism. Since p_2 is surjective, we cannot have $p_2(p_1^{-1}(U)) \subseteq \Delta$; this proves the corollary.

Henceforth we assume (as we may) that the quadric Q is given by the equation $\sum_{i=1}^4 x_i^2 = 0$. Let X denote the closure of the locus $\psi^{-1}(V \cap U)$. We know now that $M_4 \simeq X/(H \times O(4))$, although X might be neither reduced nor irreducible. However, every irreducible component of X must dominate Y.

PROPOSITION 4. *There is an $S_5 \times O(4)$-equivariant birational equivalence $X/\mu_3^5 \simeq (\mathbf{P}^3)^5$, where S_5 permutes the factors $\mathbf{P}^3$ and $O(4)$ acts in the obvious linear way on each factor $\mathbf{P}^3$.*

PROOF. Consider the action of μ_3^5 on $\mathbf{P}((\mathbf{C}^4)^5)$. This space has homogeneous coordinates $\{\alpha_{ij} | 1 \leq i \leq 5, \, i \leq j \leq 4\}$, and μ_3^5 acts via $(\lambda_1, \ldots, \lambda_5)(\alpha_{ij}) = \lambda_i \alpha_{ij}$. Then the ring of invariants is generated by the cubics in $\{\alpha_{ij}\}$ that are homogeneous in each set $\{\alpha_{1j}, \ldots, \alpha_{4j}\}$ for each $j = 1, \ldots, 5$. So we can identify $(\mathbf{P}(\mathbf{C}^4)^5)\mu_3^5$ with the space L obtained as follows: in a copy of $\mathbf{P}^{99}$, take five copies $M_1, \ldots, M_5$ of $\mathbf{P}^{19}$ embedded linearly in general position. In each M_i, take a copy D_i of the triple Veronesean embedding of $\mathbf{P}^3$. For each point $p = (p_1, \ldots, p_5) \in \prod_{i=1}^5 D_i$, take the linear space N_p in $\mathbf{P}^{99}$ spanned by $p_1, \ldots, p_5$. Then L is the union of the spaces N_p, and so is birational to a $\mathbf{P}^4$-bundle over $(\mathbf{P}^3)^5$ whose fibres are the linear varieties N_p. The equations defining X are given by the condition

$$\sum_{i=1}^4 \frac{\partial^2}{\partial x_i^2} \left(\sum_{j=1}^5 \left(\sum_{k=1}^4 \alpha_{kj} x_k \right)^3 \right) = 0,$$

and so reduce to $\sum_{i=1}^4 \sum_{j=1}^5 \alpha_{ij}^2 \cdot \alpha_{kj} = 0$ for each $k = 1, \ldots, 4$. For $r = 1, \ldots, 5$, define points $p_r \in D_r$ as follows: p_r is given by the vanishing of all cubic monomials in $\{\alpha_{ir}\}$ except α_{rr}^3 for $r \leq 4$, and p_5 by the vanishing of all cubic monomials in $\{\alpha_{i5}\}$ except α_{45}^3. Then X/μ_3^5 and the linear span $\langle p_1, \ldots, p_5 \rangle$ meet transversely in the single point defined by the conditions that $\alpha_{44}^3 + \alpha_{45}^3 = 0$ and every other cubic monomial vanish. Hence X/μ_3^5 has a unique S_5-invariant irreducible component, say Z, such that Z dominates $(\mathbf{P}^3)^5$. But $X/\mu_3^5 \rtimes S_5$ is irreducible, and so X/μ_3^5 is irreducible, and the proposition is proved.

PROOF OF THEOREM 1. It is now enough to show that $(\mathbf{P}^3)^5/S_5 \times O(4)$ is rational. Put $T = (\mathbf{C}^*)^5$, on which S_5 acts by permuting the factors, and set $N = T \rtimes S_5$. There is an obvious 5-dimensional representation U_1 of N on which S_5 acts by permutation. Let U_2 denote the irreducible four-dimensional representation of $O(4)$. Then

$$(\mathbf{P}^3)^5/S_5 \times O(4) \simeq \mathbf{P}(U_1 \otimes U_2)/N \times O(4).$$

We regard $U_1 \otimes U_2$ as the same space $(\mathbf{C}^4)^5$ as before, with coordinates α_{ij}. Let α_j denote the vector $(\alpha_{ij}, \ldots, \alpha_{4j})$, which we regard as an element of U_2, and let $(\,,\,)$ denote the bilinear form preserved by $O(4)$. Then it is a result of classical invariant theory that the ring of invariants $\mathbf{C}[\{\alpha_{ij}\}]^{O(4)}$ is generated by the terms (α_j, α_k), subject to the single relation $\det((\alpha_j, \alpha_k)) = 0$ [3]. In other words, the categorical quotient $\mathbf{P}(U_1 \otimes U_2)^{ss}/O(4)$ can be identified with the locus Θ of singular quadrics in $\mathbf{P}^4 = \mathbf{P}(U_1)$.

There is an N-equivariant rational map $\tau \colon \Theta \to \mathbf{P}^4$ given by associating to each singular quadric its vertex. Set $P = (1, 1, 1, 1, 1) \in \mathbf{P}^4$; then the N-orbit of P in $\mathbf{P}^4$ is dense, and so $\Theta/N \simeq \tau^{-1}(P)/\mathrm{Stab}(P)$. Let $\delta \subset T$ denote the diagonal subgroup; then $\mathrm{Stab}(P) = \delta \times S_5$, and since δ acts trivially we have $\Theta/N \simeq \tau^{-1}(P)/S_5$. As an S_5-space, we have $U_1 = \mathbf{1} \oplus U_3$, where $\mathbf{1}$ denotes the trivial representation, and $\tau^{-1}(P)$ can be identified with the space of quadrics in $\mathbf{P}(U_3)$ (given a quadric in $\mathbf{P}(U_3)$, take the cone over it with vertex P). So $\tau^{-1}(P)/S_5 \simeq \mathbf{P}(\mathrm{Symm}^2 U_3)/S_5$. Now $\mathrm{Symm}^2 U_3 \cong U_3 \oplus U_4$, where U_4 is a 6-dimensional S_5-space, and so the "no-name" method described in Dolgachev's talk shows that $\mathbf{P}(\mathrm{Symm}^2 U_3)/S_5$ is rational. Hence M_4 is rational.   Q.E.D.

2. THEOREM 5. *If $g \equiv 2$ (4) and $g \geq 6$, then the locus T_g of trigonal curves of genus g is rational.*

Notation. For any integer $n \geq 0$, we denote the surface $\mathbf{P}(\mathcal{O}_{\mathbf{P}^1} \oplus \mathcal{O}_{\mathbf{P}^1}(-n))$ by $\mathbf{F}_n$. If $n > 0$, then f will denote a fibre of the natural map $\pi \colon \mathbf{F}_n \to \mathbf{P}^1$ and σ the section of π with $\sigma^2 = -n$. If $n = 0$, then we let f_1, f_2 denote fibres of the two projections $\mathrm{pr}_1, \mathrm{pr}_2 \colon \mathbf{F}_0 \to \mathbf{P}^1$.

Recall that for a nonhyperelliptic trigonal curve C of genus $g \geq 4$, the canonical model $\tilde{C}$ lies on a unique rational scroll or cone $\mathbf{F}_n^*$ (the image of some $\mathbf{F}_n$ by a complete linear system $|\sigma + rf|$, $r \geq n$, or $|f_1 + rf_2|$) which is the intersection of the quadrics containing $\tilde{C}$. Also, n is even if and only if g is even; this follows from the adjunction formula and the fact that $\tilde{C} \sim 3\sigma + sf$ or $3f_1 + sf_2$, $s \geq 3n$. Finally if $n \geq 2$, then $\mathbf{F}_n^*$ can be projectively deformed in $\mathbf{P}^{g-1}$ to $\mathbf{F}_{n-2}^*$ and $\tilde{C}$ can be deformed along with it. Hence if g is even and C is a generic member of T_g, then $n = 0$. From this we deduce the following result.

PROPOSITION 6. *If g is even and $g \geq 6$, then T_g is birationally equivalent to the quotient*

$$\mathbf{P}(H^0(\mathcal{O}_{\mathbf{F}_0}(3f_1 + (\tfrac{1}{2}g + 1)f_2)))/(\mathrm{SL}_2 \times \mathrm{SL}_2).$$

PROOF OF THEOREM 5. Set $V(D) = \mathrm{Symm}^D(\mathbf{C}^2)$. By Proposition 6, it is enough to show that $\mathbf{P}(V(3) \otimes V(2K))/(\mathrm{SL}_2 \times \mathrm{PGL}_2)$ is rational. Let $f \in V(3) \otimes V(2K)$; then

$$f = \sum_{i,j} \binom{3}{i} \cdot \binom{2K}{j} \alpha_{ij} \cdot x_1^{3-i} x_2^i \otimes y_1^{2K-j} y_2^j,$$

where $\{x_1, x_2\}$ and $\{y_1, y_2\}$ are homogeneous coordinates on the two copies of $\mathbf{P}^1$. Symbolically, we can write $f = a_x^3 \otimes A_y^{2K} = b_x^3 \otimes B_y^{2k}$, where $a_x = a_1 x_1 + a_2 x_2$,

etc. So
$$\alpha_{ij} = a_1^{3-i} a_2^i A_1^{2K-j} A_2^j = b_1^{3-i} b_2^i B_1^{2K-j} B_2^j.$$

We now write down a covariant (i.e., an $\mathrm{SL}_2 \times \mathrm{PGL}_2$-equivariant map) $V(3) \otimes V(2K) \xrightarrow{\varphi} V(2) \otimes V(0)$ as follows: $\varphi(f) = (ab)^2 \cdot (AB)^{2K} \cdot a_x b_x$, where $(ab) = a_1 b_2 - a_2 b_1$, etc. Then φ induces a rational map ψ: $\mathbf{P}(V(3) \otimes V(2K)) \to \mathbf{P}(V(2) \otimes V(0))$. Let γ: $\tilde{\mathbf{P}} \to \mathbf{P}(V(3) \otimes V(2K))$ be the blow-up along the base locus of ψ and $\tilde{\psi}$: $\tilde{\mathbf{P}} \to \mathbf{P}(V(2) \otimes V(0))$ the induced morphism.

Let $T \subset \mathrm{SL}_2$ be the torus of diagonal matrices and N its normalizer in SL_2. Then the point $P = x_1 x_2 \otimes 1$ is an $(\mathrm{SL}_2 \times \mathrm{PGL}_2, N \times \mathrm{PGL}_2)$ section of $\mathbf{P}(V(2) \otimes V(0))$.

LEMMA 7. $\gamma(\tilde{\psi}^{-1}(P))$ *is irreducible of codimension* 2.

PROOF. Let Q_1 (resp. Q_2) be the coefficient of $x_1^2 \otimes 1$ (resp. $x_2^2 \otimes 1$) in $\varphi(f)$; then $\gamma(\tilde{\psi}^{-1}(P))$ is defined by the equations $Q_1 = Q_2 = 0$. Expansion of the formula defining φ shows that

$$Q_1 = \sum_{i=0}^{2K} (-1)^i \binom{2K}{i} \alpha_{0,i} \cdot \alpha_{2,2K-i} - 2 \sum_{i=0}^{2K} (-1)^i \binom{2K}{i} \alpha_{1,i} \cdot \alpha_{1,2K-i}$$

$$+ \sum (-1)^i \binom{2K}{i} \alpha_{2,i} \cdot \alpha_{0,2K-i}$$

$$= 2 \sum_{i=0}^{2K} (-1)^i \binom{2K}{i} [\alpha_{0,i} \cdot \alpha_{2,2K-i} - \alpha_{1,i} \cdot \alpha_{1,2K-i}]$$

and

$$Q_2 = 2 \sum_{i=0}^{2K} (-1)^i \binom{2K}{i} [\alpha_{3,i} \cdot \alpha_{1,2K-i} - \alpha_{2,i} \cdot \alpha_{2,2K-i}].$$

Let L be the linear subspace of $\mathbf{P}(V(3) \otimes V(2K))$ defined by the equations $\alpha_{1,i} = \alpha_{2,j} = 0 \; \forall i, j$, and M the complementary subspace defined by the equations $\alpha_{0,i} = \alpha_{3,j} = 0 \; \forall i, j$. Then projection away from L induces an $(N \times \mathrm{PGL}_2)$-equivariant rational map π: $\gamma(\tilde{\psi}^{-1}(P)) \to M$ with linear fibres. Let R_{ij} be the point defined by the equation $\alpha_{k,l} = 0 \; \forall (k,l) \neq (i,j)$; then $\pi^{-1}(R_{1,i})$ is an irreducible linear space of codimension $4K + 3$ in $\mathbf{P}(V(3) \otimes V(2K))$ if $i \neq K$, by inspection of Q_1 and Q_2. Hence $\gamma(\tilde{\psi}^{-1}(P))$ has a unique irreducible component W_1 that is mapped generically surjectively to M via π. Suppose that it has another irreducible component W_2; W_2 has codimension ≤ 2 in $\mathbf{P}(V(3) \otimes V(2K))$. Let $H \subset M$ be the linear span of the points $R_{1,0}, \ldots, R_{1,K-1}, R_{1,K+1}, \ldots, R_{1,2K}$. The torus $T \times T$ acts linearly on M with distinct weights, and so the only fixed points are the R_{ij}. Moreover, $T \times T$ acts on W_2, and so $\pi(W_2) \cap H = \varnothing$. But this is impossible, since $\dim W_2 \geq 8K + 2$. This completes the proof of Lemma 7.

We return to the proof of Theorem 5. It will suffice for us to prove that $\gamma(\tilde{\psi}^{-1}(P))/(N \times \mathrm{PGL}_2)$ is rational. Notice that $M \cong \mathbf{P}(U)$, where U is the representation $(\mathbf{C} \cdot a_1^2 a_2 \oplus \mathbf{C} \cdot a_1 a_2^2) \otimes V(2K)$ of $N \times \mathrm{PGL}_2$, which is almost free, and L, M, π are as in the proof of Lemma 7. Then projection away from π shows that $\gamma(\tilde{\psi}^{-1}(P))/(N \times \mathrm{PGL}_2)$ is ruled over $M/(N \times \mathrm{PGL}_2)$; the generic fibre has dimension $4K$. Let W be the representation $((\mathbf{C} \cdot a_1^2 a_2 \oplus \mathbf{C} \cdot a_1 a_2^2) \otimes V(0)) \oplus (V(0) \otimes V(6))$ of $N \times \mathrm{PGL}$; it is almost free and the quotient $\mathbf{P}(W)/(N \times \mathrm{PGL}_2)$ is birational to $(\mathbf{P}(\mathbf{C} \cdot a_1^2 a_2 \oplus \mathbf{C} \cdot a_1 a_2^2)/N) \times (\mathbf{P}(V(6))/\mathrm{PGL}_2) \times \mathbf{P}^1$, which is rational by the invariant theory of binary sextics. Projecting the quotient $(M \times \mathbf{P}(W))/(N \times \mathrm{PGL}_2)$ onto $M/(N \times \mathrm{PGL}_2)$ and $\mathbf{P}(W)/(N \times \mathrm{PGL}_2)$ shows that $(M/(N \times \mathrm{PGL}_2)) \times \mathbf{P}(W)$ is rational. Since $\dim \mathbf{P}(W) = 8$ and $4K \geq 8$, it follows that T_g is rational. Q.E.D.

3. THEOREM 8. *If g is odd and $g \geq 5$, then the space $T_{g,1}$ of pointed trigonal curves of genus g is rational.*

PROOF. The canonical model C of a generic trigonal curve of odd genus $g \geq 5$ lies on a rational scroll $\mathbf{F}_1$, and $C \sim 3\sigma + \frac{1}{2}(g+5)f$. Contracting σ to a point, we map C to a plane curve of degree $D = \frac{1}{2}(g+5)$ with a $(D-3)$-fold point. Fix 2 distinct points $P, Q \in \mathbf{P}^2$, and set $H = \mathrm{Stab}(P) \cap \mathrm{Stab}(Q) \subset \mathrm{PGL}(3)$. Then if W denotes the projective family of curves of degree D that pass through Q and have a $(D-3)$-fold point at P, we see that $T_{g,1} \simeq W/H$. But H is a connected soluble group, and so by Vinberg's theorem $T_{g,1}$ is rational. Q.E.D.

Appendix. In this appendix we work over a field k of characteristic $\neq 2, 3$ whose algebraic closure is $\overline{k}$. Our aim is to show that the k-variety M_4 is rational over k.

Let $Q \subset \mathbf{P}_k^3$ be a fixed quadric which is isomorphic to $\mathbf{P}_k^1 \times \mathbf{P}_k^1$ (so that if G is the k-group scheme $\mathrm{Aut}\, Q$, then $G(\overline{k}) \cong \mathrm{Aut}(Q \times \overline{k})$).

PROPOSITION 9. *M_4 is birationally equivalent to $\mathbf{P}(H^0(\mathcal{O}_Q(3)))/\mathrm{Aut}\, Q$ over k.*

PROOF. Let $\mathbf{P}_0 \subset \mathbf{P} = \mathbf{P}(H^0(\mathcal{O}_Q(3))^v)$ denote the open subscheme whose geometric points correspond to smooth cubic sections of Q. Let $\Gamma \to \mathbf{P}_0$ denote the universal family. We shall show that $M = \mathbf{P}_0/\mathrm{Aut}\, Q$ (which exists as a geometric quotient) is the coarse moduli space associated to the functor $F: Sch \to Sets$, where $F(S)$ is the set of isomorphism classes of families $C \to S$ of smooth curves of genus four whose geometric fibres have exactly two g_4^1's. We proceed to check this according to the definition of [1, Chapter 5]. Note first that $M(\overline{k})$ is certainly in bijection with $F(\overline{k})$. Define a morphism $\varphi: F \to h_M = \mathrm{Hom}_k(-, M)$ as follows: for a k-scheme S, let $C \to S$ be a (representative of) an element of $F(S)$. From the definition of F, the relative canonical model of $p: C \to S$ is a cubic section on some smooth quadric bundle $\underline{Q} \to S$ defined in $\mathbf{P}((p_*\omega_{C|S})^\vee)$ by the vanishing of $\ker(\mathrm{Symm}^2(p_*\omega_{C|S}) \to p_*\omega_{C|S}^2)$. There is an étale cover $\{T_i \to S\}$ of S such that for all i, $\underline{Q} \times_S T_i \cong Q \times_k T_i$, relative to T_i. Let S_i denote the image of T_i in S; we may assume that each T_i is Galois over S_i, with group G_i, say. Put $C_i = C \times_S T_i$; there is a Cartesian diagram

$$
\begin{array}{ccc}
C_i & \to & \Gamma \\
\downarrow & & \downarrow \\
T_i & \to & \mathbf{P}_0.
\end{array}
$$

Let $\gamma_i \colon T_i \to M$ be the composite; we need to know that it factors through $\beta_i \colon T_i \to S_i$. Note first that this is obvious if $k = \overline{k}$. Then the problem reduces to the situation where we have maps $B \hookrightarrow A \xleftarrow{\alpha} R$ of k-algebras such that $(\alpha \otimes 1)(R \otimes \overline{k}) \subseteq B \otimes \overline{k}$, and we must show that $\alpha(R) \subseteq B$. This, however, is obvious. So we have maps $f_i \colon S_i \to M$, which glue together, since the maps γ_i, and so f_i, are unique, to a morphism $f \colon S \to M$. Now define φ by $\varphi(S)([C \to S]) = f$. It remains to check that any natural transformation $\psi \colon F \to h_N$ factors through φ. Define a morphism $\chi \colon \mathbf{P}_0 \to N$ by

$$
\chi(S)(S \to \mathbf{P}_0) = \psi(S)([\Gamma \times_{\mathbf{P}_0} S \to S]),
$$

where $[X \to Y]$ denotes the isomorphism class of the morphism $X \to Y$. Then χ is Aut Q-equivariant, and so factors as $\chi = \omega \circ \pi$, where $\pi \colon \mathbf{P}_0 \to M$ is the quotient map and $\omega \colon M \to N$ is determined by χ. We have $\psi = h_\omega \circ \varphi$, and the proposition is proved.

The proof that M_4 is a rational k-variety is now very similar to that given in §1 for the case $k = \mathbf{C}$. The only difference is we must replace the group μ_3 of cube roots of unity by the group scheme $\mu_3 = \operatorname{Spec} k[t]$, where $k[t] = k[T]/(T^3 - 1)$. For each four-dimensional vector space $V_i = \sum_{j=1}^4 k \cdot \alpha_{ij}$, $i \in [1,5]$, there is a co-action $\sigma_i \colon V_i \to V_i \otimes_k k[t]$ given by $\sigma_i(v) = v \otimes t$, and so an action of μ_3 on V_i. This gives an action of μ_3^5 on $\mathbf{P}(V_1 \oplus \cdots \oplus V_5)$, and now the proof proceeds exactly as before.

REFERENCES

1. D. Mumford and J. Fogarty, *Geometric invariant theory*, 2nd ed., Springer-Verlag, 1982.
2. H. W. Richmond, *On canonical forms*, Quart. J. Math. **33** (1902), 331–340.
3. H. Weyl, *The classical groups*, 2nd ed., Princeton Univ. Press, Princeton, N.J., 1946.

COLUMBIA UNIVERSITY

Current address: University of Illinois at Chicago

Surfaces

Proceedings of Symposia in Pure Mathematics
Volume 46 (1987)

Canonical Rings and "Special" Surfaces of General Type

F. CATANESE

0. Introduction. These lecture notes are meant to survey some very recent results on surfaces of general type, focusing on what in Persson's talk has been called the botanical problem of surface geography: given a pair of integers K^2, χ in the allowed region, classify all the minimal surfaces of general type S with $K_S^2 = K^2$, $\chi(\mathcal{O}_S) = \chi$.

As pointed out in the title, this purpose can be achieved only for special values of the integers K^2, χ or by virtue of other special features of the surfaces under consideration; we are going to describe some general methods which can be used to attack the "botanical" problem, but even when the method is theoretically effective, the computations become rapidly in practice intractable for general values of (K^2, χ) (embarrassingly enough, even the classification of surfaces with $K^2 = \chi = 1$ has not yet been completely achieved!). On the other hand, as we shall see, special geometric properties of the surface under consideration (such as the existence of a fibration $f: S \to B$ onto a curve B of given genus, or with fibres of a given genus) are sometimes forced by the fact that the invariants (K^2, χ) satisfy some equalities or inequalities.

By one of the possible definitions, a surface S is of general type if there exists a positive integer m such that the rational map ϕ_m associated to the linear system $|mK_s|$ (the mth-canonical map) is a birational map onto its image $\Sigma_m \subset \mathbf{P}^{N(m)}$ (where $N(m) = P_m(S) - 1 = \chi(\mathcal{O}_S) + \frac{1}{2}m(m-1)K_S^2 - 1$); it should be rather clear, and one can see many beautiful examples in Enriques's book [**En**], that one should seek a small value of m (say $m = 1$, or 2) for which ϕ_m has a good behavior: if for instance ϕ_m is a birational morphism, then S is birational to a surface (Σ_m) of small codimension, and one has more hope to find an explicit description of S.

In fact, pre-Arcata work by Moishezon, Kodaira, and Bombieri was devoted to the fulfillment of Enriques's program, i.e., to the study of pluricanonical maps [**Sh, Ko, Bo 1, Bo 2**] and one of the achieved goals was to prove that there

1980 *Mathematics Subject Classification* (1985 *Revision*). Primary 14J10.

Author partly supported by AMS and MPI for his participation at the AMS Summer Institute on Algebraic Geometry.

exists an integer, 5, "good" for all surfaces of general type, thereby showing in particular (cf. [**B-H, Gi, B-P-V**]) that the surfaces with given invariants K^2, χ belong to a finite number of algebraic families.

Especially Bombieri's results in [**Bo 2**] were rather sharp and indicated all possible exceptions to the statements, save for the case of surfaces with $K^2 = 2, 1$ and $p_g = 0$, which seemed to deserve a special treatment (at the time only two examples of surfaces with those invariants were known).

Bombieri's paper stimulated research by several authors devoted to the study of the structure of surfaces with low values of the invariants: the interested reader may find a very good account in Chapter VII of the recent book *Complex surfaces* by Barth–Peters–Van de Ven [**B-P-V**].

We refer also to [**B-P-V**] and to the lecture notes by Barth and Persson (this volume) for the treatment of other very important progress in the theory of surfaces of general type, in particular (referring to our "botanical" theme) Horikawa's work [**Ho 2**] on surfaces with small K^2 (i.e., K^2 small with respect to χ, in particular $K^2 = 2\chi - 6$, $K^2 = 2\chi - 5$), and surfaces of positive index. An important new idea, first introduced by Mumford [**Mu 1**], and concretely exploited then by Reid (cf., e.g., [**Re 1**]), made its way in the meantime: to look at the finitely generated graded ring

$$\mathcal{R}(S) = \bigoplus_{m=0}^{\infty} H^0(S, \mathcal{O}_S(mK_S)),$$

the so-called canonical ring, as the main object of study.

In fact $X = \mathrm{Proj}(\mathcal{R}(S))$, the so-called canonical model of S, is a normal surface of which S is a minimal resolution of singularities: moreover the dualizing sheaf ω_X is invertible and the resolution morphism $\pi\colon S \to X$ is such that $\omega_S = \mathcal{O}_S(K_S) = \pi^*\omega_X$ (in particular, X has only rational double points as singularities). Hence all the pluricanonical mappings ϕ_m factor through π. X is a subvariety of a weighted projective space (cf. [**Do 2**]) and, for extremely special surfaces, it can turn out to be a (weighted) complete intersection. We shall give several simple and explicit examples of how one can describe the canonical ring, explaining also in §3 a couple of general methods (Godeaux-Reid's method of using torsion and unramified covers, our method of quasigeneric canonical projection) which can be applied with success to determine the canonical rings (hence also the minimal models) of surfaces of general type.

On the other hand, going back to the classical point of view, general results about the pluricanonical and canonical maps are not simply technical tools used to determine the mode of generation of the canonical ring.

In fact, there is an underlying philosophy that I'll try now to explain, starting to compare with the case of curves. It is well known that the bicanonical map of a curve C of genus $g \geq 3$ is an isomorphism onto its image, and the canonical map is also an isomorphism except if the curve is hyperelliptic, i.e., if C is a double cover of $\mathbf{P}^1$ branched on $2g + 2$ points. Instead, for curves of genus 2, neither the canonical nor the bicanonical map are birational. The outcome is the following: the worse the behavior of the canonical and bicanonical maps, the easier to describe the curve in very simple terms. Does the same occur for surfaces?

Already from Bombieri's paper [**Bo 2**] one sees that the exceptions to the birationality of the tricanonical map are only two: these surfaces are to be thought of as the closest analogues of genus 2 curves and, as we shall see in §1, they admit a very simple description.

What is then the analogue of hyperelliptic curves? The basic observation is that if $C \subset S$ is a curve of genus two and $C^2 = 0$, then $K \cdot C = 2$ and the restriction to C of the canonical map of S either sends C to a point, or sends C to a projective line, and in this case $\mathcal{O}_C(C) \cong \mathcal{O}_C$; furthermore, if $\mathcal{O}_C(C) \cong \mathcal{O}_C$, then also the bicanonical map of S restricts to a projection of the bicanonical map of C, which has degree at least 2.

In particular, if C belongs to a pencil of curves of genus 2, $\mathcal{O}_C(C) \cong \mathcal{O}_C$ and then the bicanonical map of S must have degree at least 2 (as we shall see in §1, the image Σ_2 of the bicanonical map ϕ_2 of S is almost always a surface). Likewise, if C belongs to a pencil of hyperelliptic curves, then the canonical map cannot be birational onto its image.

The point is (cf. §§1, 2) that, while the exceptions to the birationality of ϕ_2, apart from a finite number of families (cf. [**Bo 2, Fr**]), are due to the existence of a pencil of genus 2 curves (and then ϕ_2 yields a double cover of a ruled or rational surface), the situation for the canonical map is more complicated, sometimes nasty, and it is still dubious whether a totally clear picture will eventually emerge. In fact, when ϕ_1 is not birational, several things can happen, as has been shown by Beauville [**Be 1, Be 3**] and Xiao [**X 2, X 3**]: Σ_1 can be any surface with $p_g = 0$ (this is in some sense still an analogue of the case of curves, where the canonical image is $\mathbf{P}^1$, i.e., a curve of genus 0, if C is hyperelliptic), but moreover there are infinite examples where Σ_1 is also a canonically embedded surface.

In case Σ_1 is a curve, Xiao has proved that the genus of Σ_1 is at most 1, while Beauville had proved that the genus of the fibres of ϕ_1 is bounded (effective bounds can be given, e.g., 5 holds for $p_g \geq 20$).

But, up to now, it is totally unknown whether there can be any region in the surface geography where the canonical map is birational, and it is only known [**X 3**] that, for large values of K^2, χ, the degree of ϕ_1 is at most 6; from the point of view of surface classification, however, the case when ϕ_1 has degree at least 3 does not seem for the time being feasible to get hold and elucidate the structure of special surfaces.

As a final remark, we have mostly been trying in this exposition to present statements of theorems and concrete examples (without these last, general results about surfaces can hardly be understood). Also, in this survey, we have not tried by any means to strive for completeness; in particular, the references appear mainly when cited in the text: we apologize for the many omissions we have made.

Moreover, it is our belief that the major progress in the field done in the last ten years owes much to a lot of humble but skillful work on very special classes of surfaces: all this experimental material has been (and is) a rich humus and has allowed a better understanding of general properties of surfaces of general type.

For more broad recent surveys on surfaces of general type, we refer also to [**Ca 6**] and [**Ci 1**].

NOTATION.

S: a minimal model of a surface of general type (over **C**).

For D, C Cartier divisors, $D \equiv C$ denotes linear equivalence, $D \sim C$ denotes numerical equivalence, and $|D|$ is the linear system of effective divisors $C \equiv D$.

Furthermore, an effective divisor D is said to be m-connected if, for every decomposition $D = A + B$, where A, B are effective divisors, one has $A \cdot B \geq m$.

$K = K_S$: a canonical divisor.

$\mathcal{R} = \mathcal{R}(S) = \bigoplus_{m=0}^{\infty} H^0(\mathcal{O}_S(mK)) = \bigoplus_{m=0}^{\infty} \mathcal{R}_m$: the canonical ring of S.

$X = \mathrm{Proj}(\mathcal{R}(S))$: the canonical model of X, with $\pi: S \to X$ the canonical morphism.

$\phi_m =$ the mth canonical map (i.e., the rational map associated to the linear system $|mK_S|$) $\cdot \Sigma_m = \phi_m(S)$. For x, y points of S, $|D - x - y|$ is the linear system of divisors of sections of $\mathcal{O}_S(D - x - y) = M_x M_y \mathcal{O}_S(D)$.

$q = \dim_{\mathbf{C}} H^1(\mathcal{O}_S)$, the irregularity of S.

$p_g = \dim_{\mathbf{C}} H^2(\mathcal{O}_S)$, the geometric genus of S.

$p_a = p_g - q = \chi(\mathcal{O}_S) - 1$, the arithmetic genus of S.

P_m: the mth plurigenus of S, i.e., $\dim_{\mathbf{C}} H^0(S, \mathcal{O}_S(mK))$, equals, by [**Ko**], $\chi(\mathcal{O}_S) + (m(m-1)/2)K^2$.

For a normal C.M. variety Z of dimension n, if Z^0 is the smooth part of Z, the dualizing sheaf ω_Z is $j_*(\Omega^n_{Z^0})$, where $j: Z^0 \to Z$ is the inclusion morphism.

Since $\pi^*(\omega_X) = \omega_S = \mathcal{O}_S(K)$, the pluricanonical maps $\phi_m: S \to \Sigma_m$ factor as $\phi_m = \psi_m \circ \pi$, where ψ_m is the rational map $\psi_m: X \to \Sigma_m$ given by the sections of $(\omega_X^{\otimes m})$.

π contracts the so-called (-2) curves: irreducible curves E with $E^2 = -2$, $E \cdot K = 0$ (hence $E \cong \mathbf{P}^1$).

Tors(S): the torsion group of S, i.e., the torsion subgroup of $H_1(S, \mathbf{Z})$ (= torsion subgroup of $H^2(S, \mathbf{Z})$).

For C a divisor in S the dualizing sheaf ω_C is $\mathcal{O}_C(K+C)$ (adjunction formula).

$\mathbf{P}(e_0, \ldots, e_n)$: the weighted projective space of degrees $e_0, \ldots, e_n$ ($e_i \in \mathbf{N}$) is the quotient $\mathbf{C}^{n+1} - \{0\}/\mathbf{C}^*$, where $\lambda \in \mathbf{C}^*$ acts on $x = (x_0, \ldots, x_n)$ by the formula $\lambda x = (\lambda^{e_0} x_0, \ldots, \lambda^{e_n} x_n)$

$\pi_1^{\mathrm{alg}}(S)$: the algebraic fundamental group, the inverse limit of the quotients of the topological fundamental group $\pi_1(S)$ by normal subgroups of finite order.

The Iitaka dimension $K(S, D)$ of a divisor D is, as usual, equal to $-\infty$ if $|mD| = \varnothing \ \forall m \geq 1$, and otherwise equal to the maximum of the dimensions of the images of the rational maps associated to the linear systems $|mD|$.

1. Pluricanonical maps. At the Arcata conference the following theorem was the best available result on pluricanonical maps (cf. [**Bo 2**]).

1.1. BOMBIERI'S THEOREM. *Let X be the canonical model of a surface of general type, and let $\psi_m \colon X \to \Sigma_m$ be the mth canonical map. Then*

(i) ψ_m *is an isomorphism for $m \geq 5$, for $m = 4$ if $K^2 \geq 2$, for $m = 3$ if $K^2 \geq 6$ or $K^2 \geq 3$ and $p_g \geq 4$.*

(ii) ψ_m *is birational for $m \geq 3$ except if $K^2 = 2$, $p_g = 3$ $(m = 3)$ or if $K^2 = 1$, $p_g = 2$ $(m = 3, 4)$ and, possibly, if $K^2 = 1$, $p_g = 0$ $(m = 3, 4)$ or $K^2 = 2$, $p_g = 0$ $(m = 3)$.*

(iii) ψ_2 *is birational for $K^2 \geq 10$, $p_g \geq 6$, except if S has a pencil of curves of genus 2.*

(iv) ψ_m *is a morphism for $m \geq 4$, for $m = 3$ and $K^2 \geq 3$ (or $K^2 \geq 2$, $p_g \geq 1$), for $m = 2$ and $K^2 \geq 5$, $p_g \geq 3$ (or $p_g \geq 3$, $q = 0$).*

At the end of the section we shall summarize, for the reader's convenience, the best results available nowadays for the pluricanonical maps ψ_m $(m \geq 2)$.

1.2. REMARK. The possible exceptions mentioned in the statement (ii) of Bombieri's theorem have in fact later been shown not to occur, through work of several authors (see e.g. [**Mi, B-C, Ca 1**]).

1.3. REMARK. As we stressed in the introduction, the presence of a pencil of curves of genus 2, $f \colon S \to B$ (where f can be just a rational map if $B \cong \mathbf{P}^1$), forces ψ_2 not to be birational, since, if F is a fibre of f, then $\mathcal{O}_F(K) \cong \omega_F$. Nevertheless these surfaces are very special and amenable to a very detailed description (cf. [**X 1**]).

In fact the hyperelliptic involution which is defined for each fibre is induced by a biregular involution $i \colon S \to S$, and S/i is a ruled surface Y (this is the point of view adopted by Horikawa in [**Ho 1**]): more precisely, if S' is a blow-up of S on which f becomes a morphism (cf. [**X 1**]), $Y = \mathbf{P}(f_* \omega_{S'/B})$ is a $\mathbf{P}^1$-bundle and S' has a double cover $g' \colon S' \to Y$ (g' is not necessarily finite, due to the presence of (-2) curves), yielding a finite double cover $g \colon X \to Y$.

Since a genus 2 curve is a double cover of $\mathbf{P}^1$ branched in 6 points, it follows that the branch curve Δ of g is a reduced curve which has 6 points of intersection with the general fibre of Y. Conversely, every such double cover $g \colon X \to Y$ gives a surface with a pencil of curves of genus 2, and one of the next problems is whether the pencil of curves of genus 2 is unique for a surface of general type.

Now, $\psi_2 \colon X \to \Sigma_2$ clearly factors through $g \colon X \to Y$; hence the pencil is unique if Σ_2 is a surface and ψ_2 has degree 2. Xiao [**X 1**, 6.4, 6.5] proves that either S is a product of 2 curves of genus 2, or the pencil is unique if $K^2 \geq 5$ (he also classifies the possible exceptions).

It is curious to observe that, in spite of the simple nature of these surfaces, the only restriction to which their invariants (χ, K^2) are subject is the inequality $K^2 \leq 8\chi$ (sharper than the Bogomolov-Miyaoka-Yau inequality $K^2 \leq 9\chi$); in fact Persson [**Pe**] was able to solve the geographical question posed by Van de Ven in [**Ve**] by showing that all the invariants (χ, K^2) with $\chi, K^2 \geq 1$, $K^2 \geq 2\chi - 6$,

$K^2 \leq 8\chi - 20$, occur for some surface S carrying a pencil of curves of genus 2. This fact partly justifies the analogy set up in the introduction with hyperelliptic curves.

1.3. EXAMPLE. The surfaces which give exceptions to statement (ii) of Bombieri's theorem admit a very simple description, as we are going to show.

Surfaces with $p_g = 3$, $K^2 = 2$ belong to the I Horikawa line $K^2 = 2p_g - 4$ and as we saw in Persson's lecture (cf. [Ho 2, I]) ψ_1 has no base points; thus $\psi_1 \colon X \to \mathbf{P}^2$ is a double cover branched on a curve Δ of degree 8 (and with negligible singularities). If $f_8(x_0, x_1, x_2) = 0$ is an equation for Δ, the canonical ring $\mathcal{R} = \mathcal{R}(S)$ is $\mathbf{C}[x_0, x_1, x_2, z]/(z^2 - f_8(x_0, x_1, x_2))$; therefore X is a hypersurface of degree 8 in the 3-dimensional weighted projective space $\mathbf{P}(1, 1, 1, 4)$.

Similarly, in the case $p_g = 2$, $K^2 = 1$, one takes a basis $\{y_0, y_1\}$ of $\mathcal{R}_1 = H^0(\mathcal{O}_S(K))$, and completes the independent set $\{x_0 = y_0^2, x_1 = y_0 y_1, x_2 = y_1^2\}$ to a basis $\{x_0, x_1, x_2, x_3 = x\}$ of $\mathcal{R}_2$. One has the obvious relation $x_0 x_2 = x_1^2$ and, since one can show that $|2K|$ has no base points, the image of the bicanonical map is the quadric surface Σ_2 of equation $x_0 x_2 - x_1^2 = 0$.

Since $P_m = \dim_{\mathbf{C}} \mathcal{R}_m = 3 + m(m-1)/2$, a quick computation shows that the monomials in y_0, y_1, and x span $\mathcal{R}_3$ and $\mathcal{R}_4$, but one more generator, call it z, is needed for $\mathcal{R}_5$. We have seen that $\mathcal{R}$ contains the subring $\mathbf{C}[y_0, y_1, x]$; moreover $\mathbf{C}[y_0, y_1, x]$ and $z\mathbf{C}[y_0, y_1, x]$ are direct summands, since they belong to the $(+1)$, respectively (-1), eigenspace for the involution on $\mathcal{R}$ determined by the covering involution $i \colon S \to S$ associated to the degree 2 map $\phi_2 \colon S \to \Sigma_2$. Thus $\mathcal{R} \supset \mathbf{C}[y_0, y_1, x] \oplus z\mathbf{C}[y_0, y_1, x]$, but the graded pieces of degree m have the same dimension $\forall m \geq 1$; hence equality holds. Since z^2 belongs to the $(+1)$ eigenspace, there exists a (weighted homogeneous) polynomial $F_{10}(y_0, y_1, x)$ of degree 10 s.t.

$$\mathcal{R} = \mathbf{C}[y_0, y_1, x, z]/(z^2 - F_{10}(y_0, y_1, x)).$$

Again X is a hypersurface, of degree 10 in $\mathbf{P}(1, 1, 2, 5)$, and all such hypersurfaces with R.D.P.'s (= Rational Double Points) as singularities give rise to the minimal model of a surface with $K^2 = 1$, $p_g = 2$.

The 2 surfaces we have considered in this example belong to the I and II Horikawa lines; surfaces with $K^2 = 1$, $p_g = 2$ will be considered later in this paper also for the slow generation of their canonical ring, and are treated here in a different way than in [Ho 2]: this approach to Horikawa surfaces has been adopted by Iliev and Griffin [Il 3, Gri].

Improvements upon Bombieri's result have been recently obtained by Francia [Fr] and Reider [Rei].

1.4. FRANCIA'S THEOREM. (i) ϕ_m *is a morphism for* $m = 2$ *provided* $p_g \geq 1$, *except possibly when* $p_g = q = 1$ (*in this case, though,* $|2K|$ *has no fixed part*).

(ii) *If* $K^2 \geq 10$, ϕ_2 *is birational unless* S *has a pencil of curves of genus 2.*

(iii) $\psi_2 \colon X \to \Sigma_2$ *is an isomorphism either if*

(iiia) $p_g \geq 6$ *and divisors* $D \in |K_X|$ *are 3-connected, or if*

(iiib) $K^2 \geq 10$, $p_g \geq 6$, *unless there exists a curve* C *with either* $C^2 = 0$, $KC = 2$ *(a genus 2 curve!) or with* $C^2 = -1$, $KC = 1$ *(elliptic).*

Before giving an idea of the new ingredients employed in the proof, let's comment again on the result.

1.5. REMARK. Surfaces with $q = p_g = 1$ must have $2 \leq K^2 \leq 9$. The case $K^2 = 2$ has been completely described in [**Ca 7**], and in fact $|2K|$ is base point free; Ciliberto, Francia, and the author can prove existence of such surfaces for $K^2 = 3$, but the same ideas don't work properly for $K^2 \geq 5$. Furthermore on surfaces with $K^2 = 1$, $p_g = 0$, clearly $P_2 = 2$ and thus $|2K|$ has base points, but also it is known (cf. [**Mi 1, Bo 2**]) that $|3K|$ has base points, when the torsion group T has elements of order different from 2.

1.6. REMARK. It is clear that the existence of curves of in (iiib), as we also noticed in the introduction, forces ψ_2 not to be an isomorphism. Ciliberto (in [**Ci 2**], to show existence of surfaces for which $\mathcal{R}$ is not generated by elements of degree ≤ 2) has constructed an infinite number of families of surfaces where curves of this sort do in fact appear (clearly then K is not 3-connected).

The hypothesis $K^2 \geq 10$ is essential in statement (ii) of the theorem, as was shown by Bombieri who exhibited in [**Bo 2**, pp. 193–194] an example of a minimal surface with $K^2 = 9$, $p_g = 6$, $q = 0$, with ϕ_2 not birational and without any curve of genus 2, but with a pencil (with one base point) of hyperelliptic curves of genus 3.

Francia proved (unpublished) some results inspired by this example: for instance, if $p_g \geq 4$, $q = 0$, $K^2 \neq 8$, and ϕ_2 is not birational, S has either a genus 2 pencil, or a pencil of hyperelliptic curves of genus 3 (having a base point), and then $6 \leq K^2 \leq 9$.

Sketch of proof of (ii), 1.4. Since $H^1(\mathcal{O}_S(2K)) = 0$ by the Mumford-Ramanujam vanishing theorem [**Mu 2, Ra**], and since we have the exact sequence of sheaves

$$0 \to \mathcal{O}_S(2K - x - y) \to \mathcal{O}_S(2K) \to \mathbf{C}^2 \to 0,$$

ϕ_2 separates the two points x and y if and only if

$$H^1(\mathcal{O}_S(2K - x - y)) = 0.$$

Let $\sigma \colon \tilde{S} \to S$ be the blow-up of S at the two points x, y, and let L, M be the 2 exceptional divisors. We have

$$H^1(\mathcal{O}_S(2K - x - y)) = H^1(\mathcal{O}_{\tilde{S}}(\sigma^*(2K) - L - M)),$$

which is the Serre-dual of $H^1(\mathcal{O}_{\tilde{S}}(-(\sigma^*K) + 2L + 2M))$.

If we set $D \equiv \sigma^*K - 2L - 2M$, what is wanted is the vanishing of $H^1(\mathcal{O}_{\tilde{S}}(-D))$, which would be implied, by the Bombieri-Ramanujam vanishing theorem [**Bo 2, Ra**], by the existence of an effective 1-connected divisor in $|D|$, provided $D^2 > 0$.

To have the existence of such an effective divisor, Bombieri has to assume $p_g \geq 6$, whereas Francia can relax this assumption by using the following restatement of Miyaoka's [**Mi 2**] generalization of the Mumford-Ramanujam vanishing theorem.

VANISHING THEOREM II. *Let D be a (not necessarily effective) divisor on a surface $\tilde{S}$ such that its Iitaka dimension $K(\tilde{S}, D)$ equals 2, and such that there exists a positive integer n with $|nD|$ containing a 1-connected divisor D'. Then $H^1(\mathcal{O}_{\tilde{S}}(-D)) = 0$.*

The hypothesis $K^2 \geq 10$ applies first to ensure that $P_2 = \chi + K^2$ is at least 11: then, if x and y are mapped to the same smooth point of Σ_2, then at most 10 linear equations have to be satisfied in order that a section of $H^0(\mathcal{O}_S(2K))$ vanish of order 4 at x and y. Therefore there exists an effective divisor $D' \in |\sigma^*2K - 4L - 4M|$, and one can apply the vanishing theorem unless D' is not connected (in fact $D'^2 > 0$).

In this last case, analyzing the disconnecting decomposition of D', and using $K^2 \geq 10$, Francia concludes the proof.

Another very useful vanishing theorem, used for the proof of statement (i), is the following:

VANISHING THEOREM I. *Let $\sigma \colon \tilde{S} \to S$ be a birational morphism, and D an effective 1-connected divisor on $\tilde{S}$ with $D \cdot \sigma^*(\sigma_* D) > 0$. Then $H^1(\mathcal{O}_{\tilde{S}}(-D)) = 0$.*

1.7. REIDER'S THEOREM. *ϕ_m is a morphism for $m = 2$ and $K^2 \geq 5$, $m = 3$ and $K^2 \geq 2$, $m \geq 4$. ϕ_m is an embedding for $m = 4$ and $K^2 \geq 2$, $m = 3$ and $K^2 \geq 3$.*[*]

We reproduce Reider's beautiful and simple proof of the first assertion, noting that the last assertions follow with similar proof. Still he makes use of the connectedness property of pluricanonical divisors (used from [**Ko**] on), but there is also a new idea.

Sketch of proof of 1.7. If p is a base point of $|2K|$, by the Residue Theorem of [**G-H**] there exists a rank 2 locally free sheaf $\mathcal{E}$ on S, such that $c_1(\mathcal{E}) = K$, together with a section ξ whose scheme-theoretical zero locus is the reduced point p. Hence one has the Koszul exact sequence for $\mathcal{E}$

(1.8) $$0 \to \mathcal{O}_S \xrightarrow{\xi} \mathcal{E} \xrightarrow{\wedge \xi} \mathcal{O}_S(K - p) \to 0.$$

Since $c_1^2(\mathcal{E}) = K^2 \geq 5 > 4c_2(\mathcal{E}) = 4$, $\mathcal{E}$ is Bogomolov unstable and there exists (cf. [**Re 4**]) an extension

(1.9) $$0 \to \mathcal{O}_S(L) \to \mathcal{E} \to \mathcal{O}_S(K - L - Z) \to 0,$$

where Z is an effective 0-cycle, and L is a divisor such that $\Delta = 2L - K$ is in the positive cone of $NS(S)$. (Hence L is also a fortiori in the positive cone.) Hence

---

[*]Reider also re-proves (ii) of Francia's theorem.

it follows $(K - L) \cdot L > 0$: in fact, tensoring (1.8) and (1.9) with $\mathcal{O}_S(-L)$, we get $H^0(\mathcal{E}(-L)) \neq 0$, so that, since $H^0(\mathcal{O}_S(-L)) = 0$, $H^0(\mathcal{O}_S(K - L - p)) \neq 0$.

Thus there exists $D \in |K - L|$, and, L being positive, for $m \gg 0$, there exists also $D' \in |mL|$; now $|mK| \ni mD + D'$, and $(mD) \cdot D' = m^2 L(K - L)$, and $(K - L) \cdot L > 0$ since pluricanonical divisors are connected.

Finally, by Riemann-Roch, $(K - L) \cdot L$ is an even number, and we have a contradiction since, by (1.9), $1 = c_2(\mathcal{E}) = \deg(Z) + L(K - L)$.   Q.E.D.

1.10. REMARK. By 1.5 we see that for $m = 3$ this is the best possible result; moreover, if $K^2 = 1$ and $p_g > 0$, then $|2K|$ has no base points. (In fact, $p_g \leq 2$ by Noether's inequality, and the case $p_g = 2$ follows from [Ho 2], $p_g = 1$ from [Ca 8].)

We recapitulate the above results, also keeping track of the case $p_g = K^2 = 3$ appearing in [Ho 2, Il 1, Il 2].

1.11. THEOREM. *Let X be the canonical model of a surface of general type, and let $\psi_m\colon X \to \Sigma_m$ be the mth canonical map. Then*

(i) ψ_m *is a morphism for $m \geq 4$, $m = 3$ and $K^2 \geq 2$, for $m = 2$ if $K \geq 5$, or if $p_g \geq 1$ (except possibly if $p_g = q = 1$ and $K^2 = 3, 4$).*

(ii) ψ_m *is birational for $m \geq 3$ except if $K^2 = 2$, $p_g = 3$ ($m = 3$) or if $K^2 = 1$, $p_g = 2$ ($m = 3, 4$).*

(iii) ψ_2 *is birational for $K^2 \geq 10$ unless S has a pencil of curves of genus 2.*

(iv) ψ_m *is an isomorphism for $m \geq 5$, for $m = 4$ if $K^2 \geq 2$, for $m = 3$ if $K^2 \geq 3$, for $m = 2$ if $p_g \geq 6$, $K^2 \geq 10$ except if there are curves C with either $C^2 = 0$, $CK = 2$, or with $C^2 = -1$, $KC = 1$.*

2. The canonical map and its pathologies. We have tried in the previous section, at the risk of boring the reader, to give an idea of now necessarily complicated have to be statements concerning all surfaces of general type (and of how, though, the surfaces giving exceptions have a structure full of interesting geometry). It can be expected then that the exceptions should rapidly go out of control when one wants a good behavior of the canonical map. Therefore, after briefly surveying the general pattern which has emerged through work of several authors, we shall describe a couple of significant examples.

Let's assume $p_g \geq 3$ (this way only a finite number of families are left out), and let's consider $\psi_1\colon X \to \Sigma_1$. Two cases can a priori occur:

(A) $|K|$ is composed with a pencil (i.e., Σ_1 is a curve, whose geometric genus we shall denote by b);

(B) Σ_1 is a surface.

Let's start with case (A), which at first sight looks the more difficult to occur, and let's remark right away that there are an infinite number of families for which $|K|$ is composed of a pencil, according to classical examples of Pompilij and others (cf. [Be 1]). Nevertheless, some finiteness statements still hold true in this situation; namely, if B is the normalization of Σ_1, $f\colon S \to B$ is the pencil of which $|K|$ is composed, g denotes the genus of a general fibre of f, and b denotes the genus of B, one can give bounds for g, b.

Beauville [**Be 1**] uses the Bogomolov-Miyaoka-Yau inequality to prove the following

2.1. THEOREM. *If $|K|$ is composed of a pencil, the curves of the pencil have genus $2 \leq g \leq 5$ if $p_g \geq 20$ (moreover f is a morphism then); furthermore $g \leq 6$ if $p_g \geq 11$.*

The other bound is even nicer [**X 2**].

2.2. XIAO'S THEOREM. *If $|K|$ is composed of a pencil, then either $q(S) = b = 1$, or $b = 0$, $q(S) \leq 2$.*

As in [**X 1**] one of the basic tools employed is positivity, positivity of quotients of the locally free sheaf (of rank g) $f_*\omega_{S/B} = f_*\omega_S \otimes \omega_B^{-1}$, which follows from a theorem of Fujita [**Fu**], and positivity of $\omega_{S/B}$ (cf. [**Be 2**]).

The starting point of the argument is that the assumption that ϕ_1 factors through $f \colon S \to B$ implies the existence of a subline bundle $\mathcal{L}$ of $f_*\omega_S$ such that $H^0(B, \mathcal{L}) \cong H^0(B, f_*\omega_S)$: then positivity of the quotient bundle $f_*\omega_{S/B}/\mathcal{L} \otimes \omega_B^{-1}$ and Riemann-Roch imply $b \leq 1$. The bounds for q use the results of [**Be 2**] plus a careful analysis of two pencils existing on S, the given one $f \colon S \to B$ plus the pencil given by the Albanese map. Beyond the above bounds on b and g, there is also a "geographical" constraint in order that (A) may happen. The following results, after Castelnuovo's inequality [**C**] $K^2 \geq 3p_g - 7$, established under the assumption that ϕ_1 be an embedding, are due to the work of several authors: Horikawa, Reid, Beauville, Debarre [**Ho 2**, **Re 2**, **Be 2**, **De**].

2.3. THEOREM. *If ϕ_1 is birational, then $K^2 \geq 3p_g + q - 7$. If $K^2 < 3p_g - 7$, then ϕ_1 is of degree 2 and Σ_1 is a ruled surface. If Σ_1 is a curve, then $K^2 \geq 3p_g - 6$, and $K^2 \geq 4p_g + 4(b-1)$ if the curves of the pencil have genus g at least 3.*

As remarked in [**Be 1**] (this paper is, in our opinion, the best reference for 2.3), one sees then that surfaces with small K^2 are only in part easy to classify: in fact if $K^2 < 3p_g - 7$, S is only birational to the double cover of a geometrically ruled surface, and the study of the branch locus may become very fastidious. Beauville suggests calling "hyperelliptic" the surfaces which are birational to a double cover of a ruled surface; this is slightly confusing, since classically the hyperelliptic surfaces (bielliptic in Beauville's terminology) are the minimal elliptic surfaces with $12K \equiv 0$, $K \not\equiv 0$, and $b_1 > 0$ (in fact $b_1 = 2$), explicitly classified by Bagnera and De Franchis (see [**B-P-V**, p. 148]): nevertheless we shall temporarily adhere to this notation.

The "hyperelliptic" surfaces of general type have been studied by Xiao [**X 3**], who in particular states interesting results about unicity of the hyperelliptic pencil and about the structure of the fundamental group $\pi_1(S)$, thus confirming some conjecture of Reid. In particular Xiao proves the following.

2.4. THEOREM (XIAO). *Let S be a surface of general type with a hyper-elliptic pencil $f: S \to B$. Then, if $K^2 < 4\chi - 28$, then $q(S) = b = $ genus of B.*[†]

2.5. REMARK. In particular, when $q > 0$, the fibres of the Albanese map give the hyperelliptic pencil. The above statement confirms a conjecture of Severi to the effect that if $K^2 < 4\chi$, then the image of the Albanese map is a curve. An important step in this direction had been done earlier by Reid [**Re 5**] and Horikawa [**Ho 2**, V].

2.6. THEOREM (HORIKAWA-REID). *Let S be a minimal surface of general type with $K^2 < 3\chi$. Then, if $q(S) \geq 1$, the image of the Albanese map is a curve and the curves of the pencil thus obtained have genus 2 or 3, and indeed equal to 2 if $K^2 < \frac{8}{3}\chi$. If $q = 0$, either $\pi_1(S)$ is finite, or there exists an unramified covering $u: \tilde{S} \to S$ and a morphism $f: \tilde{S} \to B$ with fibres hyperelliptic curves of genus ≤ 5, and with $\pi_1(B) \cong \pi_1(\tilde{S})$.*

The striking fact of Theorem 2.3 above is that the topological invariants (χ, K^2) force the canonical map to have a "bad" behavior, and in some sense surfaces with "small K^2," i.e., $K^2 < 3p_g - 7$, are another surface analogue of curves of genus 2: the connection with curve theory is here more transparent, since the main methods of proof are based on the analysis of the restriction of ϕ_1 to canonical curves, and on the application of classical lemmas (such as Clifford's or Commessatti's lemma) about special systems on curves. It is somehow expected, but to our knowledge not even made explicit by some conjecture, that ϕ_1 should be birational for "K^2 very large." In view of 2.1 and 2.2, it seems that most information is missing in the case (B) where Σ_1 is a surface. Beauville proved the following [**Be 1**].

2.7. THEOREM (BEAUVILLE). *If the image Σ_1 of the canonical map is a surface, then either*

(i) $p_g(\Sigma_1) = 0$ *or*

(ii) Σ_1 *is canonically embedded (i.e., there exists a surface S' whose canonical map is birational, and with image Σ_1).*

Moreover, in case (i) $\deg \phi_1 \leq 9$ *if $\chi \leq 31$, and in fact $\deg \phi_1 \leq 4$ if Σ_1 is not ruled; in case* (ii) $\deg \phi_1 \leq 3$ *if $\chi \geq 14$.*

2.8. REMARK. Class (i) should be thought of as the rule at least when ϕ_1 is not birational. As a matter of fact, all smooth surfaces Σ with $p_g = 0$ occur as canonical images under a morphism ϕ_1 of degree 2. To see this, let's consider divisors δ, Δ on Σ such that Δ is effective, reduced, smooth and $\Delta \equiv 2\delta$. If S is the double cover of Σ branched on Δ, S being a smooth surface in the line bundle associated to the invertible sheaf $\mathcal{O}_\Sigma(\delta)$, then, in general, by the Leray spectral sequence, $H^0(\mathcal{O}_S(K_S)) \cong H^0(\mathcal{O}_\Sigma(K_\Sigma)) \oplus H^0(\mathcal{O}_\Sigma(K_\Sigma + \delta))$. By the

[†]In a more recent preprint, entitled *Fibered algebraic surfaces with low slope*, Xiao extends the result to the case where $f: S \to B$ is any pencil with $K^2 < 4\chi + 4(b-1)(g-1)$.

assumption $p_g(\Sigma) = 0$ the first summand is zero, which ensures that ϕ_1 factors through the double cover $S \xrightarrow{\pi} \Sigma$ and the map associated to the linear system $|K_\Sigma + \delta|$, which, for δ sufficiently ample, gives an embedding of Σ.

One should notice that even the classification of these simple "exceptions" is quite complicated, since a surface Σ with $p_g = 0$ can be ruled, hyperelliptic, an Enriques surface, a properly elliptic surface, a surface of general type with $p_g = 0$: thus, let alone the problem of studying the adjoint systems $|K_S + \delta|$, even the classification of the surfaces themselves with $p_g = 0$ has not been yet achieved!

On the other hand, case (ii) when $\deg \phi_1 \geq 2$ was considered for a long time to be impossible (cf. [**Ca 2**]), and in fact basically the same example (with $\deg = 2$) was found independently by Van der Geer–Zagier, Beauville, and the author [**G-Z**, **Be 2**, **Ca 2**].

One more example was found by Ciliberto, but recently Beauville [**Be 3**] found an infinite series of examples where (ii) occurs, with $\deg \phi_1 = 2$. Whether (ii) can occur with $\deg \phi_1 = 3$ is still an open problem (moreover, in this case, $K^2 > 8\chi$, and the surface S must have positive index).

2.9. *Beauville's examples.* Let A be a principally polarized abelian surface, and $f : A \to K \subset \mathbf{P}^3$ the double finite cover of its Kummer surface given by the linear system $|2\theta|$. Let $u : K \to \mathbf{P}^1$ be the rational map obtained via a generic projection of $\mathbf{P}^3$ onto $\mathbf{P}^1$. (Thus $u \circ f$ is given by two theta-functions of the second order.)

Let $v_n : \mathbf{P}^1 \to \mathbf{P}^1$ be the morphism such that $v_n(z) = z^n$, and consider the following diagram, obtained by taking Σ and X to be the respective fibre products of the pairs of morphisms (u, v_n), $(u \circ f, v_n)$.

$$
\begin{array}{ccccc}
A \times \mathbf{P}^1 & & K \times \mathbf{P}^1 & & \\
\cup & & \cup & & \\
X & \xrightarrow{\pi} & \Sigma & \to & \mathbf{P}^1 \\
\downarrow & & \downarrow & & \downarrow v_n \\
A & \xrightarrow{f} & K & \xrightarrow{u} & \mathbf{P}^1 \\
& & \cap & & \\
& & \mathbf{P}^3 & &
\end{array}
$$

Σ has $16n$ nodes, and $\pi : X \to \Sigma$ is the quotient morphism by an involution having exactly $16n$ fixed points: hence $\pi^*(\omega_\Sigma) = \omega_X$. To show that the canonical map of X factors through the canonical map of Σ, we simply compute and find $p_g(X) = p_g(\Sigma)$. In fact, since both X and Σ are hypersurfaces, by adjunction we obtain (with self-explanatory notation)

$$\omega_\Sigma = \mathcal{O}_K(1) \otimes \mathcal{O}_{\mathbf{P}^1}(n-2), \qquad \omega_X = \mathcal{O}_A(2\theta) \otimes \mathcal{O}_{\mathbf{P}^1}(n-2).$$

Therefore, for $n \geq 3$, $|\omega_\Sigma|$ clearly gives an embedding of Σ; furthermore a straightforward computation gives $p_g(X) = p_g(\Sigma) = 1 + 4(n-1) = 4n - 3$.

Where are these surfaces geographically located? We have $K_X^2 = 2K_\Sigma^2 = 2(n + 2(n-2))4 = 24n - 32$; thus $K_X^2 = 6p_g(X) - 14 > 6\chi(X) - 8$, and K^2 is not really large.

The last problem we are going to touch in this section refers to the degree of ϕ_1 in the case when $p_g(\Sigma_1) = 0$.

2.10. EXAMPLE. There is an example, due to Persson [**Per 2**, 5.8], where $\deg \phi_1 = 16$, but there is an infinite series of examples due to Beauville, where the degree is 8 (cf. Theorem 2.11, 8 is the best possible number). The following is taken from [**Be 1**]: let X be the surface $C \times \mathbf{P}^1$, C being a curve of genus 3, let $p: X \to C$ be the first projection, and F the inverse image of a point in $\mathbf{P}^1$ under the second projection. If η is a divisor on C with $\eta \not\equiv 0$, $2\eta \equiv 0$, consider the divisor δ on X such that $\delta \equiv p^*(\eta) + (a+2)F$, where a is a positive integer. Let Δ be the union of $2a + 4$ disjoint fibres F_t, and let $\pi: S \to X$ be the double cover of X branched on Δ, where the surface S is a divisor in the line bundle associated to the invertible sheaf $\mathcal{O}_X(\delta)$: as in Remark 2.8 the canonical map of S factors as $f \circ \pi$, f being the map given by the linear system $|K_X + \delta| = |p^*(K_C + \eta) + aF|$. Since C has genus 3, and $K_C + \eta$ has degree 4, but is not $\equiv K_C$, $|K_C + \eta|$ defines a covering of degree 4 of $\mathbf{P}^1$; therefore ϕ_1 has degree 8 and Σ_1 is isomorphic to $\mathbf{P}^1 \times \mathbf{P}^1$.

The program of finding restrictions upon surfaces for which the degree of ϕ_1 is large has been carried on with success by Xiao [**X 4**], using a finer analysis of the linear systems on a ruled surface Σ which give a ruling of Σ (cf. also [**Re 3**, **Di**] for a more conceptual presentation of this result), and again the consideration of the sheaf $f_* \omega_{S/B}$ associated to a given pencil $f: S \to B$.

2.11. THEOREM (XIAO). (1) *If ϕ_1 has degree 3 and $K^2 < 4p_g - 12$ (resp., 27 for $p_g = 10$), then $q(S) \leq 1$, S has a pencil of curves of genus 3, $K^2 \geq \frac{10}{3}p_g - 8$.*

(2) *If $p_g > 187$, ϕ_1 has degree ≤ 8, and if equality holds, $q(S) \leq 4$, and S has a pencil of curves of genus 5 or 6.*

(3) *If $\deg \phi_1 \geq 5$, S has bounded irregularity and a pencil of curves of bounded genus.*

3. The canonical ring $\mathcal{R}$. From the fact that there exists m s.t. $|mK|$ is free from base points Mumford's theorem (cf. [**Mu 1**], also [**B-P-V**]) that the canonical ring $\mathcal{R} = \mathcal{R}(S)$ is finitely generated follows easily. $\mathcal{R}$ is "the" birational invariant of S, since S and S' are isomorphic iff the two rings $\mathcal{R}(S)$ and $\mathcal{R}(S')$ are isomorphic, and S can be recovered from $\mathcal{R}$ since S is the minimal resolution of singularities of $X = \text{Proj}(\mathcal{R})$. (Notice that, $\mathcal{R}(S)$ being integrally closed in the field $\mathbf{C}(S)$ of rational functions of a surface S, by definition a surface of general type is just a surface for which $\mathbf{C}(S)$ equals the field of homogeneous fractions of $\mathcal{R}(S)$.) For the effective determination of $\mathcal{R}$, it is important to know an upper bound for the degree of a minimal set of homogeneous generators of $\mathcal{R}$: very

relevant to this purpose are the results mentioned in §1, and also more general results on the pluricanonical images Σ_m, such as their projective normality.

After the work of Kodaira and Bombieri, dealing mostly with the problem of projective normality, there have been better results in the direction of determining this upper bound, and also upper bounds for the relations among the generators, by Gasbarrini, Green, Ciliberto, and the present author (cf. [**Ga, Gr 1, Gr 2, Ci 2, Ca 3**]). Some of Ciliberto's results are the sharpest, e.g., he proves (we shall give here only the most general result, referring the reader to [**Ci 2**] for more precise statements)

3.1. THEOREM (CILIBERTO). $\mathcal{R}$ *is generated by (homogeneous) elements of degree* ≤ 5 *if* $p_g \geq 1$, *and of degree* ≤ 6 *if* $p_g = 0$.

The bounds in Theorem 3.1 are sharp. For instance, we have seen in Example 1.3 that a generator in degree 5 is needed for surfaces with $K^2 = 1$, $p_g = 2$.

Even if one knows some bounds about the degrees of the generators in a minimal set, one has to give some general method in order to be also able to write down the relations among them.

I *Method: the method of quasigeneric canonical projections* [**Ca 3**]. The main idea here is to understand in algebraic terms the classical picture of a generic projection, thus generalizing and systematizing the beautiful intuitions of Enriques (who often treated pluricanonical maps as being generic maps). To simplify things, a quasigeneric birational canonical projection $\Phi\colon S \to \mathbf{P}$ is defined to be (cf. [**Ca 3**]) a birational morphism of S into a 3-dimensional weighted projective space $\mathbf{P}$, given by 4 homogeneous elements y_0, y_1, y_2, y_3 of the canonical ring. The method we have introduced works for regular surfaces (i.e., with $H^1(\mathcal{O}_S) = 0$), and owes much to the work of several authors [**Ca 2, A-S, Ser, Ci 3, C-D**].

As noticed above, the results about pluricanonical maps ensure that, knowing K^2, χ, we can make an a priori choice of the degrees $e_0, \ldots, e_3$ of such elements $y_0, \ldots, y_3$. The trick is to complete $y_0, \ldots, y_3$ to a huge set of generators of $\mathcal{R}$, but in such a way that the relations become very simple and closely related to the geometry of $\Sigma = \Phi(S)$: this is done (following an old idea of Petri about canonical curves) by considering $\mathcal{R}$ as a module over the polynomial ring $\mathbf{C}[y_0, y_1, y_2, y_3]$.

Then we pick a minimal set $v_1 = 1, v_2, \ldots, v_h$ of (homogeneous) generators of $\mathcal{R}$ as a $\mathbf{C}[y_0, \ldots, y_3]$-module: the hypothesis $H^1(\mathcal{O}_S) = 0$ guarantees that the module $\mathcal{R}$ is Cohen-Macaulay; hence the (linear) relations among the v_j's have no syzygies, and they are generated by h relations

$$(3.1) \qquad \sum_{j=1}^{h} \alpha_{ij}(y)v_j = 0 \qquad (i = 1, \ldots, h).$$

Moreover, since the v_j's generate $\mathcal{R}$ as a module, and $\mathcal{R}$ is a ring, $v_i v_j$ is an element of $\mathcal{R}$; hence there exist polynomials $\lambda_{ij}^k(y)$ such that, in $\mathcal{R}$, the following holds:

$$(3.2) \qquad \sum_{k=1}^{h} \lambda_{ij}^{k}(y) v_k = v_i v_j.$$

$\mathcal{R}$ is generated by $y_0, \ldots, y_3, v_2, \ldots, v_h$ (where $\deg y_i = e_i$, $\deg v_j = l_j$), with the only relations (3.1) and (3.2), and the polynomials α_{ij}'s, λ_{ij}^{k}'s can be chosen to satisfy certain properties, stated in the following theorem: but much more important is the converse result, which gives a systematic tool to classify all the regular surfaces with given numerical invariants χ, K^2.

3.3. THEOREM [**Ca 3**]. *Let $A = (\alpha_{ij}(y))$ be a symmetric $h \times h$ matrix of weighted homogeneous polynomials in $y_0, \ldots, y_3$, such that, if $A_{\hat{i}}^{\hat{j}}$ is the minor obtained by deleting the ith row and the jth column of A, there do exist polynomials $\lambda_{ij}^{k}(y)$ with $\det(A_{\hat{i}}^{\hat{j}}) = \sum_{k=1}^{h} \lambda_{ij}^{k}(y) \det(A_{\hat{1}}^{\hat{k}})$.*

Assume further that $\det(A)$ is an irreducible polynomial, and that the surface X defined by equations (3.1) and (3.2) has only R.D.P.'s as singularities: then, if the degrees of the α_{ij}'s are suitable (there do exist integers $l_1 = 0 < l_2 \leq \cdots \leq l_h$ with $\deg(\alpha_{ij}) = (\sum_{i=0}^{3} e_i) + 1 + l_i - l_j$), X is the canonical model of a regular surface of general type.

3.4. EXAMPLE. We illustrate the previous theorem by considering surfaces with $q = 0$, $p_g = 4$, $K^2 = 6$, assuming for simplicity that ϕ_1 is a morphism and that Σ_1 is not a quadric (cf. [**Ci 3, Ca 3**]). Let $y_0, \ldots, y_3$ be a basis of $\mathcal{R}_1 = H^0(\mathcal{O}_S(K))$: then $\mathcal{R}$ is generated as a module by 1 and an element v of degree 2. The degrees of the matrix A are $\left(\begin{smallmatrix} 5 & 3 \\ 3 & 1 \end{smallmatrix}\right)$, and the above condition on the minors of A simply means that there do exist homogeneous polynomials $G(y)$, $Q(y)$, of respective degrees 4 and 2, such that $\alpha_{11}(y) = G(y)\alpha_{22}(y) + Q(y)\alpha_{12}(y)$.

The canonical ring is generated by y_0, y_1, y_2, y_3, v, and the ideal of relations among them is generated by the simple relations

$$\begin{cases} \alpha_{12}(y) + \alpha_{22}(y)v = 0, \\ G(y)\alpha_{22}(y) + \alpha_{12}(y)(Q(y) + v) = 0, \\ v^2 = G(y) + Q(y)v. \end{cases}$$

The four polynomials α_{12}, α_{22}, G, Q can be chosen generically; hence the moduli space for these surfaces is unirational. (Notice, though, that the map ϕ_1 is birational only when $\alpha_{22}(y) \not\equiv 0$.)

The following method, although applicable to a much more restricted class of surfaces, is in practice very useful for effectively enlarging the class of surfaces which have a simple algebraic description: it was basically found by Godeaux, revived by Reid who also considered nonabelian groups, extended by Barlow to the case of a nonfree action.

II *Method: Torsion, unramified coverings, and group actions.* The method is applicable when the fundamental group has a subgroup (which we may assume normal) of finite index, e.g., when the torsion group $T \subset H_1(S, \mathbf{Z})$ is nontrivial.

Then there exists an unramified Galois cover $p\colon Y \to S$ with group G, so that $S = Y/G$.

The canonical ring $\mathcal{R}(Y)$ is a representation of G and $\mathcal{R}(S)$ is (implicitly) described if one can describe $\mathcal{R}(Y)$ and the action of G. Let's give the most classical example, due to Godeaux, where Y is a hypersurface.

3.5. EXAMPLE [**Mi 1**, **Re 1**]. Assume S has $K^2 = 1$, $p_g = 0$, $T \cong \mathbf{Z}/5$. Since $q(S) = 0$, one has a $\mathbf{Z}/5$-cover Y with $p_g = 4$, $K^2 = 5$: one proves that ϕ_1 is a morphism for Y. Hence Y is a quintic surface in $\mathbf{P}^3$, and $\mathcal{R}(Y) = \mathbf{C}[y_0, \ldots, y_3]/f_5(y_0, \ldots, y_3)$. It is easy to see that $\mathcal{R}_1(Y)$ is the direct sum of the 4 nontrivial characters of $\mathbf{Z}/5$, since $p_* \mathcal{O}_Y(K_Y) = \bigoplus_{\eta \in T} \mathcal{O}_S(K + \eta)$, and $H^0(\mathcal{O}_S(K + \eta))$ has dimension 1 for $\eta \not\equiv 0$, 0 for $\eta \equiv 0$. Conversely, if $f_5(y_0, \ldots, y_3)$ is invariant for the action of $\mathbf{Z}/5$, and this action has no fixed points on Y, then $Y/(\mathbf{Z}/5) = S$ is the desired surface.

The previous example shows what has to be the strategy of the method:

(i) Describing the canonical ring of Y;

(ii) Using the representation theory of G plus the geometry of S to determine what can be the possible actions of G on $\mathcal{R}(Y)$;

(iii) Conversely, given Y and an allowed action of G on $\mathcal{R}(Y)$, checking whether there exists Y with such a *fixed-point-free* action (this is often the hardest step, at least computationally, cf. [**Bar 2**]).

3.6. EXAMPLE [**C-D**]. Let S be a surface with $K^2 = 2$, $p_g = 1$, $q = 0$, $T = \mathbf{Z}/2$, and let Y be the double cover. Step (ii) is very easy, since $\mathcal{R}(Y)$ just splits according to the two eigenspaces for $\mathbf{Z}/2$, and we write $\mathcal{R}(Y) = \mathcal{R}^+ \oplus \mathcal{R}^-$, where $\mathcal{R}^+ \cong \mathcal{R}(S)$. $\mathcal{R}(Y)$ is generated by $w \in \mathcal{R}_1^+$, $x_1, x_2 \in \mathcal{R}_1^-$, $z_3, z_4 \in \mathcal{R}_2^-$, and one proves that Y is a complete intersection of type $(4,4)$ in $\mathbf{P} = \mathbf{P}(1,1,1,2,2)$. $\mathbf{Z}/2$ acts on $\mathbf{P}$ by sending $(w, x, z) \to (w, -x, -z)$, and the 2 equations of Y must be $\mathbf{Z}/2$ invariant; with little work one can even show that the equations can be written in the very special form

$$\begin{cases} z_3^2 + wz_4 l(x_1, x_2) + G(w, x_1, x_2) = 0, \\ z_4^2 + wz_3 l'(x_1, x_2) + G'(w, x_1, x_2) = 0, \end{cases}$$

with l, l' linear forms, G, G' of degree 4 and containing only even powers of w. From this explicit form of the equations one can even show that the moduli space for our surfaces is a rational variety.

As the reader will have noticed, the method is based on the condition that one can already (step (i)) classify the surfaces Y which occur as unramified coverings, but in fact, from our geographical point of view of classifying the surfaces with given invariants K^2, χ, a preliminary question has to be answered:

(iv) what can be the torsion group T or the algebraic fundamental group $\pi_1^{\mathrm{alg}}(S)$ for a surface with given invariants, or at least is there a bound for the order of these groups?

For instance, Miyaoka and Reid [**Mi 1**, **Re 1**] proved that for a surface with $K^2 = 1$, $p_g = 0$, T has order at most five, and T cannot be isomorphic to $(\mathbf{Z}/2)^2$; then, since the representation theory of an abelian group is trivial, Reid

[**Re 1**] was able to attack problems (ii) and (iii) determining all the surfaces with $K^2 = 1$, $p_g = 0$, order of $T = 3, 4, 5$.

Regarding the "preliminary" step (iv) we remind the reader of Theorem 2.6, and mention another result of the type one is looking for ([**Re 2**], cf. also [**Be 1**]).

3.7. THEOREM (REID). *If $K^2 \neq 0$, $p_g = 0$, then the order of π_1^{alg} is at most 9.*

REMARK. In fact Reid states the incorrect bound 8, but the following nice example gives a surface with $K^2 = 2$, $p_g = 0$, $\pi_1 = (\mathbf{Z}/3)^2$ (cf. [**X 1**], anyhow we shall give the more explicit description adopted by Beauville in [**Be 3**]).

3.8. EXAMPLE (XIAO). Consider $\mathbf{P}^2 \times \mathbf{P}^2$, with homogeneous coordinates (x_0, x_1, x_2), (y_0, y_1, y_2). Let Y_λ be the complete intersection of the two hypersurfaces

$$\left\{ \sum_{i=0}^{2} x_i y_i = 0 \right\}, \qquad \left\{ \left(\sum_{i=0}^{2} x_i^3 \right) \left(\sum_{j=0}^{2} x_j^3 \right) + \lambda \prod_{i=0}^{2} x_i y_i = 0 \right\}.$$

For a general value of λ, Y_λ is smooth and $\pi_1(Y_\lambda) = 0$ by Lefschetz's theorem. On the other hand, Y_λ is clearly invariant by the group $G \cong (\mathbf{Z}/3)^2$ generated by the following two transformations g_1, g_2 such that:

$$g_1(x, y) = ((x_0, \varepsilon x_1, \varepsilon^2 x_2), (y_0, \varepsilon^2 y_1, \varepsilon y_2)) \qquad (\varepsilon = \exp(2\pi i/3)),$$
$$g_2(x, y) = ((x_1, x_2, x_0), (y_1, y_2, y_0)).$$

$K_{Y_\lambda}^2 = 18$ and it is easy to see that the action of G is free; hence the surface S has the derived invariants.

But what happens if, after prescribing a surface Y and a possible action of a group G on Y, one sees that the action is not free? This situation has been considered by Barlow [**Bar 1**, **Bar 2**], especially the "good" case where the points with a nontrivial stabilizer G_y form a finite set, and where the differential at y of a $g \in G_y$ has determinant 1. In fact in this "good" case the quotient Y/G has only R.D.P.'s as singularities; hence it is still a candidate for being the canonical model of a surface of general type.

By using this method, Barlow [**Bar 1**] was able to produce a simply connected surface of general type with $p_g = 0$ (Dolgachev, see, e.g., [**Do 1**], had already given an example of an elliptic surface with $\pi_1 = p_g = 0$, against a conjecture of Severi that such a surface should be a rational surface).

3.9. REMARK. The only drawback of Barlow's method of construction is that in this way one usually obtains only proper subvarieties of the moduli space, since the covering does not exist by topological reasons, but by the existence of singular points forming in some sense a special configuration. The reason why we have mainly been focusing on the above two methods is that often the many existing methods (among which the classical method of double multiple planes, introduced by Campedelli and recently applied again with success by several people as Burniat, Oort, Peters (see, e.g., [**O-P**, **Pet 1**]) are based on special configurations which do not exist for the general surface in the moduli space; thus they are useful methods in order to show existence of certain surfaces, but they don't solve the classification problem. Still, the "local" moduli problem can be attacked via deformation theory (cf. [**Ca 4**]), showing that a certain family of surfaces given by an explicit construction does indeed yield a Zariski open set of an irreducible component of the moduli space; but it is harder (cf. [**Ca 5**]) to see what is the closure of this Zariski open set.

3.10. EXAMPLE [**Ba-Ca**]. This is an example where first a construction was made, and then the solution of the "local" moduli problem gave confidence to attack and solve the more difficult ("global") classification problem. Let Σ be a cubic surface with 4 nodes, R the rational surface which is the double cover of Σ ramified only at the 4 nodes, and Z the double cover of Σ branched on the intersection of Σ with a quartic surface G. The fibre product $Y = Z \times_\Sigma R$ is a $(\mathbf{Z}/2)^2$ Galois cover of Σ, and the diagonal action of $\mathbf{Z}/2$ is free on Y, thus giving as quotient a smooth surface $S = Y/(\mathbf{Z}/2)$. S has $K^2 = 6$, $p_g = 3$, $\pi_1 = \mathbf{Z}/2$, and the remarkable feature that ϕ_1 has 4 base points on S, so that ϕ_1 is a (rational) double cover of $\mathbf{P}^2$.

Varying G it was first seen that one had an irreducible reduced rational component of the moduli space: then, looking at the canonical system $|K|$ and the "twisted" canonical system $|K + \eta|$, where $2\eta \equiv 0$, $\eta \not\equiv 0$, it was shown that all surfaces with $K^2 = 6$, $\chi = 4$, and with 2-torsion would occur as above (just letting Σ degenerate to a normal cubic whose equation can be written as the determinant of a symmetric 3×3 matrix of linear forms).

Added in proof. We have just received a letter from G. Xiao sketching the construction of a series of simply connected "hyperelliptic" surfaces of general type with positive index (i.e., $K^2 > 8\chi$), and with the ratio K^2/χ asymptotic to 8.726.

A previous conjecture (by Bogomolov?) to the effect that surfaces with positive index should have infinite fundamental group had previously been disproven by B. Moishezon–M. Teicher ("Simply connected algebraic surfaces with positive index," to appear), but for this we refer to Persson's lecture. Xiao's construction shows in particular also that even when K^2 is rather large, still one can have ϕ_1 not birational.

REFERENCES

[**A-S**] E. Arbarello and E. Sernesi, *Petri's approach to the study of the ideal associated to a special divisor*, Invent. Math. **49** (1978), 99–119.

[**Bar 1**] R. Barlow, *A simply connected surface of general type with $p_g = 0$*, Invent. Math. **79** (1985), 293–302.

[**Bar 2**] _____, *Some new surfaces with $p_g = 0$*, Duke Math. J. **51** (1984), 889–904.

[**Ba-Ca**] A. Bartalesi and F. Catanese, *Surfaces with $K^2 = 6$, $\chi = 4$, and with torsion*, Rend. Sem. Mat. Univ. Politec. Torino (to appear).

[**B-P-V**] W. Barth, C. Peters, and A. van de Ven, *Compact complex surfaces*, Ergeb. Math. Grenzgeb. (3), vol. 4, Springer-Verlag, 1984.

[**Be 1**] A. Beauville, *L'application canonique pour les surfaces de type general*, Invent. Math. **55** (1979), 121–140.

[Be 2] ____, *L'inégalité $p_g \geq 2q - 4$ pour les surfaces de type général*, appendix to [De].

[Be 3] ____, letters to the author of September and October 1984.

[Bo 1] E. Bombieri, *The pluricanonical map of a complex surface*, Lecture Notes in Math., vol. 155, Springer-Verlag, 1971, pp. 35–87.

[Bo 2] ____, *Canonical models of surfaces of general type*, Inst. Hautes Études Sci. Publ. Math. **42** (1973), 171–219.

[B-C] E. Bombieri and F. Catanese, *The tricanonical map of a surface with $K^2 = 2$, $p_g = 0$*, C. P. Ramanujam—A Tribute, Springer-Verlag, 1978, pp. 279–290.

[B-H] E. Bombieri and D. Husemoller, *Classification and embeddings of surfaces*, Algebraic Geometry Arcata 1974, Proc. Sympos. Pure Math., vol. 29, Amer. Math. Soc., Providence, R. I., 1975, pp. 329–420.

[C] G. Castelnuovo, *Osservazioni intorno alla geometria sopra una superficie*, Rendiconti del R. Istituto Lombardo, s. II, **24** (1891).

[Ca 1] F. Catanese, *Pluricanonical mappings of surfaces with $K^2 = 1, 2$, $q = p_g = 0$*, C.I.M.E. 1977 Algebraic Surfaces, Liguori, Napoli, 1981, pp. 249–266.

[Ca 2] ____, *Babbage's conjecture, contact of surfaces, symmetric determinantal varieties and applications*, Invent. Math. **63** (1981), 433–465.

[Ca 3] ____, *Commutative algebra methods and equations of regular surfaces*, Algebraic Geometry Bucharest 1982, Lecture Notes in Math., vol. 1056, Springer-Verlag, 1984, pp. 68–111.

[Ca 4] ____, *On the moduli spaces of surfaces of general type*, J. Differential Geom. **19** (1984), 483–515.

[Ca 5] ____, *Automorphisms of rational double points and moduli spaces of surfaces of general type*, Compositio Math. **61** (1987), 81–102.

[Ca 6] ____, *Superficie complesse compatte*, Atti Convegno GNSAGA del CNR 1984, Valetto, Torino, 1986.

[Ca 7] ____, *On a class of Surfaces of general type*, C.I.M.E. 1977 Algebraic Surfaces, Liguori, Napoli, 1981, pp. 269–284.

[Ca 8] ____, *Surfaces with $K^2 = p_g = 1$ and their period mapping*, Algebraic Geometry (Proc. Summer Meeting, Univ. Copenhagen, Copenhagen, 1978), Lecture Notes in Math., vol. 732, Springer-Verlag, 1979, pp. 1–29.

[C-D] F. Catanese and O. Debarre, *Surfaces with $K^2 = 2$, $p_g = 1$, $q = 0$*, Preprint, 1982.

[Ci 1] C. Ciliberto, *Superficie algebriche complesse: idee e metodi della classificazione* Atti del Convegno di Geometria Algebrica—Nervi 1984, Tecnopoint, Bologna, 1984, pp. 39–157.

[Ci 2] ____, *Sul grado dei generatori dell'anello canonico di una superficie di tipo generale*, Rend. Sem. Mat. Univ. Politec. Torino **41** (1983), 83–112.

[Ci 3] ____, *Canonical surfaces with $p_g = p_a = 4$ and $K^2 = 5, \ldots, 10$*, Duke Math. J. **48** (1981), 121–157.

[De] O. Debarre, *Inégalités numériques pour les surfaces de type général*, Bull. Soc. Math. France **110** (1982), 319–346.

[Di] D. Dicks, *Birational pairs according to Iitaka*, Preprint, Univ. of Warwick, 1985.

[Do 1] I. Dolgachev, *Algebraic surfaces with $p_g = q = 0$*, C.I.M.E. 1977 Algebraic Surfaces, Liguori, Napoli, 1981, pp. 97–215.

[Do 2] ____, *Weighted projective varieties*, Lecture Notes in Math., vol. 956, Springer-Verlag, 1982, 34–71.

[En] F. Enriques, *Le superficie algebriche*, Zanichelli, Bologna, 1949.

[Fr] P. Francia, *The bicanonical map for surfaces of general type*, Math. Ann. (to appear).

[Fu] T. Fujita, *On Kähler fibre spaces over curves*, J. Math. Soc. Japan **30** (1978), 779–794.

[Ga] C. Gasbarrini, Thesis, Università di Pisa, 1976.

[G-Z] G. van der Geer and D. Zagier, *The Hilbert modular group for the field $\mathbf{Q}(\sqrt{13})$*, Invent. Math. **42** (1978), 93–134.

[Gi] D. Gieseker, *Global moduli for surfaces of general type*, Invent. Math. **43** (1977), 233–282.

[Gr 1] M. Green, *The canonical ring of a variety of general type*, Duke Math. J. **49** (1982), 1087–1113.

[Gr 2] ____, *Koszul cohomology and the geometry of projective varieties*, J. Differential Geom. **19** (1984), 125–171.

[Gri] E. Griffin, *Families of quintic surfaces and curves*, Compositio Math. **55** (1985), 33–62.

[G-H] P. Griffiths and J. Harris, *Residues and zero-cycles on algebraic varieties*, Ann. of Math. (2) **108** (1978), 461–505.

[Ho 1] E. Horikawa, *On algebraic surfaces with pencils of curves of genus* 2, Complex Analysis and Algebraic Geometry, Cambridge Univ. Press, New York, 1977, 79–90.

[Ho 2] ____, *Algebraic surfaces of general type with small c_1^2*. I, Ann. of Math. (2) **104** (1976), 357–387; II, Invent. Math. **37** (1976), 121–155; III, Invent. Math. **47** (1978), 209–248; IV, Invent. Math. **50** (1979), 103–128; V, J. Fac. Sci. Univ. Tokyo, Sect. A. Math. **283** (1981), 745–755.

[Il 1] V. Iliev, *Surfaces with $p_g = 3$, $K^2 = 3$*. I, Serdica **6** (1981), 352–362; II, Serdica **7** (1982), 390–395.

[Il 2] ____, *A note on certain surfaces*, Bull. London Math. Soc. **16** (1984), 135–138.

[Il 3] ____, *Canonical rings of Horikawa surfaces*, (to appear).

[Ko] K. Kodaira, *Pluricanonical systems on algebraic surfaces of general type*, J. Math. Soc. Japan **20** (1968), 170–192.

[Mi 1] Y. Miyaoka, *Tricanonical maps of numerical Godeaux surfaces*, Invent. Math. **34** (1976), 99–111.

[Mi 2] ____, *On the Mumford-Ramanujam vanishing theorem on a surface*, Journées de Géométrie Algébrique d'Angers, Sijthoff & Noordhoff, Alphen aan den Rijn, 1980, pp. 239–247.

[Mu 1] D. Mumford, *The canonical ring of an algebraic surface*, Ann. of Math. (2) **76** (1962), 612–615.

[Mu 2] ____, *Pathologies. III*, Amer. J. Math. **89** (1967), 94–104.

[O-P] F. Oort and C. Peters, *A Campedelli surface with torsion group $\mathbf{Z}/2$*, Indag. Math. **43** (1981), 399–407.

[Pe] U. Persson, *On Chern invariants of surfaces of general type*, Compositio Math. **43** (1981), 3–58.

[Pe 2] ____, *Double coverings and surfaces of general type*, Algebraic Geometry (Proc. Sympos. Univ. Tromsø, 1977), Lecture Notes in Math., vol. 687, Springer-Verlag, 1978, pp. 168–195.

[Pet 1] C. Peters, *On certain examples of surfaces with $p_g = 0$ due to Burniat*, Nagoya Math. J. **66** (1977), 109–119.

[Ra] C. P. Ramanujam, *Remarks on the Kodaira vanishing theorem*, J. Indian Math. Soc. (1972), 41–51, suppl. ibidem **38** (1974), 121–124.

[Re 1] M. Reid, *Surfaces with $p_g = 0$, $K^2 = 1$*, J. Fac. Sci. Univ. Tokyo Sect. I A Math. **25** (1978), 75–92.

[Re 2] ____, *Surfaces with $p_g = 0$, $K^2 = 2$*, Preprint, 1979.

[Re 3] ____, *Surfaces of small degree*, preliminary version 1, 1985.

[Re 4] ____, *Bogomolov's theorem $c_1^2 \leq 4c_2$*, (Proc. Internat. Sympos. Algebraic Geom., Kyoto, 1977), Kinokuniya Book Store, Tokyo, 1977, pp. 623–642.

[Re 5] ____, *π_1 for surfaces with small c_1^2*, Algebraic Geometry (Proc. Summer Meeting, Univ. Copenhagen, Copenhagen 1978), Lecture Notes in Math., vol. 732, Springer-Verlag, 1979, pp. 534–544.

[Rei] I. Reider, letter to the author, September 1985.

[Ser] E. Sernesi, *L'unirazionalità della varietà dei moduli delle curve di genere dodici*, Ann. Scuola Norm. Sup. Pisa Cl. Sci. (4) **8** (1981), 405–439.

[Sh] I. R. Shafarevich, editor, *Algebraic surfaces*, Trudy Math. Inst. Steklov. **75** (1965); English transl. in Amer. Math. Soc. Transl. (2) **63** (1967).

[Ve] A. van de Ven, *On the Chern numbers of certain complex and almost complex manifolds*, Proc. Nat. Acad. Sci. U.S.A. **55** (1966), 1624–1627.

[X 1] G. Xiao, *Surfaces fibrées en courbes de genre deux*, Lecture Notes in Math., vol. 1137, Springer-Verlag, 1985.

[X 2] ____, *L'irrégularité des surfaces de type général dont le système canonique est composé d'un pinceau*, Compositio Math. **56** (1985), 251–257.

[X 3] ____, *Hyperelliptic surfaces of general type with $K^2 < 4\chi$*, Preprint, Shanghai, 1985.

[X 4] ____, *Algebraic surfaces with high canonical degree*, Math. Ann. **274** (1986), 473–483.

UNIVERSITY OF PISA, ITALY

Proceedings of Symposia in Pure Mathematics
Volume **46** (1987)

An Introduction to the Geography
of Surfaces of General Type

ULF PERSSON

The following exposition is intended for the benefit of the nonexpert. It neither attempts to probe any significant depths (nothing is really ever proved) nor does it aspire to be comprehensive.

Unavoidably it will overlap frequently with the accompanying contribution of Catanese, to which in fact it is meant as a complement or maybe even an introduction. The scope of Catanese's article is ostensibly more narrow, as he purports only to treat the various phenomena related to the behavior of the canonical and pluricanonical maps, but in doing so he will inevitably be led to touch upon most of the aspects of surfaces of general type considered by its devotees.

The subject of surfaces, perhaps more than any other subject of algebraic geometry (and hence mathematics?), has the distinct flavor of the study of the natural world around us.

The true practitioner, like the naturalist, is happiest when he puts his nose very close to the ground and ponders the intricate workings of all creatures, big or small, reflecting according to their abilities the almost imponderable design of their Creator, be it God, or as some Italians preferred to believe, his mischievous rival.

Thus the surface theorist shuns, or is at least weary and suspicious of, the general theory or the comprehensive fact. His ambition is not to understand surfaces, whatever that means, in order to go on to bigger and better things; but as far as it exists at all, his ambition encompasses far humbler intentions.

What his intentions really are is not clear, least of all perhaps to himself; but it would probably be not far from the truth to assert that he may be guided by a collector's bent, and like his distant cousin the naturalist, jump up and down with joy when discovering a new and maybe even unexpected specimen.

1980 *Mathematics Subject Classification* (1985 *Revision*). Primary 14J29, 14J10, 14J17.

Partly supported by the American Mathematical Society and the Swedish Research Council (NFR) for participation at this AMS Summer Institute on Algebraic Geometry.

As can be inferred it is easy to go purple, and I should be well advised not to try and rival, let alone outdo, the eloquence of one of my colleagues at this Summer Institute.

Surfaces have had their Linnaeuses, but as of yet not their Darwin. There is some formal order to the chaos, which it is the purpose of surface geography to structurize.

As a means of introduction this exposition will present a slightly unconventional definition of surfaces of general type in order to set the proper tone from the beginning. In the first section we will present the rough classification of surfaces of general type and constrast it with the simplicity curves exhibit. We also hope to indicate to the reader the significance of the recent and spectacular contributions of Freedman and Donaldson in the context of surfaces, although properly speaking those are perhaps not at first sight at the very core of our subject. Finally we will state one of the main theorems, the bound on c_1^2 of surfaces of general type, a bound due to Yau and Miyaoka, to which in all fairness Bogomolov's name should be attached as well.

In the second section we will discuss examples of surfaces with particular emphasis on the computation of their invariants. The examples will be straightforward, and the section may conceivably disappoint the reader who may have expected a rich and lush display of exotica.

The third section will be concerned with what properly is the geography of surfaces with a few token results to illustrate the kind of questions which could be asked. Activity in this field seems to have abated the last few years, leaving the most active—G. Xiao—isolated back in China.

The last section deals with surfaces of positive index occupying to some the most intriguing aspect of surface geography.

For further examples and additional insights we simply refer the reader to the previously mentioned article by Catanese, in which most of the aspects of this paper are, as mentioned before, touched upon.

Since the previous meeting at Arcata there has become available to the general public a spate of presentations and surveys devoted to surfaces. We take this to be a partial excuse for our breeziness. Rather than trying to compile a complete list of references I will limit myself to pointing out what surely is intended as a canonical source of references for the years to come, namely, the long-awaited tome of the Dutch Troika—Barth, Peters, and Van de Ven. In *Compact Complex Surfaces* (a somewhat unimaginative title appearing in the ongoing series of Ergebnisse der Mathmatik) we take it on faith that the curious reader will find references and clarification to essentially all the unproven statements of this survey.

This article is a slight elaboration of the talk and the subsequent notes I gave at the AMS Summer Institute at Bowdoin, July 1985. A first version was written in late February 1986 and circulated among my colleagues, of whom special thanks go to Arnaud Beauville and Fabrizio Catanese for many helpful remarks and for saving me from embarrassment more than once.

This version was written in late July and completed at the anniversary of my leaving Bowdoin. I want to take the opportunity to thank David Gieseker at UCLA and his colleagues Mark Green and Robert Lazarsfeld for very kindly inviting me to spend a year here in Los Angeles and whose hospitality made this write-up possible.

So let us first briefly put surfaces of general type into some kind of perspective. By a surface we will (always) mean a compact complex manifold of dimension two. Hence we will not touch upon the theory of open surfaces (surfaces with curves taken away from them) nor on finite characteristics. Furthermore all surfaces are smooth, and in case we refer to singular surfaces, we have in mind the (minimal) resolution.

We will also narrow our focus onto algebraic surfaces. Thus it is convenient to recall that for a surface X the following two statements are equivalent (see [**K 1a**]):

1. X is a projective surface (i.e., X embeddable into $\mathbf{P}^n$);

2. $\mathrm{tr.d.}_{\mathbf{C}}\mathcal{M}(X) = 2$ ($\mathcal{M}(X)$ the field of meromorphic functions).

There is a strange collection of nonprojective surfaces. (We refer to [**K 2a–c**] for details.) Two of the more accessible examples are

(a) Complex tori, $X = \mathbf{C}^2/L$ which in general have no nonconstant meromorphic functions, and

(b) Hopf surfaces $X = \mathbf{C}^2 - (0,0)/\mathbf{Z}$ for which $b_1(X) \neq 0$ (2).

The study of nonprojective surfaces is in one sense antipodal to our concern for surfaces of general type.

Among the projective surfaces there are three classes which have received particular attention. These are

(a) rational, or more generally ruled surfaces (admitting rational fibrations),

(b) elliptic surfaces (admitting elliptic fibrations), and

(c) abelian and so-called K-3 surfaces (two divergent (but on a deeper level strangely convergent) ways of generalizing the notion of an elliptic curve to surfaces).

This classification is not mutually exclusive. In fact the most tractable examples of (c) actually belong to (b) as well, and the simplest examples of (b) are in fact to be found among those in (a).

We are now ready to finally define surfaces of general type. To us the definition which most quickly comes to the core of the matter is in fact to say "none of the above." Hence,

DEFINITION. A projective surface is of general type iff it is *not* ruled, elliptic, abelian, or K-3.

This is of course not the standard definition. (And some experts may wince at my implication that surfaces that are merely birational to an abelian or a K-3 surface should be dignified by the same designation.) We have presented it here to emphasize that surfaces of general type are bagged together without having any a priori relationships to each other (very much like, say, dividing the population of this Earth into Icelanders and non-Icelanders). The disadvantage is of course that no one knows (nor may ever know) how to define surfaces of general type among all surfaces in the above vein.

A good theorem usually ends up as a mere definition. So let us reverse this trend.

THEOREM. *A projective surface is of general type iff some multiple of the canonical divisor gives a birational embedding into some projective space.*

It is convenient to recall that for surfaces in general it is hard to find divisors or linebundles; the canonical one associated to the holomorphic two forms can always be defined.

Although in general a theorem following a definition is supposed to be either trivial or false, this is truly neither.

The proof of this is essentially the life work of Enriques, referred to as the Enriques classification of surfaces (with Kodaira's name thrown in, when the context of surfaces encompasses not only the projective ones). The reader may consult any of the standard references, e.g., [**III**] or [**B 1**].

Note also that any surface satisfying the conclusion of the theorem is necessarily projective.

Finally we have to clear up one "minor" point before we are ready to continue. In contrast to the case of curves a surface is not determined by its function-field. For surfaces of general type (or more generally for nonruled surfaces) we have the equivalence of

(1) X contains no exceptional divisors of the first kind (i.e., no nonsingular rational curves with self-intersection -1), and

(2) if $\mathcal{M}(X) = \mathcal{M}(Y)$, there is a morphism $Y \to X$.

Due to the second condition such surfaces are called "minimal," and by a reprehensible abuse of language they are the "canonical" models for the function-fields.

From now on all surfaces of general type will be tacitly assumed to be minimal as well.

1. It may be illuminating to recall the case of curves before delving into surfaces. The situation is very neat. There is one fundamental discrete invariant—the genus g. We have the following theorem.

THEOREM. *If X and Y are irreducible curves, the following are equivalent:*
(i) $g(X) = g(Y)$;
(ii) *X is homeomorphic to Y; and*
(iii) *X is diffeomorphic to Y.*

In particular, g is a topological invariant; in fact $2 - 2g$ is the Euler characteristic of the curve, and it is easy to compute using any number of methods. Furthermore it can assume any nonnegative value, and the corresponding topology is well understood (the classification of real compact surfaces).

To each genus g we have a (coarse) moduli space $\mathcal{M}_g$ parametrizing the possible complex structures on a "surface" of genus g.

The basic thing to note is that each $\mathcal{M}_g$ is irreducible. This means in particular that there cannot be any more discrete invariants, if by a discrete invariant we mean something which stays locally constant.

This is not the place to give a comprehensive survey of all the known properties of the moduli spaces $\mathcal{M}_g$, so we are going to be content with listing a few salient features for future comparison with surfaces.

• $\mathcal{M}_0 = \{\mathbf{CP}^1\}$ (so there exists only one simply connected "surface," namely, S^2, which in its turn only admits one complex structure—the Riemann sphere).

• Except for $g \leq 2$, $\mathcal{M}_g$ is not explicitly known! (Everybody knows about $\mathcal{M}_1$—or should; Igusa explicitly worked out $\mathcal{M}_2$.)

• $\dim \mathcal{M}_g = 3g - 3$ if $g > 1$. (Riemann made this computation.)

• $\mathcal{M}_g$ is unirational for $g \leq 10$ and of general type if $g \geq 25$. (Note those are not sharp statements. The first is elementary; the second is considerably deeper and due to Mumford and Harris. Observe also that in a way this is a negative result; it really discourages you from getting your hands on moduli spaces of high g.)

While there is undeniable neatness in the realm of curves, surfaces exhibit chaos. (As pointed out earlier, some Italians referred to them, no doubt out of exasperation mingled with fondness, as being the progeny of Satan.)

The analog of the genus of a curve is given by the pair (c_1^2, c_2) of chern-invariants.

c_2 is simply the Euler characteristic, while c_1^2 is the self-intersection of the canonical divisor and for that reason often from now on referred to as K^2. Now due to a divisibility condition, made explicit by Noether's formula, they cannot be completely arbitrary.

In fact we have $c_1^2 + c_2 \equiv 0\ (12)$, and hence we will mostly use $c_1^2\ (= K^2)$ and $\chi\ (= c_1 + c_2)/12)$ instead and by a slight abuse of terminology refer to them as the chern-invariants (or properly speaking chern-numbers).

Those are the two fundamental discrete invariants. This claim is based on two important facts.

1. They, being chern-invariants, are easy to compute.

2. There are only a finite number of families of surfaces of general type with given chern-numbers.

One should point out that 2, as opposed to 1, is not true in general for (projective) surfaces. One may, e.g., consider elliptic surfaces and observe that their chern-numbers are invariant under logarithmic transforms (see [**K 2b**]). The validity of 2 rests on the fact that our theorem on surfaces of general type can be sharpened to

THEOREM'. *A surface is of general type iff the five-canonical map is birational.*

(For a proof see [**B**], and the lecture of Catanese.)

Statement 2 now will follow from general nonsense. Knowing K^2 and χ we can easily compute the Hilbert polynomial of the five-canonical image, and as is universally known, to each Hilbert polynomial there are only a finite number of components in the Hilbert scheme.

However there is no way to give a meaningful estimate on the number of components (not even connected components). Practical experience seems to indicate that their number increases rather rapidly with increasing numerical values of the chern-invariants. (See, e.g., [C1] for lower bounds.)

To try to remedy the situation we have to cast our nets very wide by providing a whole slew of discrete invariants.

The Italians considered the geometric genus p_g and the arithmetic genus p_a. In modern terminology $p_g = \dim H^0(\Omega^2)$ (the number of holomorphic two-forms) and $p_a = \dim H^0(\Omega^2) - \dim H^0(\Omega^1)$. We usually ignore the latter and use the so-called irregularity, denoted by q and defined as the discrepancy $p_g - p_a$. (The irregularity is hence just given by the number of holomorphic one-forms.)

More subtle invariants can be obtained by transcending the mere numerical ones and considering the intersection form on $H^2(X, \mathbf{Z})$ or the fundamental group $\pi_1(X)$.

All those invariants have in common that they are locally constant and are in fact purely topological invariants computable from a triangulation. (The complex analytic definitions of p_g, q, and K^2 should not fool the reader; in fact for (projective) surfaces $2q = b_1(X)$, $c_1^2 - 2c_2$ is the index of the intersection form on H^2, and $\chi = p_g - q + 1$ by Noether's formula.)

As we will presently see, it is rather hard to design locally constant discrete invariants which are not topological. The plurigenera P_m (m positive) are defined as $\dim H^0(X, mK)$. For m at least two they do not fit into any Hodge decomposition and one may suspect that they are truly complex analytic in character, but this is in fact not true; for *minimal* surfaces of general type both $H^1(mK)$ and $H^2(mK)$ vanish (see [**Mum 1**]). Thus P_m is expressible in the chern-numbers using Riemann-Roch. The qualification of minimal is important here; the same topological space can support both minimal and nonminimal surfaces (see below).

A truly analytic invariant, like, e.g., $\dim H^1(X, \theta_X)$ (where θ_X is the tangent sheaf) is in general not locally constant but upper semicontinuous and will jump. Typically if the component of the moduli space (cf. discussion below) is generically smooth (and its dimension given by $\dim H^1(X, \theta_X)$) the above invariant will jump at singular points (so-called obstructed deformations). To get around this one may simply consider as invariant the dimension of the moduli component, but this may in general be hard to get your hands on.

Amazingly there is a simple way of getting out of this cul de sac. The position so to speak of the canonical divisor in the intersection form is a complex invariant. We define the divisibility of a surface, denoted by $d(X)$, to be the greatest integer d such that $(1/d)K \in H^2(X, \mathbf{Z})$ (clearly there is a greatest such integer, not just maximal ones.). It turns out that $d(X)$ is in fact locally constant and not liable to jumps; in the examples we are going to look at, it will be easily computed generically, and hence everywhere, and show itself to be very useful. (We are indebted to Moishezon for pointing this out to us, and to Gieseker for reminding us of it.)

The topology of surfaces is rather complicated; in particular, there is no list of topological candidates, as in the curve case, each easily described. In fact the semigroup of topological sums is not finitely generated.

Just to give an inkling of the complexity, let us restrict to the simply connected case, which as we noted earlier reduces to a single topological space S^2 and a single complex structure $\mathbf{CP}^1$ in the case of curves.

In one direction we have a positive result. The topological invariants introduced above determine the topology.

FREEDMAN'S THEOREM. *If two simply connected surfaces have the same intersection form, then they are homeomorphic.*

(See [**F**].)

Note. The index and rank of the intersection form determine the other numerical invariants. (Recall that as X is simply connected $q = 0$.) Conversely we know that the intersection form is indefinite (e.g., Hodge index theorem), unimodular (the Poincaré duality), and hence it is determined up to parity by the topological invariants, i.e., the chern-numbers. Recall that intersection forms are either even, i.e., all squares are even, or odd (i.e., not even). An intersection form is even iff the canonical divisor is even, i.e., $K = 2D$ for some integral divisor D. (For an elementary reference see [**Srr**].)

On the other hand the topological invariants, no matter how many or how ingeniously concocted, will not suffice to determine an irreducible family.

In fact S. K. Donaldson was able to devise a diffeomorphic invariant which is not topological!

DONALDSON'S EXAMPLE. *There exist two simply connected surfaces which are homeomorphic but not diffeomorphic.*

Although Donaldson's example concerned two elliptic surfaces (a rational elliptic surface and a so-called Dolgachev surface, see [**D**] for details) there is no a priori reason why this phenomenon should not occur among surfaces of general type. (A natural pair of candidates would be simply connected surfaces with $p_g = 0$ and $K^2 = 1$; see §3)

It may in this context be worthwhile to point out that so far no two examples of diffeomorphic surfaces have been found belonging to different connected components of the moduli space. (One cannot strengthen this to conjecture that diffeomorphic surfaces belong to the same irreducible components, as the differential structure of a surface is determined by its degeneration, and a singular surface can have many different components in its versal deformation space (for the first part consult [**Psn 1**]).) It is our suspicion though that in the future such examples will be found; the difficulty of finding them is connected to the difficulty of showing that two "unrelated" surfaces are diffeomorphic.

At this point we find it expedient to discontinue the discussion before it leads us out onto water too deep for our comfort.

What about the continous analogs for surfaces? We have of course the celebrated result of David Gieseker.

GIESEKER'S THEOREM. *The moduli spaces for surfaces of general type exist and they are all quasiprojective varieties.*

(See [**G**].)

But in view of the above discussion, such an individual study of moduli spaces for surfaces is bound to be very ad hoc. Just to point out the state of affairs, I provocatively state the following (disprovable?) conjecture.

CONJECTURE. Every projective variety occurs as the compactification of some moduli space for some type of surfaces!

Note. This is rather trivially disproved for the $\mathcal{M}_g$'s! Furthermore the astute reader easily sees that due to cardinality most varieties do not at all occur as moduli spaces for surfaces (or anything else for that matter), but the same astute reader may be rather hard pressed to find a single concrete counterexample. Thus I may modify the conjecture to the statement, e.g., that every variety can be deformed to a moduli space for surfaces, or any variety defined over the integers can occur as a moduli space.

A fundamental problem is to determine what invariants can occur. Of basic importance is to be able to find the possible restrictions on the chern-invariants themselves. A more exotic question is to determine what fundamental groups can occur. One knows that such a group has to be finitely generated. (Surfaces can after all be compactly triangulated.) Serre has shown, using ideas of Godeaux, that any finite group can occur. It is still an open question whether it is possible to find X such that $\pi_{\mathrm{alg}}(X) = 0$ but $\pi_1(X) \neq 0$ (and then necessarily infinite). We will return to the fundamental group in the final section, and will concentrate from now on on the chern-invariants.

The chern-numbers are restricted by a set of inequalities made explicit by the following theorem.

THEOREM (BOGOMOLOV, MIYAOKA-YAU). *The chern-invariants c_1^2, χ of a surface of general type satisfy the inequalities*:
   (a) $c_1^2, \chi > 0$;
   (b) $c_1^2 \geq 2\chi - 6$;
   (c) $c_1^2 \leq 9\chi$.

Note. (a) and (b) were classically known. The first is part of the classification of surfaces; the second is due to M. Noether.

To indicate the proof let us consider a simplified case. Assume that $|K|$ (the canonical system) contains a smooth curve C, and let us denote by k the restriction of K to C. We have degree of $k = C^2 = K^2$ and $p_{\mathrm{g}} = h^0(k) + 1$.

As k is a special divisor, Clifford's theorem gives

$$h^0(k) \leq \tfrac{1}{2} \deg k + 1,$$

which translates into $K^2 \geq 2p_g - 4$ (which is a slightly sharper version of (b)).

In general we have to consider the free part of $|K|$ and blow up base points, but those are technicalities.

The real deep part is the existence of a sharp bound of type $c_1^2 \leq \omega\chi$. Any such bound will have the important consequence that for each χ there are only a finite number of possibilities for c_1^2 and hence only a finite number of families.

There is a sequence of such bounds, all but the first incidentally appearing publicly in 1976. The progressively smaller values of ω were given as follows:

ω

12. –	by Castelnuovo
10.667 . . .	by Van de Ven (see [**VdV**])
9.600 . . .	by Bogomolov (see [**Bog**]), which was the first break-through.
9. –	by Miyaoka (see [**M1**]), on one hand, improving the method of Bogomolov involving stability of tangentbundles etc. and by Yau (see [**Y**]), using differential geometrical methods.

The final bound is sharp. In fact equality holds for compact quotients of the unit ball in $\mathbf{C}^2$ as pointed out by Hirzebruch [**Hz**]. Yau also showed that if equality holds and provided there are no rational curves with self-intersection -2 (those by the way occur in so-called A, D and E configurations and can be viewed as resolutions of rational double points) the converse is true, i.e., the universal covering is the unit ball. This incidentally settles an old question whether a surface can be homeomorphic to $\mathbf{P}^2$ but not equal; such a surface would necessarily be of general type and satisfy the upper bound of the Yau-Miyaoka inequality, and furthermore its Picard group would be cyclic, ruling out the existence of rational curves with negative self-intersection.

The slightly annoying proviso that there exist no configurations of rational -2 curves can in fact be done away with; in fact, for surfaces possessing such samples of curves the bound be sharpened.

Finally if you do not think that progress was spectacular, translate those inequalities above in terms of c_1^2 and c_2!

2. We have now set the stage. Surfaces of general type live (and thrive) in a very well cut out region of the universe. (See Figure 1.)

So let us try to populate!

The most obvious examples are given by (smooth) hypersurfaces in $\mathbf{P}^3$. So let S_d be a smooth hypersurface of degree d. What are its invariants?

Let H ($= S_1$) be a hyperplane; by adjunction the canonical class is given by the restriction of $(d - 4)H$ to S_d. Thus, $K^2 = (d - 4)^2 d$ (by Bezout).

Furthermore counting coefficients of hypersurfaces (of degree $d - 4$) we find

$$p_g = \binom{d - 4 + 3}{3} = \frac{(d - 1)(d - 2)(d - 3)}{6}.$$

FIGURE 1. Chern-numbers of complete intersections of multiprojective spaces
$c_1^2 \leq 500$, $\chi \leq 160$.

Cheating we use the fact that $q = 0$ (in fact $\pi_1(S_d) = (1)$ by Lefschetz) to conclude $\chi = p_g + 1$.

(An alternative approach would be to use degeneration techniques degenerating S_{d+1} into $S_d \cup S_1$. Straightforward topology gives the induction formulas (see [**Psn 1**])

$$c_1^2(S_{d+1}) = c_1^2(S_d) + c_1^2(S_1) + 8(g - 1) - d(d + 1),$$
$$\chi(S_{d+1}) = \chi(S_d) + \chi(S_1) + (g - 1),$$

where $g = g(S_d \cap S_1)$ $(2g - 2 = d(d-3))$, $c_1^2(S_1) = c_1^2(\mathbf{P}^2) = 9$, $\chi(S_1) = \chi(\mathbf{P}^2) = 1$.)

Of course we can generalize this, replacing $\mathbf{P}^3$ with any three-fold on which we can do intersection theory numerically explicit.

In particular we can consider all complete intersections

$$S_{(d_1, d_2 \ldots, d_n)} = \bigcap_{i=1}^{n} H_{d_i} \subseteq \mathbf{P}^{n+1} \quad (H_d \text{ hypersurface of degree } d).$$

The formulas can still be worked out explicitly, but let us be content with the recursive formulas obtained by degeneration.

(ci)
$$c_1^2(d_1, \ldots, d_n + 1) = c_1^2(d_1, \ldots, d_{n-1}) + c_1^2(d_1, \ldots, d_{n-1}, d_n)$$
$$+ 4(D_\sigma - 2)D_\pi - D_\pi(d_{n+1}),$$
$$\chi(d_1, \ldots, d_n + 1) = \chi(d_1, \ldots, d_{n-1}) + \chi(d_1, \ldots, d_{n-1}, d_n)$$
$$+ (1/2)(D_\sigma - 2)D_\pi$$

where $D_\sigma = \sum(d_i - 1)$ and $D_\pi = \prod d_i$. (Observe that $c_1^2(d_1, \ldots, d_n) = (D_\sigma - 3)^2 D_\pi$.)

TABLE 1. Complete intersections of general type with chern-numbers $c_1^2 \le 500$ and $\chi \le 160$.

degree	c_1^2	χ	degree	c_1^2	χ	degree	c_1^2	χ
5	5	5	2,8	400	92	2,2,6	384	86
6	24	11	3,3	9	6	2,3,3	72	21
7	63	21	3,4	48	16	2,3,4	216	50
8	128	36	3,5	135	35	2,3,5	480	100
9	225	57	3,6	288	66	3,3,3	243	54
10	360	85	4,4	144	36	2,2,2,2	16	8
2,4	8	6	4,5	320	70	2,2,2,3	96	26
2,5	40	15	2,2,3	12	7	2,2,3,3	324	69
2,6	108	31	2,2,4	64	20	2,2,2,4	288	64
2,7	224	56	2,2,5	180	45	2,2,2,2,2	128	32

Just for fun we have listed those complete intersections whose chern-numbers are fairly small (see Table 1). The reader may be struck by their relative sparsity, only 30 examples spread out among 40,000 possible invariants! To these examples intrinsically defined by equations we may add complete intersections among multiprojective spaces, i.e., surfaces defined by the complete cut outs by multihomogenous forms. For amusement we have listed, somewhat bizarrely by hand, low invariants which may occur in this way among this more general class of surfaces. The diagram, although far from complete (to complete the ultimate task was just too daunting and tedious), should give the reader a first inkling.

We observe that there is indeed a great profusion of surfaces and corresponding invariants, even when we are considering such relatively simple-minded surfaces as complete intersections in multiprojective spaces. The invariants seem to follow no obvious regular pattern, but seem to be more or less randomly spread out—a feature that certainly has its charms. There seem to be empty spaces and on the other hand there are also coincidences. The observant reader may (using optical aids if necessary) detect a few binary systems and even some triplets.

Let us digress for a moment and focus on a typical triplet as given by $K^2 = 128$ and $\chi = 46$. There are at least three types of surfaces occurring in this way, either

(a) the complete intersection of two bihomogenous ternary polynomials of bidegree $(10,2)$ and $(1,1)$ in $\mathbf{P}^2 \times \mathbf{P}^2$;

(b) the hypersurface in $\mathbf{P}^1 \times \mathbf{P}^1 \times \mathbf{P}^1$ given by a trihomogenous binary polynomial of tridegree $(2,6,10)$; or

(c) the hypersurface in $\mathbf{P}^2 \times \mathbf{P}^1$ given by a polynomial $F(x_0, x_1, x_2; y_0, y_1)$ of bidegree $(4,16)$.

By Lefschetz all those examples are actually simply connected; furthermore, they are all intrinsically different as the invariant d (given by the divisibility of the canonical divisor) is easily computed (exploiting the fact that the Picard group of the generic member is simply induced from the Picard group of the ambient space) to *be respectively* 8, 4, and 1.

We see thus that (a) and (b) consist of homeomorphic surfaces, as their intersection forms are the same. It would be interesting to know whether they are even diffeomorphic. If they were, it would be very hard to prove (they cannot have common degenerations) and if not it would give an example illustrating the query at the bottom of page 201.

Another triplet is given for $K^2 = 80$ and $\chi = 25$. We simply replace the multidegrees accordingly: for (a') $(3,5)$ and $(1,1)$; for (b') $(3,4,5)$; and finally for (c') $(5,5)$. In all those cases the surfaces are, of course, simply connected and $d = 1$; hence they are all homeomorphic (odd intersection forms) so all the invariants we know of coincide. The reader may feel motivated to check whether those constructions are merely specializations of a more general construction. (If we compute the canonical divisor for a generic element in each construction, we get a different form; hence, the constructions do not coincide.)

Returning now to the general question of the spread of invariants, we may observe a few salient features which we may hope are not merely artifacts of the restricted class we have plotted.

There seem to be fault lines, invariants obediently spread along straight arrays. In the various theorems pertaining to geography of surfaces, those fault lines will play an important role. Furthermore using a construction of surfaces, say hypersurfaces in $\mathbf{P}^3$, the invariants tend to converge to certain slopes. In the case of hypersurfaces the slope is 6 (the slope being defined as the quotient of c_1^2 and χ), and for complete intersections the various slopes converge to 8.

We may at this stage pause and observe that for complete intersections the slopes may in fact never exceed 8—the magic fault line of zero index.

In fact we have

PROPOSITION. *If X is a complete intersection, then $c_1^2(X) \leq 8\chi(X)$.*

PROOF. This is obvious from the inductive formulas (ci) given after Table 1. Note also that the inequality can be slightly sharpened (although the coefficient 8 is of course the best possible) by observing that the last two terms of the upper equation of the formulas are slightly less than eight times the last term of the lower equation. Although the proposition has only been stated for complete intersections in $\mathbf{P}^n$, a corresponding bound should be expected to hold for all complete intersections in multiprojective spaces (the same kind of inductive formulas can be written down) and perhaps in even more general situations of which we are at a loss to properly formulate.

As to empty spaces there really are none, although not all invariants can be given in the simple-minded way we have just sketched. To smear out the invariants it is convenient to impose singularities on our surfaces. (We are of course (cf. Introduction) considering the minimal resolutions.) By simply specializing the surfaces, the chern-invariants will vary; in fact they will always decrease (although we know of no published proof of the fact! The interested reader may consult [**Psn 1**] to see whether the chern-invariants of the minimal models of the components of the semistable degenerate fiber always are less than those of the generic surface) and the correction terms (specialization vectors) are purely expressible in terms of local data. (This is a contradistinction to the case of more fickle invariants like p_g and q when it it necessary to take into account the global positions of the invariants, cf. the discussion below.)

To illustrate, let us consider three kinds of elliptic singularities. Of course this is not a complete classification. As the reader may suspect there is a vast literature on the subject.) They all have nonsingular elliptic curves E as resolutions and $E^2 = -1, -2$ or -3 depending on the type.

TABLE 2

ad hoc	Type mono- dromist	cusp- singularist	E^2	local equation	specialization vector
I	$\overline{E}_8$	$T_{2,3,6}$	-1	$z^2 = y(y - x^2)(y - \lambda x^2)$	$(-1, -1)$
II	$\overline{E}_7$	$T_{2,4,4}$	-2	$z^2 = xy(x - y)(x - \lambda y)$	$(-2, -1)$
III	$\overline{E}_6$	$T_{3,3,3}$	-3	$zy^2 = x(x - z)(x - \lambda z)$	$(-3, -1)$

PROPOSITION. *Let X' be a specialization of X (i.e., belonging to the same component in the Hilbert scheme) with elliptic singularities of type* I, II, *and* III. *Then*

$$c_1^2(X') = c_1^2(X) - 1\#(\text{I}) - 2\#(\text{II}) - 3\#(\text{III}),$$
$$\chi(X') = \chi(X) - 1(\#(\text{I}) + \#(\text{II}) + \#(\text{III})).$$

EXAMPLE. A quintic X with 4 type I singularities has the invariants

$$c_1^2(X) = 1, \qquad \chi(X) = 1.$$

If the singularities all lie on a plane, then $p_g = 1$ (and hence $q = 1$); otherwise $p_g = 0$ (and $q = 0$).

Now the reader expects (or should expect) examples of both those phenomena, and we have to admit that we do not know of any offhand.

This illustrates a basic technical difficulty. How do you construct a surface with a preassigned set of invariants? In particular how do we find a quintic with four type I singularities?

Thus even such simple objects as hypersurfaces are too complicated. Fortunately there is an even more elementary construction available to us, namely double covers of (ruled) surfaces.

A double covering is given by a (finite) 2:1 map $X \to Y$ branched along a (smooth) curve C.

The data is given by a surface Y along with a curve C and a square root B, i.e., a divisor B such that $\mathcal{O}(C) = \mathcal{O}(2B)$. (Thus the only condition on C is that it is so-called even. This is, however, difficult to check in general unless you master the Neron-Severi group. Hence we will from now on only consider ruled surfaces.)

It is convenient to go down to a minimal model $\overline{Y}$ and descend C to a possibly singular curve $\overline{C}$. X is then the minimal resolution of the nonfinite (in general) double cover $\overline{X} \to \overline{Y}$ branched along $\overline{C}$.

Now the singularities of $\overline{X}$ are explicitly related to the singularities of $\overline{C}$ (see, e.g., [**Psn 2**]), and the problem of imposing singularities has been reduced to a 1-dimensional problem.

The double covers of ruled surfaces play the same role as hyperelliptic curves do among curves; in fact they are characterized by being fibered by hyperelliptic curves. Unfortunately there is no good name for them as Catanese points out. (One may very well argue that the name hyperelliptic is not good to start out with.) In this article we will refer to them as hyperelliptic fibrations.

They are far too simple to be typical (or interesting?) but they are easy to manipulate. (In the business of surfaces, as in all other human endeavors, you tend to focus on what you can handle.)

But if you are asked to give an example of a curve with high genus, chances are that unless g is some very special and easily recognizable form, a hyperelliptic curve is all that you can come up with. This remark provides the justification for the following.

PROPOSITION. *If $c_1^2 \leq 8\chi$, any possible chern-invariant occurs and can be represented by some hyperelliptic fibration. (In fact even by some genus two fibration.)*

REMARK. This is essentially proved in [**Psn 2**] except for some annoying exceptions. Those are of a technical nature, and the reader may consult [**X1**] for further discussion. The ideas involved are very simple and more or less explicit in the above discussion.

In this context it is appropriate to insert two results of Xiao. First using genus-two fibrations, the proposition above cannot be improved; in fact

PROPOSITION (XIAO). *If X admits a genus-two fibration, then $c_1^2(X) \leq 8\chi$.*

REMARK. Compare this bound to that of complete intersections. Unlike that case, however, this bound is sharp, which can be seen by the curves $C_g \times C_2$ $(g(C_i) = i)$.

It comes as somewhat of a surprise, though, that this is not true for hyperelliptic fibrations in general (a guess ventured at the original delivery at Bowdoin).

EXAMPLE (XIAO). There exists (in fact infinitely many) hyperelliptic fibrations X such that $c_1^2(X) > 8\chi$.

The significance of this example will appear later and we will return to it in the final section.

We should not be too harsh on hyperelliptic fibrations because in some regions nothing else can survive.

PROPOSITION. *If $c_1^2 < 3\chi - 10$ then every surface admits a hyperelliptic fibration.*

REMARK. The key points are (see [**B 2**])

(i) if Ψ_K (the canonical map) is birational then $c_1^2 \geq 3p_g - 7$ (this is a souped-up version of $K^2 \geq 2p_g - 4$, the Noether inequality of page 202);

(ii) if Ψ_K is composite with a pencil then $c_1^2 \geq 3p_g - 6$.

From this we conclude that the canonical image is a surface and the canonical map covers it multiply. It is then straightforward to work out that the image is in fact a ruled surface and that the canonical map is two to one.

The proposition is sharp; in fact, hypersurfaces in $\mathbf{P}^2 \times \mathbf{P}^1$ given by bihomogenous polynomials $F(x_0, x_1, x_2; y_0, y_1)$ of bidegree $(4, n)$ satisfy $K^2 = 3p_g - 7$ (and can be found plotted on our diagram on page 204 along the faultline $c_1^2 = 3\chi - 10$, the line above the lowest line—the one given by the Noether inequality), and generically they have no hyperelliptic fibrations. (The natural fibrations are given by plane quartics.)

3. We have now come to a point where it is natural to state and ponder the basic questions of geography. Those are of two types.

I. If we restrict the invariants, what can we say about the surfaces?

II. If we impose some natural conditions on the surfaces (e.g., simple connect-edness, canonical embeddability, ample cotangent bundle, existence of a genus-two fibration, etc...) in what ways can we further restrict the invariants?

Carrying I to its extreme means picking out one particular pair of chern-numbers (c_1^2, χ), say, e.g., (144,36), and determining all the possible surfaces with those invariants, finding out the nature of their moduli space, etc....

Stopping short of the ultimate we may restrict ourselves to just finding the simply connected ones (ex. complete intersection of two quartics in $\mathbf{P}^4$, or the complete intersection of two hypersurfaces of bidegree (3,3) and (2,2) in $\mathbf{P}^2 \times \mathbf{P}^2$) or to determining the number of irreducible components or just bounding their dimensions.

Such an approach we call a botanical one. It requires superhuman powers of patience coupled with a penchant for a somewhat myoptic vision inspired by an inordinate fondness for detail. In general we cannot even answer the more humble questions we outlined above.

As a particular example of botanical flavor, we can consider the obsession with finding surfaces of general type with $p_g = 0$. (Such surfaces have $\chi = 1$ and hence there exists only a finite number of families, grouped together under the headings $K^2 = 1, 2, 3, \ldots, 9$.)

The first such example was found by Godeaux (see [**Gdx**]). The starting point was a nonsingular quintic and an action of $\mathbf{Z}_5$. The quintic being canonically embedded, such an action is by necessity induced by the ambient $\mathbf{P}^3$. Examples of invariant quintics with no fixed points exhibited can easily be written down (e.g., the Fermat quintic under the action $(\ldots X_i \ldots)$ goes to $(\ldots q^i X_i \ldots)$ with q a primitive fifth root of unity). The quotient surface will have the invariants $p_g = 0$ $(\chi = 1)$ and $K^2 = 1$, and will consequently correspond to the minimal chern pair invariant (1,1). Miles Reid (see [**R 1**]) made a systematic study of such Godeaux surfaces and classified the possible torsion groups (actually finite, nontrivial fundamental groups) giving examples to each (as well as extending his investigations to the next case of $K^2 = 2$ (see [**R 2**]).

But the situation is more complicated. Rebecca Barlow (see [**Baw**]) exhibited a simply connected example with $p_g = 0$, $K^2 = 1$. So far hers is the only one known, but the general consensus is that there are bound to be many more (because no one has thought of any good reason why there should not be!).

As has been mentioned before, Godeaux's construction inspired Serre to show that all finite groups can occur as fundamental groups of surfaces.

Another way of exhibiting special surfaces of the above type is to use special configurations of plane curves, say, and take the double cover. This was used by Campadelli who managed to find a configuration of plane curves adding up to degree ten and with six type I elliptic singularities. The resolution of the double covering will then have invariants $p_g = 0$ and $K^2 = 2$.

This is not the place to give a comprehensive survey of all the so far extant constructions. The list although finite continues to grow, although interest has abated lately. The ideal reference is a special chapter in the book by Barth,

Peters, and Van de Ven, where the junior author keeps tabs on the situation (see [B-P-VdV].

Suffice it to say that for some time in the early seventies only examples with low K^2 were known. (Some older ones were temporarily forgotten and rescued from oblivion.) Kuga created a minor sensation with exhibiting $K^2 = 8$ ($p_g = 0$ still!) and finally Mumford exhibited the ultimate—the "fake" projective plane ($K^2 = 9$) of which there so far is only one known example (see [Mum 2]).

The general unfeasibility of botanical classification should not discourage people completely. In a sense there are even smaller pairs than $(1,1)$, and if a class appears too big, one may always restrict it by specializing the values of p_g (or q) or impose any other natural looking condition. Both Catanese (see [C 1]) and Ciliberto (see [Ci 1, 2]) have been active in that respect, just to name a few.

Two examples, also occurring in Catanese's contribution, are actually characterized by their chern-invariants.

EXAMPLES. A surface with invariants $K^2 = 1, \chi = 3$ is given by a double cover of a quadric cone branched at the vertex and the complete intersection with a quintic.

A surface with invariants $K^2 = 2, \chi = 4$ is given by a double plane branched along an octic.

The reason those chern-invariants are small (smaller that $(1,1)$!) is that they satisfy $c_1^2 = 2\chi - 6 + k$, for small k.

So far the most complete classification of surfaces in terms of their chern-invariants has been given by Horikawa in a series of papers titled *Surfaces of general type with small $c_1^2(k+1)$*. By c_1^2 is actually meant the difference $k = c_1^2 - 2\chi + 6$, which is a nonnegative integer (cf. the lower bound on c_1^2 as in the final theorem of §1).

The simplest situation corresponds to $k = 0$, and we have:

PROPOSITION (HORIKAWA). *If X is a surface satisfying*

$$c_1^2(X) = 2\chi(X) - 6,$$

then with the exception of the double octic and the double dectic (degree ten), X is a double cover of a ruled surface along a six-section.

If $\chi \neq 3$ (4) there is only one type (the moduli space is irreducible).

If $\chi = 3$ (4) there are two types (corresponding to a connected or a disconnected branch locus).

REMARK. For the proof and more detailed statement see [H]. The proof is not difficult, and the reason we are successful in this case is because of the final proposition of the previous section. (Such surfaces all have to be double covers of ruled surfaces).

FURTHER REMARKS. Disregarding the two examples $(2,4)$ and $(8,7)$, all the surfaces possess genus-two fibrations. This may not be so surprising as we expect such surfaces to be the most special of general type (cf. Catanese's lecture).

Topologically the surfaces are simply connected, and in the case of $\chi = 3$ (4)($K^2 = 0$ (8) in Horikawa's terminology), K is always even (the second Stiefel-Whitney class vanishes, in fancier language). Thus by Freedman's theorem both types of surfaces are homeomorphic. It is an interesting open question whether they are still diffeomorphic.

Leaving botany aside one may simply ask a question of this type.

QUESTION. Does there exist a number K_2 such that if $c_1^2 < K_2\chi$ the surface necessarily has a genus-two fibration (or is deformable into one) with a possible finite number of exceptions (which may either be classified, or drowned by an estimate like $c_1^2 \le K_2\chi - L_2$)?

In general one may compute K_g (and sup K_g). So far this has not been done. As a first tentative nibble we present:

PROPOSITION. *If X is fibered over a nonrational curve, and $c_1^2 < \frac{8}{3}\chi$, this fibration is necessarily a genus-two fibration.*

REMARK. Compare this with the Horikawa-Reid result on Albanese mappings in Catanese's lecture.

The result is not difficult to prove. In view of the previously cited proposition at the end of the previous section, we know a priori that the surface is a double cover of a ruled surface. We are then reduced to estimating singularities on branch curves; this has been carried out in [**Psn 2**].

In fact, this proposition can be sharpened using slightly more sophisticated techniques.

PROPOSITION (CORNALBA-HARRIS, XIAO). *If a surface X admits a fibration over a base curve of genus b by curves of genus g then*

$$c_1^2 - 8(b-1)(g-1) \ge (4 - 4/g)(\chi - (b-1)(g-1)).$$

Consult [**X 2**, **C-H**]. Note also that the estimate is interesting for $b = 0$. (For nonminimal surfaces we do not expect any interesting bounds for $b = 0$ as even on Horikawa surfaces we can find linear systems of curves of arbitrarily high degree.)

To turn things around consider a type II question.

QUESTION. If X admits a genus-two fibration over a rational curve, find the sharpest bound on $c_1^2(X)$.

REMARK. Using a naive method of estimating singularities on a branch curve, we get that $c_1^2 < 7\chi$ (see [**Psn 2**]). This is far from being sharp, as illustrated by the difficulty of finding rational genus-two fibrations with $c_1^2 = 4\chi - 4$ (see [**Psn 2**]). It may even be so that above the slope of four only certain slopes occur and we have some kind of spectral phenomenon.

For the most up-to-date results on genus-two fibration on should consult Xiao (e.g., [**X1**]).

Let us conclude this section with two remarks of general nature.

As we have seen, the notion of a slope plays an important role in surface geography. The lower the slope, the more special the surface, and hence the more amendable to knowledge. In particular if the slope is 2, then we have essentially a Horikawa situation and there is hope of complete classification (at least if we disregard a finite number of exceptions). If the slope is at most 3, then the surface is a hyperelliptic fibration and we can clearly manipulate it very explicitly, although the complexity of possibilities should not be sneered at.

Some ten years ago Miles Reid conjectured that slope 4 is a basic fault line; below it we may hope for some unified view of some sort, above it surfaces are just hopelessly complicated. Reid's ideas can be gleaned from [**R 3**] and a further workout of the situation can be found by Xiao [**X 2**], Xiao remaining the most active geographer.

Underlying all this is the "hunch" that low slope is not typical. So what is typical?

This leads us to the second remark of even more pronounced speculative nature.

What are the chern-invariants of a typical surface? Is is possible to have a measure on the invariants? Apparently the cosmologist S. Hawkins is reputed to have some kind of intuition on this.

The basic question is how should we state this. One crude measure is the maximal dimension d of an associated moduli-space.

Catanese has shown the inequality (see [**C 2**])

$$10\chi - 2c_1^2 \le d \le 10\chi + 3c_1^2 + 108.$$

For $c_1^2 \ge 5\chi$ this estimate loses much of its appeal. The approach is also intuitively unsatisfactory. What we need is something more "probabilistic."

A first crude attempt of this is obviously our diagram on page 204. In a more systematic and serious way we observe that there are of course uniform ways of representing all surfaces. The two standard ones are:

(1) hypersurfaces in $\mathbf{P}^3$ with generic singularities;

(2) multiple coverings of $\mathbf{P}^2$, branched along curves with only nodes and cusps.

One (computer) experiment would be to plot all the invariants of say quintics and then proceed to sextics, etc., or to start with double covers of $\mathbf{P}^2$ and continue with triple covers, etc.

In this context it should be pointed out that so far nobody knows exactly what kind of surfaces may appear in the disguise of (singular) quintics. All that we basically can say is that $c_1^2 \le 5$ and $\chi \le 5$ (which as we have seen could be a lot; for surfaces of general type, see [**Ya**]).

A cautious guess would be that most surfaces have slopes between 6 and 8, just like complete intersections. This naturally leads us to the topic of the final section.

4. The line $c_1^2 = 8\chi$ has for a long time been considered somewhat of a mystery. (In fact the Old Italians suspected that beyond it, no surfaces were to be found.)

There is also topological evidence to lend credence to the superstition. The index of the intersection form of a surface turns out to be $\frac{1}{3}(c_1^2 - 2c_2)$ $(= i)$ and hence $c_1^2 > 8\chi$ iff $i > 0$.

In the mid-fifties F. Hirzebruch suggested that one should look for cocompact quotients of the unit ball in $\mathbf{C}^2$. By his proportionality theorem such surfaces should satisfy $c_1^2 = 3c_2$.

Soon thereafter Borel found some, and it was conjectured that the ratio c_1^2/c_2 was in fact always bounded by 3, a conjecture which was, as we saw in §1, vindicated some twenty years later.

Those examples are however very much removed from the projective geometric spirit, and would appear to the earth-bound classical geometer as space ships of an alien civilization lounging high up in the stratosphere. Thus a search for more elementary examples was instigated, a search that turned out to be very frustrating in the beginning and that relied upon a considerable amount of ingenuity and good luck.

Just to give an idea: surfaces with $c_1^2 = 2c_2$ are simple to produce; just consider $X = C \times C'$ (C, C' two curves).

A simple calculation yields $e(X) = e(C)e(C')$ $(= ee')$ and $K = -eC' - e'C$ (recall that in general e—the Euler characteristic—of a curve is negative); hence $K^2 = 2ee'$.

Now if we modify this construction perhaps we could slip into the mysterious region of positive index.

For example, let us take a double cover Y of X branched along a smooth curve D. Using standard formulas for double covers (see, for example, [**Psn** or **B-P-Vd V**]), one easily computes $X^2(Y) - 2e(Y) = -\frac{3}{2}D^2$; hence Y has positive index iff $D^2 < 0$.

Such curves are easy to find; just choose maps $f: D \to C, f': D \to C'$ satisfying

$$e(D) < (\deg f)e(C) + (\deg f')e(C') \quad \text{(recall } e \text{ negative)}.$$

However the big stumbling block is to ensure that D is even. (The square root of D cannot be an effective curve; Riemann-Roch cannot assure that a given even curve actually is effective; and the Neron-Severi group does not span the whole of H^2, just to mention a few obstructed alleys.)

A variation of Kodaira, which (of course) works, is to take a curve C $(g(C) \geq 3)$ with a fixed point, free involution i, and let C' be the pullback of C with regards to multiplication with m in Jac (C) (the Jacobian of C). We then get $f: C' \to C$ with $\deg f = m^{2g}$ $(g = g(C))$. Finally in $C' \times C$ consider the graphs Γ_f and Γ_{if} and let $\Gamma = \Gamma_f \cup \Gamma_{if}$ be the branch curve.

REMARK. If we can find an unramified (even) branch curve D as in the first discussion, we will actually find a complete curve (the double cover of the appropriate base curve) in $\mathcal{M}_g$ (where g is the genus of the fiber). For more details the reader is referred to [**K 3**].

In the last five to six years there have been a slew of examples of elementary nature unearthed. The obsession to find surfaces with positive index has in some sense mirrored the search for surfaces with $p_g = 0$; in both cases, ingenious ad hoc constructions have been concocted. (Mumford has, as we pointed out earlier referring to his "fake" projective plane, killed two birds with one stone, albeit with a stone not usually found in the toolbag of the average beachcomber.)

As usual I refrain from trying to be encyclopedic, and consequently let us just look at a few typical constructions.

In 1979 Miyaoka considered generic projections of embedded surfaces S onto $\mathbf{P}^2$ and considered their Galois closures. (Algebraically: Let $\mathcal{M}(X)$ be the Galois closure of the field extension $f_*\mathcal{M}(S)$ of $\mathcal{M}(\mathbf{P}^2)$ where $f: S \to \mathbf{P}^2$ is the generic projection. Geometrically: throw out the diagonal of $S \times S \times \cdots \times S$ (deg f times) and consider the surface defined by $(\dots x_i \dots)$ such that $f(x_i) = f(x_j)$.)

It turned out that with a few exceptions the Miyaoka surfaces had positive index (albeit rather small) (see [**M 2**]). No one understands why this should be so, but it indicates that positive index may be rather common. The construction of Miyaoka turned out to be very crucial in the work of Moishezon-Teicher to which we will return.

Perhaps the most spectacular examples are due to F. Hirzebruch. A few of those examples even satisfy $c_1^2 = 3c_2$. Those turned up in 1981 and are very elementary in the sense of involving configuration of lines on $\mathbf{P}^2$. What is not so classical is to take very high degree covers. As they are discussed beautifully in Barth's lecture, there is little point in dwelling on them here.

Perhaps the most elementary examples are due to Xiao. He managed to find in the summer of 1985 examples of hyperelliptic fibrations with positive index.

Before discussing it, it may be helpful to return to the example of double covers of $C \times C'$. Now let C' be $\mathbf{P}^1$ (this solves the problem of finding even branch curves) and let us look for a branch curve, not with negative self-intersection, which will prove impossible, but one with many singularities, e.g., many so-called infinitely close triple points (cf. the discussion of the Campadelli double plane). The double cover will then have many type I elliptic singularities, and if the original surface is close to the magic line (which it turns out to be if C has high genus), the low slopes of the specializations (see Table 2, bottom right) will make a crossing likely.

One possible construction would be to take nine points in a web (a two-dimensional linear system) of (2,2) curves on $\mathbf{P}^1 \times \mathbf{P}^1$ (intersections of two quadrics, recalling that $\mathbf{P}^1 \times \mathbf{P}^1$ is a quadric in $\mathbf{P}^3$). The nine points should be chosen with some care, letting them be the flexes of some nonsingular cubic. We then obtain a configuration of points such that the line through any two points contains a third point. In fact the affine plane $\mathbf{Z}_3 \times \mathbf{Z}_3$ is embeddable in $\mathbf{P}^2$. The union of the curves corresponding to the nine point of the web will then be a curve of type (18,18) and it will have $12 \times 8 = 96$ triple points (the base points of any triplet spanning a linear pencil, of which there are 12). Taking a double cover of our $\mathbf{P}^1 \times \mathbf{P}^1$ along the fibers meeting the triple points, we get a curve of type (18,36) on $\mathbf{P}^1 \times C$, with $g(C) = 47$. It is straightforward to compute the invariants of the resolution of the double cover and we obtain a surface X, fibered over a curve of genus 47, with fibers of genus 8 and with the invariants $c_1^2 = 2984$ and $\chi = 370$.

The construction of Xiao is far more ingenious. He dispenses with the simple elliptic singularities of type I, and considers ones with very high multiplicities. The nine points (the $\mathbf{Z}_3 \times \mathbf{Z}_3$ configuration) will be replaced by a finite configuration corresponding to a finite subgroup of $\mathbf{P}^3 - Q = \mathrm{PGL}(2, \mathbf{C}) \subseteq |(1,1)|$. The most spectacular result is obtained by the icosahedral group I_{60}, giving the invariants $c_1^2 = 17700, \chi = 2011$ (and slope 8.8015 as opposed to 8.065 in the simple-minded example above). For further results see [**X 3**].

The last examples illustrate some typical features of positive index constructions. The invariants are usually very high and the surfaces are usually fibered over some high genus curve. And it seems hard to "fill out" all invariants, as the constructions are so ad hoc and unstable under perturbation.

Consequently it is still an open question whether there are "holes" among the surfaces with positive index. Obviously a "hole" will be far more difficult to prove than to contradict its existence; so if any positive results will be forthcoming, they will more likely be concerned with the latter.

At some time it also looked as if not all the slopes were possible, and Hirzebruch considered the possibility of some kind of spectral phenomenon (like the possible volumes of hyperbolic spaces, etc.). Then A. Sommese came along with a very simple construction (basically just base change!), showing that all slopes up to 9 do occur (see [**Smm**]). Still if one restricts oneself to surfaces fibered over rational curves, something interesting may still be discovered.

Any discussion of the "Arctic" region of surface geography would be sadly incomplete without some mention of the Bogomolov conjecture.

BOGOMOLOV CONJECTURE. Any simply connected surface of general type has negative index.

For some years there were many (failed) attempts to prove this. The only indication of the truth of the conjecture was that so few examples had been found, and all seemed very far from being simply connected.

In the fall of 1980 Holzapfel (see [**Hoz**]) announced that there were simply connected surfaces with slopes arbitrarily close to 9. (By Yau's uniformization result, slope 9 is actually impossible.) His examples were similar to the Hilbert modular surfaces, but instead of using $H \times H$ he used the unit ball. The Hilbert modular surfaces all have negative index and are known to be simply connected (a nontrivial result). Holzapfel eventually had to retract his announcement as the determination of the fundamental groups turned out to be formidable. However certain experts believe that the examples will turn out to be simply connected, in complete analogy with the Hilbert modular surfaces, and we have to wait for an improvement of technology before Holzapfel will be vindicated.

In 1984 Moishezon and Teicher (see [**M-T**]) presented a construction of a simply connected surface of positive index. This was the first affirmed example and naturally it caused a stir. It came as the outcome of an involved theory of braid applications to surfaces which incidentally had been the preoccupation of the senior discoverer for several years, and they applied their findings to the examples suggested by Miyaoka. Naturally it was rather intimidating and definitely impenetrable to a casual observer, reinforcing the view that Bogomolov's hunch may not have been so far off anyway.

Shortly after the Bowdoin conference Xiao created another sensation. By a slight modification of the examples of hyperelliptic fibrations with positive index, he was able to make those simply connected as well. The additional high points were to take some more care in order to have a fibration over $\mathbf{P}^1$ instead of a high genus curve and to make one fiber simply connected, which naturally was done by letting a rational fiber of the rational-ruled surface be a component of the branch locus of the double covering.

BIBLIOGRAPHY

[B] E. Bombieri, *Canonical models of surfaces of general type*, Inst. Hautes Études Sci. Publ. Math. **42** (1973), 171–219.

[B 1] A. Beauville, *Surfaces algebriques complexes*, Asterisque, no. 54, Soc. Math. France, Paris, 1978.

[B 2] _____ , *L'application canonique pour les surfaces de type general*, Invent. Math. **55** (1979), 121–140.

[B-P-VdV] W. Barth, C. Peters, and A. Van de Ven, *Compact complex surfaces*, Ergeb. Math. Grenzgeb. (3) Band 4, Springer-Verlag, 1984.

[Baw] R. Barlow *A simply connected surface of general type with $p_g = 0$*, Invent. Math. **79** (1985), 293–301.

[Bog] F. Bogomolov, *Holomorphic tensors and vectorbundles on projective varieties*, Math. USSR-Izv. **13** (1979), 499–555.

[C 1] F. Catanese, *Connected components of moduli spaces*, J. Differential Geom. **24** (1986), 395–399.

[C 2] _____ , *On the moduli spaces of surfaces of general type*, J. Differential Geom. **19** (1984), 483–515.

[C-H] M. Cornalba and J. Harris, *Ample linebundles on the moduli space of curves*, (unpublished preprint–title unstable).

[Ci 1] C. Ciliberto, *Canonical surfaces with $p_g = p_a = 4$ and $K^2 = 5, \ldots, 10$*, Duke Math. J. **48** (1981), 1–37.

[Ci 2] _____ , *Canonical surfaces with $p_g = p_a = 5$ and $K^2 = 10$*, Ann. Sci. École Norm. Sup. (4) **9** (1982), 282–336.

[D] S. K. Donaldson *La topologie différentielle des surfaces complexes*, C. R. Acad. Sci. Paris Sér. I Math. **301** (1985); details to appear in *Irrationality and the h-cobordism conjecture*, J. Differential Geom.

[F] M. Freedman, *The topology of four dimensional manifolds*, J. Differential Geom. **17** (1982), 357–453.

[G] D. Gieseker, *Global moduli for surfaces of general type*, Invent. Math **43** (1977), 233–282.

[Gdx] L. Godeaux, *Sur une surface algebrique de genre zero et de bigenre deux*, Atti Accad. Naz. Lincei **14** (1931), 479–481.

[H] E. Horikawa, *Algebraic surfaces of general type with small c_1^2*. I, Ann. of Math. (2) **104** (1976), 357–387; II, Invent. Math. **37** (1976), 121–155; III, Invent. Math. **47** (1978), 209–248; IV, Invent. Math. **50** (1979), 103–128.

[Hz] F. Hirzebruch, *Automorphe Formen unde der Satz von Riemann-Roch*, Internat. Sympos. on Algebraic Topology, Univ. Nacional Autónoma de México and UNESCO, Mexico City, 1958, pp. 129–144.

[Hoz] R-P. Holzapfel, *A class of minimal surfaces in the unknown region of surface geography*, Math. Nachr. **98** (1980), 221–232.

[K 1a] K. Kodaira, *On compact complex analytic surfaces*. I, Ann. of Math. (2) **71** (1960), 111–152.

[K 2a,c] _____, *On the structure of compact complex analytic surfaces.* I, Amer. J. Math. **86** (1964), 751–798; III, Amer. J. Math. **90** (1969), 55–83.

[K 3] _____, *A certain type of irregular algebraic surfaces*, J. Analyse Math. **19** (1967), 207–215.

[M 1] Y. Miyaoka, *On the Chern numbers of surfaces of general type*, Invent. Math. **42** (1977), 225–237.

[M 2] _____, *On algebraic surfaces with positive index*, Preprint, 1980.

[Mum 1] D. Mumford, *The canonical ring of an algebraic surface*, Ann. of Math. (2) **76** (1962), 612–615.

[Mum 2] _____, *An algebraic surface with K ample, $(K^2) = 9, p_g = q = 0$*, Amer. J. Math. **101** (1979), 233–244.

[M-T] B. Moishezon and M. Teicher, *Existence of simply connected algebraic surfaces of general type with positive and zero indices*, Proc. Nat. Acad. Sci. U.S.A. **83** (1986), 6665–6666; details to appear in Invent. Math. under an almost identical title.

[Psn 1] U. Persson, *On degenerations of algebraic surfaces*, Mem. Amer. Math. Soc. No. 189 (1977).

[Psn 2] _____, *On Chern invariants of surfaces of general type*, Compositio Math. **43** (1981), 3–58.

[R 1] M. Reid, *Surfaces with $p_g = 0$, $K^2 = 1$*, J. Fac. Sci. Univ. Tokyo, Sect. 1A **25** (1978), 75–92.

[R 2] _____, *Surfaces with $p_g = 0$, $K^2 = 2$*, Preprint, 1979.

[R 3] _____, *π_1 for surfaces with small K^2*, Algebraic Geometry, Copenhagen 1978, Lecture Notes in Math., vol. 732, Springer-Verlag, 1979, pp. 534–544.

[Smm] A. Sommese, *On the density of ratios of Chern numbers of algebraic surfaces*, Math. Ann. **268** (1984), 207–221.

[Srr] J-P. Serre, *Cours d'arithmetique*, Presses Univ. de France, Paris, 1970.

[III] I. Shafarevic et al., *Algebraic surfaces*, Proc. Steklov Inst. Math. **75** (1965).

[VdV] A. Van de Van, *On the Chern numbers of surfaces of general type*, Invent. Math. **36** (1976), 285–293.

[X 1] G. Xiao, *Surfaces fibrees en courbes de genre deux*, Lecture Notes in Math., vol. 1137, Springer-Verlag, 1985.

[X 2] _____, *Hyperelliptic surfaces of general type with K^2 less than $4X$*, Preprint.

[X 3] _____, *An example of hyperelliptic surfaces with positive index*, Preprint, East China Normal University. Math P85/011.

[Y] S-T. Yau, *Calabi conjecture and some new results in algebraic geometry*, Proc. Nat. Acad. Sci. U.S.A. **74** (1977), 1798–1799.

University of Uppsala, Sweden

Threefolds

Proceedings of Symposia in Pure Mathematics
Volume **46** (1987)

Contributions to Riemann-Roch on Projective 3-folds with Only Canonical Singularities and Applications

A. R. FLETCHER

1. Introduction. The aim of this paper is to show how to apply the formula for $\chi(O_X(nK_X))$ of [**R2**, §10] to canonical 3-folds. Theorems 2.3 and 2.5 give four equivalent formulas for $\chi(O_X(nK_X))$, where X is a projective 3-fold with only canonical singularities.

The main results, Theorems 4.3 and 4.4, are that $P_{12} \geq 1$ and $P_{24} \geq 2$ for a canonical 3-fold X with $\chi(O_X) = 1$. This can be compared with surfaces of general type (see also [**W**]).

I would like to thank Miles Reid for his help and J. Kollár for suggesting ideas on how to prove Theorem 4.3. The SERC provided financial support.

2. Contributions to Riemann-Roch. Throughout §2 we assume that X is a projective 3-fold with only canonical singularities.

2.1. DEFINITION. Let r, a_1, a_2, and a_3 be coprime positive integers. Let ε be a primitive rth root of unity acting on $\mathbf{C}^3$ via

$$\varepsilon(x, y, z) = (\varepsilon^{a_1} x, \varepsilon^{a_2} y, \varepsilon^{a_3} z).$$

A singularity $Q \in X$ is of type $\frac{1}{r}(a_1, a_2, a_3)$ if (X, Q) is isomorphic on an analytic neighborhood to $(\mathbf{C}^3, 0)/\langle \varepsilon \rangle$ (see also [**R2**, 4.2] for notation).

2.2. DEFINITION. Suppose Q is a singularity of type $\frac{1}{r}(a, -a, 1)$, where r and a are coprime. Then Q is also of type $\frac{1}{r}(1, -1, b)$ where $ba \equiv 1 \mod r$ and is a canonical singularity. Define

$$\sigma(Q, n) = \sum_{k=1}^{r-1} \frac{\varepsilon^{nk}}{(1 - \varepsilon^k)(1 - \varepsilon^{ak})(1 - \varepsilon^{-ak})},$$

where ε is a primitive rth root of unity.

1980 *Mathematics Subject Classification* (1985 *Revision*). Primary 14J17, 14J30.

By [**R2**, 8.6] we have the following:

2.3. THEOREM. *For all* n,

$$\chi(O_X(nK_X)) = \frac{(2n-1)n(n-1)}{12}K_X^3 + \chi(O_X) + \frac{n\pi^*K_X \cdot c_2(Y)}{12}$$

(1)

$$+ \sum_Q \frac{1}{r}(\sigma(Q,n) - \sigma(Q,0)),$$

where the summation takes place over a basket of singularities Q *of type* $\frac{1}{r}(a, -a, 1)$, *and* $\pi \colon Y \to X$ *is a resolution.*

2.4. NOTE. The singularities $\{Q\}$ are not necessarily the singularities of X. However the singularities of X make the same contribution to $\chi(O_X(nK_X))$ as if they were those of the basket (see [**R2**, 8.2]).

This formula can be written in a number of different forms, some more useful than others depending on the application.

2.5. THEOREM. *For all* n,

$$\chi(O_X(nK_X)) = \frac{(2n-1)n(n-1)}{12}K_X^3 + \chi(O_X) + \frac{n\pi^*K_X \cdot c_2(Y)}{12}$$

(2)

$$+ \sum_Q \left(-\frac{(r^2-1)\overline{n}}{12r} + \sum_{k=1}^{\overline{n}-1} \frac{\overline{bk}(r - \overline{bk})}{2r} \right),$$

where $\overline{x}$ *denotes the smallest residue modulo* r.

$$\chi(O_X(nK_X)) = \frac{(2n-1)n(n-1)}{12}K_X^3 + (1-2n)\chi(O_X)$$

(3)

$$+ \sum_Q \left(\frac{r^2-1}{12}\left\lfloor \frac{n}{r} \right\rfloor + \sum_{k=1}^{\overline{n}-1} \frac{\overline{bk}(r - \overline{bk})}{2r} \right),$$

where $\lfloor x \rfloor$ *denotes the integral part of* x.

$$\chi(O_X(nK_X)) = \frac{(2n-1)n(n-1)}{12}K_X^3 + (1-2n)\chi(O_X)$$

(4)

$$+ \sum_Q \sum_{k=1}^{n-1} \frac{\overline{bk}(r - \overline{bk})}{2r}.$$

PROOF. By [**R2**, 8.10] we have that

$$\sigma(Q,n) = \sum_{k=1}^{\overline{n}-1} \overline{bk}(r - \overline{bk}) + \frac{r^2-1}{24}(1 - 2\overline{n}),$$

where the sum is taken to be zero if $n \equiv 0$ or $1 \bmod r$. Thus

$$\sigma(Q,n) - \sigma(Q,0) = \sum_{k=1}^{\overline{n}-1} \overline{bk}(r - \overline{bk}) - \frac{r^2-1}{12}\overline{n}.$$

So (2) follows from (1).

By an unpublished result of R. Barlow (see also [**Ka**, §2] or [**R2**, Corollary 10.3]) we have that

$$\pi^* K_X \cdot c_2(Y) = \sum_Q \frac{r^2 - 1}{r} - 24\chi(O_X).$$

This gives (3).

(4) is obtained by noticing that

$$\sum_{k=1}^{r-1} \overline{bk}(r - \overline{bk}) = \sum_{k=1}^{r-1} k(r - k) = \frac{r(r^2 - 1)}{6}$$

since r and b are coprime. □

2.6. DEFINITION. Define

$$l(Q, n) = \sum_{k=1}^{n-1} \frac{\overline{bk}(r - \overline{bk})}{2r}$$

and $l(n) = \sum_Q l(Q, n)$. This is the correction term $l(n)$ defined in [**R1**, Theorem 5.5].

2.7. NOTE. There is a closed formula for $l(Q, n)$ for a singularity Q of type $\frac{1}{r}(1, -1, 1)$:

$$l(Q, n) = \frac{\overline{n}(\overline{n} - 1)(3r + 1 - 2\overline{n})}{12r} + \frac{r^2 - 1}{12} \left\lfloor \frac{n}{r} \right\rfloor,$$

but apparently not for other types.

2.8. COROLLARY. *The correction term $l(n)$ is nonnegative for $n \geq 2$, and $l(0) = 0 = l(1)$. Moreover if X has at least one singularity of index $\neq 1$, then $l(n)$ is strictly monotonic increasing for all $n \geq 1$.*

PROOF. The correction $l(n)$ is formally equal to a sum of $l(Q, n)$ for some basket of singularities Q, each summand being strictly positive for $n \geq 2$. □

3. Technical lemmas. The following lemmas are used in the next section.

3.1. LEMMA. *For all $m \geq 0$ and $n \geq 1$*

$$l(m + 2n) \geq l(m) + nl(2)$$

with equality if and only if all the singularities are of type $\frac{1}{2}(1, 1, 1)$.

PROOF. It is enough to prove this for a single singularity Q of type $\frac{1}{r}(a, -a, 1)$. Let $ba \equiv 1 \bmod r$ and define

$$\delta_j = \overline{jb}(r - \overline{jb}) + \overline{(j+1)b}(r - \overline{(j+1)b}) - b(r - b).$$

Now
$$2r(l(Q,2n) - l(Q,m) - nl(Q,2))$$
$$= \sum_{k=m}^{m+2n-1} \overline{bk}(r - \overline{bk}) - nb(r - b)$$
$$= \sum_{j=0}^{n-1} ((2j+m)b(r - \overline{(2j+m)b}) + \overline{(2j+m+1)b}(r - \overline{(2j+m+1)b})$$
$$- b(r - b))$$
$$= \sum_{j=0}^{n-1} \delta_{2j+m} \quad \text{separating even and odd } k.$$

Consider the individual δ_j, and let $\alpha = \overline{jb}$. There are 2 cases to consider:

(i) $\alpha + b < r$. Then $\overline{(j+1)b} = \alpha + b$ and so $\delta_j = 2\alpha(r - \alpha - b) > 0$.

(ii) $\alpha + b \geq r$. Then $\overline{(j+1)b} = \alpha + b - r$ and so $\delta_j = 2(r - \alpha)(\alpha + b - r) \geq 0$.
Thus $\delta_j \geq 0$ for all $j \geq 0$. $\square$

3.2. LEMMA. *Suppose $\alpha > \beta$ are integers. Then*
$$l(\tfrac{1}{\alpha}(1,-1,1),n) \geq l(\tfrac{1}{\beta}(1,-1,1),n)$$
for all $n < \beta < \alpha$.

PROOF. This comes straight from Note 2.7. $\square$

3.3. LEMMA.
$$l(\tfrac{1}{r}(a,-a,1),n) \geq l(\tfrac{1}{r}(1,-1,1),n)$$
for all $n \leq \lfloor (r+1)/2 \rfloor$.

PROOF. For any $a, l(\tfrac{1}{r}(a,-a,1),n)$ is a sum of terms taken from the list $(r-1)/2r, 2(r-2)/2r, 3(r-3)/2r, \ldots, (r-1)/2r$, each term occurring at most twice. Suppose $t(r-t)/2r$ is such a summand occurring in $l(\tfrac{1}{r}(a,-a,1),n)$. Then there is an integer k such that $1 < k < n-1$ and either $\overline{bk} = t$ or $\overline{bk} = r - t$. So either $k = r - \overline{at}$ or $k = \overline{at}$. But only one of these solutions satisfies $k < n \leq \lfloor (r-1)/2 \rfloor$, and so each of the summands $(r-1)/2r, 2(r-2)/2r, \ldots, u(r-u)/2r$, where $u = \lfloor (r-1)/2 \rfloor$, occurs only once in $l(\tfrac{1}{r}(a,-a,1),n)$. Since
$$\frac{r-1}{2r} < \frac{2(r-2)}{2r} < \cdots < \frac{u(r-u)}{2r},$$
then
$$l\left(\frac{1}{r}(a,-a,1),n\right) \geq \sum_{k=1}^{n-1} \frac{k(r-k)}{2r} = l\left(\frac{1}{r}(1,-1,1),n\right). \quad \square$$

3.4. COROLLARY. *For all $\alpha, \beta \in \mathbf{Z}$ with $0 \leq \beta \leq \alpha$ and for all $n \leq \lfloor (\alpha+1)/2 \rfloor$, we have*
$$l(\tfrac{1}{\alpha}(a,-a,1),n) \geq l(\tfrac{1}{\beta}(1,-1,1),n).$$

3.5. DEFINITION. Define $\Delta_n(Q) = n^2 l(Q,2) + l(Q,n) - l(Q,n+1)$ and $\Delta_n = \sum_Q \Delta_n(Q)$. The purpose of this definition will become clear in Lemma 4.7 and Corollary 4.8.

3.6. LEMMA. $\Delta_n(Q)$ *is an integer for all types of singularity* Q.

PROOF. Clearly

$$\Delta_n(Q) = \frac{1}{2r}(n^2 b(r-b) - \overline{nb}(r - \overline{nb}))$$
$$= \frac{n^2 b - \overline{nb}}{2} - \frac{(nb - \overline{nb})(nb + \overline{nb})}{2r}.$$

It is easy to see that this is an integer. □

3.7. NOTE. Then

$$\Delta_2(Q) = \min(r - b, b),$$
$$\Delta_3(Q) = \min(3(r-b), r, 3b),$$
$$\Delta_4(Q) = \min(6(r-b), 3r - 2b, 2b + r, 6b),$$
$$\Delta_5(Q) = \min(10(r-b), 6r - 5b, 3r, 5b + r, 10b).$$

4. Nonzero plurigenera. In this section assume that X is a canonical 3-fold, i.e., a projective 3-fold with only canonical singularities and K_X ample. Standard use of vanishing gives

4.1. THEOREM. *For all* $n \geq 2$,

$$P_n = h^0(O_X(nK_X)) = \chi(O_X(nK_X)).$$

Thus for all $n \geq 2$,

$$(*) \qquad P_n = \frac{(2n-1)n(n-1)}{12} K_X^3 + (1 - 2n)\chi(O_X) + l(n),$$

where $l(n)$ is defined in Definition 2.6. Notice that the term involving $\chi(O_X)$ is the only one which can be negative.

4.2. THEOREM. *If* $\chi(O_X) = 0$, *then* $P_2 \geq 1$ *and* $P_4 > 2$. *If* $\chi(O_X) < 0$, *then* $P_2 \geq 4$.

PROOF. When $\chi(O_X) = 0$ then it is clear that $P_2 \geq 1$. Now

$$P_4 = 2P_2 + 6K_X^3 + l(4) - 2l(2).$$

By Lemma 3.1 $l(4) \geq 2l(2)$ and so $P_4 \geq 3$. □

By [**Ko**, Corollary 4.8] it follows that for $\chi(O_X) = 0$ the 49th pluricanonical map is birational, and when $\chi(O_X) < 0$, then the 27th pluricanonical map is birational.

The rest of this section attempts to generalize this type of result to other values of $\chi(O_X)$. J. Kollár pointed out that for any $\chi = \chi(O_X)$ there is an $n(\chi)$ such that $P_{n(\chi)} \geq 1$, but his values for $n(\chi)$ were huge. We shall calculate a reasonable (and perhaps the best possible) bound for $\chi = 1$.

4.3. THEOREM. *If $\chi(O_X) = 1$, then $P_{12} \geq 1$.*

4.4. THEOREM. *If $\chi(O_X) = 1$, then $P_{24} \geq 2$.*

By [**Ko**, Corollary 4.8] it follows that for $\chi(O_X) = 1$ the 269th pluricanonical map is birational. These 2 results will be proved later.

4.5. DEFINITION. A formal record of pluridata X is a collection $K^3 \in \mathbf{Q}$, $\chi \in \mathbf{Z}$, $p_g \in \mathbf{Z} \geq 0$, and a basket $\{Q\}$ of singularities. The plurigenera P_n of X is given by $(*)$. The pluridata of X is the data $K_X^3, \chi(O_X), p_g$, and its basket $\{Q\}$ of singularities (see Note 2.4).

4.6. EXAMPLE. Consider the formal record of pluridata X; $K^3 = 1/420$; $\chi = 1$; $p_g = 0$; and singularities: 2 of type $\frac{1}{2}(1,1,1)$, 2 of type $\frac{1}{3}(2,1,1)$, and one each of types $\frac{1}{4}(3,1,1), \frac{1}{5}(3,2,1)$, and $\frac{1}{7}(5,2,1)$. In this case the plurigenera $P_1 = P_2 = \cdots = P_{11} = 0$, $P_{12} = 1$, $P_{13} = 0$, $P_{14} = \cdots = P_{17} = 1$, $P_{18} = P_{19} = 2$, $P_{20} = P_{21} = 3$, $P_{23} = 4$, $P_{24} = 5$, and so on. It is an interesting open question to know if there exists a canonical 3-fold with this pluridata.

4.7. LEMMA. *For all formal records of pluridata (not necessarily corresponding to an X) the following are equivalent:*
(i) *P_2 is an integer;*
(ii) *P_n is an integer for $n \geq 2$;*
(iii) *$K^3 \equiv -2l(2) \mod 2\mathbf{Z}$.*

PROOF. By differencing $(*)$ and using $(*)$ with $n = 2$ to eliminate K_X^3, we get
$$P_{n+1} - P_n = n^2 P_2 + (3n^2 - 2)\chi(O_X) - \Delta_n$$
for all $n \geq 2$ and where Δ_n is given in Definition 3.5. (ii) follows from (i) by induction and Lemma 3.6. (i) and (iii) are clearly equivalent. $\square$

These difference equations give rise to 4 equalities which will be used in the proof of Theorem 4.3.

4.8. COROLLARY. (1) $P_3 - 5P_2 = 10\chi - \sum_Q \Delta_2(Q)$.
(2) $P_4 - P_3 - 9P_2 = 25\chi - \sum_Q \Delta_3(Q)$.
(3) $P_6 - P_4 - 41P_2 = 119\chi - \sum_Q(\Delta_4(Q) + \Delta_5(Q))$.
(4) $P_{12} - P_6 - 451P_2 = 1341\chi - \sum_Q(\Delta_6(Q) + \cdots + \Delta_{11}(Q))$.
So the condition $P_2 = 0 = P_3$ limits the number of singularities present.

PROOF. From the proof of the previous lemma
$$P_3 - P_2 - 4P_2 = 10\chi - \sum_Q \Delta_2(Q).$$

Similarly for (2), (3), and (4). $\square$

4.9. LEMMA. *Suppose the pluridata X contains a singularity of index*
$$r \geq s = \frac{(12\chi - 1)(24\chi - 1)}{2(6\chi - 1)}.$$

Then $P_{12\chi} \geq 1$.

PROOF. Suppose there is a singularity Q present of index $r \geq s$. Let Q' be a singularity of type $\frac{1}{s}(1, -1, 1)$. Then

$$l(12\chi) \geq l(Q, 12\chi) \geq l(Q', 12\chi) \quad \text{by Corollary 3.4}$$
$$= \frac{12\chi(12\chi - 1)(3s + 1 - 24\chi)}{12s} = (24\chi - 1)\chi.$$

So $P_{12\chi} \geq 1$. $\square$

The above corollary and lemma allow an explicit n_χ to be calculated such that $P_{n_\chi} \geq 1$ for $\chi = \chi(O_X)$.

4.10. PROOF OF THEOREM 4.3. Suppose $\chi(O_X) = 1$ and let the pluridata of X be $\chi = 1$, K_X^3, $(S_i)_{i=0,\ldots,m}$. In fact we will prove that P_{12} is nonzero for any pluridata X with $\chi = 1$ and $K^3 > 0$. So if this record of pluridata corresponds to a canonical three-fold X, then $P_{12} \geq 1$.

Suppose $\chi = 1$, $K^3 > 0$, $(S_i)_{i=0,\ldots,n}$ are pluridata such that $P_{12} = 0$. Since P_{12} is zero, so are P_2, P_3, P_4, and P_6, and hence this fixes K^3 and limits $l(n)$ by $K^3 = 2(3 - l(2)) > 0$. Now define $\Gamma_1(S_i) = \Delta_2(S_i)$, $\Gamma_2(S_i) = \Delta_3(S_i)$, $\Gamma_3(S_i) = \Delta_4(S_i) + \Delta_5(S_i)$, and $\Gamma_4(S_i) = \Delta_6(S_i) + \cdots + \Delta_{11}(S_i)$, for all $i = 0, \ldots, n$. By Corollary 4.8

(1) $\sum_{i=0}^{n} \Gamma_1(S_i) = 10$,
(2) $\sum_{i=0}^{n} \Gamma_2(S_i) = 25$,
(3) $\sum_{i=0}^{n} \Gamma_3(S_i) = 119$,
(4) $\sum_{i=0}^{n} \Gamma_4(S_i) = 1341$.

As $\Gamma_1(S) \geq 1$ for any singularity S, then there are at most 10 singularities present in the pluridata.

By Lemma 4.9, any singularity appearing in the pluridata must have index less than 26.

Hence there are only a finite number of possible combinations of singularities for this pluridata.

Appendix 1 lists the 100 singularities of index less than 26 with the corresponding values of $\Gamma_1, \Gamma_2, \Gamma_3$, and Γ_4 for each singularity.

Using the ordering of types of singularities in Appendix 1 let n_j be the number of singularities of the jth type S_j and $\Gamma_{i,j} = \Gamma_i(S_j)$. Let Γ be the 100×4 matrix $(\Gamma_{i,j})$. Then the 4 equations are given by

$$(n_1, \ldots, n_{100}) \cdot \Gamma = (10, 25, 119, 1341).$$

Column reducing Γ via the matrix E

$$E = \begin{pmatrix} 3 & -2 & -4 & -10 \\ -1 & 1 & -3 & -16 \\ 0 & 0 & 1 & -7 \\ 0 & 0 & 0 & 1 \end{pmatrix}$$

gives

$$(n_1, \ldots, n_{100}) \cdot \Gamma' = (5, 5, 4, 8),$$

where $\Gamma' = \Gamma \cdot E$. The matrix $\Gamma' = (\Gamma'_{i,j})$ is given as the last 4 columns of Appendix 1. This gives 4 new equations:

(5) $\sum_{i=0}^{n} \Gamma'_1(S_i) = 5$,

(6) $\sum_{i=0}^{n} \Gamma'_2(S_i) = 5$,

(7) $\sum_{i=0}^{n} \Gamma'_3(S_i) = 4$,

(8) $\sum_{i=0}^{n} \Gamma'_4(S_i) = 8$.

By reference to Appendix 1, there are types of singularity which have $\Gamma'_4 > 8$. These singularities can never satisfy (8) and can be deleted from the list. Likewise for those singularities with $\Gamma'_1 > 5$, $\Gamma'_2 > 5$, or $\Gamma'_3 > 4$. This reduces the number of types of singularity to exactly 36.

Suppose there is a singularity S_0 present with $\Gamma'_4 = 8$. Then there are 2 cases:

(i) $\Gamma'_3(S_0) = 2$. Thus 2 singularities S_1, S_2 of type $\frac{1}{5}(3,2,1)$ are required to satisfy (3). So $\Gamma'_2(S_0) \leq 3$ and $\Gamma'_1(S_0) \leq 3$. So S_0 is of type $\frac{1}{10}(7,3,1)$ and a further 3 singularities of type $\frac{1}{2}(1,1,1)$ are required.

(ii) $\Gamma'_3(S_0) = 3$. Then S_0 is of type $\frac{1}{5}(4,1,1)$. So in a singularity of type $\frac{1}{5}(3,2,1)$, 3 of type $\frac{1}{3}(2,1,1)$, and 4 of type $\frac{1}{2}(1,1,1)$ are required to satisfy the equations.

Suppose that $\Gamma'_4(S_0) = 7$. Now there are no types of singularity with $\Gamma'_4 = 1$ and so (8) cannot be satisfied. So any singularity S with $\Gamma'_4(S) \geq 7$ can be deleted from the list.

By considering each value of Γ'_4 in decreasing order, exactly 2 more solutions are found:

(i) 2 of type $\frac{1}{4}(3,1,1)$, 3 of type $\frac{1}{3}(2,1,1)$, and 5 of type $\frac{1}{2}(1,1,1)$;

(ii) 1 of type $\frac{1}{4}(3,1,1)$, 2 of type $\frac{1}{8}(5,3,1)$, and 3 of type $\frac{1}{2}(1,1,1)$.

In all 4 solutions $l(2) = 3$ and so $K^3 = 0$, a contradiction. So there are no pluridata with $\chi = 1$, $K^3 > 0$, and $P_{12} = 0$. $\square$

4.11. PROOF OF THEOREM 4.4. Consider pluridata with $\chi = 1$ and $K^3 > 0$. As $P_{12} \geq 1$ then $P_{24} \geq 1$. Assume that $P_{24} = 1$. Thus P_2, P_3, P_4, and P_6 are either 1 or 0. If $P_2 = 1$ then $P_{24} > 1$. So $P_2 = 0$. There are 6 cases:

(i) $P_6 = 0$, $P_4 = 0$, and $P_3 = 0$;

(ii) $P_6 = 0$, $P_4 = 1$, and $P_3 = 0$;

(iii) $P_6 = 1$, $P_4 = 0$, and $P_3 = 0$;

(iv) $P_6 = 1$, $P_4 = 1$, and $P_3 = 0$;

(v) $P_6 = 1$, $P_4 = 0$, and $P_3 = 1$;

(vi) $P_6 = 1$, $P_4 = 1$, and $P_3 = 1$.

Using Corollary 4.8, these give 4 equations for each case.

Also $P_{24} = 1$ and so $l(24) \leq 48$. Hence any singularity occurring in the pluridata has index less than 25.

Using the same techniques as in the proof of Theorem 4.3, the only pluridata with $P_{24} = 1$ are:

(i) 1 of type $\frac{1}{12}(7,5,1)$, 1 of type $\frac{1}{4}(3,1,1)$, 2 of type $\frac{1}{3}(2,1,1)$, and 2 of type $\frac{1}{2}(1,1,1)$;

(ii) 4 of type $\frac{1}{6}(5,1,1)$ and 6 of type $\frac{1}{2}(1,1,1)$;

(iii) 1 of type $\frac{1}{6}(5,1,1)$, 4 of type $\frac{1}{3}(2,1,1)$, and 5 of type $\frac{1}{2}(1,1,1)$. These solutions occur in cases (i), (ii), and (iii) respectively. Each of these solutions has $K^3 = 0$. □

4.12. NOTE. Using similar techniques it can be shown that $\chi(O_X) = 2$ implies that $P_{24} \geq 1$.

5. Appendix 1. The following table gives the values of Γ_i and Γ_i' for each type of singularity. These are used in the proof of Theorem 4.3.

No.	Singularity	Γ_1	Γ_2	Γ_3	Γ_4	Γ_1'	Γ_2'	Γ_3'	Γ_4'
1	$\frac{1}{2}(1,1,1)$	1	2	10	112	1	0	0	0
2	$\frac{1}{3}(2,1,1)$	1	3	13	149	0	1	0	0
3	$\frac{1}{4}(3,1,1)$	1	3	15	167	0	1	2	4
4	$\frac{1}{5}(4,1,1)$	1	3	16	178	0	1	3	8
5	$\frac{1}{5}(3,2,1)$	2	5	24	268	1	1	1	0
6	$\frac{1}{6}(5,1,1)$	1	3	16	185	0	1	3	15
7	$\frac{1}{7}(6,1,1)$	1	3	16	190	0	1	3	20
8	$\frac{1}{7}(5,2,1)$	3	7	34	383	2	1	1	3
9	$\frac{1}{7}(4,3,1)$	2	6	28	319	0	2	2	7
10	$\frac{1}{8}(7,1,1)$	1	3	16	194	0	1	3	24
11	$\frac{1}{8}(5,3,1)$	3	8	37	419	1	2	1	2
12	$\frac{1}{9}(8,1,1)$	1	3	16	197	0	1	3	27
13	$\frac{1}{9}(7,2,1)$	4	9	44	497	3	1	1	5
14	$\frac{1}{9}(5,4,1)$	2	6	31	346	0	2	5	13
15	$\frac{1}{10}(9,1,1)$	1	3	16	199	0	1	3	29
16	$\frac{1}{10}(7,3,1)$	3	9	41	469	0	3	2	8
17	$\frac{1}{11}(10,1,1)$	1	3	16	200	0	1	3	30
18	$\frac{1}{11}(9,2,1)$	5	11	54	610	4	1	1	6
19	$\frac{1}{11}(8,3,1)$	4	11	50	569	1	3	1	3
20	$\frac{1}{11}(7,4,1)$	3	9	43	487	0	3	4	12
21	$\frac{1}{11}(6,5,1)$	2	6	32	364	0	2	6	24
22	$\frac{1}{12}(11,1,1)$	1	3	16	200	0	1	3	30
23	$\frac{1}{12}(7,5,1)$	5	12	58	651	3	2	2	3
24	$\frac{1}{13}(12,1,1)$	1	3	16	200	0	1	3	30
25	$\frac{1}{13}(11,2,1)$	6	13	64	722	5	1	1	6
26	$\frac{1}{13}(10,3,1)$	4	12	54	618	0	4	2	8
27	$\frac{1}{13}(9,4,1)$	3	9	46	513	0	3	7	17
28	$\frac{1}{13}(8,5,1)$	5	13	61	687	2	3	2	2
29	$\frac{1}{13}(7,6,1)$	2	6	32	375	0	2	6	35
30	$\frac{1}{14}(13,1,1)$	1	3	16	200	0	1	3	30
31	$\frac{1}{14}(11,3,1)$	5	14	63	718	1	4	1	3
32	$\frac{1}{14}(9,5,1)$	3	9	47	524	0	3	8	21

No.	Singularity	Γ_1	Γ_2	Γ_3	Γ_4	Γ'_1	Γ'_2	Γ'_3	Γ'_4
33	$\frac{1}{15}(14,1,1)$	1	3	16	200	0	1	3	30
34	$\frac{1}{15}(13,2,1)$	7	15	74	834	6	1	1	6
35	$\frac{1}{15}(11,4,1)$	4	12	58	654	0	4	6	16
36	$\frac{1}{15}(8,7,1)$	2	6	32	384	0	2	6	44
37	$\frac{1}{16}(15,1,1)$	1	3	16	200	0	1	3	30
38	$\frac{1}{16}(13,3,1)$	5	15	67	767	0	5	2	8
39	$\frac{1}{16}(11,5,1)$	3	9	48	542	0	3	9	32
40	$\frac{1}{16}(9,7,1)$	7	16	78	880	5	2	2	8
41	$\frac{1}{17}(16,1,1)$	1	3	16	200	0	1	3	30
42	$\frac{1}{17}(15,2,1)$	8	17	84	946	7	1	1	6
43	$\frac{1}{17}(14,3,1)$	6	17	76	867	1	5	1	3
44	$\frac{1}{17}(13,4,1)$	4	12	61	680	0	4	9	21
45	$\frac{1}{17}(12,5,1)$	7	17	82	919	4	3	3	3
46	$\frac{1}{17}(11,6,1)$	3	9	48	549	0	3	9	39
47	$\frac{1}{17}(10,7,1)$	5	15	69	788	0	5	4	15
48	$\frac{1}{17}(9,8,1)$	2	6	32	391	0	2	6	51
49	$\frac{1}{18}(17,1,1)$	1	3	16	200	0	1	3	30
50	$\frac{1}{18}(13,5,1)$	7	18	85	955	3	4	3	2
51	$\frac{1}{18}(11,7,1)$	5	15	71	806	0	5	6	19
52	$\frac{1}{19}(18,1,1)$	1	3	16	200	0	1	3	30
53	$\frac{1}{19}(17,2,1)$	9	19	94	1058	8	1	1	6
54	$\frac{1}{19}(16,3,1)$	6	18	80	916	0	6	2	8
55	$\frac{1}{19}(15,4,1)$	5	15	73	821	0	5	8	20
56	$\frac{1}{19}(14,5,1)$	4	12	63	702	0	4	11	29
57	$\frac{1}{19}(13,6,1)$	3	9	48	560	0	3	9	50
58	$\frac{1}{19}(12,7,1)$	8	19	92	1034	5	3	3	6
59	$\frac{1}{19}(11,8,1)$	7	19	87	988	2	5	2	5
60	$\frac{1}{19}(10,9,1)$	2	6	32	396	0	2	6	56
61	$\frac{1}{20}(19,1,1)$	1	3	16	200	0	1	3	30
62	$\frac{1}{20}(17,3,1)$	7	20	89	1016	1	6	1	3
63	$\frac{1}{20}(13,7,1)$	3	9	48	565	0	3	9	55
64	$\frac{1}{20}(11,9,1)$	9	20	98	1107	7	2	2	11
65	$\frac{1}{21}(20,1,1)$	1	3	16	200	0	1	3	30
66	$\frac{1}{21}(19,2,1)$	10	21	104	1170	9	1	1	6
67	$\frac{1}{21}(17,4,1)$	5	15	76	847	0	5	11	25
68	$\frac{1}{21}(16,5,1)$	4	12	64	720	0	4	12	40
69	$\frac{1}{21}(13,8,1)$	8	21	98	1106	3	5	3	4
70	$\frac{1}{21}(11,10,1)$	2	6	32	399	0	2	6	59
71	$\frac{1}{22}(21,1,1)$	1	3	16	200	0	1	3	30

No.	Singularity	Γ_1	Γ_2	Γ_3	Γ_4	Γ_1'	Γ_2'	Γ_3'	Γ_4'
72	$\frac{1}{22}(19,3,1)$	7	21	93	1065	0	7	2	8
73	$\frac{1}{22}(17,5,1)$	9	22	106	1187	5	4	4	3
74	$\frac{1}{22}(15,7,1)$	3	9	48	574	0	3	9	64
75	$\frac{1}{22}(13,9,1)$	5	15	77	859	0	5	12	30
76	$\frac{1}{23}(22,1,1)$	1	3	16	200	0	1	3	30
77	$\frac{1}{23}(21,2,1)$	11	23	114	1282	10	1	1	6
78	$\frac{1}{23}(20,3,1)$	8	23	102	1165	1	7	1	3
79	$\frac{1}{23}(19,4,1)$	6	18	88	988	0	6	10	24
80	$\frac{1}{23}(18,5,1)$	9	23	109	1223	4	5	4	2
81	$\frac{1}{23}(17,6,1)$	4	12	64	734	0	4	12	54
82	$\frac{1}{23}(16,7,1)$	10	23	112	1263	7	3	3	11
83	$\frac{1}{23}(15,8,1)$	3	9	48	578	0	3	9	68
84	$\frac{1}{23}(14,9,1)$	5	15	78	870	0	5	13	34
85	$\frac{1}{23}(13,10,1)$	7	21	95	1087	0	7	4	16
86	$\frac{1}{23}(12,11,1)$	2	6	32	400	0	2	6	60
87	$\frac{1}{24}(23,1,1)$	1	3	16	200	0	1	3	30
88	$\frac{1}{24}(19,5,1)$	5	15	79	880	0	5	14	37
89	$\frac{1}{24}(17,7,1)$	7	21	97	1107	0	7	6	22
90	$\frac{1}{24}(13,11,1)$	11	24	118	1332	9	2	2	12
91	$\frac{1}{25}(24,1,1)$	1	3	16	200	0	1	3	30
92	$\frac{1}{25}(23,2,1)$	12	25	124	1394	11	1	1	6
93	$\frac{1}{25}(22,3,1)$	8	24	106	1214	0	8	2	8
94	$\frac{1}{25}(21,4,1)$	6	18	91	1014	0	6	13	29
95	$\frac{1}{25}(19,6,1)$	4	12	64	745	0	4	12	65
96	$\frac{1}{25}(18,7,1)$	7	21	99	1125	0	7	8	26
97	$\frac{1}{25}(17,8,1)$	3	9	48	585	0	3	9	75
98	$\frac{1}{25}(16,9,1)$	11	25	122	1377	8	3	3	13
99	$\frac{1}{25}(14,11,1)$	9	25	113	1287	2	7	2	6
100	$\frac{1}{25}(13,12,1)$	2	6	32	400	0	2	6	60

REFERENCES

[Ka] Y. Kawamata, *On the plurigenera of minimal algebraic 3-folds with $K \equiv 0$*, Preprint.

[Ko] J. Kollár, *Higher direct images of dualizing sheaves*, Ann. of Math. (2) **123** (1986), 11–42.

[R1] M. Reid, *Canonical three-folds*, Journées de Géometrie Algébrique d'Angers (Proc. Alg. Geometry, Anger, 1979), A. Beauville, editor, Sijthoff and Noordhoff, Alphen aan den Rijn, 1980, pp. 273–310.

[R2] ____ , *Young person's guide to canonical singularities*, these Proceedings.

[W] P. M. Wilson, *The pluricanonical map on varieties of general type*, Bull. London Math. Soc. **12** (1980), 103–107.

THE MATHEMATICS INSTITUTE, UNIVERSITY OF WARWICK, COVENTRY, ENGLAND

Proceedings of Symposia in Pure Mathematics
Volume 46 (1987)

Vanishing Theorems for Cohomology Groups

JÁNOS KOLLÁR

I. Introduction and classical results. The aim of this note is to survey recent results, centered around the following general question.

1. PROBLEM (TOO GENERAL FORM). Let X be an algebraic variety and let F be a sheaf on X. Find good conditions implying $H^i(X, F) = 0$ for certain given values of i.

Why should one care about this problem? Actually one doesn't, but in various situations these cohomology groups come up as intermediate objects, and understanding them helps to solve the original problem. I have two main examples in mind.

2. EXAMPLE. Maps between algebraic varieties are given by sections of line bundles. Therefore the computation of $H^0(X, L)$ is of interest. The higher cohomologies enter the picture in two ways.

(a) *Crude version.* From Riemann-Roch we can usually compute

$$\sum(-1)^i H^i(X, L).$$

Therefore any information about the groups $H^i(X, L)$ for $i > 0$ translates into some information about $H^0(X, L)$. In particular, if $H^i(X, L) = 0$ for $i > 0$, then $H^0(X, L)$ is known exactly.

(b) *Fine version.* Let Y be a subscheme of X with ideal sheaf I. Then the short exact sequence

$$0 \to I \otimes L \to L \to L \mid Y \to 0$$

gives rise to

$$H^0(X, L) \xrightarrow{r} H^0(Y, L \mid Y) \to H^1(X, I \otimes L).$$

Since Y is smaller than X, frequently $H^0(Y, L \mid Y)$ is easier to understand than $H^0(X, L)$. If $H^1(X, I \otimes L) = 0$, then r is onto, and thus one can get sections of $H^0(X, L)$.

3. EXAMPLE (DEFORMATIONS OF ALGEBRAIC VARIETIES, MAPS, SHEAVES,...). In deformation theory one would like to understand the "space

1980 *Mathematics Subject Classification* (1985 *Revision*). Primary 14-02, 14F05; Secondary 14C30, 32J25.

Partially supported by the National Science Foundation.

parameterizing" all algebraic varieties,... with given "discrete invariants." It turns out that the tangent spaces to such a "parameter space" can be identified with certain cohomology groups (usually H^0 or H^1), and local equations of the parameter space depend on the next group (H^1 or, respectively, H^2). Therefore the vanishing of the former group means that the object in question cannot be deformed (= rigid), whereas the vanishing of the latter group means that the deformations depend locally on free parameters (= unobstructed).

Since I am not an expert, I will not talk more about this. The lectures of Seshadri [**Se**] are excellent references.

From now on we restrict ourselves to char 0, mainly because very little is known in general. In studying cohomology groups one can always assume that the ground field is algebraically closed, even that it is $\mathbf{C}$. This is very important since transcendental methods play a central role in the theory.

4. DEFINITION. Let X be a projective variety and L a line bundle on X. L is said to be

(i) *very ample* if there is an embedding $f\colon X \to \mathbf{P}^?$ such that $L \cong f^*\mathcal{O}(1)$;

(ii) *ample* if L^m is very ample for some $m > 0$;

(iii) *semiample* if L^m is generated by global sections for some $m > 0$;

(iv) *numerically effective* (*nef*) if for every integral curve $C \subset X$ we have $\deg L \mid C \geq 0$;

(v) *big* if $h^0(X, L^m) > cm^{\dim X}$ if $m \gg 0$ and for some $c > 0$.
We have the implications (i)$\Rightarrow$(ii)$\Rightarrow$(iii)$\Rightarrow$ (iv) and (ii)$\Rightarrow$(v).

5. DEFINITION. Assume that X is a complex manifold and E is a holomorphic vector bundle endowed with a Hermitian metric h which has curvature form Θ (which is an End(E)-valued $(1,1)$ form). If v is a holomorphic tangent vector of X at x and $\bar{v}$ is its conjugate, then the contraction $\Theta(v, \bar{v})$ is in End(E_x).

(i) The curvature Θ at x is called positive (negative) definite if $-\sqrt{-1}\Theta(v, \bar{v})$ is a positive (negative) definite Hermitian form.

(ii) The vector bundle E is said to be positive (negative) if E can be endowed with a metric h which has everywhere positive (negative) curvature form.
An important connection between the two definitions is the following.

6. PROPOSITION. *A line bundle L on a projective manifold is ample iff it is positive.*

For the proof see [**G-H**, p. 148].
The archetype of vanishing theorems is the following.

7. THEOREM (KODAIRA [**Kod1**]). *Let X be a projective smooth algebraic variety or a compact complex manifold and L an ample or positive line bundle. Then $H^i(X, \omega_X \otimes L) = 0$ for $i > 0$, where ω_X is the canonical line bundle. Equivalently, $H^i(X, L^{-1}) = 0$ for $i < \dim X$.*

One important consequence which gives the equivalence of the two formulations is the following theorem.

8. THEOREM (KODAIRA [**Kod2**]). *A compact complex manifold X is iso-morphic to a projective variety iff it carries a positive line bundle L.*

There are two, by now classical, generalizations of this theorem. One is the following

9. THEOREM (AKIZUKI-NAKANO [**A-N**]). *For X, L as in 7 we have $H^p(X, L \otimes \Omega_X^q) = 0$ if $p + q > \dim X$.*

This can be used to prove the Lefschetz hyperplane theorem among others. The other important generalization is the following

10. THEOREM (GRAUERT-RIEMENSCHNEIDER [**G-R**]). *Let X be a pro-jective manifold and let L be a big semiample line bundle. Then $H^i(X, \omega_X \otimes L) = 0$ for $i > 0$.*

Instead of sketching the rather delicate proof I would like to point out an important local consequence of this theorem.

11. COROLLARY [**G-R**]. *Let X be a smooth projective variety and $f: X \to Y$ a generically finite map. Then $R^i f_* \omega_X = 0$ for $i > 0$.*

PROOF. Let L' be ample on Y and let $L = f^* L'$. Then L is semiample and big; hence $H^i(X, L \otimes \omega_X) = 0$ for $i > 0$. On the other hand we have a spectral sequence

$$H^i(Y, R^j f_*(L \otimes \omega_X)) = H^i(Y, L' \otimes R^j f_* \omega_X) \Rightarrow H^{i+j}(X, L \otimes \omega_X).$$

By replacing L' with a large power of L' we may assume that

$$H^i(Y, L' \otimes R^j f_* \omega_X) = 0 \quad \text{for } i > 0$$

and that $L' \otimes R^j f_* \omega_X$ is generated by global sections. Therefore the above spectral sequence gives

$$H^0(Y, L' \otimes R^j f_* \omega_X) = H^j(X, L \otimes \omega_X);$$

hence $R^j f_* \omega_X = 0$ if $j > 0$.

This proof illustrates a very strange aspect of the theory. There are local theorems (e.g., that $R^j f_* \omega_X = 0$) for which only global proofs are known or for which the global proof is much simpler.

For instance, a natural analogue of the above corollary is the following

12. PROBLEM. Let X, Y be complex spaces, X smooth, and let $f: X \to Y$ be a proper, generically finite map. Is it true that $R^i f_* \omega_X = 0$ for $i > 0$?

This problem is surprisingly difficult and was settled only recently by Take-goshi [**Tk**] building on earlier results of Nakano and Ohsawa. Recently a different proof was found by Nakayama [**N**]. His proof uses the ideas that will be discussed in 2.4.

About a year after the conference a characteristic p analog of Kodaira's vanishing theorem was proved by Raynaud. His result is the following:

THEOREM (RAYNAUD). *Let X be a smooth projective variety over a field k of characteristic p. Assume that X has a lifting over $W_2(k)$ (the ring of Witt vectors of length two). Let L be an ample line bundle on X. Then*

$$H^i(X, L \otimes \Omega_X^j) = 0 \quad if \ i + j > \max(\dim X, 2\dim X - p).$$

For the proof and for further results we refer to the forthcoming paper "Relevements modulo p^2 et decomposition du complex de de Rham" by Deligne and Illusie.

II. Recent results. In the past ten years considerable progress was achieved in improving the classical theorems. The results can be grouped together in terms of five major directions. I will discuss them one by one with emphasis on the last two groups for two reasons. Firstly these are the ones I know most about and secondly these are the ones that found significant applications in the theory of higher-dimensional algebraic varieties.

2.1. *Weaker Kodaira-Akizuki-Nakano type results.* The K-A-N vanishing theorem (9) imposes two conditions on L. It should be a line bundle and its curvature should be strictly positive everywhere. If one allows L to be a vector bundle and/or imposes various semipositivity conditions only, then one can still get various vanishing theorems for certain cohomology groups. Results in this direction are quite varied. I refer the reader to the recent book of Shiffman-Sommese [S-S] since I cannot add anything to their presentation.

2.2. *Non-Kähler manifolds.* The proof of the Kodaira vanishing theorem relies very heavily on the machinery of Kähler differential geometry. If L is a line bundle with positive curvature, then its curvature form is the Ricci form of a Kähler metric, and so the underlying manifold is automatically Kähler. If the positivity assumption is weakened to some semipositivity assumption, then one gets a degenerate Kähler metric only, and this seems to ruin the whole proof.

Recently Siu [S1, S2] developed a new approach to the problem and obtained the following

13. THEOREM (SIU [S2]). *Let X be a compact complex manifold and let L be a line bundle with a Hermitian metric whose curvature form is positive semidefinite everywhere and positive definite at some point. Then*

$$H^i(X, L \otimes \omega_X) = 0 \quad for \ i > 0.$$

As Kodaira's vanishing could be used to give a characterization of projective manifolds, this theorem can be used to give a characterization of Moishezon manifolds (= manifolds bimeromorphic to a projective variety). This was a conjecture of [G-R].

14. COROLLARY (SIU [S2]). *A compact complex manifold X is a Moishezon manifold iff it admits a Hermitian line bundle L whose curvature form is positive semidefinite everywhere and positive definite at one point.*

Unfortunately, this presentation puts the cart before the horse. The actual proof of Siu proceeds as follows. First he proves that for $i > 0$, $h^i(X, L^k) \ll k^{\dim X}$ as $k \to \infty$. This is the main part of the proof and uses very delicate estimates. Using this, Riemann-Roch implies that $h^0(X, L^k) > \varepsilon k^{\dim X}$ which easily gives Corollary 14. Now the problem of cohomology vanishing can be treated on a suitable projective X' bimeromorphic to X, and there Kähler geometry can be used to get the required vanishing. This was already done in [G-R].

Recent still unpublished results of Demailly give the above type estimates for higher cohomology groups even in the case of indefinite curvature. Unfortunately this approach falls outside the scope of the present survey.

2.3. *Singular spaces.* An attempt to generalize Kodaira's vanishing theorem to singular spaces presents considerable difficulties. First of all the two forms given in Theorem 7 are not equivalent any more. The vanishing of $H^i(X, L^{-1})$ sems to be hopelessly false. For instance, if X is a threefold with isolated non-Cohen-Macaulay singularities, then $H^2(X, L^{-1})$ is never zero. The other version is much more amenable to generalizations once one settles for a suitable definition of the "ω_X." One choice adopted in [G-R] is the following. Let $f: Y \to X$ be a resolution of singularities and set $K_X = f_* \omega_Y$. Then the following is easy to derive from Theorem 10:

15. THEOREM [G-R]. *If L is semiample and big, then $H^i(X, L \otimes K_X) = 0$ if $i > 0$ and X is projective.*

They also gave an example that this result does not hold in general if K_X is replaced by ω_X.

The generalization of Theorem 9 to singular spaces is quite complicated. First one has to observe that the reason behind Theorem 9 is that the sheaf Ω_X^p is a suitable subquotient of the de Rham complex. DuBois [D] gave a general definition of the filtered de Rham complex for singular spaces; in [G-N-P] this was used to prove the following theorem which we formulate in a very informal way due to the length of the necessary definitions.

16. THEOREM (GUILLEN-NAVARRO AZNAR-PUERTA, [G-N-P]). *K-A-N vanishing admits a good generalization to singular spaces.*

The proof of this theorem was simplified in [St], where it is derived from the following result.

17. THEOREM (STEENBRINK [St]). *Let X be a projective variety, $f: Y \to X$ a resolution of singularities, and L an ample line bundle on X. Assume that $Z \subset X$ is a subvariety containing the set of indeterminacy of f^{-1} such that $E = f^{-1}(Z)$ is a normal crossing divisor. Let $\Omega_Y^p(\log E)$ be the sheaf of p-forms with logarithmic singularities along E and let J_E be the ideal sheaf of E. Then*
  (i) $H^p(Y, J_E \Omega_Y^q(\log E) \otimes f^* L) = 0$ *for* $p + q > \dim X$,
  (ii) $R^p f_* J_E \Omega_Y^q(\log E) = 0$ *for* $p + q > \dim X$.

For the proofs we refer to the above-mentioned papers.

2.4. *Strong Kodaira-type vanishing theorems.* The original Kodaira vanishing theorem required L to be ample. The Grauert-Riemenschneider vanishing weakened this assumption to semiample and big. For the case where X is a surface, this was further weakened by Ramanujam [**R**] to L nef and big. Recently this was generalized to arbitrary dimensions.

18. THEOREM (KAWAMATA [**Ka1**], VIEHWEG [**V**]). *Let X be a smooth projective variety and let L be an nef and big line bundle. Then*

$$H^i(X, L \otimes \omega_X) = 0 \quad for \ i > 0.$$

In fact they proved a slightly more general result which I now explain. To do this we have to introduce line bundles with "rational coefficients."

We consider formal linear combinations of divisors $\sum a_i D_i$ with $a_i \in \mathbf{Q}$. We call these **Q**-divisors. If C is a curve, then $(\sum a_i D_i) \cdot C = \sum a_i (D_i \cdot C)$ is well-defined. Two **Q**-divisors D_1 and D_2 are called numerically equivalent (denoted by $D_1 \equiv D_2$) if $D_1 \cdot C = D_2 \cdot C$ for every curve $C \subset X$. If D_1 and D_2 are linearly equivalent divisors, then they are numerically equivalent. For any **Q**-divisor D there is an $m > 0$ such that mD is an integral divisor. D is called nef, big,... if mD is nef, big,.... This is independent of the choice of m.

With these definitions the promised generalization can be stated as follows:

19. THEOREM (MIYAOKA [**Mi**] FOR $\dim X = 2$, KAWAMATA [**Ka1**], VIEHWEG [**V**] IN GENERAL). *Let X be a smooth projective variety and let L be a line bundle. Assume that there exist a **Q**-divisor M, divisors E_i on X, and rational numbers a_i such that*
   (i) $L \equiv M + \sum a_i E_i$;
   (ii) M *is nef and big;*
   (iii) $\sum E_i$ *is a divisor with normal crossings only, $E_i \neq E_j$ for $i \neq j$;*
   (iv) $0 \leq a_i < 1$.
Then $H^i(X, L \otimes \omega_X) = 0$ for $i > 0$.

This theorem seems very artificial at first sight, but in fact it is a very finely tuned instrument that can be applied with great success in a variety of situations.

Instead of giving a proof I will try to clarify it with some comments.

(a) It obviously generalizes the previous result if one takes $a_i = 0$.

(b) The condition $a_i < 1$ is important. Let $X' \subset \mathbf{P}^3$ be the cone over an elliptic curve and let $f : X \to X'$ be the blowing up of the vertex, E the exceptional divisor. Then

$$H^1(X, \omega_X \otimes f^* \mathcal{O}(n) \otimes \mathcal{O}(E)) = 1 \quad \text{for every } n > 0.$$

(c) What happens if $\sum E_i$ is not a normal crossing divisor? Let $f : Y \to X$ be an embedded resolution of $\sum E_i$ and then $f^* L \equiv f^* M + f^*(\sum a_i E_i)$. $f^* M$ is still nef and big and $f^*(\sum a_i E_i)$ is a normal crossing divisor and

$$h^i(Y, \omega_Y \otimes f^* L) = h^i(X, \omega_X \otimes L).$$

$f^*(\sum a_i E_i)$ can contain some exceptional divisors of f with multiplicity larger than one. Therefore in general we get the following only:

19'. THEOREM. *With notation as in Theorem 19 assume that we have* (i), (ii), *and*

(iv') $0 \le a_i < \varepsilon$ (*depending on the singularities of* $\sum E_i$). *Then*

$$H^i(X, L \otimes \omega_X) = 0 \quad \text{for } i > 0.$$

Informally, if a line bundle is very close to being nef and big, then Kodaira vanishing still holds.

Finally, I would like to mention two situations where such formal sums of divisors naturally arise.

(d) In Mori's study of 3-folds with ω_X not numerically effective [**Mo**], he contracts "bad" divisors. This unfortunately leads to singular varieties in some cases. We consider the case (his case 3.3.5) where the exceptional divisor is $E \cong \mathbf{P}^2$ with normal bundle $\mathcal{O}(-2)$. Let $f\colon X \to Y$ be the contraction. Then Y has a singular point y and K_Y is not Cartier at that point, but $2K_Y$ is Cartier. Easy computation yields that $2K_X \sim f^*(2K_Y) + E$; therefore $K_X \equiv f^*K_Y + \frac{1}{2}E$. Similar examples abound when one studies singularities where the canonical divisor is not Cartier but some multiple of it is.

(3) Let $g\colon C \to D$ be a degree two map between smooth curves with branch locus $\{d_i\} \subseteq \mathbf{D}$. We want to understand $g_*\mathcal{O}_C$. Any degree two map is Galois, so let i be the Galois involution on C. Then $g_*\mathcal{O}_C$ splits as the sum of invariant and anti-invariant parts. The invariant part is easy to get: it is generated by the constants, so $g_*\mathcal{O}_C = \mathcal{O}_D +$ (anti-invariants).

Over $D^0 = D - \{d_i\}$ we can pick (± 1) as a 2-valued section of the anti-invariant part. What happens with this near the d_i? Let $d = d_i$ and z be a local parameter at d. Then $\sqrt{z}$ is a local parameter on C. $\sqrt{z}$ is anti-invariant and this is the local generator. Therefore our 2-valued section has a "pole of order $\frac{1}{2}$" at d, and we would like to claim that $g_*\mathcal{O}_C = \mathcal{O}_D + \mathcal{O}(\frac{1}{2}\sum d_i)$. This does not quite make sense but it is true that (anti-invariant part) $\equiv \frac{1}{2}\sum d_i$.

A similar computation gives that if $g\colon X \to Y$ is a cyclic Galois cover, X and Y smooth, then $g_*\mathcal{O}_X$ splits as a sum of line bundles and, at least numerically, these line bundles can be expressed as $\mathbf{Q}$-linear combinations of the components of the branch locus. This is the starting point of Viehweg's approach.

2.5. *Injectivity theorems.* Let us go back to Example 2 and write down one more term of the sequence we had:

$$H^0(X, L) \xrightarrow{r} H^0(Y, L \mid Y) \to H^1(X, I \otimes L) \xrightarrow{s} H^1(X, L).$$

In general we want r to be onto. This is certainly the case if $H^1(X, I \otimes L) = 0$ but the weaker assumption that s be injective still guarantees the surjectivity of r.

To fix the notation let X be a smooth projective variety and L a line bundle. Any section $s \in H^0(X, L^k)$ gives a map $\varphi_s\colon \mathcal{O} \to L^k$; hence for any sheaf F we have a map $\varphi_s(F)\colon F \to F \otimes L^k$. The induced map on cohomologies will be denoted by $H^i\varphi_s(F)\colon H^i(X, F) \to H^i(X, F \otimes L^k)$.

The first injectivity theorem was proved by Tankeev some fifteen years ago, but it went relatively unnoticed. He gave a very complicated proof of the following:

20. THEOREM (TANKEEV [**Tn**]). *If L is generated by global sections and $s \in H^0(X, L)$ is sufficiently general, then $H^1\varphi_s(L \otimes \omega_X)$ is injective.*

Recently I simplified the proof and generalized it to get

21. THEOREM [**KoI**]. *If some power of L is generated by global sections and $s \in H^0(X, L^k)$ is arbitrary, then*

$$H^i\varphi_s(L^m \otimes \omega_X) \quad \text{is injective for } i \geq 0, \ k, m > 0.$$

This form should be seen as the proper generalization of the Kodaira vanishing theorem from ample to semiample line bundles.

The Kawamata-Viehweg vanishing theorem can be generalized in a similar manner. Since the details are a little bit technical, we refer to the articles of Arapura [**A**], Kawamata [**Ka2**], Moriwaki [**Mw**], Esnault-Viehweg [**E-VII**].

We saw that the global Theorem 10 can be used to derive essentially local results about higher direct images of dualizing sheaves. Similarly Theorem 21 can be used to get the following:

22. THEOREM [**KoI**]. *Let $f: Y \to X$ be a surjective map between projective varieties, Y smooth but X can be singular. Then*
  (i) *$R^i f_* \omega_Y$ is torsionfree for every i.*
  (ii) *If L is ample on X, then $H^j(X, L \otimes R^i f_* \omega_Y) = 0$ for $j > 0$, $i \geq 0$.*

The very important special case of $f_* \omega_Y$ was proved earlier by Ohsawa [**O**] using L^2 methods.

Part (ii) of the statement can be generalized to the case of L nef and big (Esnault-Viehweg [**E-VII**]) or to the case of L semiample [**KoII**].

These results have several interesting applications to higher-dimensional geometry, but I can only give a list of them for lack of time:
  (i) characterization of Abelian varieties [**KoI**];
  (ii) pluricanonical maps of irregular 3-folds [**KoI**];
  (iii) Fano-fibrations [**KoI**];
  (iv) effective results about the Kodaira dimension of algebraic fiber spaces [**KoI**], [**E-VII**];
  (v) study of minimal models [**Ka2**].

III. Proof of the Kodaira vanishing theorem. Several of the articles mentioned earlier not only generalize the Kodaira vanishing theorem, but also provide new understanding of its proof. Here I would like to present a proof that I consider particularly simple and illuminating. Others, however, might have a different opinion.

The present proof, as all the others mentioned earlier, is not free from harmonic theory, but the necessary results can be neatly packed into a lemma which is one of the basic results of Hodge theory (cf. [**G-H**, p. 116]). For the rest of the chapter we always use the Euclidean topology of varieties.

23. LEMMA. *Let Y be a smooth projective variety over* **C**, *and let* $\mathbf{C}_Y$ *denote the constant sheaf on* Y. *The natural inclusion* $\mathbf{C}_Y \to \mathcal{O}_Y$ *induces maps* $H^i(Y, \mathbf{C}_Y) \to H^i(Y, \mathcal{O}_Y)$, *and these are all surjective.*

Now let X be an n-dimensional smooth projective variety over **C** and L an ample line bundle over X. For simplicity I assume that L is actually very ample, and then prove the following:

24 (= 7). THEOREM. *With the above notation* $H^i(X, L^{-1}) = 0$ *for* $i < n$.

PROOF. One can view L and L^2 as complex manifolds of dimension $n + 1$; let $p: L \to X$ be the natural projection. Let $S \subset L^2$ be a general section of L^2 and $D \subset X$ its divisor of zeros. Since L^2 is generated by global sections, we may assume that S and D are smooth. One has a natural map $\otimes: L \to L^2$ which is squaring in each fiber of p; let $Y = \otimes^{-1}S$. Clearly the natural projection $p: Y \to X$ is $2 : 1$ off D and $1 : 1$ on D and furthermore Y is smooth. Let τ be the natural involution of Y/X: interchanging the two sheets.

Over Y one has a natural inclusion $\mathbf{C}_Y \to \mathcal{O}_Y$, and this gives a map $p_*\mathbf{C}_Y \to p_*\mathcal{O}_Y$. τ acts on Y/X and this decomposes $p_*\mathbf{C}_Y$ and $p_*\mathcal{O}_Y$ into an invariant and an anti-invariant part. The invariant parts are easy to determine; these are $\mathbf{C}_X$, resp. $\mathcal{O}_X$. The anti-invariant part of $p_*\mathbf{C}_Y$ is a rank one local system on $X - D$ which has monodromy -1 around D. I denote this by $\mathbf{C}[-1]$.

An anti-invariant function on Y extends to a function on L which is linear on the fibers of p; these are exactly the local sections of L^{-1}. Therefore the anti-invariant part of $p_*\mathcal{O}_Y$ is exactly L^{-1}.

This gives us a map $\mathbf{C}[-1] \to L^{-1}$ and from Lemma 23 we easily conclude that the maps $H^i(X, \mathbf{C}[-1]) \to H^i(X, L^{-1})$ are all surjective. Therefore Theorem 24 follows from the next result:

25. THEOREM. *With the above notation* $H^i(X, \mathbf{C}[-1]) = 0$ *for* $i < n$.

The proof will be given in three steps.

26. *Step 1.* $H^i(X, \mathbf{C}[-1]) = H^i(X - D, \mathbf{C}[-1])$.

PROOF. Let $j: X - D \to X$ be the inclusion. One has a Leray spectral sequence
$$H^p(X, R^q j_*\mathbf{C}[-1]) \Rightarrow H^{p+q}(X - D, \mathbf{C}[-1]).$$
The computation of $R^q j_*$ is local, and since D is smooth we have locally the inclusion $(\mathbf{C} - 0) \times \mathbf{C}^{n-1} \hookrightarrow \mathbf{C} \times \mathbf{C}^{n-1}$. The $\mathbf{C}^{n-1}$ does not matter; hence we need to consider the simple situation $(\mathbf{C} - 0) \hookrightarrow \mathbf{C}$. The only possible higher direct image is R^1, but this is zero since the monodromy around 0 is nontrivial (explicit computation). Therefore $R^q j_*\mathbf{C}[-1] = \mathbf{C}[-1]$ if $q = 0$ and zero otherwise. Substituting this into the above spectral sequence gives Step 1.

27. *Step 2.* $H^i(X - D, \mathbf{C}[-1]) = 0$ if $i > n$.

PROOF. $X - D$ is Stein and therefore homotopic to a real n-dimensional space. Therefore we have no cohomology for $i > n$.

Another approach is to prove in general that if Z is affine of dimension n and F is an algebraically constructible topological sheaf, then $H^i(Z, F) = 0$ for $i > n$. Noether normalization reduces this to the case $Z = \mathbf{C}^n$. For $n = 1$ we have essentially $\mathbf{C}$ − (some points) which is homotopic to a bouquet of circles; hence $H^i(\mathbf{C}, F) = 0$ for $i > 1$. For general n one can use induction and the Leray spectral sequence for the coordinate projection $\mathbf{C}^n \to \mathbf{C}^{n-1}$.

28. *Step* 3. $H^i(X, \mathbf{C}[-1])$ and $H^{2n-i}(X, \mathbf{C}[-1])$ are dual to each other.

PROOF. On Y we have the usual Poincaré duality between $H^i(Y, \mathbf{C})$ and $H^{2n-i}(Y, \mathbf{C})$, and the cup product of a τ-anti-invariant end of a τ-invariant class is a τ-anti-invariant class. τ preserves the orientation and therefore $H^{2n}(Y, \mathbf{C}) \cong \mathbf{C}$ is τ-invariant; therefore the τ anti-invariant parts of $H^i(Y, \mathbf{C})$ and $H^{2n-i}(Y, \mathbf{C})$ are dual to each other. By what we said earlier this gives the required result.

PROOF OF 25. $H^i(X, \mathbf{C}[-1])$ is dual to $H^{2n-i}(X, \mathbf{C}[-1])$) by step 3. This is the same as $H^{2n-i}(X - D, \mathbf{C}[-1])$ by step 2, which is zero by step 1 if $i < n$.

29. REMARK. If L is only ample, then take an m such that L^m is generated by global sections. One can use the mth power map $L \to L^m$ to construct an m-sheeted cyclic cover of X, and with a little bit of care the above proof works.

30. REMARK. The central observation of the above proof is that the co-homology of $\mathcal{O}$ comes from topology, as shown by Lemma 23. There are other holomorphic sheaves whose cohomology comes from topology, and a similar argument gives various vanishing theorems for them. We refer to the articles [**E-V**], [**KoII**], and [**Sa**] for various results in this direction.

REFERENCES

[**A**] D. Arapura, *A note on Kollár's theorem*, Duke Math. J. **53** (1986), 1125–1130.

[**A-N**] Y. Akizuki and S. Nakano, *Note on Kodaira-Spencer's proof of Lefschetz theorems*, Proc. Japan Acad. **30** (1954), 266–272.

[**D**] Ph. DuBois, *Complexe de De Rham filtre d'une variete singuliere*, Bull. Soc. Math. France **109** (1981), 41–81.

[**E-VI**] H. Esnault and E. Viehweg, *Revêtements cycliques*, Algebraic Threefolds, Lecture Notes in Math., vol. 947, Springer-Verlag, 1982, pp. 241–250.

[**E-VII**] ——, *Revêtements cycliques.* II, Preprint.

[**E-V**] ——, *Logarithmic De Rham complexes and vanishing theorems*, Invent. Math. **86** (1986), 161–194.

[**G-N-P**] F. Guillen, V. Navarro Aznar, and F. Puerta, *Theorie de Hodge via schemas cubiques*, Preprint.

[**G-R**] H. Grauert and O. Riemenschneider, *Verschwindungssätze für analytische Kohomologiegruppen auf komplexen Räumen*, Invent. Math. **11** (1970), 263–292.

[**G-H**] P. Griffiths and J. Harris, *Principles of algebraic geometry*, Wiley, New York, 1978.

[**Ka1**] Y. Kawamata, *A generalization of Kodaira-Ramanujam's vanishing theorem*, Math. Ann. **261** (1982), 43–46.

[**Ka2**] ——, *Pluricanonical systems on minimal algebraic varieties*, Invent. Math. **79** (1985), 567–588.

[**Kod1**] K. Kodaira, *On a differential geometric method in the theory of analytic stacks*, Proc. Nat. Acad. Sci. U.S.A. **39** (1953), 1268–1273.

[Kod2] ____, *On deformations of complex analytic structures.* I, II, Ann. of Math. (2) **67** (1958), 328–466.

[KoI] J. Kollár, *Higher direct images of dualizing sheaves.* I, Ann. of Math. (2) **123** (1986), 11–42.

[KoII] ____, *Higher direct images of dualizing sheaves.* II, Ann. of Math. (2) **124** (1986), 171–202.

[Mi] Y. Miyaoka, *On the Mumford-Ramanujam vanishing theorem on a surface*, Journées de Géometrie Algébrique d'Angers, Juillet 1979, Sijthoff and Noordhoff, Alphen aan den Rijn, 1980, pp. 239–247.

[Mo] S. Mori, *Threefolds whose canonical bundles are not numerically effective*, Ann. of Math. (2) **116** (1982), 133–176.

[Mw] A. Moriwaki, *Some remarks on Kollár's vanishing theorem*, Preprint.

[N] N. Nakayama, *On the lower semi-continuity of the plurigenera*, Adv. Stud. Pure Math. (to appear).

[O] T. Ohsawa, *Vanishing theorems on complete Kahler manifolds*, Publ. Res. Inst. Math. Sci. **20** (1984), 21–38.

[R] C. P. Ramanujam, *Supplement to the article "Remarks on the Kodaira vanishing theorem,"* J. Indian. Math. Soc. **38** (1974), 121–124.

[Sa] M. Saito, *Hodge structures via filtered D-modules*, Preprint.

[Se] C. S. Seshadri, *Theory of moduli*, Algebraic Geometry—Arcata 1974, Proc. Sympos. Pure Math., vol. 29, Amer. Math. Soc., Providence, R.I., 1975, pp. 263–304.

[S-S] B. Shiffman and A. J. Sommese, *Vanishing theorems on complex manifolds*, Progr. Math., vol. 56, Birkhäuser, 1985.

[S1] Y.-T. Siu, *A vanishing theorem for semi-positive line bundles over non-Kahler manifolds*, J. Differential Geom. **19** (1984), 431–452.

[S2] ____, *Vanishing theorems for the semipositive case*, Lecture Notes in Math., vol. 1111, Springer-Verlag, pp. 164–192.

[St] J. H. M. Steenbrink, *Vanishing theorems on singular spaces*, Asterisque **130** (1985), 330–341.

[Tk] K. Takegoshi, *Relative vanishing theorems in analytic spaces*, Duke Math. J. **52** (1985), 273–279.

[Tn] S. G. Tankeev, *On n-dimensional canonically polarized varieties and varieties of fundamental type*, Izv. Akad. Nauk SSSR Ser. Math. **35** (1971), 31–44; English trans. in Math. USSR-Izv. **5** (1971).

[V] E. Viehweg, *Vanishing theorems*, J. Reine Angew. Math. **335** (1982), 1–8.

HARVARD UNIVERSITY

Current address: University of Utah

Proceedings of Symposia in Pure Mathematics
Volume 46 (1987)

Deformations of a Morphism
along a Foliation and Applications

YOICHI MIYAOKA

1. Introduction. In differential topology and geometry, foliations have been extensively studied because of their significant applications and their complicated structure in the large. Their local structure is, however, extremely simple. The classical theorem of Frobenius (rigorously speaking, of Frobenius-Clebsch-Deahna-Chevalley) claims that an involutive subbundle of the tangent bundle (an involutive "distribution") has a unique integral submanifold containing a given point [**C**, Chapter III, §VII].

An integral submanifold of an algebraic involutive subsheaf is not necessarily algebraic, yet it would not be apparently too unreasonable to look for a "formal" counterpart of the Frobenius theorem over rings and fields of arbitrary characteristic, since the involutiveness is purely algebraically defined. This task, though carried out without difficulty over a ring containing $\mathbf{Q}$, turns out to be rather cumbersome in a general situation. For instance, the invertibility of natural numbers is presumed in the proof of Cauchy's theorem on the existence and uniqueness of solutions of an analytic ordinary differential equation, and therefore it is impossible to mimic the original proof in a naive manner. Over a Dedekind domain of characteristic 0, the "*involutiveness*" of a "distribution" is actually insufficient to guarantee the "*integrability*" or, equivalently, the existence of the formal integral subscheme. Furthermore in characteristic p, examples show that "integral subsubschemes" would be indeterminate without liftings of the distribution to characteristic 0.

In this paper, we discuss deformations of a morphism along a smooth algebraic *foliation* (mainly in §6). The principal difficulty is how to define the "integrability" of involutive subbundles in mixed characteristics; our formulation below uses the generalized notion of *differential operators* of finite order in the sense of Grothendieck [**EGA IV**]. Once the notions are rightly defined, our argument proceeds in an exactly similar way as in [**SGA 1**, Exposé III]. The existence of the universal formal deformation along a foliation (or along one of its integrable

1980 *Mathematics Subject Classification* (1985 *Revision*). Primary 14E40; Secondary 14F10, 14M99, 14B20.

lifts to characteristic 0) will follow from the vanishing of a certain first cohomology group (Corollary 6.3). The *"formal Frobenius theorem"* (Corollary 6.4) is a special case of our result where the source variety reduces to a single point.

As an application of the deformation theory along foliations, we obtain a necessary and sufficient condition for a variety over **C** to be birationally equivalent to a conic bundle: existence of a (singular) *conic foliation* on some smooth projective birational model (Theorem 7.6). Although the assertion is valid only in characteristic 0, our proof heavily depends on S. Mori's technique to reduce the problem to positive characteristics, as was brilliantly developed in [**Mo**] to solve Hartshorne's conjecture.

From an analytic or differential-topological point of view, Theorem 7.6 shows nothing but the *"compactness and rationality of the leaves,"* a quite natural and geometric conclusion. So it would be a problem of some interest to find a differential-geometric or function-theoretic proof (e.g., construction of leaves by minimizing a suitable energy function, estimation of the growth of a solution of a special kind of algebraic differential equations, etc.)

The higher-dimensional equivalent of a conic foliation (of dimension 1) is a *"special foliation."* Its integrability on almost every reduction was shown by T. Ekedahl and O. Gabber (Theorem 8.1). A complex projective manifold is uniruled whenever it carries a special foliation (Theorem 8.5); in other words, the cotangent sheaf of a non-uniruled variety is *"generically semipositive"* (Corollary 8.6). The implications of this result will be discussed in the author's forthcoming paper [**Mi2**].

ACKNOWLEDGMENTS. The author is grateful to S. Mori, T. Fujita, Y. Kawamata, S. Iitaka, and M. Hanamura for helpful suggestions, frank criticism, and warm encouragements. One of the key results was communicated to the author by T. Ekedahl, to whom he is indebted for various remarks. Last but not least, he expresses his hearty thanks to M. Reid who gave him an opportunity to set about the present work in comfortable circumstances at the University of Warwick, England.

LIST OF NOTATIONS.

$\mathscr{I}_\Delta$: the ideal sheaf of the diagonal subscheme $X \simeq \Delta_{X/S} \subset X \times_S X$.

$\Omega^1_{X/S}$: the cotangent sheaf $\mathscr{I}_\Delta / \mathscr{I}_\Delta^2$.

$\mathscr{T}_{X/S}$: the tangent sheaf $\mathscr{H}om_{\mathscr{O}_X}(\Omega^1_{X/S}, \mathscr{O}_X)$ of X over S.

$\mathscr{P}^i_{X/S}$: $\mathscr{O}_{X \times_S X} / \mathscr{I}_\Delta^{i+1}$, the structure sheaf of the nth infinitesimal neighborhood $\Delta^{[i]}_{X/S}$ of $\Delta = \Delta_{X/S} = \Delta^{[0]}_{X/S}$.

$\mathscr{D}\!i\!f\!f^i_{X/S}$, $\mathscr{D}\!i\!f\!f_{X/S}$: the sheaf of (local) differential operators.

$\mathscr{P}\mathscr{D}\!i\!f\!f_{X/S}$, $\mathscr{P}\mathscr{D}\!i\!f\!f^i_{X/S}$: the sheaf of differential operators without terms of order 0; see §2.

$\mathscr{D}^i_{X/S}[\mathscr{F}]$, $\mathscr{D}^i_{X/S}\langle\mathscr{F}\rangle$, $\mathscr{A}^i_{X/S}(\mathscr{F})_j$: see §3.

$\mathscr{D}^i_{X/k}\langle\mathscr{F}'\rangle$, $\mathscr{A}^i_{X/k}(\mathscr{F}')_j$: see §4.

$\mathscr{F}^\perp = \mathscr{A}^1_{X/S}(\mathscr{F})_1 = \ker(\Omega^1_{X/S} \to \mathscr{F}^*)$: the annihilator of $\mathscr{F} \subset \mathscr{T}_{X/S}$.

$\partial_x = \partial/\partial x$: derivation in x.

$\det \mathscr{E}$: the determinant bundle, i.e., the double dual of the highest exterior product of a torsion free sheaf $\mathscr{E}$.

$\delta_{(f,g)}$: the differential operator representing the infinitesimal deviation of g from f, see §5.

$\mathscr{D}eform(f,B)$, $\mathscr{D}eform(f,B,\mathscr{F})$: deformation functors; see §6.

$\mathscr{H}ilb$: Hilbert scheme (-functor).

$\mathscr{RC}^{\alpha}(\mathscr{F},x)$: subfunctor of $\mathscr{H}ilb$, representing rational curves through x which are tangent to $\mathscr{F}$; see §7.

2. Basic properties of differential operators.

This section is devoted to a brief summary of Grothendieck's theory on differential operators on schemes. For a detailed treatment, we refer the reader to [**EGA IV**, §§16.7–11].

Let R be a noetherian ring with unity, S an R-scheme, and X a separated S-scheme. Denote by $\Delta^{[n]}_{X/S}$ the infinitesimal neighborhood of the image of the diagonal morphism $\Delta: X \to X \times_S X$ of order n. We have two projections $p_1, p_2: \Delta^{[n]}_{X/S} \to X$. Let $\mathscr{P}^n_{X/S}$ be the structure sheaf of $\Delta^{[n]}_{X/S}$, which is naturally identified with the sheaf $(p_1)_*(p_2)^*\mathscr{O}_X$ on X. The ring $\mathscr{P}^n_{X/S}$ has the structure of an $\mathscr{O}_X$-algebra (*canonical structure*) via the first projection p_1.

In the meantime, $\mathscr{P}^n_{X/S}$ carries a subsidiary structure of an $\mathscr{O}_X$-algebra by virtue of the ring homomorphism $d^n_{X/S} = p_2^*: \mathscr{O}_X \to \mathscr{P}^n_{X/S}$ associated with the second projection. Thus we consider $\mathscr{P}^n_{X/S}$ as an $\mathscr{O}_X$-*bimodule* on which $\mathscr{O}_X$ acts canonically from the left and via $d^n_{X/S}$ from the right. This convention is compatible with the identification of $\mathscr{P}^n_{X/S}$ with $(p_1)_*(p_2)^*\mathscr{O}_X$, which is an $\mathscr{O}_X$-bimodule in an obvious way.

Let $\mathscr{E}$ be a (left) $\mathscr{O}_X$-module. An additive map $D: \mathscr{O}_X \to \mathscr{E}$ is said to be a (relative) *differential operator of order $\leq n$* if there is a left $\mathscr{O}_X$-homomorphism $u: \mathscr{P}^n_{X/S} \to \mathscr{E}$ such that $D = u \cdot d^n_{X/S}$. Let $\mathscr{D}iff^n_{X/S}(\mathscr{E})$ denote the sheaf of local differential operators from $\mathscr{O}_X$ to $\mathscr{E}$. The natural map $\mathscr{H}om_{\mathscr{O}_X}(\mathscr{P}^n_{X/S}, \mathscr{E}) \to \mathscr{D}iff^n_{X/S}(\mathscr{E})$ is a bijection so that $\mathscr{D}iff^n_{X/S}(\mathscr{E})$ is equipped with the natural structure of an $\mathscr{O}_X$-bimodule.

For integers $m \leq n$, we have a natural projection $\mathscr{P}^n_{X/S} \to \mathscr{P}^m_{X/S}$. (In particular, the injection $p_1^*: \mathscr{O}_X \to \mathscr{P}^n_{X/S}$ together with the surjection $\mathscr{P}^n_{X/S} \to \mathscr{O}_X = \mathscr{P}^0_{X/S}$ makes $\mathscr{O}_X$ a direct summand of $\mathscr{P}^n_{X/S}$ and therefore $\mathscr{P}^n_{X/S}$ is canonically an *augmented ring* of $\mathscr{O}_X$ by $\mathscr{I}_\Delta/\mathscr{I}_\Delta^{n+1}$.) Thus, via the contravariant functor $\mathscr{H}om_{\mathscr{O}_X}(\ ,\mathscr{E})$, we get an inclusion $\mathscr{D}iff^m_{X/S}(\mathscr{E}) \subset \mathscr{D}iff^n_{X/S}(\mathscr{E})$ for $m \leq n$ which gives rise to an increasing filtration (order filtration) $\Phi = \{\mathscr{D}iff^i_{X/S}(\mathscr{E})\}$ on $\mathscr{D}iff^n_{X/S}(\mathscr{E})$. Let $\sigma^n: \mathscr{D}iff^n_{X/S}(\mathscr{E}) \to \mathrm{Gr}^n(\Phi)$ be the natural (left) $\mathscr{O}_X$-homomorphism. $\sigma^n(D)$ is called the *principal symbol* of the differential operator D. Obviously $\mathscr{D}iff^0_{X/S}(\mathscr{E})$ is the module of $\mathscr{O}_X$-homomorphisms: $\mathscr{O}_X \to \mathscr{E}$ (i.e., $\mathscr{D}iff^0_{X/S}(\mathscr{E}) \simeq \mathscr{E}$), and $\mathscr{D}iff^1_{X/S}(\mathscr{E})$ is a direct sum of $\mathscr{D}iff^0_{X/S}(\mathscr{E})$ and the left

$\mathscr{O}_X$-module of derivations: $\mathscr{O}_X \to \mathscr{E}$ over S:

$$\mathscr{D}i\!f\!f^1_{X/S}(\mathscr{E}) = \mathscr{H}om_{\mathscr{O}_X}(\mathscr{O}_X, \mathscr{E}) \oplus \mathscr{H}om_{\mathscr{O}_X}(\Omega^1_{X/S}, \mathscr{E}) \simeq \mathscr{E} \oplus \mathscr{E} \otimes_{\mathscr{O}_X} \mathscr{T}_{X/S}.$$

Let D and D' be differential operators from $\mathscr{O}_X$ to $\mathscr{O}_X$ of order $\leq m$ and n, respectively. Then the composition $D \circ D'$ is also a differential operator of order $\leq m + n$ (the proof is, though intuitively clear, rather complicated; see [**EGA IV**, §16.8]). In this way, $\mathscr{D}i\!f\!f_{X/S} = \bigcup \mathscr{D}i\!f\!f^i_{X/S}$ is a *noncommutative* associative ring, which is not an $\mathscr{O}_X$-algebra but merely an $\mathscr{O}_S$-algebra ($\mathscr{D}i\!f\!f^i_{X/S}$ denotes $\mathscr{D}i\!f\!f^i_{X/S}(\mathscr{O}_X)$).

In the following X will be always smooth over S. $\mathscr{D}i\!f\!f_{X/S}$ is an augumentation of the ring $\mathscr{O}_X$ by $\mathscr{P}\mathscr{D}i\!f\!f_{X/S}$, which is a left ideal of $\mathscr{D}i\!f\!f_{X/S}$. The graded $\mathscr{O}_X$-module associated with the order filtration Φ of $\mathscr{D}i\!f\!f_{X/S}$ is isomorphic to the commutative algebra $(\mathscr{S}\Omega^1_{X/S})^*$ (the dual of the symmetric tensor algebra of $\Omega^1_{X/S} = \mathscr{I}_\Delta/\mathscr{I}^2_\Delta$ over $\mathscr{O}_X$). There is a standard homomorphism $\mathscr{S}^n\mathscr{T}_{X/S} \to (\mathscr{S}^n\Omega^1_{X/S})^*$ defined by the formula $D_1 \otimes \cdots \otimes D_i(\omega_1 \cdots \omega_i) = \sum_\pi \{\prod_i D_i(\omega_{\pi(i)})\}$, where the indices i and π run over $\{1, \ldots, n\}$ and the symmetric group of degree n, respectively. This homomorphism is an isomorphism provided $n!$ is invertible in R, which is of course not the case in general; the ring $\mathscr{D}i\!f\!f_{X/R}$ is generated by $\mathscr{D}i\!f\!f^1_{X/R} = \mathscr{O}_X \oplus \mathscr{T}_{X/R}$ if and only if $R \supset \mathbf{Q}$.

To make things more explicit, let us fix a local parameter system $\{x_1, \ldots, x_n\}$ on X at $x \in X$ and let I be a multi-index $(i(1), \ldots, i(n))$. $\mathscr{P}^m_{X/S,x}$ is a free $\mathscr{O}_{X,x}$-module with a basis $\{dx^I\}_{|I| \leq m}$, where $dx^I = (dx_1)^{i(1)} \cdots (dx_n)^{i(n)}$ as usual. Let D_I denote the local differential operator of order $|I|$ such that $D_I(dx^J) = \delta_{IJ}$ (the Kronecker symbol). Then $\mathscr{D}i\!f\!f^m_{X/S,x}$ is also $\mathscr{O}_{X,x}$-free with a basis $\{D_I\}_{|I| \leq m}$. Put

$$\partial_i = \partial_{x_i} = \partial/\partial x_i = D_{(0, \ldots, 0, 1, 0, \ldots, 0)} \in \mathscr{T}_{X/S}.$$

Then an easy calculation shows:

$$\partial^I = \partial_1^{i(1)} \cdots \partial_n^{i(n)} = i(1)! \cdots i(n)! D_I = I! D_I.$$

Hence, in general, a differential operator of order ≥ 2 cannot be expressed as a polynomial in $\partial_1, \ldots, \partial_n$ with coefficients in $\mathscr{O}_{X,x}$. For example, in characteristic p, one can check $(\mathscr{T}_{X/S})^m \subset \mathscr{D}i\!f\!f^{np-1}_{X/S}$ for any positive integer m.

Note that $\mathscr{D}i\!f\!f_{X/S}$ and $\mathscr{P}\mathscr{D}i\!f\!f_{X/S}$ carry *Lie $\mathscr{O}_S$-algebra* structures via the *bracket product* $[\ ,\ ]$ given by:

$$[D, D'] = D \circ D' - D' \circ D.$$

Easily one shows that

$$[\mathscr{D}i\!f\!f^m_{X/S}, \mathscr{D}i\!f\!f^n_{X/S}] \subset \mathscr{D}i\!f\!f^{m+n-1}_{X/S}.$$

In particular, the tangent sheaf $\mathscr{T}_{X/S} = \mathscr{P}\mathscr{D}i\!f\!f^1_{X/S}$ is a Lie $\mathscr{O}_S$-subalgebra of $\mathscr{P}\mathscr{D}i\!f\!f_{X/S}$, while $\mathscr{D}i\!f\!f_{X/S}$ is an enveloping algebra of $\mathscr{T}_{X/S}$ (the universal one when $\mathbf{Q} \subset \mathscr{O}_X$). A coherent Lie subalgebra of $\mathscr{T}_{X/S}$ is said to be an *involutive subsheaf.* As is well known, an invertible subsheaf of $\mathscr{T}_{X/S}$ is involutive, see (4.8.1).

Assume that 2 is invertible in $\mathscr{O}_X$ so that the pairing between $\Lambda^2 \mathscr{T}_{X/S}$ and $\Omega^2_{X/S}$ is complete. Let $\mathscr{F}^\perp = \mathrm{Ker}(\Omega^1_{X/R} \to \mathscr{F}^*)$ denote the annihilator of $\mathscr{F}$ in $\Omega^1_{X/R} = \mathscr{H}om_{\mathscr{O}_X}(\mathscr{T}_{X/R}, \mathscr{O}_X)$, where $\mathscr{F}$ is a subbundle of $\mathscr{T}_{X/R}$. Then $\mathscr{F}$ is involutive if and only if $\mathscr{F}^\perp$ is a *differential ideal*, i.e.,

$$d\mathscr{F}^\perp \subset \mathscr{F}^\perp \wedge \Omega^1_{X/R} \subset \Omega^2_{X/R},$$

where d denotes the exterior derivative:

$$d(f_i dx_i) = \sum_j (\partial_j f_i)\, dx_j \wedge dx_i.$$

In fact, by direct computations, we have the identity

$$\langle d\omega, D \wedge D' \rangle = D\langle \omega, D' \rangle - D'\langle \omega, D \rangle - \langle \omega, [D, D'] \rangle$$

for $\omega \in \Omega^1_{X/S}$ and $D, D' \in \mathscr{T}_{X/S}$. Here the symbol $\langle \ , \ \rangle$ stands for the natural couplings $\Omega^1_{X/S} \times \mathscr{T}_{X/S} \to \mathscr{O}_X$, $\Omega^2_{X/S} \times \Lambda^2 \mathscr{T}_{X/S} \to \mathscr{O}_x$. Hence, if $\mathscr{F}$ is involutive, then $d\mathscr{F}^\perp$ vanishes on $\Lambda^2 \mathscr{F}$ so that $d\mathscr{F}^\perp \subset \mathscr{F}^\perp \wedge \Omega^1_{X/S}$. Conversely, if $d\mathscr{F}^\perp \subset \mathscr{F}^\perp \wedge \Omega^1_{X/S}$, then $[\mathscr{F}, \mathscr{F}]$ is annihilated by $\mathscr{F}^\perp$. Since $\mathscr{F}$ is saturated, this implies that $[\mathscr{F}, \mathscr{F}]$ is contained in $\mathscr{F}$.

3. Foliations over a Dedekind domain of characteristic 0. Let R be a Dedekind domain with field of quotients K of characteristic 0. Let X be a smooth R-scheme.

Let $\mathscr{F}$ be an *involutive* subbundle of $\mathscr{T}_{X/R}$ (i.e., $\mathscr{F}$ is a locally free Lie subalgebra of $\mathscr{T}_{X/R}$, with $\mathscr{T}_{X/R}/\mathscr{F}$ being locally free). Then $\mathscr{D}^1_{X/R}[\mathscr{F}] := \mathscr{O}_X \oplus \mathscr{F}$ generates a subring $\mathscr{D}_{X/R}[\mathscr{F}] \subset \mathscr{D}i\!f\!f_{X/R}$, which is also a left $\mathscr{O}_X$-submodule in an obvious manner. The submodules $\mathscr{D}^i_{X/R}[\mathscr{F}] = \mathscr{D}i\!f\!f^i_{X/R} \cap \mathscr{D}_{X/R}[\mathscr{F}]$ define an increasing filtration of $\mathscr{D}_{X/R}[\mathscr{F}]$ (the order filtration).

If we fix a local basis $\{\delta_1, \ldots, \delta_r\}$ of $\mathscr{F}$, an element of $\mathscr{D}^i_{X/R}[\mathscr{F}]$ is *uniquely* written in the form

$$\sum_{|M| \le i} \alpha_M \delta^M, \qquad \alpha_M \in \mathscr{O}_X,$$

by the involutiveness of $\mathscr{F}$. In particular, $\mathscr{D}^i_{X/R}[\mathscr{F}]$ is locally free of rank $\binom{r+i-1}{i}$ and the principal symbols give natural isomorphisms

$$\mathscr{D}^i_{X/R}[\mathscr{F}]/\mathscr{D}^{i-1}_{X/R}[\mathscr{F}] \simeq \mathscr{S}^i \mathscr{F}.$$

For positive integers i, j with $j \le i$, put

$$\mathscr{A}^i_{X/R}(\mathscr{F})_j = \{\omega \in \mathscr{I}^j_\Delta / \mathscr{I}^{i+1}_\Delta; \mathscr{D}^i_{X/R}[\mathscr{F}]\omega = 0\}$$
$$= \mathrm{Ker}(\mathscr{I}^j_\Delta / \mathscr{I}^{i+1}_\Delta \to (\mathscr{D}^i_{X/R}[\mathscr{F}])^*),$$
$$\mathscr{D}^i_{X/R}\langle \mathscr{F} \rangle = \{D \in \mathscr{D}i\!f\!f^i_{X/R}; D(\mathscr{A}^i_{X/S}(\mathscr{F})_0) = 0\}$$
$$= \mathrm{Ker}(\mathscr{D}i\!f\!f^i_{X/R} \to \mathscr{H}om_{\mathscr{O}_X}(\mathscr{A}^i_{X/R}(\mathscr{F})_0, \mathscr{O}_X))$$
$$= (\mathscr{D}^i_{X/R}[\mathscr{F}])^{**} = (\mathscr{D}^i_{X/R}[\mathscr{F}] \otimes_R K) \cap \mathscr{D}i\!f\!f_{X/R},$$

where the symbols $*$ and $**$ denote the dual and the double dual, respectively.

LEMMA 3.1. *Let i, j and k be natural numbers with $i \geq j \geq k$.*

(1) $\mathscr{A}^i_{X/R}(\mathscr{F})_j$ *is an ideal of* $\mathscr{P}^i_{X/R}$ *and* $\mathscr{D}_{X/R}\langle \mathscr{F} \rangle = \bigcup \mathscr{D}^i_{X/R}\langle \mathscr{F} \rangle$ *is a subring of* $\mathscr{D}\!i\!f\!f_{X/R}$.

(2) $\mathscr{I}^k_{\Delta} \mathscr{A}^i_{X/R}(\mathscr{F})_j \subset \mathscr{A}^{i+k}_{X/R}(\mathscr{F})_{j+k}$.

(3) $\mathscr{A}^i_{X/R}(\mathscr{F})_0 = \mathscr{A}^i_{X/R}(\mathscr{F})_1 \subset \mathscr{I}_{\Delta}/\mathscr{I}^{i+1}_{\Delta}$.

(4) $\mathscr{A}^i_{X/R}(\mathscr{F})_i = \mathrm{Ker}(\mathscr{S}^i \Omega^1_{X/R} \to \mathscr{S}^i \mathscr{F}^*) = \mathscr{I}^{i-1}_{\Delta} \mathscr{F}^\perp \subset \mathscr{S}^i \Omega^1_{X/R}$.

(5) $\mathscr{D}^j_{X/R}\langle \mathscr{F} \rangle = \mathscr{D}^i_{X/R}\langle \mathscr{F} \rangle \cap \mathscr{D}\!i\!f\!f^j_{X/R}$
$$= \mathrm{Ker}(\mathscr{D}\!i\!f\!f^i_{X/R} \to \mathscr{H}\!om_{\mathscr{O}_X}(\mathscr{A}^i_{X/R}(F)_1 + \mathscr{I}^j_{\Delta}\mathscr{P}^i_{X/R}, \mathscr{O}_X)).$$

(6) *If R contains the field of rational numbers $\mathbf{Q}$, then*

(a) $\mathscr{D}^i_{X/R}\langle \mathscr{F} \rangle = \mathscr{D}^i_{X/R}[\mathscr{F}]$, *and*

(b) *the canonical homomorphism* $\mathscr{A}^i_{X/R}(\mathscr{F})_j \to \mathscr{A}^j_{X/R}(\mathscr{F})_j$ *is surjective.*

PROOF. (1), (2), (3). Let $\alpha = \sum_\mu x_\mu \otimes y_\mu$ be a representative of an element of $\mathscr{A}^i_{X/R}(\mathscr{F})_j$, where $x_\mu, y_\mu \in \mathscr{O}_X$. Since α is annihilated by the constant operator 1, we have $\sum_\mu x_\mu y_\mu = 0$, which shows (3). Let $D_1, \ldots, D_s$ be tangent vectors in $\mathscr{F}$ ($s \leq i$). Then

$$D_1 \cdots D_s \alpha = \sum_\mu x_\mu D_1 \cdots D_s y_\mu = 0.$$

In order to prove that $\mathscr{A}^i_{X/R}(\mathscr{F})_j$ is an ideal, it suffices to show that $(x \otimes y)\alpha$ sits in $\mathscr{A}^i_{X/R}(\mathscr{F})_j$ for $x, y \in \mathscr{O}_X$. $(x \otimes y)\alpha$ is annihilated by $\mathscr{O}_X$ as a matter of course. On the other hand,

$$D_1 \cdots D_s \{(x \otimes y)\alpha\} = D_1 \cdots D_s \left(\sum_\mu x x_\mu \otimes y y_\mu \right)$$
$$= \sum_\mu x x_\mu D_1 \cdots D_s y y_\mu$$

is, by the Leibniz rule, written as a sum

$$\sum_\mu x x_\mu \left\{ \sum_N \binom{s}{|N|} (D_N y)(D_{N^*} y_\mu) \right\}$$
$$= \sum_n \binom{s}{|N|} x(D_N y) \left(\sum_\mu x_\mu D_{N^*} y_\mu \right)$$
$$= \sum_N \binom{s}{|N|} x(D_N y)(D_{N^*} \cdot \alpha),$$

which vanishes by our assumption. Here, D_N and D_{N^*} stand for differential operators $D_{n(1)} \cdots D_{n(|N|)}$ and $D_{n^*(1)} \cdots D_{n^*(s-|N|)}$, where $n(1) < \cdots < n(|N|)$, $n^*(1) < \cdots < n^*(s - |N|)$, $\{n(1), \ldots\} \cup \{n^*(1), \ldots\} = \{1, \ldots, s\}$. This proves the first half of (1). The proof of (2) is quite similar. The submodule

$\mathscr{D}_{X/R}[\mathscr{F}]$ is a subring of $\mathscr{D}\!i\!f\!f_{X/R}$, while $\mathscr{D}_{X/R}\langle\mathscr{F}\rangle = \mathscr{D}_{X/R}[\mathscr{F}]\otimes_R K\cap\mathscr{D}\!i\!f\!f_{X/R}$. Hence $\mathscr{D}_{X/R}\langle\mathscr{F}\rangle$ is also a subring.

(4) is a trivial consequence of our definition.

(5) follows from:

$$\mathscr{D}^i_{X/R}\langle\mathscr{F}\rangle \cap \mathscr{D}\!i\!f\!f^j_{X/R} = (\mathscr{D}^i_{X/R}[\mathscr{F}]\otimes_R K)\cap\mathscr{D}\!i\!f\!f^j_{X/R}$$
$$= (\mathscr{D}^j_{X/R}[\mathscr{F}]\otimes_R K)\cap\mathscr{D}\!i\!f\!f_{X/R} = \mathscr{D}^j_{X/R}\langle\mathscr{F}\rangle.$$

(6a) derives from the fact that the pairing between $\mathscr{S}^i\mathscr{F}$ and $\mathscr{S}^i\mathscr{F}^*$ is perfect provided $i!$ is invertible in R.

In order to show (6b), it suffices to prove that the mappings $\mathscr{A}^i_{X/R}(\mathscr{F})_j \to \mathscr{A}^{i-1}_{X/R}(\mathscr{F})_j$ are surjective. Let $\alpha \in \mathscr{P}^i_{X/R}$ be an element which is mapped into $\mathscr{A}^{i-1}_{X/R}(\mathscr{F})_j$. By definition, α is annihilated by $\mathscr{D}^{i-1}_{X/R}[\mathscr{F}]$. On the other hand, since $i!$ is invertible so that $\mathscr{S}^i\mathscr{F}^* \simeq (\mathscr{S}^i\mathscr{F})^*$, it is possible to find an element $\beta \in \mathscr{S}^i\mathscr{F}^* = \mathscr{I}_\Delta^i/\mathscr{F}^\perp\mathscr{I}_\Delta^{i-1}$ such that $\alpha+\beta$ annihilates $\mathscr{D}^i_{X/R}[\mathscr{F}]$ thanks to a *local* isomorphism $\mathscr{D}^i_{X/R}[\mathscr{F}] \simeq \mathscr{D}^{i-1}_{X/R}[\mathscr{F}]\oplus\mathscr{S}^i\mathscr{F}$. This completes the proof of the lemma.

COROLLARY 3.2. *The $\mathscr{O}_X$-modules*

$$\mathscr{D}^i_{X/R}\langle\mathscr{F}\rangle, \qquad \mathscr{D}^i_{X/R}\langle\mathscr{F}\rangle/\mathscr{D}^{i-1}_{X/R}\langle\mathscr{F}\rangle,$$
$$\mathscr{A}^i_{X/R}(\mathscr{F})_j, \quad \text{and} \quad \mathscr{A}^i_{X/R}(\mathscr{F})_j/\mathscr{A}^i_{X/R}(\mathscr{F})_{j+1}$$

are all reflexive and therefore locally free in codimension 2.

This is an immediate consequence of (3.1) and (3.3.1) below:

LEMMA 3.3. (1) *A coherent sheaf $\mathscr{F}$ on a regular scheme X is reflexive if and only if $\mathscr{F}$ is the kernel of an $\mathscr{O}_X$-homomorphism of locally free sheaves: $\mathscr{F} = \mathrm{Ker}(\mathscr{A} \to \mathscr{B})$, $\mathscr{A}$ and $\mathscr{B}$ are locally free.*

(2) *Let $f: \mathscr{F} \to \mathscr{F}'$ be an $\mathscr{O}_X$-homomorphism from a reflexive sheaf to a locally free sheaf. If f is an isomorphism in codimension 1, then f is an isomorphism on X.*

PROOF. (1) Let $\mathscr{F}$ be the kernel of $f: \mathscr{A} \to \mathscr{B}, \mathscr{A}$ and $\mathscr{B}$ being locally free. Then $\mathscr{F}^{**}$ is the kernel ${}^t({}^t f) = f: \mathscr{A} \to \mathscr{B}$. Next, suppose that $\mathscr{F}$ is reflexive. Let $\cdots \to \mathscr{E}_1 \to \mathscr{E}_0 \to \mathscr{F}^* \to 0$ be a locally free resolution of $\mathscr{F}^*$. Then $\mathscr{F}^{**} = \mathscr{F} = \mathrm{Ker}(\mathscr{E}_0^* \to \mathscr{E}_1^*)$.

(2) Let $\mathscr{F}$ be $\mathrm{Ker}(\mathscr{A} \to \mathscr{B})$ ($\mathscr{A}$ and $\mathscr{B}$ are locally free). Then we have an injection $\mathscr{F}' \to \mathscr{A}$ in codimension 1 which extends to an inclusion map on X. Now the assertion is clear.

LEMMA 3.4. (1) *There are canonical homomorphisms*

$$\mathscr{A}^i_{X/R}(\mathscr{F})_j/\mathscr{A}^i_{X/R}(\mathscr{F})_{j+1} \to \mathscr{A}^j_{X/R}(\mathscr{F})_j \simeq \mathscr{F}^\perp \cdot \mathscr{S}^{j-1}\Omega^1_{X/R},$$
$$\mathscr{D}^i_{X/R}\langle\mathscr{F}\rangle/\mathscr{D}^{i-1}_{X/R}\langle\mathscr{F}\rangle \to (\mathscr{S}^i\mathscr{F}^*)^*,$$

both of which are isomorphisms over $\mathrm{Spec}\,K$.

(2) *Assume that R is a local ring. Then there are (noncanonical) isomorphisms*

$$\det(\mathscr{A}_{X/R}^i(\mathscr{F})_j / \mathscr{A}_{X/R}^i(\mathscr{F})_{j+1}) \simeq \det(\mathscr{F}^\perp \cdot \mathscr{S}^{j-1}\Omega_{X/R}^1),$$

$$\det(\mathscr{D}_{X/R}^i\langle\mathscr{F}\rangle / \mathscr{D}_{X/R}^{i-1}\langle\mathscr{F}\rangle) \simeq \det((\mathscr{S}^i\mathscr{F}^*)^*)$$

between the determinant bundles (= the double duals of the highest exterior tensor products).

PROOF. (1) is obvious by (3.1), and therefore we have:

$$\det(\mathscr{A}_{X/R}^i(\mathscr{F})_j / \mathscr{A}_{X/R}^i(\mathscr{F})_{j+1}) = J \det(F^\perp \cdot \mathscr{S}^{j-1}\Omega_{X/R}^1),$$

$$\det(\mathscr{D}_{X/R}^i\langle\mathscr{F}\rangle / \mathscr{D}_{X/R}^{i-1}\langle\mathscr{F}\rangle) = J' \det((\mathscr{S}^i\mathscr{F}^*)^*),$$

where J and J' are *principal* ideals of R. This yields (2).

PROPOSITION 3.5. *Let m be a positive integer and $\mathscr{F} \subset \mathscr{T}_{X/R}$ an involutive subbundle. Then the following three conditions are equivalent:*

(1) *The natural map $\mathscr{A}_{X/R}^m(\mathscr{F})_1 \to \mathscr{F}^\perp = \mathscr{A}_{X/R}^1(\mathscr{F})_1$ is surjective;*

(2) *The natural map $\mathscr{A}_{X/R}^m(\mathscr{F})_j \to \mathscr{A}_{X/R}^i(\mathscr{F})_j$ is surjective for all positive integers i, j with $j \leq i \leq m$;*

(3) $\mathscr{I}_\Delta^{j-1}\mathscr{A}_{X/R}^m(\mathscr{F})_1 = \mathscr{A}_{X/R}^m(\mathscr{F})_j$ *for $j \leq m$.*

Furthermore, if $\mathscr{F}$ and m satisfy the above three conditions, then

$$\mathscr{A}_{X/R}^m(\mathscr{F})_j / \mathscr{A}_{X/R}^m(\mathscr{F})_{j+1}$$

is locally free $(j = 1, \ldots, m)$.

PROOF. We have obvious inclusion relations

$$\mathscr{I}_\Delta^{j-1}(\mathscr{A}_{X/R}^m(\mathscr{F})_1 / \mathscr{A}_{X/R}^m(\mathscr{F})_2) \subset \mathscr{A}_{X/R}^m(\mathscr{F})_j / \mathscr{A}_{X/R}^m(\mathscr{F})_{j+1} \subset \mathscr{A}_{X/R}^j(\mathscr{F})_j.$$

Condition (1) means that these three modules are equal. In fact, (1) holds if and only if

$$\mathscr{A}_{X/R}^m(\mathscr{F})_1 / \mathscr{A}_{X/R}^m(\mathscr{F})_2 = \mathscr{F}^\perp,$$

so that

(∗)
$$\mathscr{A}_{X/R}^m(\mathscr{F})_j / \mathscr{A}_{X/R}^m(\mathscr{F})_{j+1} = \mathscr{A}_{X/R}^i(\mathscr{F})_j / \mathscr{A}_{X/R}^i(\mathscr{F})_{j+1}$$
$$= \mathscr{A}_{X/R}^j(\mathscr{F})_j = \mathscr{I}_\Delta^{j-1}\mathscr{F}^\perp.$$

Thus (1) implies the surjectivity of $\mathscr{A}_{X/R}^m(\mathscr{F})_j \to \mathscr{A}_{X/R}^i(\mathscr{F})_j$ by induction on $i - j$, namely, condition (2), which is clearly stronger than (1). The identity (∗) yields (3) as well. Conversely, from (3) we derive:

$$\mathscr{I}_\Delta^{m-1}(\mathscr{A}_{X/R}^m(\mathscr{F})_1 / \mathscr{A}_{X/R}^m(\mathscr{F})_2) = \mathscr{A}_{X/R}^m(\mathscr{F})_m$$
$$= \operatorname{Ker}(\mathscr{S}^m\Omega_{X/R}^1 \to \mathscr{S}^m\mathscr{F}^*) = \mathscr{I}_\Delta^{m-1}\mathscr{A}_{X/R}^m(\mathscr{F})_1.$$

Therefore, we infer that $\mathscr{A}_{X/R}^m(\mathscr{F})_1 / \mathscr{A}_{X/R}^m(\mathscr{F})_2 = \mathscr{A}_{X/R}^1(\mathscr{F})_1$ in view of the smoothness of X over R. Now we have the equivalence of the conditions, which obviously involves the local freeness of

$$\mathscr{A}_{X/R}^m(\mathscr{F})_j / \mathscr{A}_{X/R}^m(\mathscr{F})_{j+1} \simeq \mathscr{F}^\perp \cdot \mathscr{S}^{j-1}\Omega_{X/R}^1.$$

DEFINITION 3.6. If an involutive subbundle $\mathscr{F} \subset \mathscr{T}_{X/R}$ satisfies one of the equivalent conditions above for every positive integer m, $\mathscr{F}$ is called an *integrable subbundle* of $\mathscr{T}_{X/R}$ or a *smooth foliation* on X.

COROLLARY 3.7. *Let $\mathscr{L}^{[m]}(\mathscr{F})$ denote the closed subscheme of $\Delta_{X/R}^{[m]}$ defined by the ideal sheaf $\mathscr{A}_{X/R}^m(\mathscr{F})_1$. Then $\mathscr{F}$ is integrable if and only if $\mathscr{L}^{[m]}(\mathscr{F})$ $\cap \Delta_{X/R}^{[j]} = \mathscr{L}^{[j]}(\mathscr{F})$ for all natural numbers m, j with $j \le m$. If $\mathscr{F}$ is integrable, $\mathscr{L}(\mathscr{F}) = \varinjlim \mathscr{L}^{[m]}(\mathscr{F})$ is a smooth formal subscheme $\subset X \times_R X$ of dimension $\dim_R X + \operatorname{rank} \mathscr{F}$ over R.*

EXAMPLE 3.8. Integrability actually imposes countable conditions on $\mathscr{F}$. For instance, let $R = \mathbf{Z}_p$, $X = \operatorname{Spec}[x, 1/(x-1), y]$, $\mathscr{F}_t = \mathscr{O}_X\{(x-1)\partial_x + t\partial_y\}$, $t \in \mathbf{Z}_p$. Then $\mathscr{A}_{X/R}^{p^k}(\mathscr{F}_t)_1 \to \mathscr{F}_t^\perp$ is surjective if and only if $t \in p^k \mathbf{Z}_p$. In fact, $\mathscr{A}_{X/R}^{p^k}(\mathscr{F}_t)_1 \otimes_{\mathbf{Z}_p} \mathbf{Q}_p$ is generated by

$$\delta y - t \sum_{i=1}^{p^k} \frac{(-1)^i}{i}(\delta x)^i, \qquad \delta z = p_2^* z - p_1^* z \in \mathscr{I}_\Delta.$$

4. Foliations in characteristic p.

Let k be a field of characteristic $p > 0$ and $R = W(k)$ the ring of Witt vectors with values in k. Let X be a *smooth* k-scheme, which will be always assumed to be liftable to a flat smooth R-scheme X'. Let $\mathscr{F} \subset \mathscr{T}_{X/k}$ be an involutive subbundle.

DEFINITION 4.1. $\mathscr{F}$ is said to be *L-involutive* if there exists an involutive subbundle $\mathscr{F}' \subset \mathscr{T}_{X'/R}$ which extends $\mathscr{F}$.

EXAMPLE 4.2. An involutive subbundle is not always *L*-involutive even if it is liftable to characteristic 0. Let ω be the 1-form $dx - y^{p-1} z^p \, dy + y^p z^{p-1} \, dz$ on the affine 3-space over $\mathbf{Z}$, where p is an odd prime. Let $\mathscr{F} \subset \mathscr{T}_{X/\mathbf{Z}}$ be the annihilator of ω. Since

$$\omega \wedge d\omega = 2p y^{p-1} z^{p-1} \, dx \wedge dy \wedge dz,$$

$\mathscr{F}$ is involutive over $\mathbf{F}_p$ (see §2). Let η be a 1-form $\alpha \, dx + \beta \, dy + \gamma \, dz$. Then

$$(\omega + p\eta) \wedge d(\omega + p\eta) \equiv \omega \wedge d\omega + p \, d(\eta \wedge \omega)$$
$$\equiv p(2y^{p-1} z^{p-1} - \partial_z \beta + \partial_y \gamma) \, dx \wedge dy \wedge dz \quad \bmod (p^2, \mathscr{J}^{2p-1}).$$

Here $\mathscr{J}$ denotes the ideal (x, y, z) defining the origin. Therefore $(\omega + p\eta) \wedge d(\omega + p\eta) \bmod p^2$ never vanishes for any 1-form η. In other words, $\mathscr{F} \otimes_{\mathbf{Z}} \mathbf{F}_p$ is not liftable to an involutive subbundle on the first infinitesimal neighborhood of $\mathbf{A}^3(\mathbf{F}_p)$.

Let $\mathscr{F}$ be an *L*-involutive subbundle and $\mathscr{F}'$ its involutive lifting over R. Let $\mathscr{D}_{X/k}^i\langle \mathscr{F}'\rangle$ [resp. $\mathscr{A}_{X/k}^i(\mathscr{F}')_j$] denote the natural image of $\mathscr{D}_{X'/R}^i\langle \mathscr{F}'\rangle$ in $\mathscr{D}iff_{X/k}^i$ [resp. of $\mathscr{A}_{X'/R}^i(\mathscr{F}')_j$ in $\mathscr{I}_\Delta^j / \mathscr{I}_\Delta^{i+1} = \mathscr{I}_\Delta^j \mathscr{P}_{X/k}^i$]. (Note that $R = W(k)$ is a local Dedekind domain of characteristic 0.) By the local freeness in codimension 2, the projections $\mathscr{D}_{X'/R}^i\langle\mathscr{F}'\rangle \otimes \mathscr{O}_X \to \mathscr{D}_{X/k}^i\langle\mathscr{F}'\rangle$, $\mathscr{A}_{X'/R}^i(\mathscr{F}')_j \otimes \mathscr{O}_X \to \mathscr{A}_{X/k}^i(\mathscr{F}')$ are injective outside a closed subset $\subset X$ of codimension ≥ 2. $\mathscr{D}_{X/k}\langle\mathscr{F}'\rangle = \varinjlim \mathscr{D}_{X/k}^i\langle\mathscr{F}'\rangle$ is usually called the *divided power algebra* attached to $\mathscr{F}'$.

LEMMA 4.3. $\mathscr{D}^i_{X/k}\langle\mathscr{F}'\rangle$ and $\mathscr{A}^i_{X/k}(\mathscr{F}')_j$ are independent of the choice of the lifting $\mathscr{F}'$ whenever $i < p$. In particular,

$$\mathscr{D}^1_{X/k}\langle\mathscr{F}'\rangle = \mathscr{O}_X \otimes \mathscr{F},$$

$$\mathscr{A}^1_{X/k}(\mathscr{F}')_1 = \mathscr{F}^\perp = \mathrm{Ker}(\Omega^1_{X/k} \to \mathscr{F}^*).$$

Furthermore, $\mathscr{A}^i_{X/k}(\mathscr{F}')_i \subset \mathscr{I}^i_\Delta/\mathscr{I}^{i+1}_\Delta$ is uniquely determined by $\mathscr{F}$ and equal to $\mathrm{Ker}(\mathscr{S}^i\Omega^1_{X/k} \to \mathscr{S}^i\mathscr{F}^*) \simeq \mathscr{F}^\perp \cdot \mathscr{S}^{i-1}\Omega^1_{X/k}$ for every i.

The proof is almost obvious.

EXAMPLE 4.4. For $i \geq p$, $\mathscr{D}^i_{X/k}\langle\mathscr{F}'\rangle$ and $\mathscr{A}^i_{X/k}(\mathscr{F}')_1$ do depend on the lift $\mathscr{F}'$. For example, assume that $p = 2$ and let $\mathscr{F}$ be the line bundle generated by ∂_x on $X = \mathbf{A}^2(\mathbf{F}_2)$. $\mathscr{F}' = \mathscr{O}_X(\partial_x + 2x\,\partial y)$ is an involutive lift of $\mathscr{F}$ to $\mathbf{A}^2(\mathbf{Z})$. On the other hand,

$$\{\partial_x + 2x\,\partial_y\}^2 = 2D_{xx} + 4xD_{xy} + 8x^2 D_{yy} + 2\partial_y$$

so that $\mathscr{D}^2_{X/k}\langle\mathscr{F}'\rangle = \mathscr{O}_X \oplus \mathscr{O}_X\partial_x \oplus \mathscr{O}_X(D_{xx}+\partial_y)$ is not equal to $\mathscr{D}^2_{X/k}\langle\mathscr{O}_{X'}\partial_x\rangle = \mathscr{O}_X \oplus \mathscr{O}_X\partial_x \oplus \mathscr{O}_X D_{xx}$. Here the differential operators D_{xx}, D_{xy}, and D_{yy} of order two map $(dx)^2, dx\,dy$, and $(dy)^2$ to 1, respectively.

LEMMA 4.5. Let m be a positive integer. Then the following four conditions are mutually equivalent for an involutive lifting $\mathscr{F}' \subset \mathscr{T}_{X'/R}$ of an L-involutive subbundle $\mathscr{F}$:

(1) The natural map $\mathscr{A}^m_{X/k}(\mathscr{F}')_1 \to \mathscr{F}^\perp = \mathscr{A}^1_{X/k}(\mathscr{F}')_1$ is surjective;

(2) $\mathscr{A}^m_{X/k}(\mathscr{F}')_j \to \mathscr{A}^i_{X/k}(\mathscr{F}')_j$ is surjective for $j \leq i \leq m$;

(3) $\mathscr{A}^m_{X/k}(\mathscr{F}')_j = \mathscr{I}^{j-1}_\Delta \mathscr{A}^m_{X/k}(\mathscr{F}')_1$ for every $j \leq m$;

(4) $\mathscr{F}'$ is integrable on X'.

If $\mathscr{F}'$ enjoys one of these properties, then $\mathscr{A}^m_{X/k}(\mathscr{F}')_j/\mathscr{A}^m_{X/k}(\mathscr{F}')_{j+1}$ is isomorphic to the locally free sheaf $\mathscr{F}^\perp \cdot \mathscr{S}^{j-1}\Omega^1_{X/k}$.

PROOF. An immediate consequence of (3.5) by virtue of Nakayama's lemma.

DEFINITION 4.6. An L-involutive subbundle $\mathscr{F}$ of $\mathscr{T}_{X/k}$ is said to be an integrable subbundle or a smooth foliation on X if there is an integrable lifting $\mathscr{F}'$ on X'.

We have an identity $\mathscr{D}\!i\!f\!f^j_{X'/R} \cap \mathscr{D}^i_{X'/R}\langle\mathscr{F}'\rangle = \mathscr{D}^j_{X'/R}\langle\mathscr{F}'\rangle$ on X' $(i > j)$, but this nice property is not inherited by $\mathscr{D}^i_{X/k}\langle\mathscr{F}'\rangle$ if $\mathscr{F}$ is a general L-involutive subbundle. For instance, when $k = \mathbf{F}_2$, $X' = \mathbf{A}^2$, $\mathscr{F} = \mathscr{O}_X(\partial_x + x\,\partial y)$, we have $\mathscr{D}^1_{X/k}\langle\mathscr{F}'\rangle = \mathscr{O}_X \oplus \mathscr{F}$ while $\mathscr{D}^2_{X/k}\langle\mathscr{F}'\rangle = \mathscr{O}_X \oplus \mathscr{T}_{X/k} = \mathscr{D}\!i\!f\!f^1_{X/k}$ for any lift $\mathscr{F}'$. Such phenomena are, however, ruled out for integrable bundles:

PROPOSITION 4.7. If $\mathscr{F}$ is integrable, then

$$\mathscr{D}^i_{X/k}\langle\mathscr{F}'\rangle \cap \mathscr{D}\!i\!f\!f^j_{X/k} = \mathscr{D}^j_{X/k}\langle\mathscr{F}'\rangle \quad \text{for } j \leq i.$$

PROOF. The integrability gives: $\mathscr{A}^j_{X/k}(\mathscr{F}') = \{\mathscr{A}^i_{X/k}(\mathscr{F}') + \mathscr{I}^{j+1}_\Delta\}/\mathscr{I}^{j+1}_\Delta$. Therefore, for $D \in \mathscr{D}^i_{X/k}\langle\mathscr{F}'\rangle \cap \mathscr{D}\!i\!f\!\!f^j_{X/k}$,

$$D(\mathscr{A}^j_{X/k}(\mathscr{F}')_1) = D(\{\mathscr{A}^i_{X/k}(\mathscr{F}')_1 + \mathscr{I}^{j+1}_\Delta\}/\mathscr{I}^{j+1}_\Delta)$$
$$= D(\mathscr{A}^i_{X/k}(\mathscr{F}')_1) + D(\mathscr{I}^{j+1}_\Delta/\mathscr{I}^{i+1}_\Delta) = 0.$$

On the other hand, since $\mathscr{A}^j_{X'/R}(\mathscr{F}')_1$ is saturated in $\mathscr{P}^j_{X'/R}$, we infer that $\mathscr{D}^j_{X/k}\langle\mathscr{F}'\rangle \subset \mathscr{D}\!i\!f\!\!f^j_{X/k}$ is the kernel of the projection $\mathscr{D}\!i\!f\!\!f^j_{X/k} \to (\mathscr{A}^j_{X/k}(\mathscr{F}')_1)^*$. Thus we get an inclusion relation

$$\mathscr{D}^i_{X/k}(\mathscr{F}') \cap \mathscr{D}\!i\!f\!\!f^j_{X/k} \subset \mathscr{D}^j_{X/k}(\mathscr{F}').$$

The inclusion of the converse direction is clear.

There are practical *sufficient* conditions for a subbundle $\mathscr{F}$ of $\mathscr{T}_{X/k}$ to be involutive, L-involutive, or integrable:

THEOREM 4.8. *Let $\mathscr{F}$ be a subbundle of $\mathscr{T}_{X/k}$.*

(1) *$\mathscr{F}$ is involutive if $\mathrm{Hom}_{\mathscr{O}_X}(\Lambda^2\mathscr{F}, \mathscr{T}_{X/k}/\mathscr{F}) = 0$.*

(2) *$\mathscr{F}$ is L-involutive if $\mathrm{Hom}_{\mathscr{O}_X}(\Lambda^2\mathscr{F}, \mathscr{T}_{X/k}/\mathscr{F}) = 0$ and $\mathscr{F}$ is globally liftable to a subbundle $\mathscr{F}'$ of $\mathscr{T}_{X'/W(k)}$.*

(3) *Suppose that $H^0(X, \mathscr{O}_X) = k$. Then an L-involutive subbundle $\mathscr{F}$ is integrable if*

$$\mathrm{Hom}_{\mathscr{O}_X}\left(\Lambda^n\mathscr{F}^\perp, \bigotimes_{i=2}^m \Lambda^{n_i}(\mathscr{S}^i\mathscr{F}^*)\right) = 0$$

for all integers $m \geq 2$, $n \geq 1$, and nonnegative integers n_i with $\sum n_i = n$.

(4) *When $\mathrm{Hom}_{\mathscr{O}_X}(\mathscr{F}, \mathscr{T}_{X/k}/\mathscr{F}) = 0$, a lifting $\mathscr{F}' \subset \mathscr{T}_{X'/W(k)}$ of $\mathscr{F}$ is, if any, unique. In this case, we write $\mathscr{D}_{X/k}\langle\mathscr{F}\rangle, \mathscr{A}^i_{X/k}(\mathscr{F})_j$, etc. instead of $\mathscr{D}_{X/k}\langle\mathscr{F}'\rangle$, $\mathscr{A}^i_{X/k}(\mathscr{F}')_j$, etc.*

PROOF. (1) The bracket product $[\,,\,] : \mathscr{F} \times \mathscr{F} \to \mathscr{T}_{X/k}$ induces an $\mathscr{O}_X$-bilinear map $\mathscr{F} \times \mathscr{F} \to \mathscr{T}_{X/k}/\mathscr{F}$, with $[D, D] = 0$ for every D.

(2) A global lifting $\mathscr{F}'$ remains still involutive because

$$\mathrm{Hom}_{\mathscr{O}_{X'}}(\Lambda^2\mathscr{F}', \mathscr{T}_{X'/R}) = 0$$

by the upper-semicontinuity of the dimension of cohomology groups.

(3) We prove that $\mathscr{A}^m_{X'/R}(\mathscr{F}')_1 \to \mathscr{A}^1_{X'/R}(\mathscr{F}')_1$ is surjective by induction on m. Assume that $\mathscr{A}^{m-1}_{X'/R}(\mathscr{F}')_1 \to \mathscr{A}^1_{X'/R}(\mathscr{F}')_1$ is onto. Then we have $\mathscr{I}_\Delta \cdot \mathscr{A}^{m-1}_{X'/R}(\mathscr{F}')_1 = \mathscr{A}^m_{X/k}(\mathscr{F}')_2$. In fact, $\mathscr{I}_\Delta\mathscr{A}^{m-1}_{X'/R}(\mathscr{F}')_1 \subset \mathscr{A}^m_{X'/R}(\mathscr{F}')_2$ and there are natural homomorphisms

$$\iota : \mathscr{I}_\Delta\mathscr{A}^{m-1}_{X'/R}(\mathscr{F}')_{j-1}/\mathscr{I}_\Delta\mathscr{A}^{m-1}_{X'/R}(\mathscr{F}')_j \to \mathscr{A}^m_{X'/R}(\mathscr{F}')_j/\mathscr{A}^m_{X'/R}(\mathscr{F}')_{j+1},$$

$$\pi : \mathscr{A}^m_{X'/R}(\mathscr{F}')_j/\mathscr{A}^m_{X'/R}(\mathscr{F}')_{j+1} \to \mathscr{A}^j_{X'/R}(\mathscr{F}')_j \simeq F^\perp \cdot \mathscr{S}^{j-1}\Omega^1_{X'/R}.$$

By our assumption, $\pi \cdot \iota$ is an isomorphism for $j \geq 2$. Noting that the three sheaves have the same rank, one sees that ι is an isomorphism. It follows that

$\mathscr{A}^m_{X'/R}(\mathscr{F}')_2 = \mathscr{I}_\Delta\mathscr{A}^{m-1}_{X'/R}(\mathscr{F}')_1$ and the filtration on $\mathscr{P}^m_{X'/R}$ induces a filtration on the quotient $\mathscr{P}^m_{X'/R}/\mathscr{A}^m_{X'/R}(\mathscr{F}')_2$ in an obvious way, of which the associated graded module is isomorphic to $\Omega^1_{X'/R} \oplus \sum^m_{i=2} \mathscr{S}^i\mathscr{F}^*$. The quotient

$$\mathscr{E} = \mathscr{A}^m_{X'/R}(\mathscr{F}')_1/\mathscr{A}^m_{X'/R}(\mathscr{F}')_2$$

is, in codimension 2, locally free of rank $r = \operatorname{rank}\mathscr{F}'^\perp$ with determinant $\simeq \det\mathscr{F}'^\perp$, and hence $\det(\mathscr{E} \otimes k) \simeq \det(\mathscr{F}'^\perp \otimes k)$. Moreover, the canonical map $\alpha\colon \mathscr{E} \to \Omega^1_{X'/R}$ is contained in $\mathscr{F}'^\perp$. Consider the exact sequence

$$0 \to \mathscr{G} \to \mathscr{E} \otimes k \xrightarrow{\bar{\alpha}=\alpha\otimes 1} \mathscr{F}^\perp \to \mathscr{H} \to 0$$

of coherent sheaves on X. $\mathscr{G}$ is a subsheaf of $\mathscr{I}_\Delta\mathscr{P}^m_{X/k}/\mathscr{A}^m_{X/k}(\mathscr{F}')_2$, and its naturally induced filtration gives rise to a submodule of the graded module $\bigoplus^m_{i=2}\mathscr{S}^i\mathscr{F}^*$. Hence $\det(\mathscr{H}/\text{torsion})$, being linearly equivalent to $\det\mathscr{G}-$(effective divisor), is identified with a submodule of $\bigotimes^m_{i=2}(\Lambda^{n_i}\mathscr{S}^i\mathscr{F}^*)$ for suitable integers n_i with $\sum n_i = n = \operatorname{rank}(\mathscr{H}/\text{torsion})$. Thus, unless $\mathscr{G} = 0$, we obtain a nontrivial $\mathscr{O}_X$-homomorphism

$$\Lambda^n\mathscr{A}^1_{X/k}(\mathscr{F}')_1 \to \Lambda^n(\mathscr{H}/\text{torsion})^{**} = \det(\mathscr{H}/\text{torsion}) \subset \bigotimes^m_{i=2}(\Lambda^{n_i}\mathscr{S}^i\mathscr{F}^*),$$

which contradicts our hypothesis. Therefore $\mathscr{G}$ vanishes so that $\bar{\alpha}\colon \mathscr{E} \otimes k \to \mathscr{F}^\perp$ is injective. $\Lambda^r\bar{\alpha}\colon \det(\mathscr{E} \otimes k) \to \det\mathscr{F}^\perp$ is everywhere nonvanishing since $\det(\mathscr{E} \otimes k) \simeq \det\mathscr{F}^\perp$ and $H^0(X,\mathscr{O}_x) = k$. Hence $\bar{\alpha}$ is surjective in codimension 1 and so is α by Nakayama's lemma. Now (3.3.2) applies; α is an isomorphism.

(4) Let $\mathscr{F}''$ be another lifting. $\operatorname{Hom}(\mathscr{F}''\otimes k, \mathscr{T}_{X/k}/\mathscr{F}') = \operatorname{Hom}(\mathscr{F}, \mathscr{T}_{X/k}/\mathscr{F}) = 0$, so $\operatorname{Hom}(\mathscr{F}'', \mathscr{T}_{X'/W(k)}/\mathscr{F}') = 0$ by the upper-semicontinuity theorem; in other words, $\mathscr{F}'' = \mathscr{F}'$.

COROLLARY 4.9. *Let $\mathscr{F}$ be an invertible subbundle of $\mathscr{T}_{X/k}$ which is liftable to $\mathscr{F}' \subset \mathscr{T}_{X'/R}$. If $H^0(X,\mathscr{O}_x) = k$ and*

$$\operatorname{Hom}_{\mathscr{O}_X}(\Lambda^n(\mathscr{T}_{X/k}/\mathscr{F})^*, \mathscr{F}^{*\otimes m}) \simeq H^0(X, \{\Lambda^n(\mathscr{T}_{X/k}/\mathscr{F})^*\}^* \otimes \mathscr{F}^{\otimes -m}) = 0$$

for all pairs (m,n) with $0 < n \le m$, then $\mathscr{F}$ is integrable and an integrable lift $\mathscr{F}'$ over $R = W(k)$ is unique.

5. Differential operator $\delta_{(f,g)}$. Let A be an Artinian local k-algebra with maximal ideal M. Put $S = \operatorname{Spec} A$ and let $S^{(n)}$ be the nth infinitesimal neighborhood $\operatorname{Spec} A/M^{n+1}$ of the closed point on S. For an S-scheme X, the symbol $X^{(n)}$ will denote the fibre product $X \times_S S^{(n)}$ (namely, the nth infinitesimal neighborhood of the closed fibre $X^{(0)}$). An S-morphism $f\colon Y \to X$ induces an $S^{(n)}$-morphism $f^{(n)}\colon Y^{(n)} \to X^{(n)}$ in an obvious way.

Assume that two S-morphisms f and $g\colon Y \to X$ coincide on the closed fibre $Y^{(0)}$, i.e., $f^{(0)} = g^{(0)}$. Then $(f,g)(Y) \subset X \times_S X$ is contained in $\Delta^{[m]}_{X/S}$, where m is an integer such that $M^{m+1} = 0$. Thus we get the global differential operator

$$\delta_{(f,g)} = (f,g)^*p_2^* - (f,f)^*p_2^*\colon \mathscr{O}_X \to \mathscr{O}_Y,$$

of order $\leq m$, where $\mathscr{O}_Y$ is regarded as a left $\mathscr{O}_X$-module via the first component f. The operator $\delta_{(f,g)}$ represents the (mth) *infinitesimal deviation* of g from f, and we have

$$\delta_{(f,g)} = P((f,g)^* p_2^*),$$

where $P: \mathscr{D}\!i\!f\!f^m_{X/S}(\mathscr{O}_Y) \to \mathscr{P}\mathscr{D}\!i\!f\!f^m_{X/S}(\mathscr{O}_Y)$ is the projection which kills the operators of degree 0.

Let us study some properties of the operator $\delta_{(f,g)}$.

LEMMA 5.1. (1) $f^{(j)} = g^{(j)}$ if and only if $\delta_{(f,g)} \in M^{j+1}\mathscr{D}\!i\!f\!f_{X/S}$.
(2) If $f^{(j)} = g^{(j)}$, then

$$\delta_{(f,g)} \in M^{j+1}\mathscr{P}\mathscr{D}\!i\!f\!f^i_{X/S}(\mathscr{O}_Y) + M^{(i+1)(j+1)}\mathscr{P}\mathscr{D}\!i\!f\!f_{X/S}(\mathscr{O}_Y), \qquad i = 1, 2, \ldots.$$

(3) If $g^{(j)} = h^{(j)}$, then

$$\delta_{(f,g)} - \delta_{(f,h)} \in M^{j+1}\mathscr{P}\mathscr{D}\!i\!f\!f_{X/S}(\mathscr{O}_Y).$$

(4) If $M^{m+1} = 0$ and $g^{(m-1)} = h^{(m-1)}$, then

$$\delta_{(f,g)} - \delta_{(f,h)} \in M^m f^* \mathscr{T}_{X/S} = M^m f^{(0)*}\mathscr{T}_{X/S}.$$

PROOF. (1) Clear. (2) The left $\mathscr{O}_X$-homomorphism $(f,g)^*$ is a ring homomorphism so that $(f,g)^*(\mathscr{I}^{i+1}_\Delta) \subset M^{(i+1)(j+1)}\mathscr{O}_Y$. By virtue of the equality $\delta_{(f,g)} = P((f,g)^* P_2^*)$, we obtain the assertion. (3) The proof is obvious. (4) By (3), we have $\delta_{(f,g)} - \delta_{(f,h)} \in M^m f^* \mathscr{D}\!i\!f\!f_{X/S} = M^m h^* \mathscr{D}\!i\!f\!f_{X/S}$. Therefore,

$$\delta_{(f,g)} - \delta_{(f,h)} = \delta_{(h,g)} - \delta_{(h,h)} = \delta_{(h,g)}.$$

Hence we are done, applying (2) to this situation.

Let J be an ideal of the Artinian local ring (A, M) with the property $MJ = J^2 = 0$. Assume that X is the Cartesian product $X_0 \times_k S$. Let $\mathscr{F}_0$ be an integrable subbundle of $\mathscr{T}_{X_0/k}$ with an integrable lift $\mathscr{F}_0'$ on X_0' over $R = W(k)$. Put

$$\mathscr{F} = \mathscr{F}_0 \otimes_k A \subset \mathscr{T}_{X/S},$$
$$\mathscr{D}_{X/S}\langle \mathscr{F}' \rangle = \mathscr{D}_{X_0/k}\langle \mathscr{F}_0' \rangle \otimes_k A \subset \mathscr{D}\!i\!f\!f_{X/S},$$
$$\mathscr{A}^i_{X/S}(\mathscr{F}')_j = \mathscr{A}^i_{X_0/k}(\mathscr{F}_0')_j \otimes_k A \subset \mathscr{I}^j_\Delta \mathscr{P}^i_{X/S}.$$

By the integrability condition, one shows easily that a differential operator $D \in \mathscr{D}\!i\!f\!f^i_{X/S}(\mathscr{O}_Y)$ belongs to $\mathscr{D}^i_{X/S}\langle \mathscr{F}' \rangle(\mathscr{O}_Y) + \mathscr{D}\!i\!f\!f^1_{X/S}$ if and only if $D(\mathscr{A}^i_{X/S}(\mathscr{F}')_2) = 0$, where $\mathscr{D}^i_{X/S}\langle \mathscr{F}' \rangle(\mathscr{O}_Y) = \mathscr{O}_Y \otimes_{\mathscr{O}_X} \mathscr{D}^i_{X/S}\langle \mathscr{F}' \rangle$.

LEMMA 5.2. If $\delta_{(f,g)} \in \mathscr{D}^i_{X/S}\langle \mathscr{F}' \rangle(\mathscr{O}_Y) + J\mathscr{D}\!i\!f\!f^i_{X/S}(\mathscr{O}_Y)$, then

$$\delta_{(f,g)} \in \mathscr{D}^i_{X/S}\langle \mathscr{F}' \rangle(\mathscr{O}_Y) + Jf^* \mathscr{T}_{X/Y}.$$

PROOF. It suffices to show that $\delta_{(f,g)}(\mathscr{A}^i_{X/S}(\mathscr{F}')_2) = 0$. Since $\mathscr{A}^i_{X/S}(\mathscr{F}')_2 \subset \mathscr{I}_\Delta \mathscr{P}^i_{X/S}$,

$$\delta_{(f,g)}(\mathscr{A}^i_{X/S}(\mathscr{F}')_2) = (f,g)^*(\mathscr{A}^i_{X/S}(\mathscr{F}')_2).$$

On the other hand, $\mathscr{A}^i_{X/S}(\mathscr{F}')_2 = \mathscr{I}_\Delta \mathscr{A}^i_{X/S}(\mathscr{F}')_1$ and $(f,g)^*$ is a ring homomorphism. Therefore

$$\delta_{(f,g)}(\mathscr{A}^i_{X/S}(\mathscr{F}')_2) = \{(f,g)^*(\mathscr{I}_\Delta)\}\{(f,g)^*(\mathscr{A}^i_{X/S}(\mathscr{F}')_1)\} \subset MJ = 0.$$

By definition, the following is obvious:

PROPOSITION 5.3. *Let* $\mathscr{L}^{[m]}(\mathscr{F}) \subset \Delta^{[m]}_{X/k}$ *be the closed subscheme defined by* $\mathscr{A}^m_{X/k}(\mathscr{F}')_1$. *Then* $\delta_{(f,g)} \in \mathscr{D}_{X/S}(\mathscr{F}')$ *if and only if*

$$(f,g)(Y) \subset \bigcup \mathscr{L}^{[m]}(\mathscr{F}) \times_k S.$$

EXAMPLE 5.4. As the simplest case, let X' be a smooth formal R-scheme $\operatorname{Spec} R[[x_1,\ldots,x_n]]$ and $\mathscr{F}'$ be a subbundle generated by $\partial_{x_1},\ldots,\partial_{x_r}$. Let A be an Artinian local k-algebra, $f \colon \operatorname{Spec} A \to X$ a k-morphism, and $f_0 \colon \operatorname{Spec} A \to X$ the unique constant morphism. Let us compute the differential operator $\delta_{(f_0,f)}$ and study the condition that $\delta_{(f_0,f)} \in f_0^* \mathscr{D}_{X/k}\langle\mathscr{F}'\rangle$.

Put $f_i = f^* x_i \in M$. Then $\delta_{(f_0,f)}(x^I) = f^* x^I = f^I$, where I is a multi-index $(i_1,\ldots,i_n)$. Let D_I denote the differential operator $\in \mathscr{D}\!i\!f\!f^{|I|}_{X/k}$ determined by the property $D_I x^J = \delta_{IJ}$ (the Kronecker symbol) for $|J| \le |I|$. By these operators, we can write

$$\delta_{(f_0,f)} = \sum_I f^I f_0^* D_I.$$

The subbundle $\mathscr{F}'$ is clearly involutive and integrable; $\mathscr{D}_{X/k}\langle\mathscr{F}'\rangle$ is a free $\mathscr{O}_X$-module generated by the operators $D_{I'}$, where the indices I' are of the form $(i_1,\ldots,i_r,0,\ldots,0)$, and $\mathscr{A}^i_{X/k}(\mathscr{F}') = \mathscr{P}^{i-1}_{X/k} dx_{r+1} + \cdots + \mathscr{P}^{i-1}_{X/k} dx_n$. Hence $\delta_{(f_0,f)} \in f_0^* \mathscr{D}_{X/k}\langle\mathscr{F}'\rangle$ if and only if $f_i = f^* x_i = 0$ for $i > r$, or, equivalently, if and only if the image of f is contained in the closed formal subscheme

$$\mathscr{L}(\mathscr{F}) = \operatorname{Spec} k[[x_1,\ldots,x_n]]/(x_{r+1},\ldots,x_n).$$

6. Deformations of a morphism along a foliation.

In this section, k denotes always an algebraically closed field. Let X and Y be k-schemes, $B \subset Y$ a closed k-subscheme, and $f \colon Y \to X$ a k-morphism. Let (S,s) be a *punctured k-scheme*, i.e., the pair of a k-scheme S and a k-valued point s on S. An (S,s)-*deformation of f with base subscheme B* is an S-morphism $\tilde{f} \colon Y \times_k S \to X \times_k S$ such that

$$\tilde{f}|(Y \times s \cup B \times S) = (f, \mathrm{id}_S)|(Y \times s \cup B \times S).$$

For an Artinian punctured k-scheme (S,s), an (S,s)-deformation is called an *infinitesimal deformation over* (S,s).

Let $\mathscr{D}\!e\!f\!orm(f,B) \colon (\mathscr{P}unctured\ k\text{-}schemes) \to (\mathscr{S}ets)$ denote the contravariant functor

$$(S,s) \to \{(S,s)\text{-deformation of } f \text{ with base } B\}.$$

In case X and Y are projective, it is a standard fact that the functor $\mathscr{D}\!e\!f\!orm(f,B)$ is *representable* by a k-scheme $M(f,B)$ of finite type, which is realized as a quasiprojective constructible subset of the Hilbert scheme of $X \times_k Y$ [**FGA**, no. 221].

Assume that X is smooth over k and $\mathscr{F}$ is a smooth foliation on X. (In case $\mathrm{ch}(k) > 0$, we assume that $(X, \mathscr{F})$ can be lifted to $(X', \mathscr{F}')$ over $W(k)$ and that $\mathscr{F}'$ is integrable.) Let (A, M) be an Artinian local k-algebra and let S denote the spectrum of A with a unique closed point s. An infinitesimal (S, s)-deformation of $f : Y \to X$ *along* $\mathscr{F}$ (or, more precisely, *along* $\mathscr{F}'$ when $\mathrm{ch}(k) > 0$) with base subscheme B will be an infinitesimal deformation $\tilde{f} \in \mathscr{D}\!eform(f, B)(S, s)$ such that

$$\delta_{(\tilde{f}_0, \tilde{f})} \in \tilde{f}_0^*(\mathscr{D}_{X/k}\langle\mathscr{F}'\rangle \otimes_k A) \subset \tilde{f}_0^*\mathscr{D}\!iff_{X \times_k S/S},$$

or, equivalently,

$$\mathrm{Im}(\tilde{f}_0, \tilde{f}) \subset \mathscr{L}(\mathscr{F}') \times_{X'} X \times_k S,$$

where $\tilde{f}_0$ is the trivial deformation (f, id_S) and $\mathscr{F}'$ stands for $\mathscr{F}$ if $\mathrm{ch}(k) = 0$ or an integrable lift $\mathscr{F}'$ of $\mathscr{F}$ if $\mathrm{ch}(k) > 0$.

For a general punctured k-scheme (S, s), an (S, s)-deformation is said to be *along* $\mathscr{F}$ (or *along* $\mathscr{F}'$) if, for every connected Artinian (S, s)-scheme (T, t), the induced infinitesimal deformation is along $\mathscr{F}$ (or along $\mathscr{F}'$).

By $\mathscr{D}\!eform(f, B, \mathscr{F})$ we denote the contravariant functor: $(\mathscr{P}\!unctured\ k\text{-}schemes) \to \mathscr{S}\!ets$ defined by

$$\mathscr{D}\!eform(f, B, \mathscr{F})(S, s) = \{(S, s)\text{-deformations of } f \text{ along } \mathscr{F} \text{ and with base } B\}.$$

Clearly $\mathscr{D}\!eform(f, B, \mathscr{F})$ is a subfunctor of $\mathscr{D}\!eform(f, B)$, yet it is not necessarily representable (the easiest example: $k = \mathbf{C}, X =$ an abelian variety, $Y =$ point, $B = \varnothing, \mathscr{F} = \mathscr{O}_X \xi$, where ξ is a general vector field on X).

PROPOSITION 6.1. *Let* (A, M) *be an Artinian local k-algebra and* $J \subset A$ *an ideal such that* $MJ = J^2 = 0$. *Let* S' *be the closed subscheme* $\mathrm{Spec}\, A/J$ *of* S. *Assume that* Y *is affine and that* X *is smooth over k. Let* $\mathscr{F} \subset \mathscr{T}_{X/k}$ *be a smooth foliation on X. Fix a k-morphism* $f : Y \to X$. *Then every* $\tilde{f} \in \mathscr{D}\!eform(f, B, \mathscr{F})(S', s)$ *is extendable to* $\mathscr{D}\!eform(f, B, \mathscr{F})(S, s)$. *Moreover there exists a natural bijection from the set of such prolongations to the abelian group* $J H^0(Y, \mathscr{I}_B f^* \mathscr{F})$ *once a specific extension (a "reference" extension)* $h \in \mathscr{D}\!eform(f, B, \mathscr{F})(S, s)$ *is given.*

PROOF. Since $X \times_k S$ is smooth and $Y \times_k S$ is affine over S, the morphism $\tilde{f}$ has a prolongation $g \in \mathscr{D}\!eform(f, B)(S, s)$ [**SGA 1**, Exposé III, (3.1)]. Evidently,

$$\delta_{(\tilde{f}_0, g)} \in \tilde{f}_0^*(\mathscr{D}_{X/k}\langle\mathscr{F}'\rangle \otimes_k A + J\mathscr{P}\mathscr{D}\!iff_{X \times_k S/S})$$

so that

$$\delta_{(\tilde{f}_0, g)} \in \tilde{f}_0^*(\mathscr{D}_{X/k}\langle\mathscr{F}'\rangle \otimes_k A + J\mathscr{T}_{X \times_k S/S})$$

by (5.2). The set

$$\{h \in \mathscr{D}\!eform(f, B)(S, s); h|Y \times_k S' = g|Y \times_k S' = \tilde{f}\}$$

is naturally identified with

$$JH^0(Y \times_k S, \mathscr{I}_{B \times S} g^* \mathscr{T}_{X \times S/S}) = JH^0(Y, \mathscr{I}_B f^* \mathscr{T}_{X/k})$$

by [**SGA 1**, Exposé III, (5.1)]. The identification map is nothing but the correspondence $h \to \delta_{(g,h)}$. On the other hand, by Lemma 5.2,

$$\delta_{(\tilde{f}_0, g)} - \delta_{(\tilde{f}_0, g)} = \delta_{(g,h)} - \delta_{(g,g)} = \delta_{(g,h)} \in JH^0(Y, \mathscr{I}_B f^* \mathscr{T}_X).$$

Therefore an extension h over (S, s) of $\tilde{f}$ is an infinitesimal deformation along $\mathscr{F}$ if and only if

$$\delta_{(\tilde{f}_0, g)} + \delta_{(g,h)} \in \tilde{f}_0^* (\mathscr{D}_{X/k} \langle \mathscr{F}' \rangle \otimes_k A).$$

Now the assertion is obvious. (An alternate proof of the extendability is as follows: Since $\mathscr{L}(\mathscr{F}') \otimes k$ is smooth, if $\text{Im}(\tilde{f}_0, \tilde{f})$ is contained in $\mathscr{L}(\mathscr{F}') \times_k S'$, we can extend $\tilde{f}$ to g over S such that $\text{Im}(\tilde{f}_0, g)$ is still contained in $\mathscr{L}(\mathscr{F}') \times_k S'$.)

As an immediate consequence, we get:

THEOREM 6.2. *Let A be an Artinian local k-algebra and $J \subset A$ an ideal with $J^2 = 0$. Let $S = \text{Spec} A$, $S' = \text{Spec} A/J$, and let s be the closed point. Let Y be a k-scheme, X a smooth k-scheme, and $f : Y \to X$ a k-morphism. Let $\mathscr{F} \subset \mathscr{T}_{X/k}$ be a smooth foliation and $B \subset Y$ a closed subscheme. Then the sheaf of local prolongations of a morphism $\tilde{f} \in \mathscr{D}eform(f, B, \mathscr{F})(S', s)$ to $\mathscr{D}eform(f, B, \mathscr{F})(S, s)$ is a sheaf of principal fibre bundles with sheaf of structure groups $J \mathscr{I}_{B \times S} \tilde{f}^* (\mathscr{F} \otimes_k (A/J))$ (see [**SGA 1**, Exposé III, §5] for the terminology).*

COROLLARY 6.3 (CF. [**SGA 1**, Exposé III, (5.6)]). *On the same assumption as in (6.2), suppose that $H^1(Y, \mathscr{I}_B f^* \mathscr{F}) = 0$, where Y is a complete k-variety. Then there exists a formal smooth punctured k-scheme $(\hat{M}(f, B, \mathscr{F}), o)$ of dimension $\dim_k H^0(Y, \mathscr{I}_B f^* \mathscr{F})$ which parametrizes effectively the infinitesimal deformations of f along $\mathscr{F}$, with base B. Namely, there exists a formal k-morphism*

$$\hat{f} : \hat{M}(f, B, \mathscr{F}) \to X \times_k \hat{M}(f, B, \mathscr{F}) \in \mathscr{D}eform(f, B, \mathscr{F})(\hat{M}(f, B, \mathscr{F}), o)$$

such that, for every infinitesimal deformation $\tilde{f}$ of f along $\mathscr{F}$ with base B over a connected Artinian k-scheme S, there exists a unique k-morphism $q(\tilde{f}) : (S, s) \to (\hat{M}(f, B, \mathscr{F}), o)$ inducing $\tilde{f}$.

The morphism $\hat{f}$ as above may be called the *universal formal deformation* of f along $\mathscr{F}$, with base B. In particular, in case $Y = x$ is a single point on X and B is empty, we have the *formal integral submanifold* of the foliation $\mathscr{F}$ through x. Thus we get a formal counterpart of the Frobenius theorem in differential geometry:

COROLLARY 6.4. *Let $\mathscr{F}$ be a smooth foliation on a smooth punctured k-scheme (X, x). Then $L(\mathscr{F}, x) = p_2(\mathscr{L}(\mathscr{F}') \cap p_1^{-1}(x))$ is a smooth formal k-subscheme $\subset X$ with a unique closed point x (the formal integral submanifold of $\mathscr{F}$ through x) which is characterized by the following property.*

(FIS)   *For every connected Artinian punctured k-scheme (S, s) and a k-morphism $f: (S, s) \to (X, x)$,*

$$(f, \mathrm{id}_S) \in \mathscr{D}\!\mathit{eform}(f|s, \varnothing, \mathscr{F})(S, s)$$

if and only if $f(S) \subset L(\mathscr{F}, x)$.

REMARK 6.5. In case $\mathrm{ch}(k) > 0$, $\mathscr{D}\!\mathit{eform}(f, B, \mathscr{F})$ depends on the choice of the integrable lifting $\mathscr{F}'$ and should be denoted by $\mathscr{D}\!\mathit{eform}(f, B, \mathscr{F}')$ to avoid a possible confusion. By the same reason, the integral submanifold is uniquely determined only after $\mathscr{F}'$ is specified. However, if $\mathrm{Hom}_{\mathscr{O}_X}(\mathscr{F}, \mathscr{T}_{X/k}/\mathscr{F}) = 0$, $\mathscr{F}'$ is unique so there is no ambiguity in our notation.

As a useful application of (6.3), we have the following

COROLLARY 6.6. *Let X be a smooth projective k-variety (liftable to characteristic 0) and Y a projective k-variety. Let $f: Y \to X$ be a k-morphism. Let $\mathscr{F} \subset \mathscr{T}_{X/k}$ be a subsheaf which is a smooth foliation on an open neighborhood of $f(Y)$. If $H^1(Y, \mathscr{I}_B f^* \mathscr{F}) = 0$, then the parameter space $M(f, B)$ of the universal deformation of f with base B has dimension $\geq \dim_k H^0(Y, \mathscr{I}_B f^* \mathscr{F})$ at the point (f). More generally,*

$$\dim_k M(f, B)_{(f)} \geq \dim_k H^0(Y, \mathscr{I}_B f^* \mathscr{F}) - \dim_k H^1(Y, \mathscr{I}_B f^* \mathscr{F}).$$

The proof is the same as that of Proposition 3 in [**Mo**].

Finally, note the following

PROPOSITION 6.7. *Let $X, Y, f,$ and $\mathscr{F}$ be as in (6.6). Assume that*

$$H^1(Y, \mathscr{I}_B f^* \mathscr{F}) = H^0(Y, \mathscr{I}_B f^*(\mathscr{T}_{X/k}/\mathscr{F})) = 0.$$

Let M be the irreducible component of $M(f, B)$ containing $o = (f)$ (M is uniquely determined since $\hat{M}(f, B)_o = \hat{M}(f, B, \mathscr{F})$ is smooth at o). Let $\tilde{f} \in \mathscr{D}\!\mathit{eform}(f, B)(M, o)$ be the restriction of the universal deformation. Then, for a general point $y \in Y$, $\mathscr{T}_{y \times M} \subset (\tilde{f}|y \times M)^* \mathscr{F}$ on a Zariski open subset of M. In particular, if $L(y) = \tilde{f}(y \times M)$ is reduced and if $\mathscr{F}|L(y)$ is a saturated subsheaf of $\mathscr{T}_{X/k}|L(y)$, then $\mathscr{T}_{L(y)/k} \subset \mathscr{F}|L(y)$.

PROOF. Put $f_t = \tilde{f}|Y \times t$, $t \in M$. From the upper semicontinuity of the dimension of cohomology groups, we infer that

$$H^1(Y, \mathscr{I}_B f_t^* \mathscr{F}) = H^0(Y, \mathscr{I}_B f_t^*(\mathscr{T}_{X/k}/\mathscr{F})) = 0$$

so that $\hat{M}(f_t, B, \mathscr{F})$ coincides with the formal neighborhood of (f_t) in $M(f_t, B)$ provided $t \in M$ is general. This clearly implies the proposition.

7. A characterization of conic bundles in characteristic 0. Let S be a noetherian scheme and X a proper S-scheme with irreducible geometric fibres.

A *family of irreducible curves* on X will refer to a closed flat T-subscheme $\mathscr{C} \subset X \times_S T$ of which the fibre over a general point on $T|s$ represents an irreducible curve on $X|s$, where T is an S-scheme and s is a general geometric point on S. $\mathscr{C} = \{C_t\}$ is said to be *dominant* if the projection $\mathscr{C} \to X$ is a dominant morphism or, equivalently, if the family sweeps out an open subset of X.

Let X be a proper nonsingular variety over an algebraically closed field k. A saturated subsheaf $\mathscr{F} \subset \mathscr{T}_{X/k}$ of rank 1 is called a *conic foliation* (with singularities) if there exists a dominant family of irreducible curves $\mathscr{C}$ on X (an $\mathscr{F}$-*family*) such that $\mathscr{F}$ is an ample subbundle of $\mathscr{T}_{X/k}$, with $\mathscr{F}^{\otimes -1} \otimes (\mathscr{T}_{X/k}/\mathscr{F})$ being negative, when restricted to a generic member C of $\mathscr{C}$.

EXAMPLE 7.1. Let $f: X \to W$ be a *conic bundle*; namely, X and W are proper smooth k-varieties and f is a surjective morphism whose fibres are plane conic curves. (A general fibre of f is a smooth rational curve P^1.) Then $\mathscr{T}_{X/W} = \ker(\mathscr{T}_{X/k} \to f^* \mathscr{T}_{W/k})$ is a conic foliation where an $\mathscr{F}$-family of irreducible curves is the set of fibres of f. In fact, $\mathscr{T}_{X/W} \simeq \mathscr{O}(2)$ and $f^* \mathscr{T}_W \simeq \mathscr{O}^{\oplus(n-1)}$ on the generic fibre. For further properties of conic bundles, see [**B**, Chapter II].

LEMMA 7.2. (char $p > 0$). *Let H be an ample divisor on a smooth projective variety X over k of positive characteristic. Let $\mathscr{F}$ be a conic foliation with an $\mathscr{F}$-family $\mathscr{C}$ of irreducible curves and $C \subset X$ a general member of $\mathscr{C}$. Assume that the pair $(X, \mathscr{F})$ is liftable to characteristic 0. Then there exists a rational curve L passing through a generic point of C such that*

$$(L, H) \leq \text{Max}\{2(C, H)/(C, c_1(\mathscr{F})), (C, H)\}.$$

In particular, X is a uniruled variety.

PROOF. If C is a rational curve, then the assertion trivially holds. Assume that C is not rational. Let $f: C^* \to U$ be the composite of the natural immersion $C \to U$, the normalization $C' \to C$, and the Frobenius morphism $C^* \to C'$ of degree $q = p^m$, where $U \subset X$ is an open neighborhood $X - Y$ of C on which $\mathscr{F} \subset \mathscr{T}_{X/k}$ is a subbundle ($\text{codim}_X Y \geq 2$). The coherent sheaf $\mathscr{F}$ is an invertible and L-involutive subbundle of $\mathscr{T}_{X/k}|U$. Moreover, $\mathscr{F}$ is a smooth foliation on U. In fact,

$$H^0(C, (\Lambda^n(\mathscr{T}_{X/k}/\mathscr{F})^*)^* \otimes \mathscr{F}^{\otimes -m})$$
$$= H^0(C, [\Lambda^n\{(\mathscr{T}_{X/k}/\mathscr{F})^* \otimes \mathscr{F}^{*-1}\}] \otimes \mathscr{F}^{n-m}) = 0 \qquad (m \geq n)$$

for generic C and $\mathscr{F}$ satisfies the condition in (4.9) on U. Put $d = (C, c_1(\mathscr{F}))$. By the Riemann-Roch and the Serre duality,

$$H^0(C^*, \mathscr{I}_B f^* \mathscr{F}) \neq 0 \qquad \text{if } b = \deg B < qd - g,$$
$$H^1(C^*, \mathscr{I}_B f^* \mathscr{F}) = 0 \quad \text{if } b = \deg B < qd - 2g + 2,$$

where $g = g(C) = g(C') = g(C^*)$ is the (geometric) genus. Assume that B is reduced and that the degree b satisfies the above inequality (for example, put $b = qd - 2g > 0$, where q is taken very large). Then the universal deformation space $M(f, B)$ of f with base B has positive dimension by (6.6). Now Theorem 6 in [**MM**] shows the existence of a rational curve $L \subset X$ through a general point of C such that $(L, H) \leq 2q(C, H)/b$. By letting q tend to infinity, we are done.

PROPOSITION 7.3 (char p). *Assume that* $\mathrm{ch}(k) = p > 0$ *and let the things be as in* (7.2). *Then there exists an irreducible rational curve* L *passing through a generic point of* C *which has the following two properties*:

(a) $(L, H) \leq \mathrm{Max}\{2(C, H)/(C, c_1(\mathscr{F})), (C, H)\}$;

(b) $\mathscr{T}_{L/k} \subset \mathscr{F}|L$.

PROOF. We employ the same notation as in the proof of (7.2). Since $\mathscr{F}^{-1} \otimes (\mathscr{T}_{X/k}/\mathscr{F})$ is negative on C, there exists a positive rational number ε such that

$$H^0(C^*, \mathscr{I}_B f^*(\mathscr{T}_{X/k}/\mathscr{F})) = 0 \quad \text{for } b = \deg B > q(d - \varepsilon).$$

Therefore, when q is very large and $b = qd - 2g$,

$$H^0(C^*, \mathscr{I}_B f^*(\mathscr{T}_{X/k}/\mathscr{F})) = H^1(C^*, \mathscr{I}_B f^* \mathscr{F}) = 0.$$

Hence the tangent spaces of $L(y) = \{\tilde{f}(y \times \Gamma)\}_{\mathrm{red}}$ are contained in $\mathscr{F}$ by (6.7), where $y \in C^*$ is a general point. Since L is a component of a specialization of $L(y)$, we get the assertion.

Let X be a smooth projective scheme over $S = \mathrm{Spec}\, R$ and H a relatively ample divisor on X, where R is an integral domain of characteristic 0. Let $\mathscr{F} \subset \mathscr{T}_{X/S}$ be a saturated subsheaf and $x \in X$ and S-valued point. We define a contravariant functor $\mathscr{RC}^\alpha(\mathscr{F}, x)$: ($S$-schemes) $\rightarrow$ ($\mathscr{Sets}$) by

$$\mathscr{RC}^\alpha(\mathscr{F}, x)(Z) = \{\text{proper flat family of irreducible rational curves}$$

$$L \subset X \times_S Z; \; LH \leq \alpha, \; \mathscr{T}_{L_{\mathrm{red}}/S} \subset \mathscr{F}, \; L \ni x \times_S Z\}.$$

The functor $\mathscr{RC}^\alpha(\mathscr{F}, x)$ is (represented by) a constructible (locally closed) subscheme of the Hilbert scheme of X. Moreover, it is quasiprojective since the degree α is bounded. By this functor-scheme, (7.3) will be rephrased as follows:

COROLLARY 7.4. *Assume that there exists a flat dominant family*

$$\{C_t \ni x\} \subset X \times_S T$$

of irreducible proper S-curves such that $\mathscr{F}|C_t$ is a relatively ample invertible sheaf and that $\mathscr{F}^{-1} \otimes (\mathscr{T}_{X/S}/\mathscr{F})|C_t$ is a relatively negative locally free sheaf for t lying on an open subset T' of T. Let $S' \subset S$ be the image of T' via the projection and Z a k-valued point of S', k being an algebraically closed field of positive characteristic. Then $\mathscr{RC}^\alpha(\mathscr{F}, x)(Z)$ is nonempty for some positive constant α depending only on (C, H) and $(C, c_1(\mathscr{F}))$.

COROLLARY 7.5 (char 0). *If a nonsingular projective variety X over an algebraically closed field k of characteristic 0 carries a conic foliation $\mathscr{F}$, the set $\mathscr{RC}^\alpha(\mathscr{F}, x)(k)$ is nonempty, where x is a general point on X and α is a large positive constant.*

PROOF. Let $C \ni x$ be a generic member of an $\mathscr{F}$-family $\mathscr{C}$. Without loss of generality, we may assume that X, x, C, H, and $\mathscr{F}$ are all defined over a **Z**-algebra $R \subset k$ of finite type. Hence the quasiprojective scheme $\mathscr{RC}^\alpha(\mathscr{F}, x)$ is also defined over R, if R is sufficiently localized. Put

$$\alpha = \text{Max}\{2(C, H)/(C, c_1(\mathscr{F})), (C, H)\}.$$

Since $\mathscr{RC}^\alpha(\mathscr{F}, x)$ is of finite type over $\text{Spec}\, R$ and nonempty over almost every geometric point (note that the ampleness and the negativity of vector bundles are open conditions), $\mathscr{RC}^\alpha(\mathscr{F}, x)$ is nonempty over $\text{Spec}\, k$ as well.

Thus we obtain the following characterization of conic bundles in characteristic 0:

THEOREM 7.6 (char 0). *Let k be an algebraically closed field of characteristic 0. For a k-variety X, the following two conditions are equivalent:*

(CB)   *X is birationally equivalent to a conic bundle.*

(CF)   *A suitable smooth projective model of X carries a conic foliation.*

PROOF. The implication (CB) $\Rightarrow$ (CF) has been already shown in (7.1). Assume that X has a conic foliation $\mathscr{F}$. Then, for a general k-valued point $x, \mathscr{RC}^\alpha(\mathscr{F}, x)(k) \neq \varnothing$. By definition, a general point L on some irreducible component $R_0(\mathscr{F}, x)$ of $\mathscr{RC}^\alpha(\mathscr{F}, x)$ represents an irreducible rational curve. On the other hand, $\mathscr{F}$ is of rank 1 so that $R_0(\mathscr{F}, x)(k)$ consists of a single point $r(\mathscr{F}, x)$ by the uniqueness theorem of an ordinary differential equation over a field of characteristic 0 (indeed, $L \in R_0(\mathscr{F}, x)$ is the closure of the "maximal integral variety" of $\mathscr{F}$ passing through the point x). Hence the correspondence $x \to r(\mathscr{F}, x) \in \mathscr{Hilb}(X)(k)$ determines a rational map of which a general fibre is an irreducible rational curve $L(\mathscr{F}, x) = r(\mathscr{F}, x)$. Therefore the proof is reduced to the following

LEMMA 7.7 (char 0). *A k-variety X is birationally equivalent to a conic bundle if and only if there exists a rational map $f: X \to W$ whose generic fibre is an irreducible rational curve.*

OUTLINE OF THE PROOF. By resolving the singularities of X and W, we may assume that X and W are both nonsingular and the general fibre is a smooth rational curve. Since the anticanonical divisor $-K_X$ is relatively very ample over a Zariski open subset of W, we get a birational map X into $P(f_*\mathscr{O}_X(-K_X))$ of which the image X' is fibrewise a conic curve outside a subset $V \subset W$ of codimension 2. Let $\pi: W' \to W$ be a suitable birational morphism such that the inverse image of the discriminant locus of $f': X' \to W$ becomes a divisor of simple normal crossings. Then the strict transform $X'' \subset P(\pi^*f_*\mathscr{O}_X(-K_X))$ of $X' \subset P(f_*\mathscr{O}_X(-K_X))$ is smooth, giving a conic bundle fibration $f'': X'' \to W'$. (We needed here two results in characteristic 0: the resolution of singularities [H] and Sard's theorem [S, Chapter II, §3]. The latter is actually false in positive characteristics; for instance, a smooth complete surface can carry a fibration a general fibre of which is a singular rational curve with a cusp of type $(2, q)$, where $q = p$ if p is an odd prime or an arbitrary odd natural number if $p = 2$. When

$p = 2$, $q = 3$ or $p = q = 3$, such fibrations are called *quasielliptic surfaces* since general fibres have arithmetic genus 1 [**BM**].)

8. "Generic semipositivity" of the cotangent bundles of non-uniruled varieties (characteristic 0).

Let X be a smooth variety over an algebraically closed field k of characteristic 0 and $\mathscr{F}$ a subbundle of the tangent bundle $\mathscr{T}_{X/k}$. Let $\{C_t\}$ be a family of complete irreducible curves on X which sweeps out an open dense subset of X. Assume the following conditions on the pair $(\mathscr{F}, \{C_t\})$:

(SI)   *$\mathscr{F}|C_t$ is ample for generic t;*

(SII)   *$(\mathscr{T}_{X/k}/\mathscr{F}) \otimes \mathscr{F}^*|C_t$ is negative for generic t.*

If these two conditions are satisfied for some family $\{C_t\}$, $\mathscr{F}$ is said to be a *special foliation* on X.

When X is a smooth scheme over an integral domain R of characteristic 0, a subbundle $\mathscr{F}$ of $\mathscr{T}_{X/R}$ is called a special foliation if $\mathscr{F} \otimes K$ is such on $X \otimes K$, where K denotes the algebraic closure of the field of the quotients of R.

THEOREM 8.1 (EKEDAHL-GABBER). *Let X be a smooth scheme over R with $H^0(X \otimes K, \mathscr{O}_{X \otimes K}) = K$ and $\mathscr{F}$ a special foliation on X. Then there exists an open dense subset $S \subset \operatorname{Spec} R$ such that $\mathscr{F}_s = \mathscr{F}|X_s$ is integrable on every geometric fibre X_s over $s \in S$.*

First note the following

LEMMA 8.2. *Let $\mathscr{F}$ and $\{C_t\}$ be as above. Then, for generic t, there exist a finite K-morphism $f_t: C_t' \to C_t \otimes K$ and a line bundle $\mathscr{L}_t$ of positive degree on C_t' such that $f_t^*(\mathscr{F}|C_t \otimes K) \otimes \mathscr{L}_t^{-1}$ is ample and $f_t^*\{(\mathscr{T}_{X/R}/\mathscr{F})|C_t \otimes K\} \otimes \mathscr{L}_t^{-1}$ is negative.*

PROOF. We need the theory of semistable bundles on curves in characteristic 0 and refer the reader to [**Mi2**, §§2–3]. Let $\mathscr{F}_i$ and $\mathscr{G}_j$ be the components of the graded modules associated with the semistable filtrations of $\mathscr{F}|C_t \otimes K$ and $(\mathscr{T}_{X/R}/\mathscr{F})|C_t \otimes K$, respectively. The ampleness of $\mathscr{F}|C_t \otimes K$ is equivalent to the positivity of the degrees of the **Q**-divisors

$$\delta(\mathscr{F}_i) = \det \mathscr{F}_i / \operatorname{rank} \mathscr{F}_i \in \operatorname{Pic}(C_t) \otimes \mathbf{Q},$$

while the negativity of $\mathscr{F}^* \otimes (\mathscr{T}_{X/R}/\mathscr{F})|C_t \otimes K$ yields $\deg \delta(\mathscr{G}_j) < \deg \delta(\mathscr{F}_i)$ for every i and j (for details, see [**Mi2**, §3]). Hence there is a rational number $r = m/n > 0$ such that $\deg \delta(\mathscr{F}_i) > r > \deg \delta(\mathscr{G}_j)$. Let $f_t: C_t' \to C_t$ be a finite covering of degree n and $\mathscr{L}_t$ a line bundle of degree m on C_t'. Then we have the assertion. (Note that $f_t: C_t' \to C_t$ and $\mathscr{L}_t$ can be constructed to form a family.)

(8.3) PROOF OF THEOREM 8.1. Substituting R by a suitable extension, we may assume that the families $\{C_t\}$, $\{C_t'\}$, $\{f_t\}$, and $\{\mathscr{L}_t\}$ are all defined over R.

The sheaf

$$\mathscr{H}_t = \mathscr{H}om_{\mathscr{O}_{C_t}}(\Lambda^n(\mathscr{T}_{X/R}/\mathscr{F})^*|C_t, \bigotimes_{i=2}^{m}\{\Lambda^{n_i}(\mathscr{S}^i\mathscr{F}^*)\}|C_t)$$

is *negative* over an open subset of Spec R provided t is generic ($m \geq 2$, $n \geq 1$, $\sum n_i = n$). In fact,

$$f_t^*\mathscr{H}_t \simeq [\Lambda^n\{f_t^*(\mathscr{T}_{X/R}/\mathscr{F})\}\otimes\mathscr{L}_t^{-1}]\otimes[\bigotimes\{\Lambda^{n_i}(\mathscr{S}^i(f_t^*\mathscr{F}^*\otimes\mathscr{L}_t))\}]\otimes\mathscr{L}_t^{-\Sigma in_i+n}$$

is clearly negative whenever $\mathscr{L}_t$ and $f_t^*\mathscr{F}\otimes\mathscr{L}_t^{-1}$ are ample and $f_t^*(\mathscr{T}_{X/R}/\mathscr{F})\otimes\mathscr{L}_t^{-1}$ is negative on C_t'. Therefore Condition (3) in Theorem 4.8 is verified over some open subset S of Spec R.

REMARK 8.4. T. Ekedahl kindly informed the author about Theorem 8.1, due to himself and O. Gabber. However, our proof presented above is, for the sake of the coherence of the argument, not the same as their original one, which uses explicitly the specific features of geometry in characteristic p including Frobenius morphism.

Now that a special foliation is integrable on almost every reduction, we have the following

THEOREM 8.5 (char 0). *Let X be a smooth projective variety over an algebraically closed field k of characteristic 0. If an open subset $U \subset X$ with $H^0(U, \mathscr{O}_U) = k$ carries a special foliation $\mathscr{F}$, then X is uniruled. More precisely, for a general geometric point x on U, there exists an irreducible (possibly noncomplete) rational curve $L_x \subset U$ passing through x such that $\mathscr{T}_{L_x} \subset \mathscr{F}$ at a general point of L_x.*

The proof is completely parallel to that of Theorem 7.6 so it is left to the reader.

As a direct consequence of (8.5), we have the "*generic semipositivity*" of the cotangent bundle of a non-uniruled variety in the following sense:

COROLLARY 8.6. *Let X be a normal projective variety of dimension n defined over an algebraically closed field of characteristic 0 with uncountable elements. Let $H_1, \ldots, H_{n-1}$ be ample Cartier divisors on X and $m_1, \ldots, m_{n-1}$ sufficiently large integers. Take a (sufficiently) general complete intersection curve C cut out by the complete linear systems $|m_i H_i|$. Then $\Omega^1_{X/k}|C$ is a semipositive vector bundle unless X is uniruled. ($\Omega^1_{X/k}|C$ is well defined since X is smooth in codimension 1).*

PROOF. Assume that $\Omega^1_{X/k}|C$ is not semipositive or, equivalently, that $\mathscr{T}_{X/k}|C$ is not seminegative. Let $\mathscr{F}_C \subset \mathscr{T}_{X/k}|C$ be the maximal destabilizing subsheaf. Let $\rho: X' \to X$ be a resolution and put $\mathscr{E} = \rho_*\mathscr{T}_{X'/k}$. By a theorem of Mumford-Mehta-Ramanathan [MR] (however, we need a slightly modified version; cf. [Mi2, §2]), $\mathscr{F}_C$ is a restriction of a saturated subsheaf $\mathscr{F} \subset \mathscr{E}$ defined on X. $\mathscr{F}$ is a subbundle of $\mathscr{E}$ outside a closed subset $Y \subset X$ of codimension at least 2. Let U be the open subset $X - Y - \text{Sing}(X)$. Let $\mathscr{C}$ be the family of complete intersection curves which is away from $X - U$. Since C is general, $C \in \mathscr{C}$. $\mathscr{E}|C$ is not seminegative so that $\mathscr{F}_C = \mathscr{F}|C$ is ample while $(\mathscr{E}/\mathscr{F}) \otimes \mathscr{F}^*$ is negative. Thus the pair $(\rho^*\mathscr{F}, \rho^*\mathscr{C})$ satisfies (SI) and (SII) above and $\rho^*\mathscr{F}$ is a special foliation on $\rho^{-1}(U)$.

REMARK 8.7. To the author's knowledge, Theorem 8.6 is new even in the surface case. A very partial result is found in [**Mi1**].

EXAMPLE 8.8 (EKEDAHL). An analogue of (7.6) or (8.5) is not true in positive characteristics. In fact, T. Ekedahl has found surfaces of general type which carry "conic foliations" without being uniruled. Let us sketch his construction below.

Let Y be a nonsingular surface on an abelian variety over k, $\operatorname{ch}(k) = p$. Let $\mathscr{L}$ be a very ample line bundle on Y and choose a general global section s of $\mathscr{L}^{\otimes p}$. Construct a cyclic covering $\pi \colon X \to Y$ in $P(\mathscr{O}_Y \oplus \mathscr{L})$ of degree p with branch locus along $(s)_0$. The projection π is a purely inseparable morphism of degree p and the natural homomorphism $\pi^*\Omega_Y^1 \to \Omega_X^1$ has locally free cokernel of rank 1 which is isomorphic to $\pi^*\mathscr{L}^{-1}$. The canonical divisor K_X is, by the adjunction formula, linearly equivalent to $\pi^*(K_Y + (p-1)c_1(\mathscr{L}))$ so that $\pi^*\mathscr{L} \subset \mathscr{T}_{X/k}$ is a "conic foliation" on a surface of general type. On the other hand, since X has 2-dimensional image Y on an abelian variety, X cannot be uniruled.

It should be noted that the "conic foliation" $\pi^*\mathscr{L} \subset \mathscr{T}_{X/k}$ is not liftable to characteristic 0, while X is liftable in an obvious manner as long as the pair $(Y, \mathscr{L})$ is liftable downstairs. Therefore $\pi^*\mathscr{L}$ is by no means a foliation in the sense defined in §4.

REMARK 8.9. (1) It is very likely that a special foliation on a smooth quasiprojective **C**-variety has compactifiable leaves, though the author has no proof at the moment.

(2) From the "generic semipositivity theorem" (8.6), we can derive a sort of "pseudo-effectivity" for the cycle $3c_2(X) - c_1^2(X)$ on a "minimal" projective n-fold X in the sense of S. Mori (this volume). This result has a nice application to the classification theory in dimension 3: the *nonnegativity of the Kodaira dimension of minimal threefolds*. For details, see [**Mi2**, §§6–8, **Mi3**].

REFERENCES

[**B**] A. Beauville, *Variété de Prym et Jacobian intermédiaires*, Ann. Sci. École Norm. Sup. (4) **10** (1977), 309–391.

[**BM**] E. Bombieri and D. Mumford, *Enriques classification of surfaces in char p*. III, Invent. Math. **35** (1976), 197–232.

[**C**] C. Chevalley, *Theory of Lie groups*. I, Princeton Univ. Press, Princeton, N.J., 1946.

[**EGA IV**] A. Grothendieck et al., *Éléments de géométrie algébrique*. IV. *Étude locales des schémas et des morphismes de schémas*, Inst. Hautes Études Sci. Publ. Math. No. 32, 1967.

[**FGA**] A. Grothendieck, *Fondements de géométrie algébrique*, Secrétariat Mathématique, Paris, 1962.

[**H**] H. Hironaka, *Resolution of singularities of an algebraic variety over a field of characteristic zero*, Ann. of Math. (2) **79** (1964), 109–326.

[**M1**] Y. Miyaoka, *Algebraic surfaces with positive indices*, Classification of Algebraic and Analytic Manifolds, Progr. Math., vol. 39, Birkhäuser, 1983.

[Mi2] _____ , *The Chern classes and Kodaira dimension of a minimal variety*, Algebraic Geometry Sendai, Kinokuniya-North-Holland (to appear).

[Mi3] _____ , *The Kodaira dimension of minimal threefolds*, Preprint.

[Mo] S. Mori, *Projective manifolds with ample tangent bundles*, Ann. of Math. (2) **110** (1979), 593–606.

[MM] Y. Miyaoka and S. Mori, *A numerical criterion for uniruledness*, Ann. of Math. (2) **124** (1986), 65–69.

[MR] V. B. Mehta and A. Ramanathan, *Semi-stable sheaves on projective varieties and their restriction to curves*, Math. Ann. **258** (1982), 212–224.

[S] S. Sternberg, *Lectures on differential geometry*, Prentice-Hall, Englewood Cliffs, N.J., 1964.

[SGA 1] A. Grothendieck et al., *Revêtements étales et groupe fondemental*, Lecture Notes in Math., vol. **224**, Springer-Verlag, 1971.

TOKYO METROPOLITAN UNIVERSITY, JAPAN

Proceedings of Symposia in Pure Mathematics
Volume **46** (1987)

Classification of Higher-Dimensional Varieties

SHIGEFUMI MORI

0. Introduction.

(0.1) In recent years, the classification theory of algebraic varieties programmed by Iitaka has been developed a great deal by Ueno, Viehweg, Fujita, Kawamata, Kollár (in the order they appear in the literature), and others. The theory of extremal rays introduced by Mori has also been developed by Reid, Kawamata, Shokurov, Kollár, and others, and many results of Kawamata and others showed various consequences of Minimal Model Conjectures (e.g., C_{nm}^{+}). It is from the viewpoint of the theory of extremal rays that Wilson [**W4**] and Kollár [**Ko7**] gave an overview introduction, and Kawamata, Matsuda, and Matsuki [**KMM**] gave a technically precise introduction to these theories.

My attitude in this lecture note is to review (many but not all) results in the classification theory (especially around the conjectures C_{nm}, C_{nm}^{+}, and K) which do not require conjectural existence of good minimal models. The reason is not only that there are so many important and beautiful results of this nature but also that it is, to me, the natural introduction to the theory of extremal rays. Anyway the other approach is covered well by those survey papers.

(0.2) We will work only over an algebraically closed field k of characteristic 0.

We say that two nonsingular projective varieties X and Y over k are *birationally equivalent* and write $X \sim Y$ if there is a birational mapping $X \dashrightarrow Y$, or equivalently if their function fields $Q(X)$ and $Q(Y)$ are isomorphic extension fields of k.

We can see that there are several birational invariants among which we call $h^1(X, \mathcal{O}_X)$ the *irregularity* and $h^0(X, \mathcal{O}(\nu K_X))$ the *ν-genus*, denoted by $P_\nu(x)$, where K_X is the canonical divisor of X.

The classification problem, in short, should be to classify nonsingular projective varieties up to birational equivalence. Intuitively speaking, the set of all such classes is a disjoint union of (countably?) many parameter spaces, and a slightly more specific question is to find birational invariants which separate connected components.

1980 *Mathematics Subject Classification* (1985 *Revision*). Primary 14-02, 14A10, 14E05, 14J10, 14J30, 14J40, 14M20; Secondary 14C30, 14C40.

For curves, it is known that the genus g which is equal to irregularity separates the connected components $\mathfrak{M}_g$. However for surfaces, this is already a too difficult problem, and it is not what we mean by birational classification of surfaces. We are at an intermediate stage.

(0.3) Iitaka (seel also Moishezon's paper [**Mz2**]) defined *Kodaira dimension* $\kappa(X)$ of a nonsingular projective n-fold X as κ which is an integer in $[0, n]$ or $-\infty$ such that order of growth of $P_\nu(X)$ is equal to that of ν^κ when ν is sufficiently divisible. He thus divided algebraic varieties of dimension n into $(n + 2)$ classes, among which n-folds with $\kappa = n$ are called *of general type*. In case of surfaces, these are (1) surfaces of general type ($\kappa = 2$), (2) elliptic surfaces ($\kappa = 1$), (3) abelian surfaces, hyperelliptic surfaces, $K3$ surfaces, and Enriques surfaces ($\kappa = 0$), and (4) ruled surfaces ($\kappa = -\infty$). Iitaka's philosophy is that the study of algebraic varieties is reduced to those of X with $\kappa = \dim X, 0, -\infty$ in the sense that a fiber space $f: X \to Y$ is to be viewed as a product of the base Y and a "very" general fiber F (1.11). He asked the following

$$C_{nm}: \quad \kappa(X) \geq \kappa(F) + \kappa(Y)$$

($n = \dim X$, $m = \dim Y$) as a crucial conjecture for the classification (cf. §6). One can easily see its power by applying it to the Albanese map $\mathrm{Alb}_X: X \to \mathrm{Alb}(X)$ (3.2) for X with $\kappa \leq 0$ (cf. §§10, 11).

(0.4) Ueno introduced the conjecture K to the effect that Alb_X is an algebraic fiber space birationally equivalent to an étale fiber bundle which is trivialized by a finite étale covering of $\mathrm{Alb}(X)$. The strengthened version C_{nm}^+ of C_{nm} by Viehweg (7.1) is very closely related to the conjecture K (cf. §10).

We should mention that, in this direction, 3-folds with $\kappa = 0$ and $q \geq 2$ are classified by Ueno (cf. §10) and the case $\kappa \leq 0$ and $q > 0$ by Viehweg (cf. §10, 11). However for 3-folds with $q = 0$, we need minimal models at present, and even if we assume it we still have a completely open problem of classifying special unitary 3-folds (§10).

(0.5) Another aspect is the problem of effectivity: κ is not a naive invariant in the sense that we need all the $P_\nu(X)$'s to decide, for instance, $\kappa = 0$. In this sense, Kollár's result that abelian n-folds are birationally characterized by $q = n$ and $P_{n+\alpha} = 1$ for some $\alpha > 0$ is a significant generalization of an important result of Kawamata's: $q = n$ and $\kappa = 0$. We list some of Kollár's results in Table 1.

(0.6) Finally, I would like to mention that most results on $f_*(\omega_{X/Y}^\nu)$ ($\nu > 0$) for an algebraic fiber space $f: X \to Y$ are obtained first for $\nu = 1$ by various positivity and vanishing theorems, and the general case is obtained by the mysterious covering trick (§4). Is there any direct approach?

(0.7) §1 reviews basic results on D-dimensions. Most results follow from a simple graded algebra lemma (1.2), and (1.12) is a slight refinement; (1.14) is a useful lemma of Fujita. §2 translates results in §1 to the case of canonical divisors by simple canonical bundle formulas (2.1). Viehweg's idea of discarding codimension 2 sets is used to translate (1.12) to (2.6). The first half of §3 reviews Ueno's results on varieties related to an abelian variety, and the second half

TABLE 1

κ		X		$q \geq 3$	$q \geq 4$
$-\infty$	$q = 0$			$P_{12}(X) = 0$	
	$q > 0$	uniruled			
0	$q = 0$	?		$P_{12}(X) = 1$	
	$q > 0$	$\mathrm{Alb}_X \sim$ fiber bundle			
1					Φ_{12} is the Iitaka fibration
2		nontrivial Iitaka fibration		$P_{12}(X) \geq 2$	Φ_k $(k \geq 14)$ is the Iitaka fibration
3					Φ_k $(k \geq 3)$ is generically finite

reviews more recent ones. These provide us with explicit examples of Kodaira dimensions and Iitaka fiberings. We also refer the reader to [U1] for these materials. §4 reviews several covering techniques and Viehweg's Base Change Theorem on the double dual $f_*(\omega^\nu_{X/Y})$ $(\nu > 0)$ for an algebraic fiber space $f: X \to Y$. §5 reviews various positivity and vanishing theorems. In its part II, weak positivity of $f_*(\omega^\nu_{X/Y})$ $(\nu > 0)$ of $f: X \to Y$ whose very general fiber has $\kappa = 0$ is analyzed as the sum of a nef $\mathbf{Q}$-divisor and a negligible $\mathbf{Q}$-divisor for reader's understanding. §6 reviews some history of C_{nm} with some of C^+_{nm} (which is maybe far from being complete) and studies the meaning of C_{nm} by decomposing it into 3 parts, while §7 reviews C^+_{nm} and Viehweg's question Q_{nm} whose solution implies C^+_{nm}. §8 reviews several effective results (mainly by Kollár) in classification, where (8.1) and (8.2) are modified as he mentioned in the remark in [Ko3]. We present several difficulties in generalizing the notion of minimal surfaces to 3-folds in (9.1)–(9.3.3) of §9 and make a couple of comments in (9.4.1) on definitions of canonical and minimal models. In the rest of §9, results and notions (e.g., the Abundance Conjecture and Zariski decomposition) related to the finite generation of the canonical ring

$$R(X, K_X) = \bigoplus_{\nu \geq 0} H^0(X, \mathcal{O}(\nu K_X))$$

of an algebraic variety X with $\kappa \geq 0$ are reviewed. §10 reviews several results on varieties with $\kappa = 0$ and also varieties with torsion canonical divisors. §11 is on varieties with $\kappa = -\infty$, where we take uniruledness as a natural alternative to ruledness by reviewing various difficulties around rationality and ruledness in higher dimension (11.1)–(11.2). In §12, the (standard) conjectural program to obtain a minimal model and Francia's example of directed flips are explained.

We refer the reader to Reid's beautiful introduction [**Re5**] for details of terminal and canonical singularities.

(0.8) I would like to express my deep appreciation to Professors J. Kollár and D. Morrison for the useful conversation with them and their helpful comments. I would like to express my sincere thanks to Professor Fujita for his helpful information and to Professor Nakayama for pointing out a gap in part II of §5. Finally, I would like to express my hearty thanks to members of the Columbia University Mathematics Department for their warm hospitality during the preparation of this note.

Notation. We consider varieties over a field only of characteristic 0.

For two algebraic varieties X and Y, we say that X *dominates* Y if there is given a dominating map $X \dashrightarrow Y$.

Cartier divisors, invertible sheaves, and line bundles are used interchangeably. For a coherent sheaf F on X, we write $h^i(X, F) = \dim H^i(X, F)$. When X is a nonsingular projective variety, we write $p_g(X) = h^0(X, K_X)$ (*geometric genus*) and $q(X) = h^1(X, \mathcal{O}_X)$ (*irregularity*), where K_X is the canonical divisor of X. For a Cartier divisor D on X, the *base locus* $\mathrm{Bs}\,|D|$ of the complete linear system $|D|$ is the intersection of all members of $|D|$ as a scheme. A Cartier divisor D is *base point free* if $\mathrm{Bs}\,|D| = \varnothing$. A Weil divisor D on a normal variety is **Q**-*Cartier* if there is a natural number n such that nD is a Cartier divisor. A *s.n.c. divisor* of a nonsingular algebraic variety is a union of irreducible smooth divisors intersecting transversally. A **Q**-*divisor* D on a nonsingular projective variety X is a sum $\sum a_i D_i$ of irreducible divisors D_i of X with $a_i \in \mathbf{Q}$. We use the notation

$$\lceil D \rceil = \sum_i \lceil a_i \rceil D_i, \qquad [D] = \sum_i [a_i] D_i, \qquad \langle D \rangle = D - [D],$$

where $[x]$ is the integer such that $[x] \leq x < [x] + 1$, and $\lceil x \rceil = -[-x]$. We say that D is *nef* (Reid) if $(D \cdot C) \geq 0$ for all irreducible curves C in X, that D is *negligible* if $\mathrm{Supp}\,D$ is a s.n.c. divisor and $[D] = 0$, and that D is *semiample* if some positive multiple of D is base point free.

An *algebraic fiber space* $f: X \to Y$ is a surjective morphism between nonsingular projective varieties X and Y with connected geometric fibers. As an auxiliary terminology, we say that $f^\circ: X^\circ \to Y^\circ$ is the *restriction of an algebraic fiber space to a quasiprojective base* if X°, Y° are smooth quasiprojective, f° is projective surjective with connected geometric fibers, or equivalently if f° can be extended to an algebraic fiber space $f: X \to Y$ so that X° and Y° are open dense subsets of X and Y as in the following diagram:

$$
\begin{array}{ccc}
X^\circ & \subset & X \\
f^\circ \downarrow & \circlearrowleft & f \downarrow \\
Y^\circ & \subset & Y
\end{array}
$$

Let $f: X \to Y$ be a proper surjective morphism of nonsingular algebraic varieties. We set $K_{X/Y} = K_X - f^* K_Y$, which may be used interchangeably with the relative dualizing sheaf $\omega_{X/Y}$. We say that a divisor D of X is f-*exceptional* if $\operatorname{codim}_Y f(\operatorname{Supp} D) \geq 2$. For (geometric) points y of Y, we denote the scheme-theoretic fiber of f over y by X_y. By the *branch locus* of f, we mean the subset $\{y \in Y | f$ is not smooth above $y\}$ of Y, and denote it by $\Delta(f)$. We say that f *has s.n.c. branching* if there is a s.n.c. divisor containing $\Delta(f)$.

Let $f: X \to Y$ and $g: Y' \to Y$ be projective surjective morphisms of nonsingular quasiprojective varieties. We call the irreducible components of $X \times_Y Y'$ which dominate Y' (or equivalently X) the *main components*. We note that $X \times_Y Y'$ is reduced at general points of main components. We denote by $(X \times_Y Y')_{\text{main}}$ the reduced closed subscheme of $X \times_Y Y'$ which is the union of all the main components, and call $(X \times_Y Y')_{\text{main}} \to Y'$ the morphism *induced by* (base change) g.

If furthermore f is an algebraic fiber space in the above setting, then $(X \times_Y Y')_{\text{main}}$ is irreducible and $X' \to Y'$ is called an algebraic fiber space *induced* by (base change) g, where $X' \to (X \times_Y Y')_{\text{main}}$ is an arbitrary projective resolution.

1. Basic results on D-dimensions and Iitaka fiberings. In this section, we review basic results on D-dimensions and Iitaka fiberings. We only mention a slight refinement (1.12) and useful (1.14), and the rest is quite standard.

For nonexperts, it is probably helpful to realize that the main ingredient is an elementary graded algebra lemma (1.2), except for the fundamental notions of resolution (1.5.1), elimination of indeterminacy (1.10.i), and flattening (1.10.ii). We also refer the reader to standard reference [**U1**].

Fields will be assumed to be of characteristic 0.

(1.1) Let k be a field, and let $R = \bigoplus_\nu R_\nu$ be a graded k-domain such that $R_\nu = 0$ for all $\nu < 0$, where R_ν is the degree ν part. Let $Q(R)$ be the quotient field of R, $R^{(m)}$ the graded subdomain $\bigoplus_\nu R_{\nu m}$ for $m \in \mathbf{N}$, and $R^\#$ ($\subset R$) the multiplicative subset of all nonzero homogeneous elements.

Then the quotient ring $R^{*-1}R$ is a graded k-domain, and its degree 0 part $(R^{\#-1}R)_0$ is a field which we denote by $Q((R))$. We note that $Q((R^{(m)})) = Q((R))$ for all $m \in \mathbf{N}$, and that $Q(R)$ is a rational function field over $Q((R))$ in 1 variable if $R \neq R_0$.

Let $\mathbf{N}(R) = \{\nu \in \mathbf{N} | R_\nu \neq 0\}$, and $M_{\geq n} = \{\nu \in M | \nu \geq n\}$ for a subset $M \subset \mathbf{N}$ and $n \in \mathbf{N}$. Since $\mathbf{N}(R)$ is a semigroup, we note that $\mathbf{N}(R)_{\geq n} = (d\mathbf{N})_{\geq n}$ for $n \gg 0$, where $d = \gcd \mathbf{N}(R)$.

Our arguments in this section come from the following elementary graded algebra lemma.

(1.2) LEMMA. *Let R be as above with $R \neq R_0$, and $S = \bigoplus_\nu S_\nu$ a graded k-subalgebra of R. For $\nu \geq 0$, let $S_0[S_\nu]$ denote the graded subdomain of S generated by S_0 and S_ν. Then the integral closure of S in R is graded, and*

(i) *if $Q(R)$ is finitely generated over k, then there exists $n \gg 0$ such that $Q((S_0[S_\nu])) = Q((S))$ for all $\nu \in \mathbf{N}(S)_{\geq n}$,*

(ii) *if S is integrally closed in R, then $Q((S))$ is algebraically closed in $S^{\#-1}R$, and*

(iii) *if R is integral over S, then $Q((R))$ is algebraic over $Q((S))$.*

PROOF. If $r = r_0 + \cdots + r_i + \cdots \in R$ is integral over S, then so is $r_0 + \cdots + \lambda^i r_i + \cdots$ over S for every $\lambda \in k^*$ by homogeneity of S. Thus degree i part r_i is integral over S.

(i): For $\mu, \nu \in \mathbf{N}(S)$ with $\mu|\nu$, we see $Q((S_0[S_\mu])) \subset Q((S_0[S_\nu]))$ by $S_\mu^{\nu/\mu} \subset S_\nu$. Since $Q(R)$ is finitely generated over k, so is $Q((S)) = \bigcup_{\mu \in \mathbf{N}(S)} Q((S_0[S_\mu]))$. Thus $Q((S_0[S_\mu])) = Q((S))$ for some $\mu \in \mathbf{N}(S)$. Let $n \gg 0$ be such that $\mathbf{N}(S)_{\geq n-\mu} = (d\mathbf{N})_{\geq n-\mu}$, where $d = \gcd \mathbf{N}(S)$. Then for every $\nu \in \mathbf{N}(S)_{\geq n}$, we see $\nu - \mu \in \mathbf{N}(S)$, i.e., $S_{\nu-\mu} \neq 0$, whence $Q((S_0[S_\mu])) \subset Q((S_0[S_\nu]))$ by $S_\mu \cdot S_{\nu-\mu} \subset S_\nu$.

(ii): Since S is integrally closed in R, so is $S^{\#-1}R$ in $S^{\#-1}R$, and hence $Q((S))$ is algebraically closed in $S^{\#-1}R$.

(iii): Since $(S^{\#-1}R)_0$ is integral over the field $Q((S))$, it is an algebraic field extension, and hence $Q((R)) = (S^{\#-1}R)_0$ because $Q((R))$ is a quotient ring of $(S^{\#-1}R)_0$. $\square$

(1.3) Let X be an irreducible reduced scheme over a field k, and L a Cartier divisor on it. Let

$$R(X, L) = \bigoplus_{\nu \geq 0} H^0(X, \mathcal{O}(\nu L))$$

be the graded k-domain, which may be viewed as the coordinate ring of the line bundle $\mathbf{V}(L)$ (or $\mathbf{V}(L^*)$ depending on notation) of L, and let $\mathbf{N}(X, L) = \mathbf{N}(R(X, L))$ and $Q((X, L)) = Q((R(X, L)))$.

When X is normal proper and $k = \bar{k}$, the *D-dimension* $\kappa(X, L)$ is defined as

$$\kappa(X, L) = \begin{cases} -\infty & \text{if } \mathbf{N}(X, L) = \varnothing, \\ \text{tr. } \deg_k Q((X, L)) & \text{if } \mathbf{N}(X, L) \neq \varnothing. \end{cases}$$

It is clear that $\kappa(X, L) = -\infty, 0, \ldots, \dim X$, by $Q((X, L)) \subset Q(X)$, and

$$\kappa(X, L) = -\infty \Leftrightarrow h^0(X, \mathcal{O}(\nu L)) = 0 \text{ for all } \nu > 0,$$
$$\kappa(X, L) \leq 0 \Leftrightarrow h^0(X, \mathcal{O}(\nu L)) \leq 1 \text{ for all } \nu > 0, \text{ and}$$
$$\kappa(X, L) \geq 1 \Leftrightarrow h^0(X, \mathcal{O}(\nu L)) \geq 2 \text{ for some } \nu > 0$$

follow easily by tr. $\deg_k R(X, L) = \text{tr. } \deg_k Q((X, L)) + 1$ if $\mathbf{N}(X, L) \neq \varnothing$.

(1.3.1) REMARK. If $\kappa(X, L) = 0$, then $\mathbf{N}(X, L) = c\mathbf{N}$ for some $c > 0$. Indeed since $|\nu L|$ consists of only one member L_ν for $\nu \in \mathbf{N}(X, L)$ ($h^0(X, \mathcal{O}(\nu L)) = 1$), the L_ν's must be proportional; i.e., there is a $\mathbf{Q}$-divisor D such that $L_\nu = \nu D$ for all $\nu \in \mathbf{N}(X, L)$. Thus $\mathbf{N}(X, L) = \{\nu \in \mathbf{N} | \nu L \sim \nu D\}$, which easily implies the remark.

It is easy to see that $Q((X, L)) = Q((X, \mathcal{O}(aL)))$ and $\kappa(X, L) = \kappa(X, \mathcal{O}(aL))$ for $a \in \mathbf{N}$ by $R(X, \mathcal{O}(aL)) = R(X, L)^{(a)}$.

If $\nu \in \mathbf{N}(X, L)$, then $R(X, L)_\nu = H^0(X, \mathcal{O}(\nu L))$ induces a rational map $\Phi_{|\nu L|}: X \dashrightarrow \mathbf{P}(H^0(X, \mathcal{O}(\nu L)))$. By the following (1.4), one has an equivalent definition:

$$\kappa(X,L) = \begin{cases} -\infty & \text{if } \mathbf{N}(X,L) = \varnothing, \\ \max_{\nu \in \mathbf{N}(X,L)} \dim \Phi_{|\nu L|}(X) & \text{if } \mathbf{N}(X,L) \neq \varnothing. \end{cases}$$

If X is proper irreducible reduced and nonnormal, then $\kappa(X,L)$ is defined as $\kappa(\overline{X}, \overline{L})$, where $\overline{X}$ is the normalization and $\overline{L}$ the pull back of L to $\overline{X}$.

(1.4) PROPOSITION. *Assume that X is normal proper and $k = \bar{k}$. If $\mathbf{N}(X,L) \neq \varnothing$, then there exists $n \in \mathbf{N}$ such that $Q(\Phi_{|\nu L|}(X)) = Q((X,L))$ for all $\nu \in \mathbf{N}(X,L)_{\geq n}$. Furthermore $Q((X,L))$ is algebraically closed in $Q(X)$.*

PROOF. The first assertion follows from (1.2.i). Let ξ be the $Q(X)$-scheme $\operatorname{Spec} Q(X)$, $\xi \to X$ the inclusion, and L_ξ the pull back of L. Then $R(X,L)$ is integrally closed in $R(\xi, L_\xi)$ (1.2) because elements $s \in R(\xi, L_\xi)_\nu$ integral over $R(X,L)$ ($\nu \geq 0$) are rational sections of $L^{\otimes \nu}$ with no poles and $s \in R(X,L)$ by the normality of X. So $Q((X,L))$ is algebraically closed in $R(X,L)^{\#-1}R(\xi, L_\xi)$ by (1.2.ii). Since $Q(X) = R(\xi, L_\xi)_0 \subset R(X,L)^{\#-1}R(\xi, L_\xi)$, $Q((X,L))$ is algebraically closed in $Q(X)$. $\square$

The behavior of $\kappa(X,L)$ under pull back is studied by

(1.5) PROPOSITION. *Let $f: X \to Y$ be a morphism between proper normal algebraic varieties over $k = \bar{k}$, and L, M Cartier divisors on X, Y, respectively such that $L = f^*M$. Then*
(i) *if $f_*\mathcal{O}_X = \mathcal{O}_Y$, then $R(X,L) \simeq R(Y,M)$, $Q((X,L)) \simeq Q((Y,M))$, and $\kappa(X,L) = \kappa(Y,M)$, and*
(ii) *if f is surjective, then $Q((X,L))$ is algebraic over $Q((Y,M))$ and $\kappa(X,L) = \kappa(Y,M)$.*

PROOF. (i): By $f_*\mathcal{O}_X = \mathcal{O}_Y$, one has $H^0(X, \mathcal{O}(\nu L)) = H^0(Y, \mathcal{O}(\nu M))$ for all $\nu \geq 0$, and the rest follows.

(ii): Let $g: Z \to Y$ be the Stein factorization of f. Then (i) shows that $Q((X,L)) = Q((Z, g^*M))$. Thus it is enough to show that $Q((Z, g^*M))$ is algebraic over $Q((Y,M))$. Let $h: V \to Y$ be the normalization of Y in the smallest Galois extension of $Q(Y)$ containing $Q(Z)$, and G the Galois group. Since $Q((V, h^*M)) \supset Q((Z, g^*M)) \supset Q((Y,M))$, it is enough to show that $Q((V, h^*M))$ is algebraic over $Q((Y,M))$. If $r \in R(V, h^*M)$, then $F(T) = \prod_{\gamma \in G}(T - r^\gamma) \in R(Y,M)[T]$, because

$$H^0(V, h^*\mathcal{O}(\nu M))^G = H^0(Y, h_*h^*\mathcal{O}(\nu M))^G$$
$$= H^0(Y, \mathcal{O}_Y(\nu M) \otimes h_*\mathcal{O}_V^G)$$
$$= H^0(Y, \mathcal{O}(\nu M))$$

for $\nu \geq 0$. Thus $R(V, h^*M)$ is integral over $R(Y,M)$, and $Q((V, h^*M))$ is algebraic over $Q((Y,M))$ by (1.2.iii). $\square$

(1.5.1) REMARK Let Y be a proper irreducible reduced algebraic variety over $k = \bar{k}$ and L a Cartier divisor on Y, and let Y' and L' be a normalization of Y and the pull back of L to Y', respectively. *Hironaka's resolution* [**Hi2**] says that there exists a birational morphism $f: Y'' \to Y'$ from a nonsingular projective variety Y''. Let $L'' = f^*L'$. Since $f_*\mathcal{O}_{Y''} = \mathcal{O}_{Y'}$, (1.5.i) implies $R(Y'', L'') \simeq R(Y', L')$, $Q((Y'', L'')) \simeq Q((Y', L'))$, and $\kappa(Y'', L'') = \kappa(Y', L')$. In particular $\kappa(Y, L)$ may be defined as $\kappa(Y'', L'')$ for an arbitrary resolution Y'' of Y and pull back L'' of L to Y''.

(1.6) PROPOSITION. *Let $f: X \to Y$ be a morphism between normal proper varieties such that $f_*\mathcal{O}_X = \mathcal{O}_Y$, and L a line bundle over X. Let $\bar{\eta}$ be the generic geometric point $\operatorname{Spec} \overline{Q(Y)}$ of Y. If $\kappa(X, L) \geq 0$, then $\kappa(X_{\bar{\eta}}, L_{\bar{\eta}}) \geq 0$ and $Q((X, L)) \subset Q((X_{\bar{\eta}}, L_{\bar{\eta}}))$.*

PROOF. This follows from the injection $R(X, L) \to R(X_{\bar{\eta}}, L_{\bar{\eta}})$ induced by restriction $H^0(X, \mathcal{O}(\nu L)) \to H^0(X_{\bar{\eta}}, \mathcal{O}(\nu L_{\bar{\eta}}))$ ($\nu \geq 0$). □

(1.7) COROLLARY (EASY ADDITION). *One has*

$$\kappa(X, L) \leq \kappa(X_{\bar{\eta}}, L_{\bar{\eta}}) + \dim Y.$$

We note that the case of equality is treated in (1.14).

(1.8) COROLLARY. *Under the above notation, one has*

$$Q((X, L)) \subset Q(Y) \quad \text{if } \kappa(X_{\bar{\eta}}, L_{\bar{\eta}}) = 0.$$

PROOF. One has $Q((X, L)) \subset Q((X_{\bar{\eta}}, L_{\bar{\eta}})) = k(\bar{\eta})$ (1.6). Thus $Q((X, L)) \subset k(\bar{\eta}) \cap Q(X) = Q(Y)$ since $Q(X)$ is a regular extension of $Q(Y)$. □

(1.9) PROPOSITION-DEFINITION. *Let X be a proper normal algebraic variety with generic point ξ and L a Cartier divisor on X. Then the following assertions are equivalent:*
  (i) $\kappa(X, L) = \dim X$,
  (ii) *$\Phi_{|mL|}$ is generically finite for some $m \in \mathbf{N}$, and*
  (iii) *there exists $a \in \mathbf{N}$ such that $\Phi_{|mL|}$ is birational for all $m \in \mathbf{N}(X, L)_{\geq a}$.*
  If X is projective, then one has further equivalent assertions:
  (iv) *there exist a very ample Cartier divisor H on X and $a \in \mathbf{N}$ such that $|aL - H| \neq \varnothing$, and*
  (v) *for every coherent sheaf $\mathfrak{F}$, $\mathfrak{F}(aL)$ is generically generated (i.e., $H^0(\mathfrak{F}(aL)) \otimes k(\xi) \to \mathfrak{F}(aL) \otimes k(\xi))$ for some $a \in \mathbf{N}$.*
  When these assertions hold, we say that L is big.

(1.9.1) REMARK. Under the notation of (1.9), assume that L is *nef*, that is $(L \cdot C) \geq 0$ for every irreducible curve C on X. Then it is proved by Sommese (cf. [**Ka2**, Lemma 3]) that L is big iff $(L^{\dim X}) > 0$. However, this is not so straightforward as (1.9).

PROOF. It is obvious that (iii)⇒(ii)⇒(i), and (i)⇒(iii) by (1.4). We now assume that X is projective. Then (v)⇒(iv)⇒(ii) is clear. We need to prove (ii)⇒(v). Let H be a very ample divisor such that $\mathfrak{F}(H)$ is generated by global sections. Since $\dim \Phi_{|mL|} = \dim X$ by (ii), $\Phi_{|mL|}(H) \neq \Phi_{|mL|}(X)$, and $\Phi_{|mL|}(H)$

is contained in a hypersurface section of $\Phi_{|mL|}(X)$, say of degree c. Thus $|mcL - H| \neq \varnothing$, and let $D \in |mcL - H|$. Since

$$H^0(\mathfrak{F}(H)) \otimes k(\xi) \to H^0(\mathfrak{F}(D + H)) \otimes k(\xi) \to \mathfrak{F}(D + H) \otimes k(\xi) \simeq \mathfrak{F}(H) \otimes k(\xi)$$

is surjective, (v) is proved for $D + H \sim mcL$. $\square$

(1.10) Let us recall a couple of fundamental results to study generically surjective rational mappings.

(i) Given a normal proper variety X over $k = \bar{k}$, a dominating rational map $f: X \dashrightarrow Y$ to a normal proper Y is determined by the subfield $K = Q(Y)$ of $Q(X)$ up to birational equivalence.

For such f (resp. K), Hironaka's *elimination of indeterminacy* and resolution [**Hi2**] say that there exist nonsingular projective varieties X' and Y' with birational morphism(s) $g': X' \to X$ and $h': Y' \to Y$ (resp. $Q(Y') = K$) so that the induced map $f': X' \dashrightarrow Y'$ is a morphism. We say that $f': X' \to Y'$ *represents* f (resp. K) or is *n.s. representative* of f (resp. K). We recall that f' is an algebraic fiber space iff $Q(Y)$ (resp. K) is algebraically closed in $Q(X)$.

(ii) Let $f: X \to Y$ be a proper surjective morphism of nonsingular quasiprojective varieties over $k = \bar{k}$. By the *flattening* of Hironaka [**Hi3**] or Gruson-Raynaud [**GsRa**], we can find a proper birational morphism $h'': Y'' \to Y$ from a nonsingular quasiprojective variety Y'' such that the morphism $(X \times_Y Y'')_{\text{main}} \to Y''$ induced by base change h'' is flat. We say that Y'' *flattens* f or $Y'' \to Y$ is a *flattening* of f.

(1.11) DEFINITION-THEOREM. *Let X be a proper normal algebraic variety over $k = \bar{k}$, and L a Cartier divisor on X, such that $\kappa(X, L) \geq 0$. Let $f^\flat: X^\flat \to Y^\flat$ be a n.s. representative of $Q((X, L))$, and $L^\flat$ the pull back of L to $X^\flat$. Then $f^\flat$ is an algebraic fiber space by (1.4) and we call it the Iitaka fibering of (X, L) and $Y^\flat$ the Iitaka model of (X, L), which are determined only up to birational equivalence. Then*

(i) $\dim Y^\flat = \kappa(X, L)$ *and* $\kappa(F, L^\flat|_F) = 0$, *where F is the fiber of $f^\flat$ over the generic geometric point $\bar{\eta}$ of $Y^\flat$, and*

(ii) *if a n.s. representative $f': X' \to Y'$ of a subfield $K \subset Q(X)$ is an algebraic fiber space such that L induces a Cartier divisor with $\kappa = 0$ on generic geometric fibers of f', then $K \supset Q(Y^\flat)$; i.e., $f^\flat$ factors as $X' \xrightarrow{f'} Y' \dashrightarrow Y^\flat$. Thus the two properties in* (i) *determine $f^\flat$ up to birational equivalence.*

(1.11.1) REMARK. (a) There exists m such that $Q(\Phi_{|nL|}(X)) = Q((X, L))$ for all $n \in \mathbf{N}(X, L)_{\geq m}$ (1.4). Therefore every n.s. representative of $\Phi_{|nL|}$ ($n \in \mathbf{N}(X, L)_{\geq m}$) is the Iitaka fibering.

(b) Let y be a point of $Y^{\flat\circ} = \{y \in Y^\flat | X_y^\flat \text{ is smooth}\}$. Since $\kappa(X_{\bar{\eta}}^\flat, L_{\bar{\eta}}^\flat) = 0$, one sees $\kappa(X_y^\flat, L_y^\flat) \geq 0$ and hence

$$\kappa(X_y^\flat, L_y^\flat) = 0 \Leftrightarrow h^0(X_y^\flat, L_y^{\flat \otimes \nu}) \leq 1 \text{ for all } \nu > 0.$$

This is why $\{y \in Y^{\flat\circ} | \kappa(X_y^\flat, L_y^\flat) = 0\}$ is an intersection of a countable number of Zariski open dense sets of $Y^\flat$ defined over k.

PROOF. (i): Let $H \succ 0$ be a very ample divisor on $Y^\flat$. Then

(*) CLAIM. *There exists $m > 0$ such that $f^{\flat *}\mathcal{O}(H) \subset \mathcal{O}(mL^\flat)$.*

PROOF OF CLAIM. Let $1, f_1, \ldots, f_n$ be a basis of $H^0(\mathcal{O}(H))$. By $Q((X, L)) \supset Q(Y)$, there exists $n > 0$ and $g_0, \ldots, g_n \in H^0(\mathcal{O}(mL^\flat))$ such that $g_0 \neq 0$ and $f_i = g_i/g_0$ ($i \in [1, n]$). One has

$$0 \prec (g_i) + mL^\flat = f^{\flat *}(f_i) + (g_0) + mL^\flat = f^{\flat *}((f_i) + H) + (mL^\flat + (g_0) - f^{\flat *}H).$$

Since H is very ample, $\bigcap_{1 \leq i \leq n} f^{\flat *}((f_i) + H) = \varnothing$. Thus $mL^\flat + (g_0) \succ f^{\flat *}H$, and multiplying by g_0 gives $f^{\flat *}\mathcal{O}(H) \subset \mathcal{O}(mL^\flat)$, and (*) is proved.

Let us continue (i). We only need to show $\kappa(X^\flat_{\bar{\eta}}, L^\flat_{\bar{\eta}}) = 0$. Since $\kappa(X^\flat_{\bar{\eta}} L^\flat_{\bar{\eta}}) \geq 0$ by (1.6), it is enough to show that $f^\flat_* \mathcal{O}(\nu L^\flat)$ has rank ≤ 1 for all $\nu \in \mathbf{N}(X, L)$. Since $|mL^\flat - f^{\flat *}H| \neq \varnothing$ by claim (*), there is an injection $\alpha \colon \mathcal{O}(H) \to f^\flat_* \mathcal{O}(mL^\flat)$. It induces injections $f^\flat_* \mathcal{O}(\nu L^\flat) \otimes \mathcal{O}(bH) \to f^\flat_* \mathcal{O}((\nu + mb)L^\flat)$ for all $b \in \mathbf{N}$, and one has a commutative diagram of induced maps

$$
\begin{array}{ccc}
H^0(Y^\flat, f^\flat_* \mathcal{O}(\nu L^\flat) \otimes \mathcal{O}(bH)) \otimes_k k(\bar{\eta}) & \subset & H^0(X^\flat, \mathcal{O}((\nu + mb)L^\flat)) \otimes_k k(\bar{\eta}) \\
\downarrow \beta & \circlearrowleft & \downarrow \gamma \\
f^\flat_* \mathcal{O}(\nu L^\flat) \otimes \mathcal{O}(bH) \otimes_{\mathcal{O}_{Y^\flat}} k(\bar{\eta}) & \subset & f^\flat_* \mathcal{O}((\nu + mb)L^\flat) \otimes_{\mathcal{O}_{Y^\flat}} k(\bar{\eta})
\end{array}
$$

Since $Q((X, L)) = Q(Y)$, Im γ is a $k(\bar{\eta})$-vector space of dimension ≤ 1. Since H is ample, β is surjective for $b \gg 1$. Thus $f^\flat_* \mathcal{O}(\nu L^\flat)$ has rank ≤ 1 and (i) is proved.

(ii): Since $\kappa(X'_{\bar{\eta}}, L'_{\bar{\eta}}) = 0$, we have $Q((X', L')) \subset Q(Y')$ by (1.8). We see that $Q(Y^\flat) = Q((X, L)) = Q((X', L'))$ by (1.5.i), whence the first assertion follows. If we further assume dim $Y' = \kappa(X, L)$, then $Q(Y')$ is algebraic over $Q(Y^\flat)$, and that $Q(Y') = Q(Y^\flat)$ since $Q(Y^\flat)$ is algebraically closed in $Q(X)$ (1.4). $\square$

Iitaka fibering reduces problems on Cartier divisors with $\kappa \geq 0$ to the ones on big Cartier divisors in a precise sense by the next theorem.

(1.12) THEOREM. *Let $f^\flat \colon X^\flat \to Y^\flat$ be the restriction of an algebraic fiber space to a quasiprojective base, and $L^\flat$ a Cartier divisor on $X^\flat$ with $\kappa(F, L^\flat|_F) = 0$, where F is the generic geometric fiber. Let $h' \colon Y^\# \to Y^\flat$ be a flattening of $f^\flat$, $X'' \to Y^\#$ the flat morphism induced by base change h' (1.10), $X^\#$ a resolution of X'', $f^\# \colon X^\# \to Y^\#$ the induced morphism, and $L^\#$ the induced Cartier divisor on $X^\#$. Then*

(i) *there exists $c \in \mathbf{N}(F, L^\flat|_F)$ such that $\mathcal{O}(M) = f^\#_* \mathcal{O}(cL^\#)$ is an invertible sheaf and the natural map $\mathcal{O}(\nu M) \to f^\#_* \mathcal{O}(c\nu L^\#)$ is an isomorphism for every $\nu \in \mathbf{N}$. In particular, $\kappa(Y^\#, M) = \kappa(X^\flat, L^\flat)$ if $Y^\flat$ is proper, and*

(ii) *if $f^\#_* \mathcal{O}(L^\#) \neq 0$ and fibers of $f^\#$ over codim 1 points of $Y^\#$ are all reduced, then c in (i) may be taken to be 1.*

PROOF. Let X''' the normalization of X'', and L''' the Cartier divisor on X''' induced by L. We have a commutative diagram of induced morphisms

$$\begin{array}{ccccc} X^\flat & \overset{g'}{\leftarrow} & X''' & \overset{g''}{\leftarrow} & X^\# \\ \downarrow f^\flat & & \downarrow f''' & & \\ Y^\flat & \overset{h'}{\leftarrow} & Y^\# & & \end{array}$$

with the map $f^\#$ from $X^\#$ to $Y^\#$.

Hence $\operatorname{codim}_{X'''} f'''^{-1}(y) = \operatorname{codim}_{Y^\#} y$ for all $y \in Y^\#$,

$$f_*^\# \mathcal{O}(mL^\#) = f_*''' g_*'' \mathcal{O}(mL^\#) = f_*''' \mathcal{O}(mL''')$$

has rank ≤ 1 ($m \in \mathbf{N}$) by $g_*'' \mathcal{O}_{X^\#} = \mathcal{O}_{X'''}$.

For $a \in \mathbf{N}(F, L^\flat|_F)$, choose a decomposition $aL''' \sim f'''^*(E_0) + D_1 + D_2$ such that E_0 is a Cartier divisor of $Y^\#$, and irreducible components of $D_1 \succ 0$ (resp. $D_2 \succ 0$) are dominating (resp. not dominating) over $Y^\#$. For each irreducible component E of $f'''(D_2)$, let $a_E \in \mathbf{Q}$ be the largest number such that $D_2 \succ a_E f'''^*(E)$. Let E_2 be the sum of $a_E E$ for all such E, and $b \in \mathbf{N}$ be such that bE_2 is a divisor. If E and $E' \subset Y^\#$ have no common components then so do $f'''^{-1}(E)$ and $f'''^{-1}(E')$, since $\operatorname{codim} f'''^{-1}(y) = \operatorname{codim} y$ for all $y \in Y^\#$. Thus $D_3 = bD_2 - f'''^*(bE_2)$ is effective, and there are no irreducible divisors Q of $Y^\#$ such that $\operatorname{Supp} f'''^* Q \subset \operatorname{Supp} D_3$. We note

$$abL''' = f'''^*(bE_0 + bE_2) + (bD_1 + D_3).$$

For an arbitrary $\nu \in \mathbf{N}$, an arbitrary affine open set V of $Y^\#$, and $\phi \in H^0(f'''^{-1}(V), \mathcal{O}(ab\nu L''')) = H^0(V, f_*''' \mathcal{O}(ab\nu L'''))$, we claim that

$$\phi \in H^0(V, \mathcal{O}(b\nu E_0 + b\nu E_2)).$$

Since $f_*''' \mathcal{O}(ab\nu L''')$ has rank 1 by $\kappa(F, L^\flat|_F) = 0$, one sees that $\phi \in Q(Y^\#)$, and the polar part $P \subset Y^\#$ of $\operatorname{div}_V(\phi) + (b\nu E_0 + b\nu E_2)|_V$ on V satisfies $f'''^* \prec \nu(bD_1 + D_3)_U$, where divisors are restricted to $U = f'''^{-1}(V)$. Each irreducible component of D_1 dominates $Y^\#$, whence we have $f'''^* P \prec \nu D_3$. Thus $P = 0$ by the property of D_3, and $\phi \in H^0(Y^\#, \mathcal{O}(b\nu E_0 + b\nu E_2))$ as claimed. Hence $f_*''' \mathcal{O}(ab\nu L''') = \mathcal{O}(b\nu E_0 + b\nu E_2)$. Thus $c = ab$ satisfies (i), and under the condition of (ii), we may take $a = 1$ and see $b = 1$ because $a_E \in \mathbf{Z}$ above. $\square$

Iitaka [12] originally defined D-dimension by the following estimate.

(1.13) COROLLARY. *Under the notation and assumptions of* (1.11), *there exist* $a, b, N > 0$ *such that*

$$a \cdot m^{\kappa(X,L)} \leq h^0(X, \mathcal{O}(mL)) \leq b \cdot m^{\kappa(X,L)}$$

for all $m \in \mathbf{N}(X, L)_{\geq N}$.

IDEA OF PROOF. First one can prove that L may be replaced by cL, where c is as given in (1.12.i) and we know that $\mathcal{O}(M) = f_*^\# \mathcal{O}(CL^\#)$ is a big invertible sheaf. Let H be an ample divisor on $Y^\#$. Then by (1.9), $|aH - M|, |aM - H| \neq \varnothing$ for some $a > 0$, whence $h^0(Y^\#, \mathcal{O}(mM)) = h^0(X, \mathcal{O}(mcL))$ is "comparable" with $h^0(Y, \mathcal{O}(mH))$ which is a polynomial of degree $\kappa(X, L)$ in m for $m \gg 0$. $\square$

Using Iitaka fiberings, we have a useful criterion for (1.7) to be an equality.

(1.14) PROPOSITION [**Ft1**, Proposition 1]. *Under the notation of* (1.6), *the following are equivalent*:

(i) $\kappa(X, L) = \kappa(X_{\bar{\eta}}, L_{\bar{\eta}}) + \dim Y$,

(ii) *there are* $n \in \mathbf{N}$, *an ample divisor* H *of* Y, *and an injection* $f^*\mathcal{O}(H) \subset \mathcal{O}(nL)$,

(iii) *the Iitaka model* I *of* (X, L) *dominates* Y,

(iv) *the Iitaka model* I *of* (X, L) *dominates* Y *and* $I_{\bar{\eta}}$ *is the Iitaka model of* $(X_{\bar{\eta}}, L_{\bar{\eta}})$.

In these cases, we say that L *dominates* Y.

PROOF. It is easy to see (ii) $\Rightarrow$ (iii) and (iv) $\Rightarrow$ (i). (iii) $\Rightarrow$ (ii) is proved by the argument for the claim $(*)$ in the proof of (1.11). It remains to show (i) $\Rightarrow$ (iii), (iii) $\Rightarrow$ (iv).

(i) $\Rightarrow$ (iii): By (1.6), one sees that $Q((X_{\bar{\eta}}, L_{\bar{\eta}}))$ is algebraic over $Q((X, L))$. Since $Q(Y) \subset Q((X_{\bar{\eta}}, L_{\bar{\eta}})) \cap Q(X)$ and $Q((X, L))$ is algebraically closed in $Q(X)$ (1.4), one has $Q(Y) \subset Q((X, L))$.

(iii) $\Rightarrow$ (iv): Choosing good models, we may assume that $g: X \to I$ is the Iitaka fibering such that $f = h \circ g$ for some morphism $h: I \to Y$. Since f has connected geometric fibers, so does h. By $\kappa(F, L|_F) = 0$ for the generic geometric fiber F of $g_{\bar{\eta}}: X_{\bar{\eta}} \to I_{\bar{\eta}}$ (hence, of g), $I_{\bar{\eta}}$ dominates the Iitaka model J of $(X_{\bar{\eta}}, L_{\bar{\eta}})$ (1.11.ii). By $\dim I_{\bar{\eta}} \leq \dim J$ (1.7), $I_{\bar{\eta}}$ is birational to J. $\square$

2. Kodaira dimension and Iitaka fiberings.

In this section, we simply reformulate the results in §1 for canonical divisors. All results are well known probably except for a slight refinement (2.6). We also refer the reader to standard [**U1**].

Let k be an algebraically closed field of characteristic 0. Let X be a nonsingular proper variety over k. Then the *mth plurigenus* (or *m-genus*) $P_m(X)$ is defined as $P_m(X) = h^0(X, \mathcal{O}_X(mK_X))$, and $\kappa(X) = \kappa(X, K_X)$ is called the *Kodaira dimension* of X. We also set $\kappa c(X) = \dim X - \kappa(X)$ and call it the *Kodaira codimension*. We recall the following.

(2.1) PROPOSITION. (i) *Let* $f: X \to Y$ *be a generically finite surjective morphism of nonsingular proper varieties. Then there is an effective divisor* D *on* X *such that* $K_X \sim f^*K_Y + D$ *and* $\operatorname{Supp} D = \{x \in X | f \text{ is not étale at } x\}$.

(ii) *Let* X *be a nonsingular divisor of a nonsingular variety* Y; *then* $\mathcal{O}_X(K_X) \simeq \mathcal{O}_X(K_Y + X)$.

(iii) *If* $f: X \to Y$ *is a proper algebraic fiber space, then one has* $\mathcal{O}_F(K_F) \simeq \mathcal{O}_F(K_X)$ *for every smooth fiber* F *of* f.

Using (2.1), one can reformulate results in §1 for canonical bundles.

(2.2) COROLLARY. *Let* X *and* X' *be nonsingular proper varieties over* k, *with a birational map* $f: X \dashrightarrow X'$. *Then*

(i) *if* f *is a morphism, then* $f_*\mathcal{O}(\nu K_X + B) = \mathcal{O}(\nu K_{X'})$ *for all* $\nu > 0$ *and effective* f-*exceptional divisor* B *of* X, *and*

(ii) $P_m(X) = P_m(X')$ $(m > 0)$, $R(X, K_X) \simeq R(X', K_{X'})$, $Q((X, K_X)) = Q((X', K_{X'})) \subset Q(X)$, $\mathbf{N}(X', K_{X'}) = \mathbf{N}(X, K_X)$, and $\kappa(X) = \kappa(X')$.

PROOF. By elimination of indeterminacy (1.10), (i) implies (ii). By (2.1.i), one has $\mathcal{O}(\nu K_{X'}) \subset f_* \mathcal{O}(\nu K_X + B)$ $(\nu > 0)$. Since f^{-1} is a morphism outside a closed set $B' \supset f(B)$ of codimension ≥ 2, $f_* \mathcal{O}(\nu K_X + B) = \mathcal{O}(\nu K_{X'})$ holds outside B', whence the equality holds on X' because $\mathcal{O}(K_{X'})$ is invertible. $\square$

Corollary (2.2.ii) allows us to define m-genus $P_m(X)$ $(m > 0)$ and $\kappa(X)$ for singular varieties. Indeed for an arbitrary nonsingular proper variety X' birational to X, $P_m(X) = P_m(X')$ $(m > 0)$ and $\kappa(X) = \kappa(X')$ make sense.

(2.3) COROLLARY. *Let X and Y be nonsingular projective varieties. Then*

(i) *Let $X \dashrightarrow Y$ be a dominating generically finite rational map. Then there is a natural injection $R(Y, K_Y) \to R(X, K_X)$, $\kappa(Y) \leq \kappa(X)$, and $\kappa c(Y) \geq \kappa c(X)$.*

(ii) *Let $f : X \to Y$ be a finite unramified morphism. Then $\kappa(X) = \kappa(Y)$.*

(iii) *(Easy addition) Let $f : X \to Y$ be an algebraic fiber space with generic geometric fiber F. Then $\kappa(X) \leq \kappa(F) + \dim Y$, i.e., $\kappa c(X) \geq \kappa c(F)$.*

PROOF. For (i), one may replace X, Y by their resolutions. Thus one may assume that $X \dashrightarrow Y$ is a morphism, and (2.1.i) proves (i). For (ii), one sees that $\kappa(X) = \kappa(X, f^* K_Y) = \kappa(Y)$ by $K_X \sim f^* K_Y$ and (1.5.ii). (iii) follows from (1.7) and (2.1.iii). $\square$

The following is a reformulation of (1.11) and (1.12) for the canonical divisor.

(2.4) THEOREM-DEFINITION. *Let X be a nonsingular proper algebraic variety over k with $\kappa(X) \geq 0$. Then the Iitaka fibering $f^b : X^b \to Y^b$ of (X, K_X) is called the Iitaka fibering of X. Then*

(i) $\dim Y^b = \kappa(X)$ *and* $\kappa(F^b) = 0$, *where F^b is the generic geometric fiber of f^b.*

(ii) *If a n.s. representative $f' : X' \to Y'$ (1.10) of a subfield $K \subset Q(X)$ is an algebraic fiber space such that the generic geometric fiber F' has $\kappa = 0$, then $K \supset Q(Y^b)$; i.e., f^b factors as $X' \xrightarrow{f'} Y' \dashrightarrow Y^b$. Thus the two properties in (i) determine f^b up to birational equivalence.*

(iii) *Let $f^\# : X^\# \to Y^\#$ be a n.s. representative of f^b such that $Y^\#$ flattens f^b (1.10). Then there exists $c \in \mathbf{N}(X, K_X)$ and a big Cartier divisor M on $Y^\#$ such that $\mathcal{O}(M) \subset f_*^\# \mathcal{O}(c K_{X^\#})$ and $H^0(Y^\#, \mathcal{O}(\nu M)) \simeq H^0(X^\#, \mathcal{O}(\nu c K_{X^\#}))$ for all $\nu \in \mathbf{N}$.*

(2.4.1) REMARK. It is expected that $\kappa(f^{b-1}(y))$ is constant in $y \in Y^{bo} = \{y \in Y^b | f^{b-1}(y) \text{ is smooth}\}$. If $\dim F^b \leq 2$, then this holds true by Iitaka [13]. To be precise, let $g : U \to V$ be a smooth projective morphism of nonsingular varieties. Iitaka [13] proved that plurigenera $P_n(U_v)$ are invariant for all $n \geq 0$ if $\dim U_v \leq 2$ using classification of surfaces. Levine [L2] proved the invariance of P_n if some pluricanonical system of U_V has a smooth member for all $v \in V$, and Nakayama [Ny1, Ny3] proved the invariance of P_n under the existence of a relative minimal model for $U \to V$ and a good minimal model for the generic geometric fiber of $U \to V$. However the answer is still open if $\dim F^\flat \geq 3$ or $\dim U_v \geq 3$ (cf. also [**U6**, p. 599]).

PROOF. Since $Q((X, K_X)) = Q((X^\flat, K_{X^\flat}))$ (2.2), $f^\flat$ is also the Iitaka fibering of $(X^\flat, K_{X^\flat})$. Thus (i) follows from (1.11.i) by $\kappa(F^\flat) = \kappa(F^\flat, K_{X^\flat}|_{F^\flat})$ (2.1.iii). Similarly for (ii), we see that $K \supset Q((X', K_{X'}))$ by (1.11.ii) and $\kappa(F', K_{X'}|_{F'}) = \kappa(F') = 0$ (2.1.iii), whence $K \supset Q((X, K_X)) = Q((X', K_{X'}))$ by (2.2) and (ii) is proved. Let us apply (1.12) to $f^\flat : X^\flat \to Y^\flat$ with $L^\flat = K_{X^\flat}$ and use the notation in the proof of (1.12). With M and c given in (1.12), we see that $\mathcal{O}(M) = f'''_* g'^* \mathcal{O}(cK_{X^\flat}) = f^\#_* (g' \circ g'')^* \mathcal{O}(cK_{X^\flat}) \subset f^\#_* \mathcal{O}(cK_{X^\#})$, and the rest follows from $H^0(X^\flat, \mathcal{O}(c\nu K_{X^\flat})) \simeq H^0(Y^\#, f'''_* g'^* \mathcal{O}(c\nu K_{X^\flat})) \simeq H^0(Y^\#, \mathcal{O}(\nu M))$ and (2.2). $\square$

Viehweg successfully managed to ignore f-exceptional divisors of X and codim 2 subsets of Y of an algebraic fiber space $f: X \to Y$ to study $f_*(\omega_X^{\otimes n})$. His idea is (2.2.i) and that $f_*(\omega_X^{\otimes n}(B)) \subset (f_* \omega_X^{\otimes n})^{**}$ for all effective f-exceptional divisors and the equality holds for some B, where $**$ denotes the double dual.

(2.5) PROPOSITION [**V6**, Lemma 7.3]. *Let $f: X \to Y$ be the restriction of an algebraic fiber space to a quasiprojective base, and $h: Y' \to Y$ a flattening of f (1.10.ii), $f': X' \to Y'$ the flat morphism induced by base change h, $\sigma: X'' \to X'$ the resolution of X', and let*

$$
\begin{array}{ccc}
X & \xleftarrow{\ g\ } & X'' \\
f \downarrow & & f'' \downarrow \\
Y & \xleftarrow{\ h\ } & Y'
\end{array}
$$

be the commutative diagram of induced morphisms. Then

(i) *an arbitrary effective f''-exceptional divisor B'' on X'' is g-exceptional. In particular, $H^0(\omega_{X''}^{\otimes n}(B'')) = H^0(\omega_X^{\otimes n})$ and $\kappa(\omega_{X''}^{\otimes n}(B'')) = \kappa(\omega_{X''})$ for all $n \geq 1$.*

(ii) $H^0(X'', \omega_{X''}^{\otimes n}) = H^0(Y'', (f''_* \omega_{X''}^{\otimes n})^{**})$ *for all $n \geq 1$.*

PROOF. The first part of (i) follows from the fact that such B'' is σ-exceptional by flatness of $X' \to Y'$, the rest of (i) follows from (2.2.i), and (ii) is a corollary to (i). $\square$

As (2.5.ii) shows, $(f''_* \omega_{X''}^{\otimes n})^{**}$ carries enough information, and there is essentially only one Cartier divisor to study by (2.6).

(2.6) THEOREM. *Let $f: X \to Y$ be the restriction of an algebraic fiber space to a quasiprojective base such that the generic geometric fiber F has $\kappa = 0$. Then*

(i) *there is $c \in \mathbf{N}(F, K_F)$ such that $\mathcal{O}(M) = [f_* \mathcal{O}(cK_X)]^{**}$ is an invertible sheaf and $\mathcal{O}(\nu M) \simeq [f_* \mathcal{O}(\nu c K_X)]^{**}$ for all $\nu > 0$, and*

(ii) *if furthermore $p_g(F) > 0$ and fibers of f over codim 1 points of Y are all reduced, then we may take c in (i) to be 1.*

PROOF. Deleting some codim 2 closed subset from Y, we may assume f is flat since we study the double dual of $\mathcal{O}_Y$-modules. We note $\mathbf{N}(F, K_X|_F) = \mathbf{N}(F, K_F)$ (2.1.iii). Thus we can apply (1.12) to $f: X \to Y$ and the Cartier divisor K_X with flattening id_Y. Thus there is $c \in \mathbf{N}(F, K_F)$ satisfying the required conditions. $\square$

(2.7) REMARK. Kollár has constructed an example of an algebraic fiber space $f: X \to Y$ with s.n.c. branching whose general fibers are elliptic curves such that (i) Y is a minimal surface and (ii) for an arbitrary birational morphism $Y' \to Y$ from a nonsingular projective surface, $f'_*(\omega^{5n}_{X'/Y'})$ is not locally free for any $n \in \mathbf{N}$.

3. Varieties generically finite over abelian varieties.
Kodaira dimension and Iitaka fibering are better understood for subvarieties of abelian varieties and varieties dominating and generically finite over abelian varieties. In this section, we will recall several results of this kind.

Let k be an algebraically closed field of characteristic 0.

(3.1) Let A be a nonsingular projective n-fold. We say that A is an *abelian variety* if it has a structure of a group variety [**Mf2**, §4]. When $k = \mathbf{C}$, A has a structure of a group variety iff A is a complex torus V/U, i.e., $V \simeq \mathbf{C}^n$ and U is a lattice, i.e., $U \otimes_{\mathbf{Z}} \mathbf{R} = V$, and it is known that a complex torus V/U is an abelian variety (or algebraic) iff there is a positive definite hermitian form H on V such that $\mathrm{Im}(H)$ is integral on $U \times U$ [**Mf2**, p. 35]. It is easy to see the canonical isomorphisms $V^* \otimes \mathcal{O}_A \simeq \Omega^1_A$ and $H^0(A, \Omega^p_A) \simeq \bigwedge^p H^0(A, \Omega^1_A)$ $(p \geq 0)$ for an abelian variety $A \simeq V/U$. Thus $\omega_A \simeq \mathcal{O}_A$, and abelian varieties are the typical varieties with $\kappa = 0$.

(3.2) Let X be a nonsingular projective variety. A morphism $f: X \to A$ to an abelian variety (up to translations by elements of A) induces $f^*: H^0(A, \Omega^1_A) \to H^0(X, \Omega^1_X)$, and there exists a morphism $\overline{f}: X \to \overline{A}$ such that $\overline{f}^*: H^0(\overline{A}, \Omega^1_{\overline{A}}) \xrightarrow{\sim} H^0(X, \Omega^1_X)$ and every $f: X \to A$ factors through $\overline{f}$. Indeed, one can construct $\overline{A}$ as a complex torus $H^0(X, \Omega^1_X)^*/H_1(X, \mathbf{Z})$ when $k = \mathbf{C}$, and its algebraicity follows from the Hodge structure of $H^1(X, \mathbf{Z})$. We call $\overline{A}$ the *Albanese variety* of X and $\overline{f}: X \to \overline{A}$ the *Albanese map* and denote these by $\mathrm{Alb}_X: X \to \mathrm{Alb}(X)$. From the description, we see $\dim \mathrm{Alb}(X) = q(X)$.

We prepare two easy lemmas on exterior algebras.

(3.3) Let K be an overfield of k, V a k-vector space of dimension d $(< \infty)$. Let $R = \bigoplus_i R_i$ be a k-subalgebra of the exterior K-algebra $(\bigwedge^{\cdot} V) \otimes_k K$ such that $R_0 = k$ and $\bigwedge^i V \otimes_k K \supset R_i \supset \bigwedge^i V$ for all i.

(3.3.1) LEMMA. *Assume that $R_e \subset (\bigwedge^e V) \otimes_k L$ for some $e \leq d$ and a k-vector subspace L of K. Then $R_t \subset (\bigwedge^t V) \otimes_k L$ for all $t \leq e$.*

PROOF. Let $v_1, \ldots, v_d$ be a basis of V. For $I = \{i_1, \ldots, i_j\}$ $(i_1 < \cdots < i_j)$, let $v_I = v_{i_1} \wedge \cdots \wedge v_{i_j}$. Then $R_e = \bigoplus_{|I|=e} Lv_I$. Let $w = \sum_{|J|=t} a_J v_J \in R_t$, where $a_j \in K$. Let $A \subset [1, d]$ be such that $|A| = t$. It is enough to show that $a_A \in L$. Let $A' \subset [1, d]$ be such that $A \cap A' = \varnothing$ and $|A'| = e - t$. Then from

$$w \wedge v_{A'} = \sum_{|J|=t} a_J v_J \wedge v_{A'} \in \bigoplus_{|I|=e} Lv_I,$$

one sees that $a_A \in L$. $\square$

(3.3.2) LEMMA. *Under the notation of* (3.3), *let* $v_1, \ldots, v_d$ *be a basis of* V
and $t_1, \ldots, t_d \in K$ *algebraically independent over* k, *and let* $w = t_1 v_1 + \cdots + t_d v_d$.
Let $S = \bigoplus S_i \subset (\bigwedge^{\cdot} v) \otimes_k k[t]$ *be the* k-*subalgebra generated by* $v_1, \ldots, v_d$, *and*
w. *Then*

   (i) $S_i = (\bigwedge^{i-1} V) \otimes_k kw \oplus \bigwedge^i V$ *for all* $i \le d$, *and*
   (ii) *if* $w \in R_1$ *and* $R_d = S_d$, *then* $R_i = S_i$ *for all* i.

PROOF. From the algebraic independence of $t_1, \ldots, t_d$, we see the exactness
of the Koszul complex

$$(*) \qquad\qquad \cdots \xrightarrow{\wedge w} \overset{i}{\bigwedge} V \otimes_k k[t] \xrightarrow{\wedge w} \overset{i+1}{\bigwedge} V \otimes_k k[t] \xrightarrow{\wedge w} \cdots .$$

Thus $S_i \cap \mathrm{Ker}(\bigwedge w) = S_i \cap \mathrm{Im}(\bigwedge w) \subset (\bigwedge^{i-1} V) \otimes_k k[t] \wedge w$ for all i, and hence we
see easily that $S_i = (\bigwedge^{i-1} V) \otimes_k kw \oplus \bigwedge^i V$ for all $i \le d$, which is (i). Assume
$w \in R_1$ and $R_d = S_d$. Since $S_d = (\bigwedge^d V) \otimes_k (k + kt_1 + \cdots + kt_d)$, we see
$R \subset (\bigwedge^{\cdot} V) \otimes_k (k + kt_1 + \cdots + kt_d)$ by (3.1.1). Thus $R \subset \bigwedge^{\cdot} V \otimes_k k[t]$ and
$R_i \cap \mathrm{Im}(\bigwedge w) = (\bigwedge^{i-1} V) \wedge w$, and hence $R_i = \bigwedge^i V + (\bigwedge^{i-1} V) \wedge w$ for all i, by
the exactness of $(*)$. $\square$

(3.4) COROLLARY (UENO [U1]). *Let* A *be an abelian variety of dimension*
n *and let* $f \colon X \to A$ *be a generically finite morphism from a nonsingular projec-*
tive variety X *of dimension* n. *Then* $h^{i,0}(X) \ge \binom{n}{i}$ *for* $i \in [1, n]$. *Furthermore,*
if $h^{i,0}(X) = \binom{n}{i}$ *for some* $i \in [1, n]$, *then* $h^{j,0}(X) = \binom{n}{j}$ *for all* $j \in [1, i]$.

PROOF. Since X is generically étale over A and $H^0(A, \Omega_A^{\cdot}) \otimes_k \mathcal{O}_A \simeq \Omega_A^{\cdot}$, one
has $H^0(X, \Omega_X^{\cdot}) \subset H^0(A, \Omega_A^{\cdot}) \otimes_k Q(X)$. Thus (3.4) follows from (3.3.1) in view
of

$$\overset{\cdot}{\bigwedge} H^0(A, \Omega_A^1) \subset H^0(X, \Omega_X^{\cdot}) \subset (\overset{\cdot}{\bigwedge} H^0(A, \Omega_A^1)) \otimes_k Q(X)$$

because $H^0(A, \Omega_A^{\cdot}) = \bigwedge^{\cdot} H^0(A, \Omega_A^1)$. $\square$

(3.5) COROLLARY (UENO [U1]). *Let* X *be a* d-*dimensional subvariety of*
an n-*dimensional abelian variety* A, *and* $\overline{X}$ *a resolution of* X. *Then one has*

   (0) $h^{i,0}(\overline{X}) \ge \binom{d}{i}$ *for all* $i \in [1, d]$ *and* $\kappa(X) \ge 0$. *Furthermore, the following*
are equivalent.

   (1) $h^{i,0}(\overline{X}) = \binom{d}{i}$ *for some* $i \in [1, d]$,
   (2) $p_g(\overline{X}) = 1$,
   (3) $P_m(\overline{X}) = 1$ *for some* $m > 0$,
   (4) $\kappa(X) = 0$, *and*
   (5) X *is a translate of some abelian subvariety of* A.

PROOF. Let $v_1, \ldots, v_n$ be a base of $H^0(A, \Omega_A^1)$ such that $v_1, \ldots, v_d$ form a free
base of Ω_X^1 on some open dense subset of X. Let $V = \bigoplus_{i=1}^d kv_i \subset H^0(A, \Omega_A^1)$.
Then

$$(3.5.1) \qquad\qquad \overset{\cdot}{\bigwedge} V \subset H^0(\overline{X}, \Omega_{\overline{X}}^{\cdot}) \subset (\overset{\cdot}{\bigwedge} V) \otimes_k Q(X).$$

and (0) follows from the first inclusion. From (0), it is easy to see (5) $\Rightarrow$ (4) $\Rightarrow$ (3) $\Rightarrow$ (2) $\Rightarrow$ (1), and it is enough to show (1) $\Rightarrow$ (5). Let us assume (1), that is, $H^0(\overline{X}, \Omega_{\overline{X}}^i) = \bigwedge^i V$ for some $i \in [1, d]$ (3.5.1). By applying (3.3.1) to (3.5.1), one sees that $h^{1,0}(\overline{X}) = d$. Then the Albanese variety $\mathrm{Alb}(\overline{X})$ has dimension d, and its image B in A contains some translate of X of dimension d. Hence X is a translate of B, and (5) follows. $\square$

(3.6) For an irreducible reduced closed subvariety X of an abelian variety A, the Iitaka fibering takes a very explicit form. We note that $\{a \in A | a + X \subset X\}$ is a closed set because it is $\mathrm{pr}_1(\mu^{-1}(X) \cap (A \times X))$, where $\mathrm{pr}_1, \mu : A \times A \to A$ are the projection to the first factor and the multiplication morphism, respectively. Thus the connected component $\{a \in A | a + X \subset X\}^\circ$ containing the origin is an abelian subvariety.

The following result by Ueno describes Kodaira dimension and Iitaka fibering for subvarieties of abelian varieties explicitly, and it is quite useful to analyze Albanese maps.

(3.7) THEOREM (UENO [**U1**]). *Let X be an irreducible reduced subvariety of an abelian variety A. Let $B = \{a \in A | a + X \subset X\}^\circ$. Then we have*

(i) *$X \to Y = X/B \subset A/B$ is an étale fiber bundle with fiber B and $X \times_Y Y' = Y' \times B$ for some finite étale covering Y' of Y,*

(ii) *$X \to Y$ is (birational to) the Iitaka fibering of X, and*

(iii) *Y is of general type.*

(3.7.1) REMARK. Kawamata [**Ka1**, Theorem 13] generalized (3.7) to a normal variety X dominating and finite over an irreducible subvariety of an abelian variety A using a weaker form of (3.12) (cf. also (11.5.3)).

PROOF. (i): It is clear that $X \to Y$ is a B-bundle. By Poincaré's complete reducibility [**Mf2**, p. 173], there is an abelian subvariety B' of A such that the induced map $B \times B' \to A$ is étale. Then $A \times_{(A/B)} B' \simeq B \times B'$, and hence $X \times_Y (Y \times_{(A/B)} B') \simeq B \times (Y \times_{(A/B)} B')$ because $X = A \times_{(A/B)} Y$. Thus (i) is proved.

(ii): Let $X \xleftarrow{g^\#} X^\# \xrightarrow{f^\#} Z$ be the Iitaka fibering of X (2.4). Since $\kappa(X_z^\#) = 0$ for "very" general $z \in Z$, $g^\#(X_z^\#)$ is a translate of an abelian subvariety $\ni 0$ of A by (3.5). Since abelian subvarieties $\ni 0$ of A are determined by the "sublattices," they do not have continuous moduli, and there is an abelian subvariety $C \ni 0$ of A such that $g^\#(X_z^\#)$ are translates of C. Thus $x + C \subset X$ for general $x \in X$, whence $X + C = X$ and $C \subset B$. Since general fibers of $X \to Y$ are abelian varieties $\simeq B$ (i), Y dominates Z by (2.4.ii). Thus $\dim X - \dim B = \dim Y \geq \dim Z = \dim X - \dim C$, and hence Y is birational to Z by $B \supset C$. Thus (ii) is proved.

(iii): By construction of Y (i), one has $\dim\{z \in A/B | z + Y \subset Y\} = 0$. Applying (ii) to $Y \subset A/B$, one sees that $\mathrm{id}_Y : Y \to Y$ ($\subset A/B$) is the Iitaka fibering and $\kappa(Y) = \dim Y$. $\square$

(3.8) Let X be a d-dimensional irreducible reduced subvariety of an n-dimensional abelian variety A. Then

$$\bigwedge^d H^0(A, \Omega_A^1) \otimes \mathcal{O}_X \simeq \Omega_A^d \otimes \mathcal{O}_X \to \Omega_X^d$$

induces a rational map $X \dashrightarrow \mathbf{P}(\bigwedge^d H^0(A, \Omega_A^1))$, called the *Gauss map*.

Kawamata [**KaV**, Theorem 5] proved the following when $\operatorname{codim} X = 1$.

(3.9) THEOREM (GRIFFITHS-HARRIS [**GHa**]). *If a subvariety X of an abelian variety A is of general type, then the Gauss map of X is generically finite. In particular, $|K_{\overline{X}}|$ induces a generically finite rational map, where $\overline{X}$ is a resolution of X.*

PROOF. We may assume $k = \mathbf{C}$ by the Lefschetz principle. Let $g: Z \to X$ be a resolution of the singularity and elimination of indeterminacy of the Gauss map, and let $f: Z \to Y$ be the Stein factorization of $Z \to \mathbf{P}(\bigwedge^d H^0(A, \Omega_A^1))$. Let Y_0 be an open dense subset of Y over which f is smooth. For $y \in Y_0$, $Z_y \to A$ induces a homomorphism $\alpha_y: \operatorname{Alb}(Z_y) \to A$ of abelian varieties. We claim that $\operatorname{Im} \alpha_y$ does not depend on $y \in Y_0$ and denote it by B ($\ni 0$). For the claim, it is enough to show that the kernel of

$$\alpha_y^*: H^0(A, \Omega_A^1) \to H^0(\operatorname{Alb}(Z_y), \Omega_{\operatorname{Alb}(Z_y)}^1) \simeq H^0(Z_y, \Omega_{Z_y}^1)$$

does not depend on $y \in Y_0$. It is clear that $\beta_y: H^1(A, \mathbf{Z}_A) \to H^1(Z_y, \mathbf{Z}_{Z_y})$ does not depend on $y \in Y_0$ by discreteness. Since

$$\operatorname{Ker} \alpha_y^* = \operatorname{Ker}(\beta_y \otimes \mathbf{C}) \cap H^0(A, \Omega_A^1)$$

by the Hodge decomposition, $\operatorname{Ker} \alpha_y^*$ is independent of y and the claim is proved. Since $Z_y \to A$ is generically finite, one has $\dim B \geq \dim Z_y$. Let Z_0 be the open dense subset of $f^{-1}(Y_0)$ such that $g|_{Z_0}$ is an embedding. For $z \in Z_0$, let

$$T_{X,g(z)}^{\perp} = \operatorname{Ker}[H^0(A, \Omega_A^1) \to \Omega_X^1 \otimes \mathbf{k}(g(z))].$$

Let $z' \in Z_{f(z)} \cap Z_0$. Then $T_{X,g(z)}^{\perp} = T_{X,g(z')}^{\perp}$ by the definition of Gauss map. Hence the image of $T_{X,g(z)}^{\perp}$ by $H^0(A, \Omega_A^1) \to \Omega_X^1 \otimes \mathbf{k}(g(z')) \to \Omega_{Z_{g(z)}}^1 \otimes \mathbf{k}(g(z'))$ is zero for all $z' \in Z_{f(z)} \cap Z_0$ and $\alpha_{f(z)}^*(T_{X,g(z)}^{\perp}) = 0$. Since $\alpha_{f(z)}^*$ factors as $H^0(A, \Omega_A^1) \twoheadrightarrow H^0(B, \Omega_B^1) \subset H^0(Z_{f(z)}, \Omega_{Z_{f(z)}}^1)$, $T_{X,g(z)}^{\perp}$ is sent to 0 by $H^0(A, \Omega_A^1) \to H^0(B, \Omega_B^1)$ for all $z \in Z_0$. This means $T_{X,g(z)} \supset T_{B+g(z),g(z)}$. Thus the projection $\pi: A \to A/B$ induces $X \to A/B$ whose tangent map has rank $\leq \dim X - \dim B$ at all $x \in g(Z_0)$. Thus $\dim \pi(X) \leq \dim X - \dim B$ and $X = \pi^{-1}\pi(X)$ because $\pi^{-1}\pi(X)$ is an irreducible set of dimension $\leq \dim \pi(X) + \dim B \leq \dim X$. Hence $B + X \subset X$ and $\dim B = 0$ by (3.7). Thus $\dim Z_y = 0$ for $y \in Y_0$ by $\dim Z_y \leq \dim B$. Hence the Gauss map is generically finite. $\square$

Theorem (3.9) gives us an alternate description of the Iitaka fibering of $X \subset A$.

(3.10) COROLLARY (KOLLÁR). *Under the notation of* (3.8), *the Stein factorization* $X \dashrightarrow Z$ *of the Gauss map of* X *is the Iitaka fibering of* X. *In particular, the Stein factorization of the map associated to* $|K_{\overline{X}}|$ *is the Iitaka fibering of* $\overline{X}$, *where* $\overline{X}$ *is the resolution of* X.

PROOF. Let $B = \{a \in A | a + X \subset X\}^{\circ}$ and $Y = \pi(X)$, where $\pi : A \to C = A/B$ is the projection. Let Y_0 be the open set of smooth points of Y, $X_0 = \pi^{-1}(Y_0)$, and $\pi_0 = \pi|_{X_0}$. Let $W = H^0(A, \Omega^1_{A/C})$. Then $\Omega^1_{X_0/Y_0} \simeq W \otimes_k \mathcal{O}_{X_0}$ and there is a natural commutative diagram with exact rows:

$$
\begin{array}{ccccccccc}
0 & \to & H^0(C, \Omega^1_C) \otimes_k \mathcal{O}_{X_0} & \to & H^0(A, \Omega^1_A) \otimes_k \mathcal{O}_{X_0} & \to & W \otimes_k \mathcal{O}_{X_0} & \to & 0 \\
& & \downarrow & & \downarrow & & \downarrow & & \\
0 & \to & \pi_0^* \Omega^1_{Y_0} & \to & \Omega^1_{X_0} & \to & W \otimes_k \mathcal{O}_{X_0} & \to & 0.
\end{array}
$$

Let $e = \dim Y_0$. The induced homomorphism $H^0(A, \Omega^{\cdot}_A) \otimes_k \mathcal{O}_{X_0} \to \Omega^{\cdot}_{X_0}$ of $\mathcal{O}_{X_0}$-exterior algebras sends the ideal generated by $\bigwedge^{e+1} H^0(C, \Omega^1_C) \otimes_k \mathcal{O}_{X_0}$ to 0 because $\bigwedge^{e+1} \Omega^1_{Y_0} = 0$. Thus the map $\bigwedge^d H^0(A, \Omega^1_A) \oplus_k \mathcal{O}_{X_0} \to \Omega^d_{X_0}$ factors as

$$
\bigwedge^d H^0(A, \Omega^1_A) \otimes_k \mathcal{O}_{X_0} \to \bigwedge^e H^0(C, \Omega^1_C) \otimes_k \det W \otimes_k \mathcal{O}_{X_0}
$$
$$
\to \pi_0^* \Omega^e_{Y_0} \otimes_k \det W \simeq \Omega^d_{X_0}.
$$

Hence the Gauss map $X_0 \to \mathbf{P}(\bigwedge^d H^0(A, \Omega^1_A))$ factors through the Gauss map $Y_0 \to \mathbf{P}(\bigwedge^e H^0(C, \Omega^1_C))$ of Y_0 as

$$
X_0 \to Y_0 \to \mathbf{P}(\bigwedge^e H^0(C, \Omega^1_C)) \subset \mathbf{P}(\bigwedge^d H^0(A, \Omega^1_A)).
$$

Since Y is of general type and $X \dashrightarrow Y$ is the Iitaka fibering of X (3.7), the Gauss map for Y is generically finite (3.9) and hence the Stein factorization of the Gauss map of X is the Iitaka fibering of X. □

Kawamata and Viehweg [**KaV**] used (3.9) with codim $X = 1$ to prove the following in the course of proving an important result [**KaV**, Main Theorem] which was a basis of (3.12).

(3.11) COROLLARY. [**KaV**, Theorem 1]. *Let* $\overline{X}$ *be a resolution of an irreducible reduced divisor* X *of an* n-*dimensional abelian variety* A *such that* X *is of general type. Then* $h^{i,0}(\overline{X}) \geq \binom{n}{i}$ *for all* $i \in [1, n-1]$. *If furthermore* $h^{n-1,0}(\overline{X}) = n$, *then* $h^{i,0}(\overline{X}) = \binom{n}{i}$ *for all* $i \in [1, n-1]$, *and* $\chi(\mathcal{O}_{\overline{X}}) = \pm 1$.

PROOF. Let $P \in X$ be a smooth point and let $\omega_1, \ldots, \omega_n$ be a basis of $H^0(A, \Omega^1_A)$ such that $\omega_n = t_1 \omega_1 + \cdots + t_{n-1}\omega_{n-1}$, where $t_i \in \mathcal{O}_{X,P}$. By (3.10), $t_1, \ldots, t_{n-1}$ are algebraically independent over k. Let $V = \bigoplus_{i=1}^{n-1} k v_i \subset H^0(A, \Omega^1_A)$. Then

$$
\bigwedge^{\cdot} V \subset H^0(\overline{X}, \Omega^{\cdot}_{\overline{X}}) \subset (\bigwedge^{\cdot} V) \otimes_k Q(X).
$$

By (3.3.2.i), we see $h^{i,0}(\overline{X}) \geq \binom{n}{i}$ for all $i \in [1, n-1]$. If $h^{n-1,0}(\overline{X}) = n$, then we see $h^{i,0}(\overline{X}) = \binom{n}{i}$ for all $i \in [1, n-1]$ by (3.3.2.ii), and we see $\chi(\mathcal{O}_{\overline{X}}) = (-1)^{\dim X}$. □

The following result by Kollár made results of Ueno [**U3**, Main Theorem, (1.2)] ($n = 3$) and Kawamata and Viehweg [**KaV**, Main Theorem] effective in the sense the condition $\kappa(X) = 0$ was replaced with an effective condition $P_4(X) = 1$.

(3.12) THEOREM [**Ko3**, Proposition 4.3]. *Let $f: X \to A$ be a generically finite surjective morphism from a nonsingular projective variety X with $P_4(X) = 1$ to an abelian variety A. Then X is birational to an abelian variety.*

COMMENT ON PROOF. Kollár obtained (3.12) by carefully improving the argument of Kawamata and Viehweg [**KaV**]. The idea is roughly as follows. Let $D = \bigcup_i D_i \subset A$ be the union of divisors in $\Delta(f)$. The essential point is to derive a contradiction from $D \neq \varnothing$.

(i) By $P_2(X) = 1$, one sees that D_i and D_j ($i \neq j$) are not translates of each other, and that the Iitaka model E_i of each D_i has $\chi(\mathcal{O}_{E_i}) = \pm 1$ using (3.11), and

(ii) by $P_4(X) = 1$, one can apply (i) to $X' = X \times_A A'$ with $P_2(X') = 1$, where A' is a suitable étale double cover of A. Then $D_i' = D_i \times_A A'$ is irreducible by (i), its Iitaka model E_i' is an étale double cover of E_i, and $\chi(\mathcal{O}_{E_i'}) = 2\chi(\mathcal{O}_{E_i})$, which contradicts $\chi(\mathcal{O}_{E_i'}), \chi(\mathcal{O}_{E_i}) = \pm 1$ (i).

It seems that (i) already asserts something strong. Therefore, it is possible that one can say something interesting under an even weaker assumption $P_2(X) = 1$ (cf. [**Ko3**, Remark 4.4 and Proposition 4.5]).

4. Covering constructions and base change theorems. We recall several important covering constructions in part I which are generalizations of the covering trick by [**BlG**], and recall Viehweg's base change theorems on $f_*\omega_{V/W}^n$ ($n \geq 1$) for an algebraic fiber space $f: V \to W$ in part II.

Part I. The following result by Fujita is simple but explains well the meaning of covering construction (cf. (6.3.ii)).

(4.1) LEMMA [**Ft1**, Lemma 1]. *Let Y be a nonsingular projective variety with $\kappa(Y) \geq 0$. Then there is a surjective morphism $f: X \to Y$ from a nonsingular projective variety X such that $\dim X = \dim Y$, $\kappa(X) = \kappa(Y)$, and $p_g(X) > 0$.*

A standard way to generalize a result on ω_X to $\omega_X \otimes L$ for an invertible sheaf L such that L^n is generated by global sections (for simplicity) is to identify $\omega_X \otimes L$ with a direct summand of $f_*\omega_Y$ for the cyclic n-ple cover $f: Y \to X$ of X branched along a smooth member of $|nL|$. This is a kind of "idealization" process. To be more precise, we consider the following situation.

Let X be a nonsingular quasiprojective variety, and let a **Q**-Cartier divisor D and a natural number n be such that $\mathrm{Supp}\langle D \rangle$ is a s.n.c. divisor and nD is a principal Cartier divisor (ϕ) ($\phi \in Q(X) - \{0\}$). Let $f: Y \to X$ be the normalization of X in $Q(X)(\sqrt[n]{\phi})$, and $d: Y' \to Y$ a resolution. Then

(4.2) CYCLIC COVERING LEMMA [**Es2**, §1]. *Y has only rational singularities and*

$$f_*d_*\mathcal{O}_{Y'} = \bigoplus_{i=0}^{n-1} \mathcal{O}(-\lceil iD \rceil) \quad and \quad f_*d_* \ \omega_{Y'} = \bigoplus_{i=0}^{n-1} \omega_X(\lceil iD \rceil).$$

Similarly, Kawamata has the following.

(4.3) KUMMER COVERING LEMMA [**Ka1**, Theorem 17]. *Let X be a nonsingular projective variety and D a reduced s.n.c. divisor. Let m_i be arbitrary natural numbers. Then there is a finite flat morphism $f: Y \to X$ from a nonsingular projective variety Y such that $Q(Y)/Q(X)$ is a Kummer extension, f^*D is a s.n.c. divisor, $f^*D_i = m_i E_i$ for some reduced divisors E_i.*

First applications are the unipotent reduction and the semistable reduction in codimension 1.

(4.4) Let $f: V \to W$ be an algebraic fiber space and B a divisor of W containing $\Delta(f)$. Let $W^\circ = W - B$, $V^\circ = f^{-1}(W^\circ)$, and $f^\circ = f|_{V^\circ}: V^\circ \to W^\circ$. Let k be the dimension of the general fiber of f. Then the monodromy of $R^k f_*^\circ \mathbf{C}_{V^\circ}$ around an arbitrary component of B is quasi-unipotent. We say that f has *unipotent monodromies* if these monodromies are all unipotent.

(4.5) UNIPOTENT REDUCTION LEMMA [**Ka1**, Corollary 18]. *Let $m \in \mathbf{N}$. If B is a s.n.c. divisor, then there exist a finite Kummer covering $g: W' \to W$ from a nonsingular W' and an algebraic fiber space $f': V' \to W'$ induced from f by base change g such that $g^{-1}(B)$ is a s.n.c. divisor containing $\Delta(f')$ and divisible by m, and f' has unipotent monodromies.*

Indeed we apply the Kummer covering lemma with $D = B$, $X = W$, and sufficiently divisible m_i's $\equiv 0$ (m) to get the assertion.

We note that unipotence of monodromies is independent of choice of resolution V' of $(V \times_W W')_{main}$ smooth over $W' - g^{-1}(B)$ since it is the property of $f^\circ \times_W g^{-1}(W^\circ)$.

We say that algebraic fiber space $f: V \to W$ is *semistable in codim 1* (*s.-s. in codim 1*, for short) if the fiber V_x over every codim 1 point $x \in W$ has reduced simple normal crossings. By Mumford's semistable reduction theorem, one has

(4.6) CODIM 1 S.-S. REDUCTION LEMMA [**V6**, Proposition 6.1]. *Under the notation of (4.4), let $m \in \mathbf{N}$ and assume that B is a s.n.c. divisor. Then there exist a finite Kummer covering $g: W' \to W$ from a nonsingular W' and an induced algebraic fiber space $f': V' \to W'$ such that $g^{-1}(B)$ is a s.n.c. divisor containing $\Delta(f')$ divisible by m, and f' is s.-s. in codim 1.*

The author learned from Kollár that

(4.6.1) REMARK. An algebraic fiber space with s.n.c. branching has unipotent monodromies if it is s.-s. in codim 1 by Katz [**Kz**, VII].

Let us mention the other application of (4.3).

(4.7) GALOIS COVERING LEMMA [**Ka1**, Corollary 19]. *Let* $g: Z \to W$ *be a finite morphism from a normal variety Z to a smooth quasiprojective variety W such that the branch locus B is a s.n.c. divisor. Then there is a finite Galois covering $W' \to W$ from a nonsingular W' factoring through g.*

The importance of the Codim 1 S.-S. Reduction Lemma comes from the good behavior of $f_*\omega^n_{V/W}$ for $f: V \to W$ under flat base change as explained in part II.

Part II. In later sections, we will see that $f_*(\omega^\nu_{V/W})$ plays a crucial role for C^+_{nm}, where $f: V \to W$ is an algebraic fiber space and $\nu > 0$. Our subject here is the behavior of $f_*(\omega^\nu_{V/W})$ under the base change, and we recall [**V6**, §3] in a slightly modified form.

First we note that $f_*(\omega^\nu_{V/W})$ does not depend on the choice of the nonsingular model V when the base W is fixed. To be precise:

(4.8) REMARK. If $f: V \to W$ and $f': V' \to W$ are algebraic fiber spaces such that V and V' are birational over W, then $f_*(\omega^\nu_{V/W}) \simeq f'_*(\omega^\nu_{V'/W})$ for all $\nu > 0$. Indeed if f factors through f' via a morphism $g: V' \to V$, then the remark follows from $g_*(\omega^\nu_{V/W}) \simeq \omega^\nu_{V'/W}$ (2.2.i). In the general case, we can find an algebraic fiber space $f'': V'' \to W$ factoring through both f' and f'' via birational morphisms, and the remark follows from the previous case.

To study the behavior of $f_*(\omega^\nu_{V/W})$, we start with

(4.9) LEMMA. *Let V' be an irreducible reduced Gorenstein variety and $d: V'' \to V'$ a resolution. Then for all $n > 0$, one has $\omega^n_{V'} \supset d_*\omega^n_{V''}$. If furthermore V' has only rational singularities, then one has $\omega^n_{V'} = d_*\omega^n_{V''}$ for all $n > 0$.*

(4.9.1) REMARK. We say that V' has *only rational singularities*, if V' is normal and $R^i d_*\mathcal{O}_{V''} = 0$ for all $i > 0$. The second part of (4.9) (and its proof) say that a Gorenstein singularity V' is rational iff it is canonical (cf. (9.4)).

PROOF. Let $B = \{b \in V'' | \dim d^{-1}(d(b)) > 0\}$, $V'^\circ = V' - d(B)$, $V''^\circ = d^{-1}(V'^\circ)$, and $d^\circ: V''^\circ \to V'^\circ$ the induced morphism. Then $\operatorname{codim}(V' - V'^\circ) \geq 2$. Since V' is Gorenstein, $\omega_{V'}$ (hence $\omega^n_{V'}$) is invertible and V' is Cohen-Macaulay by definition. Thus $\omega^n_{V'} = \iota_*\iota^*(\omega^n_{V'}) = \iota_*(\omega^n_{V'^\circ})$ ($\iota: V'^\circ \to V'$ is the inclusion), and it is enough to prove $\omega^n_{V'^\circ} = (d^\circ)_*\omega^n_{V''^\circ}$. Since d° is finite and birational one has $d^\circ_*\omega_{V''^\circ} \subset \omega_{V'^\circ}$ and $\omega_{V''^\circ} \subset d^{\circ*}\omega_{V'^\circ}$. Thus

$$(d^\circ)_*(\omega^n_{V''^\circ}) \subset (d^\circ)_*(\omega_{V''^\circ} \otimes d^{\circ*}\omega^{n-1}_{V'^\circ}) = (d^\circ_*\omega_{V''^\circ}) \otimes \omega^{n-1}_{V'^\circ} \subset \omega^n_{V'^\circ},$$

and the first assertion is proved. Assume that V' has only rational singularities; then one has $\omega_{V'} = d_*\omega_{V''}$ by duality. Thus $\omega_{V''} = d^*\omega_{V''}(D)$ for some effective exceptional divisor D of V'', and hence $d_*\omega^n_{V''} = \omega^n_{V'} \otimes d_*\mathcal{O}(nD) = \omega^n_{V'}$. □

(4.10) BASE CHANGE THEOREM [**V6**, §3]. *Let V, W, W' be nonsingular varieties and let $f: V \to W$ be a surjective projective morphism and $g: W' \to W$ be flat projective morphisms. Let $d: V'' \to V' = V \times_W W'$ be a resolution and* *let*

$$
\begin{array}{ccccc}
V & \xleftarrow{\ h\ } & V' & \xleftarrow{\ d\ } & V'' \\
f\downarrow & & f'\downarrow & \swarrow{\scriptstyle f''} & \\
W & \xleftarrow[\ g\ ]{} & W' & &
\end{array}
$$

be the commutative diagram of induced morphisms. Let $n > 0$. Then
  (i) *there is an inclusion*

$$
f''_*\omega^n_{V''/W'} \subset g^*[f_*(\omega^n_{V/W})].
$$

If g is smooth at $w' \in W'$ or if $g(w')$ is a codim 1 point of W such that $V_{g(w')}$ is a reduced normal crossing, then V' has only rational Gorenstein singularity at every point over w' and $f''_*\omega^n_{V''/W''} = f'_*\omega^n_{V'/W'} = g^*[f_*(\omega^n_{V/W})]$ at w'.
  (ii) *There is an inclusion*

$$
g_*f''_*\omega^n_{V''/W} \subset \{f_*(\omega^n_{V/W}) \otimes g_*(\omega^n_{W'/W})\}^{**}.
$$

*It is an equality at codim 1 point $w \in W$ if W'_w or V_w is a reduced normal crossing, where ** denotes the double dual.*

  PROOF. Let us see (i)$\Rightarrow$(ii). By (i), one has

$$
\begin{aligned}
g_*f''_*\omega^n_{V''/W} &= g_*f''_*(\omega^n_{V''/W'} \otimes f''^*\omega^n_{W'/W}) = g_*[(f''_*\omega^n_{V''/W'}) \otimes \omega^n_{W'/W}] \\
&\subset g_*[(g^*f_*\omega^n_{V/W}) \otimes \omega^n_{W'/W}] \subset [f_*(\omega^n_{V/W}) \otimes g_*(\omega^n_{W'/W})]^{**}
\end{aligned}
$$

by projection formula. Let $w \in W$ be a codim 1 point. Since (ii) is symmetric with respect to f and g, it is enough to check that the inclusion is actually an equality at w if V_w is a reduced normal crossing. Then by (i), we have the equality.

  (i): Since g is flat, V' is an irreducible reduced Gorenstein variety and $\omega_{V'/W'} = h^*\omega_{V/W}$ and hence $g^*f_*\omega^n_{V/W} = f'_*\omega^n_{V'/W'}$ by flat base change. By (4.8), one has $d_*\omega^n_{V''/W'} \subset \omega^n_{V'/W'}$ and $f''_*\omega^n_{V''/W'} \subset g^*f_*\omega^n_{V/W}$ which is the inclusion in (i). Let $w' \in W'$. In view of the argument to get the inclusion, it is enough to show V' has rational singularities along $g'^{-1}(w')$ to show that the equality holds at w'. If g is smooth at w', then V' is smooth along $f'^{-1}(w')$ and the equality is checked. Let $g''' : W''' \to W'$ be a birational morphism such that $(g \circ g''')^{-1}(g(w'))$ is normal crossing, and let

$$
\begin{array}{ccc}
V' & \xleftarrow{\ h'''\ } & V''' \qquad = V \times_W W''' \\
f'\downarrow & & f'''\downarrow \\
W' & \xrightarrow[\ g'''\ ]{} & W'''
\end{array}
$$

be the commutative diagram of induced morphisms. Then the analytic local equation of V''' at arbitrary points over $w' \in W'$ is

$$
x_1 \cdots x_n = y_1^{e_1} \cdots y_m^{e_m} \qquad (e_1, \ldots, e_m \geq 1),
$$

where $x_1, \ldots, x_n, y_1, \ldots, y_m$ are part of coordinates. It is easy to see that this is normal and toric, and hence it is a rational singularity [**KKMS**]. Thus V''' has only rational singularity along $(g''' \circ f''')^{-1}(w')$. Since f' is flat over w', one has

$$R^i h_*''' \mathcal{O}_{V'''} = f'^* R^i g_*''' \mathcal{O}_{W'''} = \begin{cases} 0 & \text{if } i > 0, \\ \mathcal{O}_{V'} & \text{if } i = 0 \end{cases}$$

over w' because W' is smooth. Thus V' has rational singularity over w' and one gets the equality at w'. $\quad\square$

Let $f \colon V \to W$ be an algebraic fiber space. Let $s \geq 1$ and let $V^s = V \times_W \cdots \times_W V$ (s times), let $V^{(s)}$ be an arbitrary resolution of the component of V^s dominating W, and $f^{(s)} \colon V^{(s)} \to W$ the induced morphism. Since $f_*^{(s)} \omega_{V^{(s)}/W}^k$ ($k \geq 1$) does not depend on the choice of resolution $V^{(s)}$ (4.8), this notation will not cause confusion.

(4.11) COROLLARY [**V6**, Lemma 3.5]. *Under the above notation, let $s, n > 0$. Then there is an injection*

$$(f_*^{(s)} \omega_{V^{(s)}/W}^n)^{**} \hookrightarrow (\overset{s}{\bigotimes} f_* \omega_{V/W}^n)^{**},$$

which is an isomorphism at a codim 1 point $w \in W$ such that V_w is a reduced normal crossing.

PROOF. Since deleting closed subsets of W of codimension ≥ 2 does not change the double dual of torsionfree sheaves, one may assume that $f^{(i)}$ ($i = 1, \ldots, s$) are flat and $f_* \omega_{V/W}^n$ is locally free on W. The previous proposition gives an injection

$$g \colon f_*^{(s)} \omega_{V^{(s)}/W}^n \to f_*(\omega_{V/W}^n) \otimes f_*^{(s-1)} \omega_{V^{(s-1)}/W}^n$$

such that g is an isomorphism at a point $w \in W$ of codimension 1 if V_w is a reduced normal crossing. This proves the assertion by induction on s. $\quad\square$

5. Positivity and vanishing theorems.

In part I, we review various results on positivity of direct images of relative dualizing sheaves and the application will be considered in later sections. In part II, we consider a special case of (5.5), where $f \colon X \to Y$ is an algebraic fiber space with generic geometric fiber of $\kappa = 0$. In this case, $f_*(\omega_{X/Y}^\nu)^{**}$ has a slightly more precise structure than being w.p.

Part I. We begin with recalling the basic notions of positivity. Weak positivity is due to Viehweg and bigness is due to Kawamata. We refer the reader to [**V7**, §3] or [**Ka10**, §5] for details.

(5.1) DEFINITION. A locally free sheaf G on a normal projective variety W is *semipositive* (*s.p.* for short) if, for each morphism g from a normal projective curve C to W, every quotient invertible sheaf of the pull back $g^* G$ has degree ≥ 0.

Let F be a torsionfree coherent sheaf on a nonsingular quasiprojective variety V. We note that there exists an open subset V° of V such that $\operatorname{codim}(V - V^\circ) \geq 2$ and $i^* F$ is locally free, where $i \colon V^\circ \to V$ is the inclusion map. Then tensor powers, symmetric powers, and determinant of such F are defined by $\hat{\otimes}^a(F) = i_* \otimes^a (i^*F)$, $\hat{S}^a(F) = i_* S^a(i^*F)$ $(a \in \mathbf{N})$, and $\det(F) = i_* \widehat{\det}(i^*F)$, where $\otimes^a$ (resp. S^a) denotes the ath tensor (resp. symmetric) power. It is easy to see that these do not depend on the choice of V°.

We say that F is *weakly positive* (or *w.p.* for short) if, for every $a > 0$ and every ample invertible sheaf H, there exists $b > 0$ such that $\hat{S}^{ab}(F) \otimes H^b$ is generically generated (1.9). We say that nonzero F is *big* if, for every ample invertible sheaf H, there exists $a > 0$ such that $\hat{S}^a(F) \otimes H^{-1}$ is w.p. It is easy to see that nonzero F is big iff, for every ample invertible sheaf H, there exists $a > 0$ such that $\hat{S}^a(F) \otimes H^{-1}$ is generically generated.

There are several different definitions of w.p. F: (i) in [V4] and [V6], $\hat{S}^{ab}(F) \otimes H^b$ is generated by global sections on an open dense subset independent of a and b, not only generically generated [V7, §3; Ka10, §5]. In this sense, Viehweg's w.p. was stronger than stated in (5.5), and (ii) in [Ka10, §5], F is nonzero coherent instead of being torsionfree [V4, V6, V7].

We also note that if V is projective and F is an invertible sheaf, then F is s.p. (resp. w.p., big in the above sense) iff F is nef (resp. pseudoeffective (11.3), big in the sense of §1).

W.p. or big sheaves behave like degenerate ample vector bundles and proofs of the following list of properties are similar to the proofs of the corresponding assertions for ample vector bundles.

(5.1.1) PROPERTIES. Let F be a nonzero torsionfree sheaf on a normal quasiprojective variety V. Then

(a) let $V^\circ \subset V$ be an open set. If F is w.p. (resp. big) then so is $F|_{V^\circ}$. If $\text{codim}(V - V^\circ) \geq 2$, then the converse also holds. (This is easy, but we may thus reduce problems to the locally free case.)

(b) If F is w.p. (resp. big) and $F \to G$ is a homomorphism which is surjective on an open dense subset, then so is G. (This again follows immediately from the definition.)

(c) If F and G are w.p. (resp. big), then so are $F \oplus G$, $F \otimes G/$ (max. torsion), and hence so are $\hat{S}^a(F), \widehat{\det}(F)$ by (b). (This is nontrivial and we refer the reader to [V7, Lemma 3.2.iii].)

(d) If $f: Y \to V$ is a finite surjective morphism, then F is w.p. (resp. big)$\Leftrightarrow$ f^*F is w.p. (resp. big) [V7, Lemma 3.2.ii].

(5.2) Let $f: X \to Y$ be an algebraic fiber space with a s.n.c. divisor B of Y such that $B \supset \Delta(f)$. Fujita [Ft3, Ft4] proved the semipositivity of $f_* \omega_{X/Y}$ when $\dim Y = 1$ by direct computation and applied it to some cases of $C_{n,1}$. Zucker [Zu] and Kawamata [Ka1] pointed out that the computation may be replaced by the theory of the variation of polarized Hodge structures which has been understood to be the main ingredient of C_{nm}^+. We should note that Fujita's result has been generalized to semipositivity (5.3) and weak positivity (5.5).

When f has unipotent monodromies, Kawamata [Ka1] proved that $f_* \omega_{X/Y}$ is the canonical extension [Sch] of $R^k f_*^\circ \mathbf{C}_{X^\circ}$ to Y to prove (5.3) for $i = 0$, where

$Y^\circ = Y - B$, $X^\circ = f^{-1}(Y^\circ)$, $f^\circ = f|_{X^\circ}$, and $k = \dim X - \dim Y$. Under the same assumption of unipotent monodromies, Kollár [Ko4] and Nakayama [Ny2] proved that $R^i f_* \omega_{X/Y}$ is the canonical extension of $R^{k+i} f^\circ \mathbf{C}_{X^\circ}$.

(5.3) THEOREM. *Let $f: X \to Y$ be an algebraic fiber space which has s.n.c. branching and unipotent monodromies. Then $R^i f_* \omega_{X/Y}$ is a s.p. locally free sheaf for all i.*

When f need not have unipotent monodromies, Viehweg [V5, Theorem 4.1] proved the local freeness (5.4) of $f_* \omega_{X/Y}$ by reducing it to (5.4) using the Unipotent Reduction Lemma (4.5). This applies to higher direct images, and [Ko4] and [Ny2] have the description of $R^i f_* \omega_{X/Y}$ as the upper canonical extension of $R^{k+i} f^\circ_* \mathbf{C}_{X^\circ}$.

(5.4) COROLLARY. *Let $f: X \to Y$ be a surjective projective morphism of quasiprojective varieties with s.n.c. branching. Then $R^i f_* \omega_{X/Y}$ is locally free for all $i > 0$.*

It should be mentioned that Moriwaki [Mwk1] generalized the above results (hence (5.3) and (5.4)) of [Ko4] and [Ny2] to the case of projective morphisms of complex spaces.

It is easy to see that s.p. and w.p. are equivalent for torsionfree sheaves on a nonsingular projective curve. This is why (5.5) is a generalization of a result of Fujita [Ft3, Ft4] ($\dim Y = 1$, $\nu = 1$). Kawamata [Ka3] proved (5.5) when $\dim Y = 1$ and Viehweg [V4] proved the general case.

(5.5) THEOREM [V4, Satz V]. *Let $f: X \to Y$ be a surjective projective morphism of quasiprojective varieties. Then $f_*(\omega^\nu_{X/Y})$ is w.p. for all $\nu > 0$ (cf. part II).*

Viehweg proved the case $\nu = 1$ by reduction to semipositivity (5.3) and reduced (5.5) to it by a mysterious argument [V6, §5] using a covering trick (4.2). The case $\nu = 1$ follows more easily from Kollár's vanishing (5.8), and Viehweg [V8, §5] gives a simpler proof of (5.5) from this viewpoint. Maehara [Mh] treated a "log version."

Kawamata and Viehweg independently generalized the Kodaira vanishing and the Ramanujam vanishing [Rm] (i.e., (5.6) with $n = 2$).

(5.6) THEOREM [Ka2, V5]. *Let D be a nef divisor on a nonsingular projective n-fold X such that $(D^n) > 0$. Then $H^i(X, \mathcal{O}(-D)) = 0$ for all $i < n$.*

IDEA OF KAWAMATA'S PROOF. By (1.9.1), there is an $m \in \mathbf{N}$ with $mD \sim H + E$ for some $E \succ 0$ and ample $H \succ 0$. He observes that (5.6) for X and D follows from the one for a blow-up (or finite flat cover) of X and the pull back of D. Thus one can reduce (5.6) to the case E is a s.n.c. divisor by blow-up, and then to the case $H = mL$ and $E = mbZ$ for some $b \in \mathbf{N}$, an ample L, and a reduced s.n.c. Z, by (4.3). One can assume $m = 1$ taking an m-ple covering. He further reduces it to case $b = 1$ observing that $H + (b - 1)Z$ is ample. Now

$H^i(X, \mathcal{O}(-H-Z)) = 0$ $(i < n)$ is a result of Norimatsu [**No**], easily reduced to the Kodaira vanishing by induction.

They further proved the "**Q**-version" of (5.6) by (4.3) (cf. (4.2)).

(5.7) THEOREM. *Let X be a nonsingular projective n-fold, and D a nef **Q**-divisor on X such that $(D^n) > 0$ and $\langle D \rangle$ is a s.n.c. divisor. Then*

$$H^i(X, \mathcal{O}([-D])) = 0 \quad \textit{for all } i < n.$$

The reader may find this a naive generalization of (5.6), but this has many important applications to problems in the theory of minimal models. We refer the reader to Kóllar's survey [**Ko6**, §2.4].

The following, (5.8) and (5.9), are the relativized vanishing theorems by Kollár. It is these that essentially made it possible to prove various problems in classification theory and the minimal model theory in a relative set up (projective morphisms to quasiprojective varieties).

(5.8) THEOREM [**Ko3**, Theorem 2.1]. *Let $f: X \to Y$ be a surjective morphism from a nonsingular projective variety X to a reduced projective variety Y with ample invertible sheaf L. Then*

(i) *$R^i f_* \omega_X$ is torsionfree for all $i \geq 0$, and*

(ii) *$H^j(Y, L \otimes R^i f_* \omega_X) = 0$ for all $j > 0$ and $i \geq 0$.*

He proved this together with the case $i = 0$ of (5.9); the proof is too complicated to outline here. The assertion (ii) was obtained earlier by Ohsawa [**O**] for Kähler X and $i = 0$. For further comments, we refer the reader to [**Ko6**, §2.5], and we only list related works [**EsV, Ka9, Ko4, Sai, Mwk1, Ny2, Ny3**].

Under the notation of (5.8), let s be a global section of M^k for an invertible sheaf M on Y and $k \in \mathbf{N}$. For any sheaf F, we have maps $\phi_s(F): F \to M^k \otimes F$, $H^i \phi_s(F): H^i(F) \to H^i(M^k \otimes F)$.

The following generalizes a result of Tankeev [**T**] which treated a general s in the case $a = k = m = j = 1$ and $i = 0$.

(5.9) THEOREM [**Ko4**, Theorem 3.6]. *If M^a is generated by global sections for some $a \in \mathbf{N}$, then*

$$H^j \phi_s(M^m \otimes R^i f_* \omega_X): H^j(M^m \otimes R^i f_* \omega_X) \to H^j(M^{m+k} \otimes R^i f_* \omega_X)$$

are injective for all $k, m > 0$ and $i, j \geq 0$.

Let us mention an important corollary of (5.8).

Let $f: X \to Y$ be a surjective morphism between nonsingular projective varieties. Then

(5.10) THEOREM [**Ko3**, Theorem 3.5]. *Let $\dim Y = k$ and let L be an invertible sheaf on Y so that $H^0(L)$ induces a generically finite rational map. Then*

(i) *$L^m \otimes \omega_Y \otimes \hat{\otimes}^s(f_* \omega_{X/Y})$ is generically generated (1.9) for all $s > 0$ if $m > k$ or if $m = k$ and Y is irrational, and*

(ii) *if Q is a torsionfree quotient of $L^m \otimes \omega_Y \otimes \hat{\otimes}^s(f_*\omega_{X/Y})$ and $p, q \in Y$ are sufficiently general points $(p \neq q)$ then*

$$H^0(Y, Q) \to H^0(\mathbf{k}(p) \otimes Q) \otimes H^0(\mathbf{k}(q) \otimes Q)$$

if $s > 0$ and $m > 1 + k$ or if $m \geq k$ and Y is irrational.

(5.10.1) REMARK. Under the notation of (5.10), let a $\mathbf{Q}$-divisor D of X and $n \in \mathbf{N}$ be such that $\mathrm{Supp}\langle D \rangle$ is a s.n.c. divisor and nD is a principal divisor. Then Esnault and Viehweg [**EsV**, Théorème 2.5] say, for instance, that the assertions (i) and (ii) remain valid after replacing $f_*\omega_{X/Y}$ with

$$f_*(\omega_{X/Y} \otimes \mathcal{O}(\lceil iD \rceil))$$

for each $i \in [0, n-1]$. The idea is to use the Cyclic Covering Lemma (4.2) to reduce the assertion to the one on a direct summand of $f_*\omega_{X/Y}$. Their result is stronger than stated here, and they use it to make the result (6.5) on C_{nm} effective [**EsV**, (2.8)] (cf. (8.4)).

We note that (ii) is vacuous if $\dim Y = 0$. To outline the proof, let us tentatively allow X and Y to be not connected when $\dim Y = 0$ so that (5.10) becomes (honestly) true when $k = 0$.

IDEA OF PROOF OF (5.10). We only need to study the case $s = 1$ by replacing X with $X^{(s)}$ (4.11). We concentrate on the essential case L is base point free and assume $m > k+1$ for simplicity. We will prove (5.10) by induction on k. Take a general $Y' \in |L|$ with smooth $X' = f^{-1}(Y')$. Then

$$0 \to L^{m-1} \otimes f_*\omega_X \to L^m \otimes f_*\omega_X \to L^{m-1} \otimes f_*\omega_{X'} \to 0$$

is exact, and $H^1(L^{m-1} \otimes f_*\omega_X) \hookrightarrow H^1(L^m \otimes f_*\omega_X)$ (5.9) by $m - 1 > 0$. Thus we have $H^0(L^m \otimes f_*\omega_X) \twoheadrightarrow H^0(L'^m \otimes f'_*\omega_{X'})$, where $f': X' \to Y'$. This is the induction step from k to $k - 1$. $\square$

The following result used in §6 follows from the above proof.

(5.11) COROLLARY. *Let $f: X \to Y$ and $g: Y \to Z$ be algebraic fiber spaces such that f has a s.n.c. branching. Let G, H be the generic geometric fibers of $g, g \circ f$ and $h: H \to G$ the induced map, and let M be a big invertible sheaf on Z generated by global sections and $L = g^*M$. Then*

(i) $H^0(Y, L^m \otimes \omega_Y \otimes (f_*^{(s)}\omega_{X^{(s)}/Y})) \twoheadrightarrow H^0(G, L^m \otimes \omega_Y \otimes (f_*^{(s)}\omega_{X^{(s)}/Y}) \otimes \mathcal{O}_G)$ *(cf. (4.11)) for all $s > 0$ if $m > \dim Z$, and*

(ii) *if $s \in \mathbf{N}$ is such that*

$$H^0(G, \omega_G \otimes h_*^{(s)}\omega_{H^{(s)}/G}) \to H^0(G, \omega_G \otimes h_*(\omega_{H/G}^s))$$

is nonzero, then $H^0(Y, L^m \otimes \omega_Y \otimes f_*(\omega_{X/Y}^s)^{**}) \neq 0$ for $m > \dim Z$.

Indeed the above proof shows (i), and (ii) follows from (i) since $\omega_Y \otimes \mathcal{O}_G = \omega_G$ and $(f_*^{(s)}\omega_{X^{(s)}/Y}) \otimes \mathcal{O}_G = h_*^{(s)}\omega_{H^{(s)}/G}$ by (2.1.iii).

Part II. We consider a special case (5.13) of (5.5), which hopefully explains the meaning of weak positivity.

(5.12) Let us consider an algebraic fiber space $f: X \to Y$ with the generic geometric fiber F of $\kappa(F) = 0$. Throughout this part II, we will always use these symbols in the above sense.

Then we study $f_*(\omega_{X/Y}^\nu)$ for $\nu > 0$ (cf. part II of §4). It is obvious that $f_*(\omega_{X/Y}^\nu) = 0$ if $\nu \notin \mathbf{N}(F, K_F)$, and that $[f_*(\omega_{X/Y}^\nu)]^{**}$ is an invertible sheaf if $\nu \in \mathbf{N}(F, K_F)$ and we denote by $D_{f,\nu}$ the associated Cartier divisor on Y. There exists $c \in \mathbf{N}(F, K_F)$ such that $D_{f,c\nu}/c\nu = D_{f,c}/c$ for all $\nu > 0$ (2.6.i).

(5.12.1) We consider the following property $(*)_f$.

PROPERTY $(*)_f$. For each $\nu \in \mathbf{N}(F, K_F)$, $\frac{1}{\nu} \cdot D_{f,\nu}$ is a sum of a nef $\mathbf{Q}$-divisor $P_{f,\nu}$ and a $\mathbf{Q}$-divisor $N_{f,\nu}$ with $[N_{f,\nu}] = 0$ such that no X_y for any codim 1 points y in $\operatorname{Supp} N_{f,\nu}$ are reduced s.n.c.

Though we do not know if f has this property, we have

(5.13) THEOREM. *There is a birational morphism $g: Y' \to Y$ from a non-singular projective Y' such that an arbitrary algebraic fiber space $f': X' \to Y'$ induced from f by base change g enjoys the property $(*)_{f'}$ (cf. (5.15.9)).*

(5.13.1) REMARK. More precisely it is shown in this part II that there is an effective divisor $B \, (\supset \Delta(f))$ of Y such that every birational morphism $g: Y' \to Y$ from a nonsingular projective Y' such that g^*B is a s.n.c. divisor works as g in Theorem (5.13).

(5.14) For a proper surjective morphism $g: Y' \to Y$ from a nonsingular Y', let $f': X' \to Y'$ be an arbitrary algebraic fiber space induced from f by base change g. Then

(5.14.1) LEMMA. *The property $(*)_{f'}$ implies $(*)_f$ if g is finite Galois and f' is semistable in codim 1.*

PROOF. Since $D_{f',\nu} = \nu \cdot P_{f',\nu}$ by $(*)_{f'}$ and is Galois group invariant, $P_{f',\nu}$ comes from a nef $\mathbf{Q}$-divisor $P_{f,\nu}$ on Y. Since Base Change Theorem (4.10.i) says $g^*D_{f,\nu} \supset D_{f',\nu}$ and X_y is not a reduced s.n.c. for any codim 1 point y in the support of $N_{f,\nu} = \frac{1}{\nu} \cdot D_{f,\nu} - P_{f,\nu} \succ 0$. It remains to show $[N_{f,\nu}] = 0$. We see that $h^*\omega_X^\nu = h^*(\omega_{X/Y}^\nu \otimes f^*\omega_Y^\nu) \subset \omega_{X'}^\nu = \omega_{X'/Y'}^\nu \otimes f'^*\omega_{Y'}^\nu$, where $h: X' \to X$ is the induced morphism. Hence we have $\frac{1}{\nu} \cdot D_{f',\nu} + K_{Y'/Y} \supset \frac{1}{\nu} \cdot D_{f,\nu}$ and $K_{Y'/Y} \supset g^*N_{f,\nu}$. We note

$$K_{Y'/Y} = g^* \sum_C \left(1 - \frac{1}{e_C}\right) C,$$

where C runs over all prime divisors on Y and e_C is the ramification index of g along an arbitrary divisor C' lying over C (which is independent of choice of C' since g is Galois). Hence $[N_{f,\nu}] = 0$. $\square$

(5.15) PROOF OF (5.13). We treat the special but essential case.

(5.15.1) LEMMA. *The property $(*)_f$ holds if $p_g(F) = 1$ and if f has a s.n.c. branching and is s.-s. in codim 1.*

PROOF. Indeed $D_{f,1}$ is a nef divisor by (5.3) and (4.6.1), and $D_{f,\nu} = \nu \cdot D_{f,1}$ for all $\nu > 0$ by (2.6.ii). $\square$

Let $f: X \to Y$ be an arbitrary algebraic fiber space as in (5.6). We start with a construction of an auxiliary algebraic fiber space over Y by taking a root of a section of an invertible sheaf (cf. (4.1)).

(5.15.2) SUBLEMMA. *There is an algebraic fiber space $h: V \to Y$ factoring through f via a generically finite surjective morphism $g: V \to X$ such that the generic geometric fiber H of h satisfies $p_g(H) = 1$ and $\kappa(H) = 0$.*

PROOF. Let $c \in \mathbf{N}$ be such that $\mathbf{N}(F, K_F) = c\mathbf{N}$ (1.3.1). Since $f_*(\omega_{X/Y}^c) \neq 0$, we may choose a very ample divisor L on Y such that

$$H^0(Y, f_*(\omega_{X/Y}^c) \otimes \mathcal{O}(cL)) = H^0(X, \mathcal{O}(K_{X/Y} + f^*L)^{\otimes c}) \neq 0.$$

Take a nonzero global section s of $\mathcal{O}(K_{X/Y} + f^*L)^{\otimes c}$, and we consider the covering of X associated to cth root $\sqrt[c]{s}$. By resolution, we get a generically finite morphism $g: V \to X$ from a nonsingular projective variety and let $h: V \to Y$ be the induced morphism. The generic geometric fiber H of h is obtained as a covering of F by taking the cth root of a nonzero section of $H^0(F, \mathcal{O}(cK_F))$. We leave the reader to see that H is connected by the choice of c and that $p_g(H) = 1$ and $\kappa(H) = 0$ (cf. (4.1)). $\square$

It is easy to see that the following claim (5.15.3) implies (5.13.1) and hence (5.13). Indeed an arbitrary effective divisor B of Y such that f and h are smooth over $Y - B$ works as the divisor B in (5.13.1) by (4.8). Hence it is enough to prove

(5.15.3) CLAIM. *If there is a reduced s.n.c. divisor B on Y such that f and h are smooth over $Y - B$, then f satisfies $(*)_f$.*

We will prove (5.15.3) in several steps.

(5.15.4) SUBLEMMA. *To prove (5.15.3), we may further assume that f is s.-s. in* codim 1.

PROOF. Let $\alpha: Y' \to Y$ be a finite surjective Galois morphism from a nonsingular variety and $f': X' \to Y'$ an algebraic fiber space induced from f by base change α such that $B' = \alpha^{-1}(B)$ is a s.n.c. divisor, f' is smooth over $Y' - B'$ and s.-s. in codim 1 (4.6). We choose $g': V' \to X'$ and $h' = f' \circ g'$ as in (5.15.3). Since $(*)_{f'}$ implies $(*)_f$ by (5.14.1), we may replace f, g, h, B by f', g', h', B'. $\square$

(5.15.5) Let $d \in \mathbf{N}$ be the least common multiple of all the ramification indices of g along divisors of V which are not g-exceptional. We consider the following commutative diagram

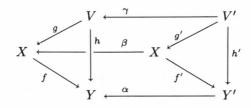

constructed as follows. Let $\alpha: Y' \to Y$ be a finite Galois morphism from a nonsingular variety and $h': V' \to Y'$ an algebraic fiber space induced from h by base change α so that $B' = \alpha^{-1}(B)$ is a s.n.c. divisor divisible by d and h' is smooth over $Y - B'$ and s.-s. in codim 1 (4.6). Let $X' = X \times_Y Y'$, and the rest are natural morphisms.

From this construction, we can observe

(5.15.6) SUBLEMMA. (i) *The singularities of X' over* codim 1 *points of Y' are at most rational Gorenstein and $\alpha^* \mathcal{O}(D_{f,\nu}) = [f'(\omega^\nu_{X'/Y'})]^{**}$ for all $\nu > 0$, and*

(ii) g' *is étale at the generic point z' of an arbitrary irreducible divisor Z' of V' which is not g'-exceptional and dominates a divisor of Y'.*

PROOF. (i) follows from (4.10.i) since f is s.-s. in codim 1 (5.15.4). (ii) is easily seen by the smoothness of $X'_{h'(z')}$ and $V'_{h'(z')}$ if $h'(Z') \not\subset B'$. If $h'(Z') \subset B'$, then (ii) follows because X' is normal at $g'(z')$ by (i) and d divides the ramification index of β at the codim 1 point $g'(z')$. $\square$

(5.15.7) Let us fix $\nu \in \mathbf{N}(F, K_F)$. Let $Y^\dagger$ be an open dense subset of Y with $\mathrm{codim}(Y' - Y^\dagger) \geq 2$ such that h' is flat over $Y^\dagger$, $h'_*(\omega^\nu_{V'/Y'}) = \mathcal{O}(D_{h',\nu})$ on $Y^\dagger$, the singularities of $f'^{-1}(Y^\dagger)$ are at most rational Gorenstein, and $\alpha^* \mathcal{O}(D_{f,\nu}) = f'_*(\omega^\nu_{X'/Y'})$ on $Y^\dagger$ (5.15.6.i). Let $X^\dagger$ be an open dense subset of $f'^{-1}(Y^\dagger)$ such that $\mathrm{codim}(f'^{-1}(Y^\dagger) - X^\dagger) \geq 2$ and g' is finite flat over $X^\dagger$. Let $V^\dagger = g'^{-1}(X^\dagger)$ and let $f^\dagger: X^\dagger \to Y^\dagger$, $g^\dagger: V^\dagger \to X^\dagger$, and $h^\dagger: V^\dagger \to Y^\dagger$ be the induced morphisms. We note that $g^{\dagger*} \omega_{X^\dagger/Y^\dagger} = \omega_{V^\dagger/Y^\dagger}(-W^\dagger)$ for an effective divisor $W^\dagger$ which is a sum of irreducible divisors dominating $Y^\dagger$ (5.15.6.ii).

Since h' has a s.n.c. branching and is s.-s. in codim 1 by the reduction steps, we see $(*)_{h'}$ by (5.15.1). Thus $(*)_f$ follows from the following.

(5.15.8) SUBLEMMA. $\alpha^* \mathcal{O}(D_{f,\nu}) = \mathcal{O}(\nu \cdot D_{h',1})$.

PROOF. It is enough to prove $f'_*(\omega^\nu_{X'/Y'}) = h'_*(\omega^\nu_{V'/Y'})$ on $Y^\dagger$ by (5.15.7), and hence it is enough to prove $f'_*(\omega^\nu_{X'/Y'}) \supset h'_*(\omega^\nu_{V'/Y'})$ on $Y^\dagger$ because $g'^* \omega_{X'/Y'} \subset \omega_{V'/Y'}$. We see that $f'_*(\omega^\nu_{X'/Y'})|_{Y^\dagger} = f^\dagger_*(\omega^\nu_{X^\dagger/Y^\dagger})$ by $\mathrm{codim}(f'^{-1}(Y^\dagger) - X^\dagger) \geq 2$, and obviously that $h'_*(\omega^\nu_{V'/Y'})|_{Y^\dagger} \subset h^\dagger_*(\omega^\nu_{V^\dagger/Y^\dagger})$ by $V' \supset V^\dagger$. Therefore we need to prove $f^\dagger_*(\omega^\nu_{X^\dagger/Y^\dagger}) \supset h^\dagger_*(\omega^\nu_{V^\dagger/Y^\dagger})$. Since $\kappa(H) = 0$, $\nu W^\dagger|_H$ is in the fixed part of $|\nu K_H|$, and

$$h^\dagger_*(\omega^\nu_{V^\dagger/Y^\dagger}) = h^\dagger_*(\omega^\nu_{V^\dagger/Y^\dagger}(-\nu W^\dagger)) = h^\dagger_* g^{\dagger*}(\omega^\nu_{X^\dagger/Y^\dagger}) = f^\dagger_* g^\dagger_* g^{\dagger*}(\omega^\nu_{X^\dagger/Y^\dagger})$$

by (5.15.7). Since $\mathcal{O}_{X^\dagger} \subset g^\dagger_* \mathcal{O}_{V^\dagger}$ globally splits, so do $\omega^\nu_{X^\dagger/Y^\dagger} \subset g^\dagger_* g^{\dagger*}(\omega^\nu_{X^\dagger/Y^\dagger})$ and $f^\dagger_*(\omega^\nu_{X^\dagger/Y^\dagger}) \subset h^\dagger_*(\omega^\nu_{V^\dagger/Y^\dagger})$. This implies $f^\dagger_*(\omega^\nu_{X^\dagger/Y^\dagger}) = h^\dagger_*(\omega^\nu_{V^\dagger/Y^\dagger})$ because both are of rank 1. $\square$

(5.15.9) REMARK. (i) We note that $D_{h',1}$ has a property slightly better than being nef (cf. (7.7.1)); therefore it may be useful to study $D_{f,\nu}$ in (5.15.3) (and hence $P_{f',\nu}$ in (5.13)).

(ii) When F is a surface or an abelian variety, H is either a $K3$ surface or an abelian variety. In this case, $D_{f,\nu}$ (and hence $P_{f',\nu}$ in (5.13)) is semiample. The following is the idea of the proof. We can blow up Y further in (5.15.3) and take a more ramified covering of Y in (5.15.4) so that h' induces a morphism of Y' to the Satake compactification $\overline{S}$ of a "fine moduli" S of H, where the boundary $\overline{S} - S$ is known to have codimension ≥ 2. There is a projective embedding of $\overline{S}$ by modular forms, and one can show that $\mathcal{O}_{\overline{S}}(1)$ is pulled back to some positive multiple of $D_{h',1}$ on Y' (cf. [U4] for the case of abelian varieties) using the fact that $\operatorname{codim}(\overline{S} - S) \geq 2$. Thus $D_{h',1}$ and hence $D_{f,\nu}$ are semiample.

6. $C_{n,m}$ with conditions on base.

(6.1) For an algebraic fiber space $f: V \to W$ with generic geometric fiber F, $\dim V = n$, and $\dim W = m$, Iitaka [I1] announced

CONJECTURE C_{nm}: $\kappa(V) \geq \kappa(W) + \kappa(F)$

to study classification of algebraic varieties; however people were rather pessimistic about C_{nm} though several special cases were known (e.g., [NU]), which was probably because the importance of algebraicity of the fiber space was not fully recognized and nonalgebraic counterexamples were known.

The tide was turned by Ueno [U2] who proved stronger $C_{2,1}^+$ (cf. §7) without using the classification of surfaces and Viehweg [V1] who proved $C_{n,n-1}^+$ (cf. §7), and both used the moduli theory of curves. Since then the vital importance of constructing the local, generic, or infinitesimal version of "moduli" for fibers (especially for C_{nm}^+) has been understood.

It should be mentioned that Ueno [U3] gave a classification of 3-folds X with $\kappa(X) = 0$ and $q(X) \geq 2$ using using $C_{n,n-1}^+$ above and $C_{.,1,0}^-$ [U3] (cf. (6.2)). (We should note that $C_{3,1}^+$ is needed for the case $q(X) = 1$.)

Fujita [Ft3, Ft4] proved the semipositivity of $f_*\omega_{V/W}$ when $\dim W = 1$ and applied it to some cases of $C_{n,1}$, which was generalized to various positivities as mentioned in §5 and to various results (6.2.c), (6.2.d), and (6.2.e) by Kawamata and Viehweg.

Since the results on C_{nm} and C_{nm}^+ are so much interrelated, let me mention three sets of results to be put together. Viehweg [V3] obtained $C_{3,.}^+$, which was necessary for the study of 3-folds with $\kappa(X) \leq 0$ and $q(X) > 0$ as a combination of $C_{n,n-1}^+$ above and $C_{3,1}^+$ [V3]. He [V6] also obtained $C_{n,n-2}$ as a combination of Kawamata's $C_{n,n-2}^+$ for the case $\kappa(F) = 0$ and $C_{n,n-2}$ for the case $\kappa(F) = 1$ [Ka4] and Viehweg's $C_{n,n-2}^+$ for the case $\kappa(F) = 2$ [V6]. This, in particular, implies $C_{4,.}$ combined with $C_{n,n-1}^+$ above and Kawamata's $C_{n,1}$ [Ka3].

In this section, we will try to understand the nature of C_{nm} by decomposing it into 3 questions, and recall the results on C_{nm} obtained under some conditions on base. Our viewpoint is different from Esnault's beautiful survey [Es1] in the sense that we do not want to strengthen C_{nm} in this section.

(6.2) Let $f: V \to W$ be an algebraic fiber space and $\bar{w}$ the generic geometric point of W. Let us formulate the following subconjectures:

$$C_{\cdot,\cdot,r} : \kappa c(W) \leq r \Rightarrow \kappa(V) \geq \kappa(V_{\bar{w}}) + \kappa(W),$$

and

$$C^-_{\cdot,\cdot,r} : \kappa(V) \geq 0, \ \kappa c(W) \leq r \Rightarrow \kappa(V) \geq \kappa(V_{\bar{w}}) + \kappa(W),$$

where $\cdot$ means an arbitrary value. The conditions on W under which the conjecture $C_{n,m}$ is proved are the following:

(6.2.a) $\kappa(W) = \dim W = 1$, $\kappa(V) \geq 0$ $(C^-_{\cdot,1,0})$ by Ueno [**U3**, 1977],

(6.2.b) $\kappa(W) = \dim W = 1$, $p_g(V_{\bar{w}}) > 0$ by Fujita [**Ft1**, 1977],

(6.2.c) $\kappa(V) \geq 0$, $\kappa c(W) = 0$ $(C^-_{\cdot,\cdot,0})$ by Kawamata [**Ka1**, 1981],

(6.2.d) $\kappa c(W) = 0$ $(C_{\cdot,\cdot,0})$ by Viehweg [**V4**, 1982].

(6.2.e) $\dim W = 1$ $(C_{\cdot,1,\cdot})$ by Kawamata [**Ka3**, 1982].

Among the three typical cases $\kappa(V) = \dim V, 0$, and $-\infty$, it is clear that $C_{n,m}$ holds true if $\kappa(V) = \dim V$. Then to understand the nature of $C_{n,m}$, let us consider the following 3 special cases of $C_{n,m}$ for $r \in \mathbf{N}$:

(6.2.1)$_r$ $\kappa(V) = -\infty$, $\kappa c(W) \leq r$? $\Rightarrow \kappa(V_{\bar{w}}) = -\infty$,

(6.2.2)$_r$ $\kappa(V) = 0$, $\kappa c(W) \leq r$? $\Rightarrow \kappa(W) = 0$, and

(6.2.3)$_r$ $\kappa(V) = \kappa(W) = 0$, $\dim W \leq r$? $\Rightarrow \kappa(V_w) = 0$.

We note that (6.2.1)$_r$ for all r is analogous to the assertion that if V is uniruled then so is W or $V_{\bar{w}}$ (cf. §11). (6.2.2)$_r$ is formulated in [**Ft2**, p. 253], and solved under the extra assumption that $dK_V \sim 0$ for some $d > 0$ [**Ft2**, Theorem 28]. Now (6.2.2)$_r$ is also easily proved under the assumption that V has a "good model" V' such that $K_{V'}$ is torsion. This is because $K_V - f^* K_W$ is pseudoeffective (11.3) by the weak positivity (5.5) of nonzero $f_* \omega^n_{V/W}$ ($n > 0$). (6.2.3)$_r$ is more delicate, but $C^+_{n,m}$ even seems to assert that f should be birational to an étale fiber bundle (cf. §§7, 10).

(6.3) REMARK. (i) It is obvious that these problems are on the "birational equivalence class" of f: if $f': V' \to W'$ is an algebraic fiber space such that there exist birational maps $V' \dashrightarrow V$ and $W' \dashrightarrow W$ fitting in the commutative diagram

$$\begin{array}{ccc} V' & \dashrightarrow & V \\ f' \downarrow & \circlearrowleft & \downarrow f \\ W' & \dashrightarrow & W \end{array}$$

then f satisfies C_{nm} (resp. (6.2.1)$_r, \ldots$) iff so does f'. This is because $\kappa(V) = \kappa(V')$, $\kappa(W) = \kappa(W')$, $\kappa(V_{\bar{w}}) = \kappa(V'_{\bar{w}'})$ (2.2.ii), where $\bar{w}'$ is the generic geometric point of w'.

(ii) For (6.2.2)$_r$ and (6.2.3)$_r$, one can replace the assumption $\kappa(V) = 0$ with "$P_t(V) = 1$ for all $t \geq 1$" by a covering trick (4.1). (6.2.1)$_r$ is hard in the sense that there seem to be no covering tricks available for reduction (except in [**Ka3**, p. 68]). In this sense, Viehweg's $Q_{n,m}$ in §7 seems to be a better formulation.

It is worthwhile to note that (6.2.1)$_r$, (6.2.2)$_r$, and (6.2.3)$_r$ for all r do imply the whole $C_{n,m}$. To be precise, one has

(6.4) THEOREM. *Let $r \in \mathbf{N}$. Assume that* $(6.2.2)_r$ *and* $(6.2.3)_r$ *hold true for all f. Then $C^-_{\cdot,\cdot,r}$ holds.*

(6.4.1) REMARK. (i) Kawamata [**Ka1**] proved $(6.2.2)_0$ first and proved $C^-_{\cdot,\cdot,0}$ (6.2.c) by essentially proving (6.4).

(ii) One merit of decomposing $C^-_{\cdot,\cdot,r}$ is the following: As $(6.2.3)_1$ is proved by Kawamata (6.2.e), $C^-_{\cdot,\cdot,1}$ follows from $(6.2.2)_1$, which is proved by Kollár by applying Kawamata's argument (iii) above and Kollár's vanishing. So $C^-_{\cdot,\cdot,1}$ holds true (cf. (6.6)).

IDEA OF PROOF. In view of (6.4.1.i), we only show steps of the proof of (6.4), and interested readers are advised to read [**Ka1**, pp. 265–266] for a full proof. For simplicity of notation, we denote by $F_{X/Y}$ the generic geometric fiber of an algebraic fiber space $X \to Y$. Let us fix an algebraic fiber space $p: X \to Y$.

Step 1. Assertion $(6.2.2)_r$ for all f implies

$(6.4.2)_r$ *Claim. If $\kappa(X) \geq 0$ and $\kappa c(Y) \leq r$, then the Iitaka model of X dominates the Iitaka model of Y.*

Step 2. Assertions $(6.2.2)_r$ and $(6.2.3)_r$ for all f imply

$(6.4.3)_r$ *Claim. If $\kappa(Y) = 0$, $\dim Y \leq r$, then $\kappa(X) \geq \kappa(F_{X/Y})$.*

Step 3. Assume that $\kappa(X) \geq 0$, $\kappa c(Y) \leq r$. Changing models, we may assume by $(6.4.2)_r$ that the Iitaka model V (resp. W) of X (resp. Y) fits in the commutative diagram of natural morphisms:

$$
\begin{array}{ccc}
X & \to & Y \\
\downarrow & & \downarrow \\
V & \to & W
\end{array}
$$

Then applying $(6.4.3)_r$ to $F_{X/W} \to F_{Y/W}$, we see $\kappa(F_{X/Y}) \leq \kappa(F_{X/W})$. Applying easy addition (2.3.iii) to $F_{X/W} \to F_{V/W}$, we see

$$\kappa(F_{X/W}) \leq \kappa(F_{X/V}) + \dim F_{V/W} = \dim F_{V/W} = \kappa(X) - \kappa(Y)$$

because $X \to V$ and $Y \to W$ are Iitaka fiberings. Hence $\kappa(F_{X/Y}) \leq \kappa(X) - \kappa(Y)$. $\square$

Viehweg proved the following result (6.2.d) by weak positivity.

(6.5) THEOREM [**V4**, Satz III]. *Let $f: V \to W$ be an algebraic fiber space such that $\kappa c(W) = 0$. Then $\kappa(V) \geq \kappa(V_{\bar{w}}) + \kappa(W)$.*

The reader will easily see that (6.5) follows from (6.5.1) whose proof is quite similar. We will show (6.5.1).

(6.5.1) PROPOSITION. *Let $p: V \to W$ be an algebraic fiber space with $\kappa(W) \geq 0$. Let $f: X \to Y$ be a n.s. representative of p such that Y flattens p (6.3.i) and such that the Iitaka fibering of W is represented by an algebraic fiber space $g: Y \to Z$. Let F, G, H be the generic geometric fibers of $f, g, g \circ f$, and $h: H \to G$ the induced algebraic fiber space. If $h_*(\omega^s_{H/G})$ is big for some $s > 0$, then ω_X dominates Y (1.14) and hence*

$$\kappa(V) = \kappa(F) + \dim W.$$

PROOF. Let M be an ample invertible sheaf on Z and $a > 0$ be such that $g^*M \subset \omega_Y^a$ (1.14), and let D be an ample invertible sheaf on Y. We note that $Q((X, \omega_X^n(B))) = Q((X, \omega_X))$ for all $n > 0$ and all effective f-exceptional divisors B of X (2.5), because Y flattens p. Since $f_*(\omega_{X/Y}^s)|_G$ is big, $D^{-1} \otimes \hat{S}^b(f_*(\omega_{X/Y}^s))|_G$ is generically generated for some $b > 0$, and therefore $g^*M^c \otimes D^{-1} \otimes \hat{S}^b(f_*(\omega_{X/Y}^s))$ is generically generated as a sheaf on Y (1.9). Since $f_*(\omega_{X/Y}^s)$ is w.p., $D^d \otimes \hat{S}^{de}(f_*(\omega_{X/Y}^s))$ is generically generated for some $e \gg 0$ and $d > 0$ depending on e. Hence we see

$$(6.5.2) \qquad f^*D \subset f^*\omega_Y^{ac} \otimes \omega_{X/Y}^{sb}(B_1),$$

$$(6.5.3) \qquad \mathcal{O}_X \subset f^*D^d \otimes \omega_{X/Y}^{sde}(B_2)$$

for some effective f-exceptional divisors B_1 and B_2 (cf. the remark before (2.5)). Thus from $(6.5.2)^{\otimes(d+1)} \otimes (6.5.3)$, we see

$$f^*D \subset f^*\omega_Y^{ac(d+1)} \otimes \omega_{X/Y}^{sb(d+1)+sde}((d+1)B_1 + B_2).$$

Now $e \gg 0$ is chosen so that $se > 2ac$; then $sb(d+1) + sde > sde > 2acd \geq ac(d+1)$, whence

$$f^*D \subset \omega_X^{sb(d+1)+sde}((d+1)B_1 + B_2)$$

and ω_X dominates Y (1.14). $\square$

We finish this section by proving $(6.2.2)_1$ by an argument similar to the one used by Kawamata to prove $(6.2.3)_1$. We will use (7.2) which however does not cause trouble since (6.6) and (6.6.1) are not used later.

(6.6) PROPOSITION (KOLLÁR, MORI). $(6.2.2)_1$ holds.

(6.6.1) COROLLARY. $C_{\cdot,\cdot,1}^-$ holds.

We only prove (6.6) since it implies (6.6.1) as in (6.4.1.ii).

PROOF. By (6.3.ii), let $f: X \to Y$ be an algebraic fiber space with $p_g(X) \geq 1$ and $\kappa c(Y) = 1$. By replacing f with its n.s. representative $f': X' \to Y'$ such that Y' flattens f, and the Iitaka fibering of Y' is represented by an algebraic fiber space $Y' \to Z'$, we may assume that there is an algebraic fiber space $g: Y \to Z$ which is the Iitaka fibering and $\kappa(\omega_X^n(B)) = \kappa(\omega_X)$ for all $n > 0$ and all effective f-exceptional divisors B on X. Let F, G, H be the generic geometric fibers of $f, g, g \circ f$ and $h: H \to G$ be the induced algebraic fiber space. Let M be an ample invertible sheaf on Z generated by global sections and $L = g^*M$. There exists $a > 0$ such that $L \subset \omega_Y^a$ (1.14). By (6.5.1), we know that $\kappa(X) \geq \kappa(F) + \kappa(Y)$ if there is $s \in \mathbf{N}$ such that $h_*(\omega_{H/G}^s)$ is big. Hence we may assume that $h_*(\omega_{H/G}^s)$ is not big for any $s > 0$. If $f_*(\omega_{H/G}^s)$ contains a big subsheaf for some $s > 0$, we can easily see that $f_*(\omega_{H/G}^s)$ contains an ample invertible sheaf for some $s \gg 0$ and $f_*(\omega_{H/G}^s)$ is big for some $s \gg 0$ by (7.2). Thus $f_*(\omega_{H/G}^s)$ does not contain big subsheaves for any $s > 0$. Since G is an elliptic curve, this means that each indecomposable component E of $f_*(\omega_{H/G}^s)$ has degree ≤ 0 (cf. [Ka3, Lemma

10]). Thus $f_*(\omega_{H/G}^s)$ is a direct sum of indecomposable vector bundles of degree 0, because $f_*(\omega_{H/G}^s)$ is s.p. (5.5). Our aim is to check the condition in (5.11.ii) for some $s \gg 0$ using Kawamata's argument in [**Ka3**, §3]. Since $p_g(X) \geq 1$ and G is an elliptic curve, $h_*\omega_{H/G}$ is a nonzero s.p. locally free sheaf (5.5). Furthermore, since $h^0(h_*\omega_{H/G}) \neq 0$, we see that $h_*\omega_{H/G}$ has a direct summand $E \simeq \mathcal{O}_G$ [**Ka3**, p. 69, lines 26–30]. Thus for all $s \gg 0$, $E^s \simeq \mathcal{O}_G \subset h_*\omega_{H/G}$ is a direct summand similarly. Under the notation of (5.11). $A^{(s)} = h_*^{(s)}\omega_{H^{(s)}/G}$ is s.p. (5.5) and hence the injection $A^{(s)} \to \otimes^s(h_*\omega_{H/G})$ is isomorphic by $\deg h_*\omega_{H/G} = 0$. Hence $A^{(s)} \supset E^s \simeq \mathcal{O}_G$ is a direct summand and $A^{(s)} \to h_*(\omega_{H/G}^s)$ induces a nonzero map $H^0(A^{(s)}) \to H^0(h_*(\omega_{H/G}^s))$. Thus the condition in (5.11.ii) is satisfied for all $s > 0$. Hence

$$H^0(X, f_*(L^m \otimes \omega_Y) \otimes \omega_{X/Y}^s(B_s)) = H^0(Y, L^m \otimes \omega_Y \otimes f_*(\omega_{X/Y}^s)^{**}) \neq 0$$

for $s \gg m > \dim Z$ and effective f-exceptional divisors B_s of X by (5.11.ii). Then $f_*\omega_Y^{ma+1} \otimes \omega_{X/Y}^s(B_s) = \omega_X^s(B_s) \otimes f^*\omega_Y^{ma+1-s}$ has a nonzero global section $(s > ma + 1)$ and $\kappa(Y) \leq \kappa(\omega_X^s(B_s)) = \kappa(X)$. $\square$

7. $C_{n,m}^+$ with conditions on fibers.

(7.1) Let $f: V \to W$ be an algebraic fiber space. Let $\bar{w}$ be the generic geometric point of W. Let $K \supset k$ be an algebraically closed field contained in $k(\bar{w})$ such that there is a nonsingular projective variety T defined over K such that $T \otimes_K k(\bar{w})$ and $V_{\bar{w}}$ are birational. The minimum of $\operatorname{tr.deg}_k K$ for all such K is called the *variation* of f and denoted by $\operatorname{Var}(f)$ (cf. [**Ko5**] for better treatment). One has $0 \leq \operatorname{Var}(f) \leq \dim W$. Viehweg strengthened $C_{n,m}$ ($n = \dim V$, $m = \dim W$) to

CONJECTURE $C_{n,m}^+$: $\kappa(V) \geq \kappa(V_w) + \operatorname{Max}\{\kappa(W), \operatorname{Var}(f)\}$ if $\kappa(W) \geq 0$.

The importance of $\omega_{V/W}$ in Viehweg's approach is summarized as follows:

(7.2) THEOREM. *Let $f: V \to W$ be an algebraic fiber space.*

(i) *The following conditions on f are equivalent.*

(a) *$\exists n > 0$ s.t. $f_*(\omega_{V/W}^{nk})$ is big $\forall k \geq 1$,*

(a') *$\exists n > 0$ s.t. $f_*(\omega_{V/W}^n)$ is big,*

(b) *$\exists n > 0$, $\exists$ big divisor H on W s.t. $\mathcal{O}(H) \subset \hat{S}^1(f_*(\omega_{V/W}^n))$,*

(c) *$\exists$ effective divisor B on V s.t. $\operatorname{codim} f(B) \geq 2$ and $\omega_{V/W}(B)$ dominates W (1.14),*

(ii) (a) *above implies the following condition.*

(d) *$\exists n > 0$ $\widehat{\det} f_*(\omega_{V/W}^n)$ is big.*

Furthermore, if the fibers of f over codim 1 points of W are all reduced simple normal crossings, then (d) $\Leftrightarrow$ (a).

(iii) *Let W' be a nonsingular projective variety over W such that $g: W' \to W$ is generically finite and dominating, V' a desingularization of $V \times_W W'$, and $f': V' \to W'$ the induced morphism. If one of the conditions* (a)–(d) *holds for f', then so does the corresponding condition for f.*

COMMENT ON PROOF. (i): It is clear that (a) $\Rightarrow$ (a') and that (b) $\Leftrightarrow$ (c) by definition of dominance of $\omega_{V/W}(B)$ over W. (a') $\Rightarrow$ (b) follows from the nonzero map $\hat{S}^m[f_*(\omega^n_{V/W})] \to \hat{S}^1[f_*(\omega^{nm}_{V/W})]$. (b) $\Rightarrow$ (a) is proved in [**V7**, (3.4)] by the technique used to prove weak positivity of $f_*\omega^n_{V/W}$ (if it is nonzero).

(ii) is proved in [**V7**, Theorem 3.5].

(iii): By (5.1.1.a), we may replace W by an open subset W° of W such that $\text{codim}(W - W^\circ) \geq 2$ and g is flat on $g^{-1}(W^\circ)$ and V, V', W' by preimages of W° by $f, g \circ f', g$. Thus we may assume that g is flat to prove (iii). Since $f'_*\omega^n_{V'/W'} \to g^*f_*\omega^n_{V/W}$ is generically surjective, (a') (resp. (d)) for f' implies the bigness of $g^*f_*\omega^n_{V/W}$ (resp. $g^*\widehat{\det}f_*\omega^n_{V/W}$). We are done by (5.1.1.d). $\square$

(7.3) Viehweg [**V7**, §3] proposed the following question $Q_{n,m}$ ($n = \dim V$, $m = \dim W$) or $Q(f)$ when f is given.

QUESTION $Q_{n,m}$ (or $(Q(f))$): Let $f: V \to W$ be an algebraic fiber space such that $\text{Var}(f) = \dim W$. Then the equivalent conditions (a)–(c) in (7.2) hold.

It is quite important that one may replace W with its generically finite covering by (7.2.iii) and hence $Q(f)$ is essentially a problem on the generic geometric fiber $V_{\bar{w}}$ of f.

To explain the importance and application of this formulation to $C^+_{n,m}$, we need the following procedure.

(7.4) Let $p: X \to Y$ be an algebraic fiber space, with $\bar{y}$ the generic geometric point of Y. Let us apply (2.5) to p to get

$$\begin{array}{ccc} X & \xleftarrow{\alpha} & V \\ p \downarrow & & f \downarrow \\ Y & \xleftarrow{\beta} & W \end{array}$$

where f is an algebraic fiber space, α and β are birational, and all f-exceptional divisors are α-exceptional. Let $\bar{w}$ be the generic geometric point of W. By definition of $\text{Var}(f)$, there is a finitely generated subfield K'' of $k(\bar{w})$ and a variety F defined over K such that $V_{\bar{w}}$ is birational to $F \otimes_{K''} k(\bar{w})$. Then for some finite extension K' ($\supset K''$) of $Q(W)$, $V \times_W \text{Spec}\, K'$ is birational to $F \otimes_{K''} K'$. Then there are nonsingular projective varieties W', W'', V', V'' such that $Q(W') = K'$, $Q(W'') = K''$, and V' (resp. V'') is an algebraic fiber space over W' (resp. W'') such that $V' \times_{W'} \text{Spec}\, Q(W') \sim V \times_W \text{Spec}\, K'$ and $V'' \times_{W''} \text{Spec}\, Q(W'') \sim F$ and fitting in the commutative diagram

$$\begin{array}{ccccccc} X & \xleftarrow{\alpha} & V & \xleftarrow{h} & V' & \xrightarrow{h'} & V'' \\ p \downarrow & & f \downarrow & & f' \downarrow & & f'' \downarrow \\ \cdot Y & \xleftarrow{\beta} & W & \xleftarrow{g} & W' & \xrightarrow{g''} & W'' \end{array}$$

Furthermore, we can replace $W' \to W''$ by a morphism induced by a base change using Viehweg's (2.5) and Codim 1 S.-S. Reduction Lemma (4.6), so that we may assume that (i) every g''-exceptional divisor of W' is g-exceptional, and (ii) g'' is s.-s. in codim 1.

(7.5) THEOREM ([**V6**, (7.4) and **Ka9**, Theorem 1.1] for refinement in (i)). *Assume that $Q(f'')$ holds true. Then one has*

(i) $\kappa(\omega_{X/Y} \otimes p^*M) \geq \kappa(X_{\bar{y}}) + \mathrm{Max}\{\mathrm{Var}(p), \kappa(Y, M)\}$ *for every Cartier divisor M on Y such that $\kappa(Y, M) \geq 0$, and*

(ii) $C^+_{n,m}$ *holds for p.*

IDEA OF PROOF. (ii) follows from (i) with $M = K_Y$, and we treat (i). Let $M = \beta^* M$. By construction, there are open subsets $W_0 \subset W$ and $W_0'' \subset W''$ such that $\mathrm{codim}(W - W_0) \geq 2$, $\mathrm{codim}(W'' - W_0'') \geq 2$, $g^{-1}W_0 \subset g''^{-1}W_0''$, f and g (resp. f'' and g'') are flat over W_0 (resp. W_0''). Let $W_0' = g^{-1}W_0$, and $g''W_0' \subset W_0''$. By Viehweg's (2.5) and the Base Change Theorem (4.10), one easily obtains

(7.5.1) SUBLEMMA. *For an arbitary effective divisor B' of V' such that $B' \subset f'^{-1}(W' - W_0')$, one has*

$$\kappa(X, \omega_{X/Y} \otimes p^*N) \geq \kappa(V, \omega_{V'/W'}(B') \otimes f'^*g^*N).$$

Thus the theorem is reduced to the following claim.

(7.5.2) CLAIM. *There is an effective divisor B' of V' such that $B' \subset f'^{-1}(W' - W_0')$ and*

$$\kappa(\omega_{V'/W'}(B') \otimes f'^*g^*N) \geq \kappa(V_{\bar{w}}) + \mathrm{Max}\{\mathrm{Var}(f), \kappa(W, N)\}.$$

PROOF OF CLAIM. Since $Q(f'')$ holds, there are $n > 0$ and an effective divisor $B'' \subset V''$ such that $B'' \subset f''^{-1}(W'' - W_0'')$ and $f''_* \omega^n_{V''/W''}(B'')$ contains a big invertible sheaf. Since

$$f'_* \omega^n_{V'/W'}|_{W_0'} \supset g''^* f''_* \omega^n_{V''/W''}(B'')|_{W_0'}$$

by (4.10.i) applied to g'' which is s.-s. in codim 1, there is an effective divisor $B' \subset f'^{-1}(W_0 - W_0')$ such that $f'_* \omega^n_{V'/W'}(B') \supset g''^* f''_* \omega^n_{V''/W''}(B'')$. Thus $\omega^n_{V'/W'}(B')$ dominates W'' and

$$\begin{aligned}
\kappa(\omega_{V'/W'}(B') \otimes f'^*g^*N) &= \kappa(\omega^n_{V'/W'}(B') \otimes f'^*g^*N) \\
&= \kappa(\omega^n_{V'/W'}(B') \otimes f'^*g^*N) \\
&= \dim W'' + \kappa(W'_{\bar{w}''} \times V''_{\bar{w}''}, \mathrm{pr}_1^* g^* N_{\bar{w}''} \otimes \mathrm{pr}_2^* \omega_{V''_{\bar{w}''}}) \\
&= \dim W'' + \kappa(W'_{\bar{w}''}, (g^*N)_{\bar{w}''}) + \kappa(V_{\bar{w}})
\end{aligned}$$

because $V'_{\bar{w}''} = W'_{\bar{w}''} \times V''_{\bar{w}'}$ and $\kappa(V''_{\bar{w}''}) = \kappa(V_{\bar{w}})$, where $\bar{w}''$ is the generic geometric point of W'' and pr_i is the ith projection. Since $\dim W'' = \mathrm{Var}(f)$ and

$$\kappa(W, N) = \kappa(W', g^*N) \leq \dim W'' + \kappa(W'_{\bar{w}''}, (g^*N)_{\bar{w}''})$$

by (1.14) and easy addition (1.7), one obtains

$$\kappa(\omega_{V'/W'}(B') \otimes f'^*g^*N) \geq \kappa(V_{\bar{w}}) + \mathrm{Max}\{\mathrm{Var}(f), \kappa(W, N)\}.$$

Thus (7.3) is proved. $\square$

(7.6) So far, $Q_{n,m}$ (and hence $C_{n,m}^+$) are proved in the following cases (however I doubt if I exhausted all the results):

(A) $\dim V_{\bar{w}} = 1$ [**V1**],

   $V_{\bar{w}}$ is an abelian variety [**U4**],

   $V_{\bar{w}}$ is of dim ≤ 2 and of general type [**V6**],

(B) $V_{\bar{w}}$ has a smooth model F with torsion K_F [**Ka4**],

   $V_{\bar{w}}$ has a smooth model F with big semiample K_F [**V7**],

   $V_{\bar{w}}$ has a good minimal model (cf. (9.7)) [**Ka10**], which in particular, implies $Q_{n,n-2}$,

(B′) $V_{\bar{w}}$ is of general type [**Ko5**].

(A)   : Results in (A) are based on the moduli theory, and thus one cannot hope to directly generalize these to solve $Q_{n,m}$.

(B)   : The approaches (B) and (B′) are based on the curvature calculation of the Hodge metric of $f_*\omega_{V/W}$. It was first done by Griffiths [**G**], and first done in this context by Fujita.

(7.6.1) THEOREM (FUJITA [**Ft3**, **Ft4**], ZUCKER [**Z**], [**Ka4**, Theorem 3]). *If $f : V \to W$ is s.-s. in codim 1 and has a s.n.c. branching, then the metric of $f_*\omega_{V/W}$ induced by Hodge structure is semipositive outside the discriminant locus D ($\subset W$) of f and, though it may degenerate (or blow up) along D, it does not affect the integral formula for $(\det f_*\omega_{V/W})^m$.*

(7.6.2) REMARK. Kollár [**Ko5**] pointed out that the original estimate of [**Ka4**] on the degeneration of a metric near D has an error, which is fixed by [**Ko5**] using [**CKS**, 5.30] and also independently by Kawamata using [**Ks**].

   The connection of bigness of $f_*\omega_{V/W}$ and the infinitesimal Torelli was first pointed out by Kawamata [**Ka4**] whose idea goes back to Griffiths [**G**].

(7.6.3) THEOREM [**Ka4**, Theorem 3]. *Assume that there is a point $x \in W$ such that f is smooth above x and the map*

$$T_{W,x} \xrightarrow{\delta_x} H^1(V_x, T_{V_x}) \xrightarrow{\lambda_x} \mathrm{Hom}(H^0(V_x, \Omega_{V_x}^{n-m}), H^1(V_x, \Omega_{V_x}^{n-m-1}))$$

is injective, where δ_x is the Kodaira-Spencer map, λ_x is the infinitesimal period map induced from the cup product, and $V_x = f^{-1}(x)$. Then $\det f_*\omega_{V/W}$ is big.

   Indeed if curvature is strictly positive at some general point of W, then $\det f_*\omega_{V/W}$ is big by (7.6.1). The injectivity of the map at x in (7.6.3) means strict positivity of curvature at x.

   The approach (B) reduces the problem to (7.6.3) by a technical but mysterious covering trick (cf. §4, part I), and a modification of the vanishing theorem in [**LWP**] was used to prove infinitesimal Torelli.

   We should note that if the minimal model conjecture (cf. (9.5)) is settled affirmatively for the generic geometric fiber of f, then $Q(f)$ holds true by [**Ka10**].

   (B′): Approach (B′) is a variation of (B) and reduces $Q(f)$ to the following instead of infinitesimal Torelli in approach (B).

(7.6.4) THEOREM [Ko5, I]. *Assume that* $\dim W = 1$, $|K_{V_{\bar{w}}}|$ *induces a birational map, and* $\deg f_* \omega_{V/W} = 0$. *Then* $\text{Var}(f) = 0$.

The theorem (7.6.4) is proved by a simple argument involving flat vector bundles using Fujita's semipositivity of $f_* \omega_{V/W}$ and Viehweg's weak positivity of $f_* \omega_{V/W}^k$ ($k \geq 2$).

Roughly speaking the problem Q in case (B′) is first reduced to a problem Q for a codim 1 s.-s. reduction $f: V \to W$ such that $|K_{V_{\bar{w}}}|$ induces a birational map by the covering trick again. Then $L = \det f_* \omega_{V/W}$ is a nef invertible sheaf on W.

We would like then to say that we could find a curve $C \subset W$ passing through a general point such that $(L \cdot C) = 0$ when $(L^m) = 0$, so that it would reduce the problem to (7.6.4). Because "moduli" of fibers should not stay constant along C if a suitable moduli theory exists, and the "generic" moduli of fibers of f is constructed in [Ko5, II] based on the idea of Q-varieties of Matsusaka.

Obviously such C does not exist as an algebraic curve, but it *does* as a local analytic curve [Ko5, Theorem 4.4], since L carries an induced semipositive metric. This is the idea of (B′).

(7.7) REMARK. (i) In view of (B′), it does not seem unrealistic to think that $\kappa_c(V_{\bar{w}})$ measures the difficulty of $Q(f)$. Therefore it seems interesting to try to prove $Q(f)$ under the assumption that the generic geometric fiber of the Iitaka fibering of $V_{\bar{w}}$ has a good minimal model.

(ii) While $f_* \omega_{V/W}$ has a beautiful interpretation in terms of Variation of Hodge structures, no direct interpretation of $f_* \omega_{V/W}^k$ ($k \geq 2$) seems available at least for the moment. To study a problem involving $f_* \omega_{V/W}^k$, we usually study the case $k = 1$ using the variation of Hodge structures, and then reduce the general case to the case of $k = 1$ by a technical yet mysterious covering trick (§4). Viehweg's weak positivity and the above approaches (B) and (B′) fall in this category. Therefore, it is highly desirable to find some interpretation of $f_* \omega_{V/W}^k$ to find a new or direct approach to the problem (if possible at all).

8. Effective results in classification of 3-folds.
In this section, we would like to review several effective results on 3-folds.

Ueno [U3] proved (8.1) and (8.2) for 3-folds X with $\kappa(X) = 0$, which was the first breakthrough in the classification of 3-folds. Kawamata and Viehweg [KaV] proved (8.1) for X with $\kappa(X) = 0$ and Kawamata [Ka1, Corollary 2] proved (8.2) for X with $\kappa(X) = 0$. These will be mentioned again in §10.

(8.1) THEOREM [Ko3, Theorem 5.1]. *Let X be a nonsingular projective n-fold with $P_{n+\alpha}(X) = 1$ for some $\alpha \geq 1$. Then the image of an arbitrary morphism $\pi: X \to A$ to an abelian variety is an abelian variety, and hence $q(X) \leq n$.*

(8.2) THEOREM [Ko3, Theorem 5.2]. *Let X be a nonsingular projective n-fold and assume that $n \geq 3$. Then X is birational to an abelian variety iff $q(X) = n$ and $P_{n+\alpha}(X) = 1$ for some $\alpha \geq 1$.*

We modified the statements following [**Ko3**, Remark 5.4].

(8.3) IDEA OF PROOF OF (8.1). We describe the steps.

Step 1. Using (3.7), it is enough to derive a contradiction assuming that $\pi(X)$ is a variety of general type and of dimension $k > 0$. (This idea goes back to Ueno [**U1**]).

Step 2. Changing models birationally, let $f: X \to Y$ be a n.s. representative of $\pi: X \to \pi(X)$ such that $P_\nu(X) = h^0(Y, \hat{S}^1 f_*(\omega_X^\nu))$ for all $\nu > 0$ (2.5). This condition is maintained if we replace X with its blow-up with a nonsingular center (4.8).

Step 3. We note $|K_Y|$ induces a generically finite morphism and in particular $P_1(Y) \geq 2$. Thus f is not generically finite (2.3.i) because $P_1(X) \leq 1 < P_1(Y)$. That is, $n > k > 0$. (The argument is the same as [**KaV**] up to here, and they [**KaV**] derived a contradiction by $C_{\cdot,\cdot,0}$ (6.5) by the assumption $\kappa(X) = 0$.)

Now we want to apply (5.10.1) to (a blow up of) f with $L = \omega_Y$, and we have to choose D. Let us make a too easy choice and see how things will go. Let $a = n + \alpha$.

Step 4. Replacing X with its blow-up, we may assume $|aK_X| \neq \varnothing$, the base locus of $|aK_X|$ is a s.n.c. divisor B, and its movable part has a smooth member H such that codim $H \cap B \geq 2$ and $H \cup B$ is a s.n.c. divisor. Let $D = K_X - (B + H)/a$. Then

$$K_{X/Y} + \lceil (a-1)D \rceil = aK_X - f^*K_Y - \left[\frac{a-1}{a}B \right] \succ H - f^*K_Y,$$

and $f_*(\omega_{X/Y} \otimes \mathcal{O}(\lceil (a-1)D \rceil)) \neq 0$. Let $\mathcal{G}$ be the image of

$$\omega_Y^m \otimes \omega_Y \otimes \hat{\otimes}^s f_*(\omega_{X/Y} \otimes \mathcal{O}(\lceil (a-1)D \rceil)) \to \omega_Y^{m+1-s} \otimes f_*\mathcal{O}(saK_X)^{**}.$$

Since Y is irrational, (5.10.ii) with Remark (5.10.1) shows $h^0(\mathcal{G}) \geq 2$ if $m \geq n$ and $s > 0$. If $s \geq m + 1$, then $\mathcal{G} \subset f_*\mathcal{O}(saK_X)^{**}$ and we obtain $P_{sa}(X) \geq 2$ by Step 2; however sa is of order $(n+1)^2$ and too big.

Let us choose a better D, using the fact that $f_*(\omega_{X/Y}^a)$ is nonzero and w.p. (5.5).

Step 5. By definition, for some $b \gg 0$ and $c = c(b) > 0$, we see that $\hat{S}^{bc}(f_*\omega_{X/Y}^a) \otimes \omega_Y^c$ is generically generated. Replacing X with its blow-up, we may assume that (i) the base locus of $|abcK_{X/Y} + cf^*K_Y + abcE|$ is a s.n.c. divisor B' and its movable part has a smooth member H' such that codim $H' \cap B' \geq 2$ and $H' \cup B'$ is a s.n.c. divisor for some effective f-exceptional divisor E, (ii) the base locus of $|aK_F|$ is a s.n.c. divisor $\overline{B}_F$, where F is the generic geometric fiber of f, and (iii) $F \cdot B' \subset bc\overline{B}_F$ by generic generatedness of $\hat{S}^{bc}(f_*\omega_{X/Y}^a) \otimes \omega_Y^c$. Setting

$$D' = K_{X/Y} + \frac{1}{a \cdot b}K_Y + E - \frac{1}{abc}(B' + H'),$$

we see

$$f_* \mathcal{O}(K_{X/Y} + \lceil (a-1)D' \rceil)^{**}$$

$$= f_* \mathcal{O}\left(aK_{X/Y} + (a-1)E + \lceil \frac{a-1}{a \cdot b}K_Y - \frac{a-1}{abc}(B' + H') \rceil \right)^{**}$$

$$= f_* \mathcal{O}\left(aK_{X/Y} + \lceil \frac{a-1}{a \cdot b}K_Y - \frac{a-1}{abc}B' \rceil \right)^{**}$$

$$\subset f_* \mathcal{O}\left(aK_{X/Y} + \lceil \frac{a-1}{a \cdot b}K_Y \rceil \right)^{**} = f_* \mathcal{O}(aK_{X/Y})^{**}$$

by $b \gg 0$. Then by (iii), we similarly obtain

$$K_F + \lceil (a-1)D'|_F \rceil = aK_F - \left[\frac{a-1}{abc}B'|_F\right] \supset aK_F - \overline{B}_F,$$

whence $f_*(\omega_{X/Y} \otimes \mathcal{O}(\lceil (a-1)D' \rceil)) \neq 0$. Let $\mathcal{G}'$ be the image of

$$\omega_Y^m \otimes \omega_Y \otimes \hat{\otimes}^1 f_*(\omega_{X/Y} \otimes \mathcal{O}(\lceil (a-1)D' \rceil)) \to \omega_Y^{m+1-a} \otimes f_* \mathcal{O}(aK_X)^{**}.$$

Setting $m = k + 1$ ($\leq n \leq a - 1$), we see $P_a(X) \geq 2$ as in Step 4. This is a contradiction. $\square$

PROOF OF (8.2). Assume $q(X) = n$ and $P_{n+\alpha}(X) = 1$ for some $\alpha > 0$. Then the Albanese map is surjective (8.1) and generically finite. By (3.4), $P_1(X) \geq 1$ and hence $P_4(X) = 1$ by $P_{n+\alpha}(X) = 1$. Thus X is birational to an abelian variety by (3.12). $\square$

The following made the result of Ueno [**U3**, Main Theorem (1), (2)] effective in the sense that $q(X)$ and $P_{12}(X)$ characterize X. (We note that $\kappa(X)$ is determined by an infinite number of $P_\nu(X)$ (cf. (1.13)), though $\kappa(X)$ is simply a number.)

(8.3) THEOREM [**Ko3**, Theorem 6.1]. *Let X be a nonsingular projective 3-fold with $q(X) \geq 3$. Then*
   (i) *X is uniruled $\Leftrightarrow \kappa(X) = -\infty \Leftrightarrow P_{12}(X) = 0$, and*
   (ii) *X is birational to an abelian variety $\Leftrightarrow \kappa(X) = 0 \Leftrightarrow P_{12}(X) = 1$.*

COMMENT ON PROOF. In general if $q(X) > 0$, one can apply Ueno's (3.7) to the Albanese map and get a surjective morphism $X \to Y$ to either (i) a variety Y of general type of dimension > 0 such that $|K_Y|$ induces a generically finite map, or (ii) the Albanese variety Y. In case (i), Kollár does not need $q(X) \geq 3$; he proves (8.3) by the vanishing theorem (5.10) when $\dim Y = 2$, and by Viehweg's weak positivity (5.5) when $\dim Y = 1$. However in case (ii), he needs $q(X) \geq 3$ to apply (8.2). For instance, if $q(X) = 1$ in case (ii), one has to analyze global monodromy when $f_*(\omega_X^{12})$ is a nonzero s.p. locally free sheaf with degree 0. Kollár [**Ko3**, Theorem 6.2] proves more when $q(X) \geq 4$ (cf. Introduction) and [**Ko3**, Remark 6.6] makes some comments when $q(X) = 1, 2$. $\square$

(8.4) Let X be a 3-fold of general type, and consider the rational map Φ_n induced by $|nK_X|$. Kollár [**Ko3**, §4] and Esnault and Viehweg [**EsV**, Corollaire (2.8)] consider which Φ_n is the Iitaka fibering (actually for higher dimensions too). One result easy to state is the following.

(8.4.1) THEOREM [**Ko3**, Corollary 4.8]. *If $P_k(X) \geq 2$ for some $k > 0$, then Φ_{7k+3} is generically finite and Φ_{11k+5} is birational.*

(8.4.2) Assume that X has a minimal model X' (cf. (9.4)), and let Φ'_n be the map induced by $|nK_{X'}|$. Benveniste obtained results on the values of n, for which Φ'_n is a birational morphism [**Ben4**, **Ben5**]. Then Benveniste [**Ben7**] proved that Φ_n is birational for $n \geq 9$ if X' is smooth, and Matsuki [**Mk**] proved that Φ_n is birational for $n \geq 8$ under the same assumption by improving the argument of [**Ben7**]. Hanamura [**Ha**] considered the case where X' is not smooth and has index > 1.

In view of (8.4.1), the work of Fletcher [**Flt**] should be mentioned (cf. [**Re5**, Chapter III]).

9. Minimal models and canonical rings.
In this section, we review the definitions of canonical and terminal singularities, canonical and minimal models, and the results on the finite generation of the *canonical ring* $R(X, K_X)$ (1.3) of a nonsingular projective variety X.

(9.1) A nonsingular projective variety X has been called an *absolute minimal model* if X does not contain a rational curve, i.e., if every morphism from $\mathbf{P}^1$ to X is a constant map. For instance, abelian varieties and hence nonsingular projective varieties finite over some abelian varieties are absolutely minimal, which is easily seen by the fact that the universal covering of an abelian variety is an affine space. The terminology is explained by the result that every rational map f from a nonsingular complete variety X to an absolute minimal model Y is a morphism, which follows easily from the elimination of indeterminacy (1.10.i).

(9.2) One of the difficulties in studying varieties of dimension > 2 is the complexity of birational geometry; there is a complete nonsingular algebraic variety of dimension 3 which is not projective (Nagata [**Na**]) in contrast to the result of Zariski that every nonsingular complete algebraic surface is projective. Furthermore, for an arbitrary nonsingular projective variety X of dimension > 2, there is a complete nonsingular algebraic variety (resp. algebraic space [**Ar**] or Moishezon space [**Mz1**]) Y birational to X which is not a projective variety (resp. an algebraic variety) (Hironaka [**Hi1**]).

(9.3) The above facts (9.2) are some of the reasons why we need the theory of extremal rays (§12), but in this subsection (9.3), we omit this subtlety and vaguely say a "nonsingular model" to mean a complete nonsingular algebraic space (or Moishezon space).

For two "nonsingular models" X and Y birational to each other, we think $X \succ Y$ when $X \dashrightarrow Y$ is a morphism as suggested by (9.1). A result by Fujiki [**Fk1**] says

(9.3.1) THEOREM. *If X is a "nonsingular model" of dimension > 2 containing a rational curve (cf. (9.1)), then there exists a "nonsingular model" Y birational to X such that $Y \nsucc X$.*

This says that, in general, there is no smallest model no matter how bad a singularity we allow on it.

(9.3.2) EXAMPLE (KOLLÁR). Let S ($\subset \mathbf{P}^3$) be a smooth surface of degree ≥ 5 such that a hyperplane section H of S is an irreducible singular rational curve, and let C be a nonsingular projective curve of genus ≥ 2. Then $X = S \times C$ is a 3-fold of general type, and there is no compact complex variety Z birational to X such that the birational map $Y \dashrightarrow Z$ from every "nonsingular model" Y birational to X is a morphism.

Indeed if there is such a Z, then $Z \prec X$ and $Z \prec Y$ (following the actual construction of (9.3.1) for $X \supset H \times P$ with $P \in C$) imply that the birational morphism $X \to Z$ collapses $H \times P$, which is however impossible because $S \times P$ ($\subset X$) cannot be collapsed by birational $X \to Z$ and because $(\mathcal{O}_S(H)^2) > 0$.

(9.3.3) Thus it seems impractical to study "minimal models" from the viewpoint of "smallest models." Deeply related is the problem of the finite generation of $R(X, K_X)$, which is an important and difficult problem, especially for varieties of general type. Finding a good model is, in the sense, a process to prove the finite generation [**Mf1**]. Reid [**R1**] ingeniously studied backward, and defined canonical singularities as "the singularities which should be on canonical models," and introduced minimal models (of general type) via canonical models (cf. (9.4.1.ii)).

(9.4) DEFINITION. Let (X, P) be a germ of a normal algebraic (or analytic) singularity. We say that X has only a *canonical* (resp. *terminal*) singularity at P iff

(i) there is an integer $r > 0$ such that rK_X is a Cartier divisor (the smallest such r is called the *index* of X at P), and

(ii) for every (or equivalently, some) resolution $f: Y \to X$ such that the exceptional set is the sum of irreducible divisors E_i, one has $rK_Y = f^*(rK_X) + \sum_i a_i E_i$, where $a_i \geq 0$ (resp. $a_i > 0$) for all i.

When a variety X has only canonical singularities, the (*global*) *index* r of X is the smallest integer > 0 such that rK_X is a Cartier divisor.

We say that a projective algebraic variety X with only canonical (resp. terminal) singularities is a *canonical* (resp. *minimal*) *model* if K_X is an ample (resp. a nef) $\mathbf{Q}$-divisor, and that a variety Y *has a canonical* (resp. *minimal*) *model* if there is a canonical (resp. minimal) model birational to Y.

(9.4.1) REMARK. (i) We should note that $\bigoplus_\nu H^0(X, \mathcal{O}(\nu K_X))$ is the canonical ring and hence $\kappa(X, K_X) = \kappa(X)$ for a projective variety X with only canonical singularities, and that a variety X of general type has a canonical model iff the canonical ring is finitely generated, because the canonical model is $\operatorname{Proj} R(X, K_X)$ by Reid's definition of canonical singularities.

(ii) Reid [**Re1**] called a birational morphism $f: X \to Y$ from a normal variety to a variety with only canonical singularities *crepant* (*not discrepant*) if $rK_X = f^*(rK_Y)$, where r is the index of Y.

The definitions of terminal singularities and minimal models are explained by the following problem (∗) and the Minimal Model Conjecture (cf. (9.5) and §12) in the sense that minimal models are the goals for both problems.

(∗) PROBLEM (REID). Given a canonical model Y, find a crepant blow-up $f: X \to Y$ such that X is a $\mathbf{Q}$-factorial minimal model.

This is answered affirmatively for 3-folds by Reid [Re2], and this follows if the Minimal Model Conjecture in the strong sense (cf. §12) holds (cf. (9.6)).

(iii) It should be mentioned that some people allow canonical singularities or only $\mathbf{Q}$-factorial terminal singularities on minimal models. At this point, it seems to be only a matter of preference which singularity to choose especially for 3-folds because Reid [Re2] showed that X has a $\mathbf{Q}$-factorial minimal model if it has a projective model Y with only canonical singularities such that K_Y is nef. We allow terminal singularities just because they were introduced as the singularities to be on "minimal models." Many results on minimal models hold for "minimal models with canonical singularities."

(iv) Francia [Fr] (cf. (12.10)) says that there is a nonsingular canonical model X^+ which is not the smallest in the sense of $\succ$ above $(Z^- \not\succ X^+)$. Thus "K_X is nef" in the definition of minimal models is a practical alternative to "being the smallest," since it is an equivalent formulation for surfaces and since we have the following.

(9.4.2) FACT. Let $f: X \to Y$ be a birational morphism between projective varieties with only terminal singularities. If f is not an isomorphism and if Y is $\mathbf{Q}$-factorial (i.e., for each Weil divisor W, there is an integer $\nu > 0$ such that νW is a Cartier divisor), then K_X is not nef.

There are also the notions of log-canonical, log-terminal, and weak log-terminal singularities, but we refer to [KMM] for these definitions. We refer to [Re5] about the details of canonical and terminal singularities.

(9.5) *The Minimal Model Conjecture*, in its simplest form, claims that every nonsingular projective variety X with $\kappa(X) \geq 0$ has a minimal model.

The importance of this conjecture is explained by

(9.6) THEOREM. *If X is a minimal model such that $\kappa c(X) = 0$, then K_X is semiample and, in particular, X has a canonical model.*

This is due to Kawamata [Ka6] and Benveniste [Ben3] when $\dim X = 3$. When $\dim X > 3$, this is proved by Kawamata as a part of the Base Point Free Theorem (12.3), which is built on the Kawamata-Viehweg Vanishing Theorem (5.6) and Shokurov's Nonvanishing Theorem (12.4).

Kawamata [Ka9] generalized (9.6) to varieties X with $\kappa c(X) > 0$. To explain it, we need some definitions.

(9.7) Let X be a projective variety of dimension n with only canonical singularities, and D a nef $\mathbf{Q}$-divisor. Then $\nu(X, D)$ is the biggest integer ν such that D^ν is not numerically trivial as a $(n-\nu)$-cycle. In general, we have $\nu(X, D) \geq \kappa(X, D)$, and we say that D is *abundant* if $\nu(X, D) = \kappa(X, D)$ and that X is a *good minimal model* if X is a minimal model and K_X is abundant.

By the uniqueness of Fujita's Zariski decomposition [**Ft7**], we can see that a minimal model X' of a nonsingular projective variety X is good iff all the minimal models X' of X are good and that $\nu(X', D)$ does not depend on the choice of the minimal model X' of X. Viehweg told me that $\nu(X, K_X)$ may be directly defined in terms of the asymptotic behavior of $h^0(nK_X + H)$ as $n \to \infty$ for some very ample divisor H without assuming the existence of a minimal model X'.

(9.8) It is known that minimal models of dimension ≤ 2 are good, and *the Abundance Conjecture* by Kawamata claims that every minimal model X is good. This conjecture is essentially reduced to the case of $\kappa(X) = 0, -\infty$ (cf. §§10,11) by the following (9.9), which is proved by the uniqueness of Zariski decomposition.

(9.9) THEOREM [**Ka9**]. *Let X be a minimal model with $\kappa(X) \geq 0$, and let $X' \to Y'$ be the Iitaka fibering of X and F the generic geometric fiber. If F has a good minimal model, then X is a good minimal model.*

Since the Abundance Conjecture holds for surfaces, every minimal model X with $\kappa c(X) \leq 2$ is good, and the following is a generalization of (9.6).

(9.10) THEOREM [**Ka9**]. *If a minimal model X is good, then K_X is semi-ample and the canonical ring of X is finitely generated.*

We should note that the semiampleness is very important for its application to C_{nm}^+ (cf. (7.6)). As for the finite generation of $R(X, K_X)$, it is easier if $\kappa(X)$ is smaller (e.g., OK if $\kappa(X) \leq 1$), while the abundance of K_X is easier if $\kappa c(X)$ is smaller (e.g., OK if $\kappa c(X) \leq 2$). Thus as far as the finite generation is concerned, showing that X has a good minimal model does not seem to be the easiest way. An alternate approach is proposed by Fujita [**Ft6**] (9.12) and generalized by Moriwaki [**Mwk3**] (9.13).

(9.11) *Zariski decomposition* [**Za**] of an effective divisor D on a surface is, in short, to write $D = P + N$, where the semipositive part P is the largest nef **Q**-divisor such that the negative part $N = D - P$ is an effective **Q**-divisor. This played a very important role in the study of open surfaces, and its generalization to higher dimensions has been tried by Benveniste [**Ben2**], Fujita [**Ft8**], Cutkosky [**Ct**], and Moriwaki [**Mwk2, Mwk3**]. So far, we do not seem to have a generally accepted candidate for the generalization, and anyway, we do not know if all effective divisors have a Zariski decomposition in any sense.

It is important to note that we have to blow up X and replace D with its total transform to get the Zariski decomposition $D = P + N$ due to the existence of a base locus of $|\nu D|$ of intermediate dimension and that we have to consider **R**-divisors P for the Zariski decomposition of an ordinary effective divisor D as shown by Cutkosky's example [**Ct**]. Their approach to the finite generation via Zariski decomposition $K_X = P + N$ is to prove that the semipositive part P is semiample (*canonical conjecture* [**Ft6**]). For varieties X of general type, they proved that the existence of the Zariski decomposition of K_X is sufficient for the finite generation of the canonical ring (necessity is done earlier by Wilson [**W2**]).

(9.11.1) REMARK. The canonical conjecture with Fujita's definition asserts more than finite generation while the conjecture with Cutkosky's is equivalent to finite generation in view of (9.12) and Wilson's result [**W2**]. Here we consider only the case of $\kappa > 0$ and the aspect of finite generation so that we do not have to pay too much attention to the difference of definitions.

Fujita [**Ft6**, **Ft8**] proved that the canonical ring of a 3-fold X with $0 < \kappa(X) < 3$ is finitely generated, and Cutkosky [**Ct**] gave another proof improving Benveniste's result [**Ben2**]. Let me outline Fujita's approach [**Ft6**].

(9.12) In general, let X be a nonsingular projective variety with $\kappa(X) > 0$, and let $f: X' \to Y$ be an algebraic fiber space with a s.n.c. branching representing the Iitaka fibering of X and $c \in \mathbf{N}$ be such that $\mathcal{O}(M) = [f_*\mathcal{O}(cK_{X'})]^{**}$ is an invertible sheaf and $\mathcal{O}(\nu M) = [f_*\mathcal{O}(\nu c K_{X'})]^{**}$ for all $\nu > 0$, and $R(X, cK_X) \simeq R(Y, M)$ by (2.5) and (2.6). Then (5.13) shows that $M/c = K_Y + \Delta + D$ after models X and Y are suitably changed, where D is a nef $\mathbf{Q}$-divisor and Δ is a negligible $\mathbf{Q}$-divisor. We note that the pull back of the semipositive part of M/c to X' "should be" the semipositive part of $K_{X'}$.

Thus the finite generation of the canonical ring of X is reduced to the finite generation of $R(Y, K_Y + \Delta + D)$ for Y of lower dimension, where $K_Y + \Delta + D$ is a big divisor. Though D is hoped to be semiample and hence $\Delta + D$ is negligible when Y is chosen properly, it is known only when the general fiber of f is an abelian variety or a surface (cf. (5.15.9.ii)). Viewing $K_Y + \Delta + D$ as a kind of "log-canonical divisor," this reduces the problem to a "log-canonical" case of lower dimension, which explains the necessity of studying "log-canonical divisors" $K_Y +$ (negligible $\mathbf{Q}$-divisor). When $\kappa(X) = 2$, Fujita proved that the semipositive part of $K_Y + \Delta + D$ is semiample and hence the canonical ring of X is finitely generated. Moriwaki [**Mwk3**] generalized the result as follows.

(9.13) THEOREM. *Let X be a nonsingular projective variety, $c \in \mathbf{N}$, and D a negligible divisor such that cD is a divisor and $\kappa(X, K_X + D) = 2$. Then $R(X, c(K_X + D))$ is finitely generated.*

10. Varieties with $\kappa = 0$.

(10.1) The first thing in studying n-folds X with $\kappa(X) = 0$ is to study the Albanese map $\mathrm{Alb}_X: X \to \mathrm{Alb}(X)$.

Theorem (8.1) says Alb_X is surjective and hence $q(X) \le n$, and Theorem (8.2) says Alb_X is birational if $q(X) = n$. In general, it is known that Alb_X is an algebraic fiber space [**Ka1**, Theorem 1]. These results were obtained under the effort to prove the following, which Ueno [**U1**] posed as a conjecture closely related to C_{nm}.

(10.2) CONJECTURE K. If X is a nonsingular projective variety with $\kappa = 0$, then the Albanese map Alb_X is an algebraic fiber space and it is birational to an étale fiber bundle over $\mathrm{Alb}(X)$ which is trivialized by a finite étale base change.

(10.3) First this conjecture was proved for 3-folds X with $q(X) \geq 2$ by Ueno [**U3**] (the first breakthrough in higher dimension) based on Viehweg's $C_{n,n-1}^+$ [**V1**] (cf. (7.6.A)), and also in the case where the generic geometric fiber is an abelian variety [**U4**] (cf. (7.6.A)), based on the study of the moduli of abelian varieties. Then Viehweg [**V3**, Satz V] (its simplified proof is in [**V6**, Lemma 9.5]) and also Kawamata [**Ka1**, a remark on p. 261] showed that C_{nm}^+ for Alb_X with $q(X) = m$ implies the conjecture K for X if $m \geq n - 2$. (The restriction on m comes from the assumption that the generic geometric fiber of Alb_X has a smooth smallest model in the sense of (9.3), cf. [**V6**, §9].) Thus the conjecture K holds for n-folds X with $q(X) = n - 1$ by [**V1**], for 3-folds X with $q(X) = 1$ [**V3**], and then for n-folds X with $q(X) = n - 2$ [**Ka10**]. It is important that X with $\kappa(X) = 0$ has thus a smooth good minimal model which is an étale fiber bundle of $\text{Alb}(X)$ if $q(X) \geq \dim X - 2$ [**V6**, Corollary 9.4]. We should remark that if $q(X) = n$, then the conjecture K holds even if the condition "$\kappa(X) = 0$" is replaced by effective "$P_{n+\alpha}(X) = 1$ for some $\alpha \geq 1$" (8.2).

(10.4) It is also an important question to find an explicit value of ν such that $P_\nu(X) = 1$ for X with $\kappa(X) = 0$.

(10.4.1) *Case* 1. X is an n-fold with $\kappa(X) = 0$ and $q(X) = n - 1$. Then X has a smooth good minimal model which is an étale fiber bundle over $\text{Alb}(X)$ whose fibers are elliptic curves. In this case, Ueno's calculation [**U3**, p. 707] (cf. also [**Ka1**, Theorem 15]) shows that $p_g(X) = 0$ and $P_{12}(X) = 1$.

(10.4.2) *Case* 2. X is a 3-fold with $\kappa(X) = 0$ and $q(X) = 1$. Then X has a smooth good minimal model which is an étale fiber bundle over $\text{Alb}(X)$ whose fibers are either abelian surfaces, hyperelliptic surfaces, $K3$ surfaces, or Enriques surfaces. Therefore Beauville [**Bea**] says that $P_\nu(X) = 1$, where

$$\nu = 2^5 \cdot 3^3 \cdot 5^2 \cdot 7 \cdot 11 \cdot 13 \cdot 17 \cdot 19$$

(cf. [**Mrr2**]). His proof gives a universal number ν such that $P_\nu(X) = 1$ for n-folds X with $\kappa(X) = 0$ and $q(X) = n - 2$.

(10.4.3) For 3-folds X with $\kappa(X) = 0$, the unsettled case $q(X) = 0$ remains, and we need to assume the existence of a good minimal model (Abundance Conjecture) to answer it at present. To be precise, Kawamata [**Ka11**] proved that there is some explicit $\nu > 0$ such that if X is a good minimal 3-fold with $\kappa = 0$, then $\nu K_X \sim 0$. D. Morrison [**Mrr2**] supplemented it and got $\nu = 120$ when $q(X) = 0$. The proof uses special properties of 3-folds (terminal lemma, etc.), and it does not seem to have immediate generalization to higher dimensions.

To go beyond this point, we have to assume the Abundance Conjecture (9.8) or, more practically, work on n-folds X such that K_X is torsion, that is, $\nu K_X \sim 0$ for some $\nu > 0$. An important result in this direction is

(10.5) THEOREM [**Ka10**, §8]. *Let X be a minimal model with numerically trivial K_X. Then*

(i) *K_X is torsion, and*

(ii) *the conjecture K holds for Alb_X.*

The assertion (i) is a cute application of C_{nm} based on Tsunoda's idea: one only needs to consider the case $q(X) > 0$, where one can apply C_{nm} (Kawamata [**Ka10**]) to $\mathrm{Alb}_X \colon X \to \mathrm{Alb}(X)$ by the induction hypothesis on the generic geometric fiber to obtain $\kappa(X) \geq 0$, which proves (i). The assertion (ii) is a generalization of Viehweg's [**V6**, Lemma 9.5], which, I heard, is also generalized by Nakayama to the following form.

(10.6) THEOREM. *Let $f \colon X \to Y$ be a surjective morphism of minimal models with connected fibers such that K_X and K_Y are torsion. Then there is a finite covering $g \colon Y' \to Y$ from a minimal model which is étale in codimension 1 such that $X \times_Y Y'$ is birational to $Y' \times F$ for some variety F.*

This suggests the possibility of generalizing Bogomolov's decomposition theorem (cf. [**Bo**] and [**Bea**]) to minimal models with torsion canonical bundles (cf. D. Morrison's problem in [**U6**, p. 614]).

In concluding this section, let me mention that even if all the conjectures (the Minimal Model Conjecture and the Abundance Conjecture) are proved for 3-folds, the problem of classification of special unitary 3-folds (cf. [**Bea**]) (possibly with terminal singularities) is completely open.

11. Varieties with $\kappa = -\infty$.

(11.1) In this section, we review several results on varieties with $\kappa = -\infty$. In case of surfaces, we had a beautiful criterion:

ruled $\Leftrightarrow$ uniruled $\Leftrightarrow \kappa = -\infty \Leftrightarrow P_{12} = 0$, and

rational $\Leftrightarrow$ unirational $\Leftrightarrow \kappa = -\infty \,\&\, q = 0 \Leftrightarrow P_2 = 0 \,\&\, q = 0$,

where we say that an n-fold X is *ruled* (resp. *uniruled*) if there exist a $(n-1)$-fold Y and a birational (resp. dominant rational) map $Y \times \mathbf{P}^1 \dashrightarrow X$, and that X is *rational* (resp. *unirational*) if there exists a birational (resp. dominant rational) map $\mathbf{P}^n \dashrightarrow X$ (we note that these are obviously birational properties).

(11.2) However Clemens and Griffiths [**CG**], Manin and Iskovskih [**IM**] gave examples of unirational 3-folds which are not rational, and Artin and Mumford [**AM**] gave examples of unirational n-folds which are not rational for all $n \geq 3$. Their criteria are all different; [**CG**] showed that the intermediate Jacobian is a Prym variety which is not a Jacobian variety, whence not rational; [**IM**] showed that the group of birational mappings is equal to the genuine automorphism group, whence not rational; [**AM**] showed that $H^3(\mathbf{Z})$ has a nonzero torsion, whence not rational. Furthermore, we can list difficulties:

(11.2.1) Segre (cf. [**IM**]) gave examples of unirational smooth quartic 3-folds in $\mathbf{P}^4$, but it is not known if general quartic 3-folds are unirational.

(11.2.2) Some smooth cubic 4-folds X in $\mathbf{P}^5$ (e.g., X's containing two linear planes in general position) are rational, but it is not known if general cubic 4-folds are rational.

(11.2.3) A result of Beauville, Colliot-Thélène, Sansuc, and Swinnerton-Dyer [**BCSS**] says that there exists a variety X which is *stably rational* (i.e., $X \times \mathbf{P}^m$ is birational to $\mathbf{P}^n$ for some $m, n \geq 0$) such that $\dim X = 3$ and X is not rational.

Thus rationality and unirationality seem too delicate to use in classification theory at present, and the difference of rationality and ruledness seem too subtle. On the other hand, we have

(11.2.4) *Fano n-folds* (nonsingular projective n-folds X with ample $-K_X$) are uniruled by Kollár, and the uniruledness is stable under small and global deformations by Fujiki and Levin [**Fk2, L1**]. We note that Fano 3-folds X are classified by Iskovskih and Shokurov [**Is1, Is2, Sho1, Sho2**] if $B_2(X) = 1$ and by Mori and Mukai [**MrMuk**] if $B_2(X) \geq 2$.

Therefore, unlike the case of $\kappa = 0$, we do not have a precise conjecture (cf. (10.2)), and the best we can hope is the uniruledness of varieties with $\kappa = -\infty$. We will introduce another notion.

(11.3) Let D be a Cartier (or $\mathbf{Q}$-Cartier) divisor on a projective variety X. We say that D is *pseudoeffective* if the following equivalent conditions are satisfied:

(i) D is a limit of effective $\mathbf{Q}$-divisors; that is, there exist sequences $\{D_i\}_{i\geq 1}$ and $\{n_i\}_{i\geq 1}$ of effective Cartier divisors and natural numbers such that $D \approx \lim_i D_i/n_i$, i.e., $(D \cdot C) = \lim_i (D_i \cdot C)/n_i$ for every irreducible curve C.

(ii) $\kappa(nD + H) \geq 0$ for all big Cartier divisors H and all natural numbers n.

(iii) There is a big Cartier divisor H such that $\kappa(nD + H) \geq 0$ for all natural numbers n.

Indeed (ii)$\Rightarrow$(iii) is obvious and (iii)$\Rightarrow$(i) follows from $D \approx \lim_n (nD + H)/n$. Assuming (i), let H be a big divisor and $n > 0$. By (1.9), there is an ample $\mathbf{Q}$-divisor $\overline{H}$ on X such that $H \succ \overline{H}$. Then there is $i \gg 0$ such that $n(D - D_i/n_i) + \overline{H}$ is ample by Kleiman's criterion [**K1**], whence $\kappa(nD + H) \geq \kappa(nD + \overline{H}) \geq \kappa(D_i) \geq 0$ (ii).

(11.4) Let X be a projective n-fold with only canonical singularities. We say that X is κ-*uniruled* if K_X is not pseudoeffective. We note that κ-uniruledness is slightly stronger than saying that the *adjunction terminates*, i.e., $|nK_X + H| = \varnothing$ for each very ample divisor H and some $n = n(H) > 0$. One can easily see that the κ-uniruledness is a birational property by the following.

(11.4.1) LEMMA. *Let $f \colon X \to Y$ be a generically finite surjective morphism of projective varieties with only canonical singularities. Then*

(i) *if X is κ-uniruled then so is Y, and*

(ii) *the converse holds if f is birational.*

Indeed (i) follows from $nK_X \succ f^*(nK_Y)$ for some $n \in \mathbf{N}$ (follows from the definition of the canonical singularity), and (ii) follows by $f_*\mathcal{O}(nK_X) = \mathcal{O}(nK_Y)$ for all $n > 0$.

(11.4.2) COROLLARY. *For a projective variety X with only canonical singularities, we have*

$$X \text{ is uniruled} \Rightarrow X \text{ is } \kappa\text{-uniruled} \Rightarrow \kappa(X) = -\infty.$$

Indeed the second implication is obvious by the definition, and for the first, it is enough to show $\mathbf{P}^1 \times Y$ is κ-uniruled by (11.4.1), which is an easy exercise.

(11.5) Now our hope is to show that the properties in (11.4.2) are all equivalent, and a classical method (if $q(X) > 0$) is to apply C_{nm} (not the full C_{nm}^+) to the algebraic fiber space $f: X \to Y$ obtained from the Albanese mapping of a nonsingular projective n-fold X with $\kappa(X) = -\infty$ and $q(X) > 0$ via Stein factorization. Since Y is generically finite over $\mathrm{Alb}(X)$, we see $\kappa(Y) \geq 0$, and C_{nm} (if proved) shows that the generic geometric fiber has $\kappa = -\infty$ and dimension $< n$. This is a reduction to lower dimensions (cf. [**V6**, Theorem 9.1]). In particular, we have the answer for 3-folds.

(11.5.1) 3-folds X with $\kappa(X) = -\infty$ and $q(X) > 0$ are uniruled by Viehweg's $C_{n,n-1}^+$ [**V1**] (cf. (7.6)) and $C_{3,1}$ [**V3**], and

(11.5.2) n-folds X with $\kappa(X) = -\infty$ and with either n-4 or $q(X) = 1$ are fibered with proper subvarieties of $\kappa = -\infty$ by Viehweg's $C_{n,n-1}^+$ [**V1**] (cf. 7.6)), Kawamata's $C_{n,1}$ [**Ka3**] (6.2.e), and $C_{n,n-2}$ which is a combination of [**Ka4**] and [**V6**] (cf. [**V6**, §9]).

Let us supplement this a little bit. We recall that Kawamata [**Ka1**, Theorem 1] proved that Alb_X is an algebraic fiber space for varieties X with $\kappa = 0$ by (3.7.1) using Kawamata's $C_{\cdot,\cdot,0}^-$ (6.2.e). This method produces

(11.5.3) PROPOSITION. *Let X be a nonsingular projective variety with $\kappa(X) = -\infty$ and $q(X) > 0$. Then we have one of the following:*

(i) *$\mathrm{Alb}_X: X \to \mathrm{Alb}(X)$ is an algebraic fiber space, or*

(ii) *there is a surjective morphism $X \to Z$ with connected fibers to a variety Z such that $\kappa(Z) = \dim Z > 0$.*

The merit in doing this is that we have a control of the dimension of the base of the algebraic fiber space in case (i), and we can apply Viehweg's $C_{\cdot,\cdot,0}$ (6.2.d) to see that the generic geometric fiber of $X \to Z$ has $\kappa = -\infty$ in case (ii). Thus we have

(11.5.4) COROLLARY. *If X is a nonsingular projective variety with $\kappa(X) = -\infty$ and $q(X) \geq \dim X - 2$, then X is fibered with proper subvarieties with $\kappa = -\infty$.*

Indeed we can use $C_{n,n-1}^+$ and $C_{n,n-2}$ as above.

If one carefully reads Viehweg's proof of weak positivity of $f_* \omega_{X/Y}^\nu$ $(\nu > 0)$ for an algebraic fiber space $f: X \to Y$ in [**V6**, §5], he can easily prove the weak positivity of $f_*(\omega_{X/Y}^\nu(H))$ for all $\nu > 0$ and all base point free Cartier divisors H. Then it is easy to see that if both Y and the generic geometric fiber of f have pseudoeffective canonical divisors, then so does X by

$$nK_X + H + f^* M = nK_{X/Y} + H + f^*(nK_Y + M),$$

where M is a very ample divisor on Y. Thus we see

(11.5.5) PROPOSITION. *Let $f: X \to Y$ be an algebraic fiber space with generic geometric fiber F. Then*

(i) *if X is uniruled, then Y or F is uniruled, and*

(ii) *if X is κ-uniruled, then Y or F is κ-uniruled.*

(11.5.6) COROLLARY. *If X is a κ-uniruled nonsingular projective variety with $q(X) > 0$, then X is fibered with κ-uniruled proper subvarieties.*

We would like to mention another new (yet conjectural) approach. We begin with a useful criterion for uniruledness by Miyaoka and Mori [**MyMr**] which generalizes the result of Kollár (cf. (11.2.4)).

(11.6) THEOREM [**MyMr**]. *Let X be a nonsingular projective n-fold with an open dense subset U with the property that for each $x \in U$ there exists an irreducible curve $C \subset X$ such that $x \in C$ and $(K_X \cdot C) < 0$. Then X is uniruled.*

The proof uses an improvement of the argument in the proof of Hartshorne's conjecture [**Mr1**], which uses an argument in char. $p > 0$. We call a projective n-fold X with only terminal (or canonical?) singularities a **Q**-*Fano n-fold* if $-K_X$ is ample. Then

(11.6.1) COROLLARY. *Every **Q**-Fano n-fold X is uniruled.*

Indeed this is obtained by applying (11.6) to a resolution of X.

(11.6.2) COROLLARY (KOLLÁR). *Let D be an irreducible reduced Cohen-Macaulay projective variety over another variety E such that D is Gorenstein in codimension 1 and $(\omega_D^{\otimes(-m)})^{**}$ is a relatively ample invertible sheaf for some $m > 0$. Then the general fibers of $D \to E$ are uniruled.*

Indeed the normalization $f: \overline{D} \to D$ of D has the property that $-mK_{\overline{D}} \succ f^*(-mK_D)$ as Weil divisors, and we can apply (11.6) to the smooth part of the general fibers of $\overline{D} \to E$.

(11.7) The conjectural approach is based on a part of the Abundance Conjecture (9.8) which claims

(11.7.1) CONJECTURE. Every minimal model X has $\kappa(X) \geq 0$.

This is solved when $\dim X = 3$ by Miyaoka [**My1**,**My2**]. The *Minimal Model Conjecture* (MMC) (cf. §12) asserts that an n-fold X with $\kappa = -\infty$ has a projective **Q**-factorial model X' with only terminal singularities which has a **Q**-*Fano fibering* $f: X' \to Y$, i.e., a surjective morphism to a normal projective variety whose generic geometric fiber is a **Q**-Fano variety. Then (11.6.1) applies and X is uniruled (modulo the above two conjectures).

We illustrate the implications as follows, where a dotted arrow means a conjectural implication.

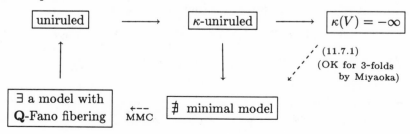

Thus if MMC holds for 3-folds then these are all equivalent (cf. also (12.12)).

(11.8) Recently Miyaoka [**My3**] constructed a general theory on uniruled varieties using Chow varieties. In particular, from (11.6.1), he proved that **Q**-Fano n-folds with Picard number 1 are *rationally connected*; that is, two general points are joined by finitely many (possibly singular) rational curves. It appears that being rationally connected is a good substitute for rationality in classification theory as uniruledness is for ruledness.

12. The theory of extremal rays. In this section, we review the theory of extremal rays only in a *simplified* form for simplicity of exposition, and we refer the reader to [**W4**] for an overall view and to [**KMM**] for details.

(12.1) Let X be a projective variety with only canonical singularities. Then the group of 1-*cycles* (formal linear combination of irreducible curves with integral coefficients) modulo *numerical equivalence* $\approx$ ($Z_1 \approx Z_2$ if $(Z_1 \cdot D) = (Z_2 \cdot D)$ for all Cartier divisors) is a finitely generated free abelian group. Its scalar extension $N(X)$ to **R** is a dual vector space of $\mathrm{NS}(X) \otimes \mathbf{R}$, the scalar extension of the Néron-Severi group $\mathrm{NS}(X)$ to **R**, whose dimension $\rho(X)$ ($< \infty$) is called the *Picard number* of X. In the finite-dimensional real vector space $N(X)$, the cone generated by the classes $[\mathbf{C}]$ of irreducible curves C is denoted by $NE(X)$ and its closure (for the metric topology) is denoted by $\overline{NE}(X)$; these are called the *cone of curves* and the *closed cone of curves*, respectively.

The idea of studying projective morphisms through these cones (in the dual form) goes back to Hironaka [**Hi1**], and the idea was taken up again by Kleiman [**Kl**]. However they did not specifically look at the half-space $\{z \in N(X)|\ (z \cdot K_X) < 0\}$; therefore the study of $\overline{NE}(X)$ was too hard. In the ideal situation, we have

(12.1.1) PROPOSITION [**Mr2**]. *Let X be a Fano n-fold. Then there exist a finite number of (possibly singular) rational curves $l_1, \ldots, l_r$ such that $(-K_X \cdot l_i) \leq n + 1$ for all i and*

$$NE(X) = \mathbf{R}_+[l_1] + \cdots + \mathbf{R}_+[l_r],$$

where $\mathbf{R}_+ = \{x \in \mathbf{R}|x \geq 0\}$.

Thus the cone of curves $NE(X)$ is rational polyhedral for such X, and we would like to call the edges the extremal rays in general. However various examples show that $\overline{NE}(X)$ is *not* always "rational polyhedral" on the closed subcone

$$\overline{NE}_K(X) = \{Z \in \overline{NE}(X)|(K_X \cdot Z) \geq 0\};$$

therefore we must keep away from $\overline{NE}_K(X)$ (in some sense) to construct a general theory. Thus we say that a half-line $R = \mathbf{R}_+ Z$ is an *extremal ray* if (i) $R \not\subset \overline{NE}_K(X)$ and (ii) if $Z_1, Z_2 \in \overline{NE}(X)$ satisfy $Z_1 + Z_2 \in R$, then $Z_1, Z_2 \in R$. Now the general form of (12.1.1) is given by

(12.2) THEOREM (CONE THEOREM). *Let X be a projective n-fold with only canonical singularities. Then the extremal rays R_i ($i \in I$) are generated by* *classes of irreducible curves, they are locally finite in* $N(X) - \overline{NE}_K(X)$ *(i.e., for each compact subset* $S \subset N(X) - \overline{NE}_K(X)$, *there exist only finitely many extremal rays* R_i *such that* $R_i \cap S \neq \varnothing$), *and*

$$\overline{NE}(X) = \overline{NE}_K(X) + \sum_i R_i.$$

The result asserts that $\overline{NE}(X)$ is rational polyhedral away from $\overline{NE}_K(X)$. (12.2) was first proved for nonsingular X by Mori [**Mr2**]. Reid [**Re4**] proved the case of 3-folds. The general case is done in a slightly weaker form (yet enough for applications) by Kawamata [**Ka8**] based on works of Reid [**Re4**] and Shokurov [**Sho3**], and supplemented by Kollár [**Ko2**] (cf. (12.3)–(12.6)).

Before going further, let me mention two of the simple technical remarks in [**Mr2**] which apply even in the general setting.

(12.2.1) (i) If L is a nef Cartier divisor on X as above, then we have the implications:

$$L^\perp \cap \overline{NE}_K(X) = 0 \leftrightarrow nL - K_X \text{ is ample} \quad \left(\begin{array}{cc} \xrightarrow[\text{vanishing}]{\text{Kodaira}} & H^i(L^{\otimes n}) = 0 \\ & i > 0, \ n \gg 0 \end{array} \right)$$
$$\text{for all } n \gg 0$$

by Kleiman's criterion [**K1**] for ampleness, where $L^\perp = \{z \in N(X) | (Z \cdot L) = 0\}$. Kawamata called such L a *supporting divisor*.

(ii) If L is a base point free supporting divisor, then $-K_X$ is relatively ample with respect to the morphism induced by $|L|$.

Kawamata first proved the following for 3-folds.

(12.3) THEOREM (BASE POINT FREE THEOREM) [**Ka7, Ka8**]. *Let* X *be a projective variety with only canonical singularities, and* D *a nef Cartier divisor* D *such that* $aD - K_X$ *is nef and big for some* $a \in \mathbf{N}$. *Then* $|mD|$ *is base point free for all* $m \gg 0$.

This proves the semiampleness of both K_X of a minimal model X of general type ($D = K_X$) and an arbitrary supporting divisor D of arbitrary X (cf. (12.2.1)). The proof of [**Ka7**] was to analyze the base locus of $|mD|$ for $m \gg 0$, which was given by the Riemann-Roch theorem. The proof worked only for 3-folds just because the Riemann-Roch formula for n-folds ($n \geq 4$) does not say much about $|mD| \neq \varnothing$ for $m \gg 0$. This difficulty was solved by Shokurov.

(12.4) THEOREM (NONVANISHING THEOREM) [**Sho3**]. *Let* X *be a projective variety with only canonical singularities, and* D *a nef Cartier divisor* D *such that* $aD - K_X$ *is nef and big for some* $a \in \mathbf{N}$. *Then* $|mD| \neq \varnothing$ *for all* $m \gg 0$.

It was Reid who found an approach to the (weak) Cone Theorem from the Base Point Free Theorem via the Rationality Theorem. He actually proved (12.5) for 3-folds and then the (weak) Cone Theorem for 3-folds.

(12.5) THEOREM (RATIONALITY THEOREM) [**Re4, Ka8**]. *Let X be a projective variety with only canonical singularities such that K_X is not nef. Then for each ample divisor H, $t(H) = \max\{t | H + tK_X$ is nef$\}$ is rational.*

The Rationality Theorem was proved in the general case by Kawamata [**Ka8**], and thus the (weak) Cone Theorem was proved. It should be noted that (12.3), (12.4), (12.5) are all proved by a similar powerful technique based on the Kawamata-Viehweg Vanishing Theroem (5.6). Even with (12.5), it was not clear that the extremal rays are locally finite in $N(X) - \overline{NE}_K(X)$. This was settled by Kollár as follows.

(12.6) THEOREM (BOUNDING DENOMINATOR) [**Ko2**]. *Under the notation of (12.5), there is a natural number d depending only on X such that $d \cdot t(H) \in \mathbf{Z}$ for all ample H.*

These all together give the full Cone Theorem (12.2). Having these theorems, we can easily see

(12.7) THEOREM (CONTRACTION THEOREM) [**Mo2, Ka7, Ka8**]. *Let X be a projective variety with only canonical singularities. For each extremal ray R, there exists a morphism $f: X \to Y$ to a normal projective variety with connected fibers such that an irreducible curve C is collapsed by f iff $[C] \in R$.*

It is easy to see that such an f is unique up to isomorphism and hence it is called the *contraction* of R. It is also easy to see that one can find a supporting divisor D such that $D^{\perp} \cap \overline{NE}(X) = R$ by the Cone Theorem (simply draw a picture), and the Base Point Free Theorem (12.3) shows that some multiple aD $(a \gg 0)$ of D is base point free and induces the contraction.

In view of these, the naivest way to get a minimal model would be to keep contracting extremal rays until we come across one.

(12.8) In case of a nonsingular projective surface X with $\kappa \geq 0$, we can find an exceptional curve E of the first kind if K_X is not nef ($\mathbf{R}_+[E]$ is an extremal ray), and we can pass to the contraction Y of E, which is again a nonsingular projective surface. Since $\rho(X) > \rho(Y) \geq 1$, we cannot have an exceptional curve of the first kind after each contraction for infinitely many times, and thus we get a minimal surface. We simply want to imitate the procedure.

(12.8.1) However, before that, one may ask "Why do I have to consider singularities when I start with a nonsingular model?" The answer (in a roundabout way) is already given in §9, but a more direct answer is in [**Mr2**], where the contractions of extremal rays for nonsingular projective 3-folds X are classified. It says, by birational contraction $f: X \to Y$ of an extremal ray, there are 3 kinds of singularities Y can get: (i) an ordinary double point, (ii) a hypersurface singularity $x^2 + y^2 + z^2 + u^3 = 0$, and (iii) a quadruple point $\mathbf{C}^3/\iota$, where $\iota: \mathbf{C}^3 \to \mathbf{C}^3$ is an involution $(x, y, z) \mapsto -(x, y, z)$. The last one is a terminal singularity of index 2 and not Gorenstein, and it is (in a sense) where the complexity of 3-dimensional birational geometry starts (cf. [**Ku, PP, Mrr1**]).

(12.9) Let us outline the conjectural method to get a minimal model, which was implicitly given in [**Mr2**] and explicitly stated by Reid [**Re4**], where he checked the Flip Conjectures I and II in the toric case. Let us start with a projective **Q**-factorial variety X of dimension $n \geq 3$ with only terminal singularities. (We saw in (12.8.1) that we may immediately get terminal singularities by the first contraction even if we start with a smooth X.) We have two cases.

(12.9.1) If K_X is nef, then X is already a minimal model, and there remains a problem of deciding whether X is good or not (*Abundance Conjecture*) (cf. (9.8)). We should note that Miyaoka [**My1**, **My2**] proved that $\kappa(X) \geq 0$ in case of 3-folds (cf. (11.7.1)), which was a part of the Abundance Conjecture.

(12.9.2) If K_X is not nef, then X has an extremal ray by the Cone Theorem (12.2), and we pick up an extremal ray at random and consider the contraction $f: X \to Y$. There are three cases.

(12.9.2.1) *Case* 1. $\dim Y < \dim X$. By (12.2.1.ii), f is a **Q**-Fano fibering, and in particular X is uniruled by (11.6). We should note that Y has only rational singularities, which was proved by Reid [**Re4**] for 3-folds and by Kollár [**Ko3**] in the general case.

(12.012 9.2.2) *Case* 2. f is birational and contracts a divisor (*good contraction*). It is easy to see that the exceptional set D is an irreducible divisor and Y is also with only **Q**-factorial terminal singularities. Thus we may replace X with Y and go back to (12.9), where we note that $\rho(X) > \rho(Y)$. (This is the case similar to the surface case.)

We should note that D here is uniruled by (11.6.2) (Kollár).

(12.9.2.3) *Case* 3. f is birational and contracts no divisors (*bad contraction*). This case really occurs, and what we can hope is the existence of the *directed flip* $X \dashrightarrow X^+$ (cf. (12.10)).

FLIP CONJECTURE I. There exists a variety X^+ over Y with only **Q**-factorial terminal singularities with a relatively ample canonical divisor over Y which is isomorphic to X outside codimension 2 subsets on both sides.

It is easy to see $\rho(X^+) = \rho(X)$, and we may replace X with X^+ and go back to (12.9). Then it is obvious that good contraction does not occur infinitely many times as we continue (admitting the conjecture), and we hope

FLIP CONJECTURE II. The bad contraction does not occur infinitely many times.

This was solved by Shokurov [**Sho3**] for 3-folds by showing that the invariant *difficulty* $d(X) \in \mathbf{Z}_+$ of X strictly decreases by the directed flip, which is generalized to the case of 4-folds in [**KMM**].

(12.10) EXAMPLE (FRANCIA [**Fr**]). Let X^+ be a smooth 3-fold containing $C \simeq \mathbf{P}^1$ with normal bundle $N_{C/X} \simeq \mathcal{O}(-1) \oplus \mathcal{O}(-2)$, and $X^+ \to B$ the analytic contraction of C. Let Z^+ be the blow-up of X^+ along C; then the exceptional divisor $D^+ \simeq F_1$ contains a negative section C' whose normal bundle in Z^+ is $\mathcal{O}(-1)^{\oplus 2}$. Thus the blow-up T of Z^+ along C' has an exceptional divisor $\simeq F_0$ which can be contracted in the other direction, whence we get $T \to Z^-$ and

$F_1 \subset T$ is mapped to $\mathbf{P}^2 \subset Z^-$ with normal bundle $\mathcal{O}(-2)$. By contracting the $\mathbf{P}^2$, we get X^- with 4-ple point P. One can actually put these in projective varieties so that X^+ is a smooth canonical (hence minimal) model. If we start with nonsingular projective Z^-, then we must have the contraction $Z^- \to X^-$ of an extremal ray, and $X^- \to B$ must be the only possible contraction at the next step. Since (B, Q) is easily seen to be not a canonical singularity (K_B is not even $\mathbf{Q}$-Cartier), we cannot get the unique minimal model X^+ by a morphism. The only way to get to X^+ is the directed flip $X^- \dashrightarrow X^+$.

(12.10.1)

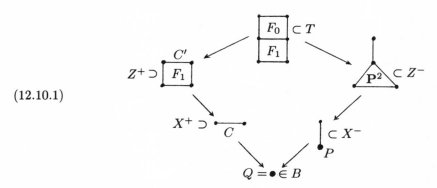

(12.11) One can formulate a log version, and also a relative version using Kollár's vanishing theorem (5.8) (cf. [**KMM**]). Let us mention two important relative cases.

(12.11.1) The process in (12.9), when applied to a birational morphism of projective varieties with only $\mathbf{Q}$-factorial terminal singularities, gives a decomposition of birational morphisms into good contractions and directed flips (modulo Flip Conjectures I and II) because of (9.4.2). Danilov [**Da**] settled the toric case.

(12.11.2) (i) The problem of finding a good model was considered in the context of 1-parameter semistable degeneration of surfaces. Kulikov [**Ku**] and Persson and Pinkham [**PP**] proved that every (sufficiently) semistable degeneration of $K3$ surfaces has a smooth good model which need not be projective, and a similar result for Enriques surfaces and hyperelliptic surfaces S with $2K_S \sim 0$ was proved by D. Morrison [**Mrr1**].

(ii) The existence of minimal models for 1-parameter projective semistable degenerations of surfaces has been announced by Tsunoda [**Ts**], Shokurov [**Sho4**], Mori, and Kawamata [**Ka12**], where the first and the last proofs are available at present in the form of preprint(s). It is easy to see that these results, when applied to the case of degeneration of surfaces S with $mK_S \sim 0$ ($m \leq 6$), find a minimal model whose global index divides m. This is why we did not come across singularities in the works of [**Ku**] and [**PP**].

(iii) Nakayama [**Ny3**] proved that this machinery is generalized to projective morphisms between complex analytic spaces.

(12.12) Based on a result in [**Ka12**], Mori recently announced a proof of the Flip Conjecture I for 3-folds [**Mr4**]. Thus only the Abundance Conjecture

$\nu(X) = \kappa(X)$ for minimal 3-folds X with $\kappa(X) = 0$ remains unsettled. (Added in proof: Miyaoka has settled the case $\nu(X) = 1$ affirmatively.)

REFERENCES

[Ab] S. Abhyankar, *Resolution of singularities of embedded algebraic surfaces*, Academic Press, 1966.

[Ar] M. Artin, *Algebraization of formal moduli*. I, Global Analysis, D. C. Spencer and S. Iyanaga, editors, Univ. of Tokyo Press, Tokyo, 1969, pp. 21–71; II. Ann. of Math. (2) **91** (1970), 88–135.

[ArMf] M. Artin and D. Mumford, *Some elementary examples of unirational varieties which are not rational*, Proc. London Math. Soc. **25** (1972), 75–95.

[Bea] A. Beauville, *Some remarks on Kähler manifolds with* $c_1 = 0$, Progr. Math., vol. 39, Birkhäuser, 1983, pp. 1–26.

[BCSS] A. Beauville, J.-L. Colliot-Thélène, J.-J. Sansuc, and P. S. Swinnerton-Dyer, *Variétés stablement rationnelles non rationnelles*, Ann. of Math. (2) **121** (1985), 283–318.

[BelFr] M. Beltrametti and P. Francia, *Threefolds with negative Kodaira dimension and positive irregularity*, Nagoya Math. J. **91** (1983), 163–172.

[Ben1] X. Benveniste, *Variétés de dimension 3 de type général tel que le systéme linéaire défini par un multiple du diviseur canonique soit sans point base*, C. R. Acad. Sci. Paris Sér. A **289** (1979), 691–694.

[Ben2] ____, *Sur la décomposition de Zariski en dimension 3*, C. R. Acad. Sci. Paris Sér. I **295** (1982), 107–110.

[Ben3] ____, *Sur l'anneau canonique de certaines variétés de dimension 3*, Invent. Math. **73** (1983), 157–164.

[Ben4] ____, *Sur les variétés de dimension 3 de type général dont le diviseur canonique est numériquement positif*, Math. Ann. **266** (1984), 479–497.

[Ben5] ____, *Sur les variétés canoniques de dimension 3 d'indice positif*, Nagoya Math. J. **97** (1985), 137–167.

[Ben6] ____, *Sur le cone des 1-cycles effectifs en dimension 3*, Math. Ann. **272** (1985), 257–265.

[Ben7] ____, *Sur les applications pluricanoniques des variétés de type trés général en dimension 3*, Preprint, Paris.

[BlG] S. Bloch and D. Gieseker, *The positivity of the Chern classes of an ample vector bundle*, Invent. Math. **12** (1971), 112–117.

[Bo] F. Bogomolov, *On the decomposition of Kähler manifolds with trivial canonical class*, Math. USSR-Sb. **22** (1974), 580–583.

[CKS] E. Cattani, A. Kaplan, and W. Schmid, *Degeneration of Hodge structures*, Ann. of Math. (2) **123** (1986), 457–535.

[CG] H. Clemens and P. A. Griffiths, *The intermediate Jacobian of the cubic threefolds*, Ann. of Math. (2) **95** (1972), 281–356.

[Ct] D. Cutkosky, *Zariski decomposition of divisors on algebraic varieties*, Duke Math. J. **53** (1986), 149–156.

[Da] V. I. Danilov, *Birational geometry of toric 3-folds*, Math. USSR-Izv. **21** (1983), 269–279.

[D-J] P. Dubois and P. Jarraud, *Une propriete de commutation de changement de base*, C. R. Acad. Sci. Paris. Sér. A **279** (1974), 745–747.

[El] R. Elkik, *Rationalité des singularités canoniques*, Invent. Math. **47** (1978), 139–147.

[Es1] H. Esnault, *Classification des variétés de dimension 3 et plus*, Sém. Bourbaki, 1980/81, Exp. 568, Lecture Notes in Math., vol. 901, Springer-Verlag, 1981, pp. 111–131.

[Es2] ____, *Fibre de Milnor d'un cóne sur une curve plane singuliére*, Invent. Math. **68** (1982), 477–496.

[EsV] H. Esnault and E. Viehweg, *Revétements cycliques*. II (*autour du théoréme d'annulation de J. Kollár*), Proceedings, La Rábida, 1984 (to appear).

[EsV2] ____, *Logarithmic De Rham complexes and vanishing theorems*, Invent. Math. **86** (1986), 161–194.

[Fln] H. Flenner, *Rational singularities*, Arch. Math. **36** (1981), 35–44.

[Flt] A. R. Fletcher, *Contributions to Riemann-Roch on projective 3-folds with only canonical singularities and applications*, these Proceedings, Part 1, pp. 221–231.

[Fr] P. Francia, *Some remarks on minimal models*. I, Compositio Math. **40** (1980), 301–313.

[FMrr] R. Friedman and D. Morrison (Editors), *The birational geometry of degenerations*, Progr. Math., vol. 29, Birkhäuser, 1983.

[Fk1] A. Fujiki, *On the minimal models of complex manifolds*, Math. Ann. **253** (1980), 111–128.

[Fk2] ____, *Deformation of uni-ruled manifolds*, Publ. Res. Inst. Math. Sci. **17** (1981), 687–702.

[Fk3] ____, *Coarse moduli space for polarized compact Kähler manifolds*, Preprint, Kyoto.

[Ft1] T. Fujita, *Some remarks on Kodaira dimensions of fiber spaces*, Proc. Japan Acad. **53** (1977), 28–30.

[Ft2] ____, *The theory of Kodaira dimension–its past, present and future*, Sûgaku **30** (1978), 243–254. (Japanese)

[Ft3] ____, *The sheaf of relative canonical forms of a Kähler fibre space over a curve*, Proc. Japan Acad. **54** (1978), 183–184.

[Ft4] ____, *On Kähler fiber spaces over curves*, J. Math. Soc. Japan **30** (1978), 779–794.

[Ft5] ____, *On L-dimension of coherent sheaves*, J. Fac. Sci. Univ. Tokyo Sect. IA Math. **28** (1981), 215–236; Correction **29** (1982), 719–720.

[Ft6] ____, *Canonical rings of algebraic varieties*, Classification of Algebraic and Analytic Manifolds (Katata, 1982), K. Ueno, editor, Progr. Math., vol. 39, Birkhäuser, 1983, pp. 65–70.

[Ft7] ____, *Fractionally logarithmic canonical rings of algebraic surfaces*, J. Fac. Sci. Univ. Tokyo Sect. IA Math. **30** (1984), 685–696.

[Ft8] ____, *Zariski decomposition and canonical rings of elliptic threefolds*, J. Math. Soc. Japan **38** (1986), 20–37.

[Ft9] ____, *A relative version of Kawamata-Viehweg's vanishing theorem*, Preprint, Tokyo.

[Ft10] ____, *Generalized adjunction mappings*, Proc. Sympos. Algebraic Geom. (Sendai, 1985) Adv. Stud. Pure Math., Vol. 10, North-Holland, 1987.

[G] P. A. Griffiths, *Periods of integrals on algebraic manifolds*. III, Inst. Hautes Études Sci. Publ. Math. **38** (1978), 779–794.

[GHa] P. A. Griffiths and J. Harris, *Algebraic geometry and local differential geometry*, Ann. Sci. École Normale Sup. (4) **12** (1979), 355–452.

[GsRa] L. Gruson and M. Raynaud, *Critères de platitude et de projectivité*, Invent. Math. **13** (1971), 1–89.

[Ha] M. Hanamura, *Pluricanonical maps of minimal 3-folds*, Proc. Japan Acad. Ser. A **61** (1985), 116–118.

[HT] T. Hayakawa and K. Takuechi, *On canonical singularities of dimension three*, Japan. J. Math. (to appear).

[Hi1] H. Hironaka, *On the theory of birational blowing-up*, Thesis, Harvard Univ., Cambridge, Mass., 1960.

[Hi2] ____, *Resolution of an algebraic variety over a field of characteristic zero*. I, II, Ann. of Math. (2) **79** (1964).

[Hi3] ____, *Flattening of analytic maps*, Manifolds—Tokyo 1973 (Proc. Internat. Conf., Tokyo, 1983), Univ. of Tokyo Press, Tokyo, 1975, pp. 313–322.

[I1] S. Iitaka, *Genera and classification of algebraic varieties*. 1, Sûgaku **24** (1972), 14–27. (Japanese)

[I2] ____, *On D-dimensions of algebraic varieties*, J. Math. Soc. Japan **23** (1971), 356–373.

[I3] ____, *Deformation of complex surfaces*. II, J. Math. Soc. Japan **22** (1976), 247–261.

[I4] ____, *Birational geometry of algebraic varieties*, Proc. Internat. Congr. Math. (Warsaw, 1983), Polish Scientific Publishers, Warsaw, 1984, pp. 727–732.

[Is1] V. A. Iskoviskih, *Fano 3-folds*. I, Math. USSR-Izv. **11** (1977), 485–527.

[Is2] ____, *Fano 3-folds*. II, Math. USSR-Izv. **12** (1978), 496–506.

[IM] V. A. Iskovskih and Yu. I. Manin, *Three-dimensional quartics and counter examples to the Lüroth problem*, Math. USSR-Sb. **15** (1971), 141–166.

[Ks] M. Kashiwara, *The asymptotic behavior of a variation of polarized Hodge structure*, Publ. Res. Inst. Math. Sci. (to appear).

[Kz] N. Katz, *The regularity theorem in algebraic geometry*, Proc. Internat. Congr. Math. (Nice, 1970), Vol. 1, Gauthier-Villars, Paris, 1971, pp. 437–443.

[Ka1] Y. Kawamata, *Characterization of abelian varieties*, Compositio Math. **43** (1981), 253–276.

[Ka2] ____, *A generalization of Kodaira-Ramanujam's vanishing theorem*, Math. Ann. **261** (1982), 57–71.

[Ka3] ____, *Kodaira dimension of algebraic fiber spaces over curves*, Invent. Math. **66** (1982), 57–71.

[Ka4] ____, *Kodaira dimension of certain algebraic fiber spaces*, J. Fac. Sci. Univ. Tokyo Sect. IA Math. **30** (1983), 1–24.

[Ka5] ____, *Hodge theory and Kodaira dimension*, Algebraic Varieties and Analytic Varieties, Adv. Stud. Pure Math., vol. 1, North-Holland, 1983, pp. 317–327.

[Ka6] ____, *On the finiteness of generators of a pluricanonical ring for a 3-fold of general type*, Amer. J. Math. **106** (1984), 1503–1512.

[Ka7] ____, *Elementary contractions of algebraic 3-folds*, Ann. of Math. (2) **119** (1984), 95–110.

[Ka8] ____, *The cone of curves of algebraic varieties*, Ann. of Math. (2) **119** (1984), 603–633.

[Ka9] ____, *Pluricanonical systems on minimal algebraic varieties*, Invent. Math. **79** (1985), 567–588.

[Ka10] ____, *Minimal models and the Kodaira dimension of algebraic fiber spaces*, J. Reine Angew. Math. **363** (1985), 1–46.

[Ka11] ____, *On the plurigenera of minimal algebraic 3-folds with $K \approx 0$*, Math. Ann. **275** (1986), 539–546.

[Ka12] ____, *The crepant blowing-ups of 3-dimensional canonical singularities and its application to degenerations of surfaces*, Ann. of Math. (2) (to appear).

[Ka13] ____, *The Zariski decomposition of log-canonical divisors*, these Proceedings, Part 1, pp. 425–433.

[KM] Y. Kawamata and K. Matsuki, *The number of the minimal models for a 3-fold of general type is finite*, Math. Ann. (to appear).

[KMM] Y. Kawamata, K. Matsuda, and K. Matsuki, *Introduction to the minimal model problem*, Proc. Sympos. Algebraic Geom., Sendai, 1985, Adv. Stud. Pure Math., vol. 10, North-Holland, 1987.

[KaV] Y. Kawamata and E. Viehweg, *On a characterization of abelian varieties in the classification theory of algebraic varieties*, Compositio Math. **41** (1980), 355–360.

[KKMS] G. Kempf, F. Knudsen, D. Mumford, and B. Saint-Donat, *Toroidal embeddings I*, Lecture Notes in Math., vol. 339, Springer-Verlag, 1973.

[Kl] S. Kleiman, *Toward a numerical theory of ampleness*, Ann. of Math. (2) **84** (1966), 293–344.

[Ko1] J. Kollár, *Toward moduli of singular varieties*, Thesis, Brandeis Univ., Waltham, Mass., 1983.

[Ko2] ____, *The cone theorem: Note to a paper of Y. Kawamata*, **119** (1984), 603–633, Ann. of Math. (2) **120** (1984), 1–5.

[Ko3] ____, *Higher direct images of dualizing sheaves*, Ann. of Math. (2) **123** (1986), 11–42.

[Ko4] ____, *Higher direct images of dualizing sheaves. II*, Ann. of Math (2) **124** (1986), 171–202.

[Ko5] ____, *Subadditivity of the Kodaira dimension: Fibers of general type*, Proc. Sympos. Algebraic Geom., Sendai, 1985, Adv. Stud. Pure Math., vol. 10, North-Holland, 1987.

[Ko6] ____, *Vanishing theorems for cohomology groups*, these Proceedings, Part 1, pp. 233–243.

[Ko7] ____, *The structure of algebraic threefolds—an introduction to Mori's program*, Bull. Amer. Math. Soc. (N.S.) (to appear).

[Ku] V. Kulikov, *Degenerations of K3 surfaces and Enriques surfaces*, Math. USSR-Izv. **11** (1977), 957–989.

[L1] M. Levin, *Deformation of uni-ruled varieties*, Duke Math. J. **48** (1981), 467–473.

[L2] ____, *Pluri-canonical divisors on Kähler manifolds*, Invent. Math. **74** (1983), 293–303.

[**LWP**] D. Lieberman, R. Wilsker, and C. Peters, *A theorem of local-Torelli type*, Math. Ann. **231** (1977), 39–45.

[**Lu**] Z. Luo, *Kodaira dimension of algebraic function fields*, Thesis, Brandeis Univ., Waltham, Mass., 1985.

[**Mh**] K. Maehara, *The weak effectivity of direct image sheaves*, Tokyo.

[**Mk**] K. Matsuki, *On pluricanonical maps for 3-folds of general type*, J. Math. Soc. Japan **38** (1986), 339–359.

[**Ms1**] T. Matsusaka, *On canonically polarized varieties. II*, Amer. J. Math. **92** (1970), 283–292.

[**Ms2**] ____, *Polarized varieties with a given Hilbert polynomial*, Amer. J. Math. **94** (1972), 1027–1077.

[**MsMf**] T. Matsusaka and D. Mumford, *Two fundamental theorems on deformations of polarized varieties*, Amer. J. Math. **86** (1964), 668–684.

[**Mn**] M. Miyanishi, *Projective degeneration of surfaces according to S. Tsunoda*, Proc. Sympos. Algebraic Geom., Sendai, 1985, Adv. Stud. Pure Math., vol. 10, North-Holland, 1987.

[**MnT**] M. Miyanishi and S. Tsunoda, *Projective degenerations of algebraic surfaces*, Proc. Sympos. Algebraic Geom., Sendai, 1985, Adv. Stud. Pure Math., vol. 10, North-Holland, 1987.

[**My1**] Y. Miyaoka, *Deformations of a morphism along a foliation*, these Proceedings, Part 1, pp. 245–268.

[**My2**] ____, *The pseudo-effectivity of $3c_2 - c_1^2$ for threefolds with numerically effective canonical classes*, Proc. Sympos. Algebraic Geom., Sendai, 1985, Adv. Stud. Pure Math., vol. 10, North-Holland, 1987.

[**My3**] ____, *On the structure of uniruled varieties*, Preprint, Tokyo.

[**MyMr**] Y. Miyaoka and S. Mori, *A numerical criterion of uniruledness*, Ann. of Math. (2) **124** (1986), 65–69.

[**Mz1**] B. G. Moishezon, *On n-dimensional compact complex varieties with n algebraically independent meromorphic functions. I, II, III*, Math. USSR Izv. **63** (1967), 51–117.

[**Mz2**] ____, *Algebraic varieties and compact complex spaces*, Proc. Internat. Congr. Math. (Nice, 1970), Vol. 2, Gauthier-Villars, Paris, 1971, pp. 643–648.

[**Mr1**] S. Mori, *Projective manifolds with ample tangent bundles*, Ann. of Math. (2) **110** (1979), 593–606.

[**Mr2**] ____, *Threefolds whose canonical bundles are not numerically effective*, Ann. of Math. (2) **116** (1982), 133–176.

[**Mr3**] ____, *On 3-dimensional terminal singularities*, Nagoya Math. J. **98** (1985), 43–66.

[**Mr4**] ____, *Flip conjecture and the existence of minimal models for 3-folds*, submitted to J. Amer. Math. Soc.

[**MrMuk**] S. Mori and S. Mukai, *Classification of Fano 3-folds with $B_2 \geq 2$*, Manuscripta Math. **36** (1981), 147–162.

[**Mwk1**] A. Moriwaki, *Torsion freeness of higher direct images of canonical bundles*, Preprint.

[**Mwk2**] ____, *Semi-ampleness of the numerically effective part of Zariski decomposition*, J. Math. Kyoto Univ. **26** (1986), 465–481.

[**Mwk3**] ____, *Several properties of Zariski decomposition*, Preprint.

[**Mrr1**] D. Morrison, *Semistable degenerations of Enriques and hyperelliptic surfaces*, Duke Math. J. **48** (1981), 197–249.

[**Mrr2**] ____, *A remark on Kawamata's paper "On the plurigenera of minimal algebraic 3-folds with $K \approx 0$,"* Math. Ann. **275** (1986), 547–553.

[**MS**] D. Morrison and G. Stevens, *Terminal quotient singularities in dimensions 3 and 4*, Proc. Amer. Math. Soc. **90** (1984), 15–20.

[**Muk**] S. Mukai, *Symplectic structure of the moduli space of sheaves on an abelian or K3 surface*, Invent. Math. **77** (1984), 101–116.

[**Mf1**] D. Mumford, *The canonical ring of an algebraic surface*, Appendix to Zariski's paper "*The theorem of Riemann-Roch for high multiples of a divisor*," Ann. of Math. (2) **76** (1962), 612–615.

[**Mf2**] ____, *Abelian varieties*, Tata Inst. Fund. Res., Bombay, and Oxford Univ. Press, 1970.

[MfF] D. Mumford and J. Fogarty, *Geometric invariant theory*, 2nd. ed., Springer-Verlag, 1982.

[Na] M. Nagata, *Existence theorems for nonprojective complete algebraic varieties*, Illinois J. Math. **2** (1958), 490–498.

[NU] I. Nakamura and K. Ueno, *An addition formula for Kodaira dimensions of analytic fibre bundles whose fibers are Moishezon manifolds*, J. Math. Soc. Japan **25** (1973), 363–371.

[Ny1] N. Nakayama, *Invariance of the plurigenera of algebraic varieties under minimal model conjectures*, Topology **25** (1986), 237–251.

[Ny2] ____, *Hodge filtrations and the higher direct images of canonical sheaves*, Invent. Math. **85** (1986), 217–221.

[Ny3] ____, *The lower semi-continuity of the plurigenera*, Proc. Sympos. Algebraic Geom., Sendai, 1985, Adv. Stud. Pure Math. vol. 10, North-Holland, 1987.

[Ny4] ____, *On Weierstrass models*, Preprint, Tokyo.

[Ny5] ____, *The singularity of the canoncial model of compact Kaehler manifold*, Preprint, Tokyo.

[No] Y. Norimatsu, *Kodaira vanishing theorem and Chern classes for ∂-manifolds*, Proc. Japan. Acad. Ser. A. Math. Sci. **54** (1978), 107–108.

[PP] U. Persson and H. Pinkham, *Degeneration of surfaces with trivial canonical divisor*, Ann. of Math. (2) **113** (1981), 45–66.

[Rm] C. P. Ramanujam, *Remarks on the Kodaira vanishing theorem*, J. Indian Math. Soc. **36** (1972), 41–51.

[Re1] M. Reid, *Canonical 3-folds*, Proc. Conf. Algebraic Geom. (Angers, 1979), Sijthoff and Nordhoff, Alphen aan den Rijn, 1980, pp. 273–310.

[Re2] ____, *Minimal models of canonical 3-folds*, Algebraic Varieties and Analytic Varieties, Adv. Stud. Pure Math., vol. 1, North-Holland, Amsterdam, and Kinokuniya Book Co., Tokyo, 1981, pp. 131–180.

[Re3] ____, *Decomposition of toric morphisms*, Arithmetic and Geometry. II, M. Artin and J. Tate, editors, Progr. Math., vol. 36, Birkhäuser, 1983, pp. 395–418.

[Re4] ____, *Projective morphisms according to Kawamata*, Preprint, Warwick.

[Re5] ____, *Young person's guide to canonical singularities*, these Proceedings, Part 1, pp. 345–414.

[Re6] ____, , *Tendencious survey of 3-folds*, these Proceedings, Part 1, pp. 333–344.

[Sai] Morihiko Saito, *Mixed Hodge modules*, Preprint, Kyoto.

[Sar] V. G. Sarkisov, *On the structure of conic bundles*, Izv. Akad. Nauk SSSR Ser. Mat. **46** (1982), 371–408.

[Sch] W. Schmid, *Variation of Hodge structure: The singularities of period mapping*, Invent. Math. **22** (1973), 211–319.

[She] N. Shepherd-Barron, *Some questions on singularities in 2 and 3 dimensions*, Thesis, Univ. of Warwick, 1980.

[Sho1] V. V. Shokurov, *Smoothness of the general anticanonical divisor on a Fano 3-fold*, Math. USSR-Izv. **14** (1980), 395–405.

[Sho2] ____, *The existence of a straight line on Fano 3-folds*, Math. USSR-Izv. **15** (1980), 173–209.

[Sho3] ____, *Theorem on non-vanishing*, Izv. Akad. Nauk SSSR Ser. Mat. **49** (1985), 635–651.

[Sho4] ____, *Letter to M. Reid dated May 24, 1985.*

[T] S. G. Tankeev, *On n-dimensional canonically polarized varieties and varieties of fundamental type*, Izv. Akad. Nauk SSSR Ser. Mat. **35** (1971), 31–44.

[Ts] S. Tsunoda, *Semi-stable degeneration of surfaces*, Proc. Sympos. Algebraic Geom., Sendai, 1985, Adv. in Math.

[U1] K. Ueno, *Classification theory of algebraic varieties and compact complex spaces*, Lecture Notes in Math., vol. 439, Springer-Verlag, 1975.

[U2] ____, *Kodaira dimension of certain fibre spaces*, Complex Analysis and Algebraic Geometry, Iwanami Shoten, Tokyo, 1977, pp. 279–292.

[U3] ____, *Classification of algebraic varieties. II: Algebraic threefolds of parabolic type*, Internat. Sympos. Algebraic Geom. (Kyoto, 1977), Kinokuniya, Tokyo, pp. 693–708.

[U4] ____, *On algebraic fibre spaces of abelian varieties*, Math. Ann. **237** (1978), 1–22.

[U5] ____, *Birational geometry of algebraic threefolds*, Géométrie Algébrique, Angers (1979), Sijthoff and Noordhoff, 1980, pp. 311–323.

[U6] ____, (Editor), *Classification of algebraic and analytic manifolds*, (Proc. Katata Sympos., 1982), Progr. Math., vol. 39, Birkhäuser, 1983.

[V1] E. Viehweg, *Canonical divisors and the additivity of the Kodaira dimension for morphisms of relative dimension one*, Compositio Math. **35** (1977), 197–233.

[V2] ____, *Rational singularities of higher dimensional schemes*, Proc. Amer. Math. Soc. **63** (1977), 6–8.

[V3] ____, *Klassifikationstheorie algebraischer Varietäten der Dimension drei*, Compositio Math. **41** (1980), 361–400.

[V4] ____, *Die Additivität der Kodaira Dimension für projektive Faserraüme über Varietäten des allgemeinen Typs*, J. Reine Angew. Math. **330** (1982), 132–142.

[V5] ____, *Vanishing theorems*, J. Reine Angew. Math. **335** (1982), 1–8.

[V6] ____, *Weak positivity and the additivity of the Kodaira dimension for certain fiber spaces*, Adv. Stud. Pure Math., vol. 1, North-Holland, 1983, pp. 329–353.

[V7] ____, *Weak positivity and the additivity of the Kodaira dimension. II: The local Torelli map*, Progr. Math., vol. 39, Birkhäuser, 1983, 567–589.

[V8] ____, *Vanishing theorems and positivity in algebraic fiber spaces*, Proc. Internat. Congr. Math. (Berkeley, Calif., 1986).

[W1] P. M. H. Wilson, *On complex algebraic varieties of general type*, Symposia Mathematica, vol. 24, Academic Press, 1981, pp. 65–73.

[W2] ____, *On the canonical ring of algebraic varieties*, Compositio Math. **43** (1981), 365–385.

[W3] ____, *On regular threefolds with $\kappa = 0$*, Invent. Math. **76** (1984), 345–355.

[W4] ____, *Towards birational classification of algebraic varieties*, Bull. London Math. Soc. **19** (1987), 1–48.

[Za] O. Zariski, *The theorem of Riemann-Roch for high multiples of an effective divisor on an algebraic surface*, Ann. of Math. (2) **76** (1962), 560–615.

[Zu] S. Zucker, *Remarks on a theorem of Fujita*, J. Math. Soc. Japan **34** (1982), 47–54.

COLUMBIA UNIVERSITY AND NAGOYA UNIVERSITY

Proceedings of Symposia in Pure Mathematics
Volume **46** (1987)

Tendencious Survey of 3-folds

MILES REID

$-\infty$. The A.M.S. Summer Institute at Bowdoin College is the first major
conference on algebraic geometry since the extinction of the dinosaurs; in the
absence of Mumford, Griffiths, and Hironaka, all kinds of creepy crawly creatures
come out from under their stones to stake out their position in the new order of
things.

DEFINITION. The *classification of varieties* is the attempt to study all algebraic varieties by dividing them into

varieties with $-K_X$ ample,

varieties with K_X numerically trivial,

varieties with K_X ample,

and fibre spaces of these.

This is a propaganda talk; its only purpose is to emphasise my conviction
that the right approach to classification is via minimal models and Mori theory.
To pursue the biological imagery: the Iitaka program goes back 20 years [**6**] and
has grown into a great school with many important results to its credit; minimal
models has at present only a handful of active practitioners. Nevertheless, I
am quite certain that within a few years the approach via minimal models will
entirely gobble up the Iitaka program. This is in no way to imply that the Iitaka
program will die out: on the contrary, it will become embedded as an important
part—perhaps the most essential part—of a much more precise and powerful
theory.

There are three preliminary things to say.

(A) *Disclaimer*. Since I may say something controversial, I would like to make
it clear that value-judgements expressed here are purely my own.

(B) *Biregular versus birational*. The theory which is advocated here is essentially biregular. Perhaps in many cases the constructions under discussion (for
example, canonical linear systems) will be birational invariants, but they will
nevertheless be studied on a particular biregular model. In this connection, I
make the following claim: there is no birational statement in the classification

1980 *Mathematics Subject Classification* (1985 *Revision*). Primary 14J10.

of surfaces which cannot be deduced as a corollary of more precise biregular statements.

(C) *Numerical versus analytic.* I propose that we work first of all with numerical properties of K_X, and only subsequently consider the analytic problem of the existence of global holomorphic differential forms. There is a very important principle at stake here: a priori, there might exist a nonsingular variety X with K_X nef, $K_X \not\approx 0$, but nevertheless $\kappa(X) = -\infty$; such an X would certainly not be uniruled, so it is clearly wrong to lump it together with the familiar $\kappa = -\infty$ varieties of the Iitaka program. (Miyaoka has recently proved [**10**] that in the 3-fold case, no such X exists, even if you allow canonical singularities.)

A classification should have two parts: first we break everything up into big classes (§§1–2), then try to say something about varieties in each class (see §4 and [**13**]). Perhaps I should advise newcomers to classification that detailed work within just one class may be the surest way into the subject—you learn to get around your own village before you set about memorising the entire motorway network. There are still plenty of good research problems in rational surfaces.

Today's talk will attempt to flesh out the definition of classification given above, mention the two big outstanding conjectures ("flip conjecture" and "abundance conjecture"), and will make some remarks about some of the individual classes. I will not make any mention of work in dimension ≥ 4.

There will be no detailed attribution of results, since just making a flow-chart of major lines of investigation would take up most of the lecture. Let me say that the major conceptual advance in minimal models is due to S. Mori. Technically, much of the machinery since the early work of Iitaka and Ueno has been put in place (in rough chronological order) by Fujita, Viehweg, Kawamata, Benveniste, Kollár, Wilson, and Miyaoka, all but one of whom are in the present audience, and by two Russians, F. A. (Fedya) Bogomolov and V. V. (Slava) Shokurov. As usual at international conferences, we keenly feel the absence of our Soviet colleagues, and the tragic damage to Soviet science resulting from it.

0. Preliminaries.

(0.1) I work with a projective variety X over the field $k = \mathbf{C}$ (geometers, GOTO (0.2)). However, any number-theorist familiar with the theory over $\mathbf{C}$ will see at once that, essentially, k could be any field of characteristic zero, provided that the cycle groups N^1 and N_1 are made up using cycles defined over k: the point is just that if a construction on X, for example, a morphism $\Phi \colon X \to \mathbf{P}^N$ is defined sufficiently intrinsically, it will automatically be defined over k. The condition on the characteristic may also be removed when the characteristic p people have sorted out vanishing.

(0.2) I assume that X has canonical singularities; the essential part of this definition is that rK_X is a Cartier divisor for some $r \in \mathbf{Z}$, $r > 0$. Notice that (up to minor technical details)

$$K_X \text{ is Cartier} \iff X \text{ is Gorenstein,}$$

which is good, since if we're lucky, X may have only hypersurface singularities; and

$$rK_X \text{ is Cartier } \Leftrightarrow \text{ locally, } X \cong (\text{Gorenstein})/(\mathbf{Z}/r).$$

(0.3) *Algebraic fibre space.* If $f: X \to Y$ is a morphism of projective varieties, with X normal, the following 3 conditions are equivalent:

(i) $f_*\mathcal{O}_X = \mathcal{O}_Y$;

(ii) Y is normal (that is, integrally closed) in the function field $k(X)$;

(iii) Y is a normal variety and f has connected general fibre.

Furthermore, the condition can always be achieved: this is Zariski's theory of projective normalisation, or Stein factorisation chez Grothendieck.

In this case, $f: X \to Y$ is a (projective, algebraic) *fibre space*. Note that the two degenerate cases $Y = \text{pt}$ and f birational are not excluded. If f is birational, then according to the point of view, f will be called a *birational contraction* of X, or a *partial resolution* of Y.

1. Case K_X not nef: Mori theory.

(1.1) *Notation.* $N^1 = N^1_{\mathbf{R}}(X)$ is the real vector spaces of Cartier divisors of X modulo $\approx$, that is, the Néron-Severi group $\otimes\mathbf{R}$; its mate N_1 is the vector space of 1-cycles with $\mathbf{R}$-coefficients modulo $\approx$. Here $\approx$ denotes numerical equivalence, which by definition is the smallest equivalence relation which makes N^1 and N_1 into dual vector spaces. $N^1_{\mathbf{Q}}, N^1_{\mathbf{Z}}, \ldots$ are the obvious rational $\ldots$ analogues. By tradition, $\rho(X) = \dim N^1(X)$.

(1.2) In Figure 1 I draw the Theorem on the Cone.

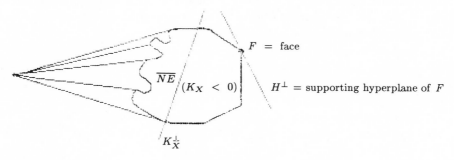

$F = \text{face}$

$H^\perp = \text{supporting hyperplane of } F$

FIGURE 1

Here $\overline{NE} \subset N_1 X$ is the closed convex cone of effective 1-cycles; I've drawn it wiggly on the left-hand side to emphasise that it's a more-or-less arbitrary closed convex cone. However, the exciting thing is the right-hand side: strictly away from the half-space $\{z | K_X z \geq 0\}$, the cone is polyhedral, with rational faces. The condition that K_X is not nef is exactly that this side is nonempty.

(1.3) Let F be a face of $\overline{NE}$ in the negative half-space; a supporting hyperplane corresponding to F is given by an element $H \in N^1_{\mathbf{Q}}$ (a linear form on N_1) such that

$$Hz \geq 0 \quad \text{for all } z \in \overline{NE}, \quad \text{and} \quad Hz = 0 \Leftrightarrow z \in F.$$

By taking a multiple, I assume H is represented by $H \in \mathrm{Pic} X$; then H is nef, and since the face F is in the negative half-space, $H - \varepsilon K_X$ is ample for $\varepsilon \in \mathbf{Q}$ with $0 < \varepsilon \ll 1$ (by the Kleiman criterion, see [23, p. 42]).

(1.4) THEOREM ON THE CONE AND CONTRACTION. K_X not nef implies that there exists a projective fibre space $f\colon X \to Y$ contracting at least one curve $\Gamma \subset X$ (this just to exclude isomorphisms) such that

$$-K_X \text{ is relatively ample for } f.$$

For surfaces, this corresponds to Castelnuovo's contractibility criterion and to the characterisation of rational and ruled surfaces. The proof of the theorem is essentially just a technical reworking of that for surfaces as given in Barth's first lecture (see the survey article [22, (3.1–2)]), although chronologically the 3-fold case came first.

It must be emphasised that the contraction morphism here and that in (2.3) below are quite different. In particular, the morphism here depends on choosing a face F of $\overline{NE}$. This choice is essential: the canonical model (if it exists) will be unique, but not the minimal model.

(1.5) If X is a 3-fold, and $\dim Y = 0, 1$, or 2, the rough division into cases is complete:

  $\dim Y = 0$: $-K_X$ is ample, X is a $\mathbf{Q}$-Fano 3-fold;
  $\dim Y = 1$: fibre space of del Pezzo surfaces;
  $\dim Y = 2$: conic bundle.

(1.6) If $\dim Y = 3$, then f is a birational contraction, and I've contracted out a piece of X on which $-K_X$ is ample; this is an exact analogue of Castelnuovo's contractibility criterion: if X were a smooth surface, f would have to be a contraction of a number of disjoint (-1)-curves.

There are two cases:

(a) Y has again canonical singularities; this happens (roughly) if f contracts the right number of divisors. Then Y has $\rho(Y) < \rho(X)$, and I can replace the study of X by that of Y.

(b) Y has worse singularities. This will happen whenever K_Y fails to be $\mathbf{Q}$-Cartier; for example, this will certainly be the case if f contracts only a finite number of curves.

At present, (b) is the main stumbling-block to further progress. Before giving the conjecture, I discuss an example due to P. Francia [4].

(1.7) EXAMPLE. Let $F \subset \mathbf{P}^4$ be the cubic scroll, that is, $\mathbf{F}_1$ embedded by $|2A + B|$, and let $Y \subset \mathbf{A}^5$ be the affine cone over F; $P \in Y$ is considered as a singularity, that is, up to local analytic isomorphism, so it could very well live on a variety of general type. Then since

$$K_F = -3A - 2B, \qquad \mathcal{O}_F(1) = 2A + B,$$

it follows that K_F is not a rational multiple of $\mathcal{O}(1)$; therefore $K_Y \notin \mathrm{Pic} Y \otimes \mathbf{Q}$.

Now take the composite of the natural projection $Y - P \to \mathbf{F}_1 \to \mathbf{P}^1$, and let $Y^+ \subset Y \times \mathbf{P}^1 \subset \mathbf{P}^4 \times \mathbf{P}^1$ be the closed graph of the resulting rational map

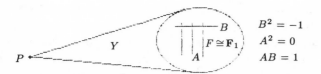

$B^2 = -1$
$A^2 = 0$
$AB = 1$

$Y \dashrightarrow \mathbf{P}^1$. Then you can see (assuming you care about it) that $g\colon Y^+ \to Y$ has the properties

(i) Y^+ is nonsingular;

(ii) $l = g^{-1}P \cong \mathbf{P}^1$, with normal bundle $\mathcal{O}(-1) \oplus \mathcal{O}(-2)$.

Therefore by the adjunction formula,

$$K_{Y^+} \cdot l = 1,$$

so K_{Y^+} is relatively ample for $g\colon Y^+ \to Y$.

It is not hard to see [4] that the singularity $P \in Y$ can be obtained as the result of a contraction of type (b) above:

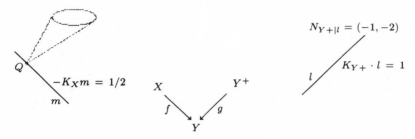

In the picture, X has a single singularity Q of type "cone over the Veronese", and f contracts a line $m \cong \mathbf{P}^1$ passing through Q with $-K_X m = \frac{1}{2}$; this is a particular case of a directed flip.

(1.8) FLIP CONJECTURE. Suppose that $f\colon X \to Y$ is a birational contraction falling into the case (1.6(b)). Then there exists a relative canonical model of Y, that is, a partial resolution $g\colon Y^+ \to Y$ such that Y^+ has canonical singularities and K_{Y^+} is relatively ample for g. The rational map $h = g^{-1} \circ f$ given as

will be called a (canonically) directed flip.

This statement, a local version of the finite generation of the global canonical ring of a 3-fold of general type, is as it stands very rough, since I have not made detailed restrictions on the singularities on X, their local class groups, or the number and kind of contracted curves and divisors.

(1.9) For the purposes of Mori theory, it would be enough to prove (1.8) under much finer assumptions (see, e.g., [15, (3.7)]): we could assume that X has only analytically $\mathbf{Q}$-factorial terminal singularities and f contracts a single curve (necessarily $\cong \mathbf{P}^1$). In this case, g will also contract only a finite number of curves. Some indirect evidence for it comes from the case of symmetric flips arising from Brieskorn's work on the simultaneous resolution of Du Val surface singularities, where the flips act in a natural way as reflections in a root system (see, e.g., [16, (8.2)]). In addition, the important particular case arising from a semistable degeneration of surfaces has been claimed recently by several authors, (see, e.g., [24]) and I expect that this case will be in place soon.

(1.10) *Minimal models.* Assuming the flip conjecture, we get the following reduction to minimal models: given any X with K_X not nef, there is a chain

$$X \dashrightarrow X_1 \dashrightarrow \cdots \dashrightarrow X_n = Y$$

such that each step is
  either (a) a birational contraction as in (1.6(a)),
  or (b) a directed flip as in (1.8),
and such that the final model Y satisfies
  either (1) K_Y is nef,
  or (2) K_Y is not nef, and has a contraction $f \colon Y \to Z$ such that $\dim Z \leq 2$ as in (1.4).

To prove this, in addition to showing that the flip exists, we have to show that we get to the final model satisfying (1) or (2) after a finite number of steps. One suggestion around for some time is that is should be possible to find a combinatorial measure of the singularities of X which decreases under each flip, and this has recently been confirmed in a very simple and elegant way [21, (2.17)].

2. Case K_X nef.

(2.1) DEFINITION. If D is a nef divisor on a projective variety X, the *characteristic dimension* of D, or the *numerical κ-dimension* of D is defined by

$$\nu(D) = \max\{k \mid D^k \not\approx O\};$$

that is, for any ample H, ν is uniquely determined by

$$D^\nu \cdot H^{3-\nu} \neq 0, \qquad D^{\nu+1} \cdot H^{3-\nu-1} = 0,$$

where $3 = \dim X$.

If the linear system $|nD|$ is free for some $n > 0$, then obviously ν is the dimension of $\Phi_{nD}(X)$, where Φ_{nD} is the morphism defined by $|nD|$.

(2.2) If K_X is nef on a projective 3-fold, then using RR and vanishing gives

$$H^0(H + nK_X) \sim n^\nu \quad \text{as } n \to +\infty,$$

where $\nu = \nu(K_X)$, for any ample H (by a suitable choice of H, it is easy to arrange that $H + nK_X$ is very ample for every $n \geq 0$). Perhaps this formula should be viewed as the true definition of $\nu(K_X)$; unfortunately, although we're promised some nice big spaces $H^0(H + nK_X)$ for $n \gg 0$, there don't seem to be too many ways of exploiting these.

(2.3) The following guess seems to be conventional wisdom:

CONJECTURE. K_X is eventually free, that is

$$|mK_X| \text{ is free for } m \text{ large and divisible.}$$

If true, then $|mK_X|$ eventually defines a fibre space $\Phi_{mK} \colon X \to Y$ with $\dim Y = \nu(X) = \kappa(X)$, and with fibres F satisfying $mK_F \sim 0$ (linear equivalence). This is at least what happens for surfaces, and the proof of it is quite mysterious and subtle (see [14] and [2]).

(2.4) DEFINITION. A nef divisor D is *abundant* if $\kappa(D) = \nu(D)$.

If $\nu(D) = \dim X = n$, that is, D is *big*, then automatically $\kappa(D) = \nu(D)$ by RR, vanishing, and a little induction on n. Abundance is intended to generalise this; the word is intended to convey the idea that D and its multiplies have "plenty of sections".

Abundance is a kind of nonvanishing statement: first of all, $\kappa(D) \neq -\infty$, that is, some $H^0(mD) \neq 0$; secondly $H^0(mD)$ contains enough elements to separate points not in the fibres of any rational map $X \dashrightarrow Y$ with $\dim Y < \nu$.

(2.5) There is an extremely important point to make about the problem of abundance of the canonical class: a key case is when $\nu(K_X) = 1$; then I'm hoping that $H^0(mK_X)$ grows with m (for sufficiently divisible m). However, a glance at RR shows that $\chi(mK_X)$ does not grow: the reason why H^0 grows is because H^1 grows. So in a sense, the onus is on me to find some nonzero cohomology from somewhere.

(2.6) THEOREM [9]. *If K_X is nef and abundant, then it is eventually free.*

You might think that getting κ right is pretty coarse—it's only asking about the dimension of the image—and that the eventual freeness of K_X is more delicate; despite this, the proof of (2.6), although technically pretty formidable, is based on the same ideas as the Contraction Theorem (1.4), with vanishing replaced by relative vanishing for the Iitaka fibration.

(2.7) The "abundance" part of Conjecture (2.3) is of course OK if $\kappa(X) \geq 1$ (for example, if $p_g \geq 2$), or if $q \geq 1$ by the methods and (hard) results of the Iitaka program: there is a nontrivial Iitaka or Albanese fibration, and off you go.

(2.8) There are however possible nonclassical approaches to the problem, for example, trying to find directly the fibres of the hoped-for contraction Φ_{mK}. This is of course much more in the spirit of the Kleiman-Mori cone: if K_X is nef and not ample, then it is a supporting function of $\overline{NE}$, and the problem is to find at least one effective curve in $K_X^\perp$. This is a subtle problem for surfaces (see [14]), and I believe that an essential ingredient lacking in our understanding of 3-folds is a proof of abundance for surfaces not relying on the Albanese fibration.

PROBLEM. S a nonsingular surface, K_S nef, $K_S^2 = 0$ but $K_S \not\sim 0$; prove that there exist curves $C \subset S$ with $K_S C = 0$.

3. Theology, biology, and art appreciation. Why do we believe in conjectures? Let me give some possible explanations.

1. *Theological.* In a famous article [20], Professor Shafarevich argues that the development of world math can best be understood by postulating divine guidance: so, for example, it is not just coincidence that the ideas of stability of tangent bundles are around just when needed to apply to the Mordell conjecture over function fields (as we heard in Barth's second lecture), but part of an overall plan. Perhaps when one of our colleagues seems to arrive at a conclusion by a process of mystic intuition rather than logical reasoning, we should bear in mind the possibility that the spirit may be speaking through him.

Although experience teaches me that there is not much future in being on the opposite side of an argument with Prof. Shafarevich, I would like to put forward an alternative view.

2. *Biological.* A useful image for contemplating the totality of all algebraic varieties is to think of them as being ALIVE; as such they will grow and degenerate and mutate and take on new forms, and will occupy every conceivable nook and cranny in the universe, including some which we will find inconceivable [1]. The origin of this imagery is Ulf Persson's notion of Geography of Surfaces: you draw all possible values for the invariants of surfaces as a map (Figure 2), and think of the surfaces as populating it. Now experience shows that wherever there is a little niche in the picture where some strange surface can get a tenuous foothold and survive by adapting its internal structure and life style, then you'll find some monstrous or exotic beast. The archetypal example is Horikawa's work on numerical quintics [5]: take the nicest possible surface of general type $S = S_5 \subset \mathbf{P}^3$; then S is canonically embedded. Straightforward arguments show that $|K_S|$ can have at worst one base point for any surface with these invariants, and sure enough, S can mutate into a surface S' with a base point, and after the mutation, S' can develop out in an entirely new direction as a surface S'' having a 2-to-1 map $\varphi\colon S'' \to Q_2 \subset \mathbf{P}^3$ to a nonsingular quadric.

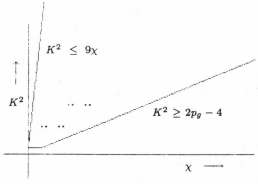

FIGURE 2

Incidentally, another intrepid Swede, Torsten Ekedahl, has considered and explored Persson's graph crossed with Spec **Z**, and returns back to report on isolated islands even off the main continent where strange new life-forms can lead an exotic and precarious existence for just a few fleeting characteristics.

Having established the nature of the problem, it is clear that we expect to be able to break varieties up into genera (as Kleiman points out, there were definite biological connotations to the word genus (Geschlecht) when it was introduced by Clebsch). Now the point is that however strange a creature may be, it will have to have intrinsic structures and general abilities in order to survive. Let me give an example: suppose that a projective variety X has unstable tangent bundle, so a subbundle $E \subset T_X$ of degree bigger than expected. Then E defines a complex foliation on X, since the Frobenius integrability obstruction is a map from $E \wedge E \to \Omega^1/E$, which must vanish for reasons of degree. There's a quite general philosophical reason to expect that E is integrable in the much stronger algebraic geometrical sense: if X is defined over a f.g. extension K of $\mathbf{Q}$, then the group $\text{Aut}(\mathbf{C}/K)$ acts on X taking the subbundle E to itself. From the time of Castelnuovo [3], there have been some successful attempts at proving this in particular cases. Recently, Miyaoka has proved (under fairly general conditions) that the leaves of the foliation are rational curves [10]. I want to interpret this as saying that X has got to be uniruled to survive in its environment.

Perhaps it should be said that of all the varieties in existence, not all will be particularly beautiful or interesting; in classification, we have to deal with every creature, no matter how dull or unpleasant it may turn out to be. Fortunately every now and then a beast of spectacular beauty will turn up; perhaps nobody who heard the talks of Barth and Klaus Hulek on the modular surface $S(5)$ and its relationship with the Horrocks-Mumford bundle on $\mathbf{P}^4$ will dispute my assertion that these constructions, put together by a dozen or so geometers of 4 or 5 different nationalities, rank together with the very greatest artistic achievements of human civilisation.

4. Individual classes of the classification. In this section I discuss some results on special classes of varieties, mainly those with $\kappa = -\infty$. We're unable to have full talks on these topics in the 3-folds seminar at Bowdoin due to lack of time, and more particularly to lack of Russians.

(4.1) *Fano 3-folds and* $\mathbf{Q}$*-Fano 3-folds.* A Fano 3-fold is a projective 3-fold X with $-K_X$ Cartier and ample; I consider mainly the case of smooth X, although some of the results can be generalised. A $\mathbf{Q}$-Fano 3-fold is a 3-fold with canonical singularities and $-K_X$ ample.

There is an important distinction between the cases $\rho(X) = 1$ and $\rho(X) \geq 2$. The point is that if $\text{Pic}X \neq \mathbf{Z}$, then $\overline{NE}$ has at least 2 extremal rays, so that Mori theory can be applied to X to get further contractions. On the other hand, if $\text{Pic}X \cong \mathbf{Z}$, then there's nothing to say except that $-K_X$ is ample. In a certain sense, passing from a variety V by contractions to a $\mathbf{Q}$-Fano 3-fold X with $\text{Pic}X \cong \mathbf{Z}$ reduces the birational classification of V to the biregular classification of X.

(4.2) By Iskovskikh's reconstruction of Fano's work, there are about 20 families of smooth Fano 3-folds with $\text{Pic}X \cong \mathbf{Z}$; see [7]. There are exactly 87 families

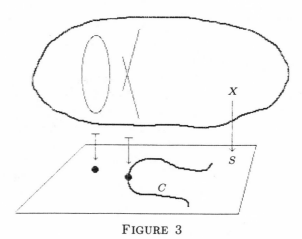

FIGURE 3

of smooth Fano 3-folds of rank ≥ 2, a result obtained by systematically applying Mori theory [**12**].

(4.3) There's not much known as yet about **Q**-Fano 3-folds, beyond the fact that they're uniruled [**11**], and that the graded ring

$$R(X, -K_X) = \bigoplus_{n \geq 0} H^0(\mathcal{O}_X(-nK_X))$$

is Gorenstein. By the general principle of stability of the tangent bundle I expect that $-K_X c_2 \geq 0$, so that using the recently proved "exact plurigenus formula" [**17**], it would follow that $H^0(-K_X) \neq 0$. Then an element $S \in |-K_X|$ can be thought of as a $K3$ surface (with arbitrary singularities, maybe nonreduced), polarised by a **Q**-Cartier Weil divisor. An example would be the weighted projective hypersurface

$$X(42) \subset \mathbf{P}(21, 14, 6, 1, 1)$$

given by

$$x^2 + y^3 + z^7 + t^{42} + u^{42} = 0;$$

the hyperplane section $(u = 0)$ is a famous example of a weighted $K3$ surface. Since $\mathcal{O}(K_X) = \mathcal{O}(42 - 21 - 14 - 6 - 1 - 1) = \mathcal{O}(-1)$, this is anticanonical. A computer search reveals exactly 95 families of these anticanonical hypersurfaces, and for each of these at least, $H^0(-K_X) \neq 0$.

(4.4) *Comic bundles.* Everyone knows the picture of Figure 3.

The interesting problems here are birational, so that I only consider a standard conic bundle ([**19**, (1.13)]): X is a smooth 3-fold fibred over a smooth surface S, so that for general $P \in S$, $f^{-1}P$ is a smooth conic; for $P \in C$, the discriminant curve, $f^{-1}P$ is a line pair (with suitable conditions at $P \in \operatorname{Sing} C$). Picking one line from each pair defines a double cover $\tilde{C} \to C$. Over $S - C$, f is a $\mathbf{P}^1$-bundle, with given degenerations over C, so that it is specified (roughly) up to a 2-torsion point in the Brauer group of S.

(4.5) THEOREM (V. G. (Volodya) Sarkisov [18, 19]). *The group* Bir$X$ *of birational transformations of* X *preserves the fibration* $f\colon X \to S$, *provided at least that* $|4K_S + C| \neq \varnothing$.

The theorem means that as soon as S or C get reasonably complicated, conic bundles are classified up to birational equivalence by the pair S and C, plus little bits of 2-torsion data. Compare this with the fact that ruled surfaces are just $C \times \mathbf{P}^1$ for some $C \in M_g$.

The theorem is proved by the method of Fano-Iskovskikh-Manin for Cremona transformations; this method is difficult, but it is a method of great power and subtlety, and it is a great disgrace to western algebraic geometers that no one outside Moscow is prepared to spend time mastering it.

(4.6) *Del Pezzo fibre spaces.* The main new result here is Kanev's computation of the intermediate Jacobian of a 3-fold fibre space $f\colon X \to B$ of del Pezzo surfaces over a base curve in terms of the curve C of lines in fibres of f [8]. There doesn't seem to be anything known yet about analogues of the results of (4.4) and (4.5) for this class of varieties.

(4.7) *Regular varieties with* $K_X \sim 0$. Here we don't know much. If we polarise X by H, a general element $S \in |H|$ is a canonically polarised surface. One could for example search for X by trying to extend a known surface S of general type, but this will usually be very difficult.

PROBLEM 1. Construct many families.

A few hundred families of examples are given by simple projective constructions, such as taking branched covers of Fano 3-folds. N. Nakayama has also pointed out that some families can be obtained as Weierstrass fibrations of elliptic curves over a rational surface S, obtained from suitable $a \in H^0(-4K_X)$, $b \in H^0(-6K_X)$.

PROBLEM 2. Connect up some of the families by degenerations.

For example, the two families of quintic hypersurfaces $X_5 \subset \mathbf{P}^4$ and double covers $Y_8 \to \mathbf{P}^3$ branched in an octic surface consist of topologically distinct varieties. However, it's easy to see that if X_5 has an ordinary triple point P, then the projection $X_5 \dashrightarrow \mathbf{P}^3$ from P presents X_5 as birational to some (singular) Y_8. I would very much like to have examples of members of two such families, say $X_5 \subset \mathbf{P}^4$ and $Y_{2,4} \subset \mathbf{P}^5$, which have only ordinary double points but are birational.

REFERENCES

1. D. Attenborough, *Life on Earth*, Collins BBC publications, 1979.

2. L. Bădescu, *Suprafeţe algebrice*, Editura Acad. Romania, Bucharest, 1981.

3. G. Castelnuovo, *Sulle superficie aventi il genere aritmetico negative*, Rend. Circ. Mat. Palermo **20** (1905), Memorie Scelte XXVII, 501–509.

4. P. Francia, *Some remarks on minimal models*. I, Compositio Math. **40** (1980), 301–313.

5. E. Horikawa, *On deformations of quintic surfaces*, Invent. Math. **31** (1975), 43–85.

6. S. Iitaka, *Genera and classification of algebraic varieties*, Sugaku **24** (1972), 14–17 (Japanese).

7. V. A. Iskovskikh, *Fano 3-folds.* I and II, I, Izv. Akad. Nauk SSSR Ser. Mat. **41** (1977), 516–562; English transl. in Math USSR Izv. **11** (1977); II, Izv. Akad. Nauk SSSR Ser. Mat. **42** (1978), 469–506; English transl. in Math USSR Izv. **11** (1977).

8. V. Kanev, *Intermediate Jacobians of threefolds with a pencil of del Pezzo surfaces and generalised Prym varieties*, C. R. Acad. Bulgare Sci. **36** (1983), 1015–1019.

9. Y. Kawamata, *Pluricanonical systems on minimal algebraic varieties*, Invent. Math. **79** (1985), 567–588.

10. Y. Miyaoka, *The Chern classes and Kodaira dimension of a minimal variety*, Algebraic Geometry, Sendai; to appear in Advanced Stud. Pure Math, Kinokuniya and North-Holland, 1987.

10'. _____, *On the Kodaira dimension of minimal threefolds*, Math. Ann. (1987) (to appear).

11. Y. Miyaoka and S. Mori, *A numerical criterion for uniruledness*, Ann. of Math. **124** (1986), 65–69.

12. S. Mori and S. Mukai, *On Fano 3-folds with $B_2 \geq 2$*, Algebraic Varieties and Analytic Varieties, S. Iitaka, editor, Advanced Stud. Pure Math. **1** (1983), 101–129.

13. S. Mori, *Classification in dimension ≥ 3*, this volume.

14. D. Mumford, *Enriques' classification of surfaces in characteristic p.* I, Global Analysis, papers in honour of K. Kodaira, Univ. of Tokyo Press, Tokyo, and Princeton Univ. Press, Princeton, N.J., 1969, pp. 325–339.

15. M. Reid, *Decomposition of toric morphisms*, Arithmetic and Geometry, papers dedicated to I. R. Shafarevich, Birkhäuser, Boston, Mass., 1983, pp. 395–418.

16. _____, *Minimal models of canonical 3-folds*, Algebraic Varieties and Analytic Varieties, S. Iitaka, editor, Advanced Stud. Pure Math. **1** (1983), 131–180.

17. _____, *Young person's guide to canonical singularities*, this volume.

18. V. G. Sarkisov, *Birational automorphisms of conic bundles*, Izv. Akad. Nauk SSSR Ser. Mat. **44** (1980), 918–945; English transl. in Math. USSR Izv. **17** (1981).

19. _____, *On the structure of conic bundles*, Izv. Akad. Nauk SSSR Ser. Mat. **46** (1982), 371–408; English transl. in Math. USSR Izv. **20** (1983).

20. I. R. Shafarevich, *On certain tendencies in the development of mathematics*, English transl. in Math. Intelligencer **3** (1980/81), no. 4, 182–184.

21. V. V. Shokurov, *The non-vanishing theorem*, Izv. Akad. Nauk SSSR Ser. Matem. **49** (1985), 635–651; English transl. in Math. USSR Izv. **27** (1986).

22. P. M. H. Wilson, *Towards birational classification of algebraic varieties*, Bull. London Math. Soc. **19** (1987), 1–48.

23. R. Hartshorne, *Ample subvarieties of algebraic varieties*, Lecture Notes in Math., vol. 156, Springer-Verlag, 1970.

24. Y. Kawamata, *Crepant blowings-up of three-dimensional canonical singularities and applications to degenerations of surfaces*, Ann. of Math. (1987) (to appear).

MATHEMATICS INSTITUTE, UNIVERSITY OF WARWICK, COVENTRY CV4 7AL, ENGLAND

Proceedings of Symposia in Pure Mathematics
Volume **46** (1987)

Young Person's Guide to Canonical Singularities

MILES REID

In memory of Oscar Zariski

This article aims to do three things: (I) to give a tutorial introduction to canonical varieties and singularities, with some of the motivating examples; (II) to provide a skeleton key to the results of my two papers on canonical singularities [**C3-f**], [**Pagoda**], and those of [**Morrison-Stevens**] and [**Mori**, Terminal singularities]; and (III) to explain the recent "exact plurigenus formula." The expository intention is reflected in explanations of some well-known standard technical points (well-known to experts but maybe not to the algebraic geometer in the street), and also worked examples, exercises, and deliberate mistakes to entertain the reader; I apologise if any secrets of the priesthood are divulged despite my best efforts.

After §4, most of the material is new: §§5, 6 and 7 contain the material of [**Morrison-Stevens**] and [**Mori**] in substantially laundered form, and the results on equivariant RR and the plurigenus formula of Chapter III appear here for the first time. The juxtaposition of these two topics reveals quite amazing relations between the cyclotomic sums appearing traditionally in connection with equivariant RR and toric geometry; of course, these are linked in a primary way by the fact that quotient singularities make contributions (for example, to $H^0(\mathbf{P}, \mathcal{O}(k))$ for a weighted projective space $\mathbf{P}$) which can be computed either by equivariant RR or as the number of lattice points of a polyhedron. However, it was something of a shock to discover how intricately cyclotomy relates to the combinatorics of the Newton polyhedron at the heart of the classification of terminal singularities.

My contribution to the subject has mainly been concerned with the study of 3-fold singularities. It should be noted that most of the recent work on varieties of dimension ≥ 4 (in particular the two circles of ideas, cone, contraction, nonvanishing theorems of Kawamata and Shokurov, and positivity of $f_*\omega$, $C^+_{n,m}$ of Fujita, Viehweg, Kawamata and Kollár) uses only the definitions of canonical and terminal singularities (and their log generalisations), together with general properties such as rationality and behaviour in codimension 2, but does not

1980 *Mathematics Subject Classification* (1985 *Revision*). Primary 14E30; Secondary 14B05.

in any essential way use specific results concerning the singularities. Indeed, counterexamples (see (3.13)) suggest that it is unlikely that we can expect any worthwhile classification of these singularities in dimension ≥ 4.

The bulk of this paper was written during a six-week visit to the NSF-sponsored special year in singularity theory and algebraic geometry at the University of North Carolina, Chapel Hill; I thank Jonathan Wahl, Jim Damon, and the other visitors for providing a stimulating environment. I am indebted to Y. Kawamata, S. Mori and D. Zagier for helpful conversations, and to A. R. Fletcher, who has repeatedly corrected false versions of the formulas of Chapter III; S. Mori has saved me from serious error in two places.

Contents

Chapter I
Overview of the Subject

1. Definitions and easy examples. Varieties are always assumed to be normal and quasiprojective, and defined over an algebraically closed field k of characteristic zero; my favourite is $k = \mathbf{C}$.

(1.1) DEFINITION. A variety X has *canonical singularities* if it satisfies the following two conditions:

(i) for some integer $r \geq 1$, the Weil divisor rK_X is Cartier;

(ii) if $f: Y \to X$ is a resolution of X and $\{E_i\}$ the family of all exceptional prime divisors of f, then

$$rK_Y = f^*(rK_X) + \sum a_i E_i, \quad \text{with } a_i \geq 0.$$

If $a_i > 0$ for every exceptional divisor E_i, then X has *terminal singularities*.

This section is devoted to explaining and motivating by examples the terms in the definition (see (1.8–9) for easy examples); the definition is ultimately justified by the fact (see (2.5)) that the canonical model of a variety of general type has canonical singularities.

Although the definition of canonical singularities is abstract, there is a kind of classification in the 3-fold case which will reduce most problems to hypersurface singularities, cyclic quotient singularities, and cyclic quotients of isolated hypersurface singularities. (It will be shown in (3.14), see also [**Pagoda**], that every terminal singularity is of this kind.) The reduction steps consist of various cyclic covers, partial resolutions, and partial smoothings, and each step involves of course some definite understanding of the "general" singularity.

Here is some more terminology: the smallest r for which rK_X is Cartier in a neighbourhood of $P \in X$ is called the *index* of the singularity P; the **Q**-divisor $\Delta = (1/r) \sum a_i E_i$ which satisfies the formal equality

$$K_Y = f^* K_X + \Delta$$

is called the *discrepancy* of f. (To remember which way round this equality goes, think of the adjunction formula for the blow-up $\sigma : Y \to X$ of a smooth point $P \in X$ of a surface: $\sigma^{-1} P = l$ is a (-1)-curve, and $K_Y = \sigma^* K_X + l$.)

(1.2) *The surface case.* Everyone knows that the ordinary double point of a surface (the singularity $X : (xz = y^2) \subset \mathbf{A}^3$) has a resolution $f : Y \to X$ for which the exceptional curve E is a (-2)-curve, that is, $E \cong \mathbf{P}^1$, $E^2 = -2$. It's easy to see by the adjunction formula (or by a direct calculation with differentials, as in (1.9) below) that $K_Y = f^* K_X$, so that this is a canonical singularity. In fact it can be proved that the surface canonical singularities are exactly nonsingular points, together with the *Du Val surface singularities*, the hypersurface singularities given by one of the equations

$$
\begin{aligned}
A_n &: x^2 + y^2 + z^{n+1} \quad &&\text{(for } n \geq 1), \\
D_n &: x^2 + y^2 z + z^{n-1} \quad &&\text{(for } n \geq 4), \\
E_6 &: x^2 + y^3 + z^4, \\
E_7 &: x^2 + y^3 + yz^3, \\
E_8 &: x^2 + y^3 + z^5.
\end{aligned}
$$

(One derivation of this list is sketched in (4.9), (3).) Among the many extraordinary properties enjoyed by these singularities is the fact that each of them has a resolution $f : Y \to X$ such that the exceptional locus of f is a bunch of (-2)-curves (forming a configuration given by the corresponding Dynkin diagram), and such that $K_Y = f^* K_X$.

It is important to realise that this harmless-looking observation is central to the theory of minimal models of surfaces and canonical models of surfaces of general type. The point is this: if X is a canonical surface (a surface with at worst Du Val singularities and ample K_X), then a minimal resolution $f : Y \to X$ is a nonsingular surface Y with K_Y nef: conversely, if Y is a nonsingular surface with K_Y nef and big (a minimal nonsingular model of a surface of general type), then the curves E with $K_Y E = 0$ form bunches of (-2)-curves, and can be contracted to Du Val singularities. Thus for surfaces of general type, it's a matter of personal preference whether you take the canonical model X, or a nonsingular minimal model Y with K_Y nef. In fact the influence of the Du Val singularities extends throughout the classification of surfaces. Now let's see how this fails in higher dimensions.

(1.3) EXAMPLE: *The Veronese cone.* The simplest example of a singularity of index > 1 is the cone over the Veronese surface; this has a resolution Y with exceptional locus E satisfying $E \cong \mathbf{P}^2$ and $\mathcal{O}_E(-E) \cong \mathcal{O}(2)$. By the adjunction formula, $\mathcal{O}_E(K_Y + E) = K_{\mathbf{P}^2} = \mathcal{O}(-3)$, so that purely formally we should have

$$K_Y = f^* K_X + \tfrac{1}{2} E.$$

We'll see presently that this is meaningful in terms of differentials. The first context in which I met a variety with these singularities was the Kummer variety of an Abelian 3-fold A; dividing out A by the involution (-1), the resulting variety $X = A/(-1)$ has 64 Veronese cone singularities (at the 64 fixed points of (-1)), and $2K_X \sim 0$. You can simply blow up these singularities to get a smooth variety Y if you wish, but then $2K_Y \sim \sum_{64} E_i$, so you've lost the good numerical properties of K_X. This is one reason why the Kummer surface does not generalise to higher dimensions (at least, not in a very simple way).

Two examples where this singularity appears on canonical models of 3-folds are given in (2.8–9); compare [**Ueno**].

(1.4) *Canonical differentials.* If V is a smooth variety, $\omega_V = \mathcal{O}_V(K_V) = \Omega_V^n$ is the invertible sheaf generated by $dx_1 \wedge \cdots \wedge dx_n$, where $x_1, \ldots, x_n$ are local coordinates. Sections of ω_V are *canonical differentials*, and sections of $\omega_V^{\otimes m}$ are *m-canonical differentials*. Canonical differentials are important for the following reasons:

(1) *Intrinsic nature.* The sheaves ω_V and $\mathcal{O}_V(rK_V)$ are part of the bundled hardware which comes free when you buy V. This is particularly important in classification theory; for example, if ω_V is ample, then there is an intrinsic way of embedding V into projective space.

(2) *Duality.* ω_V is the dualising sheaf which makes Serre duality work; that is, there is a perfect pairing $H^i(V, \mathcal{F}) \times \mathrm{Ext}_V^{n-i}(\mathcal{F}, \omega_V) \to k$.

(3) *Vanishing.* Kodaira vanishing says that $H^i(V, \mathcal{L} \otimes \omega_V) = 0$ for an ample sheaf $\mathcal{L}$ and $i > 0$.

(4) *Birational nature.* If $V \dashrightarrow W$ is a birational map between nonsingular projective varieties, then it is easy to see (for example, [**Shafarevich**, p. 167]) that regular differentials on V and W coincide, so for example

$$H^0(V, \mathcal{O}_V(rK_V)) = H^0(W, \mathcal{O}_W(rK_W)).$$

(5) *Adjunction formulas.* If two varieties X and Y are closely related, then you expect to be able to compute K_X in terms of K_Y and vice-versa; a formula of this kind is called an *adjunction formula*. In practice this means that K_X is readily computable. The following are some of the many examples of adjunction formulas.

(a) If Y is a smooth variety and $X \subset Y$ is a hypersurface, then $K_X = (K_Y + X)|_X$.

(b) The Riemann-Hurwitz formula $K_X = f^* K_Y + R_f$ for a generically finite (separable) morphism $f \colon X \to Y$ between nonsingular varieties: there is a canonical map $J \colon f^*(\Omega_Y^n) \hookrightarrow \Omega_X^n$, and the *ramification divisor* can be defined by $R_f = \operatorname{div}(J)$; this is of course just an intrinsic way of saying the determinant of the Jacobian matrix

$$J = \det \frac{\partial(x_1, \ldots, x_n)}{\partial(y_1, \ldots, y_n)}.$$

(c) If $\sigma \colon Y \to X$ is the blow-up of a nonsingular point $P \in X$ of an n-fold, and $E = \sigma^{-1} P$, then $K_Y = \sigma^* K_X + (n-1)E$.

(d) If $p \colon F \to X$ is the $\mathbf{P}^{r-1}$-bundle $F = \mathbf{P}_X(\mathcal{E})$ associated to a rank r vector bundle $\mathcal{E}$ over X, and $\mathcal{O}_F(1)$ is the tautological line bundle (that is, $\mathcal{O}_F(1)$ is $\mathcal{O}(1)$ on each fibre, and $p_* \mathcal{O}_F(1) = \mathcal{E}$), then $K_F = p^*(K_X + \det \mathcal{E}) \otimes \mathcal{O}_F(-r)$.

(e) On a deeper level, Kodaira's canonical bundle formula for an elliptic surface (see, for example, [**Barth-Peters-Van de Ven**, p. 161]) should be viewed as an adjunction formula.

(1.5) *Definition of ω_X and $\mathcal{O}_X(mK_X)$ for singular X.* Assume X is normal. Then $\Omega_{k(X)}^n$, the space of rational canonical differentials of X (more precisely, I should write $\Omega_{k(X)/k}^n$), is a 1-dimensional vector space over $k(X)$, with basis $df_1 \wedge \cdots \wedge df_n$ for any $f_1, \ldots, f_n \in k(X)$ forming a separable transcendence basis of $k(X)$ over k. Write X^0 for the nonsingular locus of X. Then for $P \in X^0$ I can choose local coordinates $x_1, \ldots, x_n$ at P, and write any $s \in \Omega_{k(X)}^n$ as

$$s = f \cdot dx_1 \wedge \cdots \wedge dx_n \quad \text{with } f \in k(X).$$

Then s is *regular* at $P \in X^0$ if f is a regular function at P. Now by definition, s is regular at $P \in X$ if there is a neighbourhood $P \in U \subset X$ such that s is regular at every $x \in U \cap X^0$. This defines a sheaf ω_X, with

$$\Gamma(U, \omega_X) = \{ s \in \Omega_{k(X)}^n \mid s \text{ is regular on } U \cap X^0 \}.$$

That is, I don't attempt to define directly a regular differential at a singular point P, but just take rational differentials which are regular on the smooth points of a neighbourhood of P. There are several traditional alternative ways of defining the same sheaf.

(a) $\omega_X = j_*(\Omega_X^n)$, where $j \colon X^0 \hookrightarrow X$ is the inclusion of the smooth locus of X;

(b) ω_X is the *double dual* of Ω_X^n.

(1.6) *Explanation.* Although the Kähler differentials Ω^1 and Ω^n have good universal properties, they are often not right for (birational) geometrical purposes; the construction (a) of ω_X in terms of rational differentials which are regular in codimension 1 is one obvious geometrical alternative, due to Zariski. To explain (b), taking the dual kills any torsion which might be present in Ω_X^n, and then taking the double dual saturates Ω_X, in the sense that any rational sections of Ω_X^n which belong to Ω_X^n in codimension 1 actually belong to ω_X; or "kills the cotorsion" in the jargon of the trade. The universal constructions of tensor product and f^* of a sheaf are often not right for geometrical purposes for similar reasons. A simple example: if $f: Y \to X$ is the blow-up of a nonsingular point of a surface, many writers who should know better write $f^* m_P$ for the ideal $m_P \cdot \mathcal{O}_Y$; in fact there are at least 3 different pull-backs of m_P, namely, the sheaf-theoretic $f^{-1} m_P$, the ringed-space construction $f^* m_P = f^{-1} m_P \otimes \mathcal{O}_Y$ (which has torsion, as you should check for yourself), and $m_P \cdot \mathcal{O}_Y = \text{Im}\{f^* m_P \to \mathcal{O}_Y\}$. In more complicated situations you might also contemplate saturating $m_P \cdot \mathcal{O}_Y$, etc.

(1.7) The sheaves

$$\mathcal{O}_X(mK_X) = \{s \in (\Omega^n_{k(X)})^{\otimes m} \mid s \text{ is regular on } X^0\} = j_*((\Omega^n_{X^0})^{\otimes m})$$

are defined in a similar way. Here the canonical divisor K_X is the Weil divisor (more precisely, divisor class) corresponding to ω_X; this means the following: take any nonzero rational differential $s \in \Omega^n_{k(X)}$, and let $K_X = \text{div}(s)$ be the divisor of zeros and poles of s. The statement that rK_X is Cartier at $P \in X$ (that is, locally principal) is equivalent to saying that $\mathcal{O}_X(rK_X)$ is invertible.

(1.8) EXAMPLES. (1) Suppose that $P \in X: (f = 0) \subset \mathbf{A}^{n+1}$ is a (normal) hypersurface singularity. Consider the expression

$$s = \frac{dx_1 \wedge \cdots \wedge dx_n}{\partial f / \partial x_0} \in \Omega^n_{k(X)},$$

where $x_0, \ldots, x_n$ are local coordinates on $\mathbf{A}^{n+1}$ in a neighbourhood of P. At any point $Q \in X$ where $(\partial f / \partial x_0)(Q) \neq 0$, X is a manifold with local coordinates $x_1, \ldots, x_n$, and $s = (\text{unit}) \cdot (dx_1 \wedge \cdots \wedge dx_n)$ is a basis on Ω^n_X. Now because of the identifications involved in the definition of Ω^1_X and in taking the wedge product, it happens that under permutation of $x_0, \ldots, x_n$, the element $s \in \Omega^n_{k(X)}$ is invariant up to ± 1. Thus x_0 does not play any particular role, and s *is a basis of Ω^n_X at any nonsingular point of X*. This means that s is regular at P. In fact s is a basis of ω_X, that is, $\omega_X = \mathcal{O}_X \cdot s$: for given any $t \in \omega_X$, I can write $t = f \cdot s$ with $f \in k(X)$; but then f must be regular at every $Q \in X^0$, and so by normality of X, f is regular on X.

(2) Let X be the quotient $X = \mathbf{A}^2 / \mu_3$ of $\mathbf{A}^2$ by the cyclic group μ_3 of cube roots of 1 acting by

$$\varepsilon: (x, y) \to (\varepsilon x, \varepsilon y) \quad \text{for all } \varepsilon \in \mu_3.$$

The ring of invariants of the action is

$$k[x^3, x^2 y, x y^2, y^3] \cong k[u_0, u_1, u_2, u_3]/(u_0 u_2 - u_1^2, u_1 u_3 - u_2^2, u_0 u_3 - u_1 u_2),$$

and the quotient X is Spec of this ring, which as you can see is the affine cone over the twisted cubic.

I now write down a basis $s \in \mathcal{O}_X(3K_X)$ as a rational 3-canonical differential on X. The idea is that upstairs on $\mathbf{A}^2$, $dx \wedge dy$ is a basis of Ω^2, but under the group action, $\varepsilon: (dx \wedge dy) \to \varepsilon^2 (dx \wedge dy)$; so $(dx \wedge dy)^{\otimes 3}$ is invariant under the group action, and should come from something on X. Now if I set

$$s = \frac{(du_0 \wedge du_1)^{\otimes 3}}{u_0^4} \in (\Omega_{k(X)}^2)^{\otimes 3},$$

then differentiating $u_0 = x^3$, $u_1 = x^2 y$ shows that

$$\pi^* s = (\text{unit}) \cdot (dx \wedge dy)^{\otimes 3},$$

(where $\pi \colon \mathbf{A}^2 \to X$ is the quotient map); since π is etale outside the origin, it is clear that s is a basis of $(\Omega_X^2)^{\otimes 3}$ everywhere on $X \setminus P$. Alternatively, note that from the equations defining X, (u_0, u_1) are local coordinates wherever $u_0 \neq 0$, and that (by direct calculation)

$$s = \frac{(du_0 \wedge du_1)^3}{u_0^4} = \frac{(du_2 \wedge du_3)^3}{u_3^4},$$

which works wherever $u_3 \neq 0$. Thus $s \in \mathcal{O}_X(3K_X)$ is a basis, and $3K_X$ is Cartier.

These two examples illustrate condition (1.1), (i). The next section tries to explain condition (ii).

(1.9) *Regularity of differentials on a resolution.* As discussed above, condition (1.1), (i) means that the sheaf $\mathcal{O}_X(rK_X)$ is invertible. Suppose that s is a local basis of $\mathcal{O}_X(rK_X)$ at a singular point $P \in X$, and that $f \colon Y \to X$ is a resolution. Then $s \in (\Omega_{k(X)}^n)^{\otimes r}$, and since $k(Y) = k(X)$, I can consider s as a rational differential on Y, and ask again whether it is regular; of course, where f is an isomorphism there is no problem, but s can perfectly well have poles along exceptional divisors of f. So condition (1.1), (ii) is the condition that s remains regular on a resolution Y.

EXAMPLES. (1) In the notation of (1.8), suppose in addition that $P \in X \subset \mathbf{A}^{n+1}$ is an ordinary point of multiplicity k, so that the projectivised tangent cone is a nonsingular hypersurface $E \subset \mathbf{P}^n$ of degree k. Then $P \in X$ is terminal if $k < n$, canonical if $k = n$, and not canonical if $k > n$.

You can see this by an explicit calculation: let $\sigma \colon Y \to X$ be the blow-up of P; then Y is nonsingular and the exceptional locus $E = \sigma^{-1} P$ is the hypersurface $E \subset \mathbf{P}^n$. I'm interested in the zeros or poles along E of the rational canonical differential

$$s = \frac{dx_1 \wedge \cdots \wedge dx_n}{\partial f / \partial x_0} \in \Omega_{k(X)}^n.$$

To calculate this, write down one affine piece of the blow-up of $\mathbf{A}^{n+1}$, which is the map $\sigma \colon \mathbf{A}^{n+1} \to \mathbf{A}^{n+1}$ given by

$$x_n = y_n, \qquad x_i = y_i y_n \quad \text{for } i = 0, 1, \ldots, n-1,$$

where $y_0, \ldots, y_n$ are coordinates in a copy of $\mathbf{A}^{n+1}$. Then

$$\sigma^* f = f(y_0 y_n, \ldots, y_n) = y_n^k \cdot g(y_0, \ldots, y_n),$$

where g is the equation of the affine piece of Y in $\mathbf{A}^{n+1}$. Now since

$$g = f(y_0 y_n, \ldots, y_n) \cdot y_n^{-k},$$

it follows that

$$\partial g/\partial y_0 = y_n^{-k+1} \cdot (\partial f/\partial x_0);$$

hence at a point $Q \in E$ where $(\partial g/\partial y_0)(Q) \neq 0$,

$$s = \frac{dx_1 \wedge dx_2 \wedge \cdots \wedge dx_n}{\partial f/\partial x_0} = y_n^{n-1} \cdot \frac{dy_1 \wedge dy_2 \wedge \cdots \wedge dy_n}{\partial f/\partial x_0}$$

$$= y_n^{n-1} \cdot \frac{dy_1 \wedge dy_2 \wedge \cdots \wedge dy_n}{y_n^{k-1} \cdot \partial g/\partial y_0} = y_n^{n-k} \cdot t,$$

where

$$t = \frac{dy_1 \wedge dy_2 \wedge \cdots \wedge dy_n}{\partial g/\partial y_0}$$

is a basis of Ω_Y^n near Q.

So the rational differential s has

a zero of order $n - k$ along E if $k < n$,

a pole of order $k - n$ along E if $k > n$,

and no zero or pole if $n = k$. More succinctly, the computation can be expressed as follows:

$$K_X = (K_{\mathbf{A}} + X)|_X \quad \text{and} \quad K_Y = (K_B + Y)|_Y$$

where $\sigma \colon B \to \mathbf{A}$ is the blow-up of P. However,

$$K_B = \sigma^* K_{\mathbf{A}} + nE \quad \text{and} \quad Y = \sigma^* X - kE,$$

where $E \subset B$ is the exceptional divisor. Adding these up gives

$$K_Y = \sigma^* K_X + (n - k)E.$$

(2) Use the notation of (1.8), (2). The quotient singularity X is resolved by a single blow-up $\sigma \colon Y \to X$, so that one affine piece of the resolution is a copy of $\mathbf{A}^2$ with coordinates (z, t), mapping to X by

$$(z, t) \mapsto (z, zt, zt^2, zt^3).$$

The exceptional curve $E = \sigma^{-1} P$ of the blow-up has $E \cong \mathbf{P}^1$, $\mathcal{O}_E(-E) \cong \mathcal{O}(3)$, and is given in the affine piece $\mathbf{A}^2$ by $z = 0$.

The rational 3-canonical differential $s = (du_0 \wedge du_1)^3/u_0^4 \in \mathcal{O}_X(3K_X)$ is a basis. Now think of s as a rational differential on the resolution $\sigma \colon \mathbf{A}^2 \to X$, by just writing $u_0 = z$, $u_1 = zt$. Then

$$s = \frac{(dz \wedge zdt)^3}{z^4} = \frac{(dz \wedge dt)^3}{z},$$

where $dz \wedge dt$ is a basis of the regular canonical differential on $\mathbf{A}^2$; so s has a pole along the exceptional curve E.

(1.10) *Exercise.* A similar calculation that the 3-fold quotient singularity $X = \mathbf{A}^3/\boldsymbol{\mu}_3$ where

$$\boldsymbol{\mu}_3 \ni \varepsilon \colon (x, y, z) \mapsto (\varepsilon x, \varepsilon y, \varepsilon^2 z)$$

is canonical of index 3. To see this, note that $X = \operatorname{Spec} A$, where

$$A = k[x^3, x^2 y, xy^2, y^3, xz, yz, z^3] = k[u_0, u_1, u_2, u_3, v_0, v_1, w]/I,$$

and the ideal I of relations between the 7 generators is generated by 10 relations of monomial type; you will enjoy checking that the projectivised tangent cone to $P \in X$ consists of $E_1 \cup E_2$, where E_1 is a plane and E_2 is a quartic scroll, and the blow-up of $P \in X$ is nonsingular. Next,

$$s = \frac{(du_0 \wedge du_1 \wedge dv_0)^{\otimes 3}}{u_0^5}$$

satisfies

$$\pi^* s = (dx \wedge dy \wedge dz)^3,$$

and so is a basis of $\mathcal{O}_X(3K_X)$. It has zeros of order 1 and 2 along the two exceptional components of the resolution.

(1.11) *Historical note.* This example was first discovered in this context (as a counterexample to my primitive idea that the index is always ≤ 2) by N. Shepherd-Barron, although it had been previously hinted at in a letter of K. Ueno. The Veronese cone singularity of (1.3) and (2.8–9) was also the main ingredient in Ueno's paper of the same period [**Ueno**]. Note that the Veronese cone of (1.3) and the quotient singularity of (1.10) are the first of a well-understood series of terminal quotient singularities (see (5.2)); the nice resolution constructed in Exercise 1.10 is generalised in (5.7).

2. Brief introduction to global canonical 3-folds.

This section is logically independent of the rest of the paper, giving some of the examples and historical motivation underlying [**C3-f**].

(2.1) *Is the canonical ring of a variety finitely generated?* Zariski's work [**Zariski**] implies that the canonical ring of a surface of general type is a f.g. k-algebra; geometric constructions of surfaces of general type since Enriques have often been closely related to a description of the canonical ring (see, for example, [**Catanese**, §1.3]). Experience has shown that many of the basic assertions in a traditional treatment of the classification of surfaces fail in higher dimensions; perhaps finite generation of the canonical ring generalises? At present this is not completely settled for 3-folds, although it looks good. Be that as it may, early examples of canonical models of 3-folds of general type for which the canonical ring is f.g. (see (2.8–11)) displayed interesting new features compared with surfaces of general type, and studying canonical 3-folds has led to some understanding of what seem to be typical features of higher-dimensional birational geometry.

(2.2) *Hilbert's 14th problem.* A standard method of constructing a graded ring: start from a nonsingular projective variety V and a divisor D on V, and set

$$R(V, D) = \bigoplus_{n \geq 0} H^0(\mathcal{O}_V(nD)).$$

Hilbert asked [**Mumford**, §3] if rings of this form are f.g. in general; this is false: the first counterexamples were given by Nagata and Zariski in the 1950s. However, Zariski gave the following sufficient condition for $R(V, D)$ to be f.g. (Graded rings appearing here are assumed to have $R_0 = k$, and f.g. means finitely generated as k-algebra.)

(2.3) THEOREM. *Suppose that the linear system $|mD|$ is free for some $m > 0$; then $R(V, D)$ is f.g.*

MODERN SKETCH PROOF. Suppose first that $m = 1$, so that $|D|$ itself is free; then $|D|$ defines a morphism $\varphi = \varphi_D \colon V \to \varphi(V) = Y \subset \mathbf{P}$ to a projective space $\mathbf{P}$ such that $\mathcal{O}_V(D) = \varphi^* \mathcal{O}_{\mathbf{P}}(1)$. If I set $\mathcal{A} = \varphi_* \mathcal{O}_V$ it follows that for any n,

$$\varphi_* \mathcal{O}_V(nD) = \mathcal{A} \otimes \mathcal{O}(n) = \mathcal{A}(n),$$

so that

$$H^0(V, \mathcal{O}_V(nD)) = H^0(\mathbf{P}, \varphi_* \mathcal{O}_V(nD)) = H^0(\mathbf{P}, \mathcal{A}(n)).$$

However, since $\mathcal{A}$ is a coherent sheaf of $\mathcal{O}_{\mathbf{P}}$-algebras, it follows easily from Serre's theorems that the ring $\bigoplus_{n \geq 0} H^0(\mathbf{P}, \mathcal{A}(n))$ is finite as a module over the homogeneous coordinate ring of $\mathbf{P}$. This gives the result in the case $m = 1$.

The more general case is similar, using the morphism $\varphi = \varphi_{mD}$ and considering the coherent sheaf of $\mathcal{O}_{\mathbf{P}}$-algebras $\mathcal{A} = \varphi_* \mathcal{O}_V$ together with the sheaves of $\mathcal{A}$-modules $\mathcal{M}_i = \varphi_* \mathcal{O}_V(iD)$ for $i = 1, \ldots, m - 1$.   Q.E.D.

(2.4) *Projective normalisation.* Suppose V and D are as in (2.3). Set

$$X = \operatorname{Proj} R(V, D),$$

and consider the morphism $\varphi_{nD} \colon V \to \varphi(V) = Y \subset \mathbf{P}$ for any n such that $|nD|$ is free. Then as is clear from the proof just given, X coincides with $\operatorname{Spec}_Y \mathcal{A}$, which in Zariski's language is the normalisation of Y in the function field of V; write

$$V \xrightarrow{\varphi} X \xrightarrow{f} Y \subset \mathbf{P}$$

for the factorisation of φ_{nD} (the Stein factorisation). Then φ is a contraction morphism corresponding to D: it is the unique morphism such that $\varphi_* \mathcal{O}_V = \mathcal{O}_X$ and for every curve $C \subset V$,

$$\varphi(C) = \mathrm{pt.} \quad \Longleftrightarrow \quad CD = 0.$$

Furthermore, since $f \colon X \to Y$ is finite, $f^* \mathcal{O}_Y(1)$ is ample on X, and some multiple is very ample. This gives:

COROLLARY. $X = \operatorname{Proj} R(V, D) \cong \varphi_{nD}(V)$ *for every sufficiently large and divisible* n.

Note that if S is a minimal surface of general type then $|mK_S|$ is free for any $m \geq 4$ (in fact for $m \geq 2$ if $K_S^2 \geq 5$, see [**Catanese**]), so that it follows that the canonical ring of S is finitely generated.

(2.5) *Canonical models.*

DEFINITION. A *canonical variety* is a projective variety X with at worst canonical singularities such that the $\mathbf{Q}$-Cartier divisor K_X is ample. If V is a variety of general type and X is a canonical variety birational to V, then X is the *canonical model* of V.

THEOREM [**C3-f**, (1.2), (II)]. *Let V be a smooth projective variety of general type. Then V has a canonical model X if and only if the canonical ring $R = R(V, K_V)$ is f.g., and then $X = \operatorname{Proj} R(V, K_V)$.*

PROOF. For a graded ring $R = \bigoplus_{k \geq 0} R_k$ and $m > 0$, the truncated ring $R^{(m)}$ is defined by $R^{(m)} = \bigoplus_{k \geq 0} R_{km}$; by the Veronese embedding, $\operatorname{Proj} R^{(m)} = \operatorname{Proj} R$.

First, let X be a canonical variety, and $m > 0$ an integer such that $\mathcal{O}(mK_X)$ is an ample Cartier divisor; then of course $R(X, \mathcal{O}_X(mK_X))$ is f.g. and $X = \operatorname{Proj} R(X, \mathcal{O}_X(mK_X))$. However, from the definition of canonical singularities and the birational invariance of $H^0(mK_V)$ it follows that

$$H^0(V, kK_V) = H^0(\mathcal{O}_X(kK_X))$$

for any nonsingular projective variety V birationally equivalent to X and any $k > 0$. Therefore $R(V, K_V)$ is finitely generated, and

$$X = \operatorname{Proj} R(X, \mathcal{O}_X(mK_X)) = \operatorname{Proj} R(V, mK_V) = \operatorname{Proj} R(V, K_V).$$

(2.6) I now prove the converse; suppose that $R(V, K_V)$ is f.g. and set $X = \operatorname{Proj} R(V, K_V)$.

It is well known that if R is a f.g. graded ring, there exists $m > 0$ such that $R^{(m)}$ is generated by elements of the smallest degree m (see, for example, EGA II, $(2.1.6)$, (v)); fix such an m. In other words, for each $k \geq 1$, $H^0(\mathcal{O}_V(kmK_V))$ is spanned as a vector space by k-fold products of elements of $H^0(\mathcal{O}_V(mK_V))$.

Let $V' \to V$ be a resolution of the base locus of $|mK_V|$; in view of the birational invariance of $H^0(mK_V)$, I can replace V by V', so assume that

$$|mK_V| = |M| + F,$$

where $|M|$ is a free linear system and F the fixed part. Because of what I just said about $H^0(kmK_V)$, I also have

$$|kmK_V| = |kM| + kF \quad \text{for } k \geq 1.$$

By (2.4) applied to M, the map $\varphi = \varphi_M : V \to \varphi_M(V) \cong X \subset \mathbf{P}$ is birational, and X is normal. By construction $\mathcal{O}_V(M) \cong \varphi^* \mathcal{O}_X(1)$.

CLAIM. *Every irreducible component Γ of F is contracted by φ to a locus $\varphi(\Gamma)$ of dimension $\leq n - 2$.*

The claim implies that X is a canonical variety: indeed, if I set

$$V^0 = V - \{\text{exceptional divisors of } \varphi\}$$

and write X^0 for the open subset of $\varphi(V^0)$ where φ^{-1} is regular, then X^0 is the complement of a subset of codimension ≥ 2 in X; and $\varphi : V^0 \xrightarrow{\sim} X^0$ is an isomorphism inducing

$$\mathcal{O}_X(1)|_{X^0} \cong \mathcal{O}_V(M)|_{V^0} \cong \mathcal{O}_V(mK_V)|_{V^0} \cong \mathcal{O}_X(mK_X)|_{X^0}.$$

So mK_X is a Cartier divisor, giving (1.1), (i); and $mK_V = \varphi^* mK_X + F$ gives (1.1), (ii).

(2.7) PROOF OF CLAIM. This is an easy result in the style of [**Zariski**]: notice first that if $\dim \varphi_M(\Gamma) = n - 1$ then $h^0(\mathcal{O}_\Gamma(kM + A)) \sim (\text{const.}) \cdot k^{n-1}$ as $k \to \infty$ for any divisor A on Γ. Now Γ is fixed in $|kM + \Gamma|$ for every k, and hence the restriction map

$$r_k \colon H^0(\mathcal{O}_V(kM + \Gamma)) \to H^0(\mathcal{O}_\Gamma(kM + \Gamma))$$

is zero. Also, one gets a bound of the form

$$h^1(\mathcal{O}_V(kM)) < (\text{const.}) \cdot k^{n-2}.$$

In fact, since $\varphi_* \mathcal{O}_V(kM) = \mathcal{O}_X(k)$ is ample, $H^i(\varphi_* \mathcal{O}_V(kM)) = 0$ for $k \gg 0$, and the Leray spectral sequence gives

$$H^1(\mathcal{O}_V(kM)) = H^0(R^1 \varphi_* \mathcal{O}_V(kM)) = H^0(R^1 \varphi_* \mathcal{O}_V \otimes \mathcal{O}_X(k)),$$

so that h^0 grows like k^d where $d = \dim \operatorname{Supp} R^1 \varphi_* \mathcal{O}_V \leq n - 2$.

This contradiction proves the claim.   Q.E.D.

(2.8) EXAMPLES. The following is an example of a canonical model of a 3-fold of general type: take the weighted projective space $\mathbf{P}(1, 1, 2, 2, 7)$ with weighted homogeneous coordinates x_1, x_2, y_1, y_2, w, and the hypersurface $X = X_{14} \subset \mathbf{P}(1, 1, 2, 2, 7)$ given by $w^2 = f_{14}(x_1, x_2, y_1, y_2)$. To explain this variety in terms of ordinary projective spaces, consider the generically 2-to-1 morphism $\pi \colon X \to \mathbf{P}(1, 1, 2, 2)$ given by omitting w. Then $\mathbf{P}(1, 1, 2, 2)$ is isomorphic in an obvious way to the quadric of rank 3, $Q \subset \mathbf{P}^4$, and π is the double covering branched in the intersection $Q \cap F_7$ of Q with a general septic, and along the vertex of Q. It is easy to see that X has 7 Veronese cone points at the intersection of the vertex with F.

This example was psychologically important, because using the easy formalism of weighted projective spaces (see [**Dolgachev**]),

$$K_X = \mathcal{O}(14 - 7 - 2 - 2 - 1 - 1) = \mathcal{O}(1)$$

is an ample **Q**-Cartier divisor satisfying $K_X^3 = 14/(2 \cdot 2 \cdot 7) = 1/2$; this number controls the growth of the plurigenera of X (that is, of a nonsingular model of X), so that

$$P_n(X) = h^0(X, \mathcal{O}(n)) \sim \left(\frac{1}{3!}\right) \cdot \left(\frac{1}{2}\right) \cdot n^3 \qquad \text{as } n \to \infty.$$

However, if X had a nonsingular model Y with K_Y nef, RR would give $P_n \sim (1/3!) \cdot K_Y^3 \cdot n^3$, with $K_Y^3 \in 2\mathbf{Z}$; so the plurigenera of this X grow a lot slower than those of any 3-fold of general type having a nonsingular model with K_Y nef. Note also that φ_{nK} cannot be birational for $n \leq 6$, so that this kind of example is analogous to Enriques' famous example $X_{10} \subset \mathbf{P}(1, 1, 2, 5)$ of a surface of general type for which φ_{4K} is not birational (compare [**Catanese**]).

(2.9) Now consider the weighted complete intersection

$$X = X_{6,6,6} \subset \mathbf{P}(2^4, 3^3).$$

This is a rare case when the theoretical idea of embedding a weighted projective space $\mathbf{P}$ by means of some $\mathcal{O}_{\mathbf{P}}(n)$ actually helps to understand it. The embedding of $\mathbf{P}(2^4, 3^3)$ by means of $\mathcal{O}(6)$ looks as follows: take a copy of $\mathbf{P}^3$ in its Veronese embedding by $\mathcal{O}(3)$, and a copy of $\mathbf{P}^2$ in its Veronese embedding by $\mathcal{O}(2)$, then take the linear join:

$$\mathbf{P} = \mathbf{P}(2^4, 3^3) = v_3(\mathbf{P}^3) * v_2(\mathbf{P}^2) \subset \mathbf{P}^{25}.$$

My 3-fold X is the intersection of $\mathbf{P}$ with 3 sufficiently general hyperplanes of $\mathbf{P}^{25}$. It's not hard to see that X has 27 Veronese cone singularities at its intersection with the 3-dimensional stratum $v_3(\mathbf{P}^3)$, and has $K_X = \mathcal{O}(1)$ and $K_X^3 = 1/2$; moreover, since there are no homogeneous polynomials of degree 1, $H^0(\mathbf{P}, \mathcal{O}(1)) = 0$, so it follows that $p_g(X) = 0$. Thus in contrast to the surface case, it's quite easy to write down 3-folds of general type with $p_g = 0$, even with the canonical ring a complete intersection.

(2.10) Since the 3-fold X of (2.9) has $p_g = 0$, it is of some interest to have an interpretation of the intermediate Jacobian JX in terms of families of 1-cycles; I am grateful to D. Ortland for permission to include a description of his beautiful solution of this. The idea is to look at the net (2-dimensional linear system) $F_\lambda = \sum \lambda_i F_i$ with $\lambda = (\lambda_1, \ldots, \lambda_3) \in \Lambda = \mathbf{P}^2$ of weighted hypersurfaces of degree 6 through X, and to note that in coordinates $y_1, \ldots, y_4$, $z_1, \ldots, z_3$ of $\mathbf{P}$, each F_λ is of the form

$$F_\lambda = c_\lambda(y_1, \ldots, y_4) + q_\lambda(z_1, \ldots, z_3)$$

with c_λ cubic and q_λ quadratic. This can be viewed as a net of cubic surfaces in $\mathbf{P}^3$ and a net of plane conics parametrised by the same base space $\Lambda = \mathbf{P}^2$.

Now the conic $Q_\lambda : (q_\lambda = 0) \subset \mathbf{P}^2$ breaks up as a line pair when λ belongs to a discriminant curve $E \subset \Lambda$ (for general X this is a nonsingular cubic curve). Also, for general $\lambda \in \Lambda$, the cubic surface $S_\lambda : (c_\lambda = 0) \subset \mathbf{P}^3$ is nonsingular and contains 27 lines. Hence the set

$$B = \{\text{pairs } (l, m) \text{ of lines} \mid \exists \lambda \in \Lambda \text{ with } l \subset Q_\lambda \text{ and } m \subset S_\lambda\}$$

is a generically 54-to-1 cover $B \to E$; each pair $b = (l, m) \in B$ corresponds to a weighted linear subspace $\Pi_b = \mathbf{P}(2, 2, 3, 3) \subset \mathbf{P}$ entirely contained in one of the hypersurfaces F_λ. It is easy to see that $C_b = \Pi_b \cap X$ is a curve of genus 2.

A general result of Lefschetz theory says that the Hodge structure on $H^3(X, \mathbf{Q})$ is irreducible for sufficiently general X, hence also the intermediate Jacobian JX, so that the family $\{C_b\}_{b \in B}$ induces an Abel-Jacobi (or cylinder) map $JB \to JX$ which must be either zero or surjective. Finally, Ortland uses methods of Clemens to interpret the derivative of $JB \to JX$ and to prove that it is nonzero.

Speculation. Note that the key to success in Ortland's example is to find some special representation of one of the defining equations: if the line $(l, m) \in B$ is given by $y_1 = y_2 = z_1 = 0$ then the corresponding F_λ is of the form

$$F_\lambda = y_1 q_1(y_1, \ldots, y_4) + y_2 q_2(y_1, \ldots, y_4) + z_1 z_2$$

(which looks almost like a quadric of rank 6); the 5-fold hypersurface $(F_\lambda = 0)$ has nontrivial 3-cycles, from which X inherits nontrivial curves.

Now there are plenty of other 3-folds of general type with $p_g = 0$ for which the defining equations do not seem to admit such nice representations. (Can you see what to do with the general quasihomogeneous polynomial $f_{18}(x, y_1, y_2, z, t)$, with $\deg x = 2$, $\deg y_i = 3$, $\deg z = 4$, $\deg t = 5$?) Here the generalised Hodge conjecture predicts a family of curves having a nontrivial Abel-Jacobi map, so that this is (to say the least) a substantial case where the Hodge conjecture has yet to be verified.

There is an analogy between these deep questions and the Bloch-Mumford conjecture on the Chow group of 0-cycles on a surface S with $p_g = 0$ (see, for example, [**Inose-Mizukami**]): in this case the traditional conjecture could also be destroyed by proving that there are no nontrivial curves on the 3-fold $S \times \mathbf{P}^1$.

(2.11) *Canonical hypersurfaces.* There are several methods of searching for canonical 3-folds which are weighted complete intersections; for example, this can be done by guessing the invariants going into the plurigenus formula of §10 (that is, an integer χ, a rational number K^3, and a basket of terminal cyclic quotient singularities), then computing the plurigenera, and determining whether or not there exists a complete intersection ring with this as its Hilbert function. This can all be done by computer, and systematic searches have been carried out by A. R. Fletcher, giving rise to many interesting families of varieties.

The following list of canonical weighted hypersurfaces was generated by a much cruder computer program. It is a complete list of $X = X_d \subset \mathbf{P}(a_1, \ldots, a_5)$ in a "well-formed" weighted projective space (that is, no 4 of the a_i have a common factor, see [**Dolgachev**, (1.3)]) such that

(i) X has terminal quotient singularities (of the type described in (5.2));

(ii) $K_X = \mathcal{O}_X(1)$;

(iii) $d \le 100$.

(Probably there are no others for any d, but the list was obtained by starting an infinite search and switching off the computer after it stopped printing out data.)

Canonical 3-fold hypersurfaces

$X_6 \subset \mathbf{P}^4$	$p_g = 5,\ K^3 = 6$
$X_7 \subset \mathbf{P}(1^4, 2)$	$p_g = 4,\ K^3 = 7/2$
$X_8 \subset \mathbf{P}(1^3, 2^2)$	$p_g = 3,\ K^3 = 2$
$X_9 \subset \mathbf{P}(1^3, 2, 3)$	$p_g = 3,\ K^3 = 3/2$
$X_{10} \subset \mathbf{P}(1^2, 2^2, 3)$	$p_g = 2,\ K^3 = 5/6$
$X_{12} \subset \mathbf{P}(1, 2^2, 3^2)$	$p_g = 1,\ K^3 = 1/3$
$X_{12} \subset \mathbf{P}(1^2, 2, 3, 4)$	$p_g = 2,\ K^3 = 1/2$
$X_{10} \subset \mathbf{P}(1^4, 5)$	$p_g = 4,\ K^3 = 2$
$X_{15} \subset \mathbf{P}(1, 2, 3^2, 5)$	$p_g = 1,\ K^3 = 1/6$
$X_{16} \subset \mathbf{P}(1, 2, 3, 4, 5)$	$p_g = 1,\ K^3 = 2/15$

$$X_{18} \subset \mathbf{P}(2, 3^2, 4, 5) \qquad p_g = 0, \ P_2 \neq 0, \ K^3 = 1/20$$

$$X_{20} \subset \mathbf{P}(2, 3, 4, 5^2) \qquad p_g = 0, \ P_2 \neq 0, \ K^3 = 1/30$$

$$X_{12} \subset \mathbf{P}(1^3, 2, 6) \qquad p_g = 3, \ K^3 = 1$$

$$X_{14} \subset \mathbf{P}(1^2, 2^2, 7) \qquad p_g = 2, \ K^3 = 1/2$$

$$X_{21} \subset \mathbf{P}(1, 3, 4, 5, 7) \qquad p_g = 1, \ K^3 = 1/20$$

$$X_{16} \subset \mathbf{P}(1^2, 2, 3, 8) \qquad p_g = 2, \ K^3 = 1/3$$

$$X_{28} \subset \mathbf{P}(3, 4, 5, 7, 8) \qquad p_g = P_2 = 0, \ P_3 \neq 0, \ K^3 = 1/120$$

$$X_{18} \subset \mathbf{P}(1, 2^2, 3, 9) \qquad p_g = 1, \ K^3 = 1/6$$

$$X_{22} \subset \mathbf{P}(1, 2, 3, 4, 11) \qquad p_g = 1, \ K^3 = 1/12$$

$$X_{28} \subset \mathbf{P}(1, 3, 4, 5, 14) \qquad p_g = 1, \ K^3 = 1/30$$

$$X_{30} \subset \mathbf{P}(2, 3, 4, 5, 15) \qquad p_g = 0, \ P_2 \neq 0, \ K^3 = 1/60$$

$$X_{40} \subset \mathbf{P}(3, 4, 5, 7, 20) \qquad p_g = P_2 = 0, \ P_3 \neq 0, \ K^3 = 1/210$$

$$X_{46} \subset \mathbf{P}(4, 5, 6, 7, 23) \qquad p_g = P_2 = P_3 = 0, \ P_4 \neq 0, \ K^3 = 1/420$$

Note that K^3 can get fairly small, although it is now known to be bounded below for canonical 3-folds with $\chi(\mathcal{O}_X) \leq 1$; see [**Fletcher**] where it is proved (following ideas of J. Kollár) that $P_{12} \neq 0$, $P_{24} \geq 2$, and hence by results of Kollár, $\varphi_{mK} \colon X \to \mathbf{P}^N$ is birational for $m \geq 269$, and so in particular, $K_X^3 \geq (1/269)^3$.

(2.12) *Exercise.* Find the singularities of some of these canonical hypersurfaces; write (x, y, z, t, u) for homogeneous coordinates on the 4-dimensional weighted projective space. Consulting [**Dolgachev**] for information on weighted projective spaces, you can prove, for example, that

(i) $X_{15} \subset \mathbf{P}(1, 2, 3^2, 5)$ has $\chi(\mathcal{O}_X) = 0$, $K^3 = 1/6$, and singularities:

$$\text{1 point of type } \tfrac{1}{2}(1, 1, 1) \text{ at } (0, 1, 0, 0, 0)$$

$$\text{and 5 points of type } \tfrac{1}{3}(2, 1, 1) \text{ along the } (z, t)\text{-axis.}$$

(ii) $X_{18} \subset \mathbf{P}(2, 3^2, 4, 5)$ has $\chi(\mathcal{O}_X) = 1$, $K^3 = 1/20$, and singularities:

$$\text{4 points of type } \tfrac{1}{2}(1, 1, 1) \text{ along the } (x, t)\text{-axis;}$$

$$\text{6 points of type } \tfrac{1}{3}(2, 1, 1) \text{ along the } (y, z)\text{-axis;}$$

$$\text{1 point of type } \tfrac{1}{4}(3, 1, 1) \text{ at } (0, 0, 0, 1, 0);$$

$$\text{and 1 point of type } \tfrac{1}{5}(3, 2, 1) \text{ at } (0, 0, 0, 0, 1).$$

(The notation for the quotient singularities is explained in (4.2).)

3. The main reduction steps.

(3.0) *Overview.* This section gives a brief run-down of the general theory of canonical singularities under the following 6 headings:

(A) Canonical $\Rightarrow$ Du Val singularities in codimension 2.

(B) Reduction to index 1 by cyclic covers.

(C) Index 1 canonical $\Rightarrow$ rational $\Rightarrow$ Cohen-Macaulay.

(D) The general section through a rational Gorenstein singularity is a rational or elliptic singularity.

(E) Reduction to cDV singularities by crepant blow-ups.

(F) Further reduction to isolated cDV singularities.

The final 7th topic

(G) Mori's detailed study of terminal singularities

will be the subject of Chapter II, §§6–7. (The material of (A)–(D) is valid in all dimensions, but (E)–(G) is restricted to $\dim X = 3$; see (3.13).)

(3.1) The overview (3.0) has introduced two new definitions:

DEFINITION. A *cDV singularity* is a 3-fold hypersurface singularity

$$P \in X \colon (F = 0) \subset \mathbf{A}^4$$

given by an equation of the form

$$F(x, y, z, t) = f(x, y, z) + tg(x, y, z, t).$$

where f is the equation of a Du Val singularity (as in (1.2)), and g is an arbitrary polynomial. So a cDV point is just a 3-fold singularity which has a Du Val surface singularity as a hyperplane section; on the other hand, a cDV point can be viewed as a 1-parameter deformation of a Du Val singularity.

DEFINITION. A birational morphism $f \colon Y \to X$ between normal varieties is *crepant* if $K_Y = f^* K_X$. If rK_X is a Cartier divisor, this means that a local basis element $s \in \mathcal{O}_X(rK_X)$ at $P \in X$ remains a local basis around $f^{-1}P$; (here and elsewhere there is an abuse of notation in writing $f^* K_X$, since K_X is not Cartier: by definition this means $(1/r)f^*(rK_X) \in \operatorname{Pic} Y \otimes \mathbf{Q}$). Note that the definitions of canonical and terminal singularities differ only in that canonical singularities are allowed to have exceptional divisors E_i appearing with multiplicity 0 in the discrepancy. A crepant partial resolution of a variety X with canonical singularities is one which pulls out only such exceptional divisors; the key example is a blow-up of a Du Val surface singularity.

(3.2) The goal of steps (A)–(F) is the following theorem, the main result of [**Pagoda**]:

THEOREM. (a) *Any terminal 3-fold point $P \in X$ is of the form $Y/\boldsymbol{\mu}_r$, where $Q \in Y$ is an isolated cDV singularity (or nonsingular), and $\boldsymbol{\mu}_r$ acts on Y freely outside Q and such that on a generator $s \in \omega_Y$,*

$$\boldsymbol{\mu}_r \ni \epsilon \colon s \mapsto \epsilon s.$$

(b) *If X is a 3-fold with canonical singularities then there exists a crepant partial resolution $f \colon Y \to X$ where Y has only terminal singularities.*

(3.3) The result (a) is a partial classification of terminal singularities. (Most of §§5–7 will be devoted to the further classification of the singularities of (a).) On the other hand (b) represents a certain reduction of all canonical singularities to terminal singularities. Compare the situation with the surface case: for surfaces, the canonical points are just the Du Val singularities; the terminal singularities are just nonsingular points. As mentioned in (1.2), the key fact is that there is a resolution $f\colon Y \to X$ such that $K_Y = f^*K_X$. (Note however that for 3-folds, the partial resolution given by (b) is not unique, so that if you've ever heard of the "absolutely minimal models" of surface theory you should do your best to forget about them in higher dimension.)

Finally, one of the key consequences of (G) (see (6.4), (A)) will be that if $P \in X$ is a terminal singularity, then it has a Q-*smoothing*, that is, a deformation X_t such that all the singular points $P_i \in X_t$ of a neighbouring fibre are terminal cyclic quotient singularities $\mathbf{A}^3/\mu_r$, where the action is

$$\mu_r \ni \varepsilon\colon (x,y,z) \mapsto (\varepsilon^a x, \varepsilon^{-a} y, \varepsilon z).$$

This reduces certain problems on 3-folds with canonical singularities to this special class of quotient singularities; in particular, this is the key to the plurigenus formula of §10 below. One can think of this as saying vaguely that a representative sample of canonical 3-folds has only this type of cyclic quotient singularities, in the same way that "most" canonical surfaces are nonsingular (so that the minimal model does not contain (-2)-curves); beware that there is definitely no theorem to this effect, even for surfaces.

I now run through the points (A)–(F) in more detail; (G) will be the main subject of §§5–7.

(3.4) (A) *Canonical* $\Rightarrow$ *Du Val singularities in codimension* 2. Canonical singularities are not necessarily isolated, but if X has canonical singularities then X is analytically of the form

$$X \cong (\text{Du Val singularity}) \times \mathbf{A}^{n-2}$$

in a neighbourhood of a general point of any codimension 2 stratum. This is what you would expect, and is easy to prove; see [**C3-f**, (1.14)].

(3.5) (B) *Reduction to index* 1 *by cyclic covers.* There are several points to make here, since cyclic covers are used in various ways. Firstly, if $P \in X$ is a canonical singularity then there is a standard local μ_r-cover $\pi\colon X' \to X$ with $K_{X'}$ a Cartier divisor and $K_{X'} = \pi^*K_X$. This construction is discussed in detail in (3.6) below. Next, in various contexts there are diagrams of the form

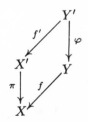

where f and f' are partial resolutions. In this set-up, I have

$$K_{X'} = \pi^*K_X \quad \text{and} \quad K_{Y'} = \varphi^*K_Y + R_\varphi,$$

where R_φ is the ramification divisor of φ. Taking f and f' to be resolutions in this diagram, it is easy to see that if $P \in X$ is canonical, then $Q \in X'$ is canonical of index 1 (where $Q = \pi^{-1}P$).

A different way of using the same kind of diagram is to take f' to be some construction made in an intrinsic way from an index 1 point, for example, a crepant blow-up. Then I can construct $f\colon Y \to X$ by taking $Y = Y'/\mu_r$ to be the quotient of Y' by the group action, and use the same kind of discrepancy calculation to show that f is also crepant. In short, this kind of argument allows me to reduce the study to the index 1 case; for details, see [**Pagoda**, §2].

(3.6) *Cyclic covering trick.* Suppose that $P \in X$ is a point of a normal variety, and D is a Weil divisor on X which is **Q**-Cartier; this means that D is a Weil divisor of X, and rD is a Cartier divisor near P for some $r \in \mathbf{Z}$, $r > 0$. Suppose that r is the smallest such r (the *index* of D). Let $s \in \mathcal{O}_X(-rD)$ be a local basis near P; by taking a smaller neighbourhood of P, I will assume that s is a basis of $\mathcal{O}_X(-rD)$ over the whole of X, and view s as giving an isomorphism $s\colon \mathcal{O}_X(rD) \xrightarrow{\sim} \mathcal{O}_X$.

PROPOSITION. *There exists a cover* $\pi\colon Y \to X$ *which is Galois with group* μ_r, *and such that the sheaves* $\mathcal{O}_X(iD)$ *are the eigensheaves of the group action on* $\pi_*\mathcal{O}_Y$, *that is,*

$$(*) \qquad \mathcal{O}_X(iD) = \{f \in \pi_*\mathcal{O}_Y \mid \varepsilon(f) = \varepsilon^i \cdot f \text{ for all } \varepsilon \in \mu_r\}.$$

Also, Y *is normal,* π *is etale over the locus* X_0 *where* D *is Cartier, and* $\pi^{-1}P = Q$ *is a single point. The* **Q**-*divisor* $\pi^*D = E$ *is a Cartier divisor on* Y.

(3.7) This is an important reduction of the problem, since working directly with the singular sheaves $\mathcal{O}_X(iD)$ is likely to be difficult. The proof can be understood as follows: over X_0, the invertible sheaf $\mathcal{O}_X(-D)$ corresponds to a line bundle $L_0 \to X_0$. A local generator $z \in \mathcal{O}_X(-D)$ is a coordinate on the fibres of L_0; now consider the locus $Y_0\colon (z^r = s) \subset L_0$. Since s is a nowhere vanishing section of $\mathcal{O}_X(-rD)$, the projection map $\pi_0\colon Y_0 \to X_0$ is etale. The idea of the proof is to extend this over the whole of X.

PROOF. Using the given section s, construct the sheaf of $\mathcal{O}_X$-algebras

$$\mathcal{A} = \mathcal{O}_X \oplus \mathcal{O}_X(D) \oplus \cdots \oplus \mathcal{O}_X((r-1)D),$$

with multiplication defined by

$$\mathcal{O}_X(aD) \otimes \mathcal{O}_X(bD) \to \mathcal{O}_X((a+b)D) \quad \text{if } a+b < r,$$

or

$$\mathcal{O}_X(aD) \otimes \mathcal{O}_X(bD) \to \mathcal{O}_X((a+b)D) \xrightarrow{s} \mathcal{O}_X((a+b-r)D) \quad \text{if } a+b \geq r.$$

There is a natural action of μ_r on $\mathcal{A}$ given by multiplication by ε^i in the summand $\mathcal{O}_X(iD)$. Then $\pi\colon Y = \operatorname{Spec}_X \mathcal{A} \to X$ is a cyclic Galois cover which is etale over X; the fact that Y is normal follows by the Serre criterion (see (3.18)). $\pi^{-1}P = Q$ is a single point, since otherwise the subgroup of μ_r stabilizing a point $Q \in \pi^{-1}P$ would be a proper subgroup, and then it is easy to get a contradiction to $r = \operatorname{index}(D)$.

To see the last sentence, suppose without loss of generality that D is an effective divisor. Then the inclusion maps $\mathcal{O}_X((i-1)D) \hookrightarrow \mathcal{O}_X(iD)$, together with the isomorphism $s\colon \mathcal{O}_X(rD) \xrightarrow{\sim} \mathcal{O}_X$ defines an $\mathcal{A}$-linear map $g\colon \mathcal{A} \to \mathcal{A}$ such that g^r is the local equation of $\pi^*(rD)$. Thinking of g as a section of $\mathcal{O}_Y$, clearly $\operatorname{div}(g) = E = \pi^*D$.

(3.8) (C) *Canonical singularities are rational.* It is known quite generally that in characteristic zero, canonical singularities are rational and therefore Cohen-Macaulay; since a quotient of a rational singularity is again rational (this is easy for a quotient $Y = X/G$ by a finite group G, essentially because if

$$
\begin{array}{ccc}
Y' & \xrightarrow{\ \pi\ } & X' \\
\downarrow & & \downarrow \\
Y & \longrightarrow & X
\end{array}
$$

where the vertical arrows are resolutions, then $\mathcal{O}_{X'}$ is a direct summand of $\pi_*\mathcal{O}_{Y'}$; see, for example, [**Pinkham**, p. 150] for details), it is in any case enough to prove this for index 1 singularities:

THEOREM [**Elkik; Flenner**, (1.3)]. *Let $P \in X$ be a canonical singularity of index 1 (that is, K_X is Cartier at P, and $f_*\omega_Y = \omega_X$ for a resolution $f\colon Y \to X$). Then for a resolution $f\colon Y \to X$,*

$$R^i f_*\mathcal{O}_Y = 0 \quad \text{for all } i > 0.$$

There are two ingredients in any proof of this: (1) vanishing and (2) duality. Let me run through Shepherd-Barron's proof in the 3-fold case, where these appear in a transparent way; although this is now a standard result, it still seems rather miraculous to me.

PROOF. Let $P \in X$ be a canonical index 1 point, and $f\colon Y \to X$ a resolution which is the minimal resolution along the Du Val locus. Grothendieck duality for the morphism f gives at once that

$$R^2 f_*\mathcal{O}_Y \overset{d}{=} \omega_X/f_*\omega_Y,$$

which is zero by assumption, so that I must prove that $R^1 f_*\mathcal{O}_Y = 0$.

Now $K_Y = f^*K_X + \Delta$, where Δ is an effective divisor with $f(\Delta) = P$. Vanishing gives

$$R^i f_*\omega_Y = 0 \quad \text{for all } i > 0.$$

However, above a neighbourhood of P, $\omega_Y = \mathcal{O}_Y(\Delta)$, so that also $R^i f_*\mathcal{O}_Y(\Delta) = 0$. Bearing this in mind, consider the cohomology long exact sequence of

$$0 \to \mathcal{O}_Y \to \mathcal{O}_Y(\Delta) \to \mathcal{O}_\Delta(\Delta) \to 0.$$

Since X is normal, $f_*\mathcal{O}_Y = f_*\mathcal{O}_Y(\Delta) = \mathcal{O}_X$, and the long exact sequence becomes

$$H^i(\Delta, \mathcal{O}_\Delta(\Delta)) = R^{i+1} f_*\mathcal{O}_Y \quad \text{for } i = 0, 1, 2.$$

Since the fibres of f have dimension ≤ 2, it follows that

$$H^2(\Delta, \mathcal{O}_\Delta(\Delta)) = R^3 f_*\mathcal{O}_Y = 0.$$

On the other hand, Δ is a Gorenstein scheme with dualising sheaf

$$\mathcal{O}_\Delta(K_Y + \Delta) = \mathcal{O}_\Delta(2\Delta),$$

and therefore Serre duality on Δ gives

$$H^0(\Delta, \mathcal{O}_\Delta(\Delta)) \overset{d}{=} H^2(\Delta, \mathcal{O}_\Delta(\Delta)) = 0.$$

This proves that $R^1 f_* \mathcal{O}_Y = 0$, so that $R^i f_* \mathcal{O}_Y = 0$ for $i > 0$, and $P \in X$ is a rational singularity.   Q.E.D.

(3.9) Rational singularities are known to be Cohen-Macaulay; the Appendix to §3 contains all you need to know about this notion, including a direct and self-contained proof of the case of the result required here: the general hypersurface section through a rational 3-fold singularity $P \in X$ is again normal (see (3.19)).

(3.10) (D) *The general section.*

THEOREM. *Let $P \in X$ be a rational Gorenstein singularity (of an n-fold X, with $n \geq 3$). Then the general hyperplane section $P \in S \subset X$ through P is a rational or elliptic Gorenstein singularity.*

Here elliptic Gorenstein means that for a resolution $f : T \to S$,

$$\varphi_* \omega_T = m_p \cdot \omega_S$$

(or equivalently $R^{n-1} f_* \mathcal{O}_T$ is 1-dimensional).

PROOF. Suppose that S runs through any linear system of sections $P \in S \subset X$ whose equations generate the maximal ideal m_P of $\mathcal{O}_{X,P}$. Then as noted in (3.19), a general element S of this linear system is normal.

Let $f : Y \to X$ be any resolution of X which dominates the blow-up of the maximal ideal m_P; by definition of the blow-up, the scheme-theoretic fibre over P is an effective divisor E such that $m_P \cdot \mathcal{O}_Y = \mathcal{O}_Y(-E)$. Hence $f^* S = T + E$, where T runs through a free linear system on Y. By Bertini's theorem, $\varphi = f|_T : T \to S$ is a resolution of S. Now I use the adjunction formula to compare K_T and $\varphi^* K_S$.

In the diagram

$$Y \supset T + E$$
$$\downarrow f \qquad \downarrow \varphi$$
$$X \supset S,$$

I have

$$K_Y = f^* K_X + \Delta \quad \text{with } \Delta \geq 0$$

and

$$T = f^* S - E,$$

so that

$$K_Y + T = f^*(K_X + S) + \Delta - E$$

and

$$K_T = (K_Y + T)_T = \varphi^* K_S + (\Delta - E)_S.$$

This just means that any $s \in \omega_S$ has at worst $(E - \Delta)_T$ as pole on T. On the other hand, since the maximal ideal $m_{S,P} \subset \mathcal{O}_{S,P}$ is the restriction to S of the maximal ideal $m_{X,P} \subset \mathcal{O}_X$ (this is where the argument uses Cohen-Macaulay in an essential way), it follows that every element of $m_{S,P}$ vanishes along $E \cap T$. Hence every element of $m_{S,P} \cdot \omega_S$ is regular on T, that is,

$$\varphi_* \omega_T \supset m_P \cdot \omega_S. \quad \text{Q.E.D.}$$

(3.11) (D) says that if $\dim X = 3$ and $P \in X$ is a canonical index 1 singularity, then *either* it is cDV, *or* a general hyperplane section through P is an elliptic Gorenstein surface singularity. From now on I will use special results on the classification of elliptic Gorenstein surface singularities, so that the remainder of the discussion is restricted to $\dim X = 3$.

(3.12) (E) *Reduction to cDV points by crepant blow-ups.* If $P \in X$ is a canonical index 1 point which is not cDV then there exists a blow-up $\sigma \colon X' \to X$ such that $K_{X'} = \sigma^* K_X$. This follows by using analogous properties of elliptic Gorenstein singularities; see [**C3-f**, (2.11–12)] for details.

COROLLARY. (i) *A rational Gorenstein 3-fold singularity $P \in X$ is terminal if and only if it is cDV.*

(ii) *Let $P \in X$ be a canonical index 1 point; then there exists a partial resolution $f \colon Y \to X$ which is crepant (that is, $K_Y = f^* K_X$) and such that Y has cDV singularities.*

EXAMPLES. (i) Suppose that $P \in X$ is the hypersurface singularity $X \colon$ $(f = 0)$ where $f = x^3 + y^3 + z^3 + t^n$ with $n \geq 3$; then the blow-up $\sigma \colon X' \to X$ of m_P is the variety $X' \colon (f' = 0)$, where $f' = x^3 + y^3 + z^3 + t^{n-3}$. Essentially the same calculation as in (1.9), (1) shows that $K_{X'} = \sigma^* K_X$.

(ii) A hypersurface double point is rather special, and in this case the required blow-up σ is not just the blow-up of m_P. For example, let $P \in X \colon (f = 0)$ where $f = x^2 + y^3 + z^6 + t^n$ with $n \geq 6$; then the required blow-up is given in one affine piece by setting

$$x = x_1 t^3, \qquad y = y_1 t^2, \qquad z = z_1 t.$$

The blown-up variety is then $X' \colon (f' = 0)$, where $f' = x_1^2 + y_1^3 + z_1^6 + t^{n-6}$. The proof that $K_{X'} = \sigma^* K_X$ in this case is similar to the calculation of (1.9), (1), and you can try it as an amusing exercise.

(3.13) It is important to understand that Corollary 3.12, (i) is proved via the classification of elliptic Gorenstein surface singularities. The statement that a terminal index 1 singularity has a rational hypersurface section is false for 4-fold singularities, as shown by the following example (one of a large class related to weighted $K3$ hypersurfaces):

$$0 \in X \colon (x^3 + y_1^4 + y_2^4 + z_1^6 + z_2^6 = 0) \subset \mathbf{A}^5;$$

here X is a terminal 4-fold singularity. This follows by the argument of Theorem 4.6 (see also [**C3-f**, (4.3)]), essentially because

$$\tfrac{1}{3} + \tfrac{1}{4} + \tfrac{1}{4} + \tfrac{1}{6} + \tfrac{1}{6} = 1 + \tfrac{2}{12} > 1 + \tfrac{1}{12}.$$

However, any hypersurface section $0 \in H \subset X$ is an irrational singularity: for example the hyperplane section $(z_2 = 0)$ is a weighted cone over a $K3$ surface.

This is one aspect of the fact that there are very many more terminal singularities in dimension ≥ 4 than in dimension 3, and it seems unrealistic to expect any useful classification.

(3.14) (F) *Further reduction to isolated cDV singularities.* The procedure of (3.12) reduces 3-fold canonical singularities of index 1 to cDV points, but the singularities are not necessarily isolated. Let X be a 3-fold with at worst cDV singularities, and suppose that Γ is an irreducible curve component of Sing X; then as mentioned in (3.4), X is analytically isomorphic to $\Gamma \times$ (Du Val singularity) in a neighbourhood of a general point of Γ (and possibly worse at a finite set of dissident points). Above this neighbourhood, the blow-up of X along Γ is just $\Gamma \times$ the blow-up of a Du Val singularity, which is crepant. The idea is to extend this crepant blow-up along the whole of Γ, so that the nonisolated singular locus of X can be reduced by a crepant morphism $f : Y \to X$.

The key to understanding this situation is the Brieskorn-Tyurina theory of simultaneous resolution of families of Du Val singularities. If $P \in X$ is a cDV point and $t \in m_P$ is such that the section $X_0 : (t = 0) \subset X$ is a Du Val singularity, then the map $t : X \to \mathbf{A}^1 = T$ represents X as a deformation of X_0 over a 1-dimensional parameter space T. Now it is well known [**Brieskorn**] that after making a base change

by a cyclic branched cover $T' \to T$ of the base, the family admits a simultaneous resolution, that is, there is a morphism $f : Y \to X'$ which fibre-by-fibre is the minimal resolution of the fibres of $X \to T$. Since we know everything about Du Val singularities and their minimal resolution, there is a lot of information available on the resolution $Y \to X'$. On the other hand $X' \to X$ is just a cyclic cover, so that the relation between singularities of X' and those of X can be studied by the methods mentioned in (3.5).

Putting together these ideas shows firstly that isolated cDV points are exactly the terminal singularities of index 1, and that nonisolated cDV singularities can be blown-up along the Du Val locus to give crepant partial resolutions; this leads to a proof of Theorem 3.2 (see [**Pagoda**] for details).

Appendix to §3. Cohen-Macaulay and all that.

(3.15) The following two properties are the main things you need to know about Cohen-Macaulay (CM):

(i) *Invariance under passing to a hyperplane section.* CM is a property of the local ring $\mathcal{O}_{X,P}$ of a point $P \in X$ of a scheme X (think of P as the generic point of an irreducible subvariety of X); if P has codimension 0 then X is CM at P by definition. Otherwise $P \in X$ is CM if and only if there exists an element

$h \in m_{X,P}$ of the maximal ideal which is a non-zerodivisor of $\mathcal{O}_{X,P}$ and such that $P \in Y$ is CM, where $Y \subset X$ is the subscheme defined locally by the principal ideal sheaf $\mathcal{I}_Y = h\mathcal{O}_X$, that is, $Y : (h = 0) \subset X$.

(ii) *The Serre criterion.* An isolated surface singularity is CM if and only if it is normal.

So by definition, a 3-fold X which is nonsingular in codimension 1 is CM if and only if there is a normal surface section through every point $P \in X$. Despite the geometric significance of this notion, young persons seem to find it hard to grasp, and I give a brief treatment.

(3.16) *The definition of depth and Serre's condition S_k.* Given a point P of a scheme X, there is a well-defined integer

$$d = \text{depth}_P \, \mathcal{O}_X$$

with the property that there exist chains of subschemes of length d

(∗) $$P \in X_d \subset X_{d-1} \subset \cdots \subset X_1 \subset X_0 = X,$$

where for each i, $X_{i+1} \subset X_i$ is the subscheme defined locally by the principal ideal sheaf $h_i \mathcal{O}_{X_i}$ with h_i a non-zerodivisor of $\mathcal{O}_{X_i,P}$, and no such chains of length $d + 1$. In this situation,

either $\dim X_d = 0$, and $P \in X$ is Cohen-Macaulay,

or $\dim X_d \geq 1$, and every element of $m_{X_d,P}$ is a zero-divisor of $\mathcal{O}_{X_d,P}$;

by elementary results in primary decomposition, the second possibility happens if and only if there exists $0 \neq f \in \mathcal{O}_{X_d,P}$ such that $m_{X_d,P} \cdot f = 0$, that is, f is a section of $\mathcal{O}_{X_d}$ whose support is just $\{P\}$.

EXAMPLE. Let $X \subset \mathbf{A}^2$ be the subscheme given by $(x^2 = xy = 0)$; then $0 \neq x \in \Gamma(\mathcal{O}_X)$ is killed by the maximal ideal $m = (x, y)$; hence m does not contain any non-zerodivisor, and $\text{depth}_0 \, \mathcal{O}_X = 0$.

This is of course just a geometrical translation of the algebraic definition of depth in terms of regular sequences. There is only one thing to be checked: that the property is independent of the choice of the chain (∗), or equivalently, that the statement of (3.15), (i) is independent of the choice of h; see for example [**Matsumura**, (16.3)].

DEFINITION. A scheme X *satisfies condition S_k* if for every point $P \in X$,

$$\text{depth}_P X \geq \min(k, \text{codim} \, P).$$

It follows at once from the above discussion that X fails to satisfy S_1 if and only if there exists a section $f \in \Gamma(U, \mathcal{O}_X)$ of $\mathcal{O}_X$ (over an open $U \subset X$) whose support has codimension ≥ 1 in X; since f is necessarily nilpotent, this can't happen for an integral scheme: a variety automatically satisfies S_1.

(3.17) Now I discuss the S_2 condition. Let X be an integral scheme and $k(X)$ its function field.

LEMMA. *Let $Q \in X$ be a point of an integral scheme; then*

$$\text{depth}_Q \, \mathcal{O}_X = 1 \quad \Longleftrightarrow \quad \exists f \in k(X) \text{ s.t. } f \notin \mathcal{O}_{X,Q} \text{ but } m_{X,Q} \cdot f \subset \mathcal{O}_{X,Q}$$

PROOF. ($\Leftarrow$) For any $0 \neq x$, $y \in m_{X,Q}$, both $xf, yf \in \mathcal{O}_{X,Q}$, but $xf \notin (x)$ (otherwise $f \in \mathcal{O}_{X,Q}$). Then y is a zero-divisor in $\mathcal{O}_{X,Q}/(x)$, since $y(xf) = x(yf)$; this proves that $\text{depth}_Q \, \mathcal{O}_X \leq 1$.

($\Rightarrow$) Let $0 \neq x \in m_{X,Q}$; x is automatically a non-zerodivisor. Then by the assumption $\text{depth}_Q \, \mathcal{O}_X = 1$, there exists $0 \neq \bar{g} \in \mathcal{O}_{X,Q}/(x)$ which is killed by $m_{X,Q}$. Let $g \in \mathcal{O}_{X,Q}$ be any lift of $\bar{g}$; then $m_{X,Q} \cdot g \subset (x)$, so that $f = g/x \in k(X)$ satisfies $f \notin \mathcal{O}_{X,Q}$ but $m_{X,Q} \cdot f \subset \mathcal{O}_{X,Q}$. Q.E.D.

EXAMPLE (Macaulay). Let $S = \text{Spec } k[x^4, x^3y, xy^3, y^4]$; this is the affine cone over an embedding $C \subset \mathbf{P}^3$ of $\mathbf{P}^1$ as a quartic in $\mathbf{P}^3$, which is not linearly normal. Then S has depth 1 only, since $x^2y^2 \notin \mathcal{O}_S$.

(3.18) *The Serre criterion: normality and S_2.* It is well known that a rational function on a normal variety X with no poles along divisors is regular on X. (In commutative algebra, the assertion is that a normal Noetherian domain is an intersection of DVR's, see [**Matsumura**, (11.5)]].)

Now let X be an affine integral scheme; say that a rational function $f \in k(X)$ is *quasiregular* if $f \in \mathcal{O}_{X,P}$ for every codimension 1 point $P \in X$.

THEOREM. *An integral scheme X satisfies S_2 if and only if*

$$\textit{quasiregular} \quad \Rightarrow \quad \textit{regular};$$

in other words, for an open set $U \subset X$,

$$\Gamma(U, \mathcal{O}_X) = \{f \in k(X) | f \in \mathcal{O}_{X,P} \ \forall \ \textit{codim. 1 points } P \in U\}.$$

In particular,

$$\textit{normal} \quad \Longleftrightarrow \quad \textit{regular (nonsingular) in codimension 1 and } S_2.$$

PROOF. Given a quasiregular element $f \in k(X)$, the set $\Sigma = \Sigma(f)$ of points $P \in X$ such that $f \notin \mathcal{O}_{X,P}$ is closed, and $\text{codim} \, \Sigma \geq 2$. By the Nullstellensatz, if Q is a generic point of a component of Σ then $(m_Q)^k \cdot f \subset \mathcal{O}_{X,Q}$, so that a suitable multiple g of f satisfies $g \notin \mathcal{O}_{X,Q}$ but $m_Q \cdot g \subset \mathcal{O}_{X,Q}$, and by the lemma, $\text{depth}_Q \, \mathcal{O}_X = 1$. This proves the first part: if X is S_2 then $\Sigma = \varnothing$, and conversely.

If X is regular in codimension 1 and S_2 then the local ring $\mathcal{O}_{X,P}$ at each prime divisor is normal, and by what I've just proved, $\Gamma(U, \mathcal{O}_X)$ is an intersection of these, hence again normal. Conversely, if X is normal then so is the local ring $\mathcal{O}_{X,P}$ at each prime divisor, and hence $\mathcal{O}_{X,P}$ is a DVR (see [**Matsumura**, (11.2)]]); this gives regular in codimension 1, and by the fact that a Noetherian normal domain is an intersection of DVRs, quasiregular implies regular, which gives S_2. Q.E.D.

Notice that the theorem must be stated in terms of scheme-theoretic points of X; for example, if S is as in Example 3.17 and $X = \mathbf{A}^1 \times S$ then the closed points of X have depth ≥ 2.

(3.19) *Rational* $\Rightarrow$ *Cohen-Macaulay*. The general proof that rational singularities are Cohen-Macaulay in characteristic 0 seems to involve vanishing and two applications of duality; the 3-fold case can be done much more simply.

THEOREM. *Let $P \in X$ be a normal 3-fold singularity and $P \in S \subset X$ a general hypersurface section. Then $R^1 f_* \mathcal{O}_Y = 0$ for a resolution $f \colon Y \to X$ implies that P is a normal point of S, and hence $P \in X$ is CM.*

PROOF. Write $m_P = m_{X,P} \subset \mathcal{O}_{X,P}$ for the maximal ideal, and $|m_P|$ for the linear system of all hypersurface sections through P. An element $S \in |m_P|$ is just a surface section $P \in S \subset X$. Then since $|m_P|$ is very ample outside P, a suitably general $S \in |m_P|$ is nonsingular outside P (by the trivial Bertini theorem). So the question is to prove that S is normal.

Let $f \colon Y \to X$ be a resolution which dominates the blow-up of m_P; then $f^* S = T + E$, where E is the scheme-theoretic fibre, and T is a surface moving in a free linear system. By Bertini (using characteristic 0), T is a resolution of S, so that $\mathcal{O}_S \subset f_* \mathcal{O}_T$ and S is normal if and only if $f_* \mathcal{O}_T = \mathcal{O}_S$. Since $\mathcal{O}_S$ is generated as a vector space by k and m_P, this is equivalent to saying that $f_* \mathcal{O}_T(-E) = m_P$. But this follows from the cohomology long exact sequence of

$$0 \to \mathcal{O}_Y(-f^* S) \to \mathcal{O}_Y(-E) \to \mathcal{O}_T(-E) \to 0.$$

Indeed, $R^1 f_* \mathcal{O}_Y(-f^* S) = (R^1 f_* \mathcal{O}_Y) \otimes \mathcal{O}_X(-S) = 0$ (by $R^1 f_* \mathcal{O}_Y = 0$); hence

$$f_* \mathcal{O}_Y(-E) \to f_* \mathcal{O}_T(-E)$$

is surjective. However, $f_* \mathcal{O}_Y(-E) = m_{X,P}$, and $m_{X,P}$ maps to $m_{S,P} \subset \mathcal{O}_S$. This proves S is normal.

Chapter II
Classification of 3-Fold Terminal Singularities

4. Toric methods for hyperquotient singularities.

(4.1) *Hyperquotient singularities in general.* This section is pure toric geometry. I put together the notation in force throughout §§4–7.

Pedantry. Recall that μ_r denotes the cyclic group of rth roots of unity in k; the choice of a primitive rth rooth of unity ε defines an isomorphism $\mathbf{Z}/r \cong \mu_r$, but I want to avoid making this choice. The point is that the action of μ_r on any k-vector space V will break it up into 1-dimensional irreducible eigenspaces, where the action is given by $\mu_r \ni \varepsilon \colon v \mapsto \varepsilon^a v$ for some $a \in \mathbf{Z}/r$; in the notation ε^a, think of the element $a \in \mathbf{Z}/r$ as a character of μ_r (the endomorphism $\mu_r \to \mu_r$ given by $\varepsilon \mapsto \varepsilon^a$). The advantage of distinguishing elements $\varepsilon \in \mu_r$ from characters $a \in \mathbf{Z}/r$ is analogous to that of distinguishing between a vector space and its dual.

Suppose that

$$Q \in Y \colon (f = 0) \subset \mathbf{A}^{n+1}$$

is a hypersurface singularity with an action of μ_r, and $P \in X = Y/\mu_r$ is the quotient; I'm interested in saying when the singularity $P \in X$ is canonical (or terminal, log canonical, etc.) in terms of the action of $\boldsymbol{\mu}_r$ and the Newton polyhedron of f. I always assume that the group action is free in codimension 1 on Y (so "no quasireflections"). The two cases

$$r = 1 \quad \text{and} \quad Y \colon (x_0 = 0) = \mathbf{A}^n \subset \mathbf{A}^{n+1}$$

are not excluded, so that this class includes both cyclic quotient singularities and hypersurface singularities. Points $P \in X$ of this kind are hereby christened *hyperquotient singularities*.

(4.2) *Type of a singularity.* Any cyclic quotient singularity is of the form $X = \mathbf{A}^n/\boldsymbol{\mu}_r$; the action $\boldsymbol{\mu}_r$ on $\mathbf{A}^n$ can be diagonalised, and is then given by

$$\boldsymbol{\mu}_r \ni \varepsilon \colon (x_1, \ldots, x_n) \mapsto (\varepsilon^{a_1} x_1, \ldots, \varepsilon^{a_n} x_n)$$

for certain $a_1, \ldots, a_n \in \mathbf{Z}/r$. The singularity is determined by a knowledge of r and $a_1, \ldots, a_n$, and I define $\frac{1}{r}(a_1, \ldots, a_n)$ to be the *type* of X; there is a reason for the fractional notation in toric geometry, although you can think of it as purely symbolic.

Now return to the set-up of (4.1); in suitable local analytic coordinates, the group action on Y extends to an action on $\mathbf{A}^{n+1}$ (it acts on the tangent space $T_{Y,Q}$), and I can assume that there, $\boldsymbol{\mu}_r$ acts diagonally by

$$\boldsymbol{\mu}_r \ni \varepsilon \colon (x_0, \ldots, x_n) \mapsto (\varepsilon^{a_0} x_0, \ldots, \varepsilon^{a_n} x_n).$$

Since Y is fixed by the action of $\boldsymbol{\mu}_r$, it follows that f is an eigenfunction, so that

$$\boldsymbol{\mu}_r \ni \varepsilon \colon f \mapsto \varepsilon^e f;$$

the symbol $\frac{1}{r}(a_0, \ldots, a_n; e)$ is the *type* of the hyperquotient singularity $P \in X$. It will be useful to note that the action of $\boldsymbol{\mu}_r$ on the standard generator

$$s = \operatorname{Res}_Y \frac{dx_0 \wedge \cdots \wedge dx_n}{f} = \frac{dx_1 \wedge \cdots \wedge dx_n}{\partial f/\partial x_0} \in \omega_Y$$

(see (1.8)) is given by

$$\boldsymbol{\mu}_r \ni \varepsilon \colon s \mapsto \varepsilon^c s, \quad \text{with } c = a_0 + \cdots + a_n - e.$$

The assumption that the group acts freely in codimension 1 on Y implies that for any divisor $d|r$,

$$\#\{i|d \text{ divides } a_i\} \le n - 1.$$

(4.3) Write $\overline{M} \cong \mathbf{Z}^{n+1}$ for the lattice of monomials on $\mathbf{A}^{n+1}$, and $\overline{N}$ for the dual lattice; then define N to be the overlattice of $\overline{N}$ given by

$$N = \overline{N} + \mathbf{Z} \cdot \tfrac{1}{r}(a_0, \ldots, a_n).$$

Thus

$$\alpha \in N \quad \Longleftrightarrow \quad \alpha \equiv \tfrac{1}{r}(ja_0, \ldots, ja_n) \mod \mathbf{Z}^{n+1} \text{ for some } j = 0, \ldots, r-1.$$

Let $M \subset \overline{M}$ be the dual sublattice. (Each of these lattices is $\cong \mathbf{Z}^{n+1}$, so it is important to give each its own name; think of M as monomials, and N as weightings or valuations of monomials.) The point of this construction is just that $\boldsymbol{\mu}_r$ acts on a monomial $x^m = x_0^{m_0} \cdots x_n^{m_n}$ by

$$\boldsymbol{\mu}_r \ni \varepsilon \colon x^m \mapsto \varepsilon^{\alpha(m)} x^m, \quad \text{with } \alpha(m) = \sum a_i m_i,$$

so that

$$x^m \text{ is invariant under } \boldsymbol{\mu}_r \quad \Longleftrightarrow \quad \alpha(m) \equiv 0 \bmod r \quad \Longleftrightarrow \quad m \in M.$$

Write σ for the positive quadrant in $N_{\mathbf{R}}$ and $\sigma^{\vee}$ for the dual quadrant in $M_{\mathbf{R}}$. Then as usual in toric geometry,

$$\mathbf{A}^{n+1} = \operatorname{Spec} k[x_0, \ldots, x_n] = \operatorname{Spec} k[\overline{M} \cap \sigma^{\vee}],$$

and the quotient, corresponding to polynomials invariant under the action of $\boldsymbol{\mu}_r$, is

$$A = \mathbf{A}^{n+1}/\boldsymbol{\mu}_r = \operatorname{Spec} k[M \cap \sigma^{\vee}].$$

Notice that $P \in X \subset A$, so that the quotient singularity still lives naturally in an ambient space, but is not necessarily a Cartier divisor there: the ideal

$$I_X \subset k[M \cap \sigma^{\vee}]$$

is the intersection of the ring of invariants $k[M \cap \sigma^{\vee}]$ with the principal ideal (f) of $k[x_0, \ldots, x_n]$, and is generated by some set $\{x^m \cdot f\}$ of invariant products of f with suitable monomials.

It's nevertheless useful to think of X as being $X \colon (f = 0) \subset A$, but beware that this can lead to error. A typical paradox of this kind is the fact that the hypersurface $X \colon (f = 0) \subset \mathbf{P}$ in a weighted projective space $\mathbf{P}$ (defined globally by a weighted homogeneous polynomial) is not necessarily defined locally by one equation.

(4.4) There is of course no purely toric method of getting a resolution of a general singularity $P \in X$ of this type (it includes all hypersurfaces). However, given a resolution $f \colon B \to A$ of the ambient space A, the proper transform $X' \subset B$ of X can be thought of as lying between X and its resolution, and the conditions

$$f_* \mathcal{O}_{X'}(rK_{X'}) = \mathcal{O}_X(rK_X)$$

for various toric resolutions $B \to A$ provide necessary conditions for $P \in X$ to be canonical; taking B related to the Newton polyhedron of X is most likely to produce useful information. If X is nondegenerate with respect to its Newton polyhedron (by definition this means that a suitable toric resolution $B \to A$ of the ambient space leads to a nonsingular X') then these conditions will also be sufficient.

(4.5) Let $\alpha = (b_0, \ldots, b_n) \in N \cap \sigma$ be a vector; this means that α is a weighting $\alpha(x_i) = b_i \in \mathbf{Q}$ on monomials such that

(i) $\alpha \in N$, that is, $\alpha \equiv \frac{1}{r}(ja_0, \ldots, ja_n) \bmod \mathbf{Z}^{n+1}$ for some $j = 0, \ldots, r-1$; and (ii) $\alpha \in \sigma$, that is, $b_i \geq 0$ for each i.

I can extend this weighting to $k[x_0, \ldots, x_n]$ in the obvious way: say that a monomial $x^m = x_0^{m_0} \cdots x_n^{m_n}$ *appears in f* (or just write $x^m \in f$) if its coefficient in f is nonzero; then define

$$\alpha(f) = \min\{\alpha(x^m) \mid x^m \in f\}.$$

In these terms, the Newton polyhedron of f is the lattice polyhedron of $M_{\mathbf{R}}$ defined by

$$\text{Newton}(f) = \{u \in M_{\mathbf{R}} \mid \alpha(u) \geq \alpha(f) \text{ for every } \alpha \in N \cap \sigma\};$$

note that the "inside" of $\text{Newton}(f)$ is the part above the polygon, except in some contexts when it is the part below.

(4.6) THEOREM. *A necessary condition for $P \in X$ to be terminal (or canonical) is that*

(*) $$\alpha(x_0 \cdots x_n) > \alpha(f) + 1$$

(respectively $\geq$) for every primitive vector $\alpha \in N \cap \sigma$. (For a quotient singularity, that is, when Y is nonsingular, the condition is $\alpha(x_1 \cdots x_n) > 1$, formally the case $f = x_0$ of (*); some of the arguments below may need minor modification to deal with this case.)

REMARK. If $r = 1$, then (*) is the condition that the point $(1, \ldots, 1) \in M$ is in the interior of $\text{Newton}(f)$. For $r > 1$, the analogous statement involves a slightly nonobvious notion of interior of a lattice polyhedron due to J. Fine which is important in other "canonical" contexts, and I discuss this in the appendix to §4.

(4.7) *The unit cube $\square$ and the weightings α_k.* I run through notation and ideas which will appear throughout the rest of this chapter. Write $\square$ for the unit cube of $N_{\mathbf{R}}$; this is the unit cell of the sublattice $\mathbf{Z}^{n+1} = \overline{N} \subset N$. Both the unit point $(1, \ldots, 1) \in N$ and the symmetry $i \colon \alpha \mapsto \alpha' = (1, \ldots, 1) - \alpha$ will appear in what follows, and obviously $\square = \sigma \cap i(\sigma)$.

(*) applies to any weighting of $N \cap \sigma$, but in practice the most important ones to consider are the points of $\overline{N}$ (which correspond to $\text{Newton}(f)$ independently of the μ_r-action), and those of $N \cap \square$.

In the case of a cyclic group action, $N = \mathbf{Z}^{n+1} + \mathbf{Z} \cdot \frac{1}{r}(a_0, \ldots, a_n)$, and it is easy to check that $N \cap \square$ consists of (the vertices of $\square$ together with) the $r - 1$ weightings

$$\alpha_k = \tfrac{1}{r}(\overline{a_0 k}, \ldots, \overline{a_n k})$$

for $k = 1, \ldots, r - 1$ (where $\bar{}$ denotes smallest residue $\bmod r$). I will usually be interested in cases where the fixed locuses of elements of μ_r have small dimension, so that most of the a_i are coprime to r.

(4.8) PROOF OF (4.6). This is a tutorial on standard toric stuff. Roughly speaking, the residue of $s = ((dx_0 \wedge \cdots \wedge dx_n)/f)^{\otimes r}$ on X bases $\mathcal{O}_X(rK_X)$ and (*) is the condition that s is regular and vanishes along an exceptional prime divisor corresponding to α. It's easy to apply the methods without understanding the proof, and some elementary applications are given in (4.9); the reader not in need of remedial instruction on these topics should GOTO (4.10).

For each primitive vector $\alpha \in N \cap \sigma$, there is toric blow-up $\varphi \colon B \to A$ of the ambient space $A = \mathbf{A}^{n+1}/\boldsymbol{\mu}_r$ with a single exceptional divisorial stratum $\Gamma \subset B$, such that the valuation of a monomial x^m (for $m \in M$) along Γ is given by

$$(**) \qquad\qquad v_\Gamma(x^m) = \alpha(m).$$

φ is given in toric geometry by the barycentric subdivision of the cone σ at α. A neighbourhood of the generic point of Γ in B is given as follows: the lattice $\alpha^\perp \subset M$ is isomorphic to $\mathbf{Z}^n$, and there is a complementary vector m_0 such that $\alpha(m_0) = 1$. The semigroup $M \cap (\alpha \geq 0)$ is then of the form $\alpha^\perp \oplus \mathbf{N} \cdot M_0$, and Spec of this is $\mathbf{G}_m^n \times \mathbf{A}^1$, with a single remaining toric stratum $\Gamma \cong \mathbf{G}_m^n$ given by $(z = 0)$, where $z = x^{m_0}$. Then for any $m \in M$, $x^m = x^{m'} \cdot z^{\alpha(m)}$ with $\alpha(m') = 0$, which proves $(**)$.

Write $X' \subset B$ for the proper transform of $X \subset A$ under the blow-up $\varphi \colon B \to A$ of the ambient space, and $\psi \colon Y \to X'$ for a resolution of X'; as usual, the pull-back of $\mathbf{Q}$-divisors is denoted by φ^*. The following proposition obviously implies (4.6).

PROPOSITION. *In the above notation,*
(I) $X' = \varphi^* X - \alpha(f)\Gamma$ *and* $K_B = \varphi^* K_A + (\alpha(x_0 \cdots x_n) - 1)\Gamma.$
(II) *Suppose that* $a = \alpha(x_0 \cdots x_n) - \alpha(f) - 1 \leq 0$, *so that*

$$K_B + X' = \varphi^*(K_A + X) + a\Gamma \qquad with \ a \leq 0;$$

then the resolution $h = \varphi \circ \psi \colon Y \to X$ *satisfies*

$$K_Y = h^* K_X - Z,$$

where every exceptional component of h lying over a component of $X' \cap \Gamma \subset B$ appears in Z with coefficient ≥ 0. In particular, X is not terminal.

PROOF OF (I). If follows from the definition of $\alpha(f)$ that $g = f^r/z^{r\alpha(f)}$ is a regular function in a neighbourhood of Γ not vanishing along Γ. Then rX' is the divisor given by $(g = 0)$ in a neighbourhood of $X' \cap \Gamma$; it follows that $g = x^m f'^r$ where $m \in \alpha^\perp$ and X' is given by $(f' = 0)$. This proves the first equality in (I).

Now to compare differentials on A and B. The following manoeuvre is the standard treatment of the canonical class in toric geometry: let $m_0, \ldots, m_n \in M$ be a $\mathbf{Q}$-basis of M, and write $z_i = x^{m_i}$ for the corresponding monomials; then write down the rational canonical differential

$$t = \frac{dz_0}{z_0} \wedge \cdots \wedge \frac{dz_n}{z_n}.$$

The point is that t is independent of the particular choice of $\mathbf{Q}$-basis of M (up to a scalar factor): to see this, consider the Jacobian of a monomial coordinate change.

Clearly t is a basis of $\mathcal{O}(K_\mathbf{T})$ over the big torus $\mathbf{T} \subset A$, with logarithmic poles along all codimension 1 strata of A. For any toric variety A, write D_A for the reduced divisor of A made up of all the codimension 1 strata of A. Then $t \in \mathcal{O}_A(K_A + D_A)$ is a basis element, essentially independent of the choice of basis $m_0, \ldots, m_n$. In particular, the toric blow-up $\varphi \colon B \to A$ satisfies

$$K_B + D_B = \varphi^*(K_A + D_A).$$

(This proves that, quite generally, a toric variety A marked with its divisor D_A has log canonical singularities.) To prove the second formula in (I), note that $\varphi^* D_A$ is a $\mathbf{Q}$-Cartier divisor which coincides with D_B outside the exceptional locus Γ of φ; on the other hand, by construction, Γ has multiplicity 1 in D_B, and it clearly has multiplicity $\alpha(x_0 \cdots x_n)$ in $\varphi^* D_A$ (since rD_A is a Cartier divisor with defining equation $(x_0 \cdots x_n)^r$).

PROOF OF (II). The statement in (II) looks obvious enough: the adjunction formula should give

$$K_{X'} \stackrel{?}{=} (K_B + X')|_{X'} = (\varphi^*(K_A + X) + a\Gamma)|_{X'} = \varphi^* K_X + a(\Gamma \cap X'),$$

so that if $a \leq 0$, the components of $\Gamma \cap X'$ make a negative contribution to the canonical class. The simple case is when the generic points of $\Gamma \cap X'$ are all contained in the nonsingular locus of B, and X' is nonsingular there; then there is no problem about using the adjunction formula, and $K_{X'} = \varphi^* K_X + a\Gamma$, where $a = \alpha(x_0 \cdots x_n) - \alpha(f) - 1$.

Unfortunately this argument doesn't work in general: the plausible-looking adjunction formula for $\mathbf{Q}$-divisors (indicated by $\stackrel{?}{=}$) is false whenever B is singular along a divisor of X'. (Consider, for example, a generator of the ordinary quadratic cone in $\mathbf{P}^3$.) The point is to see that singularities of B along a divisor of X' make a negative contribution to the canonical class.

By easy results on surface singularities, B can be resolved by a morphism $f \colon C \to B$ such that, restricting to a neighbourhood of the general point of any codimension 2 singular stratum, $K_C = f^* K_b - Z_C$ with $Z_C \geq 0$; if Y is the proper transform of X' in C then $Y = f^* X' - D$ with $D > 0$, and then, writing $g \colon Y \to B$ for the composite $Y \to X' \hookrightarrow B$, I get

$$\begin{aligned}
K_Y = (K_C + Y)|_Y &= g^*(K_B + X') - (Z_C + D)|_Y \\
&= h^* K_X + a g^* \Gamma - (Z_C + D)|_Y.
\end{aligned}$$

Thus if $a \leq 0$, any exceptional divisor lying over a component of $X' \cap \Gamma$ appears in the formula for the canonical class with coefficient ≤ 0, so that X is not terminal. (If Y is not normal, then passing to its normalisation again makes a negative contribution to the canonical class.) This proves (II).   Q.E.D.

REMARK. A component of $X' \cap \Gamma$ can easily be contained in the boundary of Γ, that is, in a codimension 2 toric stratum of B; this will happen, for example, if there is only a single monomial $x^m \in f$ for which the minimal value $\alpha(x^m) = \alpha(f)$ is achieved, since then $g = f^r / z^{r\alpha(f)}$ restricted to Γ is a monomial.

(4.9) *Elementary applications.* (1) Applying $(*)$ to the simplest possible weighting $(1, \ldots, 1)$ shows at once that a hypersurface singularity

$$P \in X \colon (f = 0) \subset \mathbf{A}^{n+1}$$

is terminal (or canonical) only if

$$\text{mult}_P f < n \quad (\text{respectively} \le n);$$

for an ordinary multiple point, the condition is also sufficient (compare (1.9), (1)).

(2) Let $X = \mathbf{A}^2/\boldsymbol{\mu}_r$ be a cyclic quotient singularity of type $\frac{1}{r}(1,a)$, for some a coprime to r; applying ($*$) to the weightings

$$\alpha_i = \tfrac{1}{r}(i, \overline{ai})$$

for $i = 1, \ldots, r - 1$ (where $\overline{}$ denotes minimal residue mod r), it is easy to see that X has a canonical singularity if and only if $a = -1$. Every quotient singularity (cyclic or otherwise) is log terminal.

The next example is important, and you won't get much further in this paper without understanding it thoroughly.

(3) Condition ($*$) gives at once Du Val's analysis of canonical points of an embedded surface $P \in X: (f = 0) \subset \mathbf{A}^3$: first, the quadratic part f_2 of f is nonzero, by ($*$) applied to $(1,1,1)$, so that by a choice of coordinates I can arrange that

$$f = f_2 + \cdots, \quad \text{where } f_2 = x^2 + y^2 + z^2 \text{ or } xy \text{ or } x^2,$$

and $\cdots$ denotes terms of degree ≥ 3. If $f_2 = xy$ then some power of z appears in f, and this gives the A_n points $f = xy + z^{n+1}$ (in suitable coordinates).

If $f_2 = x^2$ then ($*$) applied to $(2,1,1)$ implies that

$$f = x^2 + g(y,z)$$

where g has a nonzero cubic part, and in suitable coordinates,

$$f = x^2 + g_3 + \cdots \quad \text{with } g_3 = y^3 + z^3 \text{ or } y^2 z \text{ or } y^3.$$

Now $x^2 + y^2 z$ gives the D_n points $x^2 + y^2 z + z^{n-1}$.

If $f = x^2 + y^3$, then ($*$) applied to $(3,2,1)$ shows that one of the three monomials z^4, yz^3, or z^5 appears in f, which gives E_6, E_7, and E_8.

(4.10) It turns out to be pleasurable and important to consider the curious-looking question of which hyperquotient surface singularities are canonical. This means the following: take $Q \in Y: (f = 0) \subset \mathbf{A}^3$ and an action of $\boldsymbol{\mu}_r$ on Y which is free outside Q, and consider the quotient $P \in X$. If I ask for X to be canonical, it is a Du Val singularity, so the question is equivalent to asking for all the ways in which one Du Val singularity can be a cyclic unramified cover of another. This can of course be done in a number of different ways, but I particularly recommend it as an exercise in toric technique.

Exercise. In the above situation, suppose that Y is singular and $r > 1$. Prove that one of the following six cases occurs (in suitable coordinates).

r	Type	f	Description
(1) any	$\frac{1}{r}(1,-1,0;0)$	$xy + z^n$	$A_{n-1} \xrightarrow{\text{r-to-1}} A_{rn-1}$
(2) 4	$\frac{1}{4}(1,3,2;2)$	$x^2 + y^2 + z^{2n-1}$	$A_{2n} \xrightarrow{\text{4-to-1}} D_{2n+1}$
(3) 2	$\frac{1}{2}(0,1,1;0)$	$x^2 + y^2 + z^{2n}$	$A_{2n-1} \xrightarrow{\text{2-to-1}} D_{n+2}$
(4) 3	$\frac{1}{3}(0,1,2;0)$	$x^2 + y^3 + z^3$	$D_4 \xrightarrow{\text{3-to-1}} E_6$
(5) 2	$\frac{1}{2}(1,1,0;0)$	$x^2 + y^2z + z^n$	$D_{n+1} \xrightarrow{\text{2-to-1}} D_{2n}$
(6) 2	$\frac{1}{2}(1,0,1;0)$	$x^2 + y^3 + z^4$	$E_6 \xrightarrow{\text{2-to-1}} E_7$

The question may look artificial, since surface quotient singularities of this type have been well understood for more than a century, and the coverings can be classified in many other ways (and were of course, by Felix Klein and subsequently by Patrick Du Val and Coxeter). However, both the list and its derivation by toric methods seem inextricably linked to questions on terminal 3-fold singularities, and in particular to Mori's Theorem 6.1. [*Hint*: if you have trouble doing the exercise, glance at the proof of Mori's theorem in §§6–7, starting at (6.7).]

(4.11) *Canonical quotient singularities.* In the case of cyclic quotient singularities $\mathbf{A}^n/\boldsymbol{\mu}_r$ of type $\frac{1}{r}(a_1,\dots,a_n)$, condition $(*)$ of (4.6) can be rewritten in terms of the $r-1$ weightings

$$\alpha_k = \tfrac{1}{r}(\overline{ka_1},\dots,\overline{ka_n})$$

for $k = 1,\dots,r-1$ (here $^-$ denotes smallest residue mod r).

THEOREM. *A quotient singularity* $X = \mathbf{A}^n/\boldsymbol{\mu}_r$ *of type* $\frac{1}{r}(a_1,\dots,a_n)$ *is terminal* (*or canonical*) *if and only if*

$$\alpha_k(x_1 \cdots x_n) = \frac{1}{r}\sum_k \overline{ka_k} > 1 \quad \textit{for } k = 1,\dots,r-1$$

(*respectively* $\geq$).

This holds because $(*)$ of (4.6) is trivially satisfied for any lattice point α not contained in the unit cube $\square$. This theorem implies at once the Reid–Shepherd-Barron–Tai criterion [**C3-f**, (3.1); **Tai**, (3.2)]: a quotient singularity $\mathbf{A}^n/G$ by an arbitrary group G acting without quasireflections is canonical if and only if every element $g \in G$ of order r when written diagonally

$$g\colon (x_1,\dots,x_n) \mapsto (\varepsilon^{a_1}x_1,\dots,\varepsilon^{a_n}x_n)$$

in terms of any primitive $\varepsilon \in \boldsymbol{\mu}_r$ satisfies $\sum a_i \geq r$.

The combinatorics of the condition in (4.11) and some of its consequences will be studied in §5, see also [**Morrison**].

Appendix to §4. The Fine interior, plurigenera and canonical models. The ideas of this appendix are due to J. Fine (around 1980).

(4.12) *The Fine interior.* If $r = 1$, then $(*)$ of (4.6) is the condition that the point $(1, \ldots, 1) \in M_\mathbf{R}$ is in the interior of Newton(f). For $r > 1$, the statement involves the following notion of interior of a lattice polyhedron Δ (a *lattice polyhedron* is a convex polyhedron with vertices in M): define the *Fine interior* of Δ to be the closed (!) polyhedron Fine(Δ) given by

$$\text{Fine}(\Delta) = \{u \in M_\mathbf{R} \mid \alpha(u) \geq \alpha(\Delta) + 1 \text{ for every } \alpha \in N \text{ with } \alpha(\Delta) \geq 0\}.$$

To describe this in words, take each supporting hyperplane of Δ and push it into Δ until it hits another point of M (see Figure 1); now take the intersection of all of these half-spaces. It is definitely not sufficient to take just the walls of Δ; for example if $M = \{m_1, m_2 \mid m_1 + m_2 \equiv 0 \bmod 3\}$ and $\Delta = \sigma$ is the first quadrant (giving the quotient singularity X of type $\frac{1}{3}(1,1)$), the monomial xy (which gives the canonical class of X) is in the interior of σ, but not in Fine(σ) (see Figure 2); this is equivalent to the computation of (1.9), (2).

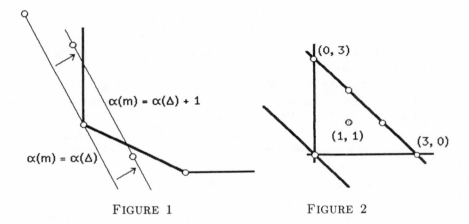

FIGURE 1   FIGURE 2

Fine(Δ) is rather tricky to calculate, because the definition involves in principle all the supporting hyperplanes of Δ; the correspondence with the canonical class in toric geometry shows how to reduce this to the finitely many vectors α involved in a toric resolution of the variety X_Δ. This proves that Fine(Δ) is a finite rational polyhedron. Note that it is not in general a lattice polyhedron.

(4.13) *Plurigenera, canonical models.* The point of the construction is that, by [**Khovanskiĭ**], lattice points in the interior of Newton(f) correspond in various set-ups to the geometric genus of the variety or singularity given by $(f = 0)$; Fine observed that the plurigenera correspond in a similar way to the lattice points of multiples of Fine(f). Clearly (4.6) just says that $P \in X$ is canonical only if (sometimes also if) $(1, \ldots, 1) \in \text{Fine}(f)$.

Quite generally, these ideas determine the plurigenera (and canonical models) of nondegenerate toric hypersurfaces in terms of the Fine interior.

EXAMPLE. Consider the hypersurface singularity $X \colon (f = 0) \subset \mathbf{A}^3$, where

$$f = x^2 + y^3 + z^k \quad \text{for } 6 \leq k \leq 11;$$

then Fine(f) is the polyhedron in $M_{\mathbf{R}} = \mathbf{R}^3$ given by

$(*)$ $l \geq 1, \quad m \geq 1, \quad n \geq 1, \quad 3l + 2m + n \geq 6 + 1 = 7.$

Of course, $\alpha = (3k, 2k, 6)$ is also a supporting hyperplane of Newton(f), but the condition

$$3kl + 2km + 6n \geq 6k + 1$$

is already implied by $(*)$ (because

$$15(l - 1) + 10(m - 1) + (11 - k)(n - 1) + (k - 5)(3l + 2m + n - 7) \geq 0).$$

This means that the pluricanonical invariants of the singularity only take account of the weighting $(\frac{1}{2}, \frac{1}{3}, \frac{1}{6})$.

In this case if $\varphi \colon Y \to X$ is a resolution of X then the pluri-adjunction ideal $\mathcal{J}_a$, defined by $\varphi_* \mathcal{O}_Y(aK_Y) = \mathcal{J}_a \cdot \mathcal{O}_X(aK_X)$, is the ideal generated by all monomials $x^l y^m z^n$ with $3l + 2m + n \geq a$. It is easy to calculate from this the (genuine) plurigenera

$$P_n(X) := \dim \mathcal{O}_X/\mathcal{J}_a = 1 + \binom{a}{2}.$$

(4.14) REMARK. There are (at least) two other notions of plurigenera in the literature on singularities.

(i) The *log plurigenera* correspond to taking the resolution Y marked with its exceptional divisor E (assumed to be a reduced normal crossing divisor) and working with $\varphi_* \mathcal{O}_Y(a(K_Y + E)) \subset \mathcal{O}_X(aK_X)$.

(ii) The L^2 *plurigenera* correspond to differentials on the nonsingular locus $X^0 \subset X$ which are square-integrable; it is known that this is the same as considering $\varphi_* \mathcal{O}_Y(aK_Y + (a - 1)E)$.

These invariants are determined by the usual interior of Newton(f) in a much simpler way than the genuine plurigenera.

(4.15) EXAMPLE. Let $M \subset \mathbf{Z}^4$ be the 3-dimensional affine lattice defined by

$$M = \{(m_1, m_2, m_3, m_4) \in \mathbf{Z}^4 \mid \sum m_i = 5 \text{ and } \sum i m_i \equiv 0 \bmod 5\},$$

and let $\Delta \subset M$ be the simplex spanned by the 4 points $(0, \ldots, 5, \ldots, 0)$, that is, the monomials $x_1^5, x_2^5, x_3^5, x_4^5$. A general polynomial f_Δ supported on Δ is the equation of a nonsingular quintic $Y \subset \mathbf{P}^3$ invariant under the μ_5-action

$$\mu_5 \ni \varepsilon \colon x_i \mapsto \varepsilon^i x_i \quad \text{for } (x_1, x_2, x_3, x_4) \in \mathbf{P}^3.$$

The toric space $\mathbf{P}_\Delta$ constructed from Δ is in this case $\mathbf{P}^3/\mu_5$, and the hypersurface $X_\Delta \colon (f_\Delta = 0) \subset \mathbf{P}_\Delta$ is the Godeaux surface $X_\Delta = Y/\mu_5$. It is a nice exercise to prove that

$$\text{Fine}(\Delta) = (1, 1, 1, 1) + \tfrac{1}{5}\Delta.$$

Problem (I. Dolgachev). It is an interesting problem to look for other examples of this phenomenon: a lattice M $(\cong \mathbf{Z}^3)$, and a lattice polyhedron Δ having no interior points (so that the toric hypersurface $X_\Delta \subset \mathbf{P}_\Delta$ has $p_g(X_\Delta) = 0$), but with Fine(Δ) $\neq$ 0 (so that $\kappa(X_\Delta) \geq 0$), or better still with Fine(Δ) of positive 3-dimensional volume (so that $\kappa(X_\Delta) = 2$). Note that J. Fine observed in 1981 that the traditional model of an Enriques surface as a space sextic passing doubly through the edges of the coordinate tetrahedron is a construction of this form (here Δ is a cube, and M the lattice generated by the vertices of Δ and the midpoints of the six faces).

5. The terminal lemma.

(5.1) *The terminal quotient singularities* $\frac{1}{r}(a, -a, 1)$. Condition (**) of (4.11) has a nice geometric interpretation: it says that all the lattice points of N contained in the cube $\square$ actually live in the middle strip (see Figure 3 for a picture). In fact (**) says that every point of $N \cap \square$ lies on or above the hyperplane $\sum y_i = 1$, and since $n \mapsto (1, \ldots, 1) - n$ is a symmetry of N (as an affine lattice), they clearly also lie below the hyperplane $\sum y_i = n - 1$ (Figure 3).

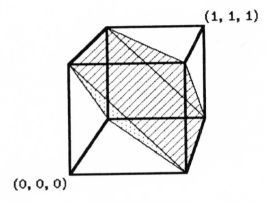

(1, 1, 1)

(0, 0, 0)

FIGURE 3

(5.2) The condition for terminal singularities is of course that the points of $N \cap \square$ live strictly in the middle strip of the cube. In the 3-dimensional case, this situation is completely understood by the following theorem of G. K. White, D. Morrison, G. Stevens, V. Danilov and M. Frumkin.

THEOREM. *A 3-fold cyclic quotient singularity $X = \mathbf{A}^3/\mu_r$ is terminal if and only if (up to permutations of (x, y, z) and symmetries of μ_r) it is of type $\frac{1}{r}(a, -a, 1)$ with a coprime to r.*

(5.3) An obvious, but nevertheless key step in the proof of (5.2) is to replace the inequality (**) by $r - 1$ equalities:

LEMMA. *Let X be of type $\frac{1}{r}(a, b, c)$, and write $d = a + b + c$. Then X is terminal if and only if*

$$\overline{ak} + \overline{bk} + \overline{ck} = \overline{dk} + r \quad \text{for } k = 1, \ldots, r - 1$$

(where $^-$ denotes smallest residue mod r).

PROOF. The points $P_k = \frac{1}{r}(\overline{ak}, \overline{bk}, \overline{ck})$ are just the points of $N \cap \square$. The two sides are congruent mod r, and equality holds if and only if the left-hand side is in the interval $(r, 2r)$.

(5.4) The lemma reduces Theorem 5.2 to the case $n = 3$, $m = 1$ of the following more general result.

THEOREM (TERMINAL LEMMA). *Let n and m be integers with $n \equiv m$ mod 2, and suppose that $\frac{1}{r}(a_1, \ldots, a_n; b_1, \ldots, b_m)$ is an $(n+m)$-tuple of rational numbers with denominator r.*

(A) *Suppose each a_i and b_j is coprime to r; then the following two conditions are equivalent:*

(i)
$$\sum_{i=1}^{n} \overline{a_i k} = \sum_{j=1}^{m} \overline{b_j k} + \frac{n-m}{2} \cdot r \quad \text{for } k = 1, \ldots, r-1.$$

(ii) *The $n+m$ elements $\{a_i, -b_j\}$ can be split up into $(n+m)/2$ disjoint pairs of the form $(a_i, a_{i'})$ or $(b_j, b_{j'})$ or $(a_i, -b_j)$ which add to 0 mod r. (That is, each a_i is either paired with another $a_{i'}$ such that $a_{i'} \equiv -a_i$ mod r, or with one of the b_j such that $b_j \equiv a_i$ mod r, and similarly for the b_j's.)*

(B) *More generally (without the coprime condition), (i) is equivalent to (ii) plus the following condition:*

(iii) *For every divisor q of r,*

$$\#\{\text{pairs}(a_i, a_{i'}) \mid q = hcf(a_i, r)\} = \#\{\text{pairs}(b_j, b_{j'}) \mid q = hcf(b_j, r)\}.$$

Note that (ii) $\Rightarrow$ (i) is trivial, since

$$\overline{ak} + \overline{(r-a)k} = \begin{cases} r & \text{if } \overline{ak} \neq 0, \\ 0 & \text{if } \overline{ak} = 0; \end{cases}$$

the implication (i) $\Rightarrow$ (iii) in (B) is similar and I leave it as an easy exercise.

(5.5) REMARK. For (5.2) I only need the case $n = 3$, $m = 1$ with a_i and b_j coprime; the case $n = 4$, $m = 2$ will be used in §6 in connection with terminal hyperquotient singularities in the form of Corollary 5.6. The tuple might more generally correspond to an action of μ_r on a complete intersection singularity $Q \in Y \subset \mathbf{A}^n$, for example with $\frac{1}{r}(a_1, \ldots, a_n)$ specifying the type of the action on the coordinates, $\frac{1}{r}(b_1, \ldots, b_{m-1})$ that on the defining equations, and $\frac{1}{r}(b_m)$ corresponding to a choice of generator of the class group of the singularity $P \in X = Y/\mu_r$ (a "polarisation" of the singularity).

(5.6) COROLLARY. *Let $\frac{1}{r}(a_1, \ldots, a_4; e, 1)$ be a 6-tuple of rational numbers with denominator r such that*

$$q = hcf(e, r) = hcf(a_4, r) \quad \text{and} \quad a_1, a_2, a_3 \text{ are coprime to } r;$$

assume that

$$\sum_{i=1}^{4} \overline{a_i k} = \overline{ek} + k + r \quad \text{for } k = 1, \ldots, r-1.$$

Then $a_4 \equiv e \bmod r$, and the remaining 4 elements can be paired together as $a_1 \equiv 1$, $a_2 + a_3 \equiv 0 \bmod r$ (or permutations).

The proof of Theorem 5.4, taken mainly from [**Morrison-Stevens**], is something of a digression from the main theme, and is left as an appendix so that the reader can skip over it.

(5.7) *Economic resolutions.* An important consequence of Theorem 5.2 is that the $r - 1$ points of $N \cap \square$ are $Q_i = \frac{1}{r}(\overline{ai}, \overline{(r-a)i}, i)$ for $i = 1, \ldots, r-1$, or equivalently,

$$P_j = \tfrac{1}{r}(j, r-j, \overline{bj}) \quad \text{for } j = 1, \ldots, r-1,$$

where $ab \equiv 1 \bmod r$; hence they all lie on the affine plane $x + y = 1$ of $\mathbf{R}^3$. It follows from this that there is a class of "economic" toric resolutions of X, due to Danilov and R. Barlow (see [**Danilov**, §4]): these correspond to subdivisions of the cone σ obtained from the picture of Figure 4 (the shaded polygonal area is to be subdivided into basic triangles).

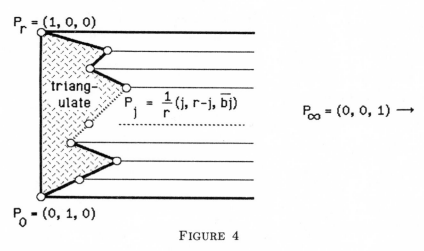

FIGURE 4

This set-up has the following nice properties.

(1) Each of the r triples (P_j, P_{j+1}, P_∞) is a basis of N, so that the subdivision leads to a toric resolution $f: Y \to X$ of $P \in X$.

(2) Each new vertex P_j of the subdivision lies in the interior of $\square$, so that $f^{-1}P = \bigcup E_j$, where E_j has fractional discrepancy a_j with $0 < a_j < 1$; these are the so-called "essential" or "semicrepant" exceptional components, which necessarily occur in any resolution of $P \in X$. In fact there is one exceptional component with each discrepancy $1/r, 2/r, \ldots, (r-1)/r$, generalising the resolution in Exercise 1.10.

(3) The shaded polygon is planar, so that triangulating it leads only to curves l with $K_Y l = 0$; in fact these curves are always $(-1, -1)$-curves, that is, $l \cong \mathbf{P}^1$ with $N_Y|_l \cong \mathcal{O}(-1) \oplus \mathcal{O}(-1)$, and there are $(r - 1)$ of them. Although the triangulation, and hence the toric resolution $Y \to X$, is not unique, you can get from one to another by a composite of flops (symmetric flips) in these curves.

Various properties of the singularities $\frac{1}{r}(a, -a, 1)$ can be read off quite conveniently from the explicit resolution given here; for example the plurigenus contributions $l(n)$ of [**C3-f**, §5] (discussed in §10 below) could in principle be calculated from it, and this is in fact a reasonable approach to calculating the main term

$$c_2 \cdot \Delta = \frac{r^2 - 1}{12r}$$

(compare (10.3) and [**Kawamata**, (2.2)]).

Exercise. If $\overline{b(j+1)} = \overline{bj} + b$ (one of the two possible cases), then

(i) the affine piece $U_j \cong \mathbf{A}^3$ of Y corresponding to $\langle P_j, P_{j+1}, P_\infty \rangle$ has coordinates

$$u = x^{-b}(xy)^{[bj/r]}, \quad v = x^r(xy)^{-j}, \quad w = x^{-r}(xy)^{j+1},$$

and in terms of these, the invariant monomials xy, z^r, etc. are given by

$$xy = vw, \quad z^r = u^r v^b (vw)^{\overline{bj}}, \quad x^r = v(vw)^j, \quad y^r = v^{-1}(vw)^{r-j}, \text{ etc.}$$

(ii) $\langle P_k, P_j, P_{j+1} \rangle$ is a basic cone contained in the shaded area of Figure 4 if and only if $[bk/r] = [bj/r] + 1$ (see Figure 5); the corresponding affine piece $V_{j,k}$ of Y then has coordinates

$$u' = u^{-1}, \quad v' = u^{k-j}x^r(xy)^{-j}, \quad w' = u^{-k+j+1}x^{-r}(xy)^{j+1},$$

and in terms of these,

$$xy = u'v'w', \quad z^r = u'^{\overline{bk}} v'^b (v'w')^{\overline{bj}}.$$

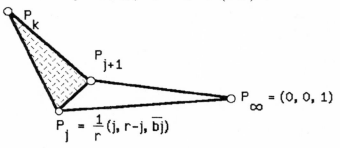

FIGURE 5

Appendix to §5. Cyclotomy and the proof of (5.4).

(5.8) Write s_a for the r-tuple

$$s_a = (\overline{ak})_{k=0,\dots,r-1} \in \mathbf{Q}^r = V;$$

then (i) of (5.4) is a certain relation between s_a for different values of a, and the implication (i) $\Rightarrow$ (ii) is deduced by proving that the s_a are as linearly independent as you could reasonably expect. As remarked above,

$$(s_a)_k + (s_{r-a})_k = \overline{ak} + \overline{(r-a)k} = \begin{cases} r & \text{if } r \nmid ak, \\ 0 & \text{if } r \mid ak; \end{cases}$$

because of this, for each $a = 0, \ldots, r - 1$, define a new r-tuple $S_a \in V$ by

$$(S_a)_k = \begin{cases} \overline{ak} - r/2 & \text{if } r \nmid ak, \\ 0 & \text{if } r \mid ak. \end{cases}$$

Then the above relation takes the form

$$(*) \qquad\qquad S_a + S_{r-a} = 0.$$

In fact only the subspace $V^- = \{(v_k) \mid v_k + v_{r-k} = 0 \ \forall k\} \subset V$ will play any part in the following argument; this is a **Q**-vector space of dimension $[(r-1)/2]$, and in due course of time I will prove that

$$\{S_a \mid \text{ for } a = 1, \ldots, [(r-1)/2]\}$$

is a basis. This easily implies (5.4).

For the purpose of proving (5.4), (A) restrict attention to a coprime to r (the more general case will be dealt with later). If all the a_i and b_j are coprime to r, then condition (i) of (5.4) is of the form

$$\sum S_{a_i} = \sum S_{b_j},$$

and the following result clearly gives the implication (i) $\Rightarrow$ (ii).

(5.9) PROPOSITION. *Consider only a coprime to r; then the relations $(*)$ are the only linear dependence relations holding between the S_a. In other words, the $\varphi(r)/2$ elements S_a with $a = 1, \ldots, [(r-1)/2]$ coprime to r are linearly independent.*

Discussion. The proof which follows, due to [**Morrison-Stevens**], is somewhat abstract. To be absolutely concrete, write out the multiplication table of the ring $\mathbf{Z}/r$ in terms of smallest residues mod r, and let

$$M_{a,k} = \{\overline{ak} - r/2 \text{ or } 0\}$$

be the $r \times r$ matrix obtained by subtracting $r/2$ from all nonzero entries; the result I am after is equivalent to saying that the top left quarter determinant of M formed by taking rows and columns $1, 2, \ldots, [(r-1)/2]$ is nonzero; for example, if $r = 11$ then

$$\begin{vmatrix} -9/2 & -7/2 & -5/2 & -3/2 & -1/2 \\ -7/2 & -3/2 & 1/2 & 5/2 & 9/2 \\ -5/2 & 1/2 & 7/2 & -9/2 & -3/2 \\ -3/2 & 5/2 & -9/2 & -1/2 & 7/2 \\ -1/2 & 9/2 & -3/2 & 7/2 & -5/2 \end{vmatrix} = -\frac{1}{2} \cdot (11)^4;$$

or if $r = 14$ then

$$\begin{vmatrix} -6 & -5 & -4 & -3 & -2 & -1 \\ -5 & -3 & -1 & 1 & 3 & 5 \\ -4 & -1 & 2 & 5 & -6 & -3 \\ -3 & 1 & 5 & -5 & -1 & 3 \\ -2 & 3 & -6 & -1 & 4 & -5 \\ -1 & 5 & -3 & 3 & -5 & 1 \end{vmatrix} = -7 \cdot (14)^4.$$

These determinants are quite fun to evaluate by row and column operations (for example, on a home computer using commercial spreadsheet software); if you difference successive rows then you see that the essential information contained in the matrix is whether

$$\overline{(a+1)k} = \overline{ak} + k \quad \text{or} \quad \overline{ak} - (r-k)$$

and whether or not $\overline{ak} = 0$ (that is, whether a "carry" occurs) for each a, k.

The nonvanishing of these determinants is equivalent to Dirichlet's theorem $L(1, \chi) \neq 0$ for odd characters χ; due to my ignorance of number theory, I can't help wondering if there isn't an elementary proof of this fact. The actual value of the determinant is presumably a power of r times a factor of number-theoretical interest, compare [**Lang**, p. 92].

(5.10) PROOF OF (5.9). I will find $\varphi(r)/2$ linearly independent vectors $w_\chi \in V \otimes \mathbf{C}$ which are linear combinations of the S_a for $a \in (\mathbf{Z}/r)^*$; this obviously proves the proposition. The vectors w_χ are constructed as eigenvectors of the natural action of $(\mathbf{Z}/r)^*$ on V with distinct characters χ^{-1} as eigenvalues, so that by a standard argument of linear algebra, to prove they are linearly independent it will be enough to prove that each $w_\chi \neq 0$.

$a \in (\mathbf{Z}/r)^*$ acts on V by $(v_k)_k \mapsto (v_{ak})_k$; note that $a(S_1) = S_a$. Given any character $\chi \colon (\mathbf{Z}/r)^* \to \mathbf{C}^*$ and any $v \in V$, the linear combination

$$w_\chi(v) = \sum_{a \in (\mathbf{Z}/r)^*} \chi(a) \cdot a(v) \in V \otimes \mathbf{C}$$

is of course an eigenvector of the action of $(\mathbf{Z}/r)^*$ on V, with eigenvalue χ^{-1}. I apply this with $v = S_1$ and χ an odd character (that is, $\chi(-1) = -1$), getting eigenvectors

$$w_\chi = w_\chi(S_1) = \sum_{a \in (\mathbf{Z}/r)^*} \chi(a) \cdot S_a;$$

there are just $\varphi(r)$ characters of $(\mathbf{Z}/r)^*$ of which $\varphi(r)/2$ are odd, so this does what I want, except that I must still prove that $w_\chi \neq 0$.

(5.11) *Proof that $w_\chi \neq 0$.* Suppose first for simplicity that χ is a primitive character mod r (that is, it does not factor through the quotient map $(\mathbf{Z}/r)^* \to (\mathbf{Z}/r')^*$ for any divisor $r'|r$), and consider the first coordinate of $w_\chi \in V$:

$$(w_\chi)_1 = \sum_{a \in (\mathbf{Z}/r)^*} \chi(a) \cdot (S_a)_1 = \sum \chi(a) \cdot \left(a - \frac{r}{2}\right) = \sum \chi(a) \cdot a$$

(the last equality uses $\sum \chi(a) = 0$). The fact that this number is nonzero now follows from results of analytic number theory:

(a) The sum for $(w_\chi)_1$ is by definition r times the generalised Bernoulli number $B_{1,\chi}$, where

$$B_{1,\chi} = \frac{1}{r} \sum_{a \in (\mathbf{Z}/r)^*} \chi(a) \cdot \left(a - \frac{r}{2}\right) = \frac{1}{r} \sum_{a \in (\mathbf{Z}/r)^*} \chi(a) \cdot a;$$

see [**Washington**, p. 30].

(b) A contour integration shows that the Dirichlet L-function $L(s, \chi)$ for χ satisfies

$$-L(0, \chi) = B_{1, \chi}$$

(see [**Washington**, p. 31]).

(c) The functional equation relating $L(s, \chi)$ and $L(1 - s, \chi^{-1})$ shows that $L(0, \chi)$ is a nonzero multiple of $L(1, \chi^{-1})$ (see [**Washington**, p. 29]).

(d) Finally, the statement that $L(1, \chi^{-1}) \neq 0$ is a famous theorem of Dirichlet (see [**Washington**, p. 33] or any book on analytic number theory).

(5.12) A suitable modification of the argument will in fact go through for any odd character χ: if $r = qf$ is a factorisation of r, and χ is induced by a primitive character $\chi \colon (\mathbf{Z}/f)^* \to \mathbf{C}^*$ mod f (that is, χ has *conductor* f) then I prove that the qth coordinate $(w_\chi)_q \neq 0$. (In fact it is true that $(w_\chi)_1 \neq 0$, as will be proved in Proposition 5.17; but the calculation is quite a lot more involved.) First, for each $a \in (\mathbf{Z}/r)^*$,

$$(S_a)_q = \overline{aq} - \frac{r}{2}$$

so that $(S_a)_q$ also only depends on a mod f. Then

$$(w_\chi)_q = \sum_{a \in (\mathbf{Z}/r)^*} \chi(a') \cdot q\left(a' - \frac{f}{2}\right),$$

where a' is the smallest residue of a mod f; now since $(\mathbf{Z}/r)^* \to (\mathbf{Z}/f)^*$ is a surjective group homomorphism, to every $a' \in (\mathbf{Z}/f)^*$ there correspond $\varphi(r)/\varphi(f)$ values of a, so that

$$(w_\chi)_q = q \cdot \frac{\varphi(r)}{\varphi(f)} \sum_{a' \in (\mathbf{Z}/f)^*} \chi(a') \cdot \left(a' - \frac{f}{2}\right) = r \cdot \frac{\varphi(r)}{\varphi(f)} \cdot B_{1, \chi}.$$

Apart from the initial factor, this is the same sum as before, now for the primitive character χ mod f, so I conclude as before.

This completes the proof of Proposition 5.9, and with it (5.4), (A).

(5.13) I now go on to prove that the only linear relations between all the S_a are given by $(*)$ of (5.8), that is, that the $[(r - 1)/2]$ elements

$$\{S_a \mid \text{for } a = 1, \ldots, [(r - 1)/2]\}$$

are linearly independent. The aim is as before: to show that the vectors $w_\chi(S_a)$ provide the right number of linear combinations of the S_a which are linearly independent. This time however I take $w_\chi(S_a)$ for different divisors $a|q$, where $q = r/f$. (Note that $w_\chi(S_a)$ is a linear combination of the vectors $S_{a'}$ for $a' \in [0, \ldots, r - 1]$ such that $\mathrm{hcf}(a', r) = a$; the divisor $a|r$ is just a natural choice of representative of this set.)

MAIN CLAIM. Let $r = fq$ be a factorisation, and $\chi \colon (\mathbf{Z}/r)^* \to \mathbf{C}^*$ an odd character which is induced from a primitive character $\chi \colon (\mathbf{Z}/f)^* \to \mathbf{C}^*$. Then the vectors $w_\chi(S_a)$ with $a|q$ are linearly independent.

The claim implies Theorem 5.4. Since there cannot be a nontrivial linear dependence relation between eigenvectors belonging to different characters, the claim gives that the set

$$\{w_\chi(S_a) \mid r = as \text{ and } \chi \text{ is an odd character} \mod s\}$$

is linearly independent; by analogy with the formula $r = \sum_{s|r} \varphi(s)$, if I write $\varphi^-(n)$ for the number of odd characters of $(\mathbf{Z}/n)^*$ then clearly

$$\left[\frac{r-1}{2}\right] = \dim V^- = \sum_{s|r} \varphi^-(s),$$

so that I have just the right number $[(r-1)/2]$ of vectors to base V^-; since they are linear combinations of the S_a for $a = 1, 2, \ldots, [(r-1)/2]$, it follows that these also base V^-. Given condition (iii), the relation (i) is (as before) just of the form

$$\sum S_{a_i} = \sum S_{b_j},$$

so that this implies (5.4), (B).

(5.14) For the proof of Claim 5.13, note first the following easy facts:

LEMMA. (i) $(w_\chi(S_a))_c$ *depends only on the product $ac \mod r$;*
(ii) *if $ac = q$ then*

$$w_\chi(S_a)_c = w_\chi(S_1)_q = r \cdot \varphi(r)/\varphi(f) \cdot B_{1,\chi} \neq 0;$$

(iii) *if $a'|r$ and $\mathrm{hcf}(a'c, r) \nmid q$ then*

$$w_\chi(S_{a'})_c = 0.$$

PROOF. If you write out the definition of $(w_\chi(S_a))_c$ then (i) is trivial, and (ii) follows by the argument of (5.12).

(iii) Write $q' = \mathrm{hcf}(a'c, r)$ and $f' = r/q'$. Arguing as in (5.12),

$$w_\chi(S_{a'})_c = \sum_{b \in (\mathbf{Z}/f)^*} \chi(b) \cdot f(b), \qquad f(b) = \begin{cases} \overline{ba'c} - \frac{r}{2} \\ 0 \quad \text{if } r|ba'c, \end{cases}$$

where the term $f(b)$ multiplying $\chi(b)$ depends only on $b \mod f'$; so it's invariant under $b \mapsto bk$ for $k \in \mathrm{Ker}\{(\mathbf{Z}/r)^* \to (\mathbf{Z}/f')^*\}$. By definition of f, this is a subgroup on which χ is nontrivial, and hence the sum is zero. Q.E.D.

(5.15) It's important to understand the distinction made in (iii); if p is a number with some factor in common with f then of course $\mathrm{hcf}(pq, r)$ is a strict multiple of q; however, if p is coprime to f then $\mathrm{hcf}(pq, r) = \mathrm{hcf}(q, r)$, and it follows that

$$pq \equiv \beta q \qquad \mod r,$$

with $\beta \in (\mathbf{Z}/r)^*$; in fact β is uniquely determined $\mod f$.

If I could replace the unpleasant second hypothesis in (iii) by the nice condition $a'c \nmid q$, then it is easy to order the divisors a and c of q in such a way that the matrix

$$\{w_\chi(S_a)_c\} \quad \text{as } a \text{ and } c \text{ run through the divisors of } q$$

is upper-triangular with nonzero diagonal entries. (Just write $a_i c_i = q$, and order the a_i such that $i < j \Rightarrow a_j \nmid a_i$.) This already proves (5.13) in the special case that every prime divisor of q also divides f. I now endeavour to modify the $w_\chi(S_a)$ to get an upper-triangular matrix in the general case. There's essentially only one way to proceed ("Möbius inversion").

(5.16) Write $\mathcal{P}$ for the set of prime divisors of q which are coprime to f; for each $p \in \mathcal{P}$ let p^α be the highest power of p dividing r, and choose $\beta_p \in (\mathbf{Z}/r)^*$ such that $\beta_p \equiv p \bmod r/p^\alpha$. (This is of course possible, and I could even require $\beta_p \equiv 1 \bmod p^\alpha$, since the numbers $p + i \cdot r/p^\alpha$ for $i = 0, \ldots, p^\alpha - 1$ take every value $\bmod\ p^\alpha$.) Then

$$pa \equiv \beta_p a \bmod r \quad \text{for every } a \text{ with } p^\alpha | a.$$

To simplify notation, if $d = \prod_{i=1}^m p_i$ is a product of distinct primes $p_i \in \mathcal{P}$, I write $\beta_d = \prod \beta_{p_i}$. Now define vectors $v_\chi(a) \in V \otimes \mathbf{C}$ for each a by the following formula:

(1)
$$v_\chi(a) = w_\chi(S_a) - \sum_{\substack{p \in \mathcal{P} \\ p | a}} w_\chi(S_{\beta_p \cdot a/p}) + \sum_{\substack{p,p' \in \mathcal{P} \\ pp' | a}} w_\chi(S_{\beta_p \beta_{p'} \cdot a/pp'}) - \cdots$$

$$= \sum_{d | a} \mu(d) w_\chi(S_{\beta_d \cdot a/d}),$$

where the sum runs over products of distinct primes in $\mathcal{P}$ dividing a, and $\mu(d)$ is the Möbius function defined on square-free integers by

$$\mu(d) = (-1)^m \quad \text{where } m = \#\{\text{distinct prime factors of } d\}.$$

This means of course that if a is not divisible by any $p \in \mathcal{P}$ then $v_\chi(a) = w_\chi(S_a)$; and for $p \in \mathcal{P}$,

$$v_\chi(p^t a) = w_\chi(S_{p^t a}) - w_\chi(S_{\beta_p p^{t-1} a}).$$

In this case, Lemma 5.14, (i) together with the definition of β_p gives

(2)
$$(v_\chi(p^t a))_c = 0 \quad \text{for all } c \text{ with } p^{\alpha-t+1} | c.$$

(5.17) PROPOSITION. *Let a, c, c' be divisors of q.*

(i) *If $ac' \nmid q$ then $(v_\chi(a))_{c'} = 0$.*

(ii) *Suppose that $ac = q$, and let $d|a$ be a product of distinct primes in $\mathcal{P}$; for each divisor $d'|d$, write $d = d'd''$. Then (sorry folks!)*

(3)
$$(w_\chi(S_{\beta_d \cdot a/d}))_c = \left\{ \sum_{d'|d} \mu(d')\chi^{-1}(d'') \right\} \cdot r \cdot \frac{\varphi(r)}{\varphi(df)} \cdot B_{1,\chi}$$

$$= \prod_{p|d} \left(\frac{\chi^{-1}(p) - 1}{p - 1} \right) \cdot r \cdot \frac{\varphi(r)}{\varphi(f)} \cdot B_{1,\chi};$$

and

(4)
$$(v_\chi(a))_c = \prod_{p|a} \left(\frac{p - \chi^{-1}(p)}{p - 1} \right) \cdot r \cdot \frac{\varphi(r)}{\varphi(f)} \cdot B_{1,\chi} \neq 0.$$

Notice that (i) and (ii) together prove Claim 5.13, and Theorem 5.4, since the matrix $\{(v_\chi(a))_c\}_{a,c}$ is upper-triangular as discussed in (5.15).

PROOF. (i) is easy: if $ac' \nmid q$ then there is some prime p appearing in ac' with higher power than in q. If p divides f then $\operatorname{hcf}(ac',r) \nmid q$, so that Lemma 5.14, (iii) gives $(w_\chi(S_a))_{c'} = 0$, and the same after dividing primes in $\mathcal{P}$ out of a, which gives the result at once. Otherwise $p \in \mathcal{P}$, and (i) comes from grouping the sum for $(v_\chi(a))_{c'}$ into pairs of terms

$$(5) \qquad \mu(d)(w_\chi(S_{\beta_d \cdot a/d}))_{c'} + \mu(pd)(w_\chi(S_{\beta_p \beta_d \cdot a/pd}))_{c'},$$

where $p \nmid d$, so that $\mu(pd) = -\mu(d)$. Under the assumption that $p^{\alpha+1} | ac'$ it follows from the definition of β_p that $\beta_p \cdot ac'/p \equiv ac' \bmod r$, so that by Lemma 5.14, (i), this pair of terms cancels out.

(5.18) The proof of (ii) breaks up into several steps. By definition

$$
\begin{aligned}
(6) \qquad w_\chi(S_{\beta_d \cdot a/d})_c &= \sum_{b \in (\mathbf{Z}/r)^*} \chi(b) \cdot \overline{\left(b\beta_d \cdot \frac{q}{d}\right)} \\
&= \frac{\varphi(r)}{\varphi(df)} \sum_{b' \in (\mathbf{Z}/f)^*} \chi(b') \sum_{\substack{b \mapsto b' \\ b \in (\mathbf{Z}/df)^*}} \overline{\left(b\beta_d \cdot \frac{q}{d}\right)}.
\end{aligned}
$$

Step 1. Consider first the range of summation of the internal sum. This can be replaced by a sum over the additive group $(\mathbf{Z}/df)$ using a Möbius inversion: for each divisor $d' | d$, write $d = d'd''$, and let

$$(7) \qquad T(d') = \sum_{\substack{b \mapsto b' \\ b \in (d'\mathbf{Z}/df\mathbf{Z})}} (\text{summand});$$

then clearly, since $(\mathbf{Z}/df) \setminus (\mathbf{Z}/df)^*$ is the union of $(d'\mathbf{Z}/df\mathbf{Z})$ for different divisors $d' > 1$, I have

$$(8) \qquad \sum_{\substack{b \mapsto b' \\ b \in (\mathbf{Z}/df)^*}} (\text{summand}) = \sum_{d' | d} \mu(d') T(d').$$

Step 2. I now make the range of summation in $T(d')$ more explicit. First, since d and f are coprime, for any b', exactly one of the integers $b' + if$ for $i = 0, \ldots, d-1$ is divisible by d, so that there exists i_0 and x with $b' + i_0 f = dx$. Since the sum $T(d')$ only involves b which are divisible by d', and the summand only depends on $b \bmod df$, I can take the range of summation to be

$$b = b' + i_0 f + jd'f \quad \text{where } j = 0, \ldots, d'' - 1.$$

Step 3. For each divisor d', I claim that

$$(9) \quad T(d') = \sum_{\substack{b \mapsto b' \\ b \in (d'\mathbf{Z}/df\mathbf{Z})}} \overline{\left(b\beta_d \cdot \frac{q}{d}\right)} = \sum_{i=0}^{d''-1} \overline{b'q + i \cdot \frac{r}{d''}} = q \cdot \overline{\overline{b'd''}} + \binom{d''}{2} \cdot \frac{r}{d''},$$

where $\overline{\overline{}}$ denotes smallest residue $\bmod f$.

PROOF. By definition,

$$(10) \qquad T(d') = \sum_{\substack{b \mapsto b' \\ b \in (d'\mathbf{Z}/df\mathbf{Z})}} \overline{\left(b\beta_d \cdot \frac{q}{d}\right)}.$$

Using Step 2, this becomes

$$(11) \qquad \sum_{j=0}^{d''-1} \overline{(b' + i_0 f + jd'f)\beta_d \cdot \frac{q}{d}};$$

now

$$(12) \quad (b' + i_0 f)\beta_d \cdot \frac{q}{d} = dx\beta_d \cdot \frac{q}{d} = x\beta_d q \equiv xdq = (b' + i_0 f)q \equiv b'q \qquad \mod r,$$

where the middle congruence uses the defining property of β_d. Gathering together the terms $d'fq/d$ into r/d'', this gives

$$(13) \qquad \sum \overline{b'q + \beta_d j \cdot \frac{r}{d''}},$$

and since $\beta_d \in (\mathbf{Z}/r)^*$, multiplication by β_d just permutes the range of summation. This proves the first part of (9).

To get the second equality, note that the numbers

$$\overline{(b'd'' + if) \cdot \frac{q}{d''}} \quad \text{as } i = 0, \dots, d'' - 1$$

take on a smallest value

$$\frac{q}{d''} \cdot \overline{\overline{b'd''}} \quad \text{when } 0 \le b'd'' + i_1 f < f,$$

and if I then change the range of summation to $i = i_1 + j$ with $j = 0, \dots, d'' - 1$, the summand simplifies, and the r.h.s. of (9) emerges after a brief struggle.

Step 4. Writing (8) for the internal sum in (6) gives

$$(14) \qquad w_\chi(s_{\beta_d \cdot a/d})_c = \frac{\varphi(r)}{\varphi(df)} \sum_{b' \in (\mathbf{Z}/f)^*} \chi(b') \sum_{d'|d} \mu(d')T(d'),$$

which by (9) is equal to

$$(15) \qquad \frac{\varphi(r)}{\varphi(df)} \sum_{b' \in (\mathbf{Z}/f)^*} \chi(b') \sum_{d'|d} \mu(d') \cdot \left\{ q \cdot \overline{\overline{b'd''}} + \binom{d''}{2} \cdot \frac{r}{d''} \right\}.$$

Now it is clear that for fixed d', summing the two terms in curly brackets against $\chi(b')$ leads respectively to $\chi^{-1}(d'') \cdot r \cdot B_{1,\chi}$ and 0. The first equality of (3) then comes out at once.

The second equality is easy: $\varphi(df) = \varphi(d) \cdot \prod(p-1)$ gives the denominator, and expanding $\prod(\chi^{-1}(p) - 1)$ the numerator.

Step 5. This is also easy: each factor on the right-hand side of (4) is

$$\frac{p - \chi^{-1}(p)}{p-1} = 1 - \frac{\chi^{-1}(p) - 1}{p-1};$$

the product of these taken over the primes $p \in \mathcal{P}$ with $p|a$ is obviously of the form

$$\sum_{d|a} \mu(d) \cdot \prod_{p|d} \left(\frac{\chi^{-1}(p) - 1}{p - 1} \right).$$

Now by (3), after multiplication by $r \cdot \varphi(r)/\varphi(f) \cdot B_{1,\chi}$, each summand becomes

$$(w_\chi(S_{\beta_d \cdot a/d}))_c,$$

and comparing with the definition of $v_\chi(a)$ in (5.16), the sum is just $(v_\chi(a))_c$.
    Amen.

(5.19) *Interpretation.* Suppose I take it into my head to write down the sum

$$B(\varepsilon) = B_r(\varepsilon) = \frac{1}{2} + \frac{1}{r} \sum_{k=1}^{r-1} k \varepsilon^k$$

for any $\varepsilon \in \boldsymbol{\mu}_r - 1$, $B(1) = 0$. Then for $a \in \mathbf{Z}$ coprime to r,

$$B(\varepsilon^a) = \frac{1}{2} + \frac{1}{r} \sum k \varepsilon^{ak} = \frac{1}{2} + \frac{1}{r} \sum_{k=1}^{r-1} \overline{a'k} \cdot \varepsilon^k,$$

where $aa' \equiv 1 \bmod r$. So the sum pulled out of a hat organizes the apparently random combinatorial data of the periodic values of $\overline{a'k}$ into a single element $B(\varepsilon) \in K = \mathbf{Q}(\boldsymbol{\mu}_r)$ of the cyclotomic field.

If I multiply the kth equality in (i) of (5.4) by ε^k for some primitive ε and sum over k, then (provided the a_i and b_j are coprime to r), I get

$$\sum_{i=1}^{n} B(\varepsilon^{a'_i}) = \sum_{j=1}^{m} B(\varepsilon^{b'_j}).$$

This reduces the proof of (i) $\Rightarrow$ (ii) in (5.4) to linear dependence relations between the $B(\varepsilon) \in K$ for primitive elements $\varepsilon \in \boldsymbol{\mu}_r$. Since

$$B_r(\varepsilon) = -\frac{1}{2} \left\{ \frac{1 \cdot \varepsilon}{1 - \varepsilon} \right\} = \frac{i}{2} \cot \frac{\pi k}{r}$$

if $\varepsilon = \exp(2\pi k i/r)$, the problem is equivalent to proving the following result:

PROPOSITION. *The $\varphi(r)/2$ numbers*

$$\cot \frac{\pi k}{r} \quad for \; k = 1, \ldots, \left[\frac{r}{2}\right] \; coprime \; to \; r$$

are linearly independent over $\mathbf{Q}$.

REMARK. The preprint version of this paper contained a false proof of this proposition.

PROOF. The vector space $\mathbf{Q}^r = V$ of (5.8) can also be thought of as $\mathbf{Q}[X]/(X^r - 1)$; now corresponding to the factorisation $X^r - 1 = \prod \Phi_d(X)$ of $X^r - 1$ into the product of cyclotomic polynomials Φ_d for divisors $d|r$, the $\mathbf{Q}$-algebra $\mathbf{Q}[X]/(X^r - 1)$ splits in a canonical way as the product of cyclotomic fields $\mathbf{Q}(\varepsilon_d)$ of degree d (where ε_d denotes a primitive dth root of 1).

Now by construction, if a is coprime to r, the projection to $\mathbf{Q}(\varepsilon_r)$ of $S_a \in V$ is

$$\sum_{k=1}^{r-1} \left(\overline{ak} - \frac{r}{2}\right) \cdot \varepsilon^k = \frac{r}{2} + \sum k \cdot \varepsilon^{a'k} = r \cdot B(\varepsilon^{a'});$$

on the other hand if $\mathrm{hcf}(a, r) > 1$, it is not hard to check that S_a projects to zero in $\mathbf{Q}(\varepsilon_r)$. It follows from this that the S_a with a coprime to r map to linearly independent elements of $\mathbf{Q}(\varepsilon_r)$.   Q.E.D.

6. Terminal 3-fold singularities according to Mori.

(6.1) The results of §§6–7 are taken essentially from [**Mori**]. Part (I) of the main theorem gives necessary conditions on the possible equations and group actions for terminal hyperquotient (= *quick*) singularities, whereas (II) is a first attempt at giving sufficient conditions.

THEOREM (S. MORI). (I) *Let* $P \in X = (Q \in Y)/\boldsymbol{\mu}_r$ *be a terminal hyper-quotient singularity, where* $r > 1$ *and* $Q \in Y$ *is singular. Then* $P \in X$ *belongs to one of the following 6 families:*

	r	Type	f	Conditions
(1)	any	$\frac{1}{r}(a, -a, 1, 0; 0)$	$xy + g(z^r, t)$	$g \in m^2$, a, r coprime
(2)	4	$\frac{1}{4}(1, 1, 3, 2; 2)$	$xy + z^2 + g(t)$	$g \in m^3$
			or $x^2 + z^2 + g(y, t)$	$g \in m^3$
(3)	2	$\frac{1}{2}(0, 1, 1, 1; 0)$	$x^2 + y^2 + g(z, t)$	$g \in m^4$
(4)	3	$\frac{1}{3}(0, 2, 1, 1; 0)$	$x^2 + y^3 + z^3 + t^3$	
			or $x^2 + y^3 + z^2t + yg(z, t) + h(z, t)$	$g \in m^4$, $h \in m^6$
			or $x^2 + y^3 + z^3 + yg(z, t) + h(z, t)$	$g \in m^4$, $h \in m^6$
(5)	2	$\frac{1}{2}(1, 0, 1, 1; 0)$	$x^2 + y^3 + yzt + g(z, t)$	$g \in m^4$
			or $x^2 + yzt + y^n + g(z, t)$	$g \in m^4$, $n \geq 4$
			or $x^2 + yz^2 + y^n + g(z, t)$	$g \in m^4$, $n \geq 3$
(6)	2	$\frac{1}{2}(1, 0, 1, 1; 0)$	$x^2 + y^3 + yg(z, t) + h(z, t)$	$g, h \in m^4$, $h_4 \neq 0$.

(II) *The general element of each of the families* (1)–(6) *is terminal.*

(6.2) REMARKS. (i) Some of the families can be tidied up into discrete normal forms using standard methods of singularity theory: for example, the second alternative case of (4) can be reduced to one of

$$x^2 + y^3 + z^2t + yt^a \text{ with } a \equiv 1 \bmod 3, \ a \geq 4$$
$$\text{or} \quad x^2 + y^3 + z^2t + t^b \text{ with } b \equiv 0 \bmod 3, \ b \geq 6$$
$$\text{or} \quad x^2 + y^3 + z^2t + \alpha yt^{2c} + \beta t^{3c} \text{ with } c \equiv 2 \bmod 3, \ c \geq 2, \ 4\alpha^3 + 27\beta^2 \neq 0.$$

More information is given in [**Mori**, (12.1), (23.1), and (25.1)].

(ii) In fact, for each of (1)–(6), every isolated singularity is terminal, as has recently been proved by [**Kollár and Shepherd-Barron**, §6] (compare (6.5), (2)). For the most important family (1), this can be proved as follows: any singularity in (1) is a Cartier divisor $P \in X: (xy + g(z^r, t)) \subset \mathbf{A}^1 \times A$, where

$\mathbf{A}^1$ corresponds to the invariant coordinate t, and A is the terminal quotient
singularity $\frac{1}{r}(a, -a, 1)$. Now let $B \to A$ be Danilov's resolution as in (5.7), and

$$\varphi \colon \mathbf{A}^1 \times B \to \mathbf{A}^1 \times A.$$

Clearly $X' = \varphi^{-1}(X) \subset \mathbf{A}^1 \times B$ is irreducible, and it can be seen that X' is
normal; $\varphi \colon X' \to X$ is totally discrepant, so that to prove that $P \in X$ is terminal
it is sufficient to prove that X' has only canonical hypersurface singularities. This
can be done (with quite a lot of pain) by a direct calculation using the explicit
coordinates for B described in Exercise 5.7.

(6.3) Part (I) of Theorem 6.1 is proved in §7 after some preliminary work at
the end of §6. The main point will be to determine what the μ_r-action looks like
by making knowledgeable use of the terminal lemma (5.4). I should emphasise
that this is merely a technical reworking of Mori's original argument, and that
each step is derived more or less directly from [**Mori**].

(6.4) *Q-smoothing and the general elephant.* (A) If $P \in X$ is a terminal
singularity it belongs to one of the families of (6.1) and is the quotient of an
isolated cDV singularity $Q \in Y \colon (f = 0) \subset \mathbf{A}^4$. It is then possible to write
down deformations of X by just varying the equation f in its eigenspace; the
singularity of X just varies inside its toric ambient space $\mathbf{A}^4/\mu_r$.

Now in every case of (6.1), one can write down a 1-parameter deformation
$\{Y_\lambda\}$ of Y compatible with the action of μ_r such that Y_λ is nonsingular for $\lambda \neq 0$
and meets the fixed locuses of μ_r transversally; this is possible because in each
case at least one of the coordinates x_i has the same eigenvalue as f. Define a
deformation of X by $X_\lambda = Y_\lambda/\mu_r$. I call this situation a *Q-smoothing* of $P \in X$:
it is a deformation $\{X_\lambda\}$ of $P \in X$ such that the general fibre has as its only
singularities a number of terminal quotient singularities $\frac{1}{r}(a, -a, 1)$.

Notice that since $\{X_\lambda\}$ is constructed as a quotient of a flat deformation $\{Y_\lambda\}$
of Y, the eigensheaves $\mathcal{L}_i$ of the action of μ_r on $\mathcal{O}_Y$ are sheaves over X_λ which
vary in a flat family with λ; this will be important for the proof of Theorem 10.2.

For example, in (1) the fixed locus of the group action is the t-axis l; suppose
that t^n is the smallest power of t appearing in f. Then $f|_l$ has 0 as a root with
multiplicity n; taking $f_\lambda = f + \lambda t$, this root splits up into n simple roots. The
picture for X is given in Figure 6.

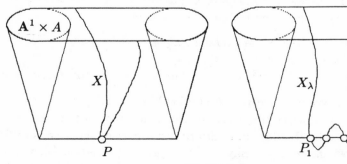

FIGURE 6

(B) If $P \in X$ is a Cohen-Macaulay point of a 3-fold then it follows by standard formalism that the general elephant $S \in |-K_X|$ has a normal Gorenstein singularity at P. The following is an interesting and important question: when does S have a Du Val singularity at P? There is some hope that under this condition, the 3-fold singularity can be treated as a kind of generalised cDV point, so that, for example, problems related to the class group and small partial resolutions of X can be dealt with by some (quite considerable) generalisation of the Brieskorn techniques for cDV points. For 3-fold terminal singularities $P \in X$, the lists of Theorem 6.1 allow me to write down an explicit $S \in |-K_X|$ with a Du Val singularity by writing down a hyperplane section of $Q \in Y$ which belongs to the eigenvalue corresponding to K_X. This shows clearly the close relationship between the list of Theorem 6.1 and that of (4.10).

	Type	Q-smoothing	General elephant	Section
(1)	$\frac{1}{r}(a, -a, 1, 0; 0)$	$f + \lambda t$	z	$A_{n-1} \xrightarrow{r-1} A_{rn-1}$
(2)	$\frac{1}{4}(1, 1, 3, 2; 2)$	$f + \lambda t$	$x - y$	$A_{2n} \xrightarrow{4-1} D_{2n+1}$
(3)	$\frac{1}{2}(0, 1, 1, 1; 0)$	$f + \lambda x$	$\lambda z + \mu t$	$A_{2n-1} \xrightarrow{2-1} D_{n+2}$
(4)	$\frac{1}{3}(0, 2, 1, 1; 0)$	$f + \lambda x$	$\lambda z + \mu t$	$D_4 \xrightarrow{3-1} E_6$
(5)	$\frac{1}{2}(1, 0, 1, 1; 0)$	$f + \lambda y$	$\lambda z + \mu t$	$D_{n+1} \xrightarrow{2-1} D_{2n}$
(6)	$\frac{1}{2}(1, 0, 1, 1; 0)$	$f + \lambda y$	$\lambda z + \mu t$	$E_6 \xrightarrow{2-1} E_7$

(6.5) *Remaining problems.* As discussed in (6.2), the converse statement Theorem 6.1, (II) is in a rather unsatisfactory state. The lists of Theorem 6.1 should perhaps be regarded as just the start of the study of these singularities.

Problems. (1) Is there a direct proof that terminal $\Rightarrow$ the general elephant $S \in |-K_X|$ is a Du Val singularity?

(2) Is there a proof of the converse statement (II) based on the fact that the general elephant is Du Val?

(3) How should one resolve these singularities? In particular, is there an analogue of the economic resolution (5.7)? It would be useful to have a partial resolution $f : Y \to X$ such that all exceptional primes have discrepancy < 1 and Y has only isolated cDV points.

For example, the equation in family (1) is of the form $xy + g(z^r, t)$; in singularity theory it is traditional to ignore quadratic factors ("Morse lemma"), so that one might look for a resolution of the 3-fold singularity in terms of that of the curve singularity $(g(Z, t) = 0)$. See also [**Kollár and Shepherd-Barron**, §6].

(6.6) The proof of (6.1) will break up into proofs in 3 disjoint cases, which are carried over to §7; I start by setting up the general framework.

Notation. Introduce the following terminology: first, $(x, y, z, t) = (x_1, \ldots, x_4)$ are local analytic coordinates on $\mathbf{A}^4$, and the group action is given by

$$\boldsymbol{\mu}_r \ni \varepsilon : (x, y, z, t; f) \mapsto (\varepsilon^a x, \varepsilon^b y, \varepsilon^c z, \varepsilon^d t; \varepsilon^e f),$$

where $(a, b, c, d) = (a_1, \ldots, a_4)$. Note that all the monomials in f belong to the same eigenvalue of the action, so that for example if $xy \in f$ then $a + b = e$ and $\alpha(f) \equiv \alpha(xy) \bmod \mathbf{Z}$ for all $\alpha \in N \cap \sigma$. The group action on the generator

$$s = \mathrm{Res}_Y \frac{dx \wedge dy \wedge dz \wedge dt}{y} \in \omega_Y$$

is given by

$$\boldsymbol{\mu}_r \ni \varepsilon \colon s \mapsto \varepsilon^{a+b+c+d-e} \cdot s,$$

and since the quotient X has index r, $a + b + c + d - e$ is coprime to r; it is therefore reasonable to fix this to be 1, which corresponds to the fact that the local class group of $P \in X$ has a canonical generator K_X.

Rules of the game.

Rule I: For every primitive $\alpha \in N \cap \sigma$,

$$\alpha(f) + 1 < \alpha(xyzt).$$

Rule II: (i) If $\mathrm{hcf}(a_i, r) \neq 1$ then a_i divides e;
(ii) $\mathrm{hcf}(a_i, a_j, r) = 1$;
(iii) $a + b + c + d - e = 1$.

Rule III: An analytic change of coordinates compatible with the group action can be used to put f in normal form with respect to its leading terms. That is, I can assume that:

(i) $$f = q(x_1, \ldots, x_k) + f'(x_{k+1}, \ldots, x_4)$$

with q a nondegenerate quadratic form in $x_1, \ldots, x_k$; moreover
(ii) if the 3-jet of f is $x^2 + y^2 z$ then

$$f = x^2 + y^2 z + yg(t) + h(z, t),$$

or if the 3-jet is $x^2 + y^3$ then

$$f = x^2 + y^3 + yg(z, t) + h(z, t).$$

Rule IV: Only one entry per household; employees of Kelloggs' or of their subsidiaries are not eligible for entry; the referees' decision is final.

Here Rule I is the condition of Theorem 4.6, and Rule II, (i) is a consequence of the fact that $\boldsymbol{\mu}_r$ acts freely on Y outside the origin: if the action of some element of $\boldsymbol{\mu}_r$ fixes the x_i-axis pointwise then some power of x_i must appear in f, hence $a_i | e$; Rule II, (ii) is similar, and (iii) has already been discussed. Rule III is a standard manipulation in singularity theory. The normal forms can be got by explicit ad hoc coordinate changes, and it is easy to see that these can be done in an equivariant way. For example, if $x^2 \in f$ then the terms in f divisible by x^2 can be gathered together to give

$$f = x^2(1 + \eta) + xg(y, z, t) + h(y, z, t)$$

with $\eta \in m$. Then $\xi = \sqrt{1 + \eta}$ is an analytic function invariant under the group action, and

$$f = x'^2 + h'(y, z, t),$$

where $x' = \xi x + (1/2)g/\xi$.

(6.7) *First division into cases.* [*Hint*: if you have trouble following the proof of Theorem 6.1, redo Exercise 4.10.]

I know from general theory that $Q \in Y$ is a cDV point, which implies f contains certain quadratic, cubic terms, etc.; as in (4.9), (3), these restrictions can of course be deduced from Rule I applied to suitable weightings $\alpha \in \mathbf{Z}^4$, but I will not spend time on this, and just assert what I need about f without proof. Write $f = f_2 + \cdots$, or $f_2 + f_3 + \cdots$, where f_2 is the quadratic part, f_3 the cubic part, and $\cdots$ indicates terms of higher degree. I write m for the maximal ideal of $k[x, y, z, t]$, so that for example $g(z, t) \in m^2$ means that $g \in (z, t)^2 \subset k[z, t]$. The following proposition deals with the quadratic part of f.

PROPOSITION. *Exactly one of the following* 5 *cases holds in suitable eigen-coordinates. To be more precise, I can make a μ_r-equivariant analytic change of coordinates (including possibly permuting the coordinates), to achieve one of the following, where the coordinate functions x, y, z and t are eigenfunctions of the μ_r-action; the 5 cases are disjoint.*

> cA *case:* $f = xy + g(z, t)$ *with* $g \in m^2$;
> *odd case:* $f = x^2 + y^2 + g(z, t)$ *with* $g \in m^3$, *and* $a \neq b$;
> cD_4 *case:* $f = x^2 + g(y, z, t)$ *with* $g \in m^3$ *and* g_3 *a reduced cubic*;
> cD_n *case:* $f = x^2 + y^2 z + g(z, t)$ *with* $g \in m^4$;
> cE *case:* $f = x^2 + y^3 + yg(z, t) + h(z, t)$ *with* $g \in m^3$ *and* $h \in m^4$.

Note that by the proposition, the proof of Theorem 6.1 breaks up into the 3 disjoint proofs of the implications

$$cA \Rightarrow (1) \text{ or } (2);$$

$$\text{odd case} \Rightarrow (2) \text{ or } (3);$$

$$cD_4, cD_n \text{ or } cE \Rightarrow (4), (5) \text{ or } (6).$$

(6.8) PROOF. If rank $f_2 \leq 1$ then there's essentially nothing to prove, since $f_2 = x^2$ with x an eigenfunction; completing the square (by Rule III) gives

$$f = x^2 + g(y, z, t) \quad \text{with } g \in m^3 \text{ and } g_3 \neq 0.$$

Now if g_3 is a reduced cubic I'm in the cD_4 case. Failing that, since the tangent cone to $(g = 0)$ is invariant under the group action, either $g_3 = y^2 z$ with y and z eigenfunctions, or $g_3 = y^3$ with y an eigenfunction.

It's convenient to deal with the case rank $f_2 \geq 2$ by means of a general result.

(6.9) LEMMA. *Let V be a vector space (over an algebraically closed field k of characteristic $\neq 2$) with a linear action of μ_r, and let $q \colon V \to k$ be a nonzero quadratic form which is an eigenform. Then there exist integers $k \geq 0$ and $l \leq 2$ with $2k + l = \text{rank } q$ and a basis of V consisting of eigenvectors*

$u_1, \ldots, u_k, v_1, \ldots, v_k, w_1, \ldots, w_{n-2k}$ *(when $n = \dim V$) such that*

(i) *q is given in the dual coordinates (x_i, y_i, z_i) by*

$$(*) \qquad\qquad q = \sum_{i=1}^{k} x_i y_i + \sum_{j=1}^{l} z_j^2;$$

and (ii) *if $l = 2$ then w_1 and w_2 have different eigenvalues.*

(6.10) This proves (6.7), since if rank $f_2 \geq 3$ then by (6.9) I can make a coordinate change to get $f_2 = xy + g_2(z, t)$, and by Rule III this reduces to the cA case; if rank $f_2 = 2$, a coordinate change will give either $f_2 = xy$, or $x^2 + y^2$ with different eigenvalues for x and y, giving the cA or odd cases.

PROOF OF (6.9). (GOTO §7 if you already know this.) This can be done as an exercise in undergraduate algebra. The main point is that if $q = \sum \lambda_i x_i^2$ is diagonal in eigencoordinates $(x_1, \ldots, x_n)$, and no two x_i's with $\lambda_i \neq 0$ have the same eigenvalue then rank $q \leq 2$; this is trivial, since the two square roots of the eigenvalue of q are the only two possibilities for the eigenvalues of x_i. On the other hand, if it is not diagonal then you can pick a term say $x_1 x_2 \in q$, and make coordinate transformations arguing on the other terms in which x_1 and x_2 appear.

(6.11) Now here's the real proof of (6.9). By choosing a μ_r-invariant complement of the kernel of q, I can assume that q is of maximal rank n. Now the existence of the normal form $(*)$ is formally equivalent to the existence of an isotropic subspace $E \subset V$ of dimension $k = (n - l)/2$ invariant under the action of μ_r. Indeed, let E be an invariant k-dimensional subspace with $q|_E \equiv 0$; then $E \subset E^{\perp}$, where $E^{\perp}$ is the orthogonal of E with respect to the associated bilinear form. Choose $F \subset V$ to be an invariant complement of $E^{\perp}$ in V, necessarily k-dimensional. By construction the quadratic form induces an isomorphism from E to the dual of F, and using this it is not hard to adjust F so that it is also isotropic. The orthogonal complement of $E \oplus F$ is then l-dimensional with $l \leq 2$, and it is easy to complete the proof.

The form q defines a nonsingular quadric $Q \subset \mathbf{P}^{n-1}$; an isotropic linear subspace $E \subset V$ of dimension k corresponds exactly to a $(k-1)$-plane in Q (that is, a linear subspace $\mathbf{P}^{k-1} \subset Q \subset \mathbf{P}^{n-1}$), and the problem is to find an invariant one. This is very well known material (see, for example [Hodge and Pedoe, vol. II, pp. 230–237]). Since $k = (\text{rank } q - l)/2$, there are lots of $(k-1)$-planes on Q; these are maximal linear subspaces of Q if $l = 0$ or 1, or one less than maximal if $l = 2$. The space parametrising them is a nonsingular variety $G_{k-1}(Q)$, and is irreducible if $l = 1$ or 2, or has two components if $l = 0$; each component is rational. Now μ_r acts on $Q \subset \mathbf{P}(V)$, and hence on $G_{k-1}(Q)$; if $n = \text{rank } q$ is even and for $l = 0$ the group action interchanges the two components of $G_{k-1}(Q)$, then there can't be any fixed point and so there's no normal form $(*)$ with $l = 0$. But in case $l = 1$ or 2, or in case $l = 0$ if the group action takes each component to itself, the action must have at least one fixed point (since the component of

$G_{k-1}(Q)$ is irreducible and rational). This corresponds to an invariant isotropic subspace of V of the required dimension.   Q.E.D.

7. Case-by-case proof of Theorem (6.1).

(7.1) *Plan of proof.* In (6.7) I divided the proof into the 3 implications

$$
\begin{array}{ll}
cA: & cA \Rightarrow (1) \text{ or } (2); \\
odd: & \text{odd case} \Rightarrow (2) \text{ or } (3); \\
cD - E: & cD_4, cD_n, \text{ or } cE \Rightarrow (4), (5) \text{ or } (6).
\end{array}
$$

Although the proofs in each case are logically disjoint, they all follow the same 4-step pattern; the first 3 steps involve only the quadratic terms xy or $x^2 + y^2$ or x^2 of f.

Step 1. Reduce to the terminal lemma. This part is analogous to Lemma 5.3 in the classification of terminal quotient singularities; it goes from the inequalities $\alpha(f) < \alpha(xyzt) - 1$ of Rule I, applied to the weightings

$$\alpha_k = \tfrac{1}{r}(\overline{ak}, \overline{bk}, \overline{ck}, \overline{dk}) \quad \text{for } k = 1, \ldots, r-1$$

to a set of $r - 1$ equalities; see (7.2).

Step 2. Coprime problem. This step discusses the possible common factors of a, b, c, d and e with r to verify the assumption of Theorem 5.4, (A) or of Corollary 5.6.

Step 3. Using the terminal lemma, write down in a mechanical way a list containing all possibilities for the type.

Step 4. Final method. The situation will be that the terms xy or $x^2 + y^2$ or x^2 of f_2 will satisfy the conditions of Rule I for one-half of all the weightings α_k: the remaining half of the weightings will then impose further monomials on f, and the condition that $f \in m^2$ will then exclude many of the cases written down in Step 3. See (7.8) for a more precise statement in the cA case.

(7.2) *How to reduce to the terminal lemma.*

PROPOSITION. (a) *Suppose that* $xy \in f$. *Then for any* $\alpha \in N \cap \square$,

$$
\begin{array}{ll}
either & \alpha(f) = \alpha(xy) < 1 \text{ and } \alpha(zt) > 1; \\
or & \alpha(f) = \alpha(xy) - 1 \text{ and } \alpha(zt) < 1.
\end{array}
$$

(b) *Suppose that* $x^2 \in f$. *Then for any* $\alpha \in N \cap \square$,

$$
\begin{array}{ll}
either & \alpha(f) = 2\alpha(x) < 1 \text{ and } \alpha(yzt) > 1 + \alpha(x); \\
or & \alpha(f) = 2\alpha(x) - 1 \text{ and } \alpha(yzt) < 1 + \alpha(x).
\end{array}
$$

In (a) *and* (b), *the alternative cases are interchanged by the symmetry*

$$\alpha \mapsto \alpha' = (1, \ldots, 1) - \alpha.$$

(c) *If either* xy *or* $x^2 \in f$, *then*

$$\overline{ak} + \overline{bk} + \overline{ck} + \overline{dk} = \overline{ek} + k + r$$

for each $k = 1, \ldots, r-1$.

(7.3) PROOF OF (a). Assume that $xy \in f$; it follows that $a + b \equiv e \mod r$, and therefore

$$\alpha(f) \equiv \alpha(xy) \quad \text{for all } \alpha \in N \cap \sigma;$$

also from Rule II, (iii), $c + d \equiv 1 \bmod r$. From this it is clear that

$$\alpha_k(zt) > 1 \quad \Longleftrightarrow \quad \alpha_{r-k}(zt) < 1.$$

In view of $\alpha(xy) \equiv \alpha(f) \bmod \mathbf{Z}$ and $0 < \alpha(f) \leq \alpha(xy) < 2$, the two cases $\alpha(f) = \alpha(xy)$ or $\alpha(xy) - 1$ seem pretty well inevitable. Also Rule I gives

$$\alpha(f) = \alpha(xy) \Rightarrow \alpha(zt) > 1.$$

So I only have to show that

$$\begin{aligned}\text{neither} \quad &\alpha(f) = \alpha(xy) \geq 1 \text{ and } \alpha(zt) > 1; \\ \text{nor} \quad &\alpha(f) = \alpha(xy) - 1 \text{ and } \alpha(zt) \geq 1\end{aligned}$$

can happen for any α. Write $\alpha' = (1, \ldots, 1) - \alpha$. Then clearly in either of the two cases

$$\alpha(xy) \geq 1 \Rightarrow \alpha'(f) \leq \alpha'(xy) \leq 1,$$

and therefore $\alpha'(f) = \alpha'(xy)$, whereas $\alpha(zt) \geq 1$ implies $\alpha'(zt) \leq 1$. This contradicts Rule I.

(7.4) PROOF OF (b). (This is word-for-word the same proof as for (a).) Assume that $x^2 \in f$. Note that since $x^2 \in f$ it follows that $2a \equiv e \bmod r$, and therefore

$$\alpha(f) \equiv 2\alpha(x) \quad \text{for all } \alpha \in N \cap \sigma;$$

also from Rule II, (iii), $b + c + d \equiv 1 + a \bmod r$. From this it is an easy computation to see that

$$\alpha_k(yzt) > 1 + \alpha(x) \quad \Longleftrightarrow \quad \alpha_{r-k}(yzt) < 1 + \alpha(x).$$

Now in view of $2\alpha(x) \equiv \alpha(f) \bmod \mathbf{Z}$ and $0 < \alpha(f) \leq 2\alpha(x) < 2$, one of the two cases $\alpha(f) = 2\alpha(x)$ or $2\alpha(x) - 1$ must hold. Also Rule I gives

$$\alpha(f) = 2\alpha(x) \Rightarrow \alpha(yzt) > 1 + \alpha(x).$$

So I only have to show that

$$\begin{aligned}\text{neither} \quad &\alpha(f) = 2\alpha(x) \geq 1 \text{ and } \alpha(yzt) > 1 + \alpha(x); \\ \text{nor} \quad &\alpha(f) = 2\alpha(x) - 1 \text{ and } \alpha(yzt) \geq 1 + \alpha(x)\end{aligned}$$

can happen for any α. Write $\alpha' = (1, \ldots, 1) - \alpha$. Then clearly in either case

$$2\alpha(x) \geq 1 \Rightarrow \alpha'(f) \leq 2\alpha'(x) \leq 1.$$

and therefore $\alpha'(f) = 2\alpha'(x)$, whereas

$$\alpha(yzt) \geq 1 + \alpha(x) \text{ implies } \alpha'(yzt) \leq 1 + \alpha(x).$$

This contradicts Rule I.

(7.5) PROOF OF (c). This is easy: if $xy \in f$ then in the two cases of (a),

$$\overline{ak} + \overline{bk} = \overline{ek} \quad \text{and} \quad \overline{ck} + \overline{dk} = k + r$$

or

$$\overline{ak} + \overline{bk} = \overline{ek} + r \quad \text{and} \quad \overline{ck} + \overline{dk} = k.$$

Similarly, if $x^2 \in f$ then in the cases of (b),

$$2\overline{ak} = \overline{ek} \quad \text{and} \quad \overline{bk} + \overline{ck} + \overline{dk} = \overline{ak} + k + r$$

or

$$2\overline{ak} = \overline{ek} + r \quad \text{and} \quad \overline{bk} + \overline{ck} + \overline{dk} = \overline{ak} + k. \quad \text{Q.E.D.}$$

(7.6) *The cA case.* The time has now come to divide up into the cases of (6.7). The case cA: $f = xy + g(z,t)$ will occupy me from now until (7.8).

cA. *Step* 2 (*coprimeness*). Lemma 7.2 gives me the conditions of the terminal lemma (5.4), (A) for the $(4+2)$-tuple $\frac{1}{r}(a,b,c,d;e,1)$, except that some of the numbers may not be coprime to r. However, this does not make too much trouble, thanks to the following argument.

Define $q = \mathrm{hcf}(e,r)$; the cases $q = 1$ and $q = r$ are not excluded in what follows.

LEMMA. *After interchanging z and t if necessary, the following hold:*
(1) $q = \mathrm{hcf}(d,r)$,
(2) a, b *and* c *are coprime to* r,
(3) $g = g(z^r, t)$.

PROOF. Since $xy \in f$, I have $e \equiv a + b \bmod r$, and it's easy to prove that a and b are coprime to r using Rule II, (ii). By Rule II, (i), any common factor of c or d and r divides q, so that I only need to prove that q divides either c or d. Note that if I set $k = r/q$ then $\overline{ek} = 0$, so I must be in the second case of the computation in (7.5), and hence

$$\overline{ck} + \overline{dk} = k \quad \text{and} \quad \overline{c(r-k)} + \overline{d(r-k)} = r - k;$$

therefore not both of

$$\overline{ck} + \overline{c(r-k)} \quad \text{and} \quad \overline{dk} + \overline{d(r-k)}$$

can be equal to r. This proves that q divides d, say.

(7.7) cA. *Step* 3. I list all the possibilities for the type; one of the following cases holds (after possibly interchanging x and y or z and t):
If $q > 1$,
    (A) $a + b \equiv 0$, $c \equiv 1$, $d \equiv e \bmod r$, that is, $\frac{1}{r}(a, -a, 1, 0; 0)$;
    (B) $a \equiv 1$, $b + c \equiv 0$, $d \equiv e \bmod r$, that is, $\frac{1}{r}(1, b, -b, b+1; b+1)$.
If $q = 1$,
    (C) $\frac{1}{r}(a, 1, -a, a+1; a+1)$;
    (D) $\frac{1}{r}(a, -a-1, -a, a+1; -1)$
with a and $a+1$ coprime to r.

PROOF. Recall that $a + b \equiv e$ and $a + b + c + d - e \equiv 1 \bmod r$.

The terminal lemma tells me that the 6 elements a, b, c, d, e and 1 must be paired off: if $q > 1$ then by Corollary 5.6 I must pair d and e, giving (A) and (B). If $q = 1$ then by Theorem 5.4 it's easy to see that (C) and (D) are the only two possibilities.

(7.8) *cA.* *Step* 4. Case (A) gives $\frac{1}{r}(a, -a, 1, 0; 0)$, which is case (1) of (6.1). To complete the proof of (6.1) in case cA, I must prove that (C) and (D) are impossible, and that (B) is only possible if either

$$\tfrac{1}{r}(a, b, c, d; e) = \tfrac{1}{4}(1, 1, 3, 2; 2) \quad \text{and} \quad z^2 \in f$$

or $\frac{1}{r}(a, b, c, d; e)$ also falls in case (A).

Method. The condition which remains to use is that, by (7.2), (a),

$$\alpha(f) = \alpha(xy) - 1 \quad \text{for every } \alpha \text{ such that } \alpha(zt) \leq 1,$$

so that g must have a monomial x^m with $\alpha(x^m) = \alpha(xy) - 1$; on the other hand, $g = g(z^r, t) \in m^2$.

To kill (C), let $k(a + 1) \equiv 1 \bmod r$ with $k < r$. Then $\alpha_k(zt) = k/r < 1$, but $\alpha_k(xy) - 1 = 1/r$, so that no monomial in m^2 stands a chance. To kill (D), choose $k = r - 1$. Then $\alpha_k(zt) = (r - 1)/r < 1$, but again $\alpha_k(xy) - 1 = 1/r$ so that again no monomial in m^2 can work.

Case (B) is $\frac{1}{r}(1, b, -b, b+1; b+1)$ with b coprime to r; if $b+1 \equiv 0 \bmod r$, this simplifies to $\frac{1}{r}(1, -1, 1, 0; 0)$ which is in (A). So assume I have

$$\tfrac{1}{r}(1, b, -b, b + 1; b + 1) \quad \text{with } b \text{ coprime to } r \text{ and } b + 1 < r.$$

Consider the weighting $\alpha = \alpha_{r-1} = \frac{1}{r}(r - 1, r - b, b, r - b - 1)$; then $\alpha(zt) = (r - 1)/r < 1$ and so by the method of this step $\alpha(g) \leq (r - b - 1)/r$. This means that there is some monomial $x^m \in (z, t)^2$ with

$$x^m \in g \quad \text{and} \quad \alpha(x^m) = \alpha(xy) - 1 = (r - b - 1)/r.$$

Obviously no multiple of t will work, so that the monomial can only be z^n for some n with $nb = r - b - 1$. Notice that since $n \geq 2$, it follows that $r \geq 3b + 1 \geq 2b + 2$.

CLAIM. $r = 4$, $b = 1$, and $n = 2$.

PROOF. I choose another weighting

$$\beta = \alpha_{r-2} = \tfrac{1}{r}(r - 2, r - 2b, 2b, r - 2b - 2);$$

since $\beta(zt) = (r - 2)/r < 1$, the method gives

$$\beta(g) = \beta(xy) - 1 = (r - 2b - 2)/r.$$

However, the same monomial z^n can't possibly work, nor can any multiple of zt; hence some power t^m must appear in g with $m \geq 2$ and

$$m(r - 2b - 2) = r - 2b - 2.$$

This implies $r = 2b + 2$, and the claim follows at once.

This completes the proof of case $cA \Rightarrow$ (1) or (2) of (6.1).

(7.9) *The odd case:* $f = x^2 + y^2 + g(z, t)$ with $g \in m^3$ and $a \neq b$; this will take until (7.11). Note that $2a \equiv 2b \equiv e$ and $a \not\equiv b \bmod r$ implies that r is even, and $b \equiv a + r/2$.

Odd. Step 2 (*coprimeness*). Either (after interchanging x and y if necessary) $a \equiv e \equiv 0 \bmod r$, and b, c, d are coprime to r, or (after interchanging z and t if necessary) I have: $q = 2 = \operatorname{hcf}(d, r)$ and a, b, c are coprime to r.

PROOF. First of all I claim that if $a \not\equiv 0 \bmod r$ then $\operatorname{hcf}(a, r) = 1$; for otherwise there exists some divisor k of r with $0 < k < r$ such that $\overline{ak} = \overline{a(r - k)} = 0$. Then by the computation in (7.5), I have

$$\overline{bk} + \overline{ck} + \overline{dk} = k + r$$

and

$$\overline{b(r - k)} + \overline{c(r - k)} + \overline{d(r - k)} = r - k;$$

(or the same with k and $(r - k)$ interchanged). Adding these together would show that not all of

$$\overline{bk} + \overline{b(r - k)}, \qquad \overline{ck} + \overline{c(r - k)} \quad \text{and} \quad \overline{dk} + \overline{d(r - k)}$$

can be r; so one of b, c, d has a common factor with a and r, contradicting Rule II, (ii).

Now since $e = 2a$ and $\operatorname{hcf}(a, r) = 1$ and r is even, it follows that $q = \operatorname{hcf}(e, r) = 2$. Then setting $k = r/2$ I get $\overline{ak} = r/2$ and $\overline{ek} = 0$, so that from (7.5),

$$\overline{bk} + \overline{ck} + \overline{dk} = r/2 + r/2,$$

and exactly one of b, c, d is even.

(7.10) *Odd. Step* 3 (*listing possible types*). I claim that the type is

$$\text{either} \quad \tfrac{1}{2}(0, 1, 1, 1; 0)$$
$$\text{or} \quad \tfrac{1}{r}(1, \tfrac{r+2}{2}, \tfrac{r-2}{2}, 2; 2) \quad \text{for some } r \text{ with } 4 | r.$$

In fact $2a \equiv 2b \equiv e \bmod r$ so that r is even and $b \equiv a + r/2$. If $a = 0$ then by Rule II, (ii), I must have $r = 2$, giving the first conclusion. Otherwise by (7.9) and (5.6), I must pair d with e, and

$$\text{either} \quad a \equiv 1, \ b + c \equiv 0 \quad \text{or} \quad a + b \equiv 0, \ c \equiv 1.$$

In the first case, $b = 1 + r/2 = (r + 2)/2$ is odd so $4 | r$, as required. It's easy to see that the second possibility gives $a = r/4$, $b = 3r/4$, and then using Rule II, (ii), necessarily $r = 4$, which gives $\tfrac{1}{4}(1, 3, 1, 2; 1)$.

(7.11) *Odd. Step* 4 (*final method*). To prove that this case implies (3) of Theorem 1, I only have to show that the case

$$\tfrac{1}{r}\left(1, \tfrac{r+2}{2}, \tfrac{r-2}{2}, 2; 2\right) \quad \text{with } 4 | r \text{ and } r > 4$$

is impossible. By (7.2), (b), for every α with $\alpha(yzt) < 1 + \alpha(x)$, there exists a monomial of weight $2\alpha(x) - 1$. So let

$$\alpha = \alpha_{r-2} = \tfrac{1}{r}(r - 2, r - 2, 2, r - 4);$$

then $\alpha(yzt) = (2r - 4)/r < \alpha(x) + 1 = (r - 2)/r + 1$. Also

$$2\alpha(x) - 1 = (r - 4)/r;$$

if $r > 4$ then it is easy to see that the only monomial in $(z,t)^2$ of weight $\leq (r-4)/r$ and in the same eigenspace as f is z^2, which is excluded by the case assumption. This proves that I am in case $\frac{1}{4}(1,3,1,2;2)$ with $f = x^2 + y^2 + g(z,t)$, and permuting y and z gives (6.1), (2).

This completes the proof of odd case $\Rightarrow$ (2) or (3) of (6.1).

(7.12) Now consider the remaining cases $f = x^2 + g(y,z,t)$ with $g \in m^3$.

cD-E. Step 2 (coprimeness). Suppose that $x^2 \in f$; then

$$\text{either} \quad a \equiv e \equiv 0 \bmod r \text{ and } b,c,d \text{ are coprime to } r,$$
$$\text{or} \quad r \text{ is odd and } a,b,c,d,e \text{ are coprime to } r,$$

or (after interchanging y, z, and t if necessary)

$$q = 2 = \text{hcf}(d,r) \text{ and } a,b,c \text{ are coprime to } r.$$

PROOF. First of all, exactly as in (7.9), if $a \not\equiv 0 \bmod r$ then $\text{hcf}(a,r) = 1$.

Now since $e = 2a$, it follows that $\text{hcf}(e,r) = 1$ if r is odd, or 2 if r is even. Also, in the first case, a,b,c, and d are coprime to r, since any common factor with r would have to divide e by Rule II, (i). On the other hand, if r is even then setting $k = r/2$ gives $\overline{ak} = r/2$ and $\overline{ek} = 0$, so that from (7.5),

$$\overline{bk} + \overline{ck} + \overline{dk} = r/2 + r/2,$$

and exactly one of b, c and d is even.   Q.E.D.

(7.13) cD-E. Step 3 (listing possible types). After possibly permuting y, z and t, the possible types are:

If $a \equiv e \equiv 0 \bmod r$, then
  (a) $\frac{1}{r}(0,b,-b,1;0)$ with b coprime to r.
If a and r are coprime and $q = 2$, then
  (b) $\frac{1}{r}(a,-a,1,2a;2a)$ with r even and a coprime to r;
or (c) $\frac{1}{r}(1,b,-b,2;2)$ with r even and b coprime to r.
If $q = 1$, then
  (d) $\frac{1}{r}((r-1)/2,-(r-1)/2,c,-c;-1)$ with r odd and c coprime to r;
or (e) $\frac{1}{r}(a,-a,2a,1;2a)$ with r odd and a coprime to r;
or (f) $\frac{1}{r}(1,b,-b,2;2)$ with r odd and b coprime to r.

PROOF. As before, this follows easily from the terminal lemma (5.4), (A) and (5.6).

(7.14) cD-E. Step 4 (Final method). First of all, in cases (b) and (c), I claim that $r = 2$; in both cases this gives $\frac{1}{2}(1,1,1,0;0)$ which implies (5) or (6) of Theorem 6.1. Indeed, suppose that r is even and $r > 2$; so choose k such that $\overline{ka} = (r+2)/2 < r$. Then in case (b),

$$\alpha_k = \frac{1}{r}\left(\frac{r+2}{2}, \frac{r-2}{2}, k, 2\right)$$

satisfies

$$\alpha_k(yzt) = \frac{1}{r}\left(\frac{r+2}{2} + k\right) < \alpha_k(x) + 1 = \frac{1}{r}\left(\frac{r+2}{2} + r\right).$$

Therefore $\alpha_k(f) = 2\alpha_k(x) - 1 = 2/r$; but nothing in m^3 has weight $\leq 2/r$, which gives a contradiction. A similar calculation also leads to a contradiction in case (c).

Next (d), (e), and (f) are impossible; indeed, choose k such that $\overline{ka} = (r+1)/2$, and consider

$$\alpha_k = \tfrac{1}{r}(\overline{ak}, \overline{bk}, \overline{ck}, \overline{dk}).$$

It is easy to check (separately in the 3 cases) that $\alpha_k(yzt) < \alpha_k(x) + 1$, so that by (7.2), $\alpha_k(f) = 2\alpha_k(x) - 1 = 1/r$. But no monomial in m^2 can have weight less that $1/r$.

(7.15) The only remaining case is (a) of (7.13), and for this I need to translate Rule I for f into a similar condition for $g(y, z, t)$. Note that $\alpha(g) \equiv 2\alpha(x) \bmod \mathbf{Z}$ for any $\alpha \in N \cap \sigma$.

LEMMA. *Assume case* (a). *Then*
(1) *For any* $\alpha \in N \cap \sigma$,

$$\text{if } \alpha(g) - 2\alpha(x) \text{ is even, then } \alpha(g) < 2\alpha(yzt) - 2;$$
$$\text{if } \alpha(g) - 2\alpha(x) \text{ is odd, then } \alpha(g) < 2\alpha(yzt) - 1.$$

(2) *The weightings*

$$\alpha_k = \tfrac{1}{r}(0, \overline{bk}, r - \overline{bk}, k) \quad \text{for } k = 1, \ldots, r-1$$

satisfy $\alpha_k(g) = 1$.

PROOF OF (1). Define $\beta = \alpha + i \cdot (1, 0, 0, 0)$, where $i = \tfrac{1}{2}(\alpha(g) - 2\alpha(x))$ or $\tfrac{1}{2}(\alpha(g) - 2\alpha(x) + 1)$ in the two cases. Then

$$\beta(f) = \min\{\alpha(g), 2\alpha(x) + 2i\} = \alpha(g).$$

Also

$$\beta(xyzt) = \alpha(x) + i + \alpha(yzt),$$

so that Rule I gives

$$\beta(f) < \beta(xyzt) - 1,$$

that is,

$$\alpha(g) < \alpha(x) + i + \alpha(yzt) - 1,$$

and writing out the definition of i gives the statement in the lemma.

PROOF OF (2). In fact $\alpha_k(g)$ even implies that

$$0 < \alpha_k(g) < 2\alpha_k(yzt) - 2 = 2k/r < 2,$$

which has no solutions, and $\alpha_k(g)$ odd implies

$$0 < \alpha_k(g) < 2\alpha_k(yzt) - 1 = (2k + r)/r < 3,$$

which has the single solution $\alpha_k(g) = 1$. Q.E.D.

(7.16) To prove that (a) $\Rightarrow$ (6.1), (4) I need to show that $r = 3$. So choose k such that $\overline{bk} = (r - 1)/2$. By interchanging y and z if necessary, I can assume that $k \geq (r - 1)/2$. Then consider

$$\alpha_k = \tfrac{1}{r}\left(0, \tfrac{r-1}{2}, \tfrac{r+1}{2}, k\right).$$

Since $\alpha_k(g) = 1$, there must be a monomial in y, z, t of weight 1, and of degree ≥ 3 (since $g \in m^3$). Note that $\alpha_k(yz) = 1$ and $\alpha_k(z^2) > 1$, so that no multiple of yz or z^2 can work. However, since $k \geq (r-1)/2$,

$$\alpha_k(y^3, y^2t, yt^2, t^3) \geq (3r - 3)/2r = 1 + (r - 3)/2r.$$

So the only way to get a monomial of weight 1 is if $r = 3$.

This shows that the type of X is $\frac{1}{3}(0, 2, 1, 1)$; it is an exercise to see that f must contain both y^3 and a monomial of degree 3 in (z, t), giving (4).

Chapter III
Contributions of Q-divisors to RR

8. Quotient singularities and equivariant RR.

(8.1) This chapter introduces a number of formulas of the type

$$\chi(X, \mathcal{L}) = (\text{RR-type expression in } D) + \sum_Q c_Q(D);$$

here X is a normal variety, and $\mathcal{L} = \mathcal{O}_X(D)$, where D is a Weil divisor which is **Q**-Cartier, and Cartier outside a finite set of points; the terms $c_Q(D)$ are contributions due to the singularities of the sheaf $\mathcal{O}_X(D)$, and are local analytic invariants of the "polarised singularity" ($Q \in X$ and $\mathcal{O}_X(D)$). Notice that this is not immediately related to the "singular Riemann-Roch theorems" in the literature, which deal in the sheaf $\mathcal{L}$, singularities and all: the formula here deals only in the **Q**-divisor class of D in $\operatorname{Pic} X \otimes \mathbf{Q}$, so involves a substantial abuse of notation.

(8.2) The existence of such a formula is not in itself particularly exciting, but in several cases of interest the computation of $c_Q(D)$ can be reduced to the contributions of a "basket"

$$\{P_\alpha \in X_\alpha \text{ and } \mathcal{O}_{X_\alpha}(D_\alpha)\}$$

of cyclic quotient singularities, which can in turn be calculated by equivariant RR. I introduce the term "basket of singularities" to emphasise the fact that the singularities $P_\alpha \in X_\alpha$ are not points of X, but only "fictitious singularities": the singularities of X and D make contributions to RR equal to those which would occur if X had these singularities. For example, X and D might in good cases deform to a variety really having these singularities.

(8.3) The notation for cyclic quotient singularities is as in (4.1).

DEFINITION. The quotient $X = \mathbf{A}^n/\boldsymbol{\mu}_r$, where $\boldsymbol{\mu}_r$ acts on $\mathbf{A}^n$ by

$$\boldsymbol{\mu}_r \ni \varepsilon \colon (x_1, \ldots, x_n) \mapsto (\varepsilon^{a_1} x_1, \ldots, \varepsilon^{a_n} x_n)$$

is a *cyclic quotient singularity of type* $\frac{1}{r}(a_1, \ldots, a_n)$. Write $\pi \colon \mathbf{A}^n \to X$ for the quotient map; then the group $\boldsymbol{\mu}_r$ acts on $\pi_* \mathcal{O}_{\mathbf{A}^n}$, and so decomposes it into r eigensheaves

$$\mathcal{L}_i = \{f \mid \varepsilon(f) = \varepsilon^i \cdot f \text{ for all } \varepsilon \in \boldsymbol{\mu}_r\}$$

for $i = 0, \ldots, r-1$. A singularity $P \in X$ with a Weil divisor D is a *cyclic quotient singularity of type* $_i(\frac{1}{r}(a_1, \ldots, a_n))$ if $P \in X$ is (locally isomorphic to) a point of type $\frac{1}{r}(a_1, \ldots, a_n)$, and $\mathcal{O}_X(D) \cong \mathcal{L}_i$.

REMARK. In writing the action $(x_1, \ldots, x_n) \mapsto (\varepsilon^{a_1} x_1, \ldots, \varepsilon^{a_n} x_n)$, I am thinking of the x_i as *coordinates* on $\mathbf{A}^n$, as usual in algebraic geometry. (This is the dual of what the topologist would write; since the tangent space is based by $\{\partial/\partial x_i\}$, the action of ε on the tangent space $T_{\mathbf{A}^n, 0}$ has eigenvalues $(\varepsilon^{-a_1}, \ldots, \varepsilon^{-a_n})$. I am repeating the bizarre pronouncement that the relation between points of $\mathbf{A}^n$ and coordinates is contravariant; sorry.)

(8.4) The following result is useful in reducing local problems concerning quotient singularities to the projective case.

PROPOSITION. *Given* $\frac{1}{r}(a_1, \ldots, a_n)$ *(with a_i coprime to r), there exists a smooth projective n-fold Y with an action of μ_r having a number N of fixed points at which μ_r acts by*

$$\mu_r \ni \varepsilon: (x_1, \ldots, x_n) \mapsto (\varepsilon^{a_1} x_1, \ldots, \varepsilon^{a_n} x_n)$$

and freely outside these points.

PROOF. This is very easy. Suppose that $k >$ maximum number of the a_i which are congruent modulo any prime p dividing r (for example, $k > n$). Consider the action of μ_r on $\mathbf{P}^{n+k}$ given in homogeneous coordinates by

$$(\underbrace{1, \ldots, 1}_{k+1 \text{ times}}, \varepsilon^{a_1}, \ldots, \varepsilon^{a_n}).$$

The action of μ_r on $\mathbf{P}^{n+k}$ has a fixed locus $\mathbf{P}^k$ where the action in the normal direction is given by $(x_1, \ldots, x_n) \mapsto (\varepsilon^{a_1} x_1, \ldots, \varepsilon^{a_n} x_n)$, and other fixed locuses of smaller dimension. Let $X \subset \mathbf{P}^{n+k}/\mu_r$ be the intersection of k general very ample divisors, and Y its inverse image under $\mathbf{P}^{n+k} \to \mathbf{P}^{n+k}/\mu_r$; then it is easy to check that Y satisfies the conditions of the proposition.   Q.E.D.

(8.5) THEOREM. *Under the conditions of (8.4), write $\pi: Y \to X$ for the quotient map, and let $\mathcal{L}_i$ be the ith eigensheaf of the action of μ_r on $\pi_* \mathcal{O}_Y$. Then for $i = 0, 1, \ldots, r-1$,*

$$\chi(X, \mathcal{L}_i) = \frac{1}{r}\chi(\mathcal{O}_Y) + \frac{N}{r}\sigma_i\left(\tfrac{1}{r}(a_1, \ldots, a_n)\right),$$

where

$$\sigma_i\left(\tfrac{1}{r}(a_1, \ldots, a_n)\right) = \sum_{\varepsilon \neq 1} \frac{\varepsilon^i}{(1 - \varepsilon^{a_1}) \cdots (1 - \varepsilon^{a_n})},$$

the sum extending over all $\varepsilon \in \mu_r - \{1\}$.

PROOF. This follows easily from equivariant RR. Let Y be a nonsingular projective n-fold and g an automorphism of Y acting with only finitely many fixed points $\{Q\}$. The Lefschetz number $L(g: \mathcal{O}_Y)$ of g acting on $\mathcal{O}_Y$ is defined by

(1) $$L(g: \mathcal{O}_Y) = \sum (-1)^j \operatorname{Tr}(g: H^j(\mathcal{O}_Y)).$$

Write $dg_Q\colon T_{Y,Q} \to T_{Y,Q}$ for the differential of the action at a fixed point Q. Then the Atiyah-Singer equivariant RR formula [**Atiyah-Segal**, p. 541] takes the simple form

$$(2) \qquad\qquad L(g\colon \mathcal{O}_Y) = \sum_Q \frac{1}{\det(1 - dg_Q)}.$$

In the present case, this can be used as follows: since

$$\pi_* \mathcal{O}_Y = \bigoplus_{i=0}^{r-1} \mathcal{L}_i,$$

it follows that

$$H^j(Y, \mathcal{O}_Y) = \bigoplus_{i=0}^{r-1} H^j(X, \mathcal{L}_i);$$

moreover, any $\varepsilon \in \boldsymbol{\mu}_r$ acts on $\mathcal{L}_i$ by multiplication by ε^i, so that,

$$\operatorname{Tr}(\varepsilon\colon H^j(Y, \mathcal{O}_Y)) = \sum_{i=0}^{r-1} h^j(X, \mathcal{L}_i) \cdot \varepsilon^i.$$

Therefore

$$(3) \qquad L(\varepsilon\colon \mathcal{O}_Y) = \sum (-1)^j \operatorname{Tr}(\varepsilon\colon H^j(Y, \mathcal{O}_Y)) = \sum_{i=0}^{r-1} \chi(X, \mathcal{L}_i) \cdot \varepsilon^i.$$

By construction, ε acts on $T_{Y,Q}$ by $(\varepsilon^{-a_1}, \ldots, \varepsilon^{-a_n})$ at each fixed point Q. Then applying (2) to $g = \varepsilon$ gives

$$(4_\varepsilon) \qquad \sum_{i=0}^{r-1} \chi(X, \mathcal{L}_i) \cdot \varepsilon^i = \frac{N}{(1 - \varepsilon^{-a_1}) \cdots (1 - \varepsilon^{-a_n})}$$

for any $\varepsilon \in \boldsymbol{\mu}_r - \{1\}$.

I consider (4_ε) as giving $r - 1$ equations for the r unknowns $\chi(X, \mathcal{L}_i)$. The final equation is

$$(4_1) \qquad\qquad \sum_{a=0}^{r-1} \chi(X, \mathcal{L}_i) = \chi(\mathcal{O}_Y).$$

I can now solve the r equations (4_ε) for $\chi(X, \mathcal{L}_i)$ by inverting a Vandermonde matrix: multiply (4_ε) through by ε^{-i} and sum over all $\varepsilon \in \boldsymbol{\mu}_r$. This gives

$$(5) \qquad\qquad \chi(\mathcal{L}_i) = \frac{1}{r} \chi(\mathcal{O}_Y) + \frac{N}{r} \sigma_i,$$

where for $i = 0, \ldots, r - 1$,

$$(6) \qquad \sigma_i\left(\tfrac{1}{r}(a_1, \ldots, a_n)\right) = \sum_{\varepsilon \neq 1} \frac{\varepsilon^{-i}}{(1 - \varepsilon^{-a_1}) \cdots (1 - \varepsilon^{-a_n})},$$

the sum extending over all $\varepsilon \in \boldsymbol{\mu}_r - \{1\}$. Since ε^{-1} runs throught $\boldsymbol{\mu}_r - \{1\}$ together with ε, I can ignore the minus signs throughout.   Q.E.D.

(8.6) It is easy to eliminate $\chi(\mathcal{O}_Y)$ from this formula, to get

$$\chi(X, \mathcal{L}_i) = \chi(\mathcal{O}_X) + \left(\frac{N}{r}\right)\{\sigma_i - \sigma_0\},$$

a prototype of the kind of formula referred to in (8.1): $\mathcal{L}_i$ is a sheaf on X with isolated singularities, and since as a $\mathbf{Q}$-divisor, $L_i = 0 \in \operatorname{Div} X \otimes \mathbf{Q}$, I can think of $\chi(\mathcal{O}_X)$ as a RR-type expression in L, and the remainder of the formula as being a sum of N local contributions $(1/r)\{\sigma_i - \sigma_0\}$ coming from the singular points of $\mathcal{L}_i$.

COROLLARY. *Let X be an n-fold having a finite set of quotient singularities $\{Q\}$, and $\mathcal{L} = \mathcal{O}_X(D)$ a divisorial sheaf on X. Then there is a formula of type*

$$\chi(X, \mathcal{L}) = (RR\text{-}type\ expression\ in\ D) + \sum_Q c_Q(D);$$

where (i) *the contributions $c_Q(X)$ are of the form*

$$c_Q(X) = \sigma_i\left(\tfrac{1}{r}(a_1, \ldots, a_n)\right) - \sigma_0\left(\tfrac{1}{r}(a_1, \ldots, a_n)\right)$$

if $\mathcal{L}$ is locally of type $_i(\tfrac{1}{r}(a_i, \ldots, a_n))$ at Q; and (ii) *the RR-type expression in D is the usual $\operatorname{ch}(D) \cdot \operatorname{Td}_X$ interpreted formally in the following ad hoc way: for the terms in which D appears, just take Td_X on the nonsingular locus of X and intersect with D (this is right because some multiple of D is a Cartier divisor, so can be moved away from the singularities); for the remaining term, just substitute $\chi(\mathcal{O}_X)$.*

SKETCH OF PROOF. The fact that a formula of this kind exists can be readily seen by comparing X and $\mathcal{L}$ with a suitable resolution; the contribution which the argument gives is a sum of local analytic invariants at each of the singularities, so can be computed on any example where only this singularity appears, for example that of (8.4–5).

Appendix to §8. Computing the σ_i.

(8.7) The sums $\sigma_i(\tfrac{1}{r}(a_1, \ldots, a_n))$ are defined by

$$\sigma_i\left(\tfrac{1}{r}(a_1, \ldots, a_n)\right) = \sum_{\varepsilon \neq 1} \frac{\varepsilon^i}{(1 - \varepsilon^{a_1}) \cdots (1 - \varepsilon^{a_n})};$$

in general one cannot expect to get a closed formula for them, but I evaluate them here in two cases of interest. The σ_i are a kind of Dedekind sum, and a lot of information on them is contained in [**Hirzebruch-Zagier**], although I prefer to work from first principles.

The sums $\sigma_i(\tfrac{1}{r}(a_1, \ldots, a_n))$ are determined recursively by the following two conditions:

(A) $$\sum_{i=0}^{r-1} \sigma_i = 0;$$

and

(B) $$(\sigma_{i+a_n} - \sigma_i)\left(\tfrac{1}{r}(a_1, \ldots, a_n)\right) = -\sigma_i\left(\tfrac{1}{r}(a_1, \ldots, a_{n-1})\right).$$

Here (A) follows from the fact that

$$\sum_{\varepsilon \in \mu_r} \varepsilon = 0,$$

and (B) is easy to check. Formula (B) is particularly useful if one or more of the a_j is ± 1.

Of course the recursion starts with

(C) $\qquad \begin{cases} \sigma_0 \left(\frac{1}{r}(\varnothing) \right) = r - 1, \\ \sigma_i \left(\frac{1}{r}(\varnothing) \right) = -1 \qquad \text{if } i = 1, \ldots, r - 1. \end{cases}$

Another useful fact is the relation

(D) $\qquad \sigma_i \left(\frac{1}{r}(a_1, \ldots, a_n) \right) = \sigma_{bi} \left(\frac{1}{r}(ba_1, \ldots, ba_n) \right)$

if b is coprime to r, which comes at once from the fact that $\varepsilon \mapsto \varepsilon^b$ is a bijection of μ_r.

(8.8) Using (A) and (B), it is easy to verify that

$$\sigma_i(\tfrac{1}{r}(-1)) = (r - 1)/2 - i \quad \text{for } i = 0, \ldots, r - 1.$$

(8.9) To calculate $\sigma_i(\tfrac{1}{r}(-1, 1))$, note that

$$\sigma_{i+1} - \sigma_i = -(r - 1)/2 + i \quad \text{for } i = 0, \ldots, r - 1,$$

and therefore

$$\sigma_i = \sigma_0 + \sum_{j=0}^{i-1} \left\{ -\frac{(r-1)}{2} + j \right\} = \sigma_0 - \frac{i(r-i)}{2}.$$

So from (A) I get

$$r\sigma_0 = \sum_{i=1}^{r-1} \frac{i(r-i)}{2} = \frac{r(r^2-1)}{12};$$

hence

$$\sigma_0 = (r^2 - 1)/12.$$

PROPOSITION. *For $i = 0, \ldots, r - 1$,*

$$\sigma_i \left(\tfrac{1}{r}(-1, 1) \right) = \frac{(r^2 - 1)}{12} - \frac{i(r-i)}{2}.$$

Using (D), it follows that for any a coprime to r, and $j = 0, \ldots, r - 1$,

$$\sigma_j \left(\tfrac{1}{r}(a, -a) \right) = \frac{(r^2 - 1)}{12} - \frac{\overline{bj}(r - \overline{bj})}{2},$$

where $ba \equiv 1 \bmod r$ and $\overline{}$ denotes smallest residue mod r.

(8.10) To calculate $\sigma_i(\tfrac{1}{r}(a, -a, 1))$ I proceed in the same way:

$$\sigma_{i+1} - \sigma_i = -\sigma_i \left(\tfrac{1}{r}(a, -a) \right) = \frac{(r^2 - 1)}{12} - \frac{\overline{bi}(r - \overline{bi})}{2},$$

where $ba \equiv 1 \bmod r$ and $^-$ denotes smallest residue $\bmod r$. This gives

$$\sigma_i = \sigma_0 - \frac{i(r^2 - 1)}{12} + \sum_{j=0}^{i-1} \frac{\overline{bj}(r - \overline{bj})}{2}.$$

As before $0 = \sum_{i=0}^{r-1} \sigma_i$, and taking account of the number of times each of the summands $\overline{bj}(r - \overline{bj})/2$ appears in the sum over i, I get

$$0 = r\sigma_0 - r(r-1) \cdot \frac{(r^2 - 1)}{24} + \sum_{j=0}^{r-1}(r - j - 1) \cdot \frac{\overline{bj}(r - \overline{bj})}{2}.$$

Now notice that the complicated sum here can be evaluated: since the factors $\overline{bj}(r - \overline{bj})/2$ in the jth and $(r - j)$th terms are the same, it simplifies to

$$\sum = \frac{(r-2)}{4} \sum_{j=0}^{r-1} \overline{bj}(r - \overline{bj}) = \frac{(r-2)}{4} \sum_{i=0}^{r-1} i(r - i) = r(r-2) \cdot \frac{(r^2 - 1)}{24};$$

this proves the following result.

PROPOSITION. $\sigma_0 = (r^2 - 1)/24$, and for $i = 0, \ldots, r - 1$,

$$\sigma_i - \sigma_0 = -i \cdot \frac{r^2 - 1}{12} + \sum_{j=0}^{i-1} \frac{\overline{bj}(r - \overline{bj})}{2}$$

(the sum is by convention 0 if $i = 0$ or 1).

The proposition was first proved by A. R. Fletcher, using results of [**Hirzebruch-Zagier**] on Dedekind sums.

(8.11) *Exercise.* Set $\sum a_i = k \bmod r$, so that the divisorial sheaf $\mathcal{O}_X(K_X)$ is of type $_k(\frac{1}{r}(a_1, \ldots, a_n))$ near Q. Give two different proofs of the proposition

$$\sigma_i = (-1)^n \sigma_{i'} \quad \text{whenever } i + i' \equiv k \bmod r;$$

(one based on Serre duality and (8.5), one by induction on n using (A)–(C)).

9. Contributions from Du Val surface singularities.

Let X be a projective surface with at worst Du Val singularities, and D a Weil divisor on X. Since D is **Q**-Cartier, the intersection numbers $D^2, K_X D \in \mathbf{Q}$ are well defined (in fact $K_X D \in \mathbf{Z}$ since K_X is Cartier).

(9.1) THEOREM. (I) *There is a formula*

$$\chi(X, \mathcal{O}_X(D)) = \chi(\mathcal{O}_X) + (1/2)(D^2 - DK_X) + \sum_Q c_Q(D)$$

where $c_Q(D) = c_Q(\mathcal{O}_X(D)) \in \mathbf{Q}$ is a contribution due to the singularity of $\mathcal{O}_X(D)$ at Q, depending only on the local analytic type of $Q \in X$ and D; the sum takes place over the singularities of D (the points $Q \in X$ at which D is not Cartier).

(II) *If $P \in X$ and D is a cyclic quotient singularity of type $_i(\frac{1}{r}(1, -1))$ then*

$$c_P(D) = -i(r - i)/2r.$$

(III) *For every Du Val singularity $Q \in X$ and Weil divisor D on X, there exists a basket of points of $\{P_\alpha \in X_\alpha$ and $D_\alpha\}$ of types $i_\alpha(\frac{1}{r_\alpha}(1, -1))$ and with i_α coprime to r_α, such that*

$$c_Q(D) = \sum_\alpha c_{P_\alpha}(D_\alpha) = -\sum_\alpha \frac{i_\alpha(r_\alpha - i_\alpha)}{2r_\alpha}.$$

REMARKS. (i) (I) is rather trivial; it is proved by comparing X and D with a resolution as in (8.6). The proof gives an expression for $c_Q(D)$ which is not very useful for computational purposes.

(ii) In (II), i does not have to be coprime to r.

(iii) The point of (III) is that the contribution $c_Q(D)$ can be expressed as a sum of a basket of contributions of the type described in (II) with i and r coprime. To prove (III), I show that there is a deformation which replaces the local analytic singularity $Q \in X$ and D with a basket $\{(P_\alpha \in X_\alpha$ and $D_\alpha)\}$.

(9.2) PROOF OF (I). It is easy to see that there is a resolution $f : Y \to X$ and a Cartier divisor E on Y such that $f^* \mathcal{O}_X(D) \to \mathcal{O}_Y(E)$ is surjective; then also $f_* \mathcal{O}_Y(E) = \mathcal{O}_X(D)$ and $R^i f_* \mathcal{O}_Y(E) = 0$, so that

$$\chi(Y, \mathcal{O}_Y(E)) = \chi(X, \mathcal{O}_X(D)).$$

(1) now follows by writing out RR for E on Y; in more detail, let $\{\Gamma_i\}$ be the exceptional curves of f. Then I can write

$$K_Y = f^* K_X + A, \quad \text{where } A = \sum a_i \Gamma_i,$$

and

$$E = f^* D + B, \quad \text{where } B = \sum b_i \Gamma_i,$$

with $a_i \in \mathbf{Z}$ and $b_i \in \mathbf{Q}$. Then (since $f^* D \Gamma_i = f^* K_X \Gamma_i = 0$ for all Γ_i),

$$\begin{aligned}
\chi(X, \mathcal{O}_X(D)) &= \chi(Y, \mathcal{O}_Y(E)) = \chi(\mathcal{O}_Y) + (1/2) E(E - K_Y) \\
&= \chi(\mathcal{O}_X) + (1/2)(f^* D + B)(f^*(D - K_X) + B - A) \\
&= \chi(\mathcal{O}_X) + (1/2)(D^2 - DK_X) + \sum_Q c_Q(D),
\end{aligned}$$

where $c_Q(D) = (1/2)(B_Q)(B_Q - A_Q)$.

(II) follows from (8.5–6), together with the result of (8.9),

$$\sigma_i - \sigma_0 = -i(r - i)/2.$$

(9.4) PROOF OF (III). This is a deformation argument; I show that given the singularity $Q \in X$ and D, there is a flat deformation $\{X_t$ and $D_t\}$ of X together with the Weil divisor class D such that X_t has only cyclic quotient singularities. The deformation family $\{X_t\}$ extends to a family of projective surfaces. Since $\chi(\mathcal{O}_X(D))$ and the invariants D^2 and DK_X are continuous in a flat family, the contribution $c_Q(D)$ is equal to the sum of the contributions from the cyclic quotient singularities of X_t.

Consider $Q \in X$ and D; let $\pi : Y \to X$ be the cyclic cover corresponding to D. Then $O \in Y$ is a Du Val surface singularity (if it's nonsingular, there's nothing to prove), together with an action of μ_r on $O \in Y$ satisfying the properties:

(i) the action is free outside O;

(ii) it acts trivially on a generator of ω_Y (corresponding to the fact that X is Gorenstein).

This is exactly the situation classified in Exercise 4.10. From the list given there, one sees that suitable Q-smoothings of X are as follows:

(1) $f + \lambda$; (-1) has two fixed points on the y-axis, given by $y^2 + \lambda = 0$.

(2) $f + \lambda z$; the subgroup $\mu_2 \subset \mu_4$ has $(2n + 1)$ fixed points on the z-axis, given by $z^{2n+1} + \lambda z = 0$.

(3) $f + \lambda$; μ_r has n fixed points on the z-axis given by $z^n + \lambda = 0$.

(4) $f + \lambda$; μ_3 has 2 fixed points on the x-axis given by $x^2 + \lambda = 0$.

(5) $f + \lambda$; (-1) has n fixed points on the z-axis given by $z^n + \lambda = 0$.

(6) $f + \lambda$; (-1) has 3 fixed points on the y-axis given by $y^3 + \lambda = 0$.

This completes the proof of Theorem 9.1.

10. The plurigenus formula. Let X be a projective 3-fold with canonical singularities, and D a Weil divisor on X such that $\mathcal{O}_X(D) \cong \mathcal{O}_X(iK_X)$ in a neighbourhood of every $P \in X$ for some i (possibly varying with P). Note that there are only finitely many points at which D is not Cartier.

(10.1) *Definition of $D \cdot c_2(X)$.* By definition,

$$D \cdot c_2(X) = (f^*D) \cdot c_2(Y),$$

where $f\colon Y \to X$ is a resolution (and as usual, f^*D refers to the pull-back of $\mathbf{Q}$-Cartier divisors). This does not depend on the resolution (if X has only 0-dimensional singular locus, $D \cdot c_2(X) = (1/r)E \cdot c_2(X)$, $E \sim rD$, where E is a divisor linearly equivalent to rD not passing through any singular points of X).

(10.2) THEOREM. (1) *There is a formula of the form*

$$\chi(X, \mathcal{O}_X(D)) = \chi(\mathcal{O}_X) + \tfrac{1}{12}D(D - K_X)(2D - K_X) + \tfrac{1}{12}D \cdot c_2(X) + \sum_Q c_Q(D),$$

where the summation takes place over the singularities of the sheaf $\mathcal{O}_X(D)$, and $c_Q(D) \in \mathbf{Q}$ is a contribution due to the singularity at Q, depending only on the local analytic type of $Q \in X$ and $\mathcal{O}_X(D)$.

(2) *If $P \in X$ is the terminal cyclic quotient singularity $X = \mathbf{A}^3/\mu_r$ of type $\frac{1}{r}(a, -a, 1)$ and $\mathcal{O}_X(D)$ is locally isomorphic to $\mathcal{O}_X(iK_X)$ (so that the pair (X and D) is of type $_i(\frac{1}{r}(a, -a, 1))$ in the terminology of (8.3)) then*

$$c_P(D) = -i \cdot \frac{(r^2 - 1)}{12r} + \sum_{j=1}^{i-1} \frac{\overline{bj}(r - \overline{bj})}{2r},$$

where b satisfies $ab \equiv 1 \bmod r$, and $\bar{\ }$ denotes smallest residue $\bmod r$ (the sum $\sum_{j=1}^{i-1}$ is zero by convention if $i = 0$ or 1).

(3) *For every 3-fold canonical singularity $Q \in X$ and Weil divisor D on X such that $\mathcal{O}_X(D) \cong \mathcal{O}_X(iK_X)$ for some i, there exists a basket of points*

$$\{P_\alpha \in X_\alpha \text{ and } D_\alpha\} \text{ of type } {}_{i_\alpha}\left(\frac{1}{r_\alpha}(a_\alpha, -a_\alpha, 1)\right)$$

such that

$$c_Q(D) = \sum c_{P_\alpha}(D_\alpha).$$

This means that the contribution from any singularity $Q \in X$ and D can be expressed as a sum of contributions from a basket of terminal cyclic quotient singularities.

PROOF. As before, (1) is proved by comparing X and D with a suitable resolution Y and E. Since X has only a finite set of dissident points, it is not hard to choose a resolution $f\colon Y \to X$ whose discrepancy Δ_f is concentrated above a finite set of X. For details of the argument, compare [**C3-f**, (5.5)].

(2) follows directly from (8.6) and (8.10).

There are two reductions in the proof of (3): the first step reduces to terminal singularities by a crepant partial resolution as described in (3.12) and (3.14), and the second to terminal quotient singularities by a flat deformation as described in (6.4), (A); the second of these is easy using (6.4), (A), and the reader should think through the details for himself.

CLAIM. *Let $P \in X$ be a canonical 3-fold singularity, and $g\colon X' \to X$ a crepant partial resolution such that X' has only terminal singularities. Then the contribution $c_P(iK_X)$ is equal to a sum of contributions $c_Q(iK_{X'})$ over the finite set of points $Q \in g^{-1}P$ at which $K_{X'}$ is not Cartier.*

This can be seen by looking more closely at the proof of (1): if I choose the resolution of X by first constructing $g\colon X' \to X$ and then resolving the singularities of X by $h\colon Y \to X'$, then obviously the discrepancy of $f\colon Y \to X$ equals that of $h\colon Y \to X'$.

Alternatively, argue as follows: there is no loss of generality in assuming that X is projective with $P \in X$ its only dissident singularity. By the fact that $X' \to X$ is crepant it follows that $g_*\mathcal{O}_{X'}(iK_{X'}) = \mathcal{O}_X(iK_X)$; also, by standard use of vanishing, $R^i g_*\mathcal{O}_{X'}(iK_{X'}) = 0$. The Leray spectral sequence then gives

$$\chi(\mathcal{O}_X(iK_X)) = \chi(\mathcal{O}_{X'}(iK_{X'}));$$

together with (1) this proves the claim.

(10.3) COROLLARY. (4) *Let X be a projective 3-fold with canonical singularities and*

$$\left\{ P_\alpha \in X_\alpha \text{ and } K_{X_\alpha}, \text{ of type } {}_1(\tfrac{1}{r_\alpha}(a_\alpha, -a_\alpha, 1)) \right\}$$

the basket for X and K_X in the sense of (3). Then $\chi(\mathcal{O}_X)$ and $K_X \cdot c_2(X)$ are related by

$$\chi(\mathcal{O}_X) = -\frac{1}{24} K_X \cdot c_2(X) + \sum_{P_\alpha} \frac{(r^2 - 1)}{24r},$$

or alternatively,

$$\frac{1}{12} K_X \cdot c_2 = -2\chi(\mathcal{O}_X) + \sum_{P_\alpha} \frac{(r^2 - 1)}{12r}.$$

(5) *Suppose that in addition H is a Cartier divisor on X; then*

$$\chi(\mathcal{O}_X(H + mK_X)) = (1 - 2m)\chi(\mathcal{O}_X)$$
$$+ \frac{1}{12}(H + mK_X)(H + (m-1)K_X)(H + (2m-1)K_X) + \frac{1}{12}H \cdot c_2(X)$$
$$+ \sum_Q \left\{ \frac{(r^2 - 1)}{12r}(m - \overline{m}) + \sum_{j=1}^{\overline{m}-1} \frac{\overline{bj}(r - \overline{bj})}{2r} \right\},$$

where inside the curly blankets, $^{-}$ *denotes smallest residue of m* mod *r and* $ab \equiv 1$ mod *r (note that r varies with Q). In particular,*

$$\chi(\mathcal{O}_X(mK_X)) = (1 - 2m)\chi(\mathcal{O}_X) + \frac{1}{12}m(m-1)(2m-1)K^3$$
$$+ \sum_Q \left\{ \frac{(r^2 - 1)}{12r}(m - \overline{m}) + \sum_{j=0}^{\overline{m}-1} \frac{\overline{bj}(r - \overline{bj})}{2r} \right\},$$

where the sum takes place over the basket of singularities for X.

Historical remark. The correct version of the formula in (5) is due to Anthony Fletcher; his paper [**Fletcher**] gives several alternative versions of the formula which are convenient in applications. The contribution in (4) has been computed by several people, probably first by Rebecca Barlow (around 1980) using her version of Danilov's economic resolution (see (5.10)); essentially her computation is given in [**Kawamata**, (2.2)].

(10.4) *Exercise.* Check that the formula gives the right values for the first few plurigenera of the 3-folds of Exercise 2.12. For example, (i) has

$$P_2 = \chi(\mathcal{O}_X(2K_X)) = \tfrac{1}{2} \cdot \tfrac{1}{6} + \{\tfrac{1}{4}\} + 5 \times \{\tfrac{1}{3}\} = 2,$$
$$P_3 = \chi(\mathcal{O}_X(3K_X)) = \tfrac{5}{2} \cdot \tfrac{1}{6} + \{\tfrac{1}{4}\} + 5 \times \{\tfrac{2}{3}\} = 4.$$

REFERENCES

[**Atiyah-Segal**] M. F. Atiyah and G. Segal, *The index of elliptic operators.* II, Ann. of Math. (2) **87** (1968), 531–545.

[**Barth-Peters-Van de Ven**] W. Barth, C. Peters, and A. Van de Ven, *Compact complex surfaces*, Springer-Verlag, 1984.

[**Brieskorn**] E. Brieskorn, *Über die Auflösung gewisser Singularitäten von holomorphen Abbildungen*, Math. Ann. **166** (1966), 76–102; *Die Auflösung der rationalen Singularitäten von holomorphen Abbildungen*, Math. Ann. **178** (1968), 255–270.

[**Catanese**] F. Catanese, *Canonical rings of surfaces*, these Proceedings, Part 1, pp. 175–194.

[**Danilov**] V. I. Danilov, *The birational geometry of toric 3-folds*, Izv. Akad. Nauk SSSR Ser. Mat. **46** (1982), 972–981; English transl. in Math USSR-Izv. **21** (1983).

[**Dolgachev**] I. Dolgachev, *Weighted projective spaces*, Group Actions and Vector Fields, Lecture Notes in Math., vol. 956, Springer-Verlag, 1982, pp. 34–71.

[**Elkik**] R. Elkik, *Rationalité des singularités canoniques*, Invent. Math. **64** (1981), 1–6.

[**Flenner**] H. Flenner, *Rational singularities*, Arch. Math. (Basel) **36** (1981), 35–44.

[**Fletcher**] A. R. Fletcher, *Contributions to Riemann-Roch on projective 3-folds with only canonical singularities and applications*, these Proceedings, Part 1, pp. 221–231.

[**Hirzebruch-Zagier**] F. Hirzebruch and D. Zagier, *The index theorem and elementary number theory*, Publish or Perish, 1974.

[Inose-Mizukami] H. Inose and M. Mizukami, *Rational equivalence of 0-cycles on surfaces of general type with $p_g = 0$*, Math. Ann. **244** (1979), 205–217.

[Kawamata] Y. Kawamata, *On the plurigenera of minimal algebraic 3-folds with $K \approx 0$*, Math. Ann. **275** (1986), 539–546.

[Khovanskiĭ] A. G. Khovanskiĭ, *Newton polyhedra and the genus of complete intersections*, Functional Anal. Appl. **12** (1978), 51–61.

[Kollár and Shepherd-Barron] J. Kollár and N. Shepherd-Barron, *Threefolds and deformations of surface singularities*, preprint.

[Lang] S. Lang, *Cyclotomic fields*, Springer-Verlag, 1978.

[Markushevich] D. G. Markushevich, *Canonical singularities of three-dimensional hypersurfaces*, Izv. Akad. Nauk SSSR Ser. Mat. **49** (1985), 334–368; English transl. in Math USSR-Izv. **26** (1986).

[Matsumura] H. Matsumura, *Commutative ring theory*, Cambridge Univ. Press, 1986.

[Mori] S. Mori, *On 3-dimensional terminal singularities*, Nagoya Math J. **98** (1985), 43–66.

[Morrison-Stevens] D. R. Morrison and G. Stevens, *Terminal quotient singularities in dimension 3 and 4*, Proc. Amer. Math Soc. **90** (1984), 15–20.

[Morrison] D. R. Morrison, *Canonical quotient singularities in dimension 3*, Proc. Amer. Math Soc. **93** (1985), 393–396.

[Mumford] D. Mumford, *Hilbert's fourteenth problem—the finite generation of subrings such as rings of invariants*, Math. Developments Arising from Hilbert Problems, Proc. Sympos. Pure Math., vol. 28, Amer. Math. Soc., Providence, R.I., 1976, pp. 431–444.

[Pinkham] H. Pinkham, *Singularités rationelles de surfaces*, Lecture Notes in Math., vol. 777, Springer-Verlag, 1980, pp. 147–168.

[C3-f] M. Reid, *Canonical 3-folds*, Journées de Géométrie Algébrique d'Angers, A. Beauville, editor, Sijthoff and Noordhoff, Alphen aan den Rijn, 1980, pp. 273–310.

[Pagoda] M. Reid, *Minimal models of canonical 3-folds*, Algebraic Varieties and Analytic Varieties, S. Iitaka and H. Morikawa, editors, Adv. Stud. Pure Math., vol. 1, Kinokuniya Book Store, Tokyo, and North Holland, Amsterdam, 1983, pp. 131–180.

[Shafarevich] I. R. Shafarevich, *Basic algebraic geometry*, Springer-Verlag, 1974.

[Tai] Y. S. Tai, *On the Kodaira dimension of the moduli space of Abelian varieties*, Invent. Math. **68** (1982), 425–439.

[Ueno] K. Ueno, *On the pluricanonical systems of algebraic manifolds*, Math. Ann. **216** (1975), 173–179.

[Washington] L. C. Washington, *Introduction to cyclotomic fields*, Springer-Verlag, 1982.

[Wilson] P. M. H. Wilson, *Towards birational classification of algebraic varieties*, Bull. London Math. Soc. **19** (1987), 1–47.

[Zariski] O. Zariski, *The theorem of Riemann-Roch for high multiples of an effective divisor on an algebraic surface*, Ann. of Math. (2) **76** (1962), 560–615.

MATHEMATICS INSTITUTE, UNIVERSITY OF WARWICK, COVENTRY CV4 7AL, ENGLAND

Affine Algebraic Geometry

Proceedings of Symposia in Pure Mathematics
Volume **46** (1987)

Classification of Noncomplete Algebraic Varieties

TAKAO FUJITA

This note is intended to be a brief survey of the classification theory of noncomplete algebraic varieties defined over the complex number field **C**. For details and proofs, consult original papers. Among others, [**I5**, Chapter 11; **My 3**; and **F4**] are recommended. We assume that the reader has some knowledge about the classification theory of complete varieties (consult, e.g., [**I5**, Chapter 10; **U**] or S. Mori's article in this book).

In §1, we define several logarithmic invariants. In §2, we describe Iitaka's classification program. §3 is devoted to the theory of Zariski decomposition on surfaces. In §4 we consider the classification of noncomplete surfaces.

Basically we employ the customary notation in algebraic geometry. Line bundles and invertible sheaves are confused occasionally and the tensor products of them are denoted by additive notation.

1. Logarithmic invariants.

(1.1) By a "*smooth completion*" we mean a pair $(\overline{V}, D)$ of a nonsingular complete variety $\overline{V}$ and an effective divisor D on $\overline{V}$ having no singularity other than normal crossings. $V = \overline{V} - D$ is called the *interior*. The pair $(\overline{V}, D)$ is said to be a *completion* of V.

Let $(\overline{V}_1, D_1)$ and $(\overline{V}_2, D_2)$ be completions as above. A holomorphic mapping $f \colon \overline{V}_1 \to \overline{V}_2$ is said to be a morphism of completions if $f^{-1}(D_2) \subset D_1$. In this case we get a morphism $f_0 \colon V_1 \to V_2$ by restriction.

(1.2) Let V be any nonsingular algebraic variety defined over **C**. Then, by virtue of Hironaka's desingularization theory, there exists a smooth completion whose interior is V. Moreover, for any morphism $f \colon V_1 \to V_2$ of smooth varieties, there exist completions $(\overline{V}_1, D_1)$, $(\overline{V}_2, D_2)$ of V_1, V_2 and a morphism $\overline{f}$ of them such that $\overline{f}_0 = f$. Such a morphism is said to be a *completion* of f.

(1.3) For a smooth completion $(\overline{V}, D)$, by $\Omega(\overline{V}, D)$ we denote the sheaf of meromorphic 1-forms on $\overline{V}$ having only logarithmic poles along D. This is a locally free sheaf of rank $n = \dim \overline{V}$.

1980 *Mathematics Subject Classification* (1985 *Revision*). Primary 14J10; Secondary 14C20, 14F45, 13F20.

For any polynomial representation ρ of $\mathrm{GL}(n; \mathbf{C})$ (here, "polynomial" means that ρ extends to a morphism of $M(n, \mathbf{C})$), let $\Omega(\overline{V}, D)^\rho$ denote the induced locally free sheaf. Then, for any morphism $\overline{f} \colon (\overline{V}_1, D_1) \to (\overline{V}_2, D_2)$ of smooth completions such that $f = \overline{f}_0$ is quasi-finite and dominant, we have a natural injective mapping $H^0(\overline{V}_2, \Omega(\overline{V}_2, D_2)^\rho) \to H^0(\overline{V}_1, \Omega(\overline{V}_1, D_1)^\rho)$. Moreover, this is bijective if f is proper and birational.

(1.4) The space $H^0(\overline{V}, \Omega(\overline{V}, D)^\rho)$ is called the *logarithmic invariant* of $(\overline{V}, D)$ of type ρ. By the above bijectivity, this is actually an invariant of the interior V. In other words, this is independent of the choice of a smooth completion of V.

For example, $h^0(\overline{V}, \Omega(\overline{V}, D))$ is called the *logarithmic irregularity* of V and is denoted by $\overline{q}(V)$. When $\rho = (\det)^m$ with m being a positive integer, we have $h^0(\overline{V}, m(K + D))$, where K is the canonical bundle of $\overline{V}$. This is called the *logarithmic m-genus* and is denoted by $\overline{P}_m(V)$. So, *logarithmic Kodaira dimension* $\kappa = \kappa(K + D, \overline{V})$ is also well defined and is denoted by $\overline{\kappa}(V)$.

(1.5) REMARK. Logarithmic invariants are *not* biholomorphic invariants of V. For example, as Serre pointed out, $\mathbf{C}^* \times \mathbf{C}^*$ can be compactified to a $\mathbf{P}^1$-bundle over an elliptic curve. We have $\overline{\kappa} < 0$ in this case, while $\overline{\kappa} = 0$ for the usual rational completion. Compare [**F3**; (1.23)] On the other hand, $H^0(\overline{V}, sK + tD)$ is a biholomorphic invariant of V for every $s > t \geq 0$ (cf. [**S**]).

2. Iitaka's classification program.

(2.1) First we recall the following

THEOREM (IITAKA [**I1**]). *Let L be a line bundle on a complete variety M such that $\kappa(L, M) \geq 0$. Then there exist a birational morphism $M' \to M$ and a surjective morphism $\Phi \colon M' \to W$ such that $\dim W = \kappa(L, M)$ and that any generic fiber F of Φ is a variety with $\kappa(L_F, F) = 0$. Moreover, such a fibration is unique up to a birational equivalence.*

Thus, Φ is sometimes called *the* Iitaka fibration.

(2.2) Let V be a smooth variety such that $\overline{\kappa}(V) \geq 0$. Taking a smooth completion $(\overline{V}, D)$ of V and applying (2.1) in case $L = K + D$, we obtain the following

THEOREM. *There exist a proper birational morphism $V' \to V$ and a surjective morphism $\Phi \colon V' \to W$ such that $\dim W = \overline{\kappa}(V)$ and that any generic fiber F of Φ is a variety with $\overline{\kappa}(F) = 0$.*

Thus, if $0 < \overline{\kappa} < \dim V$, V admits a nontrivial fibration structure. Our philosophy is that the study of the structure of such varieties can be reduced to lower-dimensional cases. So, we should first study the cases $\overline{\kappa} = \dim V$, $= 0$, and < 0.

(2.3) When $\overline{\kappa} \leq 0$ and $\overline{q} > 0$, we study logarithmic Albanese mappings in place of classical Albanese mappings.

To be precise, set

$$\overline{A}(V) = \mathrm{Coker}(H_1(V;\mathbf{Z}) \to H^0(\overline{V}, \Omega(\overline{V}, D))^{\vee}),$$

where the mapping is defined by the integration pairing

$$H_1(V;\mathbf{Z}) \times H^0(\overline{V}, \Omega(\overline{V}, D)) \to \mathbf{C}.$$

Then $\overline{A}$ gives rise to a covariant functor from the category of smooth completions to the category of abelian Lie groups. In fact, $\overline{A}(V)$ is a proper birational invariant of V, has a structure of an algebraic group, and the natural mapping $\overline{A}(V) \to \overline{A}(\overline{V}) = \mathrm{Alb}(\overline{V})$ is surjective with the kernel being isomorphic to an affine algebraic torus $\mathbf{G}_m^r$. $\overline{A}(V)$ is called the *logarithmic Albanese variety* (or quasi-Albanese variety) of V. Fixing a point o on V, one obtains a natural morphism $\overline{a}: V \to \overline{A}(V)$ defined by path-integrals, which is called the *log-Albanese mapping*. This has the usual universal property as in the case of Albanese mappings.

Using these notions, we obtain the following

(2.4) THEOREM (KAWAMATA [**Kw 3**]). *If $\overline{\kappa}(V) = 0$, the logarithmic Albanese mapping $\overline{a}$ is dominant and any general fiber of $\overline{a}$ is connected. In particular, $\overline{q}(V) \le \dim V$.*

(2.5) COROLLARY. *If $\overline{q} = \dim V$ further, then $\overline{a}$ is birational. If in addition V is affine, then $V \simeq \mathbf{G}_m^n$.*

(2.6) In order to study log-Albanese mappings, the following conjecture would be very useful:

Let $f: V \to W$ be a dominant morphism such that any general fiber F is connected. Then $\overline{\kappa}(V) \ge \overline{\kappa}(W) + \overline{\kappa}(F)$.

This is actually true if $\dim F = 1$ (cf. [**Kw 1**]). Similarly, as in the complete case, there are a couple of modified versions of this conjecture (consult, e.g., [**V**] and Viehweg's lecture in the threefold seminar). We remark that the above result (2.4) is based on a partial solution of such a modified version.

3. Zariski decomposition on surfaces.

(3.1) For a complete smooth surface S, let $\mathrm{Div}(S)$ be the group of divisors on S. An element of $\mathrm{Div}(S) \otimes \mathbf{Q}$, a $\mathbf{Q}$-linear combination of prime divisors on S, is called a $\mathbf{Q}$-*divisor*. A $\mathbf{Q}$-divisor $D = \sum \delta_i D_i$ is said to be *effective* (resp. *reduced*) if $\delta_i \ge 0$ (resp. $0 \le \delta_i \le 1$) for each coefficient δ_i. D is said to be *contractible* if $Z^2 < 0$ for any $\mathbf{Q}$-divisor Z such that $\varnothing \ne \mathrm{Supp}(Z) \subset \mathrm{Supp}(D)$.

An element of $\mathrm{Pic}(S) \otimes \mathbf{Q}$ is called a $\mathbf{Q}$-*bundle*. The $\mathbf{Q}$-bundle associated to a $\mathbf{Q}$-divisor D is denoted by $[D]$, or simply by D by abuse of notation. The intersection number $L_1 L_2 \in \mathbf{Q}$ of $\mathbf{Q}$-bundles L_1 and L_2 is naturally defined. A $\mathbf{Q}$-bundle L is said to be *nef* if $LE \ge 0$ for any effective $\mathbf{Q}$-divisor E. L is said to be *pseudo-effective* if $LH \ge 0$ for any nef $\mathbf{Q}$-bundle H.

(3.2) THEOREM (CF. [**F1, Z**]). *Let L be a pseudo-effective $\mathbf{Q}$-bundle on S. Then there is a contractible effective $\mathbf{Q}$-divisor N such that $H = L - N$ is nef and that $NH = 0$.*

The above N is determined uniquely by the numerical equivalence class of L. So it is called *the* negative part of the *Zariski decomposition* of L, while H is called the semipositive (or nef) part.

(3.3) The following fact is an important property of the Zariski decomposition.

THEOREM. *Let L, N, H be as above. For any positive integer m, let F_m be the fixed part of the linear system $|mL|$. Then $F_m - mN$ is an effective $\mathbf{Q}$-divisor.*

Thus, if mN is a usual divisor and $mL - mN = mH \in \mathrm{Pic}(S)$, then $H^0(S, mL) \simeq H^0(S, mH)$.

(3.4) COROLLARY. *Suppose in addition that $Bs|mH| = \varnothing$ for some $m > 0$ as above. Then the graded algebra $\bigoplus_{t \geq 0} H^0(S, tL)$ is finitely generated.*

For a proof, use the method in [**F4**, (1.5)].

(3.5) Our goal in this section is the following

THEOREM (CF. [**Kw 2** and **F4**]). *Let D be a reduced $\mathbf{Q}$-divisor on a smooth complete surface S such that $K + D$ is pseudo-effective for the canonical bundle K of S. Let $K + D = N + H$ be the Zariski decomposition. Then there is a positive integer m such that mD and mN are usual divisors and that $Bs|mH| = \varnothing$. (Note that $mH = mK + mD - mN$ is a line bundle.)*

4. Classification of surfaces.

(4.1) Given a smooth surface S, let $(\overline{S}, D)$ be a smooth completion of S and let K be the canonical bundle of $\overline{S}$. Applying (3.5) in this situation, we can make various observations.

First, combining (3.4) and (3.5), we get the following

THEOREM. *The logarithmic canonical ring $G = \bigoplus_{t \geq 0} H^0(\overline{S}, t(K + D))$ is a finitely generated graded algebra.*

$W = \mathrm{Proj}(G)$ is sometimes called the log-canonical model of S (especially when $\overline{\kappa}(S) = 2$).

(4.2) As for minimal models of $(\overline{S}, D)$, we have the following

THEOREM (CF. [**Kw 2, My 3**]). *Let $(\overline{S}, D)$ be a smooth completion of a surface S such that $\overline{\kappa}(S) \geq 0$. Then, there exist a smooth surface $\overline{S}'$, a birational morphism $\pi: \overline{S} \to \overline{S}'$, and a reduced $\mathbf{Q}$-divisor D' on $\overline{S}'$ such that*

(1) $\mathrm{Supp}(D') \subset \pi(\mathrm{Supp}(D))$ and $\mathrm{Supp}(D')$ has no singularity other than nodes, and

(2) $\pi^*(K' + D')$ is the semipositive part of $K + D$, where K' and K are the canonical bundles of $\overline{S}'$ and $\overline{S}$.

Such a pair $(\overline{S}', D')$ may be called *a* minimal model. However, unlike the case $D = 0$, it is not determined uniquely by $(\overline{S}, D)$. Moreover, D' and the negative part N of $K + D$ are not always usual divisors, and the contraction of N yields singularities.

There are several other notions of "*minimal model*," where we sometimes allow $\overline{S}$ to have certain mild singularities.

(4.3) When $\overline{\kappa}(S) < 0$, (3.5) implies that $K + D$ is not pseudoeffective. Hence, for any line bundle L on $\overline{S}$, $|L + t(K + D)| = \varnothing$ for any $t \gg 0$. From this one obtains the following

THEOREM (CF. [MS, My 3, My 4]). *Suppose that* $q(\overline{S}) > 0$ *or that* D *is connected. Then, if* $\overline{\kappa}(S) < 0$, S *contains an open set which is isomorphic to* $\mathbf{A}^1 \times C$ *with* C *being an affine curve.*

This was an important step of the proof of the following characterization theorem of $\mathbf{A}^2$, which in turn leads to an affirmative solution of the cancellation problem of $\mathbf{A}^2$.

(4.4) THEOREM (CF. [My 1]). *A smooth affine surface* $S = \mathrm{Spec}(A)$ *is isomorphic to* $\mathbf{A}^2$ *if and only if* $A^* = \mathbf{C}^*$, A *is factorial, and* $\overline{\kappa}(S) < 0$.

(4.5) COROLLARY (CF. [F1, Km]). *If* $S \times V \simeq \mathbf{A}^2 \times V$ *for some algebraic variety* V, *then* $S \simeq \mathbf{A}^2$.

(4.6) Furthermore, the ruling theorem (4.3) gives the following results of Lüroth type:

(1) If there is a proper surjective morphism $f \colon \mathbf{A}^2 \to S$, then $S \simeq \mathbf{A}^2$. See [My 2] or [F3, (5.8)].

(2) If there is a dominant morphism $f \colon \mathbf{A}^2 \to S$, then $\pi_1(S)$ is isomorphic to one of the following groups: cyclic group $\mathbf{Z}/k\mathbf{Z}$ ($k \geq 1$), dihedral group D_k ($k \geq 1$), symmetric group S_4, antisymmetric group A_4, A_5. See [F3, (5.15) and (4.20)].

(4.7) When $\overline{\kappa}(S) = 2$, the log-canonical model W is a normal variety and we have a birational morphism $\Phi \colon \overline{S} \to W$ such that $H = \Phi^* A$ for some ample $\mathbf{Q}$-bundle A on W. W has only rational singularities and elliptic singularities (cf. [Kw 2, My 3, F4]).

(4.8) When $\overline{\kappa}(S) = 1$, there is a morphism $\Phi \colon \overline{S} \to C$ onto a curve C such that $H = \Phi^* A$ for some ample $\mathbf{Q}$-bundle A on C. Moreover, $\Phi(\mathrm{Supp}(N))$ is a finite set. So $(K + D)F = 0$ for any general fiber F of Φ. Hence, if $DF > 0$ (this must be the case if S is affine), then $F \simeq \mathbf{P}^1$ and $DF = 2$, which implies that S admits an $\mathbf{A}^*$-fibration. For further studies, see [My 3].

(4.9) When $\overline{\kappa}(S) = 0$, H is a torsion. If $\overline{q}(S) > 0$, we study the log-Albanese mapping as in §2. As for the case $\overline{q} = 0$, much less is known.

Let us say that S is of Kummer type if $\overline{q}(\tilde{S}) > 0$ for some finite unramified covering $\tilde{S}$ of S. Such a phenomenon is impossible in case $D = 0$, but is really possible if $D \neq 0$ (for example, let S be the Kummer variety $T/\langle \iota \rangle$ deleted off the 16 double points, where ι is the (-1)-involution of an abelian surface T). In the following questions, we consider S of non-Kummer type only.

(1) How is the topological fundamental group $\pi_1(S)$? Is it a finite abelian group?

In [**F3**], we have found examples such that $\pi_1(S) \simeq \mathbf{Z}/6\mathbf{Z}, \mathbf{Z}/8\mathbf{Z}, \mathbf{Z}/9\mathbf{Z}$, and $\mathbf{Z}/4k\mathbf{Z}$ with k being any positive integer. Note that S is not always simply connected even if $\overline{p}_g(S) = 1$ (see also [**I4**]).

(2) Classify those S with $\pi_1(S) = \{1\}$.

(3) What values can be the smallest positive integer m such that $\overline{P}_m(S) > 0$? In [**F3**], we have found examples with $m = 1, 2, 3, 4$, and 6.

(4.10) Thus, we encounter several troubles which are of similar nature as those in the classification theory of higher-dimensional complete varieties.

5. Comments.

(5.1) The main trouble in positive characteristic cases is the lack of Hironaka's desingularization theory. So we consider only those varieties which admit smooth completions. Then we can define logarithmic invariants of them, study Iitaka fibrations, and transplant many results in the complex case (see [**Km, F2, F4**]). In particular, (3.5) is valid in any characteristic.

(5.2) In his papers [**I6, I7, I8**], Iitaka has developed another type of geometry on birational pairs $(D \& V)$. Although this is different from affine geometry (or noncomplete geometry), many techniques can be used in common. For details, see his original papers.

(5.3) The theory of Zariski decomposition can be generalized in higher dimensions too. The decomposition of $K + D$ is closely related to the theory of minimal models. See [**F5**].

(5.4) Mori's theory as in [**Mo**] can be generalized in our logarithmic context too. See [**KW 4**] or refer to the lecture by Miyanishi and Tsunoda.

REFERENCES

[**D**] P. Deligne, *Théorie de Hodge.* II, Inst. Hautes Études Sci. Publ. Math. **40** (1971), 5–57.

[**F1**] T. Fujita, *On Zariski problem*, Proc. Japan Acad. **55** (1979), 106–110.

[**F2**] ____, *On L-dimension of coherent sheaves*, J. Fac. Sci. Univ. Tokyo Sect. IA Math. **28** (1981), 215–236.

[**F3**] ____, *On the topology of non-complete algebraic surfaces*, J. Fac. Sci. Univ. Tokyo Sect. IA Math. **29** (1982), 503–566.

[**F4**] ____, *Fractionally logarithmic canonical ring of algebraic surfaces*, J. Fac. Sci. Univ. Tokyo Sect. IA Math. **30** (1984), 685–696.

[**F5**] ____, *Zariski decomposition and canonical rings of elliptic threefolds*, J. Math. Soc. Japan **38** (1986), 19–37.

[**I1**] S. Iitaka, *On D-dimensions of algebraic varieties*, J. Math. Soc. Japan **23** (1971), 356–373.

[**I2**] ____, *Logarithmic forms of algebraic varieties*, J. Fac. Sci. Univ. Tokyo Sect. IA Math. **23** (1976), 525–544.

[**I3**] ____, *On logarithmic Kodaira dimension of algebraic varieties*, Complex Analysis and Algebraic Geometry, Iwanami Shoten, Tokyo, 1977, pp. 175–189.

[**I4**] ____, *On logarithmic K3-surfaces*, Osaka J. Math. **16** (1979), 697–705.

[**I5**] ____, *Algebraic geometry*, Graduate Texts in Math., vol. 76, Springer-Verlag, 1982.

[**I6**] ____, *Minimal model for birational pair*, Ann. of Math. and Stat. **9** (1981), 1–12.

[**I7**] ____, *Basic structure of algebaic varieties (Part 2)*, Adv. Stud. Pure Math., vol. 1, Kinokuniya, North-Holland, 1982, pp. 303–316.

[18] ____ , *On irreducible plane curves*, Saitama Math. J. 1 (1983), 47–63.

[Km] T. Kambayashi, *On Fujita's strong cancellation theorem for the affine plane*, J. Fac. Sci. Univ. Tokyo Sect. IA Math. 27 (1980), 535–548.

[Kw 1] Y. Kawamata, *Addition formula of logarithmic Kodaira dimension for morphisms of relative dimension one*, Proc. Internat. Sympos. on Algebraic Geometry (Kyoto, 1977), Kinokuniya Book Store, 1978, pp. 207–217.

[Kw 2] ____ , *On the classification of non-complete algebraic surfaces*, Algebraic Geometry (Proc. Copenhagen, 1978), Lecture Notes in Math., vol. 732, Springer-Verlag, 1979, pp. 215–232.

[Kw 3] ____ , *Characterization of Abelian varieties*, Compositio Math. 43 (1981), 253–276.

[Kw 4] ____ , *The cone of curves of algebraic varieties*, Ann. of Math. (2) 119 (1984), 603–633.

[My 1] M. Miyanishi, *An algebraic characterization of the affine plane*, J. Math. Kyoto Univ. 15 (1975), 169–184.

[My 2] ____ , *Regular subrings of a polynomial ring*, Osaka J. Math. 17 (1980), 329–338.

[My 3] ____ , *Non-complete algebraic surfaces*, Lecture Notes in Math., vol. 857, Springer-Verlag, 1981.

[My 4] ____ , *On affine-ruled irrational surfaces*, Invent. Math. 70 (1982), 27–43.

[MS] M. Miyanishi and T. Sugie, *Affine surfaces containing cylinder-like open sets*, J. Math. Kyoto Univ. 20 (1980), 11–42.

[Mo] S. Mori, *Threefolds whose canonical bundles are not numerically effective*, Ann. of Math. (2) 116 (1982), 133–176.

[S] F. Sakai, *Kodaira dimensions of complements of divisors*, Complex Analysis and Algebraic Geometry, Iwanami Shoten, Tokyo, 1977, pp. 239–257.

[U] K. Ueno, *Classification theory of algebraic varieties and compact complex spaces*, Lecture Notes in Math., vol. 439, Springer-Verlag, 1975.

[V] E. Viehweg, *Klassificationstheorie algebraischer Varietäten der Dimension drei*, Compositio Math. 41 (1980), 361–400.

[Z] O. Zariski, *The theorem of Riemann-Roch for high multiples of an effective divisor on an algebraic surface*, Ann. of Math. (2) 76 (1962), 560–615.

UNIVERSITY OF TOKYO, JAPAN

Proceedings of Symposia in Pure Mathematics
Volume 46 (1987)

The Zariski Decomposition of Log-Canonical Divisors

YUJIRO KAWAMATA

We shall prove that the existence of the Zariski decomposition of the log-canonical divisor with real coefficients of a variety of log-general type implies the finite generatedness of the log-canonical ring. Roughly speaking, a divisor D on a variety X has the Zariski decomposition if its part defined by global sections of the mD for all $m \in \mathbf{N}$ is nef. It is easy to see that, except in the dimension 2 case as in [Z], we have at least to replace the model X by its blowing-up in order to obtain the Zariski decomposition. An example by Cutkosky [C] shows that it is necessary to consider divisors with real coefficients. The existence of the Zariski decomposition in general is an open question. The following is the main result of this paper. All varieties and morphisms are defined over $\mathbf{C}$ in this paper.

THEOREM 1. *Let $f: X \to S$ be a proper surjective morphism of normal algebraic varieties, let Δ be a $\mathbf{Q}$-divisor on X such that the pair (X, Δ) is log-terminal. Assume that $K_X + \Delta$ is f-big, i.e., $\kappa(X_\eta, K_{X_\eta} + \Delta_\eta) = \dim X_\eta$, where X_η is the generic fiber of f and $\Delta_\eta = \Delta|_{X_\eta}$, and that there exists the Zariski decomposition*

$$K_X + \Delta = P + N \quad \text{in } \operatorname{Div}(X) \otimes \mathbf{R}$$

of $K_X + \Delta$ relative to f. Then the positive part P is f-semiample, i.e., $mP \in \operatorname{Div}(X)$ and the natural homomorphism

$$f^* f_* \mathcal{O}_X(mP) \to \mathcal{O}_X(mP)$$

is surjective for some positive integer m. Thus the relative log-canonical ring

$$R(X/S, K_X + \Delta) = \sum_{m \in \mathbf{N} \cup \{0\}} f_* \mathcal{O}_X([m(K_X + \Delta)])$$

is finitely generated as an $\mathcal{O}_S$-algebra.

The proof is almost parallel to that of [K2, Theorem 3.2]. A similar result was obtained by Moriwaki [M] independently in the case where the coefficients

1980 *Mathematics Subject Classification* (1985 *Revision*). Primary 14J10, 14E30; Secondary 14C20.

in the Zariski decomposition are rational numbers. It seems that Theorem 1 was known to Benveniste [**B**] in case dim $X = 3$. We refer the reader to [**KMM**] for a general reference of the background of this paper. The author would like to thank Dr. Moriwaki for pointing out a mistake in the first version of this paper.

1. Preliminaries. Let X be a normal algebraic variety of dimension d. We denote by $Z_{d-1}(X)$ (resp. $\mathrm{Div}(X)$) the group of Weil (resp. Cartier) divisors on X. The canonical divisor K_X is defined as an element of $Z_{d-1}(X)$. We identify $\mathrm{Div}(X)$ as a subgroup of $Z_{d-1}(X)$. An **R**-*divisor* D is an element of $Z_{d-1}(X) \otimes \mathbf{R}$; i.e., $D = \sum d_j D_j$, where $d_j \in \mathbf{R}$ and the D_j are mutually distinct prime divisors on X. D is called *effective* if $d_j \geq 0$ for all j. D is said to be **R**-*Cartier* if $D \in \mathrm{Div}(X) \otimes \mathbf{R}$. We define the round-up $\ulcorner D \urcorner$, the integral part $[D]$, the fractional part $\{D\}$, and the round off $\langle D \rangle$ by

$$\ulcorner D \urcorner = \sum \ulcorner d_j \urcorner D_j, \quad [D] = \sum [d_j] D_j, \quad \{D\} = \sum \{d_j\} D_j, \quad \langle D \rangle = \sum \langle d_j \rangle D_j,$$

where $\ulcorner r \urcorner, [r]$, and $\langle r \rangle$ for $r \in \mathbf{R}$ are integers such that

$$r > \ulcorner r \urcorner \geq r \geq [r] > r - 1,$$

$$r + \tfrac{1}{2} > \langle r \rangle \geq r - \tfrac{1}{2}, \quad \text{and} \quad \{r\} = r - [r].$$

The *Iitaka dimension* of D is defined by $\kappa(X, D) = \max_{m \in \mathbf{N}} \kappa(X, [mD])$. A pair (X, Δ) for $\Delta \in Z_{d-1}(X) \otimes \mathbf{Q}$ is said to be *log-terminal* if the following conditions are satisfied.

(1) $[\Delta] = 0$ and $K_X + \Delta \in \mathrm{Div}(X) \otimes \mathbf{Q}$.

(2) There is a desingularization $\mu: Y \to X$ such that the union F of the exceptional locus of μ and the inverse image of the support of Δ is a divisor with normal crossings and

$$K_Y = \mu^*(K_X + \Delta) + \sum a_j F_j \quad \text{with } a_j > -1,$$

where the F_j are irreducible components of F.

Let $f: X \to S$ be a proper surjective morphism of normal algebraic varieties. The group of relative 1-cycles $Z_1(X/S)$ is the free abelian group generated by curves on X which are mapped to points on S. The numerical equivalence $\approx$ on $\mathrm{Div}(X) \otimes \mathbf{R}$ is defined according to the intersection pairing $(\, . \,): \mathrm{Div}(X) \times Z_1(X/S) \to \mathbf{Z}$. The closed cone of curves $\overline{NE}(X/S)$ is the closed convex cone generated by effective 1-cycles in the **R**-vector space $N_1(X/S) = Z_1(X/S)/\approx \otimes \mathbf{R}$. $D \in \mathrm{Div}(X) \otimes \mathbf{R}$ is called f-*nef* (resp. f-*ample*) if $(D \, . \, Z) \geq 0$ (resp. if f is projective and $(D \, . \, Z) > 0$) for all $Z \in \overline{NE}(X/S) - \{0\}$. The f-*base locus* $\mathrm{Bs}(X/S, D)$ for a Cartier divisor D on X is defined as the support of the cokernel of the natural homomorphism $f^* f_* \mathcal{O}_X(D) \to \mathcal{O}_X(D)$. D is called f-*free* if $\mathrm{Bs}(X/S, D) = \varnothing$. $D \in \mathrm{Div}(X) \otimes \mathbf{Q}$ is said to be f-*semiample* if mD is f-free for some positive integer m with $mD \in \mathrm{Div}(X)$. In this case D is f-nef. An **R**-Cartier divisor D on X is called f-*big* if $\kappa(X_\eta, D_\eta) = \dim X_\eta$ for the generic fiber X_η of f.

An expression $D = P + N$ of **R**-Cartier divisors $D, P,$ and N is called the *Zariski decomposition* of D relative to f if the following conditions are satisfied:

(0) D is f-big.

(1) P is f-nef.

(2) N is effective.

(3) The natural homomorphisms $f_*\mathcal{O}_X([mP]) \to f_*\mathcal{O}_X([mD])$ are bijective for all $m \in \mathbf{N}$.

P and N are called the positive part and the negative part of D, respectively. The Zariski decomposition is unique if it exists (Proposition 4), because of the condition (0). Hence our definition coincides with the original one by Zariski [**Z**] in the case $\dim X = 2$ and $S = \operatorname{Spec} k$. There is another definition by Fujita [**F**]. If $\mu: Y \to X$ is a proper birational morphism, then $\mu^* D = \mu^* P + \mu^* N$ is also the Zariski decomposition of $\mu^* D$ relative to $f \circ \mu$. The following vanishing theorem is fundamental for the proof of our theorem.

THEOREM 2. *Let $f: X \to S$ be a projective surjective morphism from a nonsingular variety X to a normal variety S, and let D be an f-ample **R**-divisor on X whose fractional part is a divisor with normal crossings. Then*

$$R^i f_* \mathcal{O}_X(\ulcorner D \urcorner + K_X) = 0 \quad \text{for all } i > 0.$$

For the proof we refer the reader to [**V, K1**, or **KMM**]. In fact, there is an f-ample **Q**-divisor D' such that $\ulcorner D'\urcorner = \ulcorner D \urcorner$.

2. Proof of Theorem 1. We begin with the following nonvanishing theorem.

THEOREM 3. *Let X be a nonsingular projective variety, let D and A be **R**-divisors on X, and let $D = \sum d_i D_i$ ($d_i \in \mathbf{R}$) be the decomposition with the irreducible components D_i of D. Assume that D is nef, the support of $\{A\}$ has normal crossings, $\ulcorner A \urcorner \geq 0$, and $t_0 D + A - K_X$ is ample for some positive real number t_0. Then there exist positive real numbers t_1 and ε such that*

$$H^0(X, \langle tD \rangle + \ulcorner A \urcorner) \neq 0$$

for all positive real numbers t satisfying

$$t \geq t_1 \quad \text{and} \quad |\langle td_i \rangle - td_i| < \varepsilon \quad \text{for all } i.$$

PROOF. We proceed by induction on $d = \dim X$. We may assume that $A \in \operatorname{Div}(X) \otimes \mathbf{Q}$. Let b be a positive integer such that $bA \in \operatorname{Div}(X)$. By assumption, there is a positive real number ε_1 such that $tD + \sum x_i D_i + A - K_X$ is ample for an arbitrary real number t satisfying $t \geq t_0$ and $|x_i| < \varepsilon_1$ for all i. We set

$$V_\alpha = \{t \in \mathbf{R}\,;\, t \geq t_0 \text{ and } |\langle td_i \rangle - td_i| < \alpha \text{ for all } i\}$$

for a real number α. We define $P(V) = \chi(X, V + \ulcorner A \urcorner)$ for $V \in \operatorname{Div}(X) \otimes \mathbf{R}$. Then by the vanishing theorem (Theorem 2), we have

$$P(\langle tD \rangle) = H^0(X, \langle tD \rangle + \ulcorner A \urcorner) \quad \text{for } t \in V_{\varepsilon_1}.$$

First, we consider the case where $D \approx 0$. Then $A - K_X$ is ample, and hence $\chi(X, \ulcorner A \urcorner) = H^0(X, \ulcorner A \urcorner) \neq 0$. Since χ is a polynomial function, there exists a real number ε with $0 < \varepsilon < \varepsilon_1$ such that

$$|P(\langle tD \rangle) - \chi(X, \ulcorner A \urcorner)| = |P(\langle tD \rangle - tD) - P(0)| < 1 \quad \text{for } t \in V_\varepsilon.$$

Therefore, the theorem is clear in this case.

Next, let us consider the case where $D \not\approx 0$. By a similar argument as in [S, Lemma 1.3] or [K2, Lemma 2.3], we find a positive number $t_2 \geq t_0$ such that if $t \geq t_2$ and $t \in V_{\varepsilon_1}$, then

$$\dim |bk(\langle tD \rangle + A - K_X)| = \frac{b^d}{d!}(\langle tD \rangle + A - K_X)^d k^d + \text{(lower terms)}$$

$$> \nu k^d \quad \text{with } \nu > \frac{b^d(d+1)^d}{d!} \text{ for } k \gg 0.$$

We fix a point $x \notin \operatorname{Supp} A$ and a positive number t_3 such that $t_3 \geq t_2$ and $t_3 \in V_{\varepsilon_1/(d+1)}$. Then for a large enough integer k, there is a member $M \in |bk(\langle t_3 D \rangle + A - K_X)|$ such that $\operatorname{mult}_x M \geq bk(d+1)$. By using Hironaka's resolution theorem [H], we construct a birational morphism $\mu: Y \to X$ from a nonsingular projective variety Y which satisfies the following conditions:

(a) μ is factorized by the blowing-up $\mu_1: X_1 \to X$ of the point x, there is a divisor $F = \sum F_j$ with normal crossings on Y, and F_1 is the strict transform of the exceptional divisor of μ_1.

(b) $K_Y = \mu^*(K_X - A) + \sum b_j F_j$ with $b_j > -1$ and $b_j \in \mathbf{Q}$.

(c) $\mu^*(\langle t_3 D \rangle + A - K_X) + (d+1)(\mu^*(t_3 D - \langle t_3 D \rangle) - \sum \delta_j F_j)$ is ample for some rational numbers δ_j with $0 < \delta_j \ll 1$.

(d) $\mu^* M = \sum r_j F_j$.

We set

$$c = \min_j \frac{b_j + 1 - \delta_j}{r_j}.$$

By changing the δ_j a little, if necessary, we may assume that the above minimum is attained at the only one value of j, say 0, which may possibly coincide with 1. We set also

$$A' = \sum_{j \neq 0}(-cr_j + b_j - \delta_j)F_j \quad \text{and} \quad B = F_0.$$

By (b), $c > 0$. Since $b_1 = d - 1$ and $r_1 \geq bk(d+1)$, we have $bck < d/(d+1)$. Therefore,

$$(1 - bck)\mu^*(\langle t_3 D \rangle + A - K_X) + \mu^*(t_3 D - \langle t_3 D \rangle) - \sum \delta_j F_j$$

$$\approx A' - B - K_Y + \mu^*(t_3 D)$$

is ample.

Let L be a very ample divisor on X such that the $L + D_i$ are also very ample for all i. Let D_i^1 and D_i^2 be general members of the $|L + D_i|$ and $|L|$, respectively. We set

$$D' = \sum d_i(D_i^1 - D_i^2).$$

Then $\langle tD \rangle \sim \langle tD' \rangle$ and $\mu^*\langle tD' \rangle = \langle t\mu^*D' \rangle$ if $t \in V_{1/2}$. Let us consider the following **Q**-divisor on Y for a real number t.

$$Q_t = \mu^*\langle tD \rangle + A' - B - K_Y$$

$$\approx \mu^*(\langle tD \rangle - \langle t_3D \rangle) + (1 - bck)\mu^*(\langle t_3D \rangle + A - K_X) - \sum \delta_j F_j.$$

By construction, there is a positive real number ε_2 with $0 < \varepsilon_2 < \varepsilon_1$ such that Q_t is ample if $t \in V_{\varepsilon_2}$ and $t \geq t_3$. By Theorem 2,

$$H^1(Y, \mu^*\langle tD \rangle + \ulcorner A' \urcorner - B) = 0 \quad \text{if } t \geq t_3 \text{ and } t \in V_{\varepsilon_2}.$$

Hence the homomorphism

$$H^0(Y, \mu^*\langle tD \rangle + \ulcorner A' \urcorner) \to H^0(B, \mu^*\langle tD \rangle + \ulcorner A' \urcorner)$$

is surjective. We have isomorphisms

$$H^0(B, \mu^*\langle tD \rangle + \ulcorner A' \urcorner) \simeq H^0(B, \mu^*\langle tD' \rangle + \ulcorner A' \urcorner)$$

$$\simeq H^0(B, \langle t\mu^*D' \rangle + \ulcorner A' \urcorner),$$

where the last cohomology group does not vanish if $t \geq t_1$ and $t \in V_\varepsilon$ for some real numbers t_1 and ε such that $t_1 \geq t_3$ and $0 < \varepsilon < \varepsilon_2$ by the induction hypothesis. Thus

$$H^0(Y, \mu^*\langle tD \rangle + \ulcorner A' \urcorner) \neq 0.$$

Since $\mu_*\ulcorner A' \urcorner \leq \ulcorner A \urcorner$, we conclude that

$$H^0(X, \langle tD \rangle + \ulcorner A \urcorner) \neq 0 \quad \text{if } t \geq t_1 \text{ and } t \in V_\varepsilon. \quad \square$$

Now we prove Theorem 1. We may assume that S is affine. Let $\mu: Y \to X$ be the desingularization as in the definition of the log-terminal pair in §1. We set $b_j = \max\{-a_j, 0\}$ and $\Delta_1 = \sum b_j F_j$. Then (Y, Δ_1) is log-terminal and $K_Y + \Delta_1 = \mu^*P + (\mu^*N + \sum(a_j + b_j)F_j)$ is the Zariski decomposition relative to $f \circ \mu$. Thus we may assume from the first that X is nonsingular and f is projective. By definition we have $\kappa(X_\eta, P_\eta) = \kappa(X_\eta, K_{X_\eta} + \Delta_\eta) = \dim X_\eta$, where $P_\eta = P|_{X_\eta}$. Hence there is an ample **R**-divisor H and an effective **Q**-divisor Γ such that $P = H + \Gamma$ by Kodaira's lemma. Then $P - \lambda\Gamma$ is also ample if $0 < \lambda \leq 1$. We take a rational number λ with $0 < \lambda \ll 1$ so that the pair $(X, \Delta + \lambda\Gamma)$ is still log-terminal, and set $A = N - \Delta - \lambda\Gamma$. Thus $tP + A - K_X$ is ample if $t \geq 2$. Let $P = \sum_{i \in I} d_i D_i$ and $A = \sum_{i \in I} a_i D_i$ be the decomposition with irreducible components D_i of P or A. Thus some of the d_i and the a_i may be zero, but since the set $\{D_i\}$ is finite, there is a positive real number ε_1 such that

$$tP + A - K_X + \sum x_i D_i$$

is ample if $t \geq 2$ and $|x_i| < \varepsilon_1$ for all i. We set

$$I' = \{i \in I \,;\, d_i \in \mathbf{Q}\} \quad \text{and} \quad I'' = \{i \in I \,;\, d_i \notin \mathbf{Q}\}.$$

Let b be a positive integer such that $b(K_X + \Delta) \in \text{Div}(X)$ and $bd_i \in \mathbf{Z}$ for all $i \in I'$. In the case where $I'' \neq \emptyset$, we set for a positive real number $\alpha > 0$ and for each $i_1 \in I''$,

$$G_{\alpha, i_1} = \{m \in b\mathbf{N} \,;\, |\langle md_i \rangle - md_i| < \alpha \text{ for all } i \in I'' \text{ and } \langle md_{i_1} \rangle - md_{i_1} > 0\}.$$

We claim here that $G_{\alpha,i_1} \neq \varnothing$. The following proof was communicated by Prof. Kazuya Kato. Let $I'' = \{i_1, \ldots, i_r\}$, and consider the subgroup $L = \mathbf{Z}^r + \mathbf{Z} \cdot b(d_{i_1}, \cdots, d_{i_r})$ of $\mathbf{R}^r$ and its closure $\overline{L}$ with respect to the real topology of $\mathbf{R}^r$. Since the maximal rank of discrete subgroups of $\mathbf{R}^r$ is r, the connected component V of the closed subgroup $\overline{L}$ at 0 is a positive-dimensional linear subspace of $\mathbf{R}^r$. We set $L_0 = \mathbf{Z}^r + \mathbf{N} \cdot b(d_{i_1}, \ldots, d_{i_r})$. Then $L = L_0 \cup (-L_0)$; hence $\overline{L} = \overline{L}_0 \cup (-\overline{L}_0)$. Thus $\overline{L}_0 \cap V$ has interior in V, and $\overline{L}_0 \supset V$. Therefore $G_{\alpha,i_1} \neq \varnothing$.

We set
$$G_\alpha = \begin{cases} \bigcup_{i \in I''} G_{\alpha,i} & \text{if } I'' \neq \varnothing, \\ b\mathbf{N} & \text{otherwise,} \end{cases}$$
and
$$M = \{m \in b\mathbf{N} \,;\, f_* \mathcal{O}_X([mP]) \neq 0\}.$$

By assumption, $M \neq \varnothing$. Since $[mP] \leq \langle mP \rangle \leq \lceil mP \rceil$ for $m \in M$,
$$\mathrm{Bs}(X/S, [mP]) \subset \mathrm{Bs}(X/S, \langle mP \rangle) \subset \mathrm{Bs}(X/S, \lceil mP \rceil),$$
and their differences are supported on the union of the D_i for $i \in I''$ by the definition of the Zariski decomposition. Let $d = \dim X$ and $\beta = \sum \lceil a_i \rceil + d$, and let us take an $m_1 \in M \cap G_{\varepsilon_1/\beta}$. Then $mP - x\langle m_1 P \rangle + A - K_X$ is ample if $m \geq \beta m_1 + 2$ and $0 \leq x \leq \beta$. By the Hironaka resolution, we can construct a projective birational morphism $\mu: Y \to X$ from a nonsingular algebraic variety Y which satisfies the following conditions:

(a) There is a divisor $F = \sum F_j$ with normal crossings on Y.

(b) $K_Y = \mu^*(K_X - A) + \sum b_j F_j$ with $b_j \in \mathbf{R}$ and $b_j > -1$.

(c) $\mu^*(mP - x\langle m_1 P \rangle + A - K_X) - \sum \delta_j F_j$ is ample for some $\delta_j \in \mathbf{Q}$ with $0 < \delta_j \ll 1$ and for all $m \geq \beta m_1 + 2$ and $0 \leq x \leq \beta$.

(d) The image of the natural homomorphism
$$\mu^* f^* f_* \mathcal{O}_X(\langle m_1 P \rangle) \to \mu^* \mathcal{O}_X(\langle m_1 P \rangle)$$
is an invertible sheaf, say $\mathcal{O}_Y(L)$, and
$$\mu^*(\langle m_1 P \rangle) = L + \sum r_j F_j$$
for some nonnegative integers r_j.

We note that $\mu(\bigcup_{r_j \neq 0} F_j) = \mathrm{Bs}(X/S, \langle m_1 P \rangle)$. In particular, if $r_j = 0$ for all j, then $\langle m_1 P \rangle$ is f-free, and hence $I'' = \varnothing$ and $\langle m_1 P \rangle = m_1 P$. Thus we assume that some of the r_j are not zero in the following. We set
$$c = \min_j \frac{b_j + 1 - \delta_j}{r_j}.$$

We may assume that the minimum is attained at $j = 0$ only. Since there is an F_1 with $r_1 \geq 1$ and $b_1 \leq \beta - 1$, we have
$$0 < c < \beta.$$

Hence there is a real number ε_2 with $0 < \varepsilon_2 < \varepsilon_1/\beta$ such that
$$\mu^* \left(mP - c\langle m_1 P \rangle + A - K_X + \sum x_i D_i \right) - \sum \delta_j F_j$$

is ample if $m \geq \beta m_1 + 2$ and $|x_i| < \varepsilon_2$ for all i. We set

$$A' = \sum_{j \neq 0}(-cr_j + b_j - \delta_j)F_j \quad \text{and} \quad B = F_0.$$

We consider the following $\mathbf{R}$-divisor for an integer m.

$$Q_m = \mu^*\langle mP \rangle + A' - B - K_Y$$

$$\approx cL + \mu^*(\langle mP \rangle - c\langle m_1 P \rangle + A - K_X) - \sum \delta_j F_j.$$

If $m \geq \beta m_1 + 2$ and $m \in G_{\varepsilon_2}$, then Q_m is ample, and hence

$$R^1(f \circ \mu)_* \mathcal{O}_Y(\mu^*\langle mP \rangle + \ulcorner A' \urcorner - B) = 0,$$

by Theorem 2. Thus we have a surjective homomorphism

$$(f \circ \mu)_* \mathcal{O}_Y(\mu^*\langle mP \rangle + \ulcorner A' \urcorner) \to (f \circ \mu)_* \mathcal{O}_B(\mu^*\langle mP \rangle + \ulcorner A' \urcorner).$$

Since $0 \leq \mu_* \ulcorner A' \urcorner \leq \ulcorner A \urcorner \leq \ulcorner N \urcorner$, there are injective homomorphisms

$$f_* \mathcal{O}_X(\langle mP \rangle) \to (f \circ \mu)_* \mathcal{O}_Y(\mu^*\langle mP \rangle + \ulcorner A' \urcorner) \to f_* \mathcal{O}_X(\langle mP \rangle + \ulcorner A \urcorner),$$

which are isomorphisms by the definition of the Zariski decomposition if m is large enough. As in the proof of Theorem 3, we take very ample divisors D_i^1 and D_i^2 so that, if we put $P' = \sum d_i(D_i^1 - D_i^2)$, then $\langle mP \rangle \sim \langle mP' \rangle$ and $\mu^*\langle mP' \rangle = \langle m\mu^*P' \rangle$ for $m \in G_{1/2}$. Then we have isomorphisms

$$(f \circ \mu)_* \mathcal{O}_B(\mu^*\langle mP \rangle + \ulcorner A' \urcorner) \simeq (f \circ \mu)_* \mathcal{O}_B(\mu^*\langle mP' \rangle + \ulcorner A' \urcorner)$$

$$\simeq (f \circ \mu)_* \mathcal{O}_B(\langle m\mu^*P' \rangle + \ulcorner A' \urcorner).$$

By applying Theorem 3 to the generic fiber of the morphism $f \circ \mu \colon B \to (f \circ \mu)(B)$, we find real numbers t_1 and ε_3 with $t_1 \geq \beta m_1 + 2$ and $0 < \varepsilon_3 < \varepsilon_2$ such that

$$(f \circ \mu)_* \mathcal{O}_B(\langle m\mu^*P' \rangle + \ulcorner A' \urcorner) \neq 0 \quad \text{if } m \geq t_1 \text{ and } m \in G_{\varepsilon_3}.$$

If $I'' \neq \varnothing$, we take a positive integral multiple $m_{2,i}$ of m_1 such that $m_{2,i} \geq t_1$ and $m_{2,i} \in G_{\varepsilon_3,i}$ for each $i \in I''$, and pick a common multiple m_2 of the $m_{2,i}$ such that $m_2 \in G_{\varepsilon_3}$. If $I'' = \varnothing$, we take a positive integral multiple m_2 of m_1 such that $m_2 \geq t_1$. The above argument shows that $\mu(B)$ is not contained in any of the $\mathrm{Bs}(X/S, \langle m_{2,i}P \rangle)$ and hence not in any of the D_i for $i \in I''$. Thus $\mu(B)$ is not contained in $\mathrm{Bs}(X/S, \ulcorner m_2 P \urcorner)$. On the other hand, $\mathrm{Bs}(X/S, \ulcorner m_1 P \urcorner) \supset \mathrm{Bs}(X/S, \ulcorner m_2 P \urcorner)$ and $\mu(B) \subset \mathrm{Bs}(X/S, \ulcorner m_1 P \urcorner)$. Therefore, $\mathrm{Bs}(X/S, \ulcorner m_1 P \urcorner) \supsetneqq \mathrm{Bs}(X/S, \ulcorner m_2 P \urcorner)$. Repeating this process, we obtain a sequence of positive integers $m_1, m_2, m_3, \ldots$ such that

$$\mathrm{Bs}(X/S, \ulcorner m_1 P \urcorner) \supsetneqq \mathrm{Bs}(X/S, \ulcorner m_2 P \urcorner) \supsetneqq \mathrm{Bs}(X/S, \ulcorner m_3 P \urcorner) \supsetneqq \cdots.$$

By the noetherian induction, we finally find a positive integer m such that $\mathrm{Bs}(X/S, \ulcorner mP \urcorner) = \varnothing$. By construction, $\ulcorner mP \urcorner \geq mP$, and hence $\ulcorner mP \urcorner = mP$. This completes the proof of Theorem 1. $\quad\square$

3. Remarks. The uniqueness of the Zariski decomposition of a relatively big divisor is a consequence of the fact that a nef and big divisor is eventually almost generated as follows.

PROPOSITION 4. *Let $f: X \to S$ be a proper surjective morphism of normal algebraic varieties and let D be an $\mathbf{R}$-Cartier divisor on X which is f-big. Then the Zariski decomposition of D is unique if it exists.*

PROOF. We may assume that S is affine and f is projective. By Kodaira's lemma, there is an ample divisor H on X and a positive integer k such that $f_*\mathcal{O}_X([kD] - H) \neq 0$. Thus we can write $D = P_0 + N_0$ for an ample $\mathbf{Q}$-Cartier divisor P_0 and an effective $\mathbf{R}$-Cartier divisor N_0 on X. Let $D = P+N = P'+N'$ be two Zariski decompositions of D relative to f. We write $P = \sum r_j D_j$ and $P_0 = \sum s_j D_j$ with mutually distinct prime divisors D_j. m being a positive integer, since $mP + P_0$ is ample, there is an ample $\mathbf{Q}$-Cartier divisor $P_m = \sum r_j(m)D_j$ such that

$$0 \leq mr_j + s_j - r_j(m) < 1/m.$$

Since P_m is f-semiample, there is a positive integer k_m such that $k_m P_m \leq [k_m(m + 1)P']$. By taking limit $m \to \infty$, we obtain $P \leq P'$, and hence the unicity. □

The Zariski decomposition is easy in the toric geometry (cf. [**TE**]). In the following proposition, we consider Zariski decompositions of not necessarily f-big divisors; we only drop the condition (0) from the definition in §1.

PROPOSITION 5. *Let $f: X \to S$ be a proper surjective toric morphism of toric varieties, and let D be an effective Cartier divisor on X. Then there is a proper birational toric morphism $\mu: Y \to X$ such that μ^*D has a Zariski decomposition relative to $f \circ \mu$ whose positive part is $f \circ \mu$-semiample.*

PROOF. Let $g: T \to T'$ be the homomorphism of tori corresponding to f. By using the action of a 1-parameter subgroup of T, we find a T-invariant effective Cartier divisor D' on X which is linearly equivalent to D. Thus we may assume that D is T-invariant. Let M (resp. M') and N (resp. N') be the groups of characters and of 1-parameter subgroups of T (resp. T'), respectively, $g_*: N_\mathbf{R} \to N'_\mathbf{R}$ the homomorphism induced by g, and let $\{\sigma_\alpha\}$ (resp. $\{\tau_\beta\}$) be the f.r.p.p. decomposition of $N_\mathbf{R}$ (resp. $N'_\mathbf{R}$) corresponding to X (resp. S). We set $|X| = \bigcup \sigma_\alpha$, $|S| = \bigcup \tau_\beta$, $I_\beta = \{\alpha; g_*(\sigma_\alpha) \subset \tau_\beta\}$, and $|X_\beta| = \bigcup_{\alpha \in I_\beta} \sigma_\alpha$. Since f is proper, $g_*^{-1}(|S|) = |X|$. Let $h: |X| \to \mathbf{R}$ be the function corresponding to $\mathcal{O}_X(D)$ by [**TE**, Theorem 9 on p. 28]. We take the convex hull of h relative to g_* in the following sense: we define a function $h^+: |X| \to \mathbf{R}$ by

$$h^+(x) = \inf\{r(x)\, ; r \in M_\mathbf{Q} \text{ and } r \geq h \text{ on } |X_\beta|\} \quad \text{for } x \in |X_\beta|.$$

We also define functions $h_m: |X| \to \mathbf{R}$ for positive integers m by

$$h_m(x) = \inf\{r(x)\, ; r \in M \text{ and } r \geq mh \text{ on } |X_\beta|\} \quad \text{for } x \in |X_\beta|.$$

Since D is effective, we have $h \leq 0$, and hence these functions are well defined. It is easy to check the following facts.

(1) The h_m satisfy conditions (*) of [**TE**, p. 27].

(2) $mh^+ \leq h_m$ for all m and the equality holds for a positive integer $m = m_0$. Let $\{\rho_\gamma(m)\}_{\gamma \in \Gamma(m)}$ be f.r.p.p. subdivisions of $\{\sigma_\alpha\}$ such that the h_m are linear on each $\rho_\gamma(m)$ for $\gamma \in \Gamma(m)$, and let $\mu_m \colon Y_m \to X$ be the corresponding proper birational morphisms. Then the Cartier divisors D_m corresponding to the h_m on the Y_m are $f \circ \mu_m$-free and the images of the natural homomorphisms $\mu_m^* f^* f_* \mathcal{O}_X(mD) \to \mu_m^* \mathcal{O}_X(mD)$ coincides with the $\mathcal{O}_{Y_m}(D_m)$ by [**TE**, Lemma on p. 47]. We note here that the inequalities $mh \leq h_m$ imply that the divisors $E_m = m\mu_m^* D - D_m$ on Y_m are effective. We set $Y = Y_{m_0}$, $\mu = \mu_{m_0}$, $P = D_{m_0}/m_0$, and $N = E_{m_0}/m_0$. Since $mh^+ \leq h_m$, the image of the homomorphism $\mu^* f^* f_* \mathcal{O}_X(mD) \to \mu^* \mathcal{O}_X(mD)$ is contained in $\mathcal{O}_X([mP])$ for an arbitrary $m \in \mathbf{N}$. Therefore, $\mu^* D = P + N$ gives the desired Zariski decomposition. $\square$

We note that a T-invariant $\mathbf{R}$-Cartier divisor A on X is f-nef if and only if the corresponding function $h_A \colon |X| \to \mathbf{R}$ is convex on each $|X_\beta|$. Thus our positive part P obtained above is the maximal f-nef part of D as well as the part given by relative global sections of the mD.

REFERENCES

[**B**] X. Benveniste, *Variétés de type général de dimension 3*, Thesis, Univ. Paris Sud, 1984.

[**C**] S. D. Cutkosky, *Zariski decomposition of divisors on algebraic varieties*, Duke Math. J. **53** (1986), 149–156.

[**F**] T. Fujita, *Zariski decomposition and canonical rings of elliptic threefolds*, J. Math. Soc. Japan **38** (1986), 19–37.

[**H**] H. Hironaka, *Resolution of singularities of an algebraic variety over a field of characteristic zero*, Ann. of Math. (2) **79** (1964), 109–326.

[**K1**] Y. Kawamata, *A generalization of Kodaira-Ramanujam's vanishing theorem*, Math. Ann. **261** (1982), 43–46.

[**K2**] _____, *The cone of curves of algebraic varieties*, Ann. of Math. (2) **119** (1984), 603–633.

[**KMM**] Y. Kawamata, M. Matsuda, and K. Matsuki, *Introduction to the minimal model problem*, Adv. Stud. Pure Math. (to appear).

[**M**] A. Moriwaki, *Semi-ampleness of the numerically effective part of Zariski decomposition*, J. Math. Kyoto Univ. **26** (1986), 465–481.

[**S**] V. V. Shokurov, *The non-vanishing theorem*, Izv. Akad. Nauk SSSR Ser. Mat. **49** (1985), 635–651; Engl. transl. in Math. USSR Izv. **26** (1986), 591–604.

[**TE**] G. Kempf, F. Knudsen, D. Mumford, and B. Saint-Donat, *Toroidal embeddings.* I, Lecture Notes in Math., vol. 339, Springer-Verlag, 1983.

[**V**] E. Viehweg, *Vanishing theorems*, J. Reine Angew. Math. **335** (1982), 1–8.

[**Z**] O. Zariski, *The theorem of Riemann-Roch for high multiples of an effective divisor on an algebraic surface*, Ann. of Math. (2) **76** (1962), 560–615.

UNIVERSITY OF TOKYO, JAPAN

Proceedings of Symposia in Pure Mathematics
Volume **46** (1987)

Open Algebraic Surfaces with Kodaira Dimension $-\infty$

M. MIYANISHI AND S. TSUNODA

1. Introduction. Let k be an algebraically closed field of characteristic p. Let X be a nonsingular algebraic surface defined over k. Then X is embedded as a Zariski open set into a nonsingular projective surface V in such a way that the complement $V - X$ is a reduced effective divisor with simple normal crossings. We set $D = V - X$. A triple (V, D, X) is called *a smooth (normal) completion of* X. The logarithmic Kodaira dimension $\bar{\kappa}(X)$ is defined as the $(D + K_V)$-dimension of V (cf. Fujita's lecture), and $\bar{\kappa}(X)$ is independent of the choice of a smooth normal completion of X. We shall discuss the structure of an open algebraic surface X with (logarithmic) Kodaira dimension $-\infty$. We say that X is *affine-ruled* if X contains a Zariski open set U which is isomorphic to $\mathbf{A}_k^1 \times U_0$, where U_0 is a curve.

In order to investigate such open surfaces X, we might as well start with a pair (V, D) of a nonsingular projective surface V and an effective reduced divisor with simple normal crossings. We then develop a theory of peeling to obtain an almost minimal model $(\tilde{V}, \tilde{D})$ of the pair (V, D). Then, if one assumes that the pair (V, D) is almost minimal, we have one of the following three cases:

(i) $X = V - D$ is affine-ruled;

(ii) the pair (V, D) is a logarithmic del Pezzo surface of rank 1 with noncontractible boundary;

(iii) the pair (V, D) is a logarithmic del Pezzo surface of rank 1 with contractible boundary.

Each of the above three cases will be discussed in detail. In the last section, we included several results concerning étale endomorphisms of an affine algebraic surface which might be related to the Jacobian conjecture. §§1–4 are based on the papers [**18, 19**].

2. Theory of peeling. Let (V, D) be a pair of nonsingular projective surface and a reduced effective divisor with simple normal crossings. Write $D = D_1 + \cdots + D_u$, where the D_i's are irreducible components of D. Let $\Gamma(D)$ be the dual

---

1980 *Mathematics Subject Classification* (1985 *Revision*). Primary 14J10, 14E30, 14J17.

Key words and phrases. Logarithmic Kodaira dimension, affine-ruled surface, del Pezzo surface, $\mathbf{C}^*$-fiber space, quotient singular point.

graph of D, which is defined in the usual manner. Let $L = D_1 + \ldots + D_r$ be a partial sum of irreducible components of D after a suitable relabeling of D_i's, and let $\{v_1, \ldots, v_r\}$ be the corresponding subgraph of $\Gamma(D)$. Define the branching number $\beta_{\Gamma(D)}(v)$ for a vertex v of $\Gamma(D)$ as the number of edges connecting v to other vertices. Suppose L satisfies a condition: $\beta_{\Gamma(D)}(v_1) = 1$, $\beta_{\Gamma(D)}(v_i) = 2$, and $(D_{i-1} \cdot D_i) = (D_i \cdot D_{i+1}) = 1$ for $2 \le i \le r - 1$. If $\beta_{\Gamma(D)}(v_r) = 1$, we call L a *rod*. If $\beta_{\Gamma(D)}(v_r) = 2$ and $(D_r \cdot D_{r+1}) = 1$ for a component D_{r+1} other than D_{r-1}, L is called a *twig*; L is a *maximal twig* if, furthermore, $\beta_{\Gamma(D)}(v_{r+1}) \ge 3$, v_{r+1} being the vertex corresponding to D_{r+1}. In general, L is called a *linear chain*. The linear chain L is *rational* (resp. *admissible*) if every component D_i is a nonsingular rational curve (resp. if $(D_i^2) \le -2$ for $1 \le i \le r$).

Let L be an admissible rational linear chain. Write L as $L = D_1 + \cdots + D_r$. Let $D' = D - L$. Noting that the intersection matrix of L, i.e., $((D_i \cdot D_j))_{1 \le i,j \le r}$, is negative-definite, we can determine rational numbers α_i $(1 \le i \le r)$ by the following condition:

$$(D' + \sum_{i=1}^{r} \alpha_i D_i + K_V \cdot D_j) = 0 \quad \text{for } 1 \le j \le r.$$

Then $0 \le \alpha_i < 1$ for every i, and $\alpha_i = 0$ for some i iff L is a rational rod

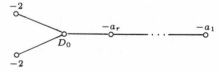

Hence $\alpha_i = 0$ for some i iff $\alpha_i = 0$ for every i. The **Q**-divisor $\mathrm{Bk}(L) = \sum_{i=1}^{r}(1 - \alpha_i)D_i$ is called *the bark of L*, and the process of subtracting $\mathrm{Bk}(L)$ from L is called *the peeling of L*.

Let $\{T_\lambda\}$ and $\{R_\mu\}$ be the sets of all admissible rational maximal twigs and all admissible rational rods, respectively. Noting that T_λ's and R_μ's are disjoint from each other, we can peel the barks of T_λ's and R_μ's independently. Let $D^* = D - \sum_\lambda \mathrm{Bk}(T_\lambda) - \sum_\mu \mathrm{Bk}(R_\mu)$, which is an effective **Q**-divisor.

A connected component F of the divisor D is called *a rational fork* if every irreducible component of F is a nonsingular rational curve and the corresponding subgraph $\Gamma(F)$ of $\Gamma(D)$ is one of the following.
(Type D)

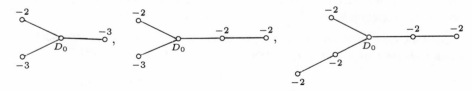

where $a_i \ge 2$ for $1 \le i \le r$,
(Type E_6)

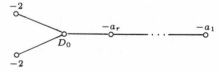

(Type E_7)

(Type E_8)

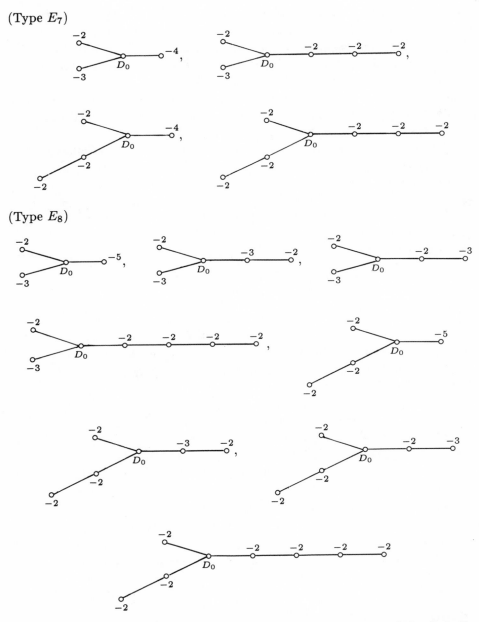

The component denoted by D_0 is called *the central component* of the fork F. The fork F is *admissible* if the intersection matrix is negative-definite, which is equivalent to saying that $(D_0^2) \leq -2$. Then we have the following:

LEMMA 1. *Let V, D, and D^* be as above. Suppose that there exists an irreducible component D_0 of D such that*

$$(D^* + K_V \cdot D_0) < 0.$$

Then the following assertions hold true:

(1) D_0 is a nonsingular rational curve;

(2) D_0 is one of the following components of D :

(i) *With an admissible rational twig* T, *which might be empty,* D_0 *forms a rational twig* $D_0 + T$;

(ii) *With admissible rational twigs* T *and* T', *which might be empty,* D_0 *forms a rational rod* $T + D_0 + T'$;

(iii) D_0 *is the central component of a rational fork* F.

Moreover, if $(D_0^2) \leq -2$, *then* D_0 *is the central component of an admissible rational fork.*

The component D in the cases (i) and (ii) or in the case (iii) with $(D_0^2) \geq -1$ is called an *irrelevant component* of a nonadmissible rational twig, rod, or fork, respectively.

Let F be an admissible rational fork, and write $F = D_1 + \cdots + D_s$. Noting that the intersection matrix of F is negative-definite, determine rational numbers $\beta_1, \ldots, \beta_s$ by the condition

$$(D - F + \sum_{i=1}^{s} \beta_i D_i + K_V \cdot D_j) = 0 \quad \text{for } 1 \leq j \leq s.$$

Then $0 \leq \beta_i < 1$ for $1 \leq i \leq s$, and $\beta_i = 0$ for some i iff every irreducible component of F has self-intersection number -2, and moreover, $\beta_i = 0$ for some i iff $\beta_i = 0$ for every i. Let $\mathrm{Bk}(F) = \sum_{i=1}^{s}(1 - \beta_i)D_i$, and call it the bark of F. If T_i $(i = 1, 2, 3)$ are three maximal twigs of the fork F, we know that $\mathrm{Bk}(F) \geq \mathrm{Bk}(T_i)$, $i = 1, 2, 3$. Let $\{F_\nu\}$ be the set of all admissible rational forks of F. We set anew $\{T_\lambda\}$ the set of all admissible rational maximal twigs which are not contained in any admissible rational forks. We set

$$D^\# = D - \sum_\lambda \mathrm{Bk}(T_\lambda) - \sum_\mu \mathrm{Bk}(R_\mu) - \sum_\nu \mathrm{Bk}(F_\nu).$$

The **Q**-divisor $D^\#$ is called *the stripped form* of D, and $\mathrm{Bk}(D) := D - D^\#$ is called *the bark of* D. By the above construction, we know that:

(1) $D^\# = D - \mathrm{Bk}(D)$ is an effective **Q**-divisor such that $\mathrm{Supp}(D)$ and $\mathrm{Supp}(D^\#)$ differ by a disjoint union of all rational rods and all rational forks whose irreducible components have all self-intersection numbers -2;

(2) $\mathrm{Bk}(D)$ has the negative-definite intersection matrix;

(3) $(D^\# + K_V \cdot Z) = 0$ for every irreducible component of T_λ's, R_μ's, and F_ν's;

(4) $(D^\# + K_V \cdot Y) \geq 0$ for every irreducible component Y of D except the irrelevant components of nonadmissible rational twigs, nonadmissible rational rods, and nonadmissible rational forks.

Suppose there exists an irreducible curve E on V such that $(D^\# + K_V \cdot E) < 0$, E is not a component of D, and the intersection matrix of $E + \mathrm{Bk}(D)$ is negative-definite. Then E is an exceptional curve of the first kind, and E meets at most two irreducible components of $\mathrm{Bk}(D)$. If E meets two components Z_1

and Z_2, then the connected components of $\mathrm{Bk}(D)$ which contain Z_i ($i = 1, 2$), the position of Z_i in the connected component, and the self-intersection number of Z_i are strongly restricted. Let $f: V \to \overline{V}$ be the composite of the contraction of E and the contractions of all possible exceptional components of $\mathrm{Bk}(D)$ (i.e., irreducible components of $\mathrm{Bk}(D)$ which become exceptional curves of the first kind after a succession of several contractions). Let $\overline{D} = f_* D$. Then we know the following:

LEMMA 2. *Let $D, f,$ and $\overline{D}$ be as above. Then the following assertions hold true:*

(1) *$\overline{D}$ is a reduced effective divisor with simple normal crossings;*

(2) *Each connected component of $f(\mathrm{Supp}(\mathrm{Bk}(D)))$ is an admissible rational twig, an admissible rational rod, or an admissible rational fork;*

(3) *$f_*(\mathrm{Bk}(D)) \le \mathrm{Bk}(\overline{D})$.*

On the other hand, suppose there exists an irreducible component E of D such that E is an exceptional curve of the first kind, $\beta_D(E)$ ($:= \beta_{\Gamma(D)}(v_E)) \le 2$, $(D^\# + K_V \cdot E) < 0$, and the intersection matrix of $E + \mathrm{Bk}(D)$ is negative-definite. Such a component E is called *a superfluous exceptional component* of D. By virtue of Lemma 1, such a component E appears only in one of the following situations:

(i) E is an isolated component;

(ii) there exists an admissible rational twig T such that T is a connected component of $\mathrm{Bk}(D)$ and $T + E$ is a twig of D, where T might be empty;

(iii) there exists admissible rational twigs T and T' such that they are connected components of $\mathrm{Bk}(D)$ and $T + E + T'$ is a rod of D, where one (or both) of T and T' might be empty.

Let $f: V \to \overline{V}$ be the contraction of E and all (subsequently) exceptional components with branching number ≤ 2 of $E + \mathrm{Bk}(D)$, i.e., irreducible components of $E + \mathrm{Bk}(D)$ which become exceptional curves of the first kind after a succession of several contractions. Let $\overline{D} = f_* D$. Then $\overline{D}$ is apparently a reduced effective divisor with simple normal crossings, and we have $f_* \mathrm{Bk}(D) \le \mathrm{Bk}(\overline{D})$. The divisor $\overline{D}$ might have superfluous exceptional components. We perform the contractions of the above kind as long as there are superfluous exceptional components. If there are no superfluous exceptional components in D, one of the following cases takes place for any irreducible curve C on V:

(i) $(D^\# + K_V \cdot C) \ge 0$;

(ii) C is an exceptional curve of the first kind such that $C \not\subset \mathrm{Supp}(D)$, $(D^\# + K_V \cdot C) < 0$ and intersection matrix of $C + \mathrm{Bk}(D)$ is negative definite;

(iii) $(D^\# + K_V \cdot C) < 0$ and the intersection matrix of $C + \mathrm{Bk}(D)$ is not negative definite.

For a $\mathbf{Q}$-divisor $Z = \sum_{i=1}^n a_i C_i$, C_i being an irreducible component, we define the integral part $[Z]$ of Z by $[Z] = \sum_{i=1}^n [a_i] C_i$, where $[a_i]$ is the Gauss symbol.

Then we have the following:

LEMMA 3. (1) $h^0(V, n(D + K_V)) = h^0(V, [n(D^\# + K_V)])$ *for every integer* $n \geq 0$.

(2) *Let* $f: V \to \overline{V}$ *be the contraction of one of the above two kinds. Then* $h^0(V, [n(D^\# + K_V)]) = h^0(\overline{V}, [n(\overline{D}^\# + K_{\overline{V}})])$ *for every integer* $n \geq 0$.

The above discussions imply the following:

THEOREM 4. *Let* (V, D) *be a pair of a nonsingular projective surface defined over* k *and a reduced effective divisor with simple normal crossings. Then there exists a birational morphism* $\mu: V \to \tilde{V}$ *onto a nonsingular projective surface* $\tilde{V}$ *such that, with* $\tilde{D} = \mu_* D$, *the following conditions are satisfied:*

(1) $h^0(V, n(D + K_V)) = h^0(\tilde{V}, n(\tilde{D} + K_{\tilde{V}}))$ *for every integer* $n \geq 0$.

(2) $\mu_* \mathrm{Bk}(D) \leq \mathrm{Bk}(\tilde{D})$ *and* $\mu_*(D^\# + K_V) \geq \tilde{D}^\# + K_{\tilde{V}}$.

(3) *For every irreducible curve* C *on* $\tilde{V}$, *we have either* $(\tilde{D}^\# + K_{\tilde{V}} \cdot C) \geq 0$ *or* $(\tilde{D}^\# + K_{\tilde{V}} \cdot C) < 0$ *and the intersection matrix of* $C + \mathrm{Bk}(\tilde{D})$ *is not negative-definite.*

3. Almost minimal models. A pair (V, D) is called *almost minimal* if, for every irreducible curve C on V, either $(D^\# + K_V \cdot C) \geq 0$ or $(D^\# + K_V \cdot C) < 0$ and the intersection matrix of $C + \mathrm{Bk}(D)$ is not negative-definite. If (V, D) is not necessarily almost minimal, the pair $(\tilde{V}, \tilde{D})$ constructed in Theorem 4 is called *an almost minimal model* of (V, D). In terms of open algebraic surfaces, a nonsingular algebraic surface X is called almost minimal if there exists a smooth normal completion (V, D, X) such that the pair (V, D) is almost minimal. Theorem 4 implies the following:

LEMMA 5. *Let* X *be a nonsingular algebraic surface defined over* k. *Then there exist an almost minimal nonsingular algebraic surface* $\tilde{X}$, *Zariski open sets* U *and* $\tilde{U}$ *of* X *and* $\tilde{X}$, *respectively and a proper birational morphism* $\mu: U \to \tilde{U}$ *such that:*

(i) $\bar{\kappa}(\tilde{X}) = \bar{\kappa}(X)$; (ii) *either* $U = X$ *or* $X - U$ *has pure codimension 1 in* X; (iii) $\mathrm{codim}_{\tilde{X}}(\tilde{X} - \tilde{U}) \geq 2$. *If* $\tilde{X}$ *is affine-ruled, so is* X.

PROOF. Let (V, D, X) be a smooth normal completion of X. Then there exists a birational morphism $\mu: V \to \tilde{V}$ onto a nonsingular projective surface $\tilde{V}$ such that, with $\tilde{D} = \mu_* D$, which is a reduced effective divisor with simple normal crossings, $(\tilde{V}, \tilde{D})$ is almost minimal. Let $\tilde{X} = \tilde{V} - \tilde{D}$. Then $\tilde{X}$ is almost minimal. Note that $\tilde{S} = \mu(\mathrm{Supp}(D)) - \mathrm{Supp}(\tilde{D})$ consists of finitely many points. Let $\tilde{U} = \tilde{X} - \tilde{S}$ and $U = \mu^{-1}(\tilde{U})$. Then U and $\tilde{U}$ are required Zariski open sets in X and $\tilde{X}$, respectively.

We assume that a pair (V, D) is almost minimal. Note that the intersection matrix of $\mathrm{Bk}(D)$ is negative-definite, that $\mathrm{Bk}(D)$ contains no exceptional curves of the first kind, and that every connected component Γ of $\mathrm{Bk}(D)$ is an admissible rational twig, an admissible rational rod, or an admissible rational fork. If

char $k = 0$, Γ is identified with the set of exceptional curves of a minimal resolution of a quotient singularity (hence a rational singularity): a cyclic quotient
singularity if Γ is either an admissible rational twig or an admissible rational
rod, and a quotient singularity $(\mathbf{A}_k^2/G, 0)$ with respect to a *small* finite subgroup
G of $\mathrm{GL}(2, k)$ if Γ is an admissible rational fork (cf. Brieskorn [**3**]). Even if
char $k > 0$, Γ contracts down to a rational singular point by virtue of Artin [**2**],
and Γ is identified with the set of exceptional curves of a minimal resolution
of this singularity. Thus, even in case char $k > 0$, we call this singularity a
"quotient singularity." Let $f: V \to \overline{V}$ be anew the contraction of all connected
components of $\mathrm{Bk}(D)$ to quotient singular points. Hence $\overline{V}$ is a normal projective surface defined over k with at worst quotient singularities. Then it is
well known that there exists an integer $N > 0$, determined by the singularities
of $\overline{V}$, such that, for every Weil divisor $\overline{A}$ on $\overline{V}$, $N\overline{A}$ is linearly equivalent to a
Cartier divisor. Hence, for Weil divisors $\overline{A}$ and $\overline{B}$ on $\overline{V}$, the intersection number
$(\overline{A} \cdot \overline{B})$ is defined as $(1/N^2)(N\overline{A} \cdot N\overline{B})$. Contrary to the intersection theory on
a nonsingular projective surface, the intersection number $(\overline{A} \cdot \overline{B})$ is a number
in $(1/N^2)\mathbf{Z}$. Then we can define the numerical equivalence, denoted by $\equiv$, in
the group $\mathrm{Pic}(\overline{V})_\mathbf{Q} = \mathrm{Pic}(\overline{V}) \otimes_\mathbf{Z} \mathbf{Q}$, where $\mathrm{Pic}(\overline{V})$ is the group of isomorphism
classes of Cartier divisors on $\overline{V}$. We denote by $NS(\overline{V})_\mathbf{Q}$ the group $\mathrm{Pic}(\overline{V})_\mathbf{Q}/(\equiv)$.
The canonical divisor $K_{\overline{V}}$ on $\overline{V}$ is defined as the unique Weil divisor such that
$K_{\overline{V}} \cap \overline{V}_0 = K_{\overline{V}_0}$, where $\overline{V}_0 = \overline{V} - \mathrm{Sing}(\overline{V})$. Let $NS(\overline{V})_\mathbf{Q} = NS(\overline{V}) \otimes_\mathbf{Z} \mathbf{Q}$. Then
we have $f^*: NS(\overline{V})_\mathbf{Q} \to NS(V)_\mathbf{Q}$, which is defined by $f^*(\overline{A}) = f^*(N\overline{A})/N$. Let
$\overline{D} = \mu_*[D^\#]$. Then we have:

LEMMA 6. (1) $K_{\overline{V}} \sim f_*(K_V)$. (2) $D^\# + K_V \equiv f^*(\overline{D} + K_{\overline{V}})$ in $NS(V)_\mathbf{Q}$.
Moreover, we have: (3) *For any irreducible curve $\overline{C}$ on $\overline{V}$, either $(\overline{D} + K_{\overline{V}} \cdot \overline{C}) \geq 0$
or $(\overline{C}^2) \geq 0$.*

On the other hand, we know the following:

LEMMA 7 (CF. TSUNODA [**24**]). *Let V be a nonsingular projective surface defined over k and let D be a reduced effective divisor with simple normal
crossings.*
(1) *Suppose $D^\# + K_V$ is numerically effective, i.e., $(D^\# + K_V \cdot C) \geq 0$ for
every irreducible curve C. Then $\bar{\kappa}(V - D) \geq 0$.*
(2) *Suppose (V, D) is almost minimal and $\bar{\kappa}(V - D) \geq 0$. Then $D^\# + K_V$ is
numerically effective and it is equal to the numerically effective part $(D + K_V)^+$
of $D + K_V$ in the Zariski decomposition*

$$D + K_V = (D + K_V)^+ + (D + K_V)^-.$$

We assume hereafter that a pair (V, D) is almost minimal and $D^\# + K_V$ is
not numerically effective. Passing to a normal projective surface $\overline{V}$, we know
that $\overline{D} + K_{\overline{V}}$ is not numerically effective. Let $N(\overline{V}) = NS(\overline{V})_\mathbf{Q} \otimes \mathbf{R}$, in which
we introduce a metric topology. Let $NE(\overline{V})_\mathbf{Q}$ be the smallest convex cone in
$NS(\overline{V})_\mathbf{Q}$ containing all irreducible curves on $\overline{V}$ and closed under multiplication

of elements of $\mathbf{Q}_+$. The cone $NE(\overline{V})_{\mathbf{R}}$ in $N(\overline{V})$ is defined in a similar fashion. Let $\overline{NE}(\overline{V})$ be the closure of $NE(\overline{V})_{\mathbf{R}}$ in $N(\overline{V})$. Let L be an ample divisor on V such that $\overline{L} = f_* L$ is an ample Cartier divisor. Since the intersection matrix of $\mathrm{Bk}(D)$ is negative-definite, $W = f^* \overline{L} - L$ is an effective divisor such that $\mathrm{Supp}(W) \subseteq \mathrm{Supp}(\mathrm{Bk}\, D)$. For a small positive number ε, we set

$$\overline{NE}_\varepsilon(\overline{D}, \overline{V}) = \{\overline{Z} \in \overline{NE}(\overline{V}); (\overline{Z} \cdot \overline{D} + K_{\overline{V}}) \geq -\varepsilon(\overline{Z} \cdot \overline{L})\}.$$

Then we have the following result which generalizes the Mori theory [2] in our context:

LEMMA 8. *For every positive number ε, there exist (possibly singular) rational curves $\overline{l}_i$, $1 \leq i \leq u$, such that*

$$\overline{NE}(\overline{V}) = \sum_{i=1}^{u} \mathbf{R}_+[\overline{l}_i] + \overline{NE}_\varepsilon(\overline{D}, \overline{V})$$

and $0 > (\overline{D} + K_{\overline{V}} \cdot \overline{l}_i) \geq -3$ for $1 \leq i \leq u$.

Since $\overline{NE}(\overline{V})$ is polyhedral on the side $\overline{D} + K_{\overline{V}} + \varepsilon\overline{L} < 0$, we can define an extremal rational curve $\overline{l}$ as follows: a half line $R = \mathbf{R}_+[Z]$ in $\overline{NE}(\overline{V})$ is an extremal ray if (i) $(\overline{D} + K_{\overline{V}} \cdot Z) < 0$ and (ii) Z_1 and Z_2 in $\overline{NE}(\overline{V})$ satisfy $Z_1, Z_2 \in R$ if $Z_1 + Z_2 \in R$; a rational curve $\overline{l}$ on $\overline{V}$ is an extremal rational curve if $0 > (\overline{D} + K_{\overline{V}} \cdot \overline{l}) \geq -3$ and $\mathbf{R}_+[\overline{l}]$ is an extremal ray. Let $\overline{l}$ be an extremal rational curve on $\overline{V}$. Then, as in Mori [21], there exists a numerically effective $\mathbf{Q}$-divisor $\overline{H}$ on $\overline{V}$ such that $\overline{H}^\perp \cap \overline{NE}(\overline{V}) = \mathbf{R}_+[\overline{l}]$, where $\overline{H}^\perp = \{Z \in \overline{NE}(\overline{V}); (\overline{H} \cdot Z) = 0\}$. Let $H = f^*(\overline{H})$. Then H is numerically effective on V. We have the following three possible cases:

(1) $(H^2) > 0$; (2) $(H^2) = 0$ and H is not numerically trivial; (3) H is numerically trival. In the case (1), the Hodge index theorem implies that the intersection matrix of $l + \mathrm{Bk}(D)$ is negative-definite, where l is the proper transform of $\overline{l}$ on V. Since $(l \cdot D^\# + K_V) = (\overline{l} \cdot \overline{D} + K_{\overline{V}}) < 0$, this contradicts the hypothesis that the pair (V, D) is almost minimal. The remaining two cases correspond to the cases (1) and (2) in the following:

LEMMA 9. *Let $\overline{l}$ be an extremal rational curve on $\overline{V}$ and let l be the proper transform of $\overline{l}$ on V. Then one of the following two cases takes place:*

(1) *The intersection matrix of $l + \mathrm{Bk}(D)$ is negative-semidefinite but not negative-definite. Moreover, $(\overline{l}^2) = 0$.*

(2) *The Picard number $\rho(\overline{V})$ equals 1, and $-(\overline{D} + K_{\overline{V}})$ is ample.*

LEMMA 10. *With the same notations and assumptions as in Lemma 9, suppose the first case occurs. Then, for sufficiently large integer n, the linear system $|nN f^*(\overline{l})|$ is composed of an irreducible pencil, free from base points, whose general members are isomorphic to $\mathbf{P}^1$. Furthermore, $V - D$ is affine-ruled. Hence $\overline{\kappa}(V - D) = -\infty$.*

In the second case of Lemma 9, we have:

LEMMA 11. *With the same notations and assumptions as in Lemma 9, suppose the second case occurs. Then the following assertions hold:*

(1) $-(D^\# + K_V)$ *is numerically effective and, for an irreducible curve C on V, $(D^\# + K_V \cdot C) = 0$ if and only if C is an irreducible component of* $\mathrm{Bk}(D)$. *Moreover,* $\bar{\kappa}(V - D) = -\infty$.

(2) *Let* $\mathrm{Supp\,Bk}(D) = \bigcup_{i=1}^r D_i$. *Then* $D - \sum_{i=1}^r D_i$ *is connected and has at most two irreducible components. Each irreducible component is a nonsingular rational curve, and if $D - \sum_{i=1}^r D_i$ has two irreducible components Y_1 and Y_2, then $(Y_1 \cdot Y_2) = 1$.*

(3) *If $D - \sum_{i=1}^r D_i$ has two irreducible components Y_1 and Y_2, then the connected component of D containing Y_1 and Y_2 has the following dual graph:*

(R$_1$)
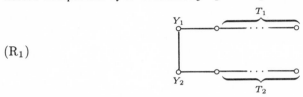

where T_i $(i = 1, 2)$ is an admissible maximal rational twig which might be empty.

(4) *Suppose $Y := D - \sum_{i=1}^r D_i$ is irreducible. Then the connected component of D containing Y is a nonadmissible rational fork or has the following dual weighted graph:*

(R$_2$)
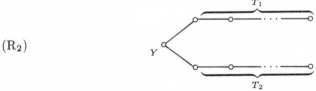

where T_i $(i = 1, 2)$ is an admissible maximal twig which might be empty.

(5) *Suppose that the connected component of D containing $D - \sum_{i=1}^r D_i$ has the dual graph of type (R$_1$) or (R$_2$). Then $V - D$ is affine-ruled.*

As a consequence of the above discussions, we have the following:

THEOREM 12. *Let (V, D) be an almost minimal pair of a nonsingular projective surface and a reduced effective divisor with simple normal crossings. Let* $\mathrm{Supp\,Bk}(D) = \bigcup_{i=1}^r D_i$. *Suppose that $\bar{\kappa}(V - D) = -\infty$ and $V - D$ is not affine-ruled. Then one of the following two cases takes place:*

(1) $D = \sum_{i=1}^r D_i$, *i.e.,* $\mathrm{Supp\,} D = \mathrm{Supp\,Bk}(D)$;

(2) $D_0 := D - \sum_{i=1}^r D_i$ *is irreducible and the connected component of D containing D_0 is a nonadmissible rational fork for which D_0 is the central component.*

4. Logarithmic del Pezzo surfaces with rank 1. Let (V, D) be an almost minimal pair consisting of a nonsingular projective surface V defined over k and a reduced effective divisor D on V with simple normal crossings. Let $f: V \to \overline{V}$ be the contraction of all connected components of the bark $\mathrm{Bk}(D)$, where $\overline{V}$ is a projective normal surface carrying at worst quotient singular points and $f: V \to \overline{V}$ is a minimal resolution of singularities on $\overline{V}$. Let $\overline{D} := f_* D^{\#}$, where $D^{\#} := D - \mathrm{Bk}(D)$. We then have $K_{\overline{V}} \sim f_* K_V$ and $D^{\#} + K_V \equiv f^*(\overline{D} + K_{\overline{V}})$. The pair (V, D) (or rather $\overline{V}$) is called a *logarithmic del Pezzo surface of rank one* if $\mathrm{Pic}(V)$ has rank one and $-(\overline{D} + K_{\overline{V}})$ is ample. Then $-(D^{\#} + K_V)$ is numerically effective. We say that (V, D) has the *noncontractible* (or *contractible*, resp.) *boundary* if $\mathrm{Supp}\, D \neq \mathrm{Supp}\, \mathrm{Bk}(D)$ (or $\mathrm{Supp}\, D = \mathrm{Supp}\, \mathrm{Bk}(D)$, resp.). If the pair (V, D) has the noncontractible boundary, there exists a unique irreducible component D_0 of D which is not contained in $\mathrm{Supp}\, \mathrm{Bk}(D)$, and the connected component of D containing D_0 is a nonadmissible rational fork for which D_0 is the central component. In particular, there are exactly three cyclic quotient singular points of $\overline{V}$ lying on $\overline{D}_0 := f(D_0)$; see Theorem 12 above. We shall look into the structures of a logarithmic del Pezzo surface of rank one with the noncontractible boundary.

Let X be a noncomplete nonsingular algebraic surface defined over k. The surface X is said to have *an $\mathbf{A}_*^1$-fiber space* structure if there exists a surjective morphism $\rho: X \to C$ from X onto a nonsingular curve C such that general fibers of ρ are isomorphic to $\mathbf{A}_*^1$, the affine line $\mathbf{A}_k^1$ with one point deleted off. A fiber $\rho^*(P)$ is said to be *singular* if either $\rho^{-1}(P)$ is reducible or $\rho^*(P) = N_P C_P$, where $N_P \geq 2$ and C_P is irreducible; in the latter case, $\rho^*(P)$ is a multiple fiber with multiplicity N_P. Given an $\mathbf{A}_*^1$-fiber space $\rho: X \to C$ with C complete, there exist a smooth completion (W, B, X) of X and a surjective morphism $\pi: W \to C$ such that general fibers of π are isomorphic to $\mathbf{P}_k^1$ and the restriction of π onto X coincides with the given morphism ρ. An $\mathbf{A}_*^1$-fiber space $\rho: X \to C$ is called *a Platonic $\mathbf{A}_*^1$-fiber space* if the following conditions are satisfied:

(1) $C \cong \mathbf{P}_k^1$;

(2) ρ has no singular fibers except three multiple fibers $\Gamma_i = \mu_i \Delta_i$, $1 \leq i \leq 3$, such that Δ_i is isomorphic to $\mathbf{A}_*^1$ and that $\{\mu_1, \mu_2, \mu_3\}$ is, up to a permutation, one of the triplets $\{2, 2, n\}$ $(n \geq 2)$, $\{2, 3, 3\}$, $\{2, 3, 4\}$, and $\{2, 3, 5\}$;

(3) There exist a smooth completion (W, B, X) of X and a surjective morphism $\pi: W \to C$ as above such that:

(i) B contains two irreducible components S_0 and S_1 which are cross-sections of π with $S_0 \cap S_1 = \varnothing$, and other irreducible components of B are contained in the fibers of π;

(ii) every fiber of π has a linear chain as its dual graph. Then we have the following:

THEOREM 13. *Suppose the characteristic of the ground field k is zero. Let (V, D) be a logarithmic del Pezzo surface of rank one with noncontractible boundary, and let $X := V - D$. Then either X is affine-ruled or X is a Platonic $\mathbf{A}_*^1$-fiber space. Moreover, if k is the complex field $\mathbf{C}$, the universal covering space of a Platonic $\mathbf{A}_*^1$-fiber space X is isomorphic to $\mathbf{A}_{\mathbf{C}}^2 - (0)$, $G := \pi_1(X)$ is a small* *finite subgroup of* $\mathrm{GL}(2, \mathbf{C})$ *which is not a cyclic group, and* X *is isomorphic to the quotient variety* $\mathbf{A}^2_{\mathbf{C}}/G$ *with the unique singular point deleted off.*

See Fujita [**4**] for relevant results on the fundamental group of a Platonic $\mathbf{A}^1_*$-fiber space.

From the above theorem, we can derive the following two results:

THEOREM 14. *Suppose the characteristic of* k *is zero. Let* X *be a nonsingular algebraic rational surface defined over* k *with* $\bar{\kappa}(X) = -\infty$. *Assume that* X *is not affine-ruled and that, for a smooth completion* (V, D, X) *of* X, *the intersection matrix of* D *is not negative-definite. Then* X *is an* $\mathbf{A}^1_*$-fiber space over $\mathbf{P}^1_k$. *Moreover,* X *is affine-uniruled. Namely, there exists a dominant quasi-finite morphism* $\pi\colon U_0 \times \mathbf{A}^1_k \to X$ *where* U_0 *is an affine curve.*

THEOREM 15 (SEE MIYANISHI [**15**] AND GURJAR-SHASTRI [**7**] FOR A TOPOLOGICAL PROOF). *Let* A *be a normal subalgebra of a polynomial ring* $\mathbf{C}[x_1, x_2]$ *such that* $\mathbf{C}[x_1, x_2]$ *is a finitely generated* A-module. *Let* $X := \operatorname{Spec} A$. *Then* X *is isomorphic to* $\mathbf{A}^2_{\mathbf{C}}$ *if* X *is nonsingular, and* X *is isomorphic to* $\mathbf{A}^2_{\mathbf{C}}/G$ *if* X *is singular, where* G *is a small finite subgroup of* $\mathrm{GL}(2, \mathbf{C})$. *In the case where* X *is singular, we have, furthermore, the following results:*

(1) A *is factorial if and only if* X *is isomorphic to a hypersurface in* $\mathbf{A}^3_{\mathbf{C}}$ *defined by* $x^2 + y^3 + z^5 = 0$.

(2) X *is affine-ruled if and only if* G *is cyclic.*

We don't know much about a logarithmic del Pezzo surface (V, D) of rank one with contractible boundary. Instead of the pair (V, D), we consider the normal projective surface $\overline{V}$ with all connected components of $\operatorname{Supp}(D) = \operatorname{Supp} \operatorname{Bk}(D)$ contracted down to quotient singular points on $\overline{V}$. In comparison with Theorem 13, the following conjecture should seem to be plausible:

CONJECTURE 16. *Let* (V, D) *(or rather* $\overline{V}$) *be a logarithmic del Pezzo surface of rank one with contractible boundary* D. *Then* $\overline{V}$ *is isomorphic to* $\mathbf{P}^2_k/G$ *for a suitable finite subgroup* G *of* $\mathrm{PGL}(2, k)$.

To test the validity of the above conjecture, we raise the following:

PROBLEM. *Let* W *be a normal projective surface defined over the complex field* $\mathbf{C}$. *Suppose there exists a finite morphism* $\psi\colon \mathbf{P}^2_{\mathbf{C}} \to W$. *Is* W *then isomorphic to* $\mathbf{P}^2_{\mathbf{C}}/G$ *for a suitable finite subgroup* G *of* $\mathrm{PGL}(2, \mathbf{C})$?

5. Affine-ruled open surfaces. Let X be an algebraic surface defined over k. We say that X is *affine-ruled* if X contains a Zariski open set U isomorphic to $U_0 \times \mathbf{A}^1_k$, where U_0 is an affine curve. If X is nonsingular and affine-ruled, then $\bar{\kappa}(X) = -\infty$ (cf. Fujita-Iitaka [**6**]). The converse is not always true as shown in Theorems 12 and 13. However, we have the following:

THEOREM 17. *Let* X *be a nonsingular algebraic surface defined over* k *with* $\bar{\kappa}(X) = -\infty$. *Suppose that* X *has a smooth completion* (V, D, X) *of* X *such* that the boundary divisor at infinity D is connected. Then X is affine-ruled. In particular, if X is affine, then X is affine-ruled.

For the relevant results, see Miyanishi-Sugie [17], Miyanishi [13], and Russell [22].

The affine-ruledness of a surface is used to characterize the affine plane $\mathbf{A}_k^2$ as an algebraic variety. Namely, we have:

THEOREM 18 [12]. Let $X = \operatorname{Spec} A$ be an affine surface defined over k. Then X is isomorphic to the affine plane $\mathbf{A}_k^2$ if and only if the following three conditions are satisfied:

(1) $A^* = k^*$, where R^* for a ring R denotes the multiplicative group of all invertible elements of R;

(2) A is factorial;

(3) X is affine-ruled.

This characterization of the affine plane no longer holds if the ground field is not an algebraically closed field, as shown by the following example due to W. Haboush: Let k be as above and let $k(t)$ be a purely transcendental extension of k. Let X be an affine surface defined over $k(t)$ by the equation $xy + z^2 = t$. Then the above three conditions are met, while $\operatorname{Pic}(X \otimes_{k(t)} k(t^{1/2})) \cong \mathbf{Z}$.

Suppose that a nonsingular affine surface X is defined over $\mathbf{C}$. Let (V, D, X) be a smooth completion of X, where D is a reduced effective divisor with simple normal crossings. Let T be a tubular neighborhood of D in V, which is suitably small, and let ∂T be its boundary. The fundamental group $\pi_1(\partial T)$ is called the fundamental group at infinity of X and denoted by $\pi_1^\infty(X)$, which is independent of the choice of a smooth completion (V, D, X). A topological characterization of the affine plane has been obtained by C. P. Ramanujam [23] and is stated as follows:

THEOREM 19 (C. P. RAMANUJAM). Let X be a nonsingular affine surface defined over $\mathbf{C}$. Then X is isomorphic to the affine plane if and only if X is topologically contractible and $\pi_1^\infty(X)$ is trivial.

A natural generalization of this result is the following:

THEOREM 20 (GURJAR-SHASTRI [8]). Let X be a normal affine surface defined over $\mathbf{C}$. Then X is isomorphic to $\mathbf{A}_{\mathbf{C}}^2/G$, where G is a small finite subgroup of $\operatorname{GL}(2, \mathbf{C})$, if and only if X is topologically contractible and $\pi_1^\infty(X)$ is a finite group.

A cancellation problem (or the Zariski problem for a polynomial ring) is:

CANCELLATION PROBLEM. Let X be an algebraic variety defined over k of dimension n. Suppose $X \times \mathbf{A}_k^m \cong \mathbf{A}_k^{n+m}$ for some positive integer m. Is X then isomorphic to $\mathbf{A}_k^n$?

A significant consequence of the theory of open surfaces is an affirmative solution of the cancellation problem in case $n = 2$; see Fujita-Iitaka [6], Fujita [4], and Miyanishi-Sugie [17].

THEOREM 21. *A cancellation holds for* $X \times \mathbf{A}_k^m \cong \mathbf{A}_k^{m+2}$, *i.e.*, X *is isomorphic to* $\mathbf{A}_k^2$.

This result was generalized to a strong cancellation theorem by Kambayashi [11].

In order to consider the above cancellation problem, it seems indispensable to obtain an algebraic (or algebro-topological) characterization of the affine n-space as an algebraic variety. An attempt in the three-dimensional case is the following:

THEOREM 22 [14]. *Let* $X = \operatorname{Spec} A$ *be a nonsingular affine threefold defined over* **C**. *Then* X *is isomorphic to the affine 3-space* $\mathbf{A}_\mathbf{C}^3$ *if and only if the following conditions are satisfied;*

(1) $A^* = \mathbf{C}^*$;

(2) *A is factorial;*

(3) $H_3(X; \mathbf{Z}) = (0)$, *where* X *is endowed with the natural structure of a complex analytic space;*

(4) X *contains a nonempty Zariski open set* U *such that* U *is isomorphic to* $U_0 \times \mathbf{A}_\mathbf{C}^2$ *with an affine curve* U_0 *and that the complement* $X - U$ *consists of nonsingular irreducible components.*

We remark that the affine-ruledness is preserved under an algebraic flat deformation. Namely, we have:

LEMMA 23 [14]. *Let* $(\mathcal{O}, t\mathcal{O})$ *be a discrete valuation ring with uniformisant* t. *Let* K *and* k *be the quotient field and the residue field of* $\mathcal{O}$, *respectively. Let* A *be a finitely generated algebra defined over* $\mathcal{O}$ *such that* A *is* $\mathcal{O}$-flat *and that* $A_K := A \otimes_\mathcal{O} K$ *and* $A_K := A \otimes_\mathcal{O} k$ *are integral domains. Assume that* k *has characteristic zero. If* $X_K := \operatorname{Spec}(A_K)$ *is affine-ruled, so is* $X_k := \operatorname{Spec} A_k$.

6. Étale endomorphisms of open surfaces.

Let X be a nonsingular, open, algebraic variety defined over k, which has characteristic zero. Let $f: X \to X$ be an étale endomorphism of X. Then f is a quasi-finite morphism. We consider the following:

PROBLEM. *Is an étale endomorphism* $f: X \to X$ *finite?*

If X is the affine plane $\mathbf{A}_k^2$, the above problem is equivalent to the Jacobian problem in dimension 2. We are not considering the Jacobian problem itself. As for the Jacobian problem in dimension two, serious efforts were made by Abhyankar and Moh. Though their methods are rather analytic by using the expansion techniques, distributions of roots, approximate roots, etc., the results obtained by them are best possible, so far; the results with some additional hypotheses are summarized in Abhyankar [1], and the Jacobian problem was affirmatively solved by Moh [20] when an étale endomorphism $f: \mathbf{A}_k^2 \to \mathbf{A}_k^2$ is given by a pair of polynomials $(\rho(x,y)), \psi(x,y))$ with $\max\{\deg_{x,y} \rho, \deg_{x,y} \psi\} \leq 100$.

Let $f: X \to X$ be as above. Then the following results hold true:

THEOREM 24 (IITAKA [9, 10]). (1) *Let X be a nonsingular algebraic variety with $\bar{\kappa}(X) = \dim X$. Let $f: X \to X$ be a quasi-finite endomorphism. Then f is an automorphism.*

(2) *Let X be a nonsingular algebraic variety with $\bar{\kappa}(X) \geq 0$. Then any dominant morphism $f: X \to X$ is an étale morphism.*

We consider, hereafter, only the case where X is an *affine surface*. By virtue of the above theorem, we have only to look into the case $\bar{\kappa}(X) \leq 1$. If $\bar{\kappa}(X) = -\infty$, we can show:

THEOREM 25 [16]. *Suppose that $\bar{\kappa}(X) = -\infty$ and that one of the following conditions is satisfied*:
(1) *X is irrational but not elliptic ruled,*
(2) *$\Gamma(X, \mathcal{O}_X) \neq k^*$ and $\mathrm{rank}(\Gamma(X, \mathcal{O}_X)^*/k^*) \geq 2$ if X is rational.*
Then an étale endomorphism $f: X \to X$ is an automorphism.

Let X be a nonsingular algebraic surface with an $\mathbf{A}^1_*$-fiber space structure given by a surjective morphism $\rho: X \to C$. Let η be the generic point of C and let X_η be the generic fiber of ρ. Then X_η has either two rational (over $k(\eta)$) places or one irrational place at infinity. In the first case, ρ is called *an $\mathbf{A}^1_*$-fibration*, and in the second case, ρ is called *a twisted $\mathbf{A}^1_*$-fibration*. Suppse X is a nonsingular *affine* surface. Then, by [13], it is known that:

(1) Either if $\bar{\kappa}(X) = 1$ or if X is an irrational surface with $\bar{\kappa}(X) = 0$, then X has an $\mathbf{A}^1_*$-fibration or a twisted $\mathbf{A}^1_*$-fibration;

(2) Conversely, if X has an $\mathbf{A}^1_*$-fibration or a twisted $\mathbf{A}^1_*$-fibration, then $\bar{\kappa}(X) \leq 1$.

Let X be a nonsingular affine surface with an $\mathbf{A}^1_*$-fiber space structure $\rho: X \to C$. Suppose S is a singular fiber of ρ. Then S is written (as a divisor) in the form $S = \Gamma + \Delta$, where

(i) $\Gamma = 0, \Gamma = \alpha\Gamma_1$ with $\alpha \geq 1$ and $\Gamma_1 \cong \mathbf{A}^1_*$, or $\Gamma = \alpha_1\Gamma_1 + \alpha_2\Gamma_2$, where $\alpha_1 \geq 1$, $\alpha_2 \geq 1$, $\Gamma_1 \cong \Gamma_2 \cong \mathbf{A}^1_k$, and Γ_1 and Γ_2 meet each other transversally in a single point;

(ii) $\Delta \geq 0$, and $\mathrm{Supp}\,\Delta$ is a disjoint union of connected components isomorphic to $\mathbf{A}^1_k$ provided $\Delta > 0$.

We have the following:

THEOREM 26 [16]. *Let X be a nonsingular affine surface with an $\mathbf{A}^1_*$-fiber space structure $\rho: X \to C$, and let $f: X \to X$ be an étale endomorphism such that $\rho \cdot f = \rho$ and $\mathrm{codim}_X(X - f(X)) \geq 2$. Then the following assertions hold*:

(1) *When $\rho: X \to C$ is an $\mathbf{A}^1_*$-fibration, f is an automorphism in each of the following cases*:

(i) *There exists a singular fiber $S = \Gamma + \Delta$ of ρ such that $\Gamma = \alpha_1\Gamma_1 + \alpha_2\Gamma_2$, where $\alpha_1 \geq 1$, $\alpha_2 \geq 1$, $\Gamma_1 \cong \Gamma_2 \cong \mathbf{A}^1_k$, and Γ_1 and Γ_2 meet each other transversally in one point.*

(ii) $\Gamma(X, \mathcal{O}_X)^* = \Gamma(C, \mathcal{O}_C)^*$.

(iii) $\Gamma(C, \mathcal{O}_C) = k$, *i.e.*, *C is complete, and there exists a singular fiber* $S = \Gamma + \Delta$ *of* ρ *such that* $\Gamma = \alpha\Gamma_1$ *with* $\alpha \geq 1$ *and* $\Gamma_1 \cong \mathbf{A}_*^1$.

When $\rho: X \to C$ *is a twisted* $\mathbf{A}_*^1$-*fibration, f is an automorphism provided C is complete and* ρ *has a singular fiber* $S = \Gamma + \Delta$ *with* $\Gamma \neq 0$.

The condition $\operatorname{codim}_X(X - f(X)) \geq 2$ is satisfied provided $\Gamma(X, \mathcal{O}_X)$ is factorial. The following result should be noteworthy:

THEOREM 27 [**16**]. *Let X be a nonsingular affine surface with an étale endomorphism* $f: X \to X$. *Suppose* $\deg f > 1$ *and* $\operatorname{codim}_X(X - f(X)) \geq 2$. *Let* $\tilde{X}$ *be the normalization of the lower X in the function field of the upper X. Suppose* $\tilde{X}$ *is nonsingular. When we regard the upper X as an open subset of* $\tilde{X}$, *then* $\tilde{X} - X$ *is a disjoint union of irreducible curves which are isomorphic to* $\mathbf{A}_k^1$.

Finally, we shall give two counterexamples:

(1) Let C be a nonsingular cubic curve on $\mathbf{P}_k^2$ and let $X := \mathbf{P}_k^2 - C$. Then X has a surjective, nonfinite, étale endomorphism $f: X \to X$ of degree 3. The surface has no $\mathbf{A}_*^1$-fibrations nor twisted $\mathbf{A}_*^1$-fibrations, while $\bar{\kappa}(X) = 0$, $\Gamma(X, \mathcal{O}_X)^* = k^*$, and $\operatorname{Pic}(X) \cong \mathbf{Z}/3\mathbf{Z}$.

(2) Let D be a nonsingular irreducible curve on $F_0 := \mathbf{P}_k^1 \times \mathbf{P}_k^1$ such that $D \sim 2M + l$, where l and M are curves of type $(1, 0)$ and $(0, 1)$, respectively, and let $X := F_0 - D$. Then X has a surjective, nonfinite, étale endomorphism $f: X \to X$ of degree 3. The surface X satisfies $\bar{\kappa}(X) = -\infty$, $\Gamma(X, \mathcal{O}_X)^* = k^*$, and $\operatorname{Pic}(X) \cong \mathbf{Z}$.

REFERENCES

1. S. S. Abhyankar, *Lectures on expansion techniques in algebraic geometry*, Tata Inst. Fund. Res., Bombay, 1977.

2. M. Artin, *Some numerical criteria for contractability of curves on algebraic surfaces*, Amer. J. Math. **84** (1962), 485–496.

3. E. Brieskorn, *Rationale Singularitäten komplexer Flächen*, Invent. Math. **4** (1968), 336–358.

4. T. Fujita, *On Zariski problem*, Proc. Japan Acad Ser. A Math. Sci. **55** (1979), 106–110.

5. _____, *On the topology of non-complete algebraic surfaces*, J. Fac. Sci. Univ. Tokyo Sec. IA Math. **29** (1982), 503–566.

6. T. Fujita and S. Iitaka, *Cancellation theorem for algebraic varieties*, J. Fac. Sci. Univ. Tokyo Sec. IA Math. **24** (1977), 123–127.

7. R. V. Gurjar and A. R. Shastri, *The fundamental group at infinity of an affine surface*, Comm. Math. Helv. **59** (1984), 459–484.

8. _____, *A topological characterization of* $\mathbf{C}^2/G$, J. Math. Kyoto Univ. **25** (1985), 767–773.

9. S. Iitaka, *On logarithmic Kodaira dimension of algebraic varieties*, Complex Analysis and Algebraic Geometry, a collection of papers dedicated to K. Kodaira, Iwanami Shoten, Tokyo, and Cambridge Univ. Press, New York, 1977, pp. 175–189.

10. _____, *Algebraic geometry*, Introduction to Birational Geometry, Graduate Texts in Math., vol. 76, Springer-Verlag, 1982.

11. T. Kambayashi, *On Fujita's strong cancellation theorem for the affine plane*, J. Fac. Sci. Univ. Tokyo Sec. IA Math. **27** (1980), 535–548.

450 MASAYOSHI MIYANISHI AND SHUICHIRO TSUNODA

12. M. Miyanishi, *Lectures on curves on rational and unirational surfaces*, Tata Inst. Fund. Res., Bombay and Springer-Verlag, 1978.

13. ———, *Non-complete algebraic surfaces*, Lecture Notes in Math., vol. 857, Springer-Verlag, 1981.

14. ———, *An algebro-topological characterization of the affine space of dimension three*, Amer. J. Math. **106** (1984), 1469–1486.

15. ———, *Normal affine subalgebras of a polynomial ring*, Algebraic and Topological Theories —to the memory of Dr. Takehiko Miyata, Kinokuniya, Tokyo, 1985, pp. 37–51.

16. ———, *Etale endomorphisms of algebraic varieties*, Osaka J. Math. **22** (1985), 345–364.

17. M. Miyanishi and T. Sugie, *Affine surfaces containing cyclinderlike open sets*, J. Math. Kyoto Univ. **20** (1980), 11–42.

18. M. Miyanishi and S. Tsunoda, *Non-complete algebraic surfaces with logarithmic Kodaira dimension* $-\infty$ *and with non-connected boundaries at infinity*, Japanese J. Math. **10** (1984), 195–242.

19. ———, *Logarithmic del Pezzo surfaces of rank one with non-contractible boundaries*, Japan. J. Math. (N.S.) **10** (1984), 271–319.

20. T. T. Moh, *On the Jacobian conjecture and the configurations of roots*, J. Reine Angew. Math. **340** (1983), 140–212.

21. S. Mori, *Threefolds whose canonical bundles are not numerically effective*, Ann. of Math. (2) **116** (1982), 133–176.

22. P. Russell, *On affine-ruled rational surfaces*, Math. Ann. **255** (1981), 287–302.

23. C. P. Ramanujam, *A topological characterization of the affine plane as an algebraic variety*, Ann. of Math. (2) **94** (1971), 69–88.

24. S. Tsunoda, *Structure of open algebraic surfaces*. I, J. Math. Kyoto Univ. **23** (1983), 95–125.

OSAKA UNIVERSITY, TOYONAKA, OSAKA 560, JAPAN

Proceedings of Symposia in Pure Mathematics
Volume **46** (1987)

Classification of Normal Surfaces

FUMIO SAKAI

Our aim is to develop an adequate theory of normal surfaces. Throughout this paper a *normal surface* will mean a normal Moishezon surface. This is equivalent to requiring that any resolution of its singularities is a smooth projective surface over **C**. We shall show a rough picture of the classification of normal surfaces. In §1 we review our method of studying divisors on normal surfaces [**Sa 3**]. In §2 we prepare some results concerning certain divisors on smooth surfaces [**Sa 2**]. In §3 we state the classification of minimal normal surfaces Y with $\kappa(Y) \le 1$ [**Sa 4**]. In §4 we study the **Q**-Gorensteinness of normal surfaces and prove the following two theorems:

THEOREM. *Let Y be a minimal normal surface. Suppose that Y is **Q**-Gorenstein. Then the Kodaira dimension $\kappa(Y)$ coincides with the numerical Kodaira dimension $\nu(Y)$.*

THEOREM. *Let Y be a canonical normal surface with $\kappa(Y) = 2$. Then the canonical ring $R(Y)$ is finitely generated if and only if Y is **Q**-Gorenstein.*

In the Appendix we consider the local properties of normal surface singularities.

This paper was written when the author stayed at the Max Planck Institut für Mathematik in Bonn (April–September 1985) and at the Institute Fourier in Grenoble (October 1985). He is grateful to both institutes for their hospitality. He would like to thank the American Mathematical Society and Surikagaku-Shinkokai for providing him the financial support to participate in the Bowdoin Summer Institute.

1. Preliminaries. Let Y be a normal surface. A *divisor* on Y will mean a Weil divisor, that is, a linear combination with coefficients in **Z** of irreducible curves on Y. Let $\mathrm{Div}(Y)$ denote the group of divisors on Y. For a divisor D let $\mathcal{O}(D)$ denote the corresponding divisorial sheaf (rank one reflexive sheaf), so that $\mathcal{O}(D) \cong j_*\mathcal{O}(D|Y_0)$ where $Y_0 \hookrightarrow Y$ is the inclusion of the smooth locus. A divisor D is *Cartier* if $\mathcal{O}(D)$ is invertible. Write $\mathrm{Div}(Y, \mathbf{Q}) = \mathrm{Div}(Y) \otimes \mathbf{Q}$: the

1980 *Mathematics Subject Classification* (1985 *Revision*). Primary 14J17, 32J15.

group of $\mathbf{Q}$-divisors on Y. For a $\mathbf{Q}$-divisor D we understand that $\mathcal{O}(D) = \mathcal{O}([D])$ where $[D]$ denotes the integral part of D. Two $\mathbf{Q}$-divisors D and D' are linearly equivalent, written $D \sim D'$, if $D - D'$ is integral and $\mathcal{O}(D) \cong \mathcal{O}(D')$.

We recall Mumford's intersection theory on a normal surface [**Mu 1**]. We first define an inverse image π^*D of a divisor D on Y by a resolution of singularities. Let $\pi: X \to Y$ be a resolution of Y, $A = \bigcup E_i$ the exceptional set. The $\mathbf{Q}$-divisor π^*D is characterized by (i) $\pi_*(\pi^*D) = D$, and (ii) $(\pi^*D)E_j = 0$ for all j. We can find it in the form: $\pi^*D = \overline{D} + \sum \alpha_i E_i, \alpha_i \in \mathbf{Q}$, where $\overline{D}$ is the strict transform of D. Since the intersection matrix $(E_i E_j)$ is negative definite, the solution exists and is unique. The *intersection number* DD' of two divisors D and D' on Y is defined to be $(\pi^*D)(\pi^*D')$. If $d = \det(E_i E_j)$, then $DD' \in (1/d)\mathbf{Z}$.

THEOREM 1.1 (CONTRACTION CRITERION). *Let $C_1, \ldots, C_n$ be irreducible curves on a normal surface Y. Then the intersection matrix $(C_i C_j)$ is negative definite if and only if the union $\bigcup C_i$ can be contracted to a finite number of normal points.*

PROOF [**Sa 3**]. Apply Grauert's contraction criterion of curves on a smooth surface [**Gr**] to those curves $\bigcup \pi^{-1}(C_i)$. $\square$

Let $f: Y \to Y'$ be a morphism between normal surfaces. It is possible to define the inverse image of a divisor D' on Y' with respect to f. There is a commutative diagram:

$$
\begin{array}{ccc}
X & \xrightarrow{\tilde{f}} & X' \\
\pi \downarrow & & \downarrow \pi' \\
Y & \xrightarrow{f} & Y'
\end{array}
$$

where π' is a resolution of Y' and π is a resolution of Y and the map $\pi'^{-1} \circ f$. Define $f^*D = \pi_*(\tilde{f}^*(\pi'^*D'))$. We now consider the birational case. By a *birational* morphism we mean a bimeromorphic morphism. Let $A_f = \bigcup C_i$ be the set of exceptional curves of f. We have the characterization: (i) $f_*(f^*D') = D'$, and (ii) $(f^*D')C_j = 0$ for all j.

THEOREM 1.2 (PROJECTION FORMULA [**Sa 3**]). *Let $f: Y \to Y'$ be a birational morphism of normal surfaces. Let D' be a divisor on Y', and Z an effective $\mathbf{Q}$-divisor supported on A_f. Then*

$$f_*\mathcal{O}(f^*D' + Z) \cong \mathcal{O}(D').$$

Two divisors D and D' on Y are *numerically equivalent*, written $D \equiv D'$, if $(D - D')C = 0$ for all irreducible curves C on Y. The *Picard number* $\rho(Y)$ is defined to be the dimension of the $\mathbf{Q}$-vector space $\{\mathrm{Div}(Y)/\equiv\} \otimes \mathbf{Q}$. We say that a divisor D on Y is *nef* if $DC \geq 0$ for all irreducible curves C on Y, and is *pseudoeffective* if $DP \geq 0$ for all nef divisors P on Y. It is known that a nef divisor is pseudoeffective. D is *ample* if some positive multiple of it is a very ample Cartier divisor in the usual sense, and is *numerically ample* if $D^2 > 0$ and $DC > 0$ for all irreducible curves C on Y.

A pair (Y, D) of a normal surface Y and a divisor D on it is called a *normal pair*. The most general problem we have in mind is to study such pairs in terms of the graded ring:

$$R(D, Y) = \bigoplus_{m \geq 0} H^0(Y, \mathcal{O}(mD)).$$

The *D-dimension* of Y is given by $\kappa(D, Y) = \text{tr. deg.}_{\mathbf{C}} R(D, Y) - 1$ but $\kappa(D, Y) = -\infty$ if $R(D, Y) = \mathbf{C}$. A *birational morphism* $f: (Y, D) \rightarrow (Y', D')$ of normal pairs is a birational morphism $f: Y \rightarrow Y'$ with the property: $f_* D = D'$. Let $A_f = \bigcup C_i$ be the exceptional set of f. Write $D = f^* D' + \sum \gamma_i C_i$. We say that f is *totally discrepant* if $A_f \neq \varnothing$ and $\gamma_i > 0$ for all i, and is *canonical* if $\gamma_i \geq 0$ for all i.

LEMMA 1.3. *Let $f: (Y, D) \rightarrow (Y', D')$ be a canonical morphism of normal surfaces. Then $R(D, Y) \cong R(Y', D')$.*

PROOF [**Sa 3**]. By definition, $D = f^* D' + Z$ where Z is an effective $\mathbf{Q}$-divisor supported on A_f. So $mD = f^*(mD') + mZ$ for every integer $m > 0$. The assertion is immediate from Theorem 1.2. $\square$

An irreducible curve C on Y is called an *exceptional curve of the first kind* on (Y, D) if $DC < 0$ and $C^2 < 0$. Given such an exceptional curve C on (Y, D), we can contract C to a normal point, $f: Y \rightarrow Y'$. Letting $D' = f_* D$, we get $D = f^* D' + ((DC)/C^2)C$. By hypothesis, f is totally discrepant. We say that (Y, D) is *minimal* if it contains no exceptional curves of the first kind. In case (Y, D) is not minimal, (Y', D') is a *minimal model* of (Y, D) if (i) (Y', D') is minimal, and (ii) there is a totally discrepant birational morphism from (Y, D) to (Y', D').

THEOREM 1.4. *Every normal pair (Y, D) has a minimal model. If D is pseudoeffective, then it admits a unique minimal model (Y', D'), and in this case D' is nef.*

PROOF [**Sa 3**]. It is sufficient to contract exceptional curves of the first kind in the above sense as many times as possible. The latter assertion can be proved logically in the same way as in the classical surface theory. $\square$

The theory of Zariski decomposition [**Z**] is related to the existence of a minimal model [**Sa 3**]. Let D be a pseudoeffective divisor on a normal surface Y. Let $(Y', D'; f)$ be its minimal model. If we put $P = f^* D'$ and $N = D - P$, then (i) N is an effective $\mathbf{Q}$-divisor and the intersection matrix of the irreducible components is negative definite unless $N = 0$, (ii) P is a nef $\mathbf{Q}$-divisor and is orthogonal to each irreducible component of N. The decomposition: $D = P + N$ is called a *Zariski decomposition*. Such a decomposition is unique.

Let D be a divisor on a normal surface Y. The *numerical D-dimension* $\nu(D, Y)$ is defined as follows [**F**, **Ka 3**]. We used instead "numerical type" in [**Sa 2**]. If D is not pseudoeffective, then we put $\nu(D, Y) = -\infty$. In case D is pseudoeffective, let $D = P + N$ be the Zariski decomposition, and define $\nu(D, Y)$ to be 0 (if $P \equiv 0$), 1 (if $P^2 = 0$, $P \not\equiv 0$), or 2 (if $P^2 > 0$).

REMARK 1.5. $\kappa(D,Y) \leq \nu(D,Y)$. Furthermore, $\kappa(D,Y) = 2$ if and only if $\nu(D,Y) = 2$. For a proof, see for instance [**Sa 2**].

In the subsequent sections we consider those pairs (Y, K_Y) where Y is a normal surface and K_Y is a canonical divisor of Y. Note that $\mathcal{O}(K_Y) \cong \omega_Y$ (the dualizing sheaf). We fix the notation:

$R(Y) = R(K_Y, Y)$,   the canonical ring,

$\kappa(Y) = \kappa(K_Y, Y)$,   the Kodaira dimension of Y as a normal surface,

$\nu(Y) = \nu(K_Y, Y)$,   the numerical Kodaira dimension of Y.

We sometimes use the anti-Kodaira dimension $\kappa^{-1}(Y) = \kappa(-K_Y, Y)$ and the numerical anti-Kodaira dimension $\nu^{-1}(Y) = \nu(-K_Y, Y)$. For a discussion of pairs $(Y, -K_Y)$, we refer to [**Sa 4**, Appendix]. In the projective geometry of Y, the pairs $(Y, K_Y + H)$ are important where H is an ample Cartier divisor on Y. See [**Sa 5**].

REMARK 1.6. Since our method is based on Mumford's intersection theory on a normal surface and Grauert's contraction criterion of curves, we can apply it to the category of 2-dimensional normal algebraic spaces in the sense of M. Artin. With slight modification (direct usage of divisorial sheaves), we can also deal with the non-Moishezon case.

2. Certain divisors on smooth surfaces. This section is devoted to considering certain divisors on smooth surfaces. First we state the classification of geometrically ruled surfaces in terms of the anticanonical divisor. To describe a geometrically ruled surface S over a smooth curve B of genus g, we employ the notation of Hartshorne's book, *Algebraic Geometry*, so that $S = \mathbf{P}(\mathcal{E})$ where $\mathcal{E}$ is a normalized rank 2 vector bundle on B. Set $\mathcal{O}(\mathfrak{e}) = \det(\mathcal{E})$, $e = -\deg \mathfrak{e}$. Let $p: S \to B$ be the projection map. There is a base section $b \in |\mathcal{O}_{\mathbf{P}(\mathcal{E})}(1)|$, and so $b^2 = -e$. If $\mathcal{E}$ is decomposable, then $e \geq 0$ and there is another section b' disjoint from b such that $b' \sim b - p^*\mathfrak{e}$. In case $e = 0$ (this condition is dropped in [**Sa 4**]), b' is unique unless $\mathfrak{e} \sim 0$. Let $\mathfrak{k}$ denote a canonical divisor of B, and $\mathfrak{t} = \mathfrak{k} + \mathfrak{e}$, $t = \deg \mathfrak{t} = 2g - 2 - e$. We obtain the following table [**Sa 4**].

TABLE 2.1

Class	$\kappa^{-1}(S)$	$\nu^{-1}(S)$	g	t	Structure
$\mathrm{I}_a, \mathrm{I}_a^*$	$-\infty$	$-\infty$	≥ 2	> 0	
I_b	$-\infty$	0	≥ 2	0	t is nontorsion in $\mathrm{Pic}^0(B)$
$_m\mathrm{II}_b$	0	0	≥ 2	0	$mt \sim 0$
II_b^*	0	0	≥ 2	0	$t \sim 0$
II_c	0	1	1	0	$\mathfrak{e}$ is nontorsion in $\mathrm{Pic}^0(B)$
II_c^*	0	1	1	0	$\mathfrak{e} \sim 0$,
$_m\mathrm{III}$	1	1	1	0	$m\mathfrak{e} \sim 0$
III^*	1	1	1	-1	
IV	2	2	≥ 0	< 0	

Here $\mathcal{E}$ is decomposable for the classes without $*$, and is indecomposable for the classes with $*$. By the symbols I, II, III, IV we mean that $\kappa^{-1}(S) = -\infty, 0, 1, 2$, respectively. For classes I, II, the subscripts a, b, c distinguish the value $\nu^{-1}(S)$.

LEMMA 2.2. *Let S be a geometrically ruled surface of class* II_c *(resp.* II_c^**). Then b and b' (resp. b) are the only (smooth) elliptic curves on S.*

PROOF. Let C be an elliptic curve on S. Suppose $C \sim nb + p^*\mathfrak{a}$. The condition $(K_S + C)C = 0$ implies that $\deg \mathfrak{a} = 0$. We have $p_*\mathcal{O}(nb + p^*\mathfrak{a}) \cong S^n\mathcal{E} \otimes \mathcal{O}(\mathfrak{a})$. For the class II_c, $S^n\mathcal{E} \otimes \mathcal{O}(\mathfrak{a}) \cong \mathcal{O}(\mathfrak{a}) \oplus \mathcal{O}(\mathfrak{e} + \mathfrak{a}) \oplus \cdots \oplus \mathcal{O}(n\mathfrak{e} + \mathfrak{a})$; hence we must have $\mathfrak{a} \sim -i\mathfrak{e}$ for some i with $0 \le i \le n$. Thus $C \sim (n - i)b + ib'$, and so $C = (n - i)b + ib'$. It follows that $C = b$ or b'. For the class II_c^*, we know from Atiyah's theorem that $S^n\mathcal{E} \cong \mathcal{F}_{n+1}$ (with his notation) and that $H^0(B, \mathcal{F}_{n+1} \otimes \mathcal{O}(\mathfrak{a})) \ne 0$ if and only if $\mathfrak{a} \sim 0$. Therefore we must have $n = 1$, $\mathfrak{a} \sim 0$ and so $C = b$. $\square$

A smooth rational surface S is called a *degenerate del Pezzo surface* if $-K_S$ is nef but not ample.

LEMMA 2.3 [Sa 4, Proposition 3.3]. *Let S be a degenerate del Pezzo surface with $K_S^2 = 0$. Then there is an indecomposable curve of canonical type b belonging to the anticanonical system $|-K_S|$. Furthermore, $\kappa^{-1}(S) = 1$ or 0, according as the normal sheaf $\mathcal{O}(b) \otimes \mathcal{O}_b$ is a torsion element or not in $\mathrm{Pic}^0(B)$.*

REMARK 2.4. A curve $b = \sum n_i C_i$ with C_i irreducible on a smooth surface S is *indecomposable* if $\mathrm{Supp}(b)$ is connected and g.c.d.$(n_i) = 1$. We say that b is *of canonical type* if $K_S C_j = bC_j = 0$ for all j [Mu 2].

THEOREM 2.5 [Sa 4, Proposition 6.1]. *Let C be a nef curve on a smooth projective surface S. Suppose that (i) $C^2 = 0$, (ii) $K_S C = 0$, (iii) S contains no (-1)-curves E with $CE = 0$. Then the pair (S, C) is classified as follows:*

$\kappa(C, S)$	S	C
0	geometrically ruled surface : II_c	$nb + n'b'$
	geometrically ruled surface : II_c^*	nb
	degenerate del Pezzo surface : $\kappa^{-1}(S) = 0$	nb
1	minimal elliptic surface	sum of fibres

PROOF. Decompose C into connected components $C = \sum n_i b_i$ where the b_i are indecomposable. We show that all b_i are of canonical type. Clearly, b_i is nef and $b_i^2 = 0$. It follows that b_i is orthogonal to each irreducible component of b_i. Next we see that $K_S b_i = 0$ for all i. If not, by (ii), $K_S b_j < 0$ for some j. By the Riemann-Roch theorem, some multiple of b_j gives a ruled fibration on S. We infer that some multiple of b_i is a fibre and hence $K_S b_i < 0$ for all i, and so $K_S C < 0$, which is a contradiction. We can easily deduce from the condition (iii) that K_S is orthogonal to each irreducible component of b_i. Thus all b_i are indecomposable and of canonical type. In case $\kappa(C, S) = 1$, $|mC|$ for some

$m > 0$ defines a minimal elliptic fibration and C is a sum of fibres. The case $\kappa(C, S) = 0$ is subtle. We know [**Mu 2**] that $p_g(S) = 0$, and by a covering technique we get $\kappa(S) = -\infty$. It is easy to see that S cannot have a ruled fibration over a curve of genus ≥ 2. Therefore, two cases remain: (1) S is a rational surface, and (2) S has a ruled fibration $p: S \to B$ over an elliptic curve B and every b_i is an elliptic curve. For case (1), we refer to [**Sa 2**, p. 111]. For case (2), we give a simpler argument than that in [**Sa 2**, 4].

To classify case (2), we first show that S is geometrically ruled. If we can find an effective divisor D numerically equivalent to $m(K_S + qC)$ for some $m, q > 0$, $q \in \mathbf{Q}$, then we are done. Indeed, if so, D is nef, for otherise there must be an irreducible curve E with $DE < 0$, and hence $E^2 < 0$, so E would be a (-1)-curve with $CE = 0$, which contradicts (iii). In particular, $D^2 \geq 0$. It follows from this that $K_S^2 \geq 0$. We infer that S is geometrically ruled. Now we prove the existence of such D. The addition formula [**Sa 2**] tells us that $\kappa(K_S + C, S) \geq 0$ unless $C = nb$ where the b is a section for the ruled fibration. We consider this exceptional case. Set $\mathcal{O}(b) \otimes \mathcal{O}_b = \mathcal{O}(\mathfrak{a})$; then $\deg \mathfrak{a} = 0$. Since b is a section, by identifying b with B, we can regard $\mathfrak{a}$ as a divisor on B. Let $\mathcal{L}_k = \mathcal{O}(K_S + kb - p^*\mathfrak{a})$. We show that $H^0(S, \mathcal{L}_k) \neq 0$ for $k = 1$, or 2. We have $D \equiv K_S + (k/n)C$, for $D \in [\mathcal{L}_k]$. Suppose $H^0(S, \mathcal{L}_1) = 0$. Since $\chi(\mathcal{L}_1) = 0$ and $H^2(S, \mathcal{L}_1) \cong H^0(S, \mathcal{O}(-b + p^*\mathfrak{a})) = 0$, we get $H^1(S, \mathcal{L}_1) = 0$. By definition, $\mathcal{L}_2 \otimes \mathcal{O}_b \cong \omega_b \cong \mathcal{O}_b$, and so we have an exact sequence: $0 \to \mathcal{L}_1 \to \mathcal{L}_2 \to \mathcal{O}_b \to 0$. From the long exact sequence

$$0 = H^0(S, \mathcal{L}_1) \to H^0(S, \mathcal{L}_2) \to H^0(b, \mathcal{O}_b) \to H^1(S, \mathcal{L}_1) = 0$$

we deduce that $H^0(S, \mathcal{L}_2) \neq 0$. Thus we may assume that S is geometrically ruled. We must have $\kappa^{-1}(S) = 0$. Indeed, if $\kappa^{-1}(S) = 1$, then $|-mK_S|$ for some $m > 0$ defines an elliptic fibration on S. Since $K_S C = 0$, some multiple of C is a sum of fibres and hence $\kappa(C, S) = 1$, which is a contradiction. In view of Table 2.1, S must be of class II_c or II_c^*. We complete the proof by Lemma 2.2. $\square$

3. Minimal normal surfaces. Let Y be a normal surface. Hereafter we consider Y from the view point of the pair (Y, K_Y). Let $\pi: X \to Y$ be the minimal resolution of Y, and $A = \bigcup E_i$ the exceptional set. If K_X is a canonical divisor of X, then $K_Y = \pi_* K_X$ becomes a canonical divisor of Y. There is an effective $\mathbf{Q}$-divisor Δ supported on A such that $K_X = \pi^* K_Y + \Delta$. With the notation of the Appendix, $\Delta = \sum \Delta_y$, $y \in \mathrm{Sing}(Y)$. We refer to the triple (X, π, Δ) as the minimal resolution of Y. Note that $\Delta = 0$ if and only if Y has only RDP's (rational double points). An irreducible curve C on Y is an *exceptional curve of the first kind* if $K_Y C < 0$ and $C^2 < 0$. If Y contains no such exceptional curves, then Y is said to be *minimal*. By successive contractions of exceptional curves of the first kind, we obtain a minimal model (cf. §1). We know that the canonical ring is invariant in the process of going to a minimal model.

PROBLEM 3.1. Describe exceptional curves of the first kind. For the Gorenstein case, see [**Sa 4**]. Cf. [**Mor**].

We consider a minimal normal surface Y. In view of Theorem 1.4, we have $\nu(Y) \geq 0$ if and only if K_Y is nef.

THEOREM 3.2 [**Sa 4**. Theorem 4.9]. *Let Y be a minimal normal surface. Suppose that $\nu(Y) = -\infty$. Then either* (i) $-K_Y$ *is numerically ample and $\rho(Y) = 1$, or* (ii) Y *has a minimal ruled fibration, i.e., a fibration $p: Y \to B$ onto a smooth curve B such that the general fibre is $\mathbf{P}^1$ and every fibre contains no exceptional curves of the first kind.*

PROBLEM 3.3. Classify those normal surfaces in the class (i). Let $f: \tilde{Y} \to Y$ be a finite covering of normal surfaces. If $\tilde{Y}$ belongs to the class (i), then so does Y. Indeed, we know that $K_{\tilde{Y}} = f^* K_Y + R$ with $R \geq 0$ (cf. §1 for the definition of f^*). We infer from the condition $\rho(\tilde{Y}) = 1$ that $\rho(Y) = 1$ and that $f^*(-K_Y) = -K_{\tilde{Y}} + R$ is numerically ample, and so is $-K_Y$.

REMARK 3.4. If Y belongs to the class (ii), then $\rho(Y) = 2$ and Y has only rational singularities [**Sa 4**], and furthermore, we have the inequality [**Sa 5**]: $K_Y^2 \leq 8\chi(\mathcal{O}_Y)$.

To describe normal surfaces Y with $\nu(Y) = 0$ and 1, we introduce the following notion. A *smooth pair* (S, D) is a pair of a smooth surface S and an effective $\mathbf{Q}$-divisor D on it. A (-1)-curve E on S is said to be (adjointly) *redundant* if $(K_S + D)E = 0$. Let $\varphi: S \to S'$ be the contraction of such E. Letting $D' = \varphi_* D$, we get

$$(K_S + D) = \varphi^*(K_{S'} + D').$$

We say that (S, D) is (adjointly) *irredundant* if it contains no redundant (-1)-curves. By successive contractions of redundant (-1)-curves, we arrive at an irredundant smooth pair (S', D'). Given a normal surface Y, we say that (S, D) *is associated to Y* if (S, D) is obtained from the pair (X, Δ) by successive contractions of redundant (-1)-curves, where (X, π, Δ) denotes the minimal resolution of Y. We have

$$\pi^* K_Y = K_X + \Delta = \varphi^*(K_S + D)$$

and

$$R(Y) \cong R(K_X + \Delta, X) \cong R(K_S + D, S).$$

PROBLEM 3.5. When is a smooth pair (S, D) associated to a normal surface? A necessary condition is that $K_S + D$ is orthogonal to each component of D, for (X, Δ) satisfies this property. We answered [**Sa 4**] this question for the case in which D is a disjoint sum of indecomposable curves of canonical type.

We state the structure theorems. We omit the case in which Y has only RDP's, for the classification reduces to that of X.

THEOREM 3.6 [**Sa 4**, Theorem 5.3]. *Let Y be a minimal normal surface having at least one non-RDP singularity. Suppose that $\nu(Y) = 0$. Let (S, D) be an associated irredundant smooth pair. Then S is either $\mathbf{P}^2$ or a geometrically* *ruled surface. In the former case, $\kappa(Y) = 0$. In the latter case, (S, D) is classified as follows:*

$\kappa(Y)$	class of S	D	smallest n with $nK_Y \sim 0$
$-\infty$	I_b	$2b$	
	II_c	$2b, 2b'$	
	IV $(g \geq 1)$	$\alpha b + D'$ $(\alpha > 1)$	
0	$_m\mathrm{II}_b$	$2b$	m
	II_b^*	$2b$	1
	II_c	$b + b'$	1
	II_c^*	$2b$	1
	$_m\mathrm{III}$	$2b, 2b'$	m
		$b + b'$	1
	$_2\mathrm{III}$	2-section	2
	$_1\mathrm{III}$	two sections	1
	III^*	2-section	2
	IV		

PROBLEM 3.7. In case $\kappa(Y) = 0$, Y can be obtained as a cyclic quotient of a normal Gorenstein surface $\tilde{Y}$ with $K_{\tilde{Y}} \sim 0$. Describe $\tilde{Y}$ in detail. See [**W**] for the local observation.

THEOREM 3.8 [**Sa 4**, Theorem 6.3]. *Let Y be a minimal normal surface having at least one non-RDP singularity. Suppose that $\nu(Y) = 1$. Then the associated irredundant smooth pair (S, D) is classified as follows:*

$\kappa(Y)$	S	D
$-\infty$	*geometrically ruled surface* : II_c	$\alpha b, \alpha' b'$ $(\alpha, \alpha' > 2)$
0	*geometrically ruled surface* : II_c	$\alpha b + \alpha' b'$ $(\alpha, \alpha' \geq 1,\ \alpha + \alpha' > 2)$
	geometrically ruled surface : II_c^*	αb $(\alpha > 2)$
	geometrically del Pezzo surface :	αb $(\alpha > 1)$
	$\kappa^{-1}(S) = 0$	
1	*minimal elliptic surface*	*sum of fibres*

PROOF. Let d be a positive integer such that dD is integral, and write $P = d(K_S + D)$. By hypothesis, P is nef, $P^2 = 0$, and $P \not\equiv 0$. We find also that $K_S P = 0$. In case $\kappa(P, S) \geq 0$, any curve $C \in |nP|$ for some $n > 0$ satisfies the assumptions in Theorem 2.5. So we can use the classification given there. In case $\kappa(P, S) = -\infty$, of course $\kappa(S) = -\infty$, and S is never rational [**Sa 2**], so there is a ruled fibration $p: S \to B$. It can be shown that $|P + p^*\mathfrak{a}| \neq \varnothing$ for a degree zero divisor $\mathfrak{a}$ on B. Now apply Theorem 2.5 to a curve $C \in |P + p^*\mathfrak{a}|$. □

EXAMPLE 3.9. We give simple examples with $\kappa(Y) \neq \nu(Y)$. Let S be a geometrically ruled surface of class II_c. Namely, $S = \mathbf{P}(\mathcal{O} \oplus \mathcal{O}(\mathfrak{e}))$ over an elliptic curve B where $\mathfrak{e}$ is a degree zero, nontorsion divisor on B. Note that $-K_S \sim b + b'$.

(i) $\kappa(Y) = -\infty$, $\nu(Y) = 0$. Let $D = 2b$. Blow up at a point $x_0 \in b$, and then blow up at a general point x_1 on the exceptional curve E_1 over x_0. By general we mean that it is not the point where E_1 meets the strict transform of b. Let X be the resulting surface, and define Δ in the following way:

$$(X, \Delta) = (S_2, D_2) \xrightarrow{\varphi_2} (S_1, D_1) \xrightarrow{\varphi_1} (S_0, D_0) = (S, D),$$

where $D_i = \varphi_i^* D_{i-1} - E_i$. We have $\Delta = 2\bar{b} + \overline{E}_1$ where the bar means the strict transform on X. Let $\pi : X \to Y$ be the contraction of $\mathrm{Supp}(\Delta)$. We have $\pi^* K_Y = \varphi^*(p^* \mathfrak{e})$ where $\varphi = \varphi_1 \circ \varphi_2$, and so $\kappa(Y) = -\infty$.

(ii) $\kappa(Y) = -\infty$, $\nu(Y) = 1$. Let $D = 3b$. Blow up at a point $x_0 \in b$, and blow up at a general point x_1 on the exceptional curve E, and again at a general point x_2 on the exceptional curve E_2 over x_1. Let (X, Δ) be the resulting pair. In this case $\Delta = 3\bar{b} + 2\overline{E}_1 + \overline{E}_2$. By contracting $\mathrm{Supp}(\Delta)$, we get a normal surface Y with $\pi^* K_Y = \varphi^*(b + p^* \mathfrak{e})$, and hence $\kappa(Y) = -\infty$.

(iii) $\kappa(Y) = 0$, $\nu(Y) = 1$. Let $D = 2b + b'$. First blow up b as in the case (i), then blow up at a point x_0' on b'. In this case, $\Delta = 2\bar{b} + \overline{E}_1 + \bar{b}'$. By contracting $\mathrm{Supp}(\Delta)$, we get Y with $\pi^* K_Y = \varphi^* b$, and hence $\kappa(Y) = 0$.

4. Q-Gorensteinness. A normal surface Y is **Q**-*Gorenstein* if K_Y is **Q**-Cartier, that is, some multiple of K_Y is a Cartier divisor.

THEOREM 4.1. *Let Y be a minimal normal surface. Suppose that Y is **Q**-Gorenstein. Then*

$$\kappa(Y) = \nu(Y).$$

PROOF. We first show the following:

LEMMA 4.2. *Let (S, D) be a smooth pair associated to a normal surface Y. Suppose that there exists a curve b such that $D \geq b$. If mK_Y is a Cartier divisor for some $m > 0$, then*

(4.3) $$\mathcal{O}(m(K_S + D)) \otimes \mathcal{O}_b \cong \mathcal{O}_b.$$

PROOF. Let (X, π, Δ) be the minimal resolution of Y. Since (S, D) is associated to Y, there is a sequence of smooth pairs:

$$(X, \Delta) = (S_n, D_n) \to \cdots \to (S_0, D_0) = (S, D),$$

where $\varphi_i : (S_i, D_i) \to (S_{i-1}, D_{i-1})$ is a contraction of a redundant (-1)-curve E_i to a point $x_{i-1} \in D_{i-1}$. Note that $D_i = \varphi_i^* D_{i-1} - E_i$. Define $b_i \subset S_i$ as follows: $b_0 = b$, $b_i = \varphi_i^* b_{i-1} - E_i$ (if $x_{i-1} \in b_{i-1}$), $\varphi_i^* b_{i-1}$ (if $x_{i-1} \notin b_{i-1}$). It is easy to see that b_i has integral coefficients and that $D_i \geq b_i$. Let $\mathcal{L}_i$ denote the invertible sheaf $\mathcal{O}(m(K_{S_i} + D_i))$. We infer from the hypothesis that mD_n is integral and that $\mathcal{L}_n \otimes \mathcal{O}_{b_n} \cong \mathcal{O}_{b_n}$. It follows that mD_i is integral and $\mathcal{L}_i \cong \varphi_i^* \mathcal{L}_{i-1}$ for every i. We prove the assertion by induction. Assume that $\mathcal{L}_i \otimes \mathcal{O}_{b_i} \cong \mathcal{O}_{b_i}$. Taking the direct image of the exact sequence:

$$0 \to \mathcal{L}_i \otimes \mathcal{O}(-b_i) \to \mathcal{L}_i \to \mathcal{L}_i \otimes \mathcal{O}_{b_i} \to 0,$$

we get the exact sequence:

$$0 \to \mathcal{L}_{i-1} \otimes \varphi_{i*}\mathcal{O}(-b_i) \to \mathcal{L}_{i-1} \to \varphi_{i*}\mathcal{O}_{b_i} \to \mathcal{L}_{i-1} \otimes R^1\varphi_{i*}\mathcal{O}(-b_i).$$

But $\varphi_{i*}\mathcal{O}(-b_i) \cong \mathcal{O}(-b_{i-1}), R^1\varphi_{i*}\mathcal{O}(-b_i) = 0$. Therefore, $\mathcal{L}_{i-1} \otimes \mathcal{O}_{b_{i-1}} \cong \varphi_{i*}\mathcal{O}_{b_i}$. On the other hand, by taking the direct image sheaves of the exact sequence: $0 \to \mathcal{O}(-b_i) \to \mathcal{O} \to \mathcal{O}_{b_i} \to 0$, we get $\varphi_{i*}\mathcal{O}_{b_i} \cong \mathcal{O}_{b_{i-1}}$. Thus we obtain the desired isomorphism: $\mathcal{L}_{i-1} \otimes \mathcal{O}_{b_{i-1}} \cong \mathcal{O}_{b_{i-1}}$. □

PROOF OF THEOREM 4.1. We have remarked that $\kappa(Y) = 2$ if and only if $\nu(Y) = 2$. See Remark 1.5. If Y has only RDP's, then $\kappa(X) = \kappa(Y)$, $\nu(X) = \nu(Y)$, and $\kappa(X) = \nu(X)$ is a consequence of the Enriques classification of smooth projective surfaces [**Mu 2**]. Now assume that Y has at least one non-RDP singularity. Looking in the list of minimal normal surfaces with $\kappa(Y) < \nu(Y)$ in Theorems 3.6 and 3.7, we can find a curve b on each such Y with $D \geq b$. Furthermore, it does not satisfy the property (4.3) for any m such that mD is integral. So Y is never **Q**-Gorenstein. □

REMARK 4.4. The Gorenstein case has been discussed previously in [**Sa 1**]. Unfortunately, the **Q**-Gorensteinness is not necessarily preserved in the process of going to the minimal model. This was first pointed out to the author by Y. Kawamata. The following are variants of his example.

EXAMPLE 4.5. Let S be a geometrically ruled surface of class II_c. Take a point $x_0 \in b$, and the fibre F passing through x_0. Blow up X at x_0, then blow up at the point x_1 where the strict transform of F' meets the exceptional curve E_1 over x_0. Let X be the resulting surface, $\pi: X \to Y$ the contraction of $A = \bar{b} \cup \overline{E}_1$ to a point y. Then Y is a normal surface constructed in Example 3.9, (i). So $\kappa(Y) = -\infty$, $\nu(Y) = 0$. We have just seen that Y is not **Q**-Gorenstein. Take another point $x_0' \in b$ and the fibre F' passing through x_0'. Blow up X at x_0', and again at the point x_1' where the strict transform of F' meets the exceptional curve E_1' over x_0', and again at the point x_2' where the strict transform of F' meets the exceptional curve E_2'. Let $\tilde{X}$ be the resulting surface, and $\tilde{\pi}: \tilde{X} \to \tilde{Y}$ the contraction of $\tilde{A} = \bar{b} \cup \overline{E}_1 \cup \overline{E}_1' \cup \overline{E}_2'$ where the bar means the strict transform on $\tilde{X}$. By a calculation (cf. the Appendix), we have

$$\tilde{\Delta} = \tfrac{1}{5}(12\bar{b} + 6\overline{E}_1 + 8\overline{E}_1' + 4\overline{E}_2').$$

Let $\tilde{C}$ be the image of the last exceptional curve E_3' by $\tilde{\pi}$. We have $K_{\tilde{Y}}\tilde{C} < 0$ and $\tilde{C}^2 < 0$, so $\tilde{C}$ is an exceptional curve of the first kind on $\tilde{Y}$. Hence Y is the minimal model of $\tilde{Y}$. We have a commutative diagram:

$$
\begin{array}{ccc}
\tilde{X} & \xrightarrow{\quad\tilde{\varphi}\quad} X \longrightarrow S \\
{\scriptstyle\tilde{\pi}}\downarrow & \quad\downarrow{\scriptstyle\pi} \\
\tilde{Y} & \xrightarrow{\quad f \quad} Y
\end{array}
$$

We have $\tilde{\varphi}^*F = \overline{F} + \overline{E}_1 + 2\overline{E}_2$, $\tilde{\varphi}^*F' = \overline{F}' + \overline{E}_1' + 2\overline{E}_2' + 3E_3'$. We regard $\mathfrak{a} = x_0 - x_0'$ as a divisor on B by identifying b with B. Then

$$\tilde{\varphi}^*(p^*\mathfrak{a}) \sim \overline{E}_1 + 2\overline{E}_2 - \overline{E}_1' - 2\overline{E}_2' - 3E_3' \quad \text{near } \tilde{A}.$$

Since b' is disjoint from b, $\mathcal{O}(b')$ is trivial near b and so $\tilde{\varphi}^* b \sim \tilde{\varphi}^*(p^* \mathfrak{e})$ near $\tilde{A}$. On the other hand,

$$K_{\tilde{X}} = \tilde{\varphi}^* K_S + \overline{E}_1 + 2\overline{E}_2 + \overline{E}'_1 + 2\overline{E}'_2 + 3E'_3.$$

Combining these together, we have $5(K_{\tilde{X}} + \tilde{\Delta}) \sim \tilde{\varphi}^*(p^*(7\mathfrak{e} - \mathfrak{a}))$ near $\tilde{A}$. We deduce from this that $\tilde{Y}$ is $\mathbf{Q}$-Gorenstein if and only if $7\mathfrak{e} - \mathfrak{a}$ is a torsion element in $\mathrm{Pic}^0(B)$. Choosing x_0, x'_0 suitably, we can make $\tilde{Y}$ $\mathbf{Q}$-Gorenstein. As a consequence, Theorem 4.1 does not hold without the minimality hypothesis.

EXAMPLE 4.6. Let S be a geometrically ruled surface of class I_a, so that $S = \mathbf{P}(\mathcal{O} \oplus \mathcal{O}(\mathfrak{e}))$, $g \geq 2$ and $t = 2g - 2 - e > 0$. Suppose $g \geq 3$. Choose $\mathfrak{e}$ as $e > t$. In particular, $b^2 < 0$. Let $\pi: X = S \to Y$ be the contraction of b to a point y. We have $\pi^* K_Y = (t/e)b + p^*(\mathfrak{l} + \mathfrak{e})$. It turns out that $\kappa(Y) = 2$. Since $\rho(Y) = 1$, Y must be minimal. It is easy to see that Y is $\mathbf{Q}$-Gorenstein if and only if $(2g - 2)\mathfrak{e} + e\mathfrak{l}$ is a torsion element in $\mathrm{Pic}^0(B)$. Now blow up X in the following way. Take a point $x_0 \in b$, and the fibre F passing through x_0. Blow up at x_0, then at the point x_1 where the strict transform of F meets the exceptional curve E_1 over x_0, and again at the point x_2 where the further strict transform of F meets the exceptional curve E_2. Let $\tilde{X}$ be the resulting surface. Let $\tilde{\pi}: \tilde{X} \to \tilde{Y}$ be the contraction of $\tilde{A} = \bar{b} \cup \overline{E}_1 \cup \overline{E}_2$. Let $\tilde{C}$ be the image of the last exceptional curve E_3 by $\tilde{\pi}$. We find that $\tilde{C}$ is an exceptional curve of the first kind. Here $t > e$ is used. Let $f: \tilde{Y} \to Y$ be the contraction of $\tilde{C}$ to y, so that Y is the minimal model of $\tilde{Y}$. In a similar manner to that in Example 4.5, we can show that $\tilde{Y}$ is $\mathbf{Q}$-Gorenstein if and only if $(3e + 1)\mathfrak{l} + (6g - 4)\mathfrak{e} + (e - t)\mathfrak{d}$ is a torsion element in $\mathrm{Pic}^0(B)$, where we regard $\mathfrak{d} = x_0$ as a divisor on B by identifying b with B. Therefore, the $\mathbf{Q}$-Gorensteinness of $\tilde{Y}$ has nothing to do with that of Y.

We have mainly studied normal surfaces Y with $\kappa(Y) \leq 1$. Very little is known for the case in which $\kappa(Y) = 2$. To answer the first question, whether the canonical ring $R(Y)$ is finitely generated or not, we introduce the notion of canonical model. A normal surface Y is *canonical* if it contains no irreducible curves C with $K_Y C \leq 0$, $C^2 < 0$. A canonical normal surface Y'' is a *canonical model* of Y if there is a canonical morphism $g: Y \to Y''$, that is, $K_Y \geq g^* K_{Y''}$. The canonical model exists but is not unique even if $\kappa(Y) \geq 0$. However, for the case $\kappa(Y) = 2$, it is unique [**Sa 3**]. Note that Y with $\kappa(Y) = 2$ is canonical if and only if K_Y is numerically ample.

THEOREM 4.7. *Let Y be a canonical normal surface with $\kappa(Y) = 2$. Then $R(Y)$ is finitely generated if and only if Y is $\mathbf{Q}$-Gorenstein.*

PROOF. The if part is standard. We show the only if part. Suppose that $R(Y)$ is finitely generated. Let (X, π, Δ) be the minimal resolution of Y, and $A = \bigcup E_i$ the exceptional set. Note that $R(Y) \cong R(K_X + \Delta, X)$. Since $K_X + \Delta$ is nef, according to a theorem of Zariski [**Z, F**], there is a positive integer m such that (i) $m\Delta$ is integral, (ii) $\mathcal{O}(m(K_X + \Delta))$ is generated by global sections. Write $\mathcal{L} = \mathcal{O}(m(K_X + \Delta))$ for such m. Since $\mathcal{L}$ is generated by global sections, and

$\mathcal{L}E_i = 0$ for all i, we can easily show that $\mathcal{L}$ is trivial near A. By the projection formula, $\pi_*\mathcal{L} \cong \mathcal{O}(mK_Y)$. Hence $\mathcal{O}(mK_Y)$ is invertible. □

REMARK 4.8. In [Sa 3] we considered a canonical model for a normal pair. The above theorem is stated there for this general setting. Meanwhile, a proof has been given by Bădescu [B 2], who also gave attention to the above particular case. In Example 4.6, Y is also the canonical model of $\tilde{Y}$ because $\rho(Y) = 1$. Therefore, we cannot tell the finitely generatedness of $R(\tilde{Y})$ by the Q-Gorensteinness of $\tilde{Y}$.

CONCLUDING REMARK 4.9. We finally mention other aspects of normal surfaces. We refer to Brenton [Br] for the projectivity criteria. In our situation, he proved that if $p_g(Y) = 0$, then Y is always projective. Normal surfaces often appear as the compactification of quotients of the ball or the product of two discs. See [H, Hö]. Miyaoka [Mi] and Kobayashi [Ko] established an inequality of Chern numbers for normal surfaces having quotient singularities. See [I] for its applications. Normal surfaces with $\mathbf{C}^*$-action were extensively studied. Degenerations of smooth surfaces to normal surfaces are discussed in [B 1, C, L, Wi]. Intersection theory on a normal surface has been used by Giraud [G], Morales [Mo], and Saito [S]. We refer to [K+N] for the classification of quartic surfaces with RDP's.

Appendix. Normal surface singularities. Let (V, y) be a normal surface singularity. Assume V is Stein. Let $\pi: U \to V$ be a resolution of V, $A = \bigcup E_i$ the exceptional set. If K_U is a canonical divisor of U, then $K_V = \pi_* K_U$ is a canonical divisor of V. Now define a Q-divisor $\Delta = \sum \delta_i E_i$ by the equations $K_U E_i + \sum \delta_i E_i E_j = 0$ for all j. We infer from the definition in §1 that $\pi^* K_V = K_U + \Delta$. When π is the minimal resolution in the sense that there are no (-1)-curves in A, we have $\Delta \geq 0$. Furthermore, $\Delta = 0$ if and only if y is an RDP, and otherwise $\mathrm{Supp}(\Delta) = A$. This type of result follows from Zariski's lemma on curves with negative definite intersection matrix. We also write $\Delta = \Delta_y$.

EXAMPLE A.1. Suppose A is a chain of $\mathbf{P}^1$'s.

$$\underset{-a_1}{\overset{E_1}{\circ}} \underline{\quad\quad} \underset{-a_2}{\overset{E_2}{\circ}} \underline{\quad} \cdots \underline{\quad} \underset{-a_n}{\overset{E_n}{\circ}} \qquad (a_i \geq 2)$$

The calculation of Δ is given by Knöller [Kn] as follows. Define the continued fraction $d/e = [a_1, \ldots, a_n]$. Consider the linear equations: $X_{i+1} = a_i X_i - X_{i-1}$ with indeterminates X_i, $0 \leq i \leq n + 1$. Let $\{c_i\}, \{c_i'\}$ be two solutions with $c_0 = d$, $c_1 = e$ and with $c_0' = 0$, $c_1' = 1$, respectively. Then $c_n = 1$, $c_{n+1} = 0$ and $c_n' = e'$, $c_{n+1}' = d$ ($ee' = 1 \bmod d$). Note that $c_i > 0, c_i' > 0$ for $1 \leq i \leq n$. The formula for Δ is given as follows:

$$\Delta = \sum \left(1 - \frac{1}{d}(c_i + c_i')\right) E_i.$$

EXAMPLE A.2. Let us take the case in which the dual graph of A is a star. Suppose in addition that the central curve E_0 has genus g and that all other curves are $\mathbf{P}^1$'s. We have a finite number of branches of chains of $\mathbf{P}^1$'s,

TABLE A.4 (log canonical singularities).

(1)   smooth points ⎫
                            ⎬ log terminal
(1)*   quotient singularities ⎭

(2)   cusp singularities

(2)*

(3)   simple elliptic singularities

(3)*

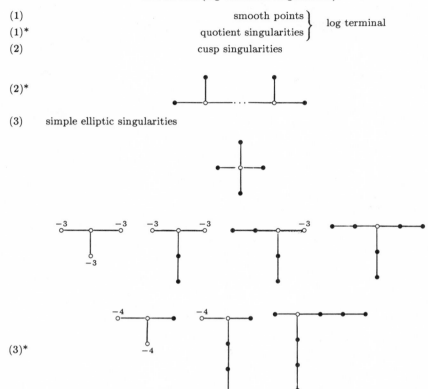

E_{ki}, $i = 1, \ldots, n_k$. Let $E_0^2 = -a_0$, $E_{ki}^2 = -a_{ki}$ $(a_{ki} \geq 2)$. Define $d_k/e_k = [a_{k1}, \ldots, a_{kn_k}]$, $\{c_{ki}\}$, $\{c'_{ki}\}$ as above. The negative definiteness of the intersection matrix is equivalent to: $a_0 - \sum e_k/d_k > 0$. With this notation we get

$$\Delta = \delta_0 E_0 + \sum \left(1 - \frac{1}{d_k}((1 - \delta_0)c_{ki} + c'_{ki}) \right) E_{ki},$$

where

$$\delta_0 = 1 + \frac{2g - 2 + \sum(1 - 1/d_k)}{a_0 - \sum e_k/d_k}.$$

REMARK A.3 [**Wa**]. Suppose that π is minimal. Let $A' \subset A$ be a proper subset, and Δ' the corresponding **Q**-divisor supported on A'. Then $\Delta > \Delta'$ unless y is an RDP. Indeed, $(\Delta - \Delta')E_j = 0$ (if $E_j \subset A'$), ≤ 0 (if $E_j \not\subset A'$). It follows from Zariski's lemma that $\Delta \geq \Delta'$. By the connectedness of A, one can complete the proof.

We now consider log canonical singularities. Assume that the resolution π is the minimal good resolution in the sense that every E_i is smooth, A has normal crossings, and every (-1)-curve in A meets the other components in more than two points. Note that Δ is not necessarily effective. The singularity y having Δ with $\delta_i \leq 1$ (resp. $\delta_i < 1$) for all i is called *log canonical* (resp. *log terminal*). In [**Ka 2**] the **Q**-Gorensteinness is in addition assumed, but it follows from the classification below. The classification of log canonical singularities is implicit in Kawamata [**Ka 1**] and in Watanabe [**Wa**]. In Table A.4 we give the list. Here ∘ denotes a smooth rational curve and • denotes a (-2)-curve.

We see this classification quickly, as follows. The exceptional set A consists of only $\mathbf{P}^1$'s unless it is a single elliptic curve, and π is the minimal resolution except for the case ⊂◯⊃ -1. To see this, let (U_0, π_0, Δ_0) be the minimal resolution of (V, y). Since π factors through π_0 by $\varphi: U \to U_0$, $\Delta = \varphi^*\Delta_0 - G$ with $G \geq 0$ and so every coefficient of $\Delta_0 \leq 1$. With the help of Remark A.3, it is easy to check that A_0 is either a single elliptic curve, a rational curve with a node, or a collection of $\mathbf{P}^1$'s having normal crossings. If the graph of A has at least two vertices with more than two branches, then it must have the shape $(2)^*$. If the graph of A has a cycle, then it coincides with this cycle (cf. [**W**]). If A is a chain of $\mathbf{P}^1$'s, then every coefficient of Δ is less than 1 (cf. Example A.1). It remains the case in which the graph of A is a star of $\mathbf{P}^1$'s with a central curve. The formula in Example A.2 says that $\delta_0 \leq 1$ if and only if $\sum(1 - 1/d_k) \leq 2$. If so, the other coefficients are less than 1. The above inequality has finite solutions: $(2, 2, d), (2, 3, 3), (2, 3, 4), (2, 3, 5)$ (log terminal); $(2, 2, 2, 2), (2, 3, 6), (2, 4, 4), (3, 3, 3)$. The log terminal singularities exactly correspond to the quotient singularities. Singularities of types (3) and $(3)^*$ have appeared as ball cusp singularities [**Ho**, **Y + H**]. Note that singularities of types $(1)^*, (2)^*, (3)^*$ are rational.

REFERENCES

[**B 1**] L. Bădescu, *Normal projective degenerations of rational and ruled surfaces*, J. Reine Angew. Math. **367** (1986), 76–89.

[**B 2**] ____ , *Some remarks on a result of Zariski*, Preprint.

[**Br**] L. Brenton, *Some algebraicity criteria for singular surfaces*, Invent. Math. **41** (1977), 129–147.

[**C**] F. Catanese, *Automorphisms of rational double points and moduli spaces of surfaces of general type*, Preprint.

[**F**] T. Fujita, *Semipositive line bundles*, J. Fac. Sci. Univ. Tokyo Ser. I A **30** (1983), 353–378.

[**G**] J. Giraud, *Surfaces d'Hilbert-Blumenthal*, Lecture Notes in Math., vol. 868, Springer-Verlag, 1981, pp. 35–57.

[**Gr**] H. Grauert, *Über Modifikationen und exzeptionelle analytische Mengen*, Math. Ann. **146** (1962), 331–368.

[H] F. Hirzebruch, *Arrangements of lines and algebraic surfaces*, Arithmetic and Geometry. II, Progr. Math., vol. 36, Birkhäuser, 1983, pp. 113–140.

[Ho] R. Holzapfel, *Ball cusp singularities*, Nova Acta Leopoldina (N.F.) **52** (1981), 109–117.

[Hö] T. Höfer, *Ballquotienten als verzweigte Überlagerungen der projectiven Ebene*, Dissertation, Bonn, 1985.

[I] K. Ivinskis, *Normale Fläche und die Miyaoka-Kobayashi Ungleichung*, Diplomarbeit, Bonn, 1985.

[K+N] M. Kato and I. Naruki, *On rational double points on quartic surfaces*, Preprint.

[Ka 1] Y. Kawamata, *On the classification of non-complete algebraic surfaces*, Lecture Notes in Math., vol. 732, Springer-Verlag, 1979, pp. 215–232.

[Ka 2] ____, *The cone of curves of algebraic varieties*, Ann. of Math. (2) **119** (1984), 603–633.

[Ka 3] ____, *Pluricanonical systems on minimal algebraic varieties*, Invent. Math. **79** (1985), 567–588.

[Kn] F. W. Knöller, *Zweidimensionale Singularitäten und Differentialformen*, Math. Ann. **206** (1973), 205–213.

[Ko] R. Kobayashi, *Einstein-Kähler V-metrics on open Satake V-surfaces with isolated quotient singularities*, Math. Ann. **272** (1985), 385–398.

[L] E. Looijenga, *Rational surfaces with effective anticanonical divisor*, Ann. of Math. (2) **114** (1981), 267–322.

[M+T] M. Miyanishi and S. Tsunoda, *Open algebraic surfaces with Kodaira dimension* −∞ *and logarithmic del Pezzo surfaces*, this volume.

[Mi] Y. Miyaoka, *The maximal number of quotient singularities on surfaces with given numerical invariants*, Math. Ann. **26** (1984), 159–171.

[Mo] M. Morales, *Une propriété asymptotique des puissances symboliques d'un idéal*, Ann. Inst. Fourier (Grenoble) **32** (1982), 219–228.

[Mor] D. Morrison, *The birational geometry of surfaces with rational double points*, Math. Ann. **271** (1985), 415–438.

[Mu 1] D. Mumford, *The topology of normal singularities of an algebraic surface and a criterion for simplicity*, Inst. Hautes Études Sci. Publ. Math. **9** (1961), 5–22.

[Mu 2] ____, *Enriques classification of surfaces in char. p*, Global Analysis, Univ. of Tokyo Press, Tokyo, 1969, pp. 325–340.

[S] K. Saito, *A new relation among Cartan matrix and Coxeter matrix*, Preprint.

[Sa 1] F. Sakai, *Enriques classification of normal Gorenstein surfaces*, Amer. J. Math. **104** (1982), 1233–1241.

[Sa 2] ____, *D-dimensions of algebraic surfaces and numerically effective divisors*, Compositio Math. **48** (1983), 101–118.

[Sa 3] ____, *Weil divisors on normal surfaces*, Duke Math. J. **51** (1984), 877–887.

[Sa 4] ____, *The structure of normal surfaces*, Duke Math. J. **52** (1985), 627–648.

[Sa 5] ____, *Ample Cartier divisors on normal surfaces*, J. Reine Angew. Math. **336** (1986), 121–128.

[W] J. Wahl, *Equations defining rational singularities*, Ann. Sci. École Norm. Sup. (4) **10** (1977), 231–264.

[Wa] K. Watanabe, *On plurigenera of normal isolated singularities. I*, Math. Ann. **253** (1980), 241–262.

[Wi] P. M. H. Wilson, *The behaviour of the plurigenera of surfaces under algebraic smooth deformations*, Invent. Math. **47** (1978), 289–299.

[Y+H] M. Yoshida and S. Hattori, *Local theory of Fuchsian systems with certain discrete monodromy groups. III*, Funkcional. Ekvac. **22** (1978), 1–49.

[Z] O. Zariski, *The theorem of Riemann-Roch for high multiples of an effective divisor on an algebraic surface*, Ann. of Math. (2) **76** (1962), 560–615.

SAITAMA UNIVERSITY, JAPAN

Proceedings of Symposia in Pure Mathematics
Volume **46** (1987)

Divisors with Finite Local Fundamental Group
on a Surface

A. R. SHASTRI

Abstract. We expound the view that the local fundamental group of a
divisor on a surface determines the combinatorics of the divisor. A classification of the divisors with finite local fundamental group on a smooth
projective surface is obtained, and some applications to the study of affine
surfaces are indicated.

0. Introduction. We consider pairs (X, F), where X is a smooth projective
surface/$\mathbf{C}$ and F is a finite connected system of algebraic curves on X. Two
pairs (X, F) and (X', F') are said to be equivalent if there is a sequence $(X, F) =
(X_0, F_0), \ldots, (X_n, F_n) = (X', F')$, where (X_{i+1}, F_{i+1}) is obtained from (X_i, F_i)
by either blowing-up a point in F_i or blowing-down an exceptional curve of the
1st kind in F_i. By blowing up if necessary one can assume that F consists of
nonsingular curves with normal crossings only. If $N = N_\varepsilon(F)$ is a small tubular
neighborhood of F in X, then $\pi_1(N - F)$ does not depend upon N and is defined
as the local fundamental group at F, which we shall denote by $\pi(F)$. It is well
known that $\pi(F)$ is an invariant of the above equivalence. In this paper we
give a classification of equivalence classes of F with $\pi(F)$ finite. Since N can
be deformed to F, finiteness of $\pi(F)$ would imply that all components of F are
simply connected and that the dual graph of F is a tree (see §1 for defintions). It
also follows that the components of F are independent in the Neron-Severi vector
space, and the intersection form restricted to the subspace generated by these
components defines a nonsingular bilinear form $B(F)$. When F is obtained by
resolving an isolated singularity, $B(F)$ is negative definite. This case was studied
in [**7**] when $\pi(F) = (1)$. Now the case of quotient singularities, viz., when
$\pi(F)$ is finite, is well understood. When one considers the compactification of
affine surfaces, we come across with (X, F) where $B(F)$ is indefinite. This was
considered for the first time in [**8**], again for the case when $\pi(F) = (1)$. In [**2**],
the case when $\pi(F)$ is the binary icosahedral group was considered. In all these,
a group-theoretic idea due to Mumford plays an important role.

1980 *Mathematics Subject Classification* (1985 *Revision*). Primary 14C20, 14J17, 14J99,
14M20.

In §1 we fix up the notations and conventions, most of which can be found in the earlier works [1, 7, 9, 6, 3] and then state the classification theorem (Theorem 1). In §2 we shall prove a rationality result answering a question of Van de Ven partially. Also we shall summarize results from [3] and [4] and indicate the role of Theorem 1 in proving these results. In §3 we prove a lemma by a simple application of Dehn's lemma which does not need much expertise in 3-dimensional topology. In §4 we complete the proof of Theorem 1 by arguing more or less as in [9].

Walter Neumann has obtained a most general classification of plumbing graphs, and it seems easier to deduce our Theorem 1 from that of 4.1 of [8]. However, some of the "moves" described by Neumann are not geometric, and so his result is not ready-made for geometric use. One has to show that the nongeometric moves are not necessary in our context, and this itself involves arguments used herein; this doesn't cut down the number of pages.

1. Preliminaries and the statement of the classification theorem.
Recall some terminologies about weighted graphs: T denotes a graph with vertices v_i (u_i, etc.), having weights $\Omega_{v_i} \in \mathbf{Z}$. The links (or edges) of T will be denoted by $[u; v]$, where u and v are vertices linked in T. $T - \{v\}$ will denote the subgraph of T obtained by removing v and all the links $[v; u]$ of T, at v. A graph is a tree if it is connected and has no proper loops. For a tree T, the components of $T - \{v\}$ are called *branches* of T at v. A vertex v is called *free* (*linear* or a *branch* point) if T has at most one branch (at most two or at least three branches, respectively) at v. A branch Γ of T at v is *simple* if it does not have any branch points of T. An *extremal branch point* is a branch point at which at most one branch is not simple. Clearly a finite tree always has an extremal branch point. A tree is linear if all its vertices are linear. For instance, a simple branch is necessarily linear.

Associated to T is the *bilinear form* $B(T)$, on the real vector space spanned by the vertex set of T, defined as follows:

$$v_i \cdot v_i = \Omega_{v_i} \quad \forall i,$$

$$v_i \cdot v_j = \begin{cases} 1 & \text{if } [v_i; v_j] \text{ is a link in } T, \\ 0 & \text{otherwise, for } i \neq j. \end{cases}$$

The *discriminant* of this form is denoted by $d(T)$.

Now let T be a finite tree. Fix an ordering of the vertices. Then the local fundamental group $\pi(T)$ of T is defined as the quotient of the free group over $\{v_i\}$ by the relations

(a) $[v_i, v_j] = e$ if $[v_i; v_j]$ is a link in T, $i \neq j$.

(b) For each vertex v, if $\{v_{i_1}, \ldots, v_{i_k}\}$ is the set of vertices linked to v, $i_1 < i_2 < \cdots < i_k$, then $v_{i_1} \cdots v_{i_k} \cdot v^{\Omega_v} = e$.

It can be checked that the isomorphism class of $\pi(T)$ does not depend upon the choice of ordering of vertices. The abelianized group $\mathrm{ab}\,\pi(T)$ is of finite order if and only if $d(T) \neq 0$ and then $\mathrm{ord}(\mathrm{ab}\,\pi(T)) = d(T)$.

We say T is spherical (or cyclic of order $\leq n$) if $\pi(T) = (e)$ (or is cyclic of order $\leq n$), etc.

Now for a pair (X, F), as in the introduction, where all components of F are rational, the dual weighted graph $T = T(X, F)$ is defined as follows: The vertices of T are the lines in F; the weight Ω_C is the self-intersection number (C^2); two vertices C and D are linked in T if and only if $(C.D) \neq 0$. If we assume $\pi(F)$ is finite, then it follows that $T = T(X, F)$ is a tree. Moreover, we have $\pi(F) \simeq \pi(T)$ (see [7] or [9]). Also it is easily seen that $B(F) \simeq B(T)$ under the obvious identification of basic elements.

Definition of "blow-up" and "blow-down." Let $[u; v]$ be a link in T. By "blow-up at $[u; v]$" we mean to obtain a new tree T' as follows: Introduce a new vertex w in T, delete the link $[u; v]$, and introduce links $[u; w]$ and $[w; v]$. Define the new weights Ω' by

$$\Omega'_x = \begin{cases} \Omega_x, & \text{if } x \neq u, v, w, \\ \Omega_x - 1, & \text{if } x = u \text{ or } v, \\ -1 & \text{if } x = w. \end{cases}$$

Now let v be a free vertex in T. By "blow-up at v" we mean to obtain a new tree T' as follows: Introduce a new vertex w and a new link $[v; w]$, and define the new weights Ω' by

$$\Omega'_x = \begin{cases} \Omega_x, & \text{if } x \neq v, w, \\ \Omega_v - 1, & \text{if } x = v, \\ -1 & \text{if } x = w. \end{cases}$$

"Blow-down" is described precisely as the inverse process of blow-up and, as such, we need to have a linear vertex w with $\Omega_w = -1$ to perform the blow-down on a given tree T.

Two trees are equivalent if there is a finite chain of blow-ups and blow-downs to obtain one from the other. A tree is minimal if it has no linear vertex v with $\Omega_v = -1$. Every finite tree is equivalent to a minimal one, which may be empty. There is no uniqueness, in general, though if $B(T)$ is negative definite, then there is a unique minimal tree in the equivalence class of T. If T' is obtained from T by blowing-up once, then obviously $B(T') \simeq B(T) \perp (-1)$.

We say T satisfies the hypothesis (E) if every positive semidefinite subspace W of $B(T)$ is of real dimension ≤ 1. We say T satisfies (R) if no tree equivalent to T has a subtree of the form

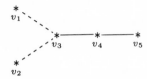

with $\Omega_{v_3} = -1$ and $\Omega_{v_5} \geq 0$.

REMARK. Clearly if T satisfies (E), then every subtree of T also satisfies (E) and every T' equivalent T also satisfies (E). Further, there can be at most two vertices with nonnegative weights, and if there are two of them, then these two vertices are linked in T and one of the weights is actually zero.

For any tree T and a vertex $v \in T$, let $T^{(v)}$ denote the tree obtained from T by adding two more vertices x and y, say, with $\Omega_x = 0 = \Omega_y$, and two more links, viz., $[v; x]$ and $[x; y]$. It is easily seen that $\pi(T) \simeq \pi(T^{(v)})$ and $B(T^{(v)}) \simeq B(T) \perp \left(\begin{smallmatrix} 0 & 1 \\ 1 & 0 \end{smallmatrix} \right)$.

LEMMA 1.1. (i) $T^{(v)}$ satisfies (E) if and only if T is negative definite.
(ii) $T^{(v)}$ is minimal and satisfies (R) implies T is minimal.

PROOF. (i) follows from the remark that $B(T^{(v)}) \simeq B(T) \perp \left(\begin{smallmatrix} 0 & 1 \\ 1 & 0 \end{smallmatrix} \right)$. To see (ii), note that T may fail to be minimal only if v is linear in T and $\Omega_v = -1$. Since $T^{(v)}$ is minimal, v is not free in T. Hence $T^{(v)}$ will have a subtree of the form

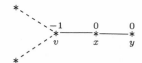

contradicting (R).

LEMMA 1.2. Suppose T has a subtree of the form $* \!\!-\!\!-\!\!-\!\! * \!\!-\!\!-\!\!-\!\! *$, where v is

linear and $\Omega_v = 0$. Then T is equivalent to a tree T' which is obtained from T by merely changing the weights at u and w to $\Omega_u + 1$ and $\Omega_w - 1$ respectively.

PROOF. Blow-up $[v; w]$ to obtain $* \!\!-\!\!-\!\!-\!\! * \!\!-\!\!-\!\!-\!\! * \!\!-\!\!-\!\!-\!\! *$ with weights Ω_u, $-1, -1$, and $\Omega_w - 1$ respectively. Now blow-down the vertex v_1 (and rename v_2 as v).

LEMMA 1.2'. Suppose $v \in T$ is a free vertex linked to $u \in T$, and $\Omega_v = 0$. Then T is equivalent to a tree T' which is obtained by changing the weight at u to any number.

PROOF. Blow-up at the free vertex v to obtain $* \!\!-\!\!-\!\!-\!\! * \!\!-\!\!-\!\!-\!\! *$ with weights $\Omega_u, -1, -1$ respectively. Now blow-down v_1 and rename v_2 as v. The result is that the weight at u has increased by 1. Repeat it as often as needed. To decrease the weight reverse the process.

THEOREM 1.3. If $T = T(X, F)$, where X is a nonsingular, projective surface/$\mathbf{C}$, then T satisfies (R).

PROOF. First, suppose X contains two lines C_1 and C_2 such that $(C_1^2) \geq 0$ and $(C_1.C_2) = 1$. Then we claim that $\pi_1(X) = (e)$. Since monoidal transformations do not change $\pi_1(X)$, using the lemma above, we can assume that either $(C_1^2) > 0$ or $(C_2^2) > 0$. Now apply Lefschetz's argument to conclude that $\pi_1(C_i) \to \pi_1(X)$ is surjective where $(C_i^2) > 0$. Hence $\pi_1(X) = (e)$. Now, to prove the lemma, if T violates (R), then without loss of generality we can assume that T itself contains a subtree of the form

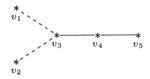

with $\Omega_{v_3} = -1$ and $\Omega_{v_5} \geq 0$. By the above observation this implies $\pi_1(X) = (e)$, and hence $H^1(X, \mathcal{O}_X) = 0$. But Lemma 6 of [8] precisely states that if $H^1(X, \mathcal{O}_X) = 0$, then T satisfies (R). This proves the lemma.

We now recall some notations from [1]. For any positive integers, $0 < \lambda < n$, such that $(n, \lambda) = 1$ let $\langle n, \lambda \rangle$ denote the negative definite linear tree $\overset{-a_1}{*}\!\!\!\rule[0.5ex]{1.5em}{0.4pt}\!\!\!\overset{-a_2}{*}\!\cdots\overset{-a_{k-1}}{*}\!\!\!\rule[0.5ex]{1.5em}{0.4pt}\!\!\!\overset{-a_k}{*}$ where $a_i \geq 2$ are integers defined by

$$n/\lambda = a_1 - \cfrac{1}{a_2 - \cfrac{}{\ddots - \cfrac{1}{a_k}}}$$

Let $\langle\langle n, \lambda \rangle\rangle$ denote the tree $\overset{0}{*}\!\!\!\rule[0.5ex]{1.5em}{0.4pt}\!\!\!\overset{0}{*}\!\!\!\rule[0.5ex]{1.5em}{0.4pt}\!\!\!\overset{-a_1}{*}\!\cdots\overset{-a_k}{*}$. Note that if $\lambda\lambda' \equiv 1 \pmod{n}$, then $\langle n, \lambda \rangle = \langle n, \lambda' \rangle$ and applying Lemma 1.2 successively, it can be seen that $\langle\langle n, \lambda \rangle\rangle$ is equivalent to $\langle\langle n, \lambda' \rangle\rangle$. For n_j and λ_j as above, define a_{ji} similarly, and for any $a \in \mathbf{Z}$ let $\langle a; n_1, \lambda_1; n_2, \lambda_2; n_3, \lambda_3 \rangle$ denote the tree

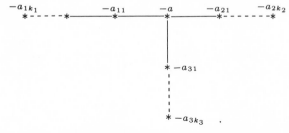

THEOREM 1. *Let T be a finite weighted tree satisfying* (E) *and* (R). *Then $\pi(T)$ is finite if and only if T is equivalent to one of the following trees:*

(i) *the empty tree $\varnothing$ or $\overset{0}{*}\!\!\!\rule[0.5ex]{1.5em}{0.4pt}\!\!\!\overset{0}{*}$;*

(ii) $\langle n, \lambda \rangle$ *for $0 < \lambda < n$, $(n, \lambda) = 1$;*

(iii) $\langle a; 2, 1; n_2, \lambda_2; n_3, \lambda_3 \rangle$, *where $\{n_2, n_3\}$ is one of the pairs $\{3, 3\}$, $\{3, 4\}$, $\{3, 5\}$, or $\{2, n\}$, for some $n \geq 2$ and $0 < \lambda_i < n_i$, $(n_i, \lambda_i) = 1$, and $a \geq 2$;*

(iv) *the trees as mentioned in* (iii) *except that $a \leq 1$;*

(v) *the trees $T_1^{(v)}$, where T_1 is one of the trees in* (ii) *or* (iii) *and $v \in T_1$ is any vertex.*

It is easily checked that these trees satisfy (E), in general. Those which are negative definite satisfy (R), trivially. Also, the trees $\langle\langle n, \lambda \rangle\rangle$ and those listed in (iv) satisfy (R). It is not clear whether $T_1^{(v)}$ mentioned in (v) satisfy (R) or not. In fact, some of them do not arise geometrically. However, it is possible to determine all those which arise geometrically. We shall not go into this. The proof of this theorem will be completed in §3 and §4.

2. Some applications.

THEOREM 2. *Let X be a smooth projective surface/$\mathbb{C}$, and let F be a connected system of curves on X with the local fundamental group $\pi(F)$ being finite. Suppose there is a divisor D supported on F such that $D^2 > 0$. Then X is rational.*

Van de Ven [10, p. 197] asks whether all algebraic nonsingular compactifications of a homology 2-cell V are rational. Whereas the above theorem answers this partially, viz., when $\pi_1^\infty(V) < \infty$, it is more general, in that it does not assume any homology restriction on V itself.

To prove the above theorem we need the following elementary fact:

LEMMA 2.1. *Let X be a smooth projective surface, and let C be an irreducible curve on X with $(C^2) \geq 0$. Suppose $(C.K) < 0$. Then X is birationally a ruled surface.*

Here K denotes the canonical divisor. The lemma is obvious if X is minimal. If not, blow-down as often as needed to make X minimal and see that the image of C satisfies the same hypothesis.

PROOF OF THE THEOREM. By blowing-up if necessary, we can assume that F is a connected system of smooth curves with normal crossings (and is minimal with respect to this property). Now $\pi(F)$ is finite and so all the components are rational and the dual graph is a tree. Further the fact that $\pi(F)$ is finite implies the discriminant of $B(F)$ is nonzero, and hence F supports a divisor D with $(D^2) > 0$. By the Hodge index theorem, it follows that F satisfies (E). By Lemma 1.3, F satisfies (R). Hence we can assume that the dual tree T is as in (iv) or (v) of Theorem 1. In the latter case we have a line $C \in F$ such that $(C^2) = 0$. Hence $(C.K) = -2$. By Lemma 2.1, X is birationally a ruled surface. In the former case T has a subtree of the form

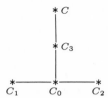

with $(C_0^2) = -1$, $(C_1^2) = -2$, $(C_2^2) = -a$, $(C_3^2) = -b$ say, where $(a, b) = (2, n), (3, 3), (3, 4)$ or $(3, 5)$. Blow-down C_0 and then the image of C_1; denote the images of C_2 and C_3 by the same letters. Then $(C_2^2) = -a + 2$, $(C_3^2) = -b + 2$. If $a = 2$, then we have $(C_2^2) = 0$ and $(C_2.K) = -2$ as before. Otherwise $(C_2^2) = -3 + 2 = -1$. Note that $(C_2.C_3) = 2$. Hence, when we blow-down C_2, the image C_3' of C_3 is a rational curve with an ordinary cusp of multiplicity 2 and $(C_3)'^2 = -b + 2 + 4 = -b + 6$. Again by the adjunction formula $(C_3' \cdot K) = -(C_3'^2) = b - 6 < 0$. Thus in any case X is birationally a ruled surface. As before by a Lefschetz-type argument, X is simply connected also. Hence X is rational.

REMARK. R. V. Gurjar has obtained an alternative proof of this theorem.

THEOREM 3 (MIYANISHI). *Let V be a normal affine surface/$\mathbf{C}$ and let $\varphi\colon \mathbf{C}^2 \to V$ be a proper morphism onto V. Then*
(i) *$V \simeq \mathbf{C}^2$ as an affine variety if V is nonsingular;*
(ii) *in general, $V \simeq \mathbf{C}^2/G$, where G is a small, finite subgroup of $\mathrm{GL}(2, \mathbf{C})$.*

More generally, in [4] one has

THEOREM 4 (CHARACTERIZATION OF $\mathbf{C}^2/G$ AS AN AFFINE SURFACE). *Let V be an affine normal surface/$\mathbf{C}$ which is topologically contractible. Assume further that the fundamental group at infinity of V, $\pi_1^\infty(V)$, is finite. Then $V \simeq \mathbf{C}^2/G$, where G is a small subgroup of $\mathrm{GL}(2, \mathbf{C})$ isomorphic to $\pi_1^\infty(V)$. In particular, V has at most one singular point.*

Under the hypothesis of Theorem 3 it is proved in [2] that V is contractible. Further the proper map $\varphi\colon \mathbf{C}^2 \to V$ also tells us that $\pi_1^\infty(V)$ is finite. Then Theorem 3 is an immediate consequence of Theorem 4.

The proof of Theorem 4 can be summarized as follows: Let P be the set of singularities of V. By a Lefschetz-type argument, observe that $\pi_1(V - P)$ is a quotient of $\pi_1^\infty(V)$ and hence finite. Let $\pi'\colon \overline{V}' \to V - P$ be the universal covering projection, and let $\overline{V}$ be an affine normal surface containing $\overline{V}'$ such that $\overline{V} - \overline{V}' = \overline{P}$ consists of a finite number of points. Extend π' to a proper morphism $\pi\colon \overline{V} \to V$. It suffices to show that $\overline{V} \simeq \mathbf{C}^2$. One first shows that $\overline{V}$ is also contractible. $\pi_1^\infty(\overline{V})$, being a subgroup of $\pi_1^\infty(V)$, is finite. Using Poincaré duality and the fact that $\pi_1(\overline{V}') = (1)$ one shows that $\pi_1^\infty(\overline{V})$ is perfect. Hence it is either trivial or isomorphic to the binary icosahedral group.

Let Y be a projective normal surface containing $\overline{V}$ such that $F_\infty = Y - \overline{V}$ has smooth components with normal crossings, and let Y be smooth near F_∞. Let $X \to Y$ be the minimal resolution. By definition $\pi_1^\infty(\overline{V})$ is isomorphic to the local fundamental group at F_∞. Thus by Theorem 1, we may assume that the dual tree corresponding to F_∞ is either $\overset{0}{*}\!\!-\!\!-\!\!\overset{0}{*}$ or $\langle 1; 2, 1; 3, 1; 5, 1\rangle$ respectively. Using the theory of elliptic fibrations, one shows that the latter case cannot occur. In the former case one obtains X as a ruled surface over $\mathbf{P}^1$ and shows that Y itself is smooth. Now it is easily seen that $\overline{V} \simeq \mathbf{C}^2$.

3. Some auxiliary results. For the study of trees with branch points we need a stronger version of a group-theoretic result due to Mumford. Let $G_1, \ldots, G_n$ be any nontrivial groups, and $a_i \in G_i$, $i = 1, \ldots, n$, any elements. Let $\tau(G_1, \ldots, G_n)$ denote the quotient group of the free product $G_1 * \cdots * G_n$ by the single relation $a_1 * \cdots * a_n = e$. For $n = 3$, and $G_1 \simeq \mathbf{Z}/(\lambda_i)$ and $a_i \in \mathbf{Z}/(\lambda_i)$ being generators, $\tau(G_1, G_2, G_3)$ is denoted by $\tau(\lambda_1, \lambda_2, \lambda_3)$, where $\lambda_i \geq 2$ are integers. These are classically known as triangle groups, all nontrivial, noncyclic, and those which are finite among them are all known. We sum up all these into:

LEMMA 3.1. *Let* $G_1, \ldots, G_n$ *be any nontrivial groups, and* $a_i \in G_i$ *any elements. Then*

(i) *for* $n \geq 4$, $\tau(G_1, \ldots, G_n)$ *is infinite.*

(ii) $\tau(G_1, \ldots, G_n)$ *is nontrivial, noncyclic, for* $n \geq 3$.

(iii) $\tau(G_1, G_2, G_3)$ *is finite* $\Leftrightarrow$ G_i *are all cyclic groups generated by* a_i, *say* $G_i = \mathbf{Z}/(\lambda_i)$ *with* $(\lambda_1, \lambda_2, \lambda_3) = (2, 2, k)$ $(k \geq 2), (2, 3, 3), (2, 3, 4)$, *or* $(2, 3, 5)$.

REMARK. For a proof of (i) and (ii), see [3] or [7]. (iii) is classical. Weaker versions of this result have been employed in [7, 9, 3] in a typical way as in the following:

LEMMA 3.2. *Let* T *be a minimal tree and* v *be any vertex in* T. *If* $\pi(T)$ *is cyclic, then there are most two nonspherical branches at* v. *If* $\pi(T)$ *is finite, then there can be at most three nonspherical branches at* v; *further if there are three nonspherical branches, then they are all finite cyclic.*

PROOF. Let $T_1, \ldots, T_n$ be the branches of T at v. From the standard presentation of $\pi(T)$, one obtains

$$\pi(T)/(v) \simeq \tau(\pi(T_1), \ldots, \pi(T_n)) = \pi(T_1) * \cdots * \pi(T_n)/(v_1 \cdots v_n),$$

where $v_i \in T_i$ is linked to v. Now apply the above lemma.

We now bring in a topological concept which will help us to conclude that certain trees T with two branch points have $\pi(T)$ infinite. For details about plumbing 2-disc bundles over S^2 along a tree T, see [6, Chapter 8]. We are interested in the following situation: Let T be a finite tree, and $v_0 \in T$ a linear vertex such that T_1 and T_2 are the two nonempty branches of T. Let Δ, δ_1, and δ_2 denote the discriminants of T, T_1, and T_2 respectively. Let $M = \partial P(T)$, where $P(T)$ is the 4-dimensional real manifold obtained by plumbing 2-disc bundles over S^2 along T. Let $S_{v_0}^2$ denote the 2-sphere at the base corresponding to the vertex v_0 in the plumbing data; w_0 is a fibre of the principal S^1-bundle, $\xi_{(m_0)} \to S_{v_0}^2$, such that $w_0 \subset M$. For a sufficiently small open 2-disc $D^2 \subset S_{v_0}^2$, we can assume that $\xi_{(m_0)}/D^2 \subset M$ is a tubular neighborhood of w_0 in M. Using some trivialization of $\xi_{(m_0)}/D^2$ we identify it with $D^2 \times w_0$. Let $W = M - D^2 \times w_0$.

LEMMA 3.3. *Suppose* $\pi(T)$ *is cyclic,* $\delta_1 \cdot \delta_2 \neq 0$, *and* $\pi(T_i) \neq (e)$ *for* $i = 1, 2$. *Suppose further that* $\partial P(T)$ *is either a lens space or* $S^2 \times S^1$. *Then* W *is incompressible in* W, *i.e., the inclusion map* $\iota: \partial W \to W$ *induces monomorphism in* π_1.

PROOF. Assuming on the contrary by the Loop theorem and Dehn's lemma (see [5, Chapter 4]) it follows that there is a properly imbedded 2-disc A in W with A representing a nontrivial element of $\pi_1(\partial W)$. Let U_1 be a small collar of ∂W, and U_2 a small thickening of A in W. Then $\partial(U_1 \cup U_2) = \partial W \amalg B$, say, where B is a 2-sphere in $W \subset M$. Clearly B separates M, and since $M = \partial P(T)$ is either a lens space or $S^2 \times S^1$, B bounds a 3-disc D in M. Clearly either $D \subset W$ or $D^2 \times w_0 \subset D$.

Case 1. Suppose $D \subset W$. It follows that W is a solid torus. Now consider a tree Γ obtained by joining a new vertex u with $\Omega_u = -m$, $m \geq 2$, at the vertex v_0 to T. Then $\partial P(\Gamma)$, being the union of two solid tori, is a lens space or $S^2 \times S^1$, and hence $\pi_1 \partial P(\Gamma) = \pi(\Gamma)$ is cyclic. Using the standard presentation of $\pi(\Gamma)$, it follows that $\pi(\Gamma)/(v_0) \simeq \tau(\pi(T_1), \pi(T_2), \mathbf{Z}/(m))$. By Lemma 3.1, either $\pi(T_1)$ or $\pi(T_2)$ has to be trivial, which is a contradiction.

Case 2. Suppose $D^2 \times w_0 \subset D$. It follows that w_0 is an unknotted loop in D, and hence D is the union of two solid tori $D^2 \times w_0$ and $S^1 \times D_0^2$ say, identified along the boundaries by a diffeomorphism $h: \partial(D^2 \times w_0) \to \partial(S^1 \times D_0^2)$ reversing the orientations, and represented by a matrix $\begin{pmatrix} -0 & 0 \\ \mu & 1 \end{pmatrix}$ for some integer μ. It follows that $\partial P(\Gamma)$, where Γ is the tree considered in case 1, is a connected sum $L \,\#\, \partial P(T)$ for some lens space $L = D^2 \times S^1 \,\mathrm{II}_g\, S^1 \times D_0^2$, where now g will be represented by the product matrix

$$\begin{pmatrix} -1 & 0 \\ \mu & 1 \end{pmatrix} \begin{pmatrix} 0 & 1 \\ 1 & 0 \end{pmatrix} \begin{pmatrix} -1 & 0 \\ -m & 1 \end{pmatrix} \begin{pmatrix} -m & -1 \\ -\mu - 1 & \mu \end{pmatrix}.$$

In particular $H_1(L) = \mathbf{Z}/(m)$, and hence $H_1(\partial P(\Gamma))$ is of order $|m\Delta|$. On the other hand, the discriminant $d(\Gamma)$ is given by $d(\Gamma) = m\Delta - \delta_1 \cdot \delta_2$, and hence $\pm m\Delta = m\Delta - \delta_1 \cdot \delta_2$. This is true for all $m \geq 2$. Hence $\delta_1 \cdot \delta_2 = 0$, which is a contradiction. This completes the proof of Lemma 3.3.

As a consequence of this we derive:

LEMMA 3.4. *Let T be a tree of the form*

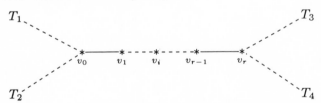

where $\gamma_1 = T_1 \text{---}*_{v_0}\text{---} T_2$ and $\gamma_r = T_3 \text{---}*_{v_r}\text{---} T_4$ satisfy the conditions of Lemma 3.3 and all $v_1, \ldots, v_{r-1}$ are linear in T, $r \geq 1$. Then $\pi(T)$ contains a subgroup isomorphic to $\mathbf{Z} \oplus \mathbf{Z}$.

PROOF. Let $M_i = \partial P(\gamma_i), i = 0, \ldots, r$. Let w_i denote the fibres of the S^1-bundles over S^2 corresponding to the vertices v_i, contained in $M_i, i = 0, \ldots, r$, respectively. Then by Lemma 3.3

$$\iota_\#: \pi_1(\partial D^2 \times w_i) \to \pi_1(M_i - \overset{\circ}{D}{}^2 \times w_i)$$

are monomorphisms for $i = 0, \ldots, r$. Let h_i be an orientation-reversing diffeomorphism of $S^1 \times S^1$, represented by the matrix

$$\begin{pmatrix} -1 & 0 \\ n_i & 1 \end{pmatrix},$$

where $\Omega_{v_i} = n_i$. Then as in [**6**, pp. 67–68] it follows that

$$\partial P(T) = (M_0 - \overset{\circ}{D}{}^2 \times w_0)\,\mathrm{II}_h\,(M_r - \overset{\circ}{D}{}^2 \times w_r)$$

with $h = h_r \circ t \circ h_{r-1} \circ t \circ \cdots \circ h_0$, where t is the involution of $S^1 \times S^1$ given by $(x, y) \to (y, x)$. By Van Kampen's theorem it follows that $\pi_i(\partial P(T))$ is given by the pushout diagram

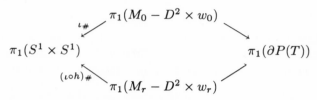

which is actually an amalgamated free product of groups. In particular $\pi_1(S^1 \times S^1)$ is a subgroup of $\pi_1(\partial P(T))$. This completes the proof of Lemma 3.4.

4. Completion of the proof of Theorem 1. We begin with:

LEMMA 4.1. *Let T be a minimal tree, $u \in T$ a vertex, and γ a simple branch at u with some nonnegative weights. Suppose $\pi(\gamma)$ is finite. Then T is equivalent to a minimal tree T_1 obtained by replacing the branch γ by an equivalent branch*

$$\gamma = *\!-\!-\!-\!-\overset{0}{\underset{x}{*}}\!-\!-\!-\!-\overset{0}{\underset{y}{*}},$$

where the vertex y is free in T_1, and possibly changing the weight at u.

PROOF. Let $v \in \gamma$ be such that $\Omega_v \geq 0$. Let v' be the vertex in γ linked to v and "nearer" to u (or if such a vertex does not exist, then let $v' = u$). Blow-up $[v'; u]$ if necessary, successively, to make $\Omega_v = 0$. Then using Lemma 1.2 to shift the weights across v, we can assume that the free vertex $y \in \gamma$ has $\Omega_y = 0$. Since $\pi(\gamma)$ is finite, at this stage $\gamma \neq \overset{0}{*}$ and hence there is a vertex $x \in \gamma$, linked to y in γ, and $x \neq y$. If there is any vertex v with $\Omega_v = -1$ in γ, $v \neq x$, blow-down, so that $\gamma - \{x, y\}$ is minimal. Now apply Lemma 1.2 to make $\Omega_x = 0$. At some stage it is possible that we might have altered the weight at u, but the rest of the tree T is unaltered. Hence the lemma.

From the above lemma it follows that if T is a linear tree satisfying (E) and $\pi(T)$ finite, then $T \simeq \langle n, \lambda \rangle$ or $\langle\langle n, \lambda \rangle\rangle$ or $\overset{0}{*}\!\!-\!\!-\!\!-\!\!\overset{0}{*}$. We next deal with trees with one branch point:

LEMMA 4.2. *Let T be a minimal tree satisfying (E) and (R), and let $\pi(T)$ be finite. Then at any branch point v, no branch can be a single vertex u with $\Omega_u = 0$.*

PROOF. Assuming on the contrary if $T_1, \ldots, T_n$ are the other branches of T at v, it follows from the standard presentation that $\pi(T) \simeq \pi(T_1) * \cdots * \pi(T_n)$. Hence one of the T_i is spherical, and also T_i is negative definite. Hence by

Mumford's result (see [77]) T_i is not minimal. That means there is $w \in T_i$, such that $\Omega_w = -1$, $[w, v]$ is a link in T, and w is a branch point in T. So T contains a subtree

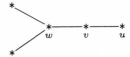

contradicting (R).

THEOREM 4.3. *Let T be a tree with exactly one branch point v, and having weights on each branch ≤ -2. Then $w(T)$ is finite if and only if $T = \langle a; 2, 1; n_2, \lambda_2; n_3, \lambda_3 \rangle$ with the restrictions stated in* (iii) *and* (iv) *of Theorem 1.*

PROOF. $\pi(T) < \infty$ implies, by Lemma 4.2, that there are precisely three branches of T at v, say $\langle n_i, \lambda_i \rangle$, $i = 1, 2, 3$, with $\pi(\langle n_i, \lambda_i \rangle) \simeq \mathbf{Z}/(n_i)$ and then $\pi(T)/(v) \simeq \tau(\mathbf{Z}/(n_1), \mathbf{Z}/(n_2), \mathbf{Z}/(n_3))$, and hence by Lemma 4.1 (iii), $n_1 = 2$, and n_2, n_3 are as stated in Theorem 1. Conversely, we have a central extension

$$(e) \to (v) \to \pi(T) \to \tau(2, n_2, n_3) \to (e),$$

where $\tau(2, n_2, n_3)$ is finite; (v) is the central subgroup generated by $v \in \pi(T)$. It suffices to show that (v) is finite. If $(v) \simeq \mathbf{Z}$, by a standard result about central extensions, it follows that $\mathrm{ab}\,\pi(T)$ is infinite; equivalently the discriminant $d(T) = 0$. But then $d(T) = (-2a + 1)n_2 n_3 + 2\lambda_2 n_3 + 2\lambda_3 n_2$ is easily seen to be nonzero for the specific values of $\{n_2, n_3\}$ as mentioned in Theorem 1. This proves the lemma.

For a tree T satisfying (E) and (R) and having one branch point with $\pi(T) \simeq \mathbf{Z}/(n)$, it follows from Lemma 3.2 that one of the branches is spherical, and hence by Lemma 4.1, we have $T \simeq T_1^{(v)}$ for some $v \in T_1 = \langle n, \lambda \rangle$. In view of Lemma 4.1, Theorem 1 would follow from:

PROPOSITION. *Let T be a minimal tree satisfying* (E) *and* (R), $2 \leq k =$ *number of branch points of T. If all simple branches of T carry negative weights, then $\pi(T)$ is noncyclic and infinite.*

We shall prove this proposition by induction on k; so denote the statement for trees with k number of branch points by P_k. Consider the following statement $(k \geq 2)$; $\overline{\mathrm{P}}_k$: "Let T be a minimal tree satisfying (E) and (R) and having no spherical simple branches, and number of branch points $= k \geq 2$. Then $\pi(T)$ is noncyclic and infinite."

Obviously, in view of Lemma 4.1, $\mathrm{P}_k \Rightarrow \overline{\mathrm{P}}_k$. Now to prove P_k we shall assume that $\pi(T)$ is cyclic or finite and arrive at some contradiction, for $k = 2, 3$, and ≥ 4, separately.

Case $k = 2$. Let u and v be the branch points of T. We first claim that there are exactly two simple branches of T at u and v. If possible suppose at v, T has more than two simple branches. Let γ be the nonsimple branch at u. Then γ is minimal, has one branch point, and has more than two nonspherical branches.

But $\pi(T)/(u)$ is cyclic or finite, and hence $\pi(\gamma)$ has to be cyclic, contradicting Lemma 3.2.

Thus we can apply Lemma 3.4, with $v_0 = v$, $v_r = u$ say; T_i are all linear, nonspherical, and so γ_1 and γ_2 are linear. Thus $P(\gamma_i)$ are either lens spaces or $S^2 \times S^1$ (see [6]). All the conditions of Lemma 3.4 are satisfied, and hence $\pi(T)$ contains $\mathbf{Z} \oplus \mathbf{Z}$, which is a contradiction. Hence P_2 and thereby $\overline{P}_2$ are proved.

Now assume $k \geq 3$. We shall first claim

LEMMA 4.4. *Under the induction hypothesis that* P_l *holds for* $2 \leq l \leq k-1$, *at any extremal branch point* v *of* T, T *looks like*

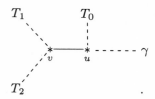

with T_0, T_1, T_2 *being simple and* $\Omega_u = -1$.

PROOF OF THE LEMMA. Let $T_1, \ldots, T_n$, $n \geq 2$, be the simple branches and γ the nonsimple branch at v. $\pi(T_i)$ are nontrivial, and hence $\pi(\gamma)$ has to be finite cyclic. If $l = $ no. of branch points of γ, then $1 \leq k - 2 \leq l \leq k - 1$. If $l = k - 1$, then γ is minimal. Hence γ should have a spherical simple branch, say $\gamma = \gamma_1^{(v)}$, for some negative definite minimal γ_1 with $k - 2$ branch points. $\pi(\gamma) \simeq \pi(\gamma_1)$ is finite cyclic. So by the induction hypothesis $k - 2 \leq 1$, and hence $k - 2 = 0$ by Lemma 4.2, which is a contradiction. So $l = k - 2$; then there is a branch point u of T, linear in γ, which is linked to v. If $\Omega_u \neq -1$, then γ is minimal. Also now no simple branch of γ can be spherical. Hence by the induction hypothesis $l = 1$, again contradicting Lemma 3.2. So $\Omega_u = -1$. Further since u is the only vertex in γ with $\Omega_u = -1$, it follows that one of the branches, say T_0, at u is linear. It remains to show that $n = 2$. If $n \geq 3$, then looking at $\pi(T)/(u)$, one concludes that γ is spherical. This immediately implies that γ' is not minimal and so there is a vertex w, linear in γ, with $\Omega_w = -1$ which is linked to u. Further one of the branches, say T_0', of γ' at w should be linear:

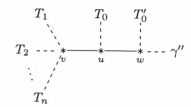

Looking back at γ, we conclude that all the weights on T_0 should be equal to -2. If r is the no. of vertices on T_0, then γ is blown-down successively to:

with the weight at w changed to $\Omega_w + r + 1 = r \geq 1$. Hence $\pi(\gamma)$ is nontrivial. Again since $n \geq 3$, $\pi(T)/(v)$ cannot be finite or cyclic. This contradiction shows that $n = 2$ completing the proof of the lemma.

We shall now consider the case $k = 3$ separately. From the above Lemma 4.4 it follows that T looks like

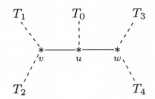

with all the branches T_i being simple and $\Omega_u = -1$, and all the weights on T_i being ≤ -2. Let $\gamma = T_1 \text{-}\text{-}\text{-}*\text{-}\text{-}\text{-}T_2$ and $\gamma' = T_3 \text{-}\text{-}\text{-}*\text{-}\text{-}\text{-}T_4$. Let Γ be the nonsimple branch at v and Γ'' the nonsimple branch at w. As before, it follows that $\pi(\Gamma)$ is finite cyclic and hence $T_0 \text{-}\text{-}\text{-}*$ is spherical. Hence all the weights on T_0 should be equal to -2 and if r is the number of vertices on T_0, Γ is blown-down to $T_3 \text{-}\text{-}\text{-}*\text{-}\text{-}\text{-}T_4$ with the weight at v changed to $\Omega_w + r + 1$. Similarly Γ' is blown-down to $T_1 \text{-}\text{-}\text{-}*\text{-}\text{-}\text{-}T_2$ with the weight at v changed to $\Omega_v + r + 1$. If $d(\gamma) \neq 0$ and $\pi(\gamma) \neq (e)$, then we could apply Lemma 3.4 to Γ' and γ' in place of γ_1 and γ_r of Lemma 3.4, to conclude that $\pi(T)$ contains $\mathbf{Z} \oplus \mathbf{Z}$. Hence $d(\gamma) = 0$ or $\pi(\gamma) = (e)$. Similarly $d(\gamma') = 0$ or $\pi(\gamma') = (e)$.

In any case it follows that $\Omega_v = \Omega_w = -1$. Now looking at $\pi(T)/(v)$, it follows that $\pi(\gamma)$ or $\pi(\gamma')$ is finite, say $\pi(\gamma)$ is finite, i.e., $d(\gamma) = 0$. Hence $\pi(\gamma) = (e)$. On the other hand, since Γ' is equivalent to $T_3 \text{-}\text{-}\text{-}*\text{-}\text{-}\text{-}T_4$ with the weight at w being $\Omega_w + r + 1 \geq r \geq 1$, one easily checks that the order of $\pi(\gamma') \geq 6$. Now $\pi(T)/(u) \simeq \tau(\pi(T_1), \pi(T_2), \pi(\Gamma'))$. Hence $\pi(T_1) \simeq \mathbf{Z}/(2), \simeq \pi(T_2)$, i.e., $T_1 = \overset{-2}{*} = T_2$. This implies $\gamma = \overset{-2}{*}\!\!\text{---}\!\!\overset{-1}{*}\!\!\text{---}\!\!\overset{-2}{*}$, which is equivalent to $\overset{0}{*}$, i.e., $\pi(\gamma) \simeq \mathbf{Z}$. This contradiction completes the proof of P_3, and hence that of $\overline{P}_3$.

Finally let $k \geq 4$. Consider the case where for all extremal branch points v, $\Omega_v = -1$. Let v_1 and v_2 be two extremal branch points (which exist, because $k \geq 4$). By Lemma 4.4, there are two vertices u_1 and u_2 with $\Omega_{u_1} = -1$ and links $[v_i; u_i]$, $i = 1, 2$. Since $k \geq 4$, it also follows that v_1 is not linked to v_2 or u_2, and v_2 is not linked to u_1. In particular $u_1 \neq u_2$. We also claim that u_1 is linked to u_2. For otherwise $v_1 + u_1$ and $v_2 + u_2$ will span a two-dimensional positive semidefinite subspace of $B(T)$, contradicting (E). Hence $[u_1; u_2]$ is a link. To sum up, T looks like:

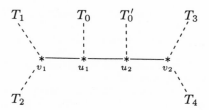

with T_i nonempty simple branches, $\Omega_{u_i} = -1$, $i = 1, 2$. If γ is the nonsimple branch at v_1, by $\overline{P}_3$, it follows that $T_0\text{---}\underset{u_1}{*}$ is spherical and hence all the weights on T_0 should be equal to -2. As before γ is then blow-down to a minimal tree

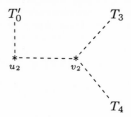

with the weight at u_2 changed to $\Omega_{u_2} + r + 1 = r \geq 1$. This contradicts the cyclicity of $\pi(\gamma)$.

Hence there should exist an extremal branch point v with $\Omega_v \neq 1$, in T. In particular, $T_1\text{---}\underset{v}{*}\text{---}T_2$ is not spherical. Hence looking at $\pi(T)/(u)$ (as in Lemma 4.4) we obtain that $\pi(\gamma')$ is cyclic. By the induction hypothesis applied to γ', it follows that there is a vertex w in γ' linked to u in T, linear in γ' with $\Omega_w = -1$. T looks like

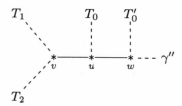

with $\Omega_u = -1 = \Omega_w$. Since $k \geq 4$, γ'' is nonsimple. But now the nonsimple branch at v blows down to Γ:

with the weight at w changed to $\Omega_w + r + 1 = r \geq 1$. Γ has at least one branch point, and all its simple branches are nonspherical. But $\pi(\Gamma) \simeq \pi(\gamma)$ is cyclic, contradicting the induction hypothesis.

This completes the proof of the proposition, and thereby, the proof of Theorem 1.

References

1. E. Brieskorn, *Rationale Singularitäten komplexer Flächen*, Invent. Math. **4** (1968), 336–358.

2. R. V. Gurjar, *Affine varieties dominated by* C^2, Comment Math. Helv. **55** (1980), 378–389.

3. R. V. Gurjar and A. R. Shastri, *The fundamental group at infinity of affine surfaces*, Comment. Math. Helv. **59** (1984), 459–484.

4. ――, *A topological characterization of* C^2/G, J. Math. Kyoto Univ. **25** (1985), 767–773.

5. J. Hempel, *3-manifolds*, Ann. of Math. Studies, no. 86, Univ. of Tokyo Press, Tokyo, 1976.

6. F. Hirzebruch and W. D. Neumann, *Differentiable manifolds and quadratic forms*, Marcel Dekker, New York, 1971.

7. D. Mumford, *The topology of normal singularities of an algebraic surface and a criterion for simplicity*, Inst. Hautes Études Sci. Publ. Math. **9** (1961), 5–22.

8. W. D. Neumann, *A calculus for plumbing applied to the topology of complex surface singularities and degenerating complex curves*, Trans. Amer. Math. Soc. **268** (1981), 299–344.

9. C. P. Ramanujam, *A topological characterization of the affine plane as an algebraic variety*, Ann. of Math. (2) **94** (1971), 69–88.

10. A. Van de Ven, *Analytic compactifications of complex homology cells*, Math. Ann. **147** (1962), 189–204.

11. P. Wagreich, *Singularities of complex surfaces with solvable fundamental group*, Topology **11** (1971), 87–117.

Tata Institute of Fundamental Research, India